Chitin in Nature and Technology

Chitin in Nature and Technology

Edited by

Riccardo Muzzarelli
University of Ancona
Ancona, Italy

Charles Jeuniaux
University of Liège
Liège, Belgium

and

Graham W. Gooday
University of Aberdeen
Aberdeen, Scotland

PLENUM PRESS • NEW YORK AND LONDON

Library of Congress Cataloging in Publication Data

International Conference on Chitin and Chitosan (3rd: 1985: Ancona, Italy)
Chitin in nature and technology.

"Proceedings of the Third International Conference on Chitin and Chitosan, held April 1–4, 1985, in Senigallia, Italy"—Verso of t.p.
Includes bibliographies and index.
1. Chitin—Congresses. 2. Chitosan–Congresses. I. Muzzarelli, Riccardo A. A., 1937– . 2. Jeuniaux, Charles. III. Gooday, G. W., 1942– . IV. Title.
QP702.C5I58 1985 574.19′2482 85-28088
ISBN 0-306-42211-5

Proceedings of the Third International Conference on Chitin and Chitosan, held April 1–4, 1985, in Senigallia, Italy

Printed in the United States of America

FOREWORD

Exegi monumentum aere perennius.
The monument I have built will last longer than bronze.
Horace

My previous book, "Chitin", (1977) was listed by the publisher, as a "key research book", among the most requested books by libraries. It received favorable comments from each of the journals which reviewed it, Science, 198, 28 Oct. 1977, Physiological Entomology, 2(4), Dec. 1977, The Canadian Institute of Food Science and Technology Journal, April 1978, The Quarterly Review of Biology, 53:361, 1978, Oceanographic Abstracts, 15:182, 1979, Annales de Zoologie-Ecologie Animale, 11:127, 1979, and Enzyme & Microbial Technology, 2, 1980. The variety of these journals testifies to the interdisciplinary character of chitin studies. "Chitin" has really been a landmark, to use the definition given by Science, because it stimulated interest in the less known polysaccharides and in modified chitins, besides chitin itself, to the point that three International Conferences on Chitin / Chitosan were convened (Boston, U.S.A. 1977, Sapporo, Japan 1982 and Senigallia, Italy 1985).

In convening the 3rd International Conference on Chitin / Chitosan (1-4 April 1985), one of the main objectives was the preparation of the present book. While the proceedings of the previous two Conferences were very valuable, they did not appear in any book catalogs and this severely limited their distribution. Therefore, in view of the need for a full-coverage book accessible world-wide, and in consideration of the large attendance (238 participants from 16 Countries), Plenum Press agreed to publish the present book which wants to be more than a mere proceeding. Its spirit is that of a "Chitin" book for the eighties, of lasting interest for the expert, and of passionate reading for the cultured man.

While offering the essential background, the present book is devoted mostly to advanced aspects of pure and applied research. I think that "Chitin" will remain a major reference for those who seek more basic information.

It is through the competence of the participants who supplied their manuscripts without delay, the help offered by the coeditors, Prof. Charles Jeuniaux and Prof. Graham W. Gooday, who also provided some of the "editor's reports" on interesting material that would have otherwise been inadequate for publication, and the understanding shown by the Government of Regione Marche, Ancona, Italy, which offered a special grant, that the final typescript has been produced, and it is a merit of Plenum Press that the book has been published so shortly after the Conference.

I wish to express my heartful and sincere thanks to all participants in the 3rd International Conference on Chitin / Chitosan, that I had the honor to chair; its success is also a merit of Maria G. Weckx, general manager, Irene Muzzarelli, secretary, Barbara B. Muzzarelli, social activities and Domenico Mencarelli, social relations. I thank Frank Salvi and Catherine Lough who assisted me in editing and producing the final typescript, as well as the staff of Chito-Bios s.r.l., Ancona, including among others Pamela La Marca and Lorella Falcetelli. All of them join me in expressing best wishes to the readers of this book, for fruitful studies and gratifying research activities.

Ancona, June 5th, 1985

Riccardo A. A. Muzzarelli

Professor of Chemistry and Biochemistry
Faculty of Medicine
University of Ancona
I-60100 Ancona, Italy

CONTENTS

Section III

CHITIN SYNTHESIS INHIBITION

Section IV

MOLECULAR RECOGNITION AND BIODEGRADATION

Section V

CHEMISTRY OF CHITIN

Section VI

MEDICAL USES OF CHITIN

Section VII

CHITIN ECOLOGY

PROGRESS TOWARDS SOLVING THE STRUCTURES OF POLYSACCHARIDES

K. D. PARKER

Astbury Department of Biophysics, The University of Leeds
Leeds, LS2 9JT, U.K.

STRUCTURE AT LOW RESOLUTION

The art of determining the structure of a regular fibrous molecule is well illustrated by the example of α-chitin. In 1950 Darmon and Rudall obtained good X-ray fibre diffraction patterns, good polarised infra-red spectra and built analogue molecular models of the stick and ball kind. Their conclusions are now considered to be wrong whilst Carlstrom in 1957 using almost the same data deduced a structure which is widely held to be substantially correct. It appears to us now that both these attempts were based on data which is insufficient to generate a uniquely correct solution. Darmon and Rudall took account of some fine detail in the X-ray and I.R. results and were led astray. Carlstrom, more selective, minimised or put aside these details, used superior model building and gave an acceptable solution. These discarded details still await a satisfactory explanation.

Many polysaccharide structures have been studied since then by similar methods with similar certainty and uncertainty in the results. Progress in protein and nucleic acid structures has tended to lead and inform parallel work on polysaccharides, and whilst the practical techniques have improved over 30 years the general approach has changed very little. Alginic acid in our laboratory provides a good illustration. Astbury in 1945 was hopelessly handicapped by lack of chemical information at that time. Subsequently in 1973 we were able to deduce crystal structures for the two components of alginic acid, D-mannuronic acid and D-guluronic acid, studied separately. The basic conformations of these molecular chains each in a crystalline state is thus known with some confidence. We can also surmise that the manner of packing of chains in the crystal is energetically favourable. We were also able to make tentative suggestions about the dispositions of some of the side groups, but it is at this level that hard evidence weakens and personal judgement or guesswork enters.

In 1950 to study structure was an end in itself; now we do so with the further aim of increasing our knowledge of the physical and chemical potentialities of the substance examined. With nucleic acid the mere outline structure has been sufficient to generate new ideas of molecular biology. With alginic acid very limited structural information has stimulated speculation on the mechanism of gel formation which may be technologically helpful. In both these examples the contribution of structure to the argument is so general that its value is difficult to assess. Trying to be more specific, from a more detailed knowledge of polymer structures we might hope to understand and predict their detailed physical chemistry, particularly their solubility, and their interaction with enzymes. Some of these topics will be discussed at this meeting. Two less general problems of chitin spring to mind. The helicoidal layer structure of insect cuticle is an ultrastructural phenomenon which may well originate from helicity at the molecular level a subtlety of structure probably beyond the sensitivity of current techniques to demonstrate. The difficulty of explaining the biogenesis of the anti-parallel structure of α-chitin by a mechanism at a cellular level has caused some workers to doubt the uniqueness of this interpretation of the data. It is thus still necessary to establish beyond all doubt that α-chitin is in fact an anti-parallel structure.

THE PURSUIT OF STRUCTURE REFINEMENT

Remarkably these outline structures of polysaccharides have been found without the direct application of computing. Much work has been done to improve results by the use of computing; the gains though useful have not been dramatic.

The X-ray diffraction data has been the subject of rigid-body refinement procedures and difference Fourier maps have been studied to enable the utmost information to be extracted from fibre diagrams. Owing to the limitations of the diffraction data these have proved to be of limited value and have completely resolved very few points. Three independently conducted refinements of α-chitin all came to the conclusion that less symmetrical structures than that of Carlstrom should be considered, probably with statistically distributed disorder, but the data did not allow a more precise specification. Personal preference based on model building must then take over. Blackwell and Gardener found that for cellulose I a parallel chain model agreed more closely with the X-ray data than the best anti-parallel-chain model tested by them. The difference in agreement was relatively small and although considered to be signficant by them one is still left wishing for further confirmation of the conclusion reached.

Model building is being refined by computing structures of least free energy. Such conformational analysis will almost certainly become of great use in future when the techniques are perfected but so far it seems merely to provide circumstantial approval or disapproval to structures which have already been decided upon.

Infra-red spectroscopy (to which Raman spectroscopy has been added) has been taken to its limits with attempts at a full normal coordinate analysis of the vibrational modes of cellulose. However the imperfect data on the force fields made a direct comparison of the observed and calculated frequencies meaningless, whilst the calculation of the intensities was even more rudimentary. This work was valuable in a limited way, in that it gave insight into the general types of the modes to be encountered in a polysaccharide vibrational spectrum and must certainly be taken into account by anyone using these methods.

Generalising the position with respect to techniques we note that the X-ray method is not directed at any particular detail in a structure so that in the case of fibres, lacking the extreme accuracy possible with single crystal studies, it cannot give detailed information at the atomic level. Moreover X-ray diffraction is not particularly well suited to studying either static or dynamic variability of structure. Vibrational spectroscopy gives information which is in part localised and in part non-localised but the theory of interpreting spectra in all but the simplest cases is very difficult. High resolution nuclear magnetic resonance which is by far the most informative spectroscopic tool could not previously be applied to solid specimens.

THE WAY AHEAD?

Progress towards a more complete understanding of polysaccharides through their structure has thus been a difficult and piecemeal process. Whilst a number of hopeful technical developments have recently appeared it is worth remembering that fibre structure determinations have often in the past been advanced by preparing or by discovering in nature improved specimens for examination. Thus Loligo pen gives a β-chitin of modest crystallinity but the same structure has been discovered in a highly crystalline form in Pogonophore tubes and in diatom spines. Similarly crab tendon gives an α-chitin of good crystallinity but more recently Rudall has discovered that the grasping spines of Sagitta are α-chitin of a very high degree of crystallinity. The latter has not yet been fully exploited because of technical difficulties but its very existence is reassuring. This kind of research contains too large a sporting element for most researchers, but with patience, skill or luck better specimens can be obtained _in vitro_ as shown by the improvements obtained by annealing films and fibres cast from solutions of mannuronic acid.

The determination of its crystal structure, however accurate, gives just one configuration of a molecule in one particular environment. From a physical point of view this represents just one point on a multi-dimensional energy graph. To understand properties under different conditions (in solution perhaps) we should like to have information on more such points. The α, β (and γ) forms of chitin and the modifications of cellulose do give us limited variation for these substances but the opening of research into the structures of their oligomers could give us much more information about the force fields which are operating. This approach is promoted on the one hand by improved methods of preparing and purifying the oligo-saccharides and on the other hand by the availability of fast direct methods for single crystal structure determination. Structures related to but different in detail from the polymer may thus be studied with reasonable speed and with full single crystal accuracy.

Although N.M.R. techniques have been improving steadily since their inception the application of N.M.R. at high resolution to polysaccharides has increased rapidly in significance only during the last few years. (Notice the paucity of N.M.R. references in the Sapporo meeting in 1982.) In solution (where applicable) improved resolution and sensitivity, double resonance methods and the use of other nuclei have all been exploited. ^{13}C spectroscopy using cross polarisation and magic angle spinning allows high resolution spectra of solids to be observed for the first time thus linking up with the infra-red observations of 35 years ago. Moreover magnetic resonance methods are capable of giving information about the motion of molecules. In the fields of molecular enzymology and nucleic acids the possibility of variations of structure within limits is now seen to be vital to their functions. Perfectly static material is dead. Whilst it would be stretching analogy too far to argue for equally important effects in polysaccharides it may be that some of their properties and some of the effects by which we observe them will turn out to be influenced by fluctuations of their structure.

In conclusion the role of the structure of biological molecules in determining their properties remains paramount as originally postulated by Astbury but the concept of structure has now to be extended beyond the static and highly symmetrical arrangements with which he was chiefly occupied.

Section I

CHITIN IN SKELETAL STRUCTURES OF ANIMALS

CHITINOPROTEIC COMPLEXES AND MINERALIZATION IN MOLLUSK SKELETAL STRUCTURES

M. POULICEK, M. F. VOSS-FOUCART and Ch. JEUNIAUX

Laboratory of Animal Morphology, Systematics and Ecology
Zoological Institute, 22, Quai Van Beneden
B-4020 Liège, Belgium

STRUCTURE AND CHEMICAL COMPOSITION OF CHITINOPROTEIC MATRICES OF MOLLUSK SHELLS

The deposition of shell material by Mollusks is one of the more fully studied processes of extracellular calcification, and a large amount of data has been collected on shell structure, formation and regeneration, minerology, trace elements and organic compounds biochemistry. Like in many other invertebrates, the nucleation and growth of mineral crystallites occur on a well defined structure of organic components, these components being more or less "frozen" within developing crystals : this constitutes the "organic matrix template theory".

Both vertebrate and invertebrate organic matrices seem to be built on the same general scheme: a core of structural macromolecule (polysaccharide or protein) sheathed by one or several envelopes of acidic proteins, glycoproteins and mucopolysaccharides. All these components may be stabilized by structural bonding, mostly quinone-tanning and sulfur bonds.

The organic matrix isolated through decalcification of calcified layers of Mollusk shells (the so-called "conchyolins") is very complex in composition and structure. Proteins are the main component and account for 50 to 80% of matrix dry weight (1). Chitin was demonstrated in the shells of several species (2,3) and recent investigations made obvious its presence in every calcified layer of Mollusk shell so far examined (ca. 150 species). Its amount, estimated by the enzymatic method of Jeuniaux (2,4), is quite variable : from 0.01 to 40 % of the matrix dry weight (1 - 3).

The conformation and orientation of the matrix macromolecules were relatively poorly understood until the studies of Weiner and co-workers (5,6, 7,8). As stressed by Weiner (8), the main difficulty was the fact that X-ray and electron diffraction studies need to remove the minerals and, during this operation, some hydrophilic components are lost. The use of non-calcified material (for example, *Loligo* pen) may lead to misinterpretation (9). X-ray and electron diffraction studies (5) of the insoluble components of the matrix reveal that there is an association of chitin in its β-form (parallel chains crystalline form) with proteins that, similarly, adopt an antiparallel β-sheet conformation (8). The chitin polymer would be oriented approximately perpendicular to the protein-polypeptide chains (5,6,7), so that this crossed construction presumably contributes to the mechanical strength of the matrix (6).

This theory is consistent with transmission electron microscopy results (shadowcast preparations, ultrathin sections) (10 - 12). The representative ultrastructure of the matrix is a micro-meshwork made of dense grains (2.3 - 4.0 nm average diameter) united by short, straight and thin organic connections (10). The sheets of the matrix are composed of several (up to five) different layers. The two surface layers are composed mainly of more soluble, acidic constituents. The core comprises a thin layer of chitin sandwiched between two thicker layers of proteins (12,13).

Bearing in mind those structural results, the calcophillic matrix of Mollusk shells appears, from a chemical point of view, as composed of two structural units :

1. An acidic polypeptide fraction with strong affinity to Ca^{++} ions, mostly soluble in decalcifying reagents (1, 14, 15, 16, 17). Its most probable arrangement involves a spiraled peptide chain (18). It is generally called "Mineralization Matrix"(MM).

2. A high molecular weight chitinoproteic complex with no affinity to calcium, arranged in the form of sheets and layers (1, 18), called "Carrier Protein" (CP).

The attachment of the soluble "mineralization matrix" to the "carrier protein" complex will activate the mineralizing substrate, leading to epitaxial $CaCO_3$ deposition (18). A "sandwich" structure thus appears with the chitinoproteic CP embedded between two sheets of the mineralization matrix, as presented on Figure 1. Subsequently, quinone-tanning can stabilize this whole molecular framework (1).

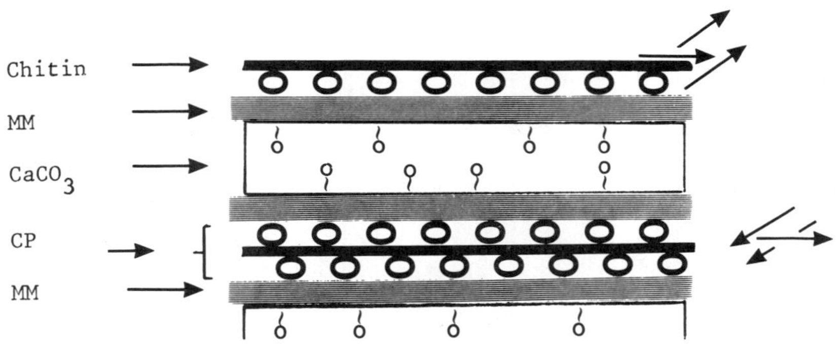

Fig. 1 : Schematic interpretation of the structure of the organic matrix in Mollusk shells (not drawn to scale).

CHITIN-PROTEIN BOND IN MOLLUSK SHELLS

The nature of the chitin-protein association in Mollusk calcified shells is still poorly understood, particularly as far as linkage is concerned. As chitin always appears accompanied by proteins, it is obvious that there is/are site/s in the polysaccharide polymer through which covalent binding to protein occurs. This proteoglycan may be accompanied by some other proteins or glycoproteins, specifically or not specifically associated with it by weak forces (9, 19).

According to the generalized high aspartic acid content of the insoluble protein complex (1), Hackman datas on chitin-protein relations may tentatively be extended to calcified Mollusk shells.

N-acetylglucosamine and chitin can react with α-aminoacids and peptides to give stable complexes and it seems that chitin can form more or less stable glycoproteins, probably through aspartyl and histidyl residues(20,21). Similarly, Brine and Austin (22) showed that predominant aminoacids in the residual chitin after vigorous alkaline hydrolysis were aspartic acid, serine and glycine, which may be involved in the chitin-protein linkage. It should be remembered that those three aminoacids represent 35 to 50 % of aminoacid residues in the soluble matrix of Mollusk shells (1). In a similar perspective, Gottschalk proposed a linkage in the form of glycosidic ester between acetylglucosamine residues and β-carboxyl groups of aspartyl residues (23). According to these ideas, Stegeman reported that, in the non-calcified pen of Cephalopods, the proteins nost firmly bound to chitin contain large amounts of aspartic acid (24).

In contrast, Atwood and Zola (25) found that, after prolonged alkali treatment, there is no predominance of aspartic acid and histidine, nor indeed, of any other aminoacid in the chitin isolated from *Loligo* pen. According to these authors, these results do not indicate the lack of protein-chitin covalent linkages, but only that the use of this method is inadequate (25). The results of hydroxylamine and lithium borohydride treatments on the *Loligo* pen chitinoproteic complex show that ester linkages are probably absent. If present, the protein/s joined to chitin by these links is/are also joined to chitin through another type of bond (25).

Hydroxylamine cleaves ester and amide linkages and has been used extensively in studies of protein-polysaccharide linkages (26). Lithium borohydride cleaves esters but has less action on amides (25).

These controversial results clearly show that existing evidences for covalent binding between chitin and proteins must be reexamined, particularly in the case of calcified matrix.

CHITIN AND CALCIFICATION IN MOLLUSK SKELETAL STRUCTURES

It seems well established that $CaCO_3$ nucleation occurs through a Ca^{++} high affinity soluble proteic complex, associated with an insoluble chitinoproteic complex. Chitin does not appear to be directly related to calcification process, but physiological and evolutive evidence indicates indirect relationship of chitin with $CaCO_3$ metabolism. Three kinds of indirect arguments were recently proposed.

A) Correlative chitin and $CaCO_3$ content in slug shells during annual cycles

The internal shells of the slugs *Agriolimax reticulatus*, *Milax rusticus*, *Limax maximus*, *Limax cinereoniger* and *Lehmania marginata* do contain chitin, and an important part of this chitin (60 %) is not firmly bound to the proteins of the insoluble complex (27).

$CaCO_3$ and chitin estimations in shells taken at different periods of the year (under natural conditions) show important cyclic fluctuations (fig.2). An obvious decrease of the amount of chitin in the organic matrix of the shell is observed during hibernation of the slugs (October to March) and this is parallel with the decrease of the amount of $CaCO_3$ during the same period. During the period of activity (March to October) the internal shell simultaneously restore the $CaCO_3$ content and the amount of chitin of the organic matrix (maximum just before hibernation). During shell resorption, histoenzymological methods make obvious a chitinolytic activity from the whole palleal epithelium (28).

Correlatively, a high amount of hydrophillic acidsoluble proteins (20% of total shell proteins, with an aminoacid pattern similar to other mineralization matrices (27)) should be referred to an increased calcium metabolism of the structure. The simultaneous metabolic variations of chitin and $CaCO_3$ can be simulated, under laboratory conditions, by starvation of the slugs or sexual maturity induction (28).

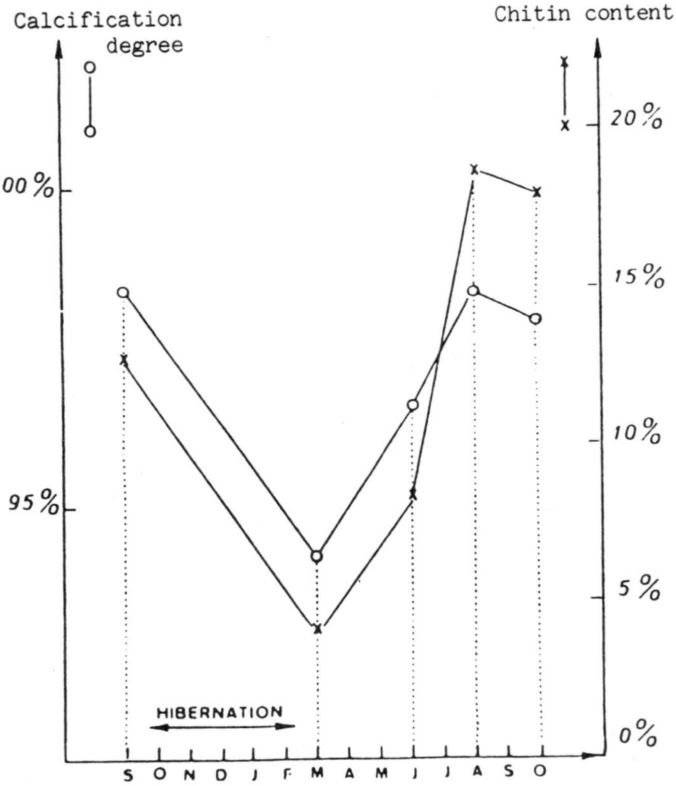

Fig. 2 : Cyclic fluctuations of calcification degree and chitin content in the shell of *Agriolimax* (from (27), slightly modified).

B) <u>Chitin content of Gastropod operculi</u>

The operculi of 114 species of Gastropods representing 38 families with wide ranging morphological, ecological and geographical differences have been examined for the presence of chitin (29,30). In all the species so far analysed, calcified operculi were found relatively rich in chitin ; the amount of chitin was similar in operculi of neighbouring species. No chitin was found in the non-calcified, corneous operculi. The level of calcification was the unique character found to be correlated to the presence of chitin and no other parameter showed any reciprocity.

The organic matrix of calcified operculi is built of two layers, a quinone tanned proteic sheet, and a calcified glycoproteic matrix. The tanned sheet of all operculi tested is totally devoid of chitin. Mainly proteic, its aminoacid composition is similar with the composition of the whole typical corneous operculum. The calcophillic glycoproteic matrix (insoluble complex) does contain chitin, the amount of which depends directly on the calcification degree of the operculum (31). Aminoacid composition of this matrix appears very close to that of insoluble shell proteins.

C) <u>Evolution of the chitin content of Mollusk shells</u>

From an evolutive point of view, chitin content and calcification degree of Mollusk shells appear to vary concommitantly, but in opposite directions. During the evolution of the external shells of Conchifera (Gastropods,

Cephalopods, Bivalvia, Scaphopods and Monoplacophora), the less calcified chitin rich microstructural layers (mother-of-pearl) are replaced by highly calcified crossed-lamellar layers, much poorer in chitin (1).

On the contrary, in the evolutive lines where the shells become more and more embedded within dorsal palleal expansions (Cephalopods, Opisthobranchs, Pulmonates, Placophora,...), the increase of the chitin content is connected with a decrease of the calcification degree (1).

In both instances, the mineralization matrix remains remarkably stable in its aminoacid composition pattern. The proteins associated with chitin to form the insoluble "carrier protein" complex, show only slight modifications, with no general obvious tendency.

CONCLUSIONS

The main results on chitin-protein complexes in Mollusk shells were here summarized and focused on the problems linked with calcification, one of the most tremendous features of Mollusk shells. It is suggested that chitin may behave as a molecular skeleton onto which structural proteins and mineralization proteins are bound. This is consistent with electron microscopy results (10,11,12,13) and X-ray diffraction data (5,6,7,8). It is likely that the binding of proteins to chitin reveals some active site used for nucleation and growth of $CaCO_3$ crystals.

According to this preliminary hypothesis of indirect influence of chitin on mineralization of Mollusk shells, the nature of chitin-protein bond and fine molecular organization of chitin-protein complexes in calcified shells will certainly need further research.

REFERENCES

1. M. Poulicek, "Coquilles et autres structures squelettiques des Mollusques. Composition chimique, biomasse et biodégradation en milieu marin", Ph. D. Thesis, University of Liège, p. 180 (1982).
2. Ch. Jeuniaux, "Chitine et Chitinolyse", Masson ed., Paris (1963).
3. G. Goffinet and Ch. Jeuniaux, Distribution et importance quantitative de la chitine dans les coquilles de Mollusques, Cah.Biol.Mar., 20 : 341 (1979).
4. Ch. Jeuniaux, Chitine et phylogénie : application d'une méthode enzymatique de dosage de la chitine, Bull.Soc.Chim.Biol., 47 : 2267 (1965).
5. S. Weiner and W. Traub, X-ray diffraction study of the insoluble organic matrix of mollusk shells, Fed.Eur.Biochem.Soc., 111 : 311 (1980).
6. S. Weiner and W. Traub, Macromolecules in mollusc shells and their functions in biomineralization, Phil.Trans.R.Soc.Lond.,B, 304: 425 (1984).
7. S. Weiner, Y. Talmon and W. Traub, Electron diffraction of mollusk shell organic matrices and their relationship to the mineral phase, Int.J. Biol.Macromol., 5 : 325 (1983).
8. S. Weiner, Organization of organic matrix components in mineralized tissues, Amer.Zool., 24 : 945 (1984).
9. S. Hunt and M. Nixon, A comparative study of protein composition in the chitin-protein complexes of the beak, pen, sucker disc, radula and oesophageal cuticle of cephalopods, Comp.Biochem.Physiol., 68B : 535 (1981).
10. G. Goffinet, Ch. Grégoire and M.F. Voss-Foucart, On ultrastructure of the trabeculae in the interlamellar membranes of nacre conchyolin of the *Nautilus* shell, Arch.Internat.Physiol.Bioch., 85 : 849 (1977).
11. G. Bevelander and H. Nakahara, Compartment and envelope formation in the process of biological mineralization, in : "The mechanisms of biomin-

eralization in Invertebrates and Plants", M. Omori and N. Watabe ed., Tokai University Press, Tokyo (1980).

12. H. Nakahara, G. Bevelander and M. Kakei, Electron microscopic and amino acid studies on the outer and inner shell layers of *Haliotis rufescens* Venus, 39 : 205 (1982).

13. H. Nakahara, Calcification of Gastropod nacre, in : "Biomineralization and biological metal accumulation", P. Westbroek and E.W. de Jong ed., Reidel, Dordrecht (1983).

14. E.M. Greenfield, D.C. Courtney and M.A. Crenshaw, Ionotropic nucleation of calcium carbonate by molluscan matrix, Amer.Zool.,24 : 925 (1984).

15. S. Weiner and L.H. Hood, Soluble proteins of the organic matrix of mollusc shells : a potential template for shell formation, Science, 190 : 987 (1975).

16. M.A. Crenshaw, The soluble matrix from *Mercenaria mercenaria* shell, Biomineralisation, 6 : 6 (1972).

17. T. Samata and G. Krampitz, Calcium binding proteins in oyster shells, Malacologia, 22 : 225 (1982).

18. E.T. Degens, Carbonate, Phosphate and Silica Deposition in the living cell, in "Topics in Current Chemistry 64", F. Boschke ed., Springer Verlag Berlin (1979).

19. K.M. Rudall and W. Kenchington, The chitin system, Biol.Rev., 48 : 597 (1973).

20. R.H. Hackman, Chitin 2.-Reaction of N-acetyl-D-glucosamine with α-amino acids, peptides and proteins, Austr.J.Biol.Sci., 8 : 83 (1955).

21. R.H. Hackman, Chitin 4.-The occurence of complexes in which chitin and protein are covalently linked, Austr.J.Biol.Sci., 13 : 568 (1960).

22. C.J. Brine and P.R. Austin, Chitin isolates : species variation in residual amino acids, Comp.Biochem.Physiol.,70B : 173 (1981).

23. A. Gottschalk, W.H. Murphy and E.R.B. Graham, Carbohydrates-peptide linkages in glycoproteins and methods for their elucidation, Nature, 194 : 1051 (1962).

24. H. Stegeman, Protein (conchagen) and chitin in the supporting tissue of the cuttlefish, Z.Physiol.chem., 331 : 269 (1963).

25. M.M. Attwood and H. Zola, The association between chitin and protein in some chitinous tissues, Comp.Biochem.Physiol., 20 : 993 (1967).

26. E.R.B. Graham, W.H. Murphy and A. Gottschalk, Studies on mucoproteins -IX.On the suceptibility to alkali and to hydroxylamine of the predominant carbohydrate-peptide linkage in ovine-submaxillary-gland glycoprotein, Biochem.biophys.Acta, 74 : 222 (1963).

27. M. Poulicek and M.F. Voss-Foucart, Variations saisonnières de la composition chimique de la coquille d'*Agriolimax reticulatus* (Müller,1774) (Gastropode, Limacidae), Arch.Zool.exp.gen., 121 : 77 (1980).

28. M. Poulicek and M.F. Jaspar-Versali, Essai d'interprétation d'un cycle saisonnier de la limacelle chez quelques Pulmonés Limacidae, Malacologia, 22 : 241 (1982).

29. M. Poulicek, Chitin in Gastropod operculi, Bioch.Syst.Ecol., 11 : 47 (1983).

30. M. Grasset, M. Poulicek, M. Truchet and J. Vovelle, Composants accessoires de l'opercule protéique tanné de *Büccinum undatum* (L.), Haliotis, 13 : 115 (1983).

31. M. Poulicek, La matrice organique des opercules calcifiés, Malacologia, 22 : 235 (1982).

CHARACTERIZATION OF THE INSECT CUTICLE LAYERS BY LASER RAMAN SPECTROMETRY

M. TRUCHET and B. MAUCHAMP*

Laboratoire d'Histophysiologie Fondamentale et Appliquée
U.A. 680 C.N.R.S., Université Pierre et Marie Curie
F-75005 Paris, France

*Laboratoire de Phytopharmacie, I.N.R.A., Etoile de Choisy
F-78000 Versailles, France

INTRODUCTION

Raman spectrometry allows to characterize all kinds of molecules by vibrational spectra. Laser is used as a light excitation source and a monochromator and recording system analyse scattered light for frequency and intensity. The theoretical basis of Raman vibrational spectra is photon energy exchange with the molecules vibrational states, or inelastic light scattering. Two kinds of informations are given by Raman spectra :
a. The peak positions are a property of the vibrational levels of the ground electronic state of the molecule.
b. The peak intensities depends upon ground and excited electronic states[1].

To our knowledge, except tne referenced paper, very few works have been published, dealing with Raman spectrometry of chitin and chitosans, whereas this powerful method is widely used in other fields, such as nucleic acids research, or chromophores biochemistry[1,2].

In "classical" Raman spectrometry, scattered light is collected through a lens system, from samples usually in the range of 0.1-1g (millimetric scale). Since 1976, it has been possible to record Raman spectra directly from samples as small as 10-11 g, by photon microscope, This Raman microprobe (Fig.1) of Delhaye and Dhamelincourt (3) uses a photon microscope, and a laser focused in a 1µm probe. It enables us to determine molecular composition in volumes of a few cubic micrometers and has been successfully applied to biological samples for various studies at cellular level (reviewed 4).

We found that to characterize *in situ* insect cuticle components, in natural conditions and after experimentation of pesticides[5] could be a potentially important application.

As a first step, in order to check the feasibility of our project, we will deal, in this paper, with recording of micro-Raman spectra of pure chitin and cuticle.

MATERIAL AND METHODS

Insect cuticle was obtained from freshly sectioned abdomen of locust,

Locusta migratoria. After fixation in Carnoy's, dehydration, inclusion in paraffin, the abdomen was sectioned at 7 µm and spread on glass microscope slides. After drying, the sections were deparaffined with toluene and absolute alcohol purified to avoid fluorescence. Raman analyses were made without any treatment (4). Standard chitin was obtained from crab shell, purified and powdered (6). Raman microanalysis was performed with a Jobin-Yvon MOLE 77 (Longjumeau, France), in the following analytical conditions: laser beam (Spectra-Physics, U.S.A.), 514.5 nm, 100 mW; objective, x 100 (N.A. 0.90); probe, 1 µm, approximately (the laser power was about 5 mW in the volume of the probe focus); monochromator slit width, 700 µm (spectral resolution of about 7 cm^{-1}). Recording of Raman spectra were done from standard fragments of approximately 20 µm; in the cuticle, the laser was focused in two different layers : an upper layer at about 5 µm under the top, called exocuticle, and a lower layer at about 5 µm above the epidermis, called endocuticle. The first two spectra were recorded by simple scanning, and short photon counting time; the spectra of endocuticle were recorded by multiple scanning and computer-treated to obtain better peak-to-noise ratios i.e. "accumulated" (for laser localization, see Fig. 2). The assignments of the Raman bands, were drawn from the work of Galat and Popowicz (2).

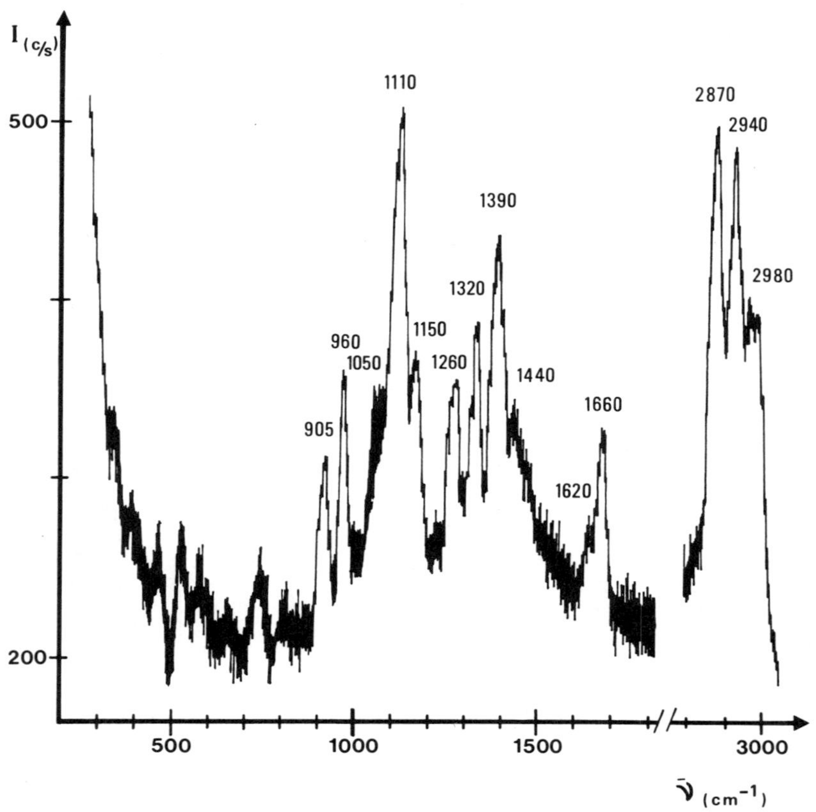

Fig. 3. *Raman spectrum of purified chitin. Size of the sample : approximately 20 µm. Laser beam : 514.5 nm, 100 mW ; focus : 1 µm, approximately ; power at the focus : 5mW. Aperture diaphragm : 5 µm, approximately. Slit width : 700 µm. Photon counting :* $10^2 c.s^{-1}$ *; 4 s. Recorder : 500 mV, 20* $cm^{-1}.mn^{-1}$*, 100* $cm^{-1}.cm^{-1}$.

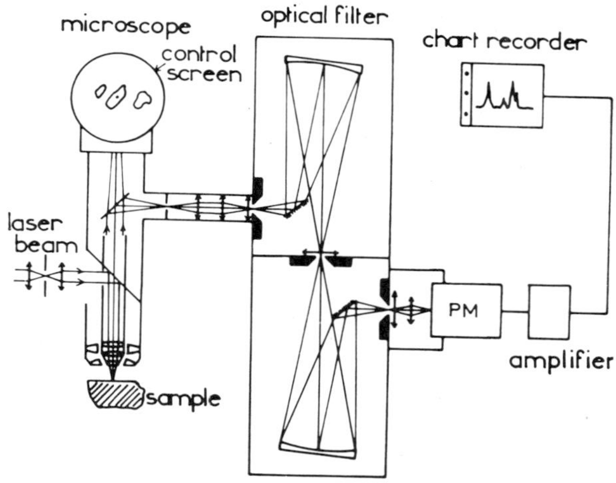

Fig. 1. *Schematic representation of the Raman Laser Microprobe (from Delhaye and Dhamelincourt[3]).*

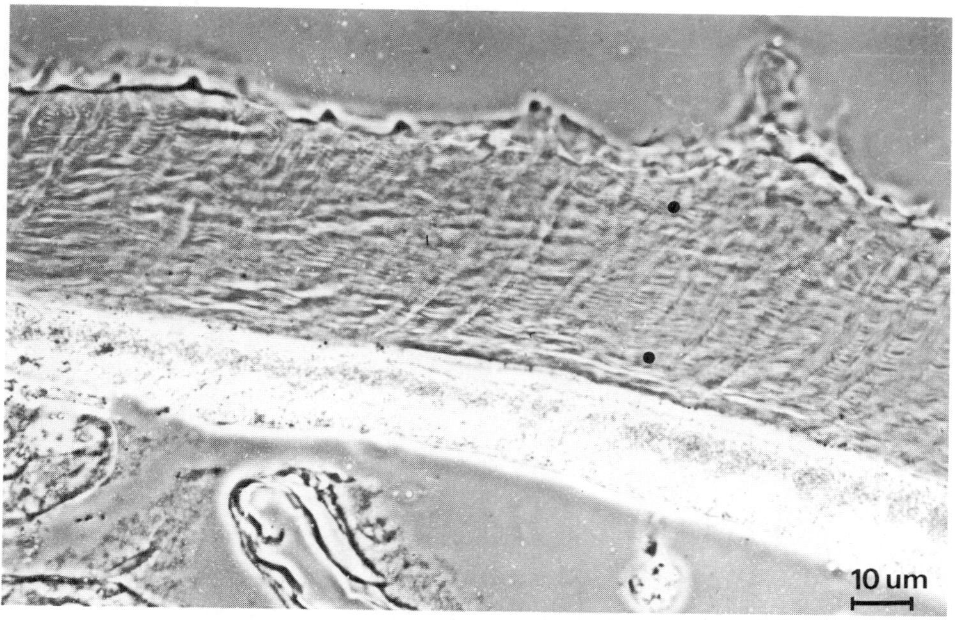

Fig. 2. *Section of an insect cuticle. The points indicate the positions and sizes of laser impacts.*

RESULTS AND DISCUSSION

In each case, three to seven spectra were recorded, at different time intervals, and from different points. Examples of good records, manually re-drawn, are given in figures 3 to 5. Table 1 lists the Raman peaks in frequencies and gives the relative intensities, by normalisation to the intensity of the 2930 cm^{-1} response, taken as unity.

Our results demonstrate that it is possible to obtain Raman spectra from different insect cuticle layers, with volumes of a few cubic microns (approximately 10^{-11}g) and to obtain the chemical composition of cuticular sub-structures, optically determined. However, the material is easily burned by the laser, and little power should be used. Thus, peak-to-noise ratios are low.

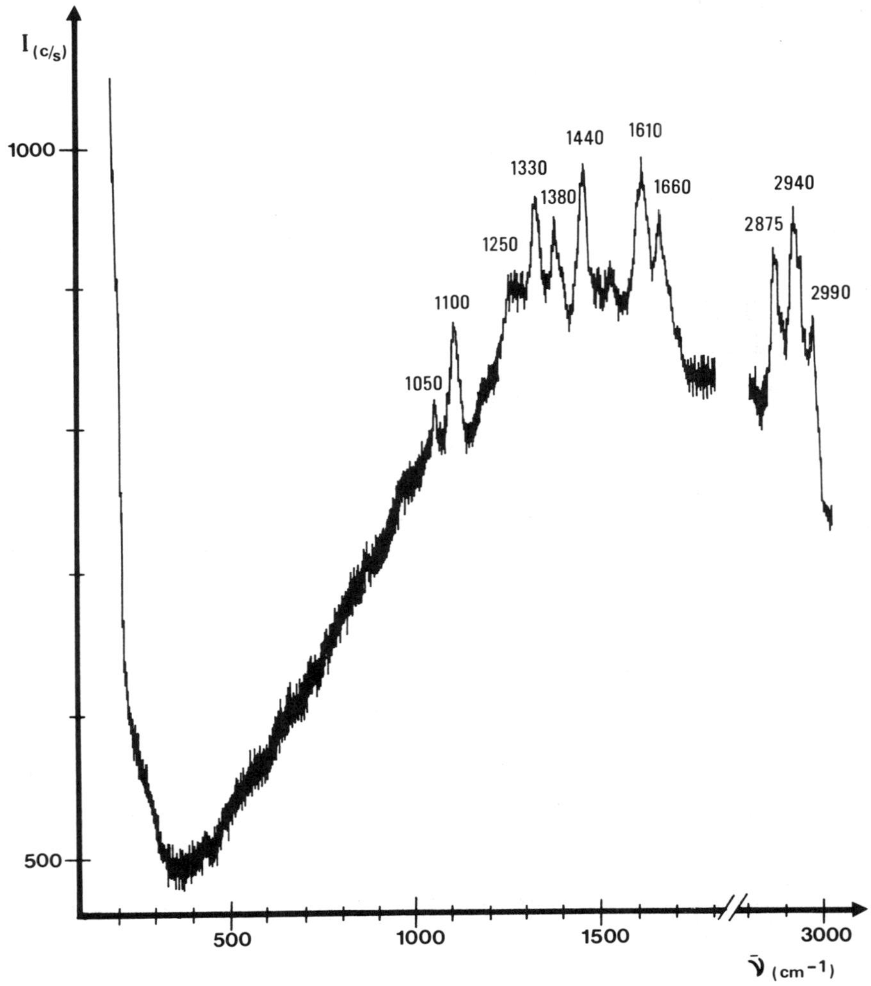

Fig. 4. *Raman spectrum of exocuticular layer. The instrumental conditions are the same as for fig. 3, except : recorder, 200 mV.*

TABLE 1. *Comparison of Raman frequencies.*

A pure chitin		B Exocuticle		C Endocuticle	
2980	0.7			2970	0.8
2940	1	2940	1	2925	1
2875		2875	0.9	2865	0.25
1660	0.3	1660	0.8	1660	0.3
1620	0.09	1620	1	1610	0.2
1440	0.25	1440	0.9	1440	0.25
1380	0.6	1390	0.5	1350	0.2
1320	0.4	1320	0.8	1310	0.2
1260	0.3			1250	0.25
1150	0.4				
1100	0.9	1100	0.5	1000	0.5
1050	0.3	1060	0.2	1050	0.3
950	0.4			940	0.25
900	0.3			890	0.2

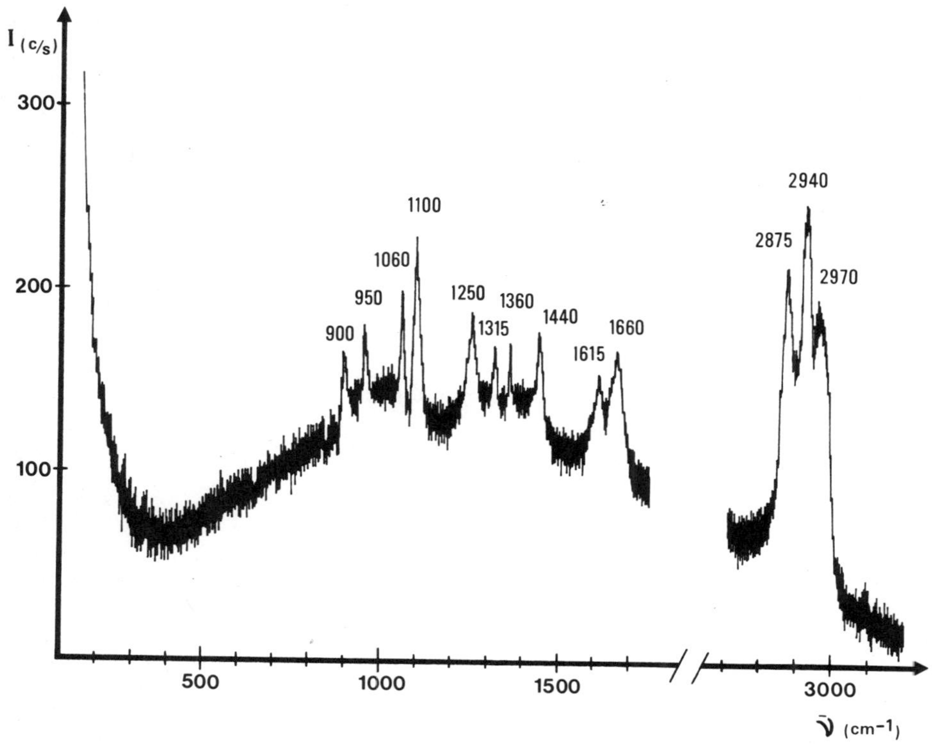

Fig. 5. *Raman spectrum of endocuticular layer. The instrumental conditions are the same as fig. 3 and 4, except : recorder, 100 mV ; accumulations : 9.*

In the purified chitin, and in the exocuticular layers, we encountered a fluorescence (see Fig. 3 and 4, Y-axis) which makes difficult but not impossible to record spectra. On the other hand, in the endocuticular layer, despite the fact that the fluorescence was not stronger, it was quite impossible to obtain a spectrum in the same analytical conditions. We obtained satisfying results only by accumulation. This technique of signal treatment is very time-consuming, but the long time punctual illumination decreases considerably the flourescence (see Fig. 4, Y-axis).

Our spectra of pure chitin appears almost similar to the description of Galat and Popowicz, both in frequencies and relative intensities except for the peak at 1100 cm^{-1}, which is intense in our spectra, and given as medium by the above authors [2].

This spectrum is also roughly similar to that of cuticle, but with several differences. We tentatively interpret them as follows:

In the CH stretching region, the peak at 2875 cm^{-1} is more intense in chitin than in cuticle. We attribute this difference to the chemical fixation. In the amide I region, in the chitin, the peak at 1620 cm^{-1}, very weak, is a shoulder of the intense 1660 cm^{-1}. In the cuticle, both peaks are well represented, and the 1620 cm^{-1} can be more intense than the 1660 cm^{-1} (exocuticle). We suggest that this region may be a superimposition of amide bands of the glucidic polymer and of the accompanying proteins.

It is the same for the 1440 cm^{-1} peak, more intense in both cuticle layers than in the standard of chitin.

The peaks at 1390 cm^{-1} and 1320 cm^{-1} are reduced in cuticle layers, especially in endocuticle. At 1250 cm^{-1}, an intense Raman peak, likely to correspond to the amide III mode, is observed in both standard and endocuticle, whereas it is very weak in the exocuticular layer. We suggest that the reduction in intensity of this mode may be an effect of the hardening processing, although reduction in 1390 cm^{-1} and 1320 cm^{-1} intensities result of linkage between chitin and associated proteins.

The 1150 cm^{-1} peak of the pure chitin is not observed in the cuticle; likewise for the peak at 2870 cm-1, we attribute this modification to the chemical fixation.

The intense response at 1100 cm^{-1}, attributed to C-C stretching, is well represented in the three cases.

The peak at 1050 cm^{-1}, attributed to C-C and C-N stretching modes, is present in the cuticular layer spectra, but it is a simple shoulder in our pure chitin standard. Like for the 1440 cm^{-1} and 1620 cm^{-1} peaks, we suggest a contribution of the accompanying proteins for this mode.

Above 1000 cm^{-1}, numerous peaks are described by Galat and Popowicz [2] and are also observed in some of our spectra. However, our peak-to-noise ratios are not sufficient to support analytical interpretations, and we do not take them into account for this study.

In our spectra, only the reduction of intensity of the peak at 2870 cm^{-1} and the disappearance of the peak at 1150 cm^{-1} can be attributed to the chemical fixation ; among the other modifications, the enhancement of the peaks at 1440 cm^{-1} and 1620 cm^{-1} are attributed to the accompanying proteins. They are very difficult to extract completely during the chitin purification which makes possible a slight contribution of associated proteins in the standard spectra; this difficulty has also been pointed out by Galat and Popowicz (2). Thus, the spectral differences generated by the proteins are not "all or nothing", but observed only as differences in intensities, which complicates the spectral interpretation.

Regarding the two cuticular layers, the differences between the spectra which may be attributed to the hardening are even more discrete (diminution of the intensity of the 1250 cm^{-1} peak). This may be due to the slight difference in hardening between the two layers in the locust abdomen. To better determine differences due to hardening, it would be more suitable to use samples more sclerotized, such as pupal exuviae, as done by Peter et al. (7) (see also Peter et al., this Volume).

In fact, our purpose was essentially to check the feasibility of microscopic Raman analysis of cuticle; from this point of view, the results presented here are stimulating. However, to reach our main target, the intracellular secretions of the cuticular material, improvements are required in the analytical processing. We have undertaken a computer-treatment to enhance the peak-to-noise ratios; it is also possible to make more numerous accumulations. However this technique is so time-consuming, that this step would be feasible only with the second generation Raman microsprobe, with multichannel detection (8). With this technology, the different elements of the spectra are acquired simultaneously, and not sequentially, which allows greater integration time and more numerous accumulations in a shorter time.

ACKNOWLEDGMENTS

Samples preparation and typescript were made by H. Moysan. Raman analyses were performed in the Laboratoire de Spectrochimie Infrarouge et Raman, L.P. n° 002631 du C.N.R.S., Thiais, France.

REFERENCES

1. P.R. Carey, Resonance Raman spectroscopy in biochemistry and biology. *Quart. Rev. Biophys.*, II (3): 309-370 (1978).
2. A. Galat and J. Popowicz, Study of the Raman Scattering Spectra of Chitins. *Bull. Acad. Pol. Sci.*, (Cl. II), 26 (8) : 519-524 (1978).
3. M. Delhaye and P. Dhamelincourt, Raman Microprobe and Microscope with Laser excitation. *J. Raman Spectrosc.*, 3 : 33-43 (1975).
4. M. Truchet, Applications biologiques de la microsonde Raman. *J. Microsc. Spectrosc. Electron.*, 9 : 117-120 (1984).
5. B. Mauchamp, Biosynthesis of chitin by insect epidermal cells : effects of diflubenzuron. *(this volume).*
6. R.A.A. Muzzarelli, Chitin. *Pergamon Press, Oxford, New-York.* (1977).
7. M.G. Peter, L. Grün and H. Förster, CP/MAS-^{13}C-NMR Spectra of Sclerotized Insect Cuticle and of Chitin. *Angew. Chem. Int. Ed. Engl.*, 23 (8) : 638-639 (1984).
8. J. Barbillat, Etude d'une seconde génération de microsonde optique a effet Raman mettant à profit les avantages de la détection multicanale. *Thèse Doctorat d'Etat*, Université des Sciences et Techniques de Lille (France), n° 605 (1983).

STRUCTURAL STUDIES ON SCLEROTIZED INSECT CUTICLE

M. G. PETER*, G. KEGEL and R. KELLER

*Institut für Organische Chemie und Biochemie and
Institut für Zoophysiologie der Rheinischen Friedrich-Wilhelms
Universität Bonn
D-5300 Bonn 1, F.R. Germany

INTRODUCTION

The integument of insects in composed mainly of chitin and proteins[1,2]. Maintenance of normal life functions requires that cuticular structures are sclerotized, either in smaller parts, such as mandibles, or in larger areas for the construction of the exoskeleton. Sclerotization of the insect cuticle actually results from the incorporation of diphenolic compounds into the outer parts of the integument. Due to the classical work of Karlson and his school [3,4], it is well established that N-acetyldopamine is an essential component of the sclerotization system in many insects. More recently, N-β-alanyldopamine has been discovered as another sclerotization agent [5].

Very little is known on the molecular mechanism of sclerotization, and, despite many efforts to analyze the chemical details, this important process is a matter of great controversy. Since a number of excellent recent reviews [6-8] are available, only the major hypothesis will be briefly summarized here. The sclerotization process is initiated in the cuticle by an enzymatic oxidation of the diphenolic substrate. It is thought that this oxidation yields the corresponding *o*-quinone and/or *p*-quinonemethide as a reactive intermediate. According to the conclusions drawn from chemical model reactions, crosslinking of cuticle proteins results then from Michael type conjugate addition and Schiff's base formation with free peptidic amino groups. It has been proposed [8] that also chitin may be involved in the *o*-quinone or *p*-quinonemethide mediated formation of crosslinks, either *via* unacetylated amino groups or hydroxy oxygen. On the other hand, non-covalent interactions between oxidation products of the sclerotization agents and polypeptides and/or chitin may contribute significantly, if not predominantly to the stability of sclerotized insect cuticle [9-11].

Earlier studies have shown that in vitro oxidation of N-acetyldopamine yields complex product mixtures, containing mostly polymeric material [11]. In order to investigate the mode of binding of these polymers to chitin, we have now prepared [U-^{14}C]-chitin by in vivo biosynthesis and studied its interaction with [β-^{3}H]-N-acetyldopamine under oxidative conditions. We also report on the investigation of polymeric N-acetyldopamine oxidation products by ^{13}C-NMR spectroscopy in solution and in the solid state.

PREPARATION AND ANALYSIS OF [U-^{14}C]-CHITIN

The method of the preparation of radiolabelled chitin makes use of the fact that in early postmoult of decapod crustaceans carbohydrate precursors are incorporated into chitin of the new endocuticle at a remarkably high rate [12]. Soft shelled crayfish, Orconectes limosus, in stage A of the moulting cycle [13] were used within 24 h after ecdysis. [U-^{14}C]-glucose (spec. activity 283 mCi·mmol^{-1}) was injected in aqueous solution into one of the pereiopods which was ligated around the needle to prevent loss of label. Doses were 1.5-6 µCi·g^{-1} live weight in a volume not exceeding 50 µl per animal. After a 24 h incubation, animals were sacrificed, most of the tissues removed, and the integuments boiled and stored frozen. Chitin was prepared by an established procedure [14] which was used in modified form. The integument was disintegrated in 2N KOH (250 ml per 4 animals) with an Ultra Turrax homogenizer, and protein was extracted by stirring the suspension for 24 h at 100°C. Centrifugation and washes with water were followed by resuspending the sediment in hydrochloric acid (2 N, 250 ml) at 20°C for 15 h for decalcification. The next steps consisted in washing with water, stirring in potassium hydroxide (1 N, 250 ml) for 2 h at 100°C, washing with water, extracting with acetone (100 ml) and then with chloroform and methanol (1:1) and drying at 90°C. The remaining chitin was dissolved in 3 portions of concentrated hydrodhloric acid (100 ml each) and stirred for 2 h at 4°C. The combined supernatants were slowly poured through glass wool into ice cold water (2 l) under stirring. The colloidal chitin was collected by centrifugation, washed and dialyzed against water. Aliquots were taken for gravimetric determination of chitin content and radioactivity. The suspension was stored at 4°C after addition of a few drops of toluene. Several batches of [U-^{14}C]-chitin were prepared. The highest spec. radioactivity obtained was 215 nCi·mg^{-1}. Up to 30% of the injected [U-^{14}C]-glucose were incorporated into chitin.

Enzymatic hydrolysis of [U-^{14}C]-chitin was performed with varying amounts of chitinase from Streptomyces antibioticus (Calbiochem) (Fig. 1). It seems that the enzymatic reaction proceeds more rapidly in the first time interval shown in Fig. 1, whereas a rather linear time course is observed later.

A sample of 1.05 mg [U-^{14}C]-chitin was subjected to hydrolysis in 2 ml 6N HCl at 110°C for 20 h and the hydrolysate analyzed on a Biotronik

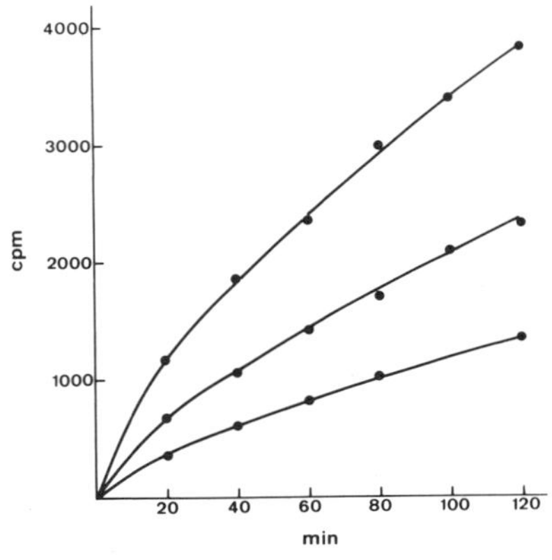

Fig. 1. Liberation of soluble radiolabelled products from [U-^{14}C]-chitin. Mixtures containing a suspension of 110 µg [U-^{14}C]-chitin (spec. activity 215 nCi·mg^{-1}) and 12.5 or 25 or 50 µg chitinase in 50 µl 50 mM Na-acetate buffer pH 5.5 with 0.15 M NaCl, were stirred at 25°C. Aliquots were counted after centrifugation at 10000 g for 2 min

LC 6000 E amino acid analyzer. It was found that the chitin contained the following residual amino acids: aspartic acid, threonine, serine, glutamic acid, valine, leucine and histidine: between 10 and 15 nmol·mg^{-1}; glycine and alanine: between 24 and 25 nmol·mg^{-1}; isoleucine and arginine: 7.5 nmol·mg^{-1}; lysine was exceptionally abundant with 113 nmol·mg; proline and methionine were not detected. Tyrosine and phenylalanine were not separated from glucosamine. Two major, unidentified components eluted between glucosamine and lysine, and between histidine and arginine, respectively.

^{13}C-NMR SPECTROSCOPY OF SCLEROTIZED INSECT CUTICLE, CHITIN, AND OXIDATION PRODUCTS OF N-ACETYLDOPAMINE

Indirect structural information on sclerotized insect cuticle has been obtained so far from the analysis of products that were isolated after complete or partial destruction of biological samples by chemical or limited enzymatic degradation [6,8]. We have recently applied a rather new, non-destructive methodology, namely solid state CP/MAS-^{13}C-NMR spectroscopy to the analysis of native sclerotized insect cuticle (Fig. 2) [15]. A solid state ^{13}C-NMR spectrum of chitin from larvae of the tobacco hornworm, *Manduca sexta*, is shown in Fig 3. The signals of the carbon atoms in N-acetyl-glucosamine moieties are easily identified in the spectrum of the cuticle, and a full assignment of the spectra has been published [15]. Here, we shall focus on the spectral region of aromatic carbon atoms between 110 and 160 ppm in more detail.

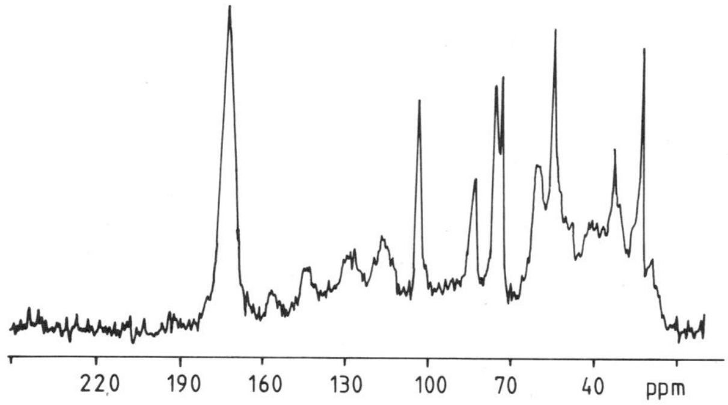

Fig. 2. CP/MAS-^{13}C-NMR spectrum of exuviae from *M. sexta* pupae (15).

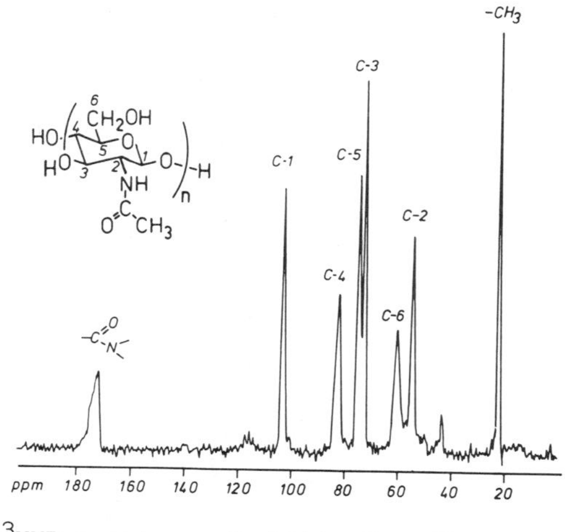

Fig. 3. CP/MAS-^{13}NMR-spectrum of chitin from *M. sexta* larvae (15).

Structural elements containing aromatic carbon atoms are derived in sclerotized insect cuticle from aromatic amino acids and from sclerotization agents. ^{13}C-NMR signals arising from the former appear at 155 and 130-115 ppm, as has been demonstrated by a corresponding ^{13}C-NMR spectrum of bovine serum albumin [15]. The ^{13}C resonances of N-acetyldopamine ring carbon atoms are shown in Fig. 4, where the two diphenolic phenoxy carbon atoms appear at 145.0 and 144.2 ppm. When, in a model system, N-acetyldopamine is oxidized by means of mushroom tyrosinase (monophenol monooxigenase, EC 1.14.18.1) in the presence of β-alanine methyl ester as a potential amino group donor, a complex water soluble product mixture is formed. Partial resolution by chromatography on a Sephadex LH-20 column yields fractions of polymeric mixtures that account for ca. 85% of total products [11]. Fig. 5 shows the region of aromatic and carbonyl carbon atoms in a ^{13}C-NMR spectrum of such polymers in solution. There are notable similarities with the corresponding region in the solid state ^{13}C-NMR spectrum of sclerotized insect cuticle shown in Fig. 2. The presence of diphenolic phenoxy carbon atoms is proven by the group of signals between 140 and 150 ppm in both, sclerotized cuticle and model system. We thus may reasonably assume that the major oxidation products are principally the same in vitro and in vivo. It should be noted that no evidence is obtained for the presence of Schiff's base type partial structures, neither in the cuticle nor in the model system.

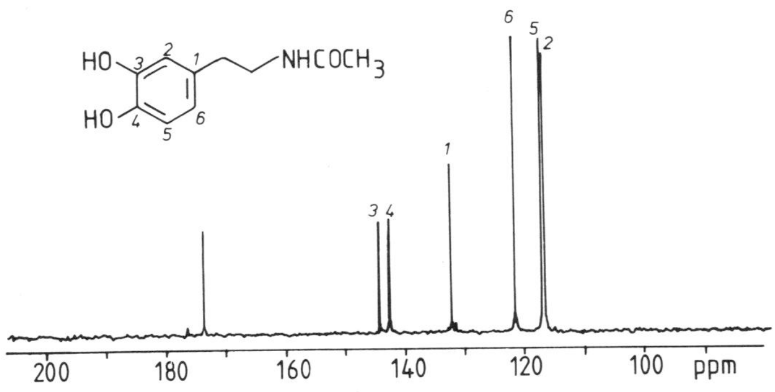

Fig. 4. ^{13}C-NMR spectrum of N-acetyldopamine in D_2O-5% CD_3OD. Only aromatic and carbonyl carbon atoms are shown.

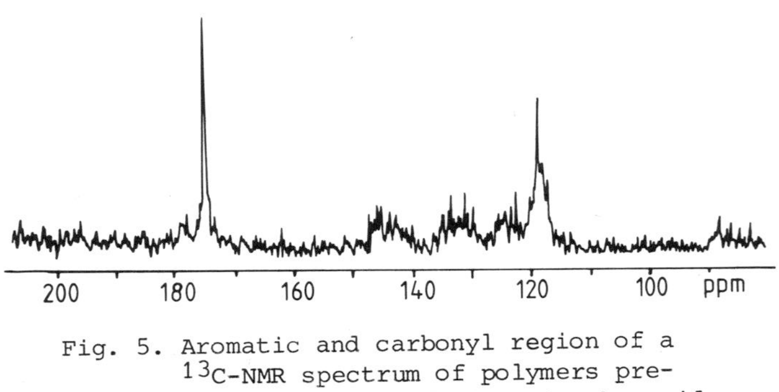

Fig. 5. Aromatic and carbonyl region of a ^{13}C-NMR spectrum of polymers prepared from N-acetyldopamine by oxidation with tyrosinase and separation on Sephadex LH-20. See [11] for details.

Table 1. Binding of oxidation products of N-acetyldopamine to chitin and to sclerotized cuticle from M. sexta pupae.

initial concentration of N-acetyldopamine (mM)	incorporation of N-acetyl-dopamine into bound products ($nmol \cdot mg^{-1}$)	
chitin from crab shells		
0.6	85 ± 8	(± S.D., n=3)
1.1	120	
2.0	218 ± 19	(± S.D., n=3)
M. sexta cuticle		
0.08	24 ± 0.5	(± range, n=2)
0.33	81 ± 5.0	(± range, n=2)
0.82	125 ± 10	(± range, n=2)
9.4	840	

Chitin from crab shells (5.7 - 12.6 mg) or cuticle from M. sexta pupae [15] (15.3 mg) was incubated with appropriate amounts of $|\beta-{}^{3}H, \alpha-{}^{14}C|$-N-acetyldopamine in 10 or 15 ml 50 mM PO_4 buffer pH 6.8. The reactions were started in the experiments with chitin by addition of 0.1 mg (100 U) tyrosinase. Sclerotized insect cuticle contains intrinsic diphenoloxidase activity. The mixtures were stirred at 22° C for 18 h, and counted after centrifugation.

BINDING OF OXIDATION PRODUCTS OF N-ACETYLDOPAMINE TO CHITIN

When N-acetyldopamine is oxidized by means of mushroom tyrosinase in the presence of chitin, the insoluble polysaccharide acquires initially a reddish colour which changes gradually into brown. A quantitation of the binding of the oxidation products has been made by using radiolabelled N-acetyldopamine, and measuring the decrease in the concentration of soluble radiolabelled material by liquid scintillation counting.

Table 1 shows the results from experiments with commercially available chitin from crab shells. For comparison, analogous data from the binding of N-acetyldopamine oxidation products to sclerotized cuticle of M. sexta pupae are included in Table 1. The time course of the reaction reveals a rapid initial binding of 50-70% of the total within 20 min after addition of tyrosinase, followed by a gradual increase up to the final amount (data not shown). The specific incorporation, calculated as nmol N-acetyldopamine bound in the form of oxidation products per mg chitin or cuticle, respectively, is roughly proportional to the initial concentration of N-acetyldopamine. It is interesting to note that we have also recorded a solid state ^{13}C-NMR spectrum of chitin from crab shells, and found that it is indistinguishable from that of chitin from M. sexta larvae, shown in Fig. 3.

N-acetyldopamine does not bind detectably to chitin in the absence of tyrosinase, nor do preformed polymeric oxidation products of the diphenol. Therefore, binding requires the generation of reactive intermediates in situ. However, the following experiments show that the association of chitin with the in situ generated oxidation products is essentially non-covalent.

Doubly labelled polysaccharide-polyphenol complexes were prepared from $[\beta-{}^{3}H]$-N-acetyldopamine and $[U-{}^{14}C]$-chitin in principally the same way as described in Table 1. However, binding of polyphenol was calculated from

Table 2. Binding of oxidation products of [β-^{3}H]-N-acetyldopamine to [U-^{14}C]-chitin

initial concentration of N-acetyldopamine (mM)	incorporation of N-acetyl-dopamine into bound products (nmol·mg^{-1})
0.74	31
1.48	47
2.95	58

the amount of ^{3}H label remaining associated with the chitin after extensive washing with water. Digestion with chitinase then liberated ^{3}H and ^{14}C radioactivity from the doubly labeled complexes into solution, and the stoichiometry of binding was calculated from these data. The results are shown in Table 2. The binding to [U-^{14}C]-chitin is lower than that to chitin from crab shells by a factor of ca. 4 which may be due to the removal of adsorbed polyphenol by the washing procedure and to the different chitin preparations.

The time course of the liberation of ^{3}H and ^{14}C radioactivity into solution is depicted in Fig. 6. Initially, the appearance of soluble ^{3}H labelled material is faster than that of ^{14}C from chitin hydrolysis. After this initial period, the reaction shows a linear course for both labels. Hydrolysis of chitin was 88.2 ± 3.2 % of the expected total after 10 h. The kinetics of chitinase catalyzed hydrolysis was the same with all doubly labelled complexes, and also with a sample of [U-^{14}C]-chitin that had not been treated with N-acetyldopamine under oxidative conditions. Thus, binding of polyphenols to chitin does not result in an inhibition of chitinase in these particular cases.

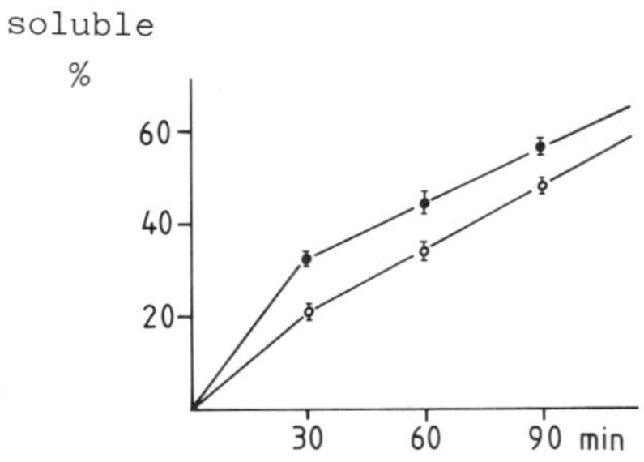

Fig. 6. Chitinase catalyzed hydrolysis of doubly labelled polyphenol-chitin complexes. Filled circles: ^{3}H from polyphenol; open circles: ^{14}C from chitin. Incubations contained the preparations listed in table 2, and 5.5 mg chitinase in 2 ml 50 mM KPO_4 buffer pH 5.0. The mixtures were stirred magnetically at 22°C. Soluble radioactivity was determined by counting aliquots of the centrifuged mixtures.

If covalent bonds between oxidation products of N-acetyldopamine and chitin formed, doubly labelled products would be expected after hydrolysis with chitinase. Therefore, an aliquot containing 173000 dpm ^3H and 85500 dpm ^{14}C was subjected to chromatography on a silica gel column (fig. 7). Recovery of ^3H was 28160 dpm (16.3%) and of ^{14}C 85100 dpm (99.6%). A large quantity of polyphenols was apparently strongly adsorbed on the silica gel, whereas the chitin degradation products were recovered nearly quantitatively. Furthermore, since the peak center of ^3H was separated from that of ^{14}C, it is estimated that maximally 5% of the bound polyphenols might be associated covalently with chitin. This figure is probably an overestimation, since the products eluting from the column are still impure. Similar results were obtained from a chromatographic analysis on a µ-Bondapak AX ion exchange column, and further separations were not attempted so far.

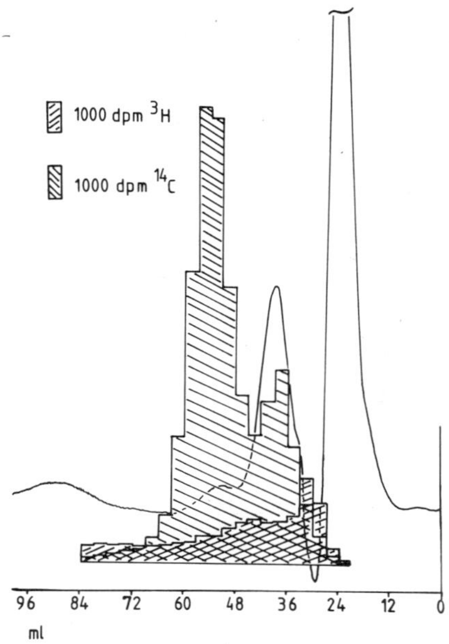

Fig. 7. Chromatography of products from digestion of doubly labelled polyphenol-chitin complex with chitinase. Conditions: 0.7 x 62 cm SiO_2 column (10-40 µm; Macherey-Nagel); mobile phase: ethanol-H_2O-CH_3CN-27% NH_3 = 27-11-61-0.1; flow 3 ml·min^{-1}. The line shows the response of a differential refractive index detector. Fractions of 3 ml were collected and counted.

CONCLUSIONS

The results of this study show that <u>in situ</u> generated oxidation products of N-acetyldopamine bind to chitin by noncovalent interactions. The complex of polyphenol and chitin is obviously stable only as a macromolecular assembly. We may conclude that the same type of binding applies to the interaction of oxidation products of sclerotization agents with chitin in the stabilization of the exoskeleton of insects. Covalent bonds between polyphenols and free amino or hydroxy groups in proteins or chitin are not formed in detectable amounts. There is, however, the possibility that a copolymerization of diphenols and tyrosine residues of proteins occurs, and that this type of reaction is of fundamental importance [16,17]. It becomes increasingly clear that the model describing sclerotization as a result of crosslinking of cuticle proteins *via* free amino groups does not really apply to the facts.

n interesting application of a more practical importance is the introduction of solid state NMR spectroscopy as an analytical tool for the investigation of a complex insoluble matrix. This non-destructive method should allow a rapid and rather simple quantitative determination of chitin in biological samples.

ACKNOWLEDGEMENT

This work was supported by grants of the Deutsche Forschungsgemeinschaft to M.G.P. (Pe 264/4-1) and to R.K.

REFERENCES

1. A.C. Neville, "Biology of the Arthropod Cuticle", Springer Verlag, Berlin (1975).
2. R.H. Hackman, Chemistry of the Insect Cuticle, in: M. Rockstein, ed., "The Physiology of Insecta", Vol. 6, p. 215, Academic Press, New York (1974).
3. P. Karlson and C.E. Sekeris, N-Acetyldopamine as Sclerotization Agent of the Insect Cuticle, Nature 195: 183 (1962).
4. P. Karlson and C.E. Sekeris, Control of Tyrosine Metabolism and Cuticle Sclerotization by Ecdysone, in: H.R. Hepburn, ed., "The Insect Integument", p. 145, Elsevier, Amsterdam (1976).
5. T.L. Hopkins, T.D. Morgan, Y. Aso, and K.J. Kramer, N-β-Alanyldopamine: Major Role in Insect Cuticle Tanning, Science 217: 364 (1982).
6. S.O. Andersen, Biochemistry of Insect Cuticle, Ann. Rev. Entomol. 24: 29 (1979).
7. P.C.J. Brunet, The Metabolism of Aromatic Amino Acids Concerned in the Crosslinking of Insect Cuticle, Insect Biochem. 10: 467 (1980).
8. H. Lipke, M. Sugumaran, and W. Henzel, Mechanism of Sclerotization in Diptera, Advan. Insect Physiol. 17: 1 (1983).
9. R.H. Hackman and M. Goldberg, Molecular Crosslinks in Cuticles, Insect Biochem. 7: 175 (1977).
10. J.F.V. Vincent and J.E. Hillerton, The Tanning of Insect Cuticle: A Critical Review and a Revised Mechanism, J. Insect Physiol. 25: 653 (1978).
11. M.G. Peter, Products of in vitro Oxidation of N-Acetyldopamine as Possible Components in the Sclerotization of Insect Cuticle, Insect Biochem. 10: 221 (1980).
12. R. Lang, Chitinsynthese bei dem Flußkrebs Orconectes limosus: Aktivität der Phosphoglucosamin Isomerase und Einbau von [U-^{14}C]-Glucose in Chitin, Z. Vergl. Physiol. 73: 305 (1971).
13. M. Lunt and P.W. Kent, A Chitinase System from Carcinus maenas, Biochim. Biophys. Acta 44: 371 (1960).
14. A. Willig and R. Keller, Molting Hormone Content, Cuticle Growth and Gastrolith Growth in the Molt Cycle of the Crayfish, Orconectes limosus, J. Comp. Physiol. 86: 377 (1973).
15. M.G. Peter, L. Grün and H. Förster, CP/MAS-^{13}C-NMR Spectra of Sclerotized Insect Cuticle and of Chitin, Angew. Chem. Int. Ed. Engl. 23: 638 (1984).
16. L. Grün and M.G. Peter, Incorporation of Radiolabelled Tyrosine, N-Acetyldopamine, N-β-Alanyldopamine, and the Arylphorin Manducin into the Sclerotized Cuticle of Tobacco Hornworm (Manduca sexta) Pupae, Z. Naturforsch. C 36: 1066 (1984).
17. L. Grün and M.G. Peter, Selective Crosslinking of Tyrosine Rich Larval Serum Proteins and of Soluble Manduca sexta Cuticle Proteins by Nascent N-Acetyldopamine Quinone and N-β-Alanydopamine Quinone, in: K. Scheller, ed., "The Larval Serum Proteins of Insects", p. 102, Thieme Verlag, Stuttgart (1983).

CHITIN-PROTEIN MOLECULAR ORGANIZATION IN ARTHROPOD

M. M. GIRAUD-GUILLE and Y. BOULIGAND

E.P.H.E. & C.N.R.S.,
62, rue Maurice Günsbourg
F-94200 Ivry-sur-Seine, France

INTRODUCTION

Many fibrous biological structures show a supramolecular organization geometrically comparable to a "twisted plywood"[1]. This arrangement is similar to that of molecules in cholesteric* liquid crystals, but the liquid character is often abolished by the presence of molecular crosslinks. Liquid crystals were first described in cholesterol esters, and this explains the term cholesteric given to this particular state of matter.

The best example of such cholesteric liquid crystal analogues is the chitin-protein matrix of arthropod cuticles[1]. However examples of this state of matter are actually known in other biological structures made of quite different polymers, such as nucleic acids[2], collagen[3], cellulose[4].

The question therefore arises of the origin of the cholesteric twist. Biological structures showing the geometry of continuously twisted plywoods are often composed of one major polymer having a helical configuration, as DNA in certain procaryote chromosomes[5], or collagen in many extracellular matrices[3]. In this case a model due to Rudall[6] can be considered, that shows how the packing of helical polymers in concentrated solution gives rise to an architecture comparable to a twisted plywood.

In arthropod cuticles two different kinds of polymers coexist in the matrix, a polysaccharide and proteins. To understand the origin of the cholesteric twist present in these skeletal structures the mutual arrangement of chitin and protein must be known. We present here ultrastructural observations obtained on the decalcified crab cuticle of *Carcinus maenas* that lead us to discuss possible models of the chitin-protein molecular organization in arthropod cuticles.

ULTRASTRUCTURAL OBSERVATIONS

The fully grown crab cuticle is composed of four distinct superimposed layers. The outer epicuticle, a lipoproteic coating, will not be considered in this study. The three following layers form the procuticle[7] whose general structure is stratified. As already said the organic fibrils forming the procuticle are composed of chitin crystallites and proteins. The epicuticle and the pigmented layer are preecdysial layers; the principal layer and the membranous layer are post-ecdysial layers. Calcification occurs after ecdysis by impregnation of calcite in the pre- and post-ecdysial cuticle, but we will consider in the following observations the organic matrix after demineralization.

Various aspects at low resolution

All observations are made on specimens routinely processed for electron microscopy. Thin sections are contrasted by uranyl acetate and lead citrate.

The three layers of the procuticle when viewed in oblique section show superimposed series of nested arcs (Fig. 1 : a, c, e, g); this means that the cholesteric spatial organization is present in all the chitin-protein layers. However differences appear due to variations in the lamina thickness, and variations in the grouping of microfibrils[8]. We call microfibril the unit composed of a clear rod and the adjacent opaque material as observed at high magnification in the electron microscope.

The thickness of one lamina is the half-helicoïdal pitch of the structure and results from a 180° rotation of the microfibril direction. This pitch varies from 0.1 to 5 µm in small crabs[8]. The chitin-protein system presents various textures at different levels of the procuticle. The microfibrils can be grouped into fibrils of different sizes; in the upper half of the exocuticle the fibril diameters range from 10 to 25 nm (Fig. 1a), and in the endocuticle from 30 to 100 nm (Fig. 1e). A progressive change from dispersed microfibrils to a reticulate structure appears in the lower half of the exocuticle (Fig. 1c). In the membranous layer the structure is dense with few free spaces (Fig. 1g). This last situation resembles that of insect cuticles.

Uniform alveolar pattern at high resolution

Perfectly transverse ultrathin sections show at high resolution a uniform alveolar pattern at all levels of the crab procuticle[8] (Fig. 1 : b, d, f, h). This pattern consists of rods transparent to electrons, surrounded by a dark matrix. The average diameter of the lucent rods is 3 nm and the mean distance between two rod axes is 4,6 nm. The contrast and regularity of this alveolar structure, forming more or less extended hexagonal arrays, is however better in the preecdysial layers of the crab than in the post-ecdysial ones. The alveolar pattern is limited to small domains that further extend in opposite images of dark filaments. Control by focus series shows that this inversion of images is not due to phase contrast reversal. This transformation is the result of the progressive tilting of microfibrils in the section plane, and is therefore a consequence of the twisted fibrous organization.

This pattern, where lucent rods in cross sections form semi-hexagonal arrays, was often observed in arthropod exocuticles. The clear rods are classically assumed to represent the chitin crystallite, and the dark phase proteins.

INTERPRETATION OF THE ALVEOLAR PATTERN

Anterior works

Rudall[9] first proposed from the study in electron microscopy of the ovipositor of Megarhyssa, that clear rods were made of chitin, whereas the dark cement represented the associated proteins.

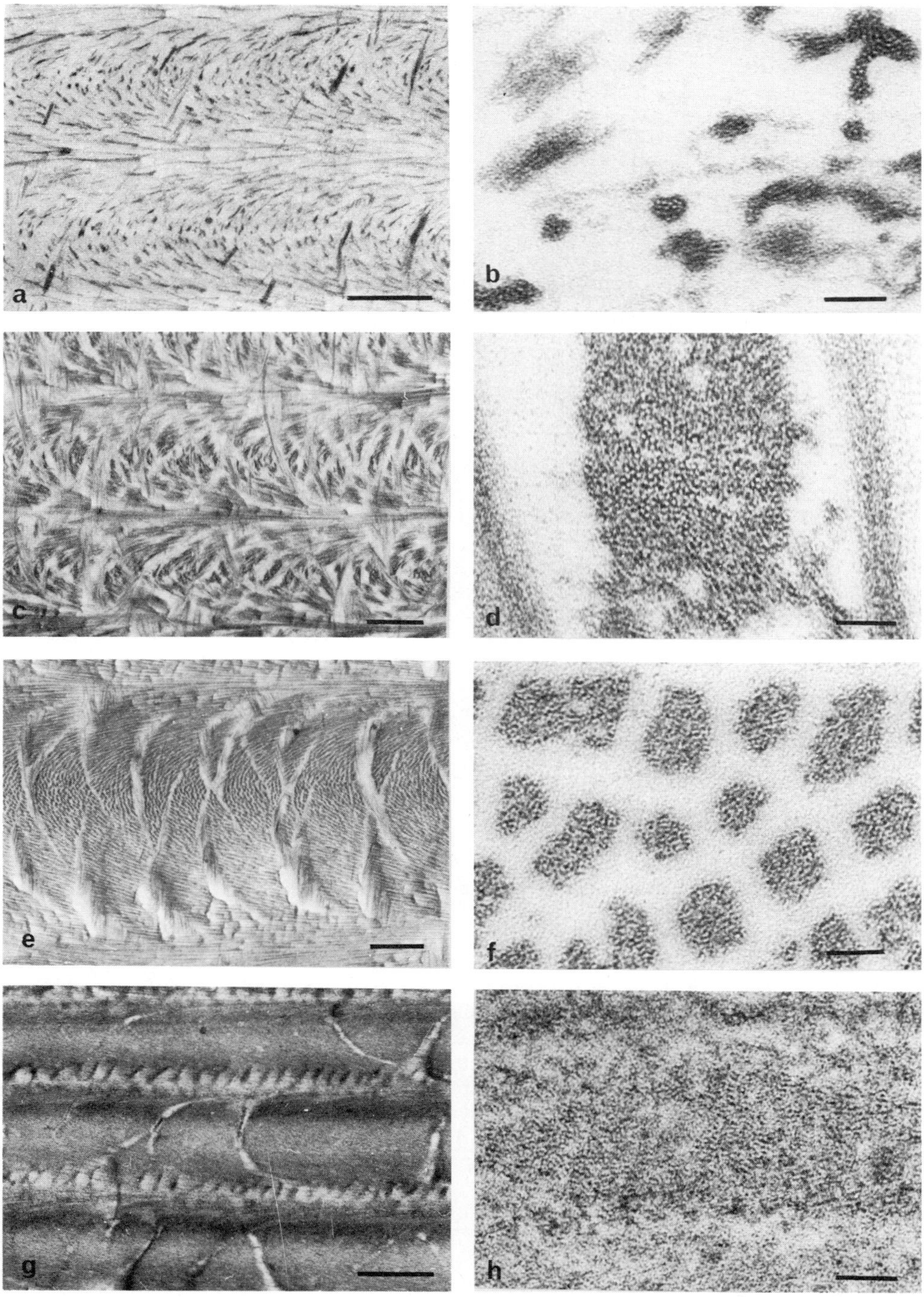

Fig. 1. Ultrastructural observations. At low resolution in transmission electron microscopy (on the left), four different levels of the crab procuticle show a constant cholesteric organization, but various groupings of microfibrils. At high resolution (on the right), the corresponding regions show a uniform alveolar pattern. a, b : upper pigmented layer. c, d : lower pigmented layer. e, f : principal layer. g, h : membranous layer. a, c, e, g : bar = 1 µm. b, d, f, h : bar = 40 nm.

Neville[10] generalized this interpretation to arthropod cuticles arguing that chitin is a dense crystallite and a neutral polymer, therefore not stainable for electron microscopy. He also indicated that proportions of dark and clear materials in the examined micrographs were in good agreement with the chitin-protein ratio.

Filshie[11] suspected this interpretation on the view of tick cuticles presenting the same alveola structure when the chitin content is lower than 4 %.

We presented similar objections from the study of the crab cuticle[8]; large changes in the chitin-protein estimations are found between the different layers[12], whereas a constant alveolar pattern is seen at all levels of the procuticle. Possible extraction of proteins by decalcifying agents, as suggested by Hackman[13], does not modify the fact that the postecdysial layers contain 30 % more chitin than the preecdysial ones.

Staining of chitin

Many questions remain about the ability of stains to differentiate chitin and proteins in the fine structure of arthropod cuticles.

Chitin appears electron dense after uranyl acetate and lead citrate staining in thin sections of deproteinized crab cuticle[8] and in the periostracum of a whelk[14]. The removal of proteins by hot alkali could partially deacetylate chitin and confer charges to the polymer, but *in vitro* thin sections of pellets of purified chitin also appear electron dense[8]. Conversely, cuticular proteins can appear very poorly contrasted after classical staining, for example, resilin in the pleural cuticle of a flea[15].

Hexagonal order and cholesteric packing

In certain situations the cholesteric twist is not present in the cuticle and all chitin crystallites lie parallel. It is the case of *Megarhyssa* ovipositor which provides the best conditions for X-ray diffraction and ultrastructural studies. The chitin-protein system observed in this material shows a very clear hexagonal pattern, the diameter of the clear rods reaching 6.6 nm.

In regularly twisted materials, as in the crab cuticle, this order is much more difficult to observe because perfectly transverse sections of microfibrils are limited to very small regions. Furthermore liquid crystalline cholesteric order in polymers is compatible only with a short range hexagonal order[16]. Therefore the continuously twisted plywood structure will tend to disorganize the hexagonal arrangement.

New hypothesis

We believe that the classical interpretation of the chitin nature of transparent rods is questionnable. The principal argument, beyond the problem of the chitin crystallite staining, is the contradiction between the presence of an alveolar pattern in different species and different regions of arthropod cuticles where the chitin content varies greatly.

We propose a different interpretation of the two phase hexagonal pattern observed in cross sections of arthropod cuticles. Two phase systems often result from the segregation of incompatible molecules in solutions. X-ray diffraction studies in liquid crystalline systems have shown that, among other arrangements, hexagonal networks can arise from such a phase segregation, for example in lipid-water systems[17,18] or in block copolymers[19,20]. Thin sections of such polymerized systems show hexagonal

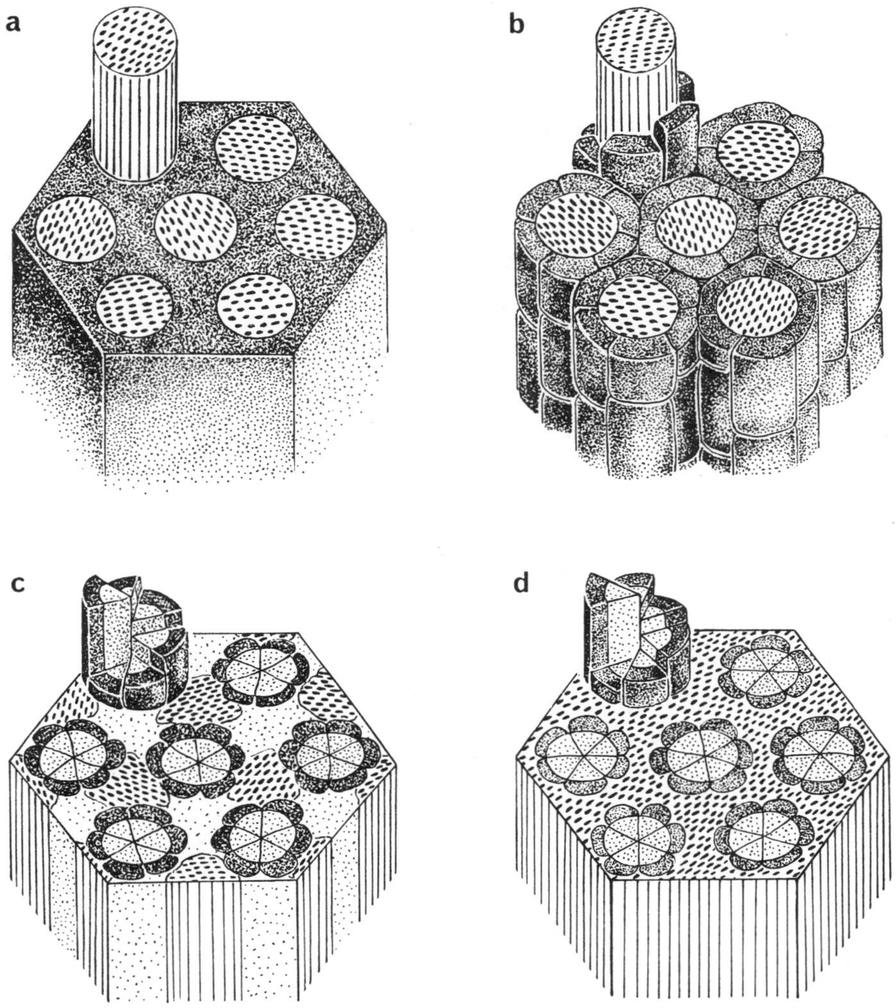

Fig. 2. Different cuticle models. a) Parallel chitin crystallites cemented with proteins; b) parallel chitin crystallites surrounded by helically arranged proteins[21]; c) polar and non polar parts of proteins segregate; clear rods are supposed to be formed by polar aminoacids and electron dense alveolar walls are supposed to contain polar aminoacids and separated chitin crystallites; d) model slightly different from that proposed in c. The chitin lattice is connected and more or less regularly ordered.

arrays of clear rods embedded in a dark matrix (or inversely).

In arthropod cuticles, which are liquid crystalline analogues, such phase segregations could also occur. The proteins made of polar and non polar amino acids may well separate between different regions, the hydrophilic parts being contrasted by electronic stains, and the hydrophobic parts appearing clear. The remaining question in this model is the position of the chitin crystallites.

POSSIBLE CUTICLE MODELS

Fig. 2a corresponds to the model proposed by Neville[10] from ultrastructural observations of insect cuticles. The clear rods are assumed to represent the chitin crystallites surrounded by disordered proteins, opaque to electrons.

Fig. 2b corresponds to the model proposed by Blackwell and Weih[21] from X-ray diffraction studies on Megarhyssa ovipositor. Two interesting facts can be ascertained from the fiber diagrams of intact and deproteinized cuticles; small angle reflections on the equator are indexed by an hexagonal network, and reflections attributed to proteins are suggestive of a helical arrangement. The authors propose a chitin-protein microfibril structure consisting of a crystallite chitin core surrounded by a sheath of ordered proteins. In this model they adopt the classical interpretation of the chitin nature of transparent rods.

We have made the hypothesis that the hexagonal array observed on electron micrographs is due to a phase segregation between hydrophylic and hydrophobic parts of cuticular proteins. Covalent links between chitin and proteins occur via aspartic acid[13], a polar amino-acid, and this suggests that chitin is incorporated in the dark phase. We propose two possible models. In Fig. 2c the chitin crystallites are independant whereas in Fig. 2d the chitin lattice is connected. Equatorial reflections visible on the X-ray diffraction patterns of chitin are sharp enough to consider that eventually the crystallites form a connected network. Chitin is generally considered as forming clear rods embedded in a dark protein matrix, but this is not easily compatible with strong variations of the chitin content, and a geometrically invariable alveolar pattern. A new hypothesis is proposed and is shown to be compatible with X-rays diffraction data. The contrast would be due mainly to a segregation of polar and non polar residues of amino-acids. Chitin and polar parts of proteins form the dark honeycomb pattern. More studies are necessary to know whether or not the chitin crystallites are separated units or belong to a connected lattice.

ACKNOWLEDGMENT

We thank Dr. A.M. Levelut for stimulating discussions.

REFERENCES

1. Y. Bouligand, Twisted fibrous arrangements in biological materials, and cholesteric mesophases, Tissue & Cell, 4:189 (1972).
2. Y. Bouligand, M. O. Soyer, and S. Puiseux-Dao, La structure fibrillaire et l'orientation des chromosomes chez les Dinoflagellés, Chromosoma, 24:251 (1968).
3. Y. Bouligand and M. M. Giraud-Guille, Spatial organization of collagen fibrils in skeletal tissues : Analogies with liquid crystals, in : "Biology of Invertebrate and lower vertebrate collagens", A. Bairati and R. Garrone ed., Plenum publishing corp., New-York (in press).
4. J. C. Roland and B. Vian, The wall of the growing plant cell : its three dimensional organization, Int. Rev. Cytol., 61:129 (1979).
5. F. Livolant and Y. Bouligand, New observations on the twisted arrangement of Dinoflagellate chromosomes, Chromosoma, 68:21 (1978).
6. K. M. Rudall, Protein ribbons and sheets, Lectures on the scientific basis of medicine, V:217 (1956).
7. A. G. Richards, in : "The Integument of Arthropoda", Univ. Minessota press, ed. Minneapolis (1951).
8. M. M. Giraud-Guille, Fine structure of the chitin-protein system

in the crab cuticle, *Tissue & Cell*, 16:75 (1984).

9. K. M. Rudall, Skeletal structures in insects, *Biochem. Soc. Symp.*, 25:83 (1965).
10. A. C. Neville, *in* : "Biology of the Arthropod Cuticle", D. S. Farner ed., Springer-Verlag, Berlin (1975).
11. B. K. Filshie, Fine structure of the cuticle of insects and other arthropods, *in* : "Insect Ultrastructure", R. C. King and M. Akai ed., Plenum publishing corp., New-York, 1:281 (1982).
12. M. Lafon, Nouvelles recherches biochimiques et physiologiques sur le squelette tégumentaire des crustacés, *Bull. Inst. Océan.*, 939:1 (1948).
13. R. H. Hackman, Cuticle : Biochemistry, *in* : "Biology of the Integument", Invertebrates, J. Bereiter-Hahn et al. ed., Springer-Verlag, Berlin, 1:583 (1984).
14. S. Hunt and K. Oates, Chitin helicoïds accompany protein helicoïds in the periostracum of a whelk *Buccinum*, *Tissue & Cell*, 16:565 (1984).
15. A. C. Neville, Cuticle : organization, *in* : "Biology of the Integument", Invertebrate, J. Bereiter-Hahn et al. ed., Springer-Verlag, Berlin, 1:611 (1984).
16. E. T. Samulski and A. V. Tobolsky, Cholesteric liquid crystals formed by certain polypeptides with organic solvents, *in* : "Liquid crystals and plastic crystals", G. W. Gray and P. A. Winsor ed., Halsted press, Chichester, 1:175 (1974).
17. V. Luzzati and F. Husson, The structure of the liquid crystalline phases of lipid-water systems, *J. Cell Biol.*, 12:207 (1962).
18. W. Stoeckenius, Some electron microscopical observations on liquid-crystalline phases in lipid-water systems, *J. Cell Biol.*, 12: 221 (1962).
19. J. Dlugosz, A. Keller, and E. Pedemonte, Electron microscope evidence of a macroscopic "single crystal" from a three block copolymer, *Kolloid. Z. Polymere*, 242:1125 (1970).
20. B. Gallot, Liquid crystalline structure of block copolymers, *in* : "Liquid crystalline order in polymers", A. Blumstein ed., Acad. Press, New-York, 191 (1978).
21. J. Blackwell and M. A. Weih, Structure of chitin-protein complexes ovipositor of the ichneumon fly *Megarhyssa*, *J. Molec. Biol.*, 137:49 (1980).

PORE CANALS AND ORGANIZATION OF CHITINOPROTEINS IN THE CUTICLE OF THE CRAB CARCINUS MAENAS

G. GOFFINET and P. COMPERE

Laboratory of Animal Morphology, University of Liége
22, quai Van Beneden
B-4020 Liége, Belgium

STRUCTURE AND COMPOSITION OF THE CUTICLE OF THE DECAPOD CRUSTACEANS

The structure of the calcified exoskeleton of decapod crustaceans consists of four distinct layers according to their sequence of deposition and their chemical composition (1).

Typically the procuticle (pigmented layer, principal layer and membranous layer) is formed of stacked laminae composed of fiber beds in which the chitin-protein microfibrils run parallel to the surface and undergo a slight and progressive change in orientation from one bed to the next (twisted plywood organization of Bouligand (2)). Microfibrils constitute the basic structural units of the procuticle and are made up of a core of chitin cristallite embedded in a protein sheath (3,4). The chitinous fibrils arise at the surface of plasma membrane dense plaques. The latter are located at the tip of short microvilli of the apical membrane of the epidermis and are considered to be responsible for the changing orientation of the chitin-protein microfibrils that result in laminae (5). At a higher level of organization, these microfibrils are associated into macrofibers in the principal layer, into reticulate structures in the pigmented layer (6). In the membranous layer (as well as in the non-mineralized procuticles) they are homogeneously distributed.

The laminae of the cuticle are crossed by vertical fibers closely associated to cuticular canals, or pore canals, which extend from the apical plasma membrane of the epidermis to the inner epicuticle. The organization of this canalicular system, that must be distinguished from the pore canal filaments (or muscle attachment fibers), varies to a precise schedule during the moult/intermoult cycle.

ULTRASTRUCTURE AND EVOLUTION OF THE PORE CANAL SYSTEM OF THE CUTICLE OF CARCINUS MAENAS DURING THE MOULT/INTERMOULT CYCLE

When the edification of a new cuticle takes place, the pore canals are composed of two distinct elements : microvilli and vertical fibers. The evolution of these two components during the moult cycle is quite different.

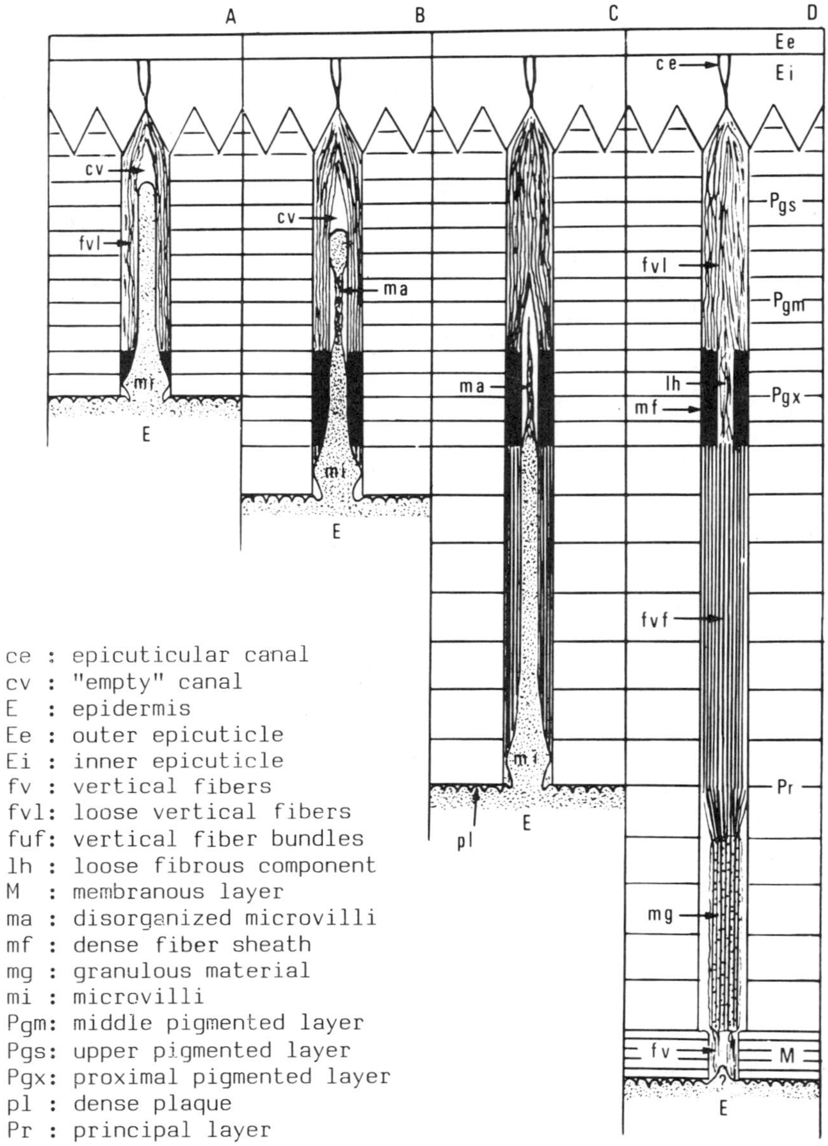

Fig. 1. Diagrammatic section of Carcinus maenas cuticle showing the evolution of the cuticular canals during the moult/intermoult cycle.

A. Cuticle at the end of the preecdysial period (stage D2). The microvilli reach the upper laminae of the pigmented layer.

B. Beginning of the postecdysial period (stage A2). The microvilli retract from the upper laminae of the pigmented layer while they become loose in the upper area.

C. Stage C1. The disorganization of the microvilli progresses through the pigmented layer up to the principal layer. Vertical fibers originating from the sheath are taking up the canal lumen at the middle and upper levels of the pigmented layer.

D. End of the postecdysial period (stage 3C). The deposition of the membranous layer occurs. Vertical fibers fill the overall cuticular canals that are devoid of any cytoplasmic extension.

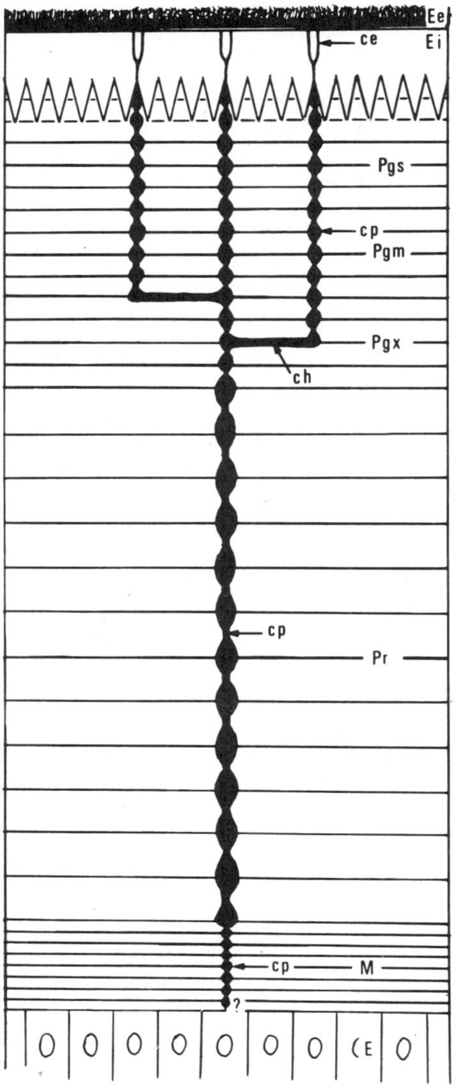

Fig. 2. Diagrammatic view of the organization of the canalicular system in the outer wall of the branchiostegal cuticle of Carcinus maenas at stage C4.

ce : epicuticular canal
ch : horizontal procuticular canal
cp : vertical procuticular canal
E : epidermis
Ee : outer epicuticle
Ei : inner epicuticle
M : membranous layer
Pg : pigmented layer
Pr : principal layer

A. Microvilli

As soon as the first preecdysial laminae are laid down, the procuticle is penetrated up to the basis of the inner epicuticle by tubular cytoplasmic extensions 0.2 to 0.3 μm in diameter originating from the epidermis (fig.3). While the deposition of the preecdysial procuticle is going on (stage D2-D3) (fig. 1A), some microvilli retract from the most distal laminae, leaving short hollow canals facing the epicuticle. In agreement with observations in Orconectes limosus (7) and Uca pugilator (8), these canals enter the innermost region of the epicuticle and extend up to the cuticulin by narrow epicuticular canals. The latter, 25 nm in diameter, are often dichotomized. The disorganization of the microvilli begins just after ecdysis, at the middle part level of the pigmented layer. Their retraction and dislocation progress during the further postecdysial stages so that, at the end of stage C3, when the membranous layer is laid down, the procuticle is devoid of any cytoplasmic material (fig. 1B, C and D).

2. Vertical fibers

Whatever the pre- or postecdysial stage considered (except for C3 and C4), the vertical fibers arise at first at the basis of the microvilli (Fig. 3). The plasma membrane of the latter seems to be closely concerned with the deposition of the vertical fibrous components but it never exhibits dense plaques. As soon as they are deposited, the fibers take up a helical orientation and constitute progressively a twisted sheath surrounding the axial microvilli, a profile superimposable on the helical pore canal model described in insect cuticles (9, 10). Following the retraction and the disorganization of the microvilli, some fibers separate from the sheath and gradually invade the lumen of the canals (Fig. 1). The "empty" pattern of a pore canal is thus transitory. At the end of the postecdysial period (C3) and during the intermoult stage (C4), the pore canals appear as lacunar formations, the lumen of which is more or less filled with loose fibrous vertical bundles (Fig. 4).

The presence of vertical fibers in other decapod (7, 8) and in amphipod (11) cuticles suggests that they are a constant feature of the crustacean cuticle organization. Their chemical composition remains however hypothetical. Some observations support the view that they are of chitin-protein composition. At high magnification, when they are perfectly cross-sectioned and after staining with classical contrasting agents, they appear as composite fibrils formed by an electron lucent core about 3 nm in diameter embedded in a dark matrix. On the other hand, they are degraded in the old cuticle during the preecdysial period just at the time they are in contact with the ecdysial fluid which is known to contain chitinases and proteases. Even if the vertical fibers are chitin-protein elements, which remains to be formally demonstrated, the mechanism involved in their deposition seems however to be different than that which regulates the orientation of the horizontal chitinous fibrils, since the plamsa membrane of microvilli are devoid of dense plaques (Fig. 3).

3. Horizontal pore canals

Contrary to the classical opinion, the pore canals project lateral branches (Figs.4 and 6). These horizontal canals, mainly located in the middle and proximal regions of the pigmented layer, run parallel to the chitin protein fibers of the laminae. During the preecdysial period, they are occupied by cellular extensions. After the regression of the microvilli, they are progressively invaded by loose fiber material such as in vertical canals.

4. Spatial organization of the pore canal system

In the principal layer of the branchiostegal cuticle (outer wall) of *Carcinus maenas*, pore canals are roughly 600,000 per mm^2. In the upper laminae of the pigmented layer, this number increases up to 2 millions per mm^2, the mean surface of a canal cross-section decreasing according to an inverse ratio (from 1.9 to 0.51 μm^2). Taking into consideration that pore canals are branched out at the middle and proximal laminae of the pigmented layer and that epicuticular canals are regarded as fine extensions of procuticular canals, the pore canal system of *Carcinus* can be compared to a stag horn-shaped system joining the epidermis with the outermost layers of the cuticle (Fig. 2).

DEGRADATION OF THE OLD CUTICLE COMPONENTS DURING THE PREECDYSIAL PERIOD IN *CARCINUS MAENAS*

Morphologically, the first step (stage Do) of the moulting process is revealed by the detachment of the apical epithelial membrane(apolysis)

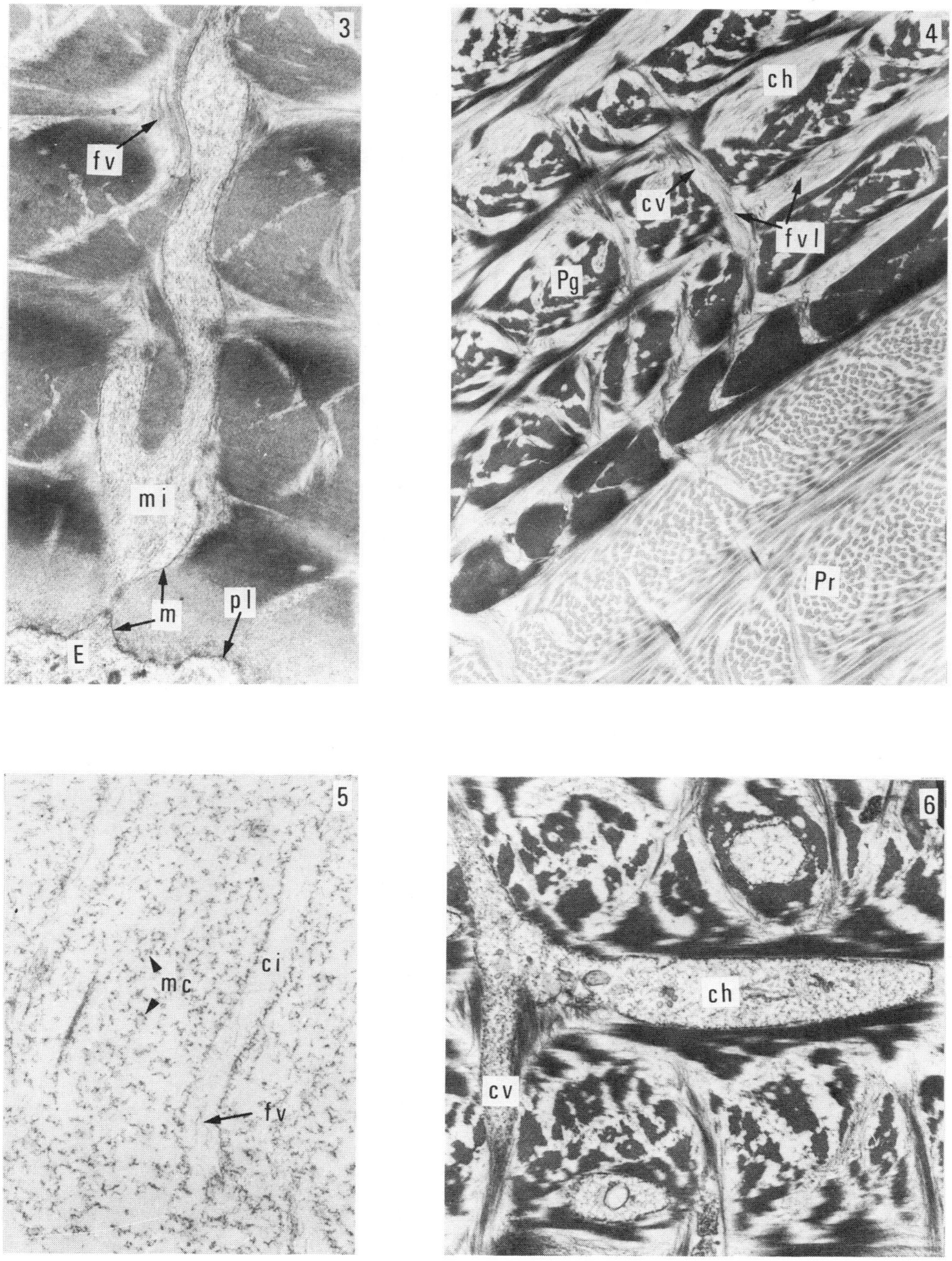

Figs. 3 - 6. Outer wall of the branchiostegal cuticle of *Carcinus maenas* in cross section. Fig. 3. Proximal pigmented layer (stage D2). Microvilli (mi) and vertical fibres (fv) crossing the lamellae. Fig. 4. Proximal pigmented layer (Pg) and principal layer (Pr) (stage C4). Vertical (cv) and horinzontal (ch) canals are filled with loose fibrous component (fvl). Fig. 5. Section across the middle region of the principal layer in the exuvium. Macrofibers (mc) and vertical fibers (fv) are altered. Fig. 6. Middle region of the pigmented layer (stage A2). Horizontal canal (ch) branching from a vertical canal (cv). ci: procuticular canal; E: epidermis; m: cellular membrane; pl: dense plaque.

and the temporary disappearance of the apical dense plaques(5). Before the edification of the new cuticle takes place (stage D1) the epidermis secretes the exuvial fluid containing chitinolytic and proteolytic enzymes. The membranous layer is transformed into a hygroscopic glycoproteic gel which keeps hydrolytic enzymes in store during the preecdysial period (12, 13) . It is accepted that the digestion products are absorbed by the epithelial cells and used for the biosynthesis of components which will be incorporated into the new cuticle. The results referring to the degradation of *Carcinus maenas* cuticles (table 1) suggest that about 50% of chitin and 60% of HCl-soluble proteins are probably recovered by the animal. In contrast, the NaOH-soluble proteins and calcite are not concerned with the hydrolytic activity of the ecdysial fluid. Thus they appear definitively lost with the rejection of the exuvium. These conclusions agree with previous observations on *Cancer pagurus* and *Hemigrapsus nudus* (14).

At the ultrastructural level, the membranous layer displays sinuous membrane-like formations in which neither the lamellate organization, nor the pore canal system are recognizable. In the proximal laminae of the principal layer the macrofibres as well as the vertical fibres are disorganized and locally destroyed (Fig. 5). No obvious changes in the organization of the epicuticle and of the pigmented layer are detected.

Table 1. Degradation of the cuticular components of the carapace of *Carcinus maenas* during the preecdysial period (values expressed in mg/cm^2 of cuticle) (15).

Stages : Component	C4	D2	Exuvium
Ca *	14.2 (± 2.4)	14.7 (± 3.1)	14.7 (± 3.7)
HCl(0.5 N) soluble proteins	0.60(± 0.1)	0.41(± 0.1)	0.24(± 0.1)
NaOH(0.5 N) soluble proteins	1.13(± 0.2)	1.34(± 0.4)	1.00(± 0.3)
Chitin **	3.27(± 1.30)	3.42(± 0.41)	1.76(± 0.64)

(1) established by atomic absorption spectrophotometry
(2) dosed by enzymatic method

CONCLUSIONS

The three-dimensional organization of the chitin protein fibers according to a twisted plywood system confers to cuticles a high resistance against mechanical distortions. The vertical fibers, probably composed of chitino-proteins, associated with the pore canals extend from the basis of the procuticle up to the inner epicuticle. The fact that they are moreover closely imbricated to the lamellate chitin-protein fibrous components leads to believe that they contribute to insure the cohesion between the successive fiber beds of the stacked laminae.

From a physiological point of view, pore canals can be regarded as a continuous connection system between the cuticle and the epidermis, at least during the moulting-time. In *Carcinus* they are probably involved in the cuticle hardening processes (transport of Ca) (16, 17) and in the stiffening of the epicuticle and of the outermost procuticle laminae (transport of polyphenols - Compère and Goffinet, unpubl.). At the beginning of the preecdysial period, the pore canals (microvilli) are in contact with the epicuticle. Thus they constitute routes through which digestion products from the old cuticle can be absorbed by the epithelial cells. In the old cuticle, before ecdysis, they constitute also preferential ways through which hydrolases of the ecdysial fluid can gain access to the procuticle laminae.

ACKNOWLEDGEMENT

P. Compère is reasearch assistant of the "Institut pour l'Encouragement de la Recherche Scientifique dans l'Industrie et l'Agriculture".

REFERENCES

1. P. Drach, Mue et cycle d'intermue chez les Crustacés Décapodes, Ann.Inst.Océan., 19 : 103 (1939).
2. Y.Bouligand, Twisted fibrous arrangements in biological materials and cholesteric mesophases, Tissue Cell, 4(2) : 189 (1972).
3. A.C. Neville, D.A. Parry and J. Woodhead-Galloway, The chitin crystallite in Arthropod cuticle, Tissue Cell, 21 : 73 (1976).
4. J. Blackwell and M.A. Weih, Structure of chitin-protein complexes : ovopositor of the Ichneumon fly Megarhyssa, J. Molec. Biol., 137 : 49 (1980).
5. M. Locke, The role of plasma membrane plaques and Golgi complex vesicles in cuticle deposition during the moult/intermoult cycle, in : "The Insect Integument", H.R. Hepburn edit., Elsevier, 237 (1976).
6. M.M. Giraud-Guille, Fine structure of the chitin-protein system in the crab cuticle, Tissue Cell, 16 (1) : 75 (1984).
7. G. Kümmel, H. Claasen and R. Keller, Zur Feinstruktur von Cuticula und Epidermis beim Flusskrebs Orconectes limosus während eines Häutungszyklus, Z. Zellforsch. Microsk. Anat., 109 : 517 (1970).
8. J.P. Green and M.R. Neff, A survey of the fine structure of the integument of the fiddler crab, Tissue Cell, 4 : 137 (1972).
9. A.C. Neville, M.G. Thomas and Zelany, Pore canal shape related to molecular architecture of Arthropod cuticle, Tissue Cell, 1 : 183 (1969).
10. A.C. Neville and C.W. Berg, Cuticle ultrastructure of a Jurassic Crustacean (Eryma stricklandi), Paleontology, 14 : 201 (1971).
11. K. Halcrow, Modified pore canals in the cuticle of Gammarus (Crustacea Amphipoda). A study by scanning and TEM, Tissue Cell, 169 : 267 (1976).
12. Ch. Jeuniaux, Sur la gélification de la couche membraneuse chez les crabes en mue, Arch.Intern.Physiol. Bioch., 67 : 516 (1959).
13. Ch. Jeuniaux , "Chitine et Chitinolyse", Masson ed., Paris (1963).
14. L.A. Hohnke, Enzymes of chitin metabolism in the Decapod Hemigrapsus nudus, Comp. Biochem. Physiol., 40B : 757 (1971).
15. G. Goffinet, D. Raick and M.F. Voss-Foucart, Degradation of cuticular material in the exoskeleton of Carcinus maenas during the preecdysial period, in preparation.
16. D.R. Roer, Mechanisms of resorption and deposition of calcium in the carapace of the crab Carcinus maenas, J. Exp. Biol., 88 : 205 (1980).
17. M.M. Giraud, Carbonic anhydrase activity in the integument of the crab Carcinus maenas during the intermolt cycle, Comp. Biochem. Physiol., 69 : 381 (1981).

LIPID-LINKED INTERMEDIATES IN CRUSTACEAN CHITIN SYNTHESIS

M. N. HORST

Division of Basic Medical Science
Mercer University School of Medicine
Macon, GA 31207 USA

INTRODUCTION

In the arthropods, chitin is associated with protein [1,2]. Although a structural role for the protein seems clear, a biosynthetic function is also possible, e.g. the protein may serve in the initial formation of a primer molecule. Regarding the origin of chitin synthesis primers, Cabib and co-workers found that partially purified chitin synthetase from *Saccharomyces* does not require an exogenous primer for chitin synthesis to occur.[3] Similar results have been reported by Cohen[4] who studied the *in vitro* synthesis of insect chitin. In both studies, the chitin synthetase preparation appears to contain an endogenous primer. Previous studies in my laboratory have shown that crustacean chitin synthetase will utilize both oligosaccharides of chitin and macromolecular chitin as substrates.[5] The crustacean enzyme is similar to its fungal counterpart in several respects, e.g., antibiotic sensitivity, but is clearly different in lacking protease activation properties and not being stimulated by GlcNAc.

We have formulated the following hypothesis regarding the pathway of chitin synthesis in crustaceans: (1) synthesis of a lipid-linked chitin oligosaccharide occurs on a polyprenol intermediate; (2) the oligosaccharide is transferred *en bloc* to a protein acceptor; (3) the resultant primer is extended by sequential addition of N-acetylglucosamine (GlcNAc) residues via chitin synthetase; and (4) the final product is deposited at the growing cuticle site. This paper will focus on the first step in the pathway, i.e., the formation of lipid-linked intermediates. One might ask, are lipid-linked intermediates involved in other chitin synthesis systems? In fungal systems the answer is no, based upon the studies of Cabib and co-workers.[6] Studies by Quesada-Allue[7] suggest that lipid-linked intermediates are involved in insect glycoprotein and chitin synthesis. Lipid intermediates are also implicated in the biogenesis of other polysaccharide polymers, e.g., glycogen[8] and cellulose.[9]

Previous work on lipid-linked intermediates in crustacean chitin synthesis. We have described an *in vitro* assay for chitin synthesis using crude microsomes prepared from 72h larvae of *Artemia salina*.[10]

Upon incubation of microsomes with UDP-^{3}H GlcNAc, radioactivity is incorporated into 4 fractions: chloroform:methanol (2:1) soluble, chloroform:methanol:water (10:10:3) soluble, 9.5 M urea (or 1% SDS) soluble and the final insoluble residue containing macromolecular chitin. Preliminary experiments show that 70% of the ^{3}H GlcNAc in the urea (or SDS) soluble fraction is cleaved by chitinase; the ^{3}H GlcNAc appears to be associated with a chitoprotein. A major difficulty associated with this system is the concurrent incorporation of labeled substrate into four different fractions; thus, the rate of one reaction may affect levels of substrate available for other reactions. Analysis of C:M (2:1) soluble material by TLC (Fig. 1A) shows two major components which co-chromatograph with dolichol pyrophosphate N-acetylglucosamine (Dol-P-P-GlcNAc) and Dol-P-P-N-acetylchitobiose. Analysis of the C:M:W (10:10:3) soluble extract (Fig. 1B) shows this fraction contains material which remains at the origin of the TLC plate, i.e., it contains oligosaccharide-lipids with three or more carbohydrate units. In a related series of experiments, ^{3}H dolichol phosphate (Dol-P) was prepared and incubated with *Artemia* microsomes plus unlabeled UDP-GlcNAc; samples were extracted and lipid fractions were analyzed by TLC. The results of this experiment were similar to those shown in Fig. 1, indicating that dolichol is glycosylated by *Artemia* microsomes, i.e., dolichol is most likely the carrier lipid. When *in vitro* labeled oligosaccharide lipids soluble in C:M:W (10:10:3) were chromatographed on DEAE-Sephacel, all radiolabeled material was bound to the column and was eluted by 0.1 M ammonium formate; this chromatographic behavior suggests the presence of a pyrophosphate linkage. Mild acid hydrolysis of C:M:W (10:10:3) soluble material allowed quantitative release of radiolabeled material in a water soluble form. Following chromatography on Biogel P-4, three size classes of oligosaccharides were observed, corresponding to octa-, penta- and disaccharides of chitin (Fig. 2A). Upon digestion with affinity purified *Streptomyces* chitinase, all radioactivity migrated as a disaccharide (Fig. 2B). The starting material was resistant to α-mannosidase digestion. These results show that the C:M:W (10:10:3) fraction contains oligosaccharides of chitin attached to dolichol, with a maximum length of 8 GlcNAc residues.

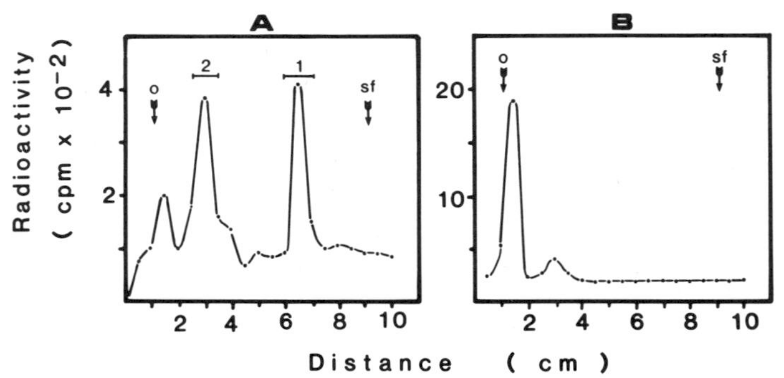

Fig. 1. TLC of ^{3}H GlcNAc-labeled oligosaccharide-lipids on silica gel in chloroform:methanol water (65:35:4). A: Chloroform:methanol (2:1) soluble, B: Chloroform: methanol:water (10:10:3) soluble. Position of standard Dol-P-P-GlcNAc ("1") and Dol-P-P-N-acetyl-chitobiose ("2") are indicated by bars; o:origin; s:solvent front.

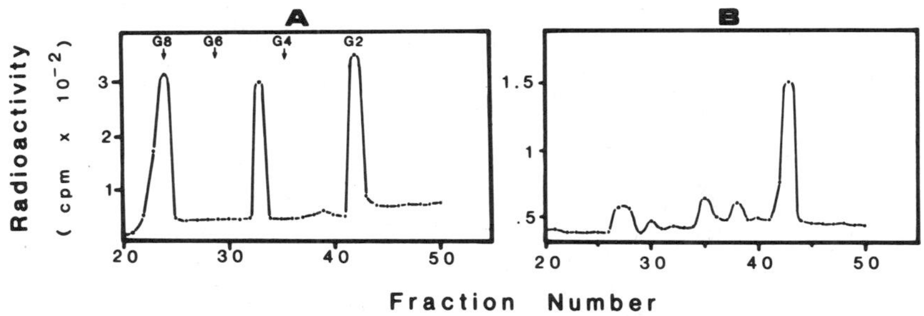

Fig. 2. Gel chromatography of oligosaccharides from C:M:W (10:10:3) soluble material following mild acid hydrolysis. Control ("A") and _Streptomyces_ chitinase digested samples ("B") were chromatographed on a Biogel P-4 column (1x50cm) equilibrated with 25mM acetic acid. Positions of standard chitin oligosaccharides (G_2 through G_8) are indicated.

Recent studies on inhibitors of lipid intermediate and chitin synthesis. We wished to examine differential effects of antibiotics on the _Artemia_ system. Antibiotics were selected based upon published modes of action, e.g., tunicamycin is a nucleoside antibiotic which inhibits formation of Dol-P-P-GlcNAc.[11] Amphotericin B, which blocks chitin synthesis in yeast[12], is an ionophore which binds to cholesterol. Nikkomycin inhibits chitin synthesis in both fungal[13] and insect systems.[14] Both nikkomycin and another antibiotic, polyoxin D, are UDP-GlcNAc analogs. Polyoxin D inhibits chitin synthesis in fungi[15] and insects.[16] There is no information on the effects of amphotericin B, nikkomycin or polyoxin D on the biosynthesis of lipid intermediates.

Effects of antibiotics on in vivo synthesis of lipid intermediates and chitin products. When _Artemia_ larvae were incubated with ^{3}H-GlcN in the presence of various antibiotics (Table I), several compounds inhibited the synthesis of lipid-linked intermediates. Tunicamycin

Table I. _In Vivo_ inhibition of chitin synthesis by antibiotics

	Radioactivity (Percent of Control Values)			
Antibiotic	CM (2:1)	CMW (10:10:3)	Urea Soluble	Residue (Chitin)
Nikkomycin	27%	50%	93%	42%
Amphotericin B	72%	64%	67%	23%
Dimilin	68%	100%	67%	58%
Tunicamycin*	60%	37%	62%	48%

Larvae (10 ml packed volume) were incubated with ^{3}H GlcN (0.5μCi/ml) plus antibiotics (20μg/ml) for 1h (*4h tunicamycin), extracted and radioactivity in each fraction measured.

blocks synthesis of both C:M (2:1) and C:M:W (10:10:3) material; a significant decrease in synthesis of chitin products was also observed. Similar results were obtained by Quesada-Allue,[17] who studied *in vivo* effects of tunicamycin on ^{14}C glucose incorporation into insect chitin. Others have reported that tunicamycin has no effect on chitin synthesis in insects.[18] Nikkomycin blocks *in vivo* synthesis of lipid-linked intermediates in *Artemia* larvae from 50-75%; synthesis of urea insoluble chitin was also reduced 60%, consistent with previous reports on chitin synthesis in fungi and insects. Amphotericin B causes a slight decrease in lipid intermediate synthesis and a dramatic reduction of formation of urea insoluble chitin. Dimilin (diflubenzuron) causes a decrease in synthesis of Dol-P-P-GlcNAc and also decreases synthesis of chitin.

In vitro effects of antibiotics on synthesis of lipid intermediates and chitin products. Various antibiotics were tested using the *in vitro* assay (Table II). We observed variable rates of inhibition when tunicamycin was simply mixed with microsomes; since this preparation consists entirely of vesicles,[5] it appeared possible that tunicamycin was not penetrating the vesicle bilayers to reach the target enzyme. When microsomes were sonicated in the presence of tunicamycin (Table II), incorporation of radioactivity into the C:M (2:1) fraction was totally inhibited. Incorporation of radioactivity into C:M:W (10:10:3) soluble material was inhibited by 50%, indicating that addition of ^{3}H-GlcNAc residues to endogenous lipid-bound oligosaccharides was taking place. Radiolabeling of SDS soluble and insoluble chitin products was reduced by only 20%, indicating that 20% of the labeling in these fractions is via lipid intermediates and 80% is due to direct transfer of ^{3}H GlcNAc to existing primer chains. When tested in the *in vitro* assay, polyoxin D caused a 75% decrease in synthesis of SDS insoluble chitin; however, synthesis of SDS soluble chitin product was unaffected by polyoxin, suggesting that different enzymes are involved in the synthesis of these two products. Polyoxin D caused only a 20% decrease in the synthesis of C:M (2:1) soluble material. Nikkomycin inhibits the *in vitro* synthesis of C:M:W (10:10:3) soluble oligosaccharide-lipids (70%) and SDS insoluble chitin product (85%). Amphotericin B inhibits the *in vitro* synthesis of both classes of lipid-linked intermediates (30-40%) but has less effect on chitin product synthesis (20%). In summary, these data suggest that several distinct enzymes in the chitin synthesis pathway are being assayed and that these activities are differentially sensitive to antibiotic action. This may allow the various enzymes to be studied individually by proper selection of antibiotics.

Table II. Effect of antibiotics on *in vitro* synthesis of lipid intermediates and chitin products.*

	Radioactivity (Percent of Control)			
Antibiotic	CM (2:1)	CMW (10:10:3)	SDS Sol.	Residue (Chitin)
Polyoxin D (40μg/ml)	80%	100%	100%	25%
Amphotericin B (40μg/ml)	68%	58%	81%	86%
Nikkomycin (40μg/ml)	78%	30%	75%	15%
Tunicamycin (sonicated; 50μg/ml)	8%	55%	75%	80%

*Incorporation into lipid intermediates and products during *in vitro* incubation of microsomes and UDP ^{3}H GlcNAc.

Recent studies on solubilized enzymes of the chitin-lipid intermediate pathway. In order to obtain more information on the formation of lipid-linked oligosaccharides, experiments were undertaken to study the first enzyme in the pathway, the dolichol phosphate:UDP-N-acetylglucosamine:N-acetylglucosamine-1-phosphate transferase (GlcNAc-1-P transferase). This enzyme has previously been described from protozoan and mammalian sources.[19-20]

Extraction and assay of the GlcNAc-1-P transferase. Brine shrimp microsomes were extracted with 0.5% lubrol PX and centrifuged (100 K x g/1h). The insoluble material (containing 20% of the total protein and 80% of the enzymatic activity) was extracted with 0.5% Triton X-100 which solubilized all the remaining enzymatic activity. After ultracentrifugation, the soluble fraction was treated with Amberlite XAD-2 to remove detergent. The resulting preparation, containing 85% of the solubilized protein, could be stored at -70°C for up to 5 weeks with no loss of enzymatic activity. For assay of transferase activity, the XAD-treated enzyme (0.2 to 2 mg protein) was incubated in 10mM KPO_4 buffer, containing 15 mM $MgCl_2$ and UDP ^{3}H-GlcNAc (3 pmol) in a final volume of 0.5 ml. After incubation at 30°C for 30 min, samples were extracted with C:M (2:1), and the organic layer was washed with methanol:0.7% NaCl:chloroform (45:48:3) and with water. The washed organic layer was dried and counted. In some experiments, liposomes containing phosphatidyl choline were added to the incubations and bath sonicated for 30 sec before addition of radiolabeled substrate.

Identification of reaction product. When the C:M (2:1) soluble material was analyzed by TLC (Figure 3), a single radiolabeled peak was observed which co-chromatographed with authentic Dol-P-P-GlcNAc. After mild acid hydrolysis of the C:M (2:1) soluble material (0.1 N HCl in 50% n-propanol/100°/15'), the water soluble radioactivity was analyzed by descending paper chromatography (Fig. 4); one radioactive component was observed which co-chromatographed with authentic GlcNAc. These results support the notion that the product of the assay is Dol-P-P-GlcNAc.

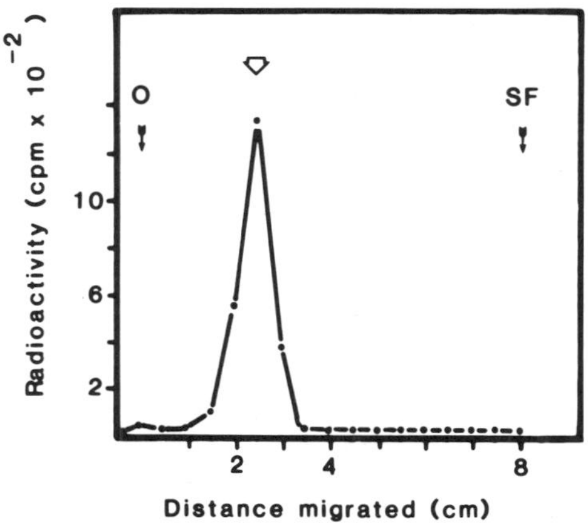

Fig. 3. TLC of radiolabeled product formed by GlcNAc-1-P transferase, soluble in C:M (2:1) using diisobutylketone:acetic acid: water (6:4:3) as solvent. Arrow shows position of standard Dol-P-P-GlcNAc.

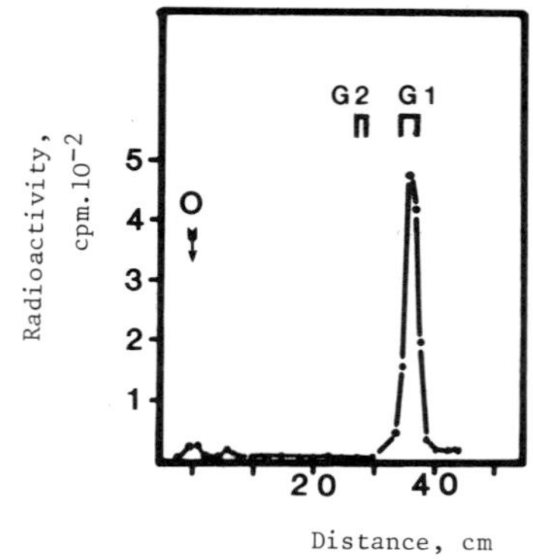

Fig. 4. Paper chromatography of mild acid hydrolyzed product from GlcNAc-1-P transferase reaction. Sample soluble in C:M (2:1) was hydrolyzed (0.1M HCl in 50% n-propanol/100°C/15 min) and subjected to paper chromatography in butanol:pyridine:water (6:4:3) for 40h. The position of standard GlcNAc (G1) and N-acetylchitobiose (G2) is indicated.

<u>Properties of GlcNAc-1-P transferase reaction</u>. The incorporation of ^{3}H GlcNAC into C:M (2:1) soluble material is linear with respect to (a) time of incubation (up to 40 min) and (b) added enzyme protein (up to 2 mg). Incorporation of radioactivity into product is stimulated up to 6-fold by exogenous dolichol phosphate. The approximate K_m for Dol-P is 3.3μM. This value is in good agreement with values reported for protozoan[19] and mammalian[20] GlcNAc-1-P transferases. The approximate K_m for UDP-GlcNAc is 0.17μM. The *Artemia* GlcNAC-1-P transferase requires 15mM $MgCl_2$ for maximal activity; the enzyme is totally inhibited by 100mM EDTA. Inclusion of 500μM UMP in the incubation reduces GlcNAc incorporation by 95%. The enzyme is stimulated up to 3-fold by exogenous phospholipids added as liposomes; maximal stimulation is achieved with phosphatidyl glycerol while phosphatidyl choline is nearly as active. The XAD treated enzyme was stable when stored at 4°C for up to 24h. When the enzyme was stored in 0.1% Triton X-100, an irreversible loss of activity was observed; similar inactivation was observed with digitonin and deoxycholate. On the other hand, the GlcNAc-1-P transferase appeared quite stable to storage in the presence of Lubrol PX (0.1%).

<u>Inhibition of GlcNAc-1-P transferase by antibiotics</u>. The XAD-treated enzyme was sonicated in the presence of various antibiotics and then tested for activity (Table III). Tunicamycin causes a 98% decrease in activity; this result was expected, since tunicamycin inhibits the formation of Dol-P-P-GlcNAc. Diumycin causes a 30% decrease in activity; the product synthetized in the presence of diumycin is Dol-P-P-GlcNAc; this result is in agreement with previous data on the protozoan enzyme,[19] which showed that diumycin only inhibits synthesis of Dol-P-P-$(GlcNAc)_2$. The *Artemia* GlcNAc-1-P transferase is also inhibited by amphomycin; this finding is consistent with results in other systems.[22] Other antibiotics, including nikkomycin, polyoxin D and amphotericin B, showed no effect on the solubilized GlcNAc-1-P transferase.

Liposome nature of the XAD-2 treated enzyme. When XAD-2 treated enzyme was chromatographed on Sepharose 4B, most of the applied protein and the majority of the enzyme was obtained in the voided volume. When the preparation was chromatographed on Sepharose 4B in the presence of 0.1% Triton X-100, most of the applied protein was obtained as a distinct peak in the included volume. These results suggest that the

Table III. Effect of antibiotics on *Artemia* GlcNAc-1-P transferase.*

Antibiotic		Radioactivity (Percent of Control)
Control	(12,419 cpm)	100%
Amphomycin	(50μg/ml)	39%
Diumycin	(50μg/ml)	67%
Nikkomycin	(20μg/ml)	83%
Polyoxin D	(20μg/ml)	71%
Tunicamycin	(50μg/ml)	2%

*XAD-2 treated enzyme was sonicated with antiobiotics, incubated with UDP-^{3}H GlcNAc and extracted.

XAD-treated enzyme is associated with a phospholipid liposome; such liposomes are obtained in the voided volume of Sepharose 4B. After treatment with Triton X-100, the liposome integrity is destroyed and the protein now migrates as a smaller molecular weight component, perhaps associated with a detergent micelle; similar results have been observed with other liposome-associated enzymes.[23]

Future experiments. Purification of the GlcNAc-1-P transferase is now underway using affinity and gel chromatography. When purifed, the physical properties of the *Artemia* enzyme will be examined and compared to the enzyme from protozoan and mammalian sources. We have achieved successful extention of ^{3}H GlcNAc labeled oligosaccharide lipid prepared from scaled-up incubations of GlcNAc-1-P transferase. When added to fresh microsomes in the presence of 500μM UDP-GlcNAc, nearly 40% of the material was converted to a form soluble in C:M:W (10:10:3), i.e., the oligosaccharide contains 3 or more GlcNAc residues. We plan to study the physical properties and kinetics of this enzyme. Finally, we intend to study the oligosaccharidyl transferase which transfers the chitin oligosaccharide from dolichol to a protein acceptor. This enzyme has been detected in *Artemia* microsomes in preliminary experiments and we are now optimizing conditions for its assay.

Acknowledgements. This work was supported by Grant GM-30952 from the National Institute of Health and was conducted with the technical assistance of Ms. Lissa Jackson and Ms. Catherine Carter. I thank Ms. Sonja Randall for typing the manuscript.

References

1. S. Hunt, "Polysaccharide-Protein Complexes in Invertebrates". Academic Press, New York, N.Y. (1970).
2. P.R. Austin, C.J. Brine, J.E. Castle, and J.P. Zikakis, Chitin: New facts of research. Science 212:749 (1980).
3. M.S. Kang, N. Elango, E. Mattia, J. Au-Young, P.W. Robbins and E. Cabib, Isolation of chitin synthetase from Saccharomyces cerevisiae. Purification of an enzyme by entrapment in the reaction product. J. Biol. Chem. 259:14966 (1984).
4. E. Cohen, In vitro chitin synthesis in an insect: formation and structure of microfibrils. Eur. J. Cell Biol. 16:289 (1982).
5. M.N. Horst, The biosynthesis of crustacean chitin by a microsomal enzyme from larval brine shrimp. J. Biol. Chem. 256:1412 (1981).
6. E. Cabib, B. Bowers and R.L. Roberts, Vectorial synthesis of a polysaccharide by isolated plasma membranes. Proc. Nat. Acad. Sci. USA 80:3318 (1983).
7. L.A. Quesada Allue and E. Belocopitow, Lipid-bound oligosaccharides in insects. Eur. J. Bioch. 88:529 (1978).
8. C.R. Krisman and R. Barengo, A precursor of glycogen biosynthesis: α-1,4-glucan-protein. Eur. J. Bioch. 52:117 (1975).
9. H.E. Hopp, P.A. Romero, G.R. Daleo and R. Pont Lezica, Synthesis of cellulose precursors. The involvement of lipid-linked sugars. Eur. J. Bioch. 84:561 (1978).
10. M.N. Horst, The biosynthesis of crustacean chitin. Isolation and characterization of polyprenol-linked intermediates from brine shrimp microsomes. Arch. Biochem. Biophys. 223:254 (1983).
11. A. Elbein, The tunicamycins - useful tools for studies on glycoproteins. Trends in Biochem. Sci. August, 1981:219 (1981).
12. D.M. Rast and S. Bartnicki-Garcia. Effects of amphotericin B, nystatin and other polyene antibiotics on chitin synthetase. Proc. Nat. Acad. Sci. USA 78:1233 (1981).
13. V. Dahn, H. Hagenmaier, H. Hohne, W. Konig, G. Wolf and H. Zahner. Stoffwechselprodukte von microorganismen. 154 Mitteilung. Nikkomycin, ein neuer Hemmstoff der Chitin synthese bei Pilzen, Arch. Microbiol. 107:143 (1976).
14. E. Cohen and J. Casida, Properties and inhibition of insect integumental chitin synthetase. Pestic. Biochem. Physiol. 17:301 (1982).
15. N. Ohta, K. Kakiki and T. Misato, Studies on the mode of action of polyoxin D. Part II. Effect of polyoxin D on synthesis of fungal cell wall chitin. Agr. Biol. Chem. 34:1224 (1970).
16. T. Leighton, E. Marks and F. Leighton, Pesticides: Insecticides and fungicides are chitin synthesis inhibitors. Science 213:905 (1981).
17. L.A. Quesada Allue, The inhibition of insect chitin synthesis by tunicamycin. Biochem. Biophys. Resh. Commun. 105:312 (1982).
18. R.T. Mayer, A.C. Chen and J.R. DeLoach. Chitin synthesis inhibiting insect growth regulators do not inhibit chitin synthetase. Experientia 37:337 (1981).
19. C.L. Villemez and P.L. Carlo, Properties of a soluble polyprenol phosphate UDP-GlcNAc: N-acetylglucosamine-1-phosphate transferase. J. Biol. Chem. 255:8174 (1980).
20. R.K. Keller, D.Y. Boon and F.C. Crum, N-acetylglucosamine-1-phosphate transferase from hen oviduct: solubilization, characterization and inhibition by tunicamycin. Biochemistry 18:3946 (1979).
22. M.S. Kang, J.P. Spencer and A.D. Elbein, Amphomycin inhibition of mannose and GlcNAc incorporation into lipid-linked saccharides. J. Biol. Chem. 253:8860 (1978).
23. G. Sessa and G. Weissman, Incorporation of lysozyme into liposomes. J. Biol. Chem. 245:3295 (1970).

Section II

CHITIN IN SKELETAL STRUCTURES OF MICROBES

THE ROLE OF CHITIN IN FUNGAL GROWTH AND MORPHOGENESIS

E. LOPEZ-ROMERO and J. RUIZ-HERRERA

Facultad de Quimica, Universidad de Guanajuato

Departamento de Génetica y Biologia Molecular, CINVESTAV
Instituto Politécnico Nacional
Mexico

INTRODUCTION

The cell wall plays a prominent role in the structure and behaviour of fungi. The presence of the wall defines the shape and strength of the fungal cell and is a critical determinant for the morphological development of these eucaryots. Functions such as vegetative growth, reproduction, dispersal, survival and cell-cell interaction, depend to a great extent in the mode of cell wall construction. This means that, in symple terms, fungal differentiation and morphogenesis can be explained on the basis of cell wall morphogenesis.

In general, the fungal cell wall consists of 80 to 90 % polysaccharides with protein and lipids accounting for the remaining portion. Physically, it can be considered as a mesh of microfibrillar polymers imbedded in a matrix of amorphous or cementing components. Except for the lower fungi and most true yeasts, which have cellulose and glucans in the cell wall, respectively, the most important structural components of the fungal cell wall is chitin. Proteins and polysaccharides such as mannans, glucans, galactans and some heteropolysaccharides represent the cementing material which, by joining the skeletal components together, provide the macromolecular dimension of the cell wall.

CHITIN IN THE FUNGAL WALL

Chemical considerations

Fungal chitin is exclusively confined to the cell wall, except for some Oomycetes where it has been found as a cytoplasmic inclusion complexed with glucan in the form of cellulin granules[1]. Chitin has the same chemical structure in fungi and animals. In all fungi studied thus far it exists in the crystalline form denominated α-chitin, which is the most widely distributed polymorph in Nature. In this crystalline form the neighboring chains of N-acetylglucosamine are antiparallel and are organized in an orthorombic unit cell with dimensions: a= 4.74 Å, b= 18.86 Å and c= 10.32 Å[2].

Physical and structural considerations

Electron microscopy techniques combined with cytochemical, chemical and enzymatic analyses indicate that a certain degree of stratification exists in the walls of fungi. It is a common observation that the outer surface of the wall is made up of amorphous material, whereas microfibrils are mainly confined to the innermost layer of the wall adjacent to the plasmalemma.

Two types of cell walls can be recognized in the filamentous fungi: a thin, rather plastic, primary wall, which is formed at the growing zone of the hyphae and a more rigid, secondary or mature wall which is formed by thickening of the former. In both of them chitin is the main structural component. Chitin microfibrils usually have a diameter of 10-25 nm and may be up to 2-3 um long[3], values closely similar to those estimated for the microfibrils synthesized in vitro by chitosomes from *Mucor rouxii*[4]. Poly-GlcNAc chains of these last microfibrils are formed by about 2000 units (C. Calvo and J. Ruiz-Herrera, unpublished results). Observations made using fluorescent conjugated wheat germ agglutinin[5], the fluorescent brightener calcofluor[6] and chitinase[7] have shown that chitin is more accessible to these agents in the primary wall than in the mature wall. These results speak of a different architectural organization between primary and secondary walls.

In most fungal cell walls chitin microfibrils are randomly oriented with the exception of septa where they can be seen organized in a circular fashion. For the sporangiophore of *Phycomyces blakesleeanus*, it has been suggested[8,9] that the microfibrils in the inner wall of the growing zone are preferentially organized in the direction of a right-handed spiral or flatly transverse, whereas in the non-growing zone they lie parallel to the longitudinal axis. The implications of this arrangement to the spiral mode of growth of the sporangiophore are described below.

There is almost general agreement that the primary wall present in the growing region of fungal cells behaves as a two-phase system composed of rigid chitin microfibrils embedded in the Newtonian viscous fluid of the cementing material (for a review, see ref 10). In the nongrowing zone,the viscoelastic properties of the cell wall decrease and the secondary wall becomes rigidified. Reasons for this change in the physicochemical characteristics of the wall remain unknown, but Burnett[3] has suggested that it may be due to late increase in microfibril diameter and packing. The possibility of crosslinking of chitin with other polymers was studied[11] (cfr. Sietsma et al., this Volume).

ROLE OF CHITIN AS STRUCTURAL COMPONENT

Several lines of evidence define the function of chitin in keeping the wall shape and rigidity and consequently its crucial role in fungal morphogenesis. For instance, *Aspergillus* cell walls retain their original shape upon removal of amorphous cell components under conditions where only chitin remains unaffected [12]. The organization of the chitin microfibrillar mesh has been studied after removal of the other wall components by enzymatic digestion or by chemical oxidation[3]. These treatments do not cause much spatial distortion in the orientation of the microfibrils, indicating that once formed, the interwoven skeletal structure exists as a rigid, three-dimensional net with the ability to retain its shape even after removal of the embedding components. In this sense chitin microfibrils play a genuine structural role[3]. This role is also illustrated by the behavior of chitin-less cells obtained either by mutation or by inhibition of its synthesis with drugs. Mutants of *Aspergillus nidulans* which lack chitin produce swollen hyphae and are osmotically fragile[13]. The best known example of a wall less-mutant is the "slime" variant of *Neurospora crassa*[14]. This mucoid mutant lacks the ability to synthesize both chitin and ß-glucan *in vivo*[15], and its growth and development is sustained only in osmotically-protected media where it produces stable protoplasts of variable sizes. Polyoxin D[16] and nikkomycin[17], which are strong competitive inhibitors of chitin synthetase, inhibit the growth of fungi[18,21], induce morphological alterations[20,22], and inhibit cell wall regeneration of stable *N. crassa* os-1 protoplasts[23].

CHITIN AND CELL WALL GROWTH

Structural rearrangements in the growing cell wall

Structure of the cell wall is of paramount importance in defining the mechanism of its expansion. As mentioned above primary wall in the growing zone has a viscoelastic structure, whereas the secondary, nongrowing wall is mostly rigid. In this regards it is important the observation that walls of dead *Phycomyces* sporangiophores are rigid[24]. Comparison of the mechanical extensibility between the cell wall of the nongrowing stage II and the growing stage IVb of the sporangiophore of *Phycomyces* revealed an increase in extensibility accompanied by a transition from a basically elastic structure to a more plastic one[25]. This same change occurs after stimulation of growth by a light pulse[26] and it is apparently due to breakage of a few glycosidic bonds in the chitin chains[27] which induces a plastification of the cell wall. This splitting process has been suggested to be responsible for maintaining the plastic properties characteristic of a growing wall allowing the intersusception of incoming microfibrils[28], although alternative possibilities have been entertained[11]. Chitinases have been invoked as elicitors of this mechanism (cfr. Gooday et al., this Volume) to explain[29] the increase in growth rate of the sporangiophore of *Phycomyces* after a light stimulus[27]. In this sense, the presence of chitinases in several fungal species seems relevant[30-33]. Thus, wall expansion may be envisaged as a dynamic process involving the turgor pressure of the cell acting on a viscoelastic structure whose rigidity depends on the length and stiffness of the chitin microfibrils, which in turn are under enzymatic control.

Reorientation of chitin microfibrils has also been considered important in determining the mode of wall growth. Thus, this mechanism appears to be responsible for the spiral growth of the *Phycomyces* sporangiophore[34]. This reorientation occurs as chitin microfibrils move down the growing zone of the sporangiophore.

Chitin biosynthesis

Since the first description of chitin biosynthesis *in vitro* using cell-free extracts from *N. crassa*[35], numerous studies dealing with the biochemical properties of chitin synthetase in a great number of fungi have appeared in the literature. Most of these have been compiled in a few reviews[10, 36, 37].

Studies on chitin synthesis by fungal extracts have been performed mostly with crude particulate preparations which catalyze the conversion of small amounts of the soluble precursor UDP-N-acetylglucosamine (UDP-GlcNAc) into an insoluble polymer characterized as chitin mainly by enzymatic and/or chemical criteria. However, the physical characterization of the product remained unresolved until 1974 when chitin microfibrils were synthesized for the first time in the assay tube using a high-speed, membrane-free supernatant of the fungus *Mucor* as the source of the synthetase[38]. The product was characterized as α-chitin by X-ray diffraction, the only valid criterium to ascertain the crystalline nature of a polysaccharide.

A survey of the many reports on fungal chitin synthetase reveals that the enzyme system from these organisms shares common properties. Some of these are outlined in Table 1. These characteristics are emphasized on the basis of their possible participation in the mechanism of enzyme regulation *in vivo*.

Cytological aspects of chitin biosynthesis and the role of chitosomes

Cytological studies of fungal hypha have revealed that the marking difference between the apical dome which corresponds to the growing zone, and their distal portion is the accumulation of cytoplasmic vesicles at the hyphal tip[39,40]. These vesicle clusters have been considered to be equivalent with the so-called "Spitzenkörper", originally described as a dark-staining body by Brunswick[41].

Table 1.Some biochemical properties of fungal chitin synthetase

1. The enzyme is activated by limited, unspecific proteolysis.
2. The enzyme requires a divalent cation (Mg^{++} or Mn^{++}) for activity.
3. N-Acetylglucosamine is a strong allosteric activator of the enzyme, mimicking the effect of substrate at the allosteric site. The aminosugar itself is not incorporated into chitin.
4. Chitin synthetase has unusual kinetics. Km values for UDP-GlcNAc are in the mM range.
5. Chitin synthesis occurs in the absence of an exogenous primer.
6. Stability of chitin synthetase is high in purified preparations but poor in crude fractions.
7. Chitin synthetase is competitively inhibited by UDP, a product of the reaction.
8. There is no high energy lipid intermediate involved in chitin biosynthesis.

It is generally recognized that these apical vesicles are responsible for the polarized growth in filamentous fungi, even though how this mechanism is accomplished is not completely understood.

Two vesicle sizes have been recognized in the growing hyphal tip: large and small. The former have a diameter ranging from 100 to 400 nm while the small ones measure 30-100 nm. The small vesicles are comparable in size, membrane width and staining properties with microvesicles isolated from several fungal species, and which are able to synthesize chitin microfibrils *in vitro*. These microvesicles named chitosomes, have been considered to be the carriers of chitin synthetase from its site of origin to the cell surface[42-46]. They accumulate at the hyphal tip[27] and contain practically all of the chitin synthetase activity of the fungal cells[47], which has been found to be located in the cytoplasm[48]. General properties of chitosomes are shown in Table 2.

Table 2. Some physical and biochemical characteristics of chitosomes from yeast cells of *Mucor rouxii*[42-47]

Physical

Chitosomes are microvesicles with a mean diameter of 40-70 nm delimited by a thin (ca. 7.5 nm) unit membrane.

M.W. is $5X10^6$ daltons (by gel filtration), their sedimentation coefficient of 105 S, and their specific gravity is 1.145 g/cm^3

Chitosomes are dissociated by digitonin into 16S subunits measuring 7-12 nm with M.W. ca. $5X10^5$ daltons. They reassociate into chitosome-like structures upon removal of the saponin.

Chitosomes synthesize α-chitin microfibrils *in vitro* when incubated with UDPGlcNAc and activators.

Biochemical

Chitin synthetase in chitosomes is inactive. It is activated *in vitro* by limited proteolysis.

Chitosomal chitin synthetase is stable.

Chitosomes are made of two thirds protein and one third lipid. Lipids contain about equal proportion of polar and neutral components. Eight major polypeptides with M.W. ranging between 17-55 Kdaltons are recognized.

Cell wall growth in fungi depends on the operation of fine temporal and spatial regulatory systems. Temporal control requires a mechanism to maintain chitin synthetase inactive in the cytoplasmic milieu full of substrate. In this sense, it is known that chitosomal chitin synthetase is inactive and requires proteolytic activation[44]. Spatial regulation demands a driving mechanism to mobilize chitosomes to the sites of chitin microfibril deposition. A role for microtubules to fulfill this role has been both sustained[49] and denied[50]. Recent observations from our group suggest that Ca^{++} ionic currents may be the leading force for chitosome movilization in the cytoplasm[51]. Once microfibril biosynthesis has occurred, a mechanism must operate to stop further enzymatic synthesis to guarantee the observed kinetics of hyphal growth. As a matter of example, N. crassa hypha having a growing zone of 29 nm exhibit an extension rate of 38 nm min^{-1}. Accordingly it may be expected that half-life of active chitin synthetase in the apical zone would be of only 0.38 min. The observed instability of chitin synthetase in most fungal systems[10,36,37] suggests that post-synthesizing arrest of the enzyme activity involves its irreversible modification.

According to these concepts fungal growth and morphogenesis are the resultant of the temporal control and the spatial distribution of chitosomes in the cell. Thus, synthesis of chitin in a cylindrical hyphae takes place apically, whereas in a spherical cell chitin biosynthesis occurs isodiametrically.

REFERENCES

1. H.Y. Lee and J.M. Aronson, Composition of cellulin, the unique chitin-glucan granules of the fungus, Apodachlya sp, Arch. Microbiol. 102:203 (1975).
2. R. Minke and J. Blackwell, The structure of α-chitin, J. Mol. Biol. 120: 167 (1978).
3. J.H. Burnett, Aspects of the structure and growth of hyphal walls, in: "Fungal Walls and Hyphal Growth", J.H. Burnett and A.P.J. Trinci, eds., Cambridge University Press, Cambridge (1979)
4. J. Ruiz-Herrera, V.O. Sing, W.J. van der Woude, and S. Bartnicki-Garcia, Microfibril assembly by granules of chitin synthetase, Proc. Nat. Acad. Sci. USA 72: 2706 (1975).
5. M. Galun, A. Braun, A. Frensdorff, and E. Galun, Hyphal walls of isolated lichen fungi. Autoradiographic localization of precursor incorporation and binding of fluorescein-conjugated lectins, Arch. Microbiol. 108: 9 (1976).
6. K. Gull and A.P.J. Trinci, Detection of areas of wall differentiation in fungi using fluorescent staining, Arch. Mikrobiol. 96: 53 (1974).
7. C.A. Vermeulen and J.G.H. Wessels, Ultrastructural differences between wall apices of growing and non-growing hyphae of Schizophyllum commune, Protoplasma 120: 123 (1984).
8. P.A. Roelofsen and A.L. Houwink, Architecture and growth of the primary cell wall in some plant hairs and in the Phycomyces sporangiophore, Acta Bot. Neerl. 2: 218 (1953).
9. P.A. Roelofsen, Ultrastructure of the wall in growing cells and its relation to the direction of growth, in: "Advances in Botanical Research", vol. 2, R.D. Preston, ed., Academic Press, New York (1965).
10. G.W. Gooday and A.P.J. Trinci, Wall structure and biosynthesis in fungi, in: "The Eucaryotic Microbial Cell", G.W. Gooday, D. Lloyd, and A.P.J. Trinci, eds., Cambridge University Press, Cambridge (1980).
11. J.G.H. Wessels, Apical hyphal extension. Do lytic enzymes play a role? in: "Microbial Cell Wall Synthesis and Autolysis", C. Nombela, ed., Elsevier Science Publishers, Amsterdam (1984).
12. J. Ruiz-Herrera, Chemical components of the cell wall of Aspergillus species, Arch. Biochem. Biophys. 122: 118 (1967).
13. D. Katz and R.F. Rosenberger, Lysis of Aspergillus nidulans mutant blocked in chitin synthesis and its relation to wall assembly and wall metabolism, Arch. Mikrobiol. 30: 284 (1971).

14. S. Emerson, Slime, a plasmodioid variant of Neurospora crassa, Genetica 34: 162 (1963).
15. C.A. Leal-Morales and J. Ruiz-Herrera, Alterations in the biosynthesis of chitin and glucan in the slime mutant of Neurospora crassa, Exp. Mycol. In press.
16. A. Endo, K. Kakiki, and T. Misato, Mechanism of action of the antifungal agent polyoxin D, J. Bacteriol. 104: 189 (1970).
17. U. Dahn, H. Hagenmaier, H. Hohne, W.A. Konig, G. Wolf, and H. Zahner, Stoffwechselprodukte von mikroorganismen. 154. Mitteilung nikkomycin, ein neuer hemmstoff der chitinsynthese bei pilzen, Arch. Microbiol. 107: 143 (1976).
18. K. Isono, J. Nagatsu, Y. Kawashima, and S. Susuki, Studies on polyoxins, antifungal antibiotics. Part I. Isolation and characterization of polyoxins A and B, Agr. Biol. Chem. 29: 848 (1965).
19. K. Isono, J. Nagatsu, K. Kobinata, K. Sasaki, and S. Susuki, Studies on polyoxins, antifungal antibiotics. Part V. Isolation and characterization of polyoxins C, D, E, F, G, H and I, Agr. Biol. Chem. 31: 190 (1967).
20. S. Bartnicki-García and E. Lippman, Inhibition of Mucor rouxii by polyoxin D: Effects on chitin synthetase and morphological development, J. Gen. Microbiol. 71: 301 (1972).
21. H. Muller, R. Furter, H. Zahner, and D.M. Rast, Inhibition of chitosomal chitin synthetase and growth of Mucor rouxii by nikkomycin Z, nikkomycin X, and polyoxin A: comparison, Arch. Microbiol. 130: 195 (1981).
22. T. Benitez, T.G. Villa, and I. García Acha, Effects of polyoxin D on germination, morphological development and biosynthesis of the cell wall of Trichoderma viride, Arch. Microbiol. 108: 183 (1976).
23. C.P. Selitrennikoff, Use of temperature-sensitive protoplasts of the osmotic-1 variant of Neurospora crassa for the detection of cell wall acting antifungal compounds, in: "Microbial Cell Wall Synthesis and Autolysis", C. Nombela, ed., Elsevier Science Publishers, Amsterdam (1984).
24. J.K.E. Ortega, Mechanical and Structural Dynamics of Cell Wall Growth, Ph D Thesis, University of Colorado, Boulder, USA (1976).
25. C.N. Anliquist and R.I. Gamow, Phycomyces: mechanical behavior of stage II and stage IV, Plant Physiol. 51: 586 (1973).
26. J.K.E. Ortega, R.I. Gamow, and C.N. Ahlquist, Phycomyces: a change in mechanical properties after a light stimulus, Plant Physiol. 55: 333 (1980).
27. L. Herrera-Estrella and J. Ruiz-Herrera, Light response in Phycomyces blakesleeanus: evidence for roles of chitin biosynthesis and breakdown, Exp. Mycol. 7: 362 (1983).
28. S. Bartnicki-García, Fundamental aspects of hyphal morphogenesis, in: "Microbial Differentiation", J.M. Ashworth and J.E. Smith, eds., Cambridge University Press, Cambridge (1973).
29. S. Bartnicki-García and E. Lippman, The busting tendency of hyphal tips of fungi: presumptive evidence for a delicate balance between wall synthesis and wall lysis in apical growth, J. Gen. Microbiol. 73: 487 (1972).
30. R.J. Cohen, Some properties of chitinase from Phycomyces blakesleeanus, Life Sciences 15: 289 (1974).
31. K. Soderhall, E. Svensson, and T. Unestam, Chitinase and protease activities in germinating zoospore cysts of a parasitic fungus, Aphanomyces astaci, oomycetes, Mycopathologia 64: 9 (1978).
32. E. López-Romero, J. Ruiz-Herrera, and S. Bartnicki-García, The inhibitory protein of chitin synthetase from Mucor rouxii

is a chitinase, Biochim. Biophys. Acta 702: 233 (1982).
33. K. Barrett-Bee and M. Hamilton, The detection and analysis of chitinase activity from the yeast form of Candida albicans, J. Gen. Microbiol. 130: 1857 (1984).
34. R.I. Gamow and B. Bottger, Phycomyces: modification of spiral growth after mechanical conditioning of the cell wall, Science 203: 268 (1979).
35. L. Glaser and D.H. Brown, The synthesis of chitin in cell-free extracts of Neurospora crassa, J. Biol. Chem. 228: 729 (1957).
36. V. Farkas, Biosynthesis of cell walls of fungi, Microbiol, Rev. 43: 117 (1979).
37. J. Ruiz-Herrera, La síntesis de quitina por los hongos,Ciencia 32: 89 (1981).
38. J. Ruiz-Herrera and S. Bartnicki-García, Synthesis of wall microfibrils in vitro by a "soluble" chitin synthetase from Mucor rouxii, Science 186: 357 (1974).
39. S.N. Grove and Bracker, C.E. Protoplasmic organization of hyphal tips among fungi: vesicles and spitzenkorper, J. Bacteriol. 104: 989 (1970).
40. S.N. Grove, The cytology of hyphal tip growth, in: "The Filamentous Fungi", vol. 3, J.E. Smith and D.R. Berry, eds., Edward Arnold Publishers, London (1978).
41. H. Brunswick, Untersuchungen uber geschlechts und kernverhaltnisse bei der Hymenomyzetegattung Coprinus, in: "Botanische Abhandlungen", vol. 5. K. Goebel, ed., Gustav Fischer, Jena (1924).
42. J. Ruiz-Herrera, E. López-Romero, and S. Bartnicki-García, Properties of chitin synthetase in isolated chitosomes from yeast cells of Mucor rouxii, J. Biol. Chem. 252: 3338 (1977).
43. S. Bartnicki-García, Chitosomes and the origin of chitin microfibrils, in: "Advances in Protoplast Research", L. Ferenczy and G.L. Farkas, eds., Pergamon Press, London (1980).
44. S. Bartnicki-García, Unique properties of chitosomes, in: "Microbial Cell Wall Synthesis and Autolysis", C. Nombela, ed., Elsevier Science Publishers, Amsterdam (1984).
45. S. Bartnicki-García, C.E. Bracker, E. Reyes, and J. Ruiz-Herrera, Isolation of chitosomes from taxonomically diverse fungi and synthesis of chitin microfibrils in vitro, Exp. Mycol. 2: 173 (1978).
46. S. Bartnicki-García, J. Ruiz-Herrera, and C.E. Bracker, Chitosomes and chitin synthesis, in: "Fungal Walls and Hyphal Growth", J.H. Burnett and A.P.J. Trinci, eds., Cambridge University Press, Cambridge (1979).
47. J. Ruiz-Herrera, C.E. Bracker, and S. Bartnicki-García, Sedimentation properties of chitosomes from Mucor rouxii, Protoplasma 122: 178 (1984).
48. R. Sentandreu, A. Martínez-Ramón, and J. Ruiz-Herrera, Localization of chitin synthase in Mucor rouxii by an autoradiographic method, J. Gen. Microbiol. 130: 1193 (1984).
49. R.J. Howard and J.R. Aist, Cytoplasmic microtubules and fungal morphogenesis, Ultrastructural effects of methyl benzimidazole-2-ylcarbamate determined by freeze-substitution of hyphal tip cells, Cell Biol. 87: 55 (1980).
50. F.B. Herr and M.C. Heath, The effect of antimicrotubule agents on organelle positioning in the cowpea rust fungus, Uromyces phaseoli var. vignae, Exp. Mycol. 6: 15 (1982).
51. M.C. Valenzuela and J. Ruiz-Herrera, Posible papel de las corrientes de calcio en la síntesis de la pared celular de Phycomyces, XV Congress of the Mexican Society of Biochemistry, Proc. p 79 (1984).

THE ROLE OF CHITIN IN HYPHAL MORPHOGENESIS

J. H. SIETSMA, C. A. VERMEULEN and J. G. H. WESSELS

Department of Plant Physiology
University of Groningen, Biological Centre
Haren, The Netherlands

INTRODUCTION

Chitin is a structural component present in the wall of all fungi except the Oomycetes where cellulose plays a similar role. It is usually considered to be present as microfibrils, consisting of crystalline chitin embedded in an amorphous matrix (1), because most electronmicroscopical and X-ray diffraction studies show microfibrillar and crystalline chitin. However these results were usually obtained after drastic chemical treatments of the wall and induction of crystallisation and microfibril formation by these treatments remains a possibility (2).

Recently it has been shown that chitin may not always solely occur as a microfibrillar, microcrystalline homopolymer of N-acetyl glucosamine in fungal walls; in Ascomycetes and Basidiomycetes chitin chains may be linked to β-glucan (3,4) and in Zygomycetes it is partly deacetylated to chitosan (5) and linked to peptides (6). These findings suggest that chitin is probably not the only component which is important in generating the hyphal form. In order to obtain a better understanding of the process of hyphal morphogenesis a more detailed knowledge of the mature hyphal wall is necessary.

STRUCTURE OF THE WALL

Although many fungal walls have been analysed chemically or microscopically, in only a few cases has a comprehensive analysis been conducted (2). In a long-term study we have gathered a large number of data on the hyphal wall of the basidiomycete *Schizophyllum commune* and a model of its structure has been proposed (4,7). As the major morphogenetic entity of the wall a covalently linked complex between chitin and the (1→3)-β/(1→6)-β-glucan was envisaged. An alkali-soluble (1→3)-α-glucan, mainly located at the outside of the wall as crystalline material, probably has little morphogenetic significance because mutations that block its synthesis or 2-deoxyglucose that inhibits its synthesis do not have a major effect on hyphal morphogenesis. The glucan-chitin complex is highly insoluble and resistant to alkali presumably because the glucan chains are linked to chitin through their reducing ends via amino acids, particularly lysine. The β-glucan chains are solubilized when chitin is depolymerised by chitinase or by treatment with nitrous acid after deacetylation. Apart from the linkage of the two polymers to each other, the wall model also assumes that both chitin and β-glucan chains are hydrogen bonded among themselves . In the case of chitin this would mean that certain chains must be unsubstituted and still can form microfibrils. The (1→3)-β-glucan and C-6 substituted (1→3)-β-glucan are known to form triple helices (8,9) and by doing so would crosslink the chitin microfibrils into a rigid network. However the crystallites formed must be rather imperfect because X-ray diffraction of native walls give no or very weak interference lines of crystalline chitin or hydroglucan (10,11). Only after treatment with hot dilute mineral acid crystallinity is revealed. Such a treatment destroys the linkage between chitin and β-glucan and also removes (1→6)-β-linked side chains thus favouring the release of homopolymers which can crystallize further. In this view at least some of the microfibrils and the good crystallinity of polymers as observed in wall preparations after extractions (11) could thus be induced by the treatments to which the walls were subjected.

SYNTHESIS OF THE GLUCAN-CHITIN COMPLEX

The proposed structure of the chitin-glucan complex immediately suggests that its synthesis must occur in at least two distinct phases. First the individual chitin and β-glucan must be polymerised and deposited outside the plasma membrane. Only then the coupling of the β-glucan chains to the chitin can proceed. This coupling probably competes with the tendency of the chitin chains to crystallise, which readily occurs in the absence of such modifying activity.

The following observations support the occurrence of the suggested sequence of events. By light microscopic autoradiography it could be shown that both chitin and the glucan are synthesized in the hyphal apex but that only label from [^{3}H]-N-acetylglucosamine immediately appears in an alkali-insoluble form. Label from [^{3}H]-glucose is primarily incorporated at the apex into water- and alkali-soluble β-glucans, becoming alkali-insoluble (interpreted as linkage to chitin) during a chase period (12). According to these observations not only the β-glucan but also the chitin is modified after being polymerised and deposited into the wall. Support for this chain of events also comes from studies on wall regeneration by protoplasts of *Schizophyllum commune* (13). When in regenerating protoplasts chitin synthesis is blocked by polyoxin-D the β-glucan accumulates as a water-soluble material, indicating that this β-glucan only becomes alkali-insoluble when it can be attached to insoluble chitin. On the other hand, when β-glucan synthesis is blocked by cycloheximide chitin synthesis continues and chitin is deposited as a microfibrillar crystalline component. Also in the absence of inhibitors, in the first hours of their regeneration, the protoplasts fail to synthesize the chitin-glucan complex but form highly crystalline chitin microfibrils and water and DMSO-soluble β-glucans. Only after the formation of the chitin-glucan complex commences do these protoplasts produce hyphae (13,14). This suggests that the linkage between chitin and glucan is a prerequisite for the hyphal mode of growth.

CHITIN SYNTHASE LOCALISATION AND ACTIVATION

In vivo chitin is always deposited at the outside of the plasma membrane. Electron microscopic autoradiography has shown that the polymer is synthesized very close to the plasma membrane (15). Cell fractionation studies have shown that membrane preparations contain chitin synthase activity. Of particular value are fractionation studies in which ConA-stabilized plasma membranes were isolated free from other membranes (16, 17). Such plasma membranes contain chitin synthase in an inactive form, as in the case of the yeast *Saccharomyces cerevisiae* (16) or in an active form, as in the case of *Schizophyllum commune* (17). The case of *S. cerevisiae* is special because the wall contains very little chitin except in the septum that separates mother and daughter cell (18). The inactive enzyme in the plasma membrane preparations from this organism can, however be activated by limited proteolytic digestion (18) and localized proteolytic activation in vivo has been proposed as a regulatory mechanism for the initiation of septum synthesis (18). In the case of *S. commune*, more representative for chitin-containing filamentous fungi, the plasma membrane bound enzyme was largely in an active state but high speed centrifugation of the post-membrane supernatant sedimented an inactive chitin synthase preparation that could be activated by trypsin digestion (17).

High-speed sedimentable particles containing inactive chitin synthase have been isolated from a variety of fungi and have been called chitosomes (19,20). Such chitosomes contain protein and lipids which differ in composition from those in the membranes (21). Upon activation with proteolytic enzymes they produce crystalline chitin (20) and in *Mucor rouxii* most of the chitin synthase was localized in the cytoplasm in a zymogenic form by using cells permeabilized by toluene/ethanol (22). The idea that chitosomes are particles or vesicles which are extruded through the plasma-membrane into the wall and there function as chitin-synthesizing entities is still entertained (23). However it is more probable that chitosomes are the conveyers of chitin synthase to the plasma-membrane, ensuring that no chitin is synthesized inside the cell (see previous Chapter).

The activity of chitin synthase is clearly influenced by lipids and lipophilic substances. The effects of polyene antibiotics and digitonin on chitosomal activity has been attributed to interactions of chitin synthase with sterols (24). The enzyme is also stimulated by primary alcohols such as ethanol and methanol, particularly when proteolytic activation is omitted (25). Delipification of membrane-derived chitin synthase decreased its activity, while membrane-derived chitin synthase dissociated by digitonin could be activated by the addition of certain phospholipids but not by sterols (26,27). In addition, Arrhenius plots of the temperature dependence of the membrane-associated chitin synthase showed deflection from linearity that are characteristic for lipid-effected membrane-bound enzymes (27).

This lipid-dependence of chitin synthase suggests that in the membrane the specific lipid environment of the enzyme may be crucial to its activity. It even seems possible that incorporation of the inactive chitosome-bound enzyme in the plasma-membrane results in lipid enzyme interactions which bring about its activity. These lipid-enzyme interactions could also result in a positioning of the enzyme in the membrane in such a way that substrate can be accepted at the inside while chitin chains are deposited at the outside of the membrane. Such vectorial synthesis of chitin has been indicated experimentally (28). Whether proteolytic activation is operative in vivo remains to be determined. Even in the case of yeast septum synthesis in which proteolytic activation was first proposed as a regulatory mechanism (18) analysis of protease mutants has failed to corroborate this scheme (29,30).

CHITIN SYNTHESIS AT THE HYPHAL APEX

In actively growing mycelium chitin is continuously synthesized at ever extending apices of the hyphae. Synthesis is maximal at the extreme tip and declines subapically according to a gradient required for hyphal morphogenesis (1). Therefore it appears as if a chitin synthesizing entity maximally active at the extreme tip lost most of its activity a few minutes later when it is in juxtaposition to the wall which has left the apical dome.

One hypothesis which has been proposed depends on the postulated proteolytic activation of chitin synthase (31). After being transported to the extreme tip, the enzyme would be proteolytically activated. Continued proteolysis would then slowly inactivate the enzyme, accounting for the subapical decrease in activity. This scheme would imply a very high turnover rate of the enzyme. However, this is at variance with an apparent high stability of the enzyme in the presence of cycloheximide at concentrations which completely block protein synthesis. Chitin synthesis went on undisturbed for hours, both in regenerating protoplasts (13) and in mycelium (25). Since the presence of cycloheximide has been shown by autoradiography to obliterate apical chitin synthesis but to generate subapical chitin synthesis (32), this has been interpreted as evidence for the presence of inactive chitin synthase along the lateral wall. Consequently a substance has been sought that could reversibly inactivate chitin synthase rather than destroying the enzyme as in the case of proteolytic digestion. Such a substance has indeed been found (33,34). However, it has been recently noted that this inhibitor protein does not inhibit chitin synthase activity but actually displays chitinase activity destroying the native chitin chains made by the enzyme (35 and see below). Because this hypothesis of sequential activation and inactivation is thus difficult to reconcile with experimental data we would like to propose an alternative hypothesis based on the reasonable assumption that the enzyme is an integral membrane protein.

If chitin synthase at the hyphal apex is an integral membrane protein it could probably move laterally in the plane of the membrane. While new membrane material is being inserted at the hyphal apex chitin synthase could retain its position at the apex by lateral movement. The force that drives this movement could be generated by the synthesis of chitin itself, as has been proposed for the cellulose-synthesizing enzyme in plant-membranes (36). The positioning of chitin synthase in the apex could also be mediated by proton currents that traverse hyphal tips (37). We have attempted to show that an electrical current can move chitin synthase in the plane of the membrane by placing protoplasts of *S. commune* in an electric field and then localizing the enzyme by autoradiography after a pulse with [^{3}H]-N-acetylglucosamine (38). Although we failed to demonstrate any redistribution of the enzyme over the surface of the protoplast, the current did determine the site of emergence of the hyphal apex towards the anode. However, the relationship of chitin synthase with the currents that traverse the apex may be much more indirect. Evidence is accumulating that the apex of tip-growing cells in fungi and plants contains a highly structured cytoskeleton (39.40). If chitin synthase in the membrane were connected to this cytoskeleton, the whole assemblage of chitin synthase molecules could be pushed forward through the plane of the membrane while other membrane components are inserted during extension growth. In this view the spatial distribution of chitin synthase activity at the apex is thus regarded as a consequence of the organisation of the underlying cytoplasm. Any disturbance of this organisation, e.g. by cycloheximide, could thus lead to displacement of chitin synthase in the membrane.

PROPERTIES OF NEWLY SYNTHESIZED CHITIN

It was shown that the product of chitin synthase can be subject to further enzymatic modifications in the hyphal wall, i.e. coupling to β-glucan chains in Ascomycetes and Basidiomycetes (3) and deacetylation to chitosan in the case of Zygomycetes (5). If chitin synthesis is allowed to proceed in the absence of these modifying reactions as during synthesis in vitro using chitosomes (20) or membranes (17), the product is a fully crystallized microfibrillar material with a characteristic X-ray diffraction pattern. This we presume is also what happens in the early stages of protoplast regeneration when the modifying system is not yet operative (14) and it may also occur during synthesis of the chitinous core of the septum of S. commune. This crystalline chitin is very resistant to chemical and enzymatic degradation and probably not further accessible for modifying reactions. This would suggest that chitin after synthesis exists for a short period in an amorphous state during which modifying reactions can occur. The most obvious location to find chitin in this amorphous state is at the hyphal apex for it was shown that at this site chitin is not yet linked to β-glucan.

It was indeed shown that newly synthesized chitin of S. commune is highly susceptible to chitinase degradation and solubilization by dilute mineral acid. With time this chitin became gradually more resistant probably due to crystallisation and/or coupling to glucan chains (41). By autoradiography of shadowed preparations it was shown that this susceptible chitin is only present in growing hyphal apices but not in nongrowing ones where chitin is already linked to glucan or is crystallised and therefore more resistant to chemical and enzymatic treatments (41). Similarly, chitin synthesised in vitro was shown to be susceptible to chitinase for a short period (1,35).

Calcofluor was reported to interfere with the crystallisation of both β-chitin and cellulose (42). We found that also the crystallisation of α-chitin can be prevented by calcofluor for chitin synthesized in the presence of this substance remains sensitive to chitinase and X-ray diffraction analysis revealed no distinct deflection lines of crystalline chitin as long as the material was kept in a moist condition. As soon as the material became dry the characteristic deflection lines of crystalline α-chitin appeared.

CONCLUSION

According to our model (7,12,41) the newly synthesized wall at the hyphal apex has plastic properties and is extensible by turgor pressure. The chitin of this extensible wall is continuously synthesized probably at the plasma membrane according to a gradient ensuring maximal synthesis at the extreme tip and low synthesis at the base of the extension zone. How this gradient in chitin synthase activity arises is still unknown but the gradient may be intimately associated with the polarized cytoplasmic structure at the apex. It appears that the newly synthesized chitin chains at the apex exist for some time in an amorphous (non-crystalline) condition, making them accessible for reactions which attach β-glucan chains or peptides or just deacetylate the chains. Subsequent crystallisation may involve unmodified chains, producing perfect crystallites, or partly modified chitin, which can only form hydrogen bonds in limited regions. In effect, one could conjecture that microfibrils are formed with a rather crystalline core surrounded by less perfectly crystallized chitin chains which are covalently linked to the matrix of the wall. Together with hydrogen-bonding among glucan chains this could result in a cross-linked network which confers rigidity to the wall. Such a model would not be unlike that proposed for primary walls of plants on the basis of different constituent polymers. In the case of apical wall synthesis in fungi we presume that the process of rigidification of the wall at the base of the extension zone has progressed to such an extent that the wall no longer yields to the turgor pressure and the tubular form of the hypha is attained.

REFERENCES

1. G. W. Gooday and A. P. J. Trinci, Wall structure and biosynthesis in fungi, in: "The eukaryotic microbial cell . Soc. Gen. Microbiol. Symp. 30", G. W. Gooday, D. Lloyd and A. P. J. Trinci, eds., Cambridge University Press, Cambridge (1980).
2. J. G. H. Wessels and J. H. Sietsma, Fungal cell walls: A survey, in: "Encyclopedia of Plant Physiology, Vol 13B", W. Tanner and F. A. Loewus. eds., Springer Verlag, Berlin (1981).
3. J. H. Sietsma and J. G. H. Wessels, Solubility of (1-3)-β-D/(1-6)-β-D-glucan in fungal walls: Importance of presumed linkage between glucan and chitin. J. Gen. Microbiol. 125:209 (1981).
4. J. H. Sietsma and J. G. H. Wessels, Evidence for covalent linkages between chitin and β-glucan in a fungal wall. J. Gen. Microbiol. 114:99 (1979).
5. L. L. Davis and S. Bartnicki-Garcia. Chitosan synthesis by the tandem action of chitin synthetase and chitin deacetylase from Mucor rouxii. Biochem. 23:1065 (1984).
6. R. Datema, J. G. H. Wessels and H. van der Ende, The hyphal wall of Mucor mucedo. Eur. J. Biochem. 80:621 (1977).
7. J. G. H. Wessels and J. H. Sietsma, Cell wall synthesis and hyphal morphogenesis: A new model for apical growth. In: "Cell walls '81", D. G. Robinson and H. Quader, eds., Wissenschaftliche Verlagsgesellschaft, Stuttgart (1981).
8. J. Jelsma and D. R. Kreger. Ultrastructural observations on (1-3)-β-D-glucan from fungal cell walls. Carbohydr. Res. 43:200 (1975).
9. R. H. Marchessault and Y. Deslandes, Fine structure of (1-3)-β-D-glucans: curdlan and paramylon. Carbohydr. Res. 75:231 (1979).
10. J. H. Sietsma and J. G. H. Wessels, Chemical analysis of the hyphal wall of Schizophyllum commune. Biochim. Biophys. Acta 496:225 (1977).
11. P. van der Valk, R. Marchant and J. G. H. Wessels, Ultrastructural localization of polysaccharides in the wall and septum of the basidiomycete Schizophyllum commune. Exp. Mycol. 1:69 (1977).
12. J. G. H. Wessels, J. H. Sietsma and A. S. M. Sonnenberg, Wall synthesis and assembly during hyphal morphogenesis in Schizophyllum commune. J. Gen. Microbiol. 129:1607 (1983).
13. A. S. M. Sonnenberg, J. H. Sietsma and J. G. H. Wessels, Biosynthesis of alkali-insoluble cell wall glucan in Schizophyllum commune protoplasts. J. Gen. Microbiol. 128:2667 (1982).
14. P. van der Valk and J. G. H. Wessels, Ultrastructure and localization of wall polymers during regeneration and reversion of protoplasts of Schizophyllum commune. Protoplasma 90:65 (1976).
15. P. van der Valk and J. G. H. Wessels, Light and electron microscopic autoradiography of cell wall regeneration by Schizophyllum commune protoplasts. Acta Bot. Neerl. 26:43 (1977).
16. A. Duran, B. Bowers and E. Cabib, Chitin synthetase zymogen is attached to the yeast plasma membrane. Proc. Nat. Acad. Sci. 72:3952 (1975).
17. C. A. Vermeulen, M. B. J. M. Raeven and J. G. H. Wessels. Localization of chitin synthase activity in subcellular fractions of Schizophyllum commune protoplasts. J. Gen. Microbiol. 114:87 (1979).
18. E. Cabib, A. Duran and B. Bowers, Localized activation of chitin synthetase in the initiation of yeast septum formation. In: "Fungal walls and hyphal growth", J. H. Burnett and A. P. J. Trinci, eds., Cambridge University Press, Cambridge (1979).
19. S. Bartnicki-Garcia, C. E. Bracker, E. Reyes and J. Ruiz-Herrera, Isolation of chitosomes from taxonomically diverse fungi and synthesis of chitin microfibrils in vitro. Exp. Mycol. 2:173 (1978).
20. C. E. Bracker, J. Ruiz-Herrera and S. Bartnicki-Garcia, Structure and transformation of chitin synthase particles (chitosomes) during

microfibril synthesis in vitro.Proc.Nat.Acad.Sci. 73:4570 (1976).

21. J. Hernandez, E. Lopez-Romero, J. Cerbon, J. Ruiz-Herrera, Lipid analysis of chitosomes, chitin synthesizing micro vesicles from Mucor rouxii. Exp. Mycol. 5:349 (1981).

22. R. Sentandreu, A. Martinez-Ramon and J. Ruiz-Herrera, Localization of chitin synthase in Mucor rouxii by an autoradiographic method. J. Gen. Microbiol. 130:1193 (1984).

23. S. Bartnicki-Garcia and C. E. Bracker, Unique properties of chitosomes, in: "Microbial cell wall synthesis and autolysis", C. Nombela, ed., Elsevier Science Publishers, Amsterdam (1984).

24. D. M. Rast and S. Bartnicki-Garcia, Effects of amphotericin B, nystatin and other polyene antibiotics on chitin synthase. Proc. Nat. Acad. Sci. 78:1233 (1981).

25. M. B. J. M. Raeven, Biosynthesis of chitin in fungi. Thesis University of Groningen (1981).

26. A. Duran and E. Cabib, Solubilization and parial purification of yeast chitin synthetase. Conformation of the zymogenic nature of the enzyme. J. Biol. Chem. 253:4419 (1978).

27. C. A. Vermeulen and J. G. H. Wessels, Evidence for a phospholipid requirement of chitin synthase in Schizophyllum commune. Curr. Microbiol. 8:67 (1983).

28. E. Cabib, B. Bowers and R. L. Roberts, Vectorial synthesis of a polysaccharide by isolated plasma membranes. Proc. Nat. Acad. Sci. 80: 3318 (1983).

29. D. H. Wolf and C. Ehmann, Studies on a proteinase B mutant of yeast. Eur. J. Biochem. 98:375 (1979).

30. G. S. Zubenko, A. P. Mitchell and E. W. Jones, Septum formation, cell division and sporulation in mutants of yeast deficient in proteinase B. Proc. Nat. Acad. Sci. 76:2395 (1979).

31. J. Ruiz-Herrera and S. Bartnicki-Garcia, Proteolytic activation and inactivation of chitin synthetase from Mucor rouxii. J. Gen. Microbiol. 97:241 (1976).

32. D. Katz and R. F. Rosenberger, Hyphal wall synthesis in Aspergillus nidulans: effect of protein synthesis inhibitor and osmotic shock on chitin insertion and morphogenesis. J. Bact. 108:184 (1971).

33. E. Lopez-Romero, S. Bartnicki-Garcia and J. Ruiz-Herrera. Purification of inhibitory protein of chitin synthase from Mucor rouxii. Biochim. Biophys. Acta 525:338 (1978).

34. G. D. Craig, J. M. Campbell and J. F. Peberdy, Endogenous chitin synthase inhibitor in Aspergillus nidulans. Trans. Br. Mycol. Soc. 7:579 (1981).

35. E. Lopez-Romero, J. Ruiz-Herrera and S. Bartnicki-Garcia, The inhibitory protein of chitin synthetase from Mucor rouxii is a chitinase. Biochim. Biophys. Acta 702:233 (1982).

36. R. M. Brown Jr. and D. Montezinos, Cellulose microfibrils: Visualization of biosynthetic and orienting complexes in association with the plasma membrane. Proc. Nat. Acad. Sci. 73:143 (1976).

37. N. A. R. Gow. Transhyphal electrical currents in fungi. J. Gen. Microbiol. 130:3313 (1984).

38. S. C. de Vries and J. G. H. Wessels, Polarized outgrowth of hyphae by constant electrical fields during reversion of Schizophyllum commune protoplasts. Exp. Mycol. 6:95 (1982).

39. J. M. Picton and M. W. Steere, A model for the mechanism of tip extension in pollen tubes. J. Theor. Biol. 98:15 (1982).

40. R. J. Howard and J. R. Aist. Cytoplasmic microtubules and fungal morphogenesis. J. Cell Biol. 87:55 (1980).

41. C. A. Vermeulen and J. G. H. Wessels. Ultrastructural differences between wall apices of growing and non-growing hyphae of Schizophyllum commune. Protoplasma 120:123 (1984).

42. W. Herth. Calcofluor white and congo red inhibit chitin microfibril assembly of Poterioochromonas. J. Cell Biol. 87:442 (1980).

CHITIN SYNTHESIS, DIMORPHISM AND VIRULENCE IN THE PATHOGENIC FUNGUS CANDIDA ALBICANS

A. CASSONE

Medical Mycology Unit, Istituto Superiore di Sanità
00100 Rome, Italy

INTRODUCTION

Candida albicans is the most frequent aetiologic agent of human and animal candidosis. It is a polymorphic organism which, depending on temperature and the available nutritional source may grow and differentiate into yeast (Y), the unicellular budding form, pseudomycelial or true mycelial form (M), the hypha growing by apical extension, and chlamydospores[1]. C. albicans is present, under the saprobic Y form in the gastroenteric tract of all healthy individuals without causing disease. On occasions, the microorganism is given the opportunity to colonize and invade host's tissues and this usually happens when either serious underlying disease (such as diabetes of haemopoietic malignancy) or simply a digression from the normal physiological state (such as pregnancy or parenteral nutrition) is present[2].

In initiating the disease, the Y form often converts to M form, a process called dimorphism or dimorphic transition[3]. Although associated with the disease, the exact role of Y-M conversion in the pathogenesis of candidosis is still obscure and equally obscure are the mechanisms by which the Y form transforms to the M form "in vivo". It is known, however, that, both "in vitro" and "in vivo", the Y-M conversion occurs through the intermediary formation of a germ-tube[4] (Fig.1). Germ-tube formation may be induced by a number of factors with a mandatory temperature in the range 33-37°C[5]. Among all the numerous inducers of the process, the acetylated hexosamine N-acetyl-D-glucosamine (GlcNAc) is a most simple and reliable one[6,7]. As this inducer is both the natural precursor of chitin and a powerful allosteric activator of chitin synthase[8], it has been logical to wonder whether 1. chitin synthase was activated during germ-tube formation; 2. there were modulations in chitin amount during the different stages of germ-tube formation; 3. there was any relationship between the mechanism(s) through which GlcNAc induces Y-M transition and a possible activation of chitin synthase.

This paper synthetically reviews the experimental evidence about these fundamental questions that may contribute to a better understanding of the enzymatic, morphological and pathogenic events correlated with germ-tube formation and virulence of C.albicans.

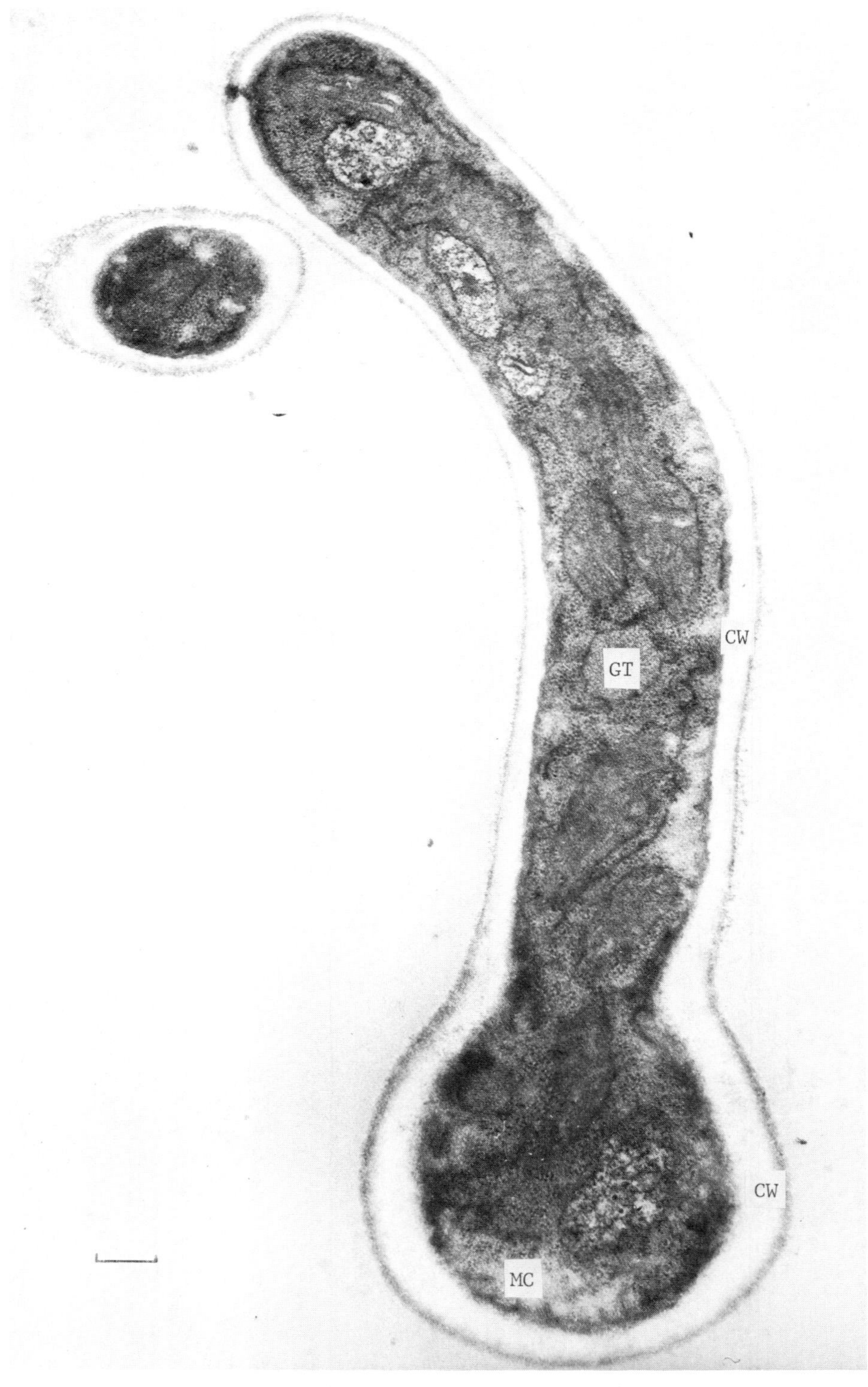

Fig.1. A germ-tube of a high-responder strain of Candida albicans after 300 min in GlcNAc medium. MC, mother yeast cell;GT,germ-tube; CW,cell wall. The magnification bar corresponds to 0.25 microns.

RESULTS

1. Activation of chitin synthase during germ-tube formation

The enzyme chitin synthase of *C.albicans* is essentially located in the cytoplasmic membrane, under inactive zymogenic form that can be activated by proteinases. Some activity is, however, expressed "in vivo" without apparent need of exogenous proteinase activation. Total enzyme activity is therefore the experimental sum of the expressed plus the zymogenic chitin synthase activity [9]. Assay of the enzyme therefore involves preparation of plasmamembrane and incorporation of labelled GlcNAc from UDP-GlcNAc precursor into a product having the characteristics of chitin (insolubility, chemical composition, etc).

Using this or comparable approaches, there have been numerous reports showing that during germ-tube formation induced by either GlcNAc or other compounds, the expressed activity of a crude preparation of the enzyme increases significantly, therefore suggesting "in vivo" physiological activation. In particular, Chiew et al.[9] showed that during germ-tube formation the specific activity of total chitin synthase increased by about 50% in the initial stages, then returned to basal (yeast) level. The expressed activity of the enzyme was calculated to increase of about 5 times. Braun and Calderone[10] also detected augmented expression of enzyme activity during germ-tube formation, although this occurred in a medium different from the one containing GlcNAc. Further, these authors showed by autoradiography that the increased chitin synthase activity was localized at the growing tip of the germling.

In general these studies did not include non-responder strains of *C. albicans* i.e. strains which do not form germ-tubes after stimulation with GlcNAc or other triggers and can be useful as controls. We have recently isolated variants which are either unable to respond to GlcNAc with germ-tubes or responded very poorly to induction (less than 20% germ-tubes formed in 300 min)[11]. These strains contrast with some others which vigourously germinate (practically 100% germlings in 300 min at 37°C) upon simple exposure to GlcNAc in a basal saline. The presence of such variants allowed a comparison of their response with activation of chitin synthase under various experimental conditions. The experiments along this line were performed recently by Sullivan and collaborators (P.A.Sullivan, *personal communication*) by the use of a mixed membrane preparation of the enzyme. Altogether, the results showed that when the cells of a high responder strain were incubated under conditions favoring mycelial development in GlcNAc, the expressed chitin synthase activity increased significantly with respect to that of starting Y cells. No such an increase was noted when yeast cells were incubated without the inducer or at 28°C, a germ-tube non-permissive temperature (Table I). There were conditions, however, under which marked activation of chitin synthesis occurred even in non-germinative strains and independently upon exposure to inducer.

2. Chitin content during germ-tube formation

The importance of changes in chitin content for the construction of the specific shape of mycelial wall dates from Chattaway's studies[12] showing that the cell wall of the hyphal cells of *C.albicans* contained in its alkali

Table I. Expressed activity (nmol/min/mg protein) of chitin synthase in a high-responder strain of *C.albicans*.[a)]

Condition	Activity	% Increase
No inducer, 37°C	0.04	- -
No inducer, 28°C	0.06	50
Inducer, 28°C	0.07	75
Inducer, 37°C	0.14	350

a) for methodology details, see Ref.9.

insoluble fraction 3 times as much chitin as that present in the analogous fraction of yeast cell wall. Moreover, by the use of special cytochemical techniques, Cassone et al.[13] showed that the inner wall of the emerging germ-tube was enriched with beta 1-3 glucan and chitin and it is now of interest that the former polymer is the predominant form of insoluble glucan synthesized by the germling[14]. The earlier reports of increased chitin content of the mature hyphal wall in systems which were germinated in complex media have been largely confirmed and somewhat extended for the cell wall of germ-tube formed in presence of GlcNAc. Table 2 compares two sets of five independent determinations obtained with two different, high responder strains of *C.albicans*. It is easily observed that there is a net increase in the specific chitin content of the cell wall, ranging from 4 to about 6 times and that the increase is not due to new overall cell wall synthesis which accompanies germination, as shown by total polysaccharide determination.

In fact, the two other principal cell wall polysaccharides,i.e. mannan and glucan, while increasing their total content, did not change significantly either their proportion or their specific content over the yeast form (Table 3). Of interest is also that most of the relative increase of chitin content of the mycelial form respect to yeast form occurred relatively early in morhogenesis (Table 3). Note that, in assessing the significance of analytical determinations of chitin, the large variability of the individual determinations should be appreciated that might explain the different results obtained in different laboratories using comparable methods. Overall, there seems to be little doubt that the mycelial wall of *C.albicans* has more chitin than the yeast wall and that this increased chitin content is amplified when germ-tubes are induced in GlcNAc-medium.

3. Chitin visualization

Most of the experimental approaches used to study chitin content and synthesis during morphogenesis in *C.albicans* have some strong,inherent li-

Table 2. Chitin content during germ-tube formation in two high-responder strains of C.albicans[a)]

Form of growth	Chitin A72	Chitin BP	Total saccharides A72	Total saccharides BP
Yeast, 0 min	0.95(0.7)	1.6(1.0)	82(7.4)	74(6.0)
Germ-tube, 300 min	3.7(1.4)	8.5(3.1)	87(9.4)	89(8.5)

a) Values expressed as mg % dry cell walls (SD) as from 5 independent determinations. For the analytical methods, see Refs. 9, 11, 12.

Table 3. Principal cell wall polysaccharides : ratio of their content between mycelial and yeast form[a)]

Time	Chitin	Mannan	Sol.Glucan	Ins.Glucan
30 min, no germlings	3.8	1.06	1.3	0.8
120 min, early germ-tubes	4.6	1.0	1.4	0.9
300 min, germ-tubes well formed	5.9	0.99	1.6	0.9

a) From a representative experiment (single determination) with the strain BP germinated in GlcNAc. For the analytical determinations, see Ref. as in Table 2.

mitations. For instance, what is the true control of chitin content in the yeast form if this content fluctuates during growth and in the mother cell even during germ-tube formation? Also,if chitin has any role in mycelial development it should be played locally, i.e. by regulation of growth and cell wall physiology at the hyphal apex, as aptly exemplified by Cabib's work on septum formation[15]. Measurements of both chitin content and enzymic determinations do not take into account these simple questions. Cytochemical visualization of chitin under the electron microscope may help resolving these particular points. It has been already shown[10] by autoradiography that increased chitin synthase is detectable at the growing tip of the germ-

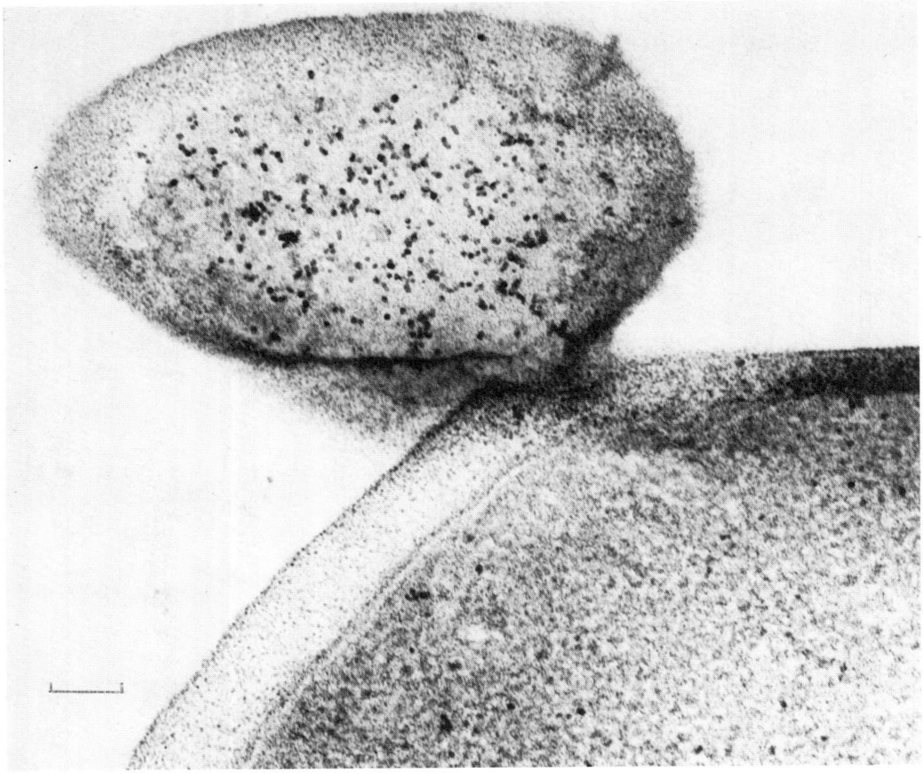

Fig.2. Cytochemical visualization of chitin by WGA-gold markers. For details, see text. Magnification bar :0.25 microns.

tube. Recently, electron-microscopic visualization of chitin in thin sections of chemically-fixed material has been made possible by the use of the specific marker wheat germ agglutinin (WGA) which can be conveniently conjugated with dense markers such as colloidal gold particles[16,17]. We have adapted this technique to chitin detection during growth and germ-tube formation in several variants of C.albicans. As expected, much of the marker was seen to be deposited at the growing tip of the germ-tube (Carruba et al.,manuscript in preparation) but some was also seen in vacuoles and scattered in the cell wall. In grazing sections through the wall of the emerging germling, WGA-gold marker was specially concentrated in the innermost layers (Fig.2). Although no quantitative estimation is possible by this technique, the results are indirect confirmation that germ-tube formation in C.albicans is charcterized by increased chitin synthesis and deposition at places in the cell wall where chitin may decisively contribute to the new cell wall building and physiology.

CONCLUSIONS

Within the limitations already indicated, it seems sufficiently clear that 1. chitin content significantly increases during germ-tube formation, in general, and specifically when the process is induced by GlcNAc. This increase, possibly accompanied by new beta 1-3 glucan[14], is not attributable to the overall, new cell wall synthesis that occurs prominently in this model of induction[11]; 2. the expressed chitin synthase activity augments during morphogenic induction in high responder strains, suggesting that chitin synthesis is significantly activated when cell undergoes yeast-mycelial conversion; 3. although preliminary, the evidence so far accumulated points out to a localization of activated synthesis of chitin at the emerging area of the germ-tube and in the innermost cell wall layers. These studies do certainly support a role for chitin as a determinant in the construction of specific mycelial shape and, therefore, in direct terms, as participant in the invasiveness of the fungus. The morphogenic compound N-acetylglucosamine may be a direct or indirect element in the activation of chitin synthase but further experiments are required to identify the specific function of this aminosugar in the induction of morphogenesis. However, this model of morphogenesis remains the most suitable in future studies on the relationship between chitin modulation and morphology, hence virulence in C.albicans.

ACKNOWLEDGEMENT

This work was in part carried out under contract N°83.02916.52, Special Project "Control of Infectious Diseases" of the Italian National Research Council, Rome.

REFERENCES

1. A. Cassone, P. A. Sullivan, and M. G. Shepherd, N-Acetyl-D-glucosamine induced morphogenesis in Candida albicans, Microbiologica, 8:85 (1985).
2. F. C. Odds, "Candida and Candidosis", Leicester University, Leicester (1979).
3. S. Bartnicki-Garcia, Cell wall chemistry, morphogenesis and taxonomy of fungi, Ann. Rev. Microbiol., 22:87 (1968).
4. A. Cassone, N. Simonetti, and V. Strippoli, Ultrastructural changes in the wall during germ-tube formation from blastospores of Candida albicans, J. Gen. Microbiol., 77:417 (1973).
5. E. Mattia, and A. Cassone, Inducibility of germ-tube formation in Candida albicans at different phases of yeast growth, J. Gen. Microbiol., 113:439 (1979).
6. N. Simonetti, V. Strippoli, and A. Cassone, Yeast-mycelial conversion induced by N-acetyl-D-glucosamine in Candida albicans, Nature, 250:344 (1974).
7. M. G. Shepherd, Y. Y. Chiew, S. P. Ram, and P. A. Sullivan, Germ-tube induction in Candida albicans, Can. J. Microbiol., 26:21 (1980).
8. G. W. Gooday, Biosynthesis of the fungal wall: mechanisms and implications, J. Gen. Microbiol., 99:1.
9. Y. Y. Chiew, M. G. Shepherd, and P. A. Sullivan, Regulation of chitin synthesis during germ-tube formation in Candida albicans, Arch. Microbiol., 125:97 (1980).
10. P. C. Braun, and R. A. Calderone, Chitin synthesis in Candida albicans: comparison of yeast and hyphal forms, J. Bacteriol., 133:1427 (1978).
11. E. Mattia, G. Carruba, L. Angiolella, and A. Cassone, Induction of germ-tube formation by N-acetyl-D-glucosamine in Candida albicans: uptake of inducer and germinative response, J. Bacteriol., 152:555 (1982).
12. F. W. Chattaway, M. R. Holmes, and A. J. E. Barlow, Cell wall composi-

tion of yeast and mycelial forms of Candida albicans, J. Gen. Microbiol., 51:367 (1968).
13. N. Simonetti, and A. Cassone, Dinamica della formazione di tubi germinativi in Candida albicans, Atti XV Congresso Nazionale di Microbiologia, Torino-Saint Vincent, 2:150 (1969).
14. M. G. Shepherd, and P. A. Sullivan, The control of morphogenesis in Candida albicans, J. Dent. Res., 76:42 (1984).
15. E. Cabib, R. Roberts, and B. Bowers, Synthesis of the yeast cell and its regulation, Ann. Rev. Biochem., 51:763 (1982).
16. D. Poulain, G. Tronchin, Dubremetz, and J. Biguet, Ultrastructure of the cell wall of Candida albicans blastospores: study of its layers, Ann. Microbiol., 129:141 (1978).
17. G. Tronchin, D. Poulain, and J. Biguet, Etude chitochimique et ultrastructurales de la paroi de Candida albicans, Arch. Microbiol., 123:245 (1979).

EFFECT OF CONGO RED ON YEAST MORPHOGENESIS

G. L. VANNINI, S. PANCALDI and F. POLI

Institute of Botany, University of Ferrara
Corso Porta Mare, 2
I-44100 Ferrara, Italy

INTRODUCTION

Valuable information about fungal morphogenesis can be gained by treating the cells with agents that vary the mode in which the wall components are synthesized and/or deposited. In yeasts, chitin, a structural polysaccharide responsible for fundamental morphogenetic events, is synthesized by enzyme complexes inserted in the plasma membrane (1). However, this structure, when studied by means of freeze etching and freeze fracturing, did not reveal aggregates of particles interpretable as the site of chitin synthesis (2). In any case, it is generally assumed that each enzyme complex is composed of as many units as are necessary to form the chains that, in the periplasmic space, stack into a crystalline microfibril (3).

Are synthesis and crystallization separable or inseparable events? Does the lateral aggregation of the chains operate as a control over the synthase complexes and as a means to maintain their integrity? In the attempt to furnish some indications on these topics, *Saccharomyces cerevisiae* and *Rhodotorula glutinis* were treated with Congo red (CR). The stain specifically binds to β-linked glucosides, particularly cellulose and chitin, and prevents their crystallization without blocking the synthetic process of the polymers (4). In *Saccharomyces*, chitin is confined to the primary septum and to the bud scars, whereas in *Rhodotorula*, the polymer is present in the entire wall and in the septum.

RESULTS

When observed under u.v. light immediately after CR addition, the two yeasts revealed a fluorescence pattern corresponding to chitin localization. Probably the fluorescence emission, in the red range, is dichroic since the unlinked CR was not fluorescent.

During an 18 h exposure to 0.25 mg/ml of CR, *S. cervisiae* continued to bud, but the separation between mother and daughter cells did not take place. In this way, multicellular aggregates of different size and complexity arose in which several aberrations of wall septa were evident (Fig. 1a). Precisely, diffuse materials were inserted within the original wall and the septa were thicker than normal and irregularly developed (Fig. 1b). The newly apposed constituents, fluorescent under u.v. light in both living cells and semithin sections, were unstained after periodic acid-thiocarbohydrazide-silver

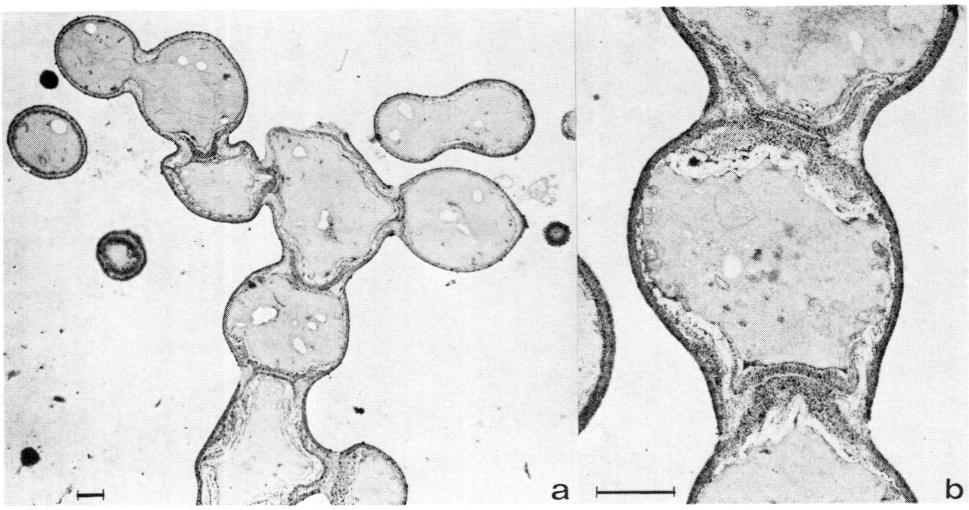

Fig. 1. Saccharomyces cerevisiae treated with Congo Red. a) Multicellular aggregate showing aberrant walls and septa. b) PATAg technique. Abnormally structured septa with large amounts of unstained components. In all micrographs the bar corresponds to 1 μm.

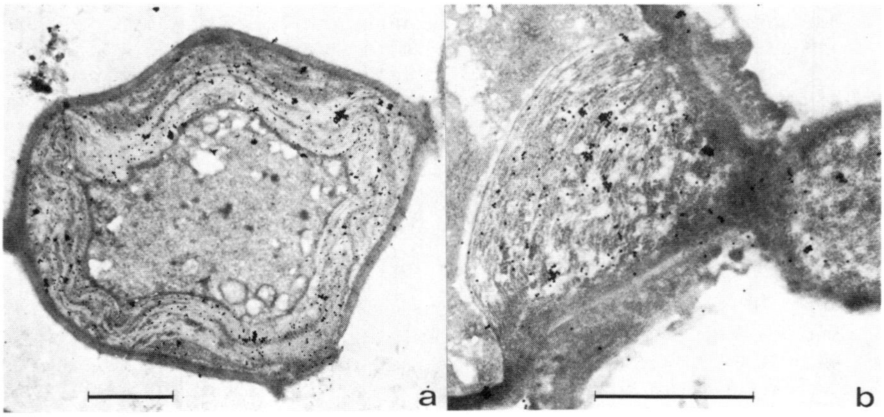

Fig. 2. WGA-Au labeling of a) a thickened cell wall in Saccharomyces cerevisiae and of b) an aberrant septum of Rhodotorula glutinis exposed to Congo Red.

proteinate (PATAg) treatment (Fig. 1b) and were labeled on thin sections by colloidal gold coupled with wheat germ agglutinin (WGA-Au) (Fig. 2a). The lectin binding to chitin was favoured by pretreating the plastic sections with a KOH saturated solution. Thus, it can be inferred that the aberrant materials were formed by irregularly arranged chains of N-acetylaglucosamine.

In some cases the wall broke, suggesting that the unassembled chains are available to lytic enzyme. At the cytoplasmic level, numerous vesicles of different size appeared. They are probably of a secretory type since the limiting membrane was stained by the PATAg technique and phosphotungstic acid similarly to the plasma membrane.

When treated with 0.10 mg/ml of CR, Rhodotorula glutinis almost immediately stopped multiplying. During the entire period of treatment (18 h), no multicellular aggregate was noted. This behaviour is in accordance with the fact that, in this basidiomycetous yeast, new chitin fibrils are indispensable for bud growth. Under the electron microscope, it was evident that the wall structure and the septum area were modified by the insertion of aberrant materials (Fig. 2b) which, for cytochemical reactivity and fluorescence emission, were similar to those appearing in Saccharomyces.

DISCUSSION

The results suggest that, in yeasts, chitin synthesis and crystallization can be artificially uncoupled and that the first process continues outside spatial and temporal controls when the second is prevented. The oversynthesis of materials with a random distribution in the periplasmic space suggests that the enzyme complexes are spread over the plasma membrane. We propose that the integrity of the chitin-synthesizing complexes depends on the ordered assembly of their products and vice versa. Precisely, the polymers could react with CR as soon as they form; consequently they could assume a disordered disposition that forces the synthase complexes to dissociate into single units which move freely and randomly along the plasma membrane. Thus, too many chains are produced and wall zones normally not involved in chitin synthesis become accumulation sites of amorphous materials. The defects in chitin crystallization disturb cell separation but not bud development in S. cerevisiae. In R. glutinis, instead, the main steps in wall morphogenesis are altered.

REFERENCES

1. E. Cabib, R. Roberts and B. Bowers, Synthesis of the yeast cell wall and its regulation, Ann. Rev. Biochem. 5:763 (1982).
2. A. Maurer and K. Muhlethaler, Isolation and characterization of paracrystalline arrays of the plasma membrane of baker's yeast Saccharomyces cerevisiae, Biochem. J. 105:189 (1982).
3. S. Bartnicki-Garcia, J. Ruiz-Herrera and C.E. Bracker, Chitosmoses and chitin synthesis, in: "Fungal Walls and Hyphal Growth", J. H. Burnett and A. P. J. Trinci, eds., Cambridge University Press, Cambridge (1979).
4. H. Quarter, D. G. Robinson and R. Van Kempen, Cell wall development in Oocystis solitaria in the presence of polysaccharide binding dyes, Planta 157:317 (1983).

ROLES OF CHITINASES IN FUNGAL GROWTH

G. W. GOODAY, A. M. HUMPHREYS and W. H. McINTOSH

Department of Microbiology, Marischal College, University of Aberdeen
Aberdeen AB9 1AS, United Kingdom

INTRODUCTION

There are three roles for chitinases in fungi:

(a) Most spectacularly, they are involved in the gross autolysis associated with the release of spores in some basidiomycete fruit bodies. Examples include the maturation of puffballs, *Lycoperdon* species[1], and the autodigestion of gill tissue following spore release in the ink-caps, *Coprinus* species[2].

(b) They have a nutritional role. In the case of soil saprophytes such as *Aspergillus* species[3,4] and *Trichoderma* species[5], chitinase enables them to utilise chitinous debris from dead invertebrates and fungi as food sources. In the case of pathogens of crustacea, insects and fungi, chitinases also enable them to penetrate their hosts. Examples include the crayfish pathogen *Aphanomyces astaci*[6], insect pathogens such as *Beauvaria* and *Cordyceps* species[7,8]. There is however no convincing evidence that there is any appreciable re-cycling of the chitin in cell walls[9,10] even though chitinase accumulates to a marked extent in old cultures [11,12], i.e. there does not seem to be any appreciable autophagic utilisation of chitin in starving cultures.

(c) They have a morphogenetic role in the growth and differentiation of all chitin-containing fungi. This, the most fundamental of the three roles, has however proved the most difficult to obtain evidence for. Indeed some authors question whether chitinase and other lytic enzymes are involved in hyphal apical growth. Burnett[13] suggests that "teleologically speaking, the apex would seem to be a most dangerous location for a lytic entity!", and Wessels[14] proposes a model for hyphal growth that does not require wall lytic enzymes. The unitary model for hyphal growth proposed by Bartnicki-Garcia[15] does however see the control of apical wall growth as being the result of a "delicate balance between wall synthesis and wall lysis", with some of the vesicles being transported to the apex containing lytic enzymes, to keep the wall at the apex in a plastic and extensible condition. Certainly in every case investigated chitinase activities can be detected in actively growing chitinous fungi. Examples include *Mucor* and *Phycomyces* species[16-20], *Neurospora crassa*[21,22] *Aspergillus nidulans*[9,23], *Saccharomyces cerevisiae*[24] and *Candida albicans* [25].

The rest of this article asks the question "is chitinase involved in the vegetative growth of hyphae?"

CHITINASE SYSTEMS IN *MUCOR MUCEDO*

When *Mucor mucedo* is grown on agar, three sites of chitinase activity can be distinguished:

(i) High-speed centrifugation of cell homogenates yields a *supernatant activity*[17-19]. From 18 hour cultures this is the major activity. It has not yet been purified, and may be heterogeneous.

(ii) This centrifugation also yields a *microsomal activity*. Quantitatively this is only a minor activity, but it has important properties: it is membrane-bound, showing a bi-phasic Arrhenius plot and requiring phospholipids; it can be solubilised by mild detergents such as Triton X-100 or Zwittergent-14;it is partially zymogenic, being activated by trypsin; and as we shall show below, it appears to be physically associated with chitin synthase.

(iii) There is *extracellular activity*. This has not been characterised further.

We have no further information on the sub-cellular localisations of the supernatant or microsomal activities, or on any relationships between the enzymes in these three sites. In *Saccharomyces cerevisiae* about half of the chitinase was in intracellular vesicles, and the other half was in the periplasm, clearly suggesting a process of storage and secretion[24].

Cellular chitinase during growth

Using methods described before[17-19], the microsomal and supernatant chitinase activities were assayed during the growth of *Mucor mucedo* strain Z46 (+) on agar (Table 1). Under these conditions, germ tubes formed at about 12 hours, the first branches at about 15 hours, exponential growth ceased at about 30 hours, and sporulation occurred at about 48 hours. The sporangiospores had detectable chitinase activities. The ratio of microsomal:supernatant chitinase was much higher in spores and in early germinating spores than in the resultant mycelial cultures. The ratio of native microsomal chitinase activity to trypsin-treated activity increased dramatically during the first 5 minutes of germination. The specific activity of the supernatant chitinase showed little change from spores and then throughout mycelial growth and differentiation.

Association between microsomal chitinase and chitin synthase

In these cultures of *Mucor mucedo*, the microsomal chitinase and chitin synthase have been shown to have the following properties in common: (i) they are both membrane-bound, requiring phospholipids for activity[17-19] ,and (ii) they are both partially zymogenic, being activatable by endogenous or exogenous proteases[17-19] . From the results in Table 1 it is seen that the activities of the two enzymes follow very similar trends during growth of a culture. Both have high specific activities in spores, and both show a very marked increase in ratio of native to trypsin-activated activity after 5 minutes germination.

Evidence for a physical association between the two enzymes comes from two types of experiment; assays for net chitin synthesis *in vitro* and *in situ*.

Table 1. Chitinase and chitin synthase activities during growth of cultures of *Mucor mucedo* on agar

	Time (hours)						
	0	0.1	16	24	48	72	96
Growth phase*	G	G	E	E	S	S	S
Dry wt (mg)	0.6	0.6	12	88	474	480	468
Chitinase act.⁺							
Supernatant	0.60	0.79	2.9	21.8	121.9	183.3	129.8
	(46.7)	(42.0)	(40.5)	(47.4)	(41.9)	(62.9)	(22.3)
Native micro.	0.06	0.20	0.04	0.07	0.51	0.45	0.19
	(5.1)	(10.5)	(0.55)	(0.15)	(0.18)	(0.18)	(0.04)
Trypsin + micro.	0.13	0.12	0.12	0.28	0.94	0.98	0.50
	(9.8)	(6.1)	(1.4)	(0.62)	(0.33)	(0.38)	(0.11)
Ratio micro:sup.	0.22	0.15	0.04	0.013	0.008	0.005	0.004
Chitin synthase @							
Native	0.79	0.84	2.7	16.2	65.0	59.2	90.5
	(2.2)	(4.2)	(0.42)	(0.59)	(0.31)	(0.30)	(0.45)
Trypsinized	1.47	0.55	6.3	20.9	102.6	67.1	113.1
	(4.1)	(2.8)	(0.87)	(0.76)	(0.50)	(0.34)	(0.57)

*G, germination: E, exponential; S, stationary.

⁺Prepared and assayed as described in ref. 17. Units are total activity from two 9 cm Petri dishes: nmol diacetylchitobiose. min $^{-1}$; and in parentheses (specific activity): nmol. min^{-1}. (g fresh wt)$^{-1}$. Ratio is total trypsin-treated microsomal activity: total supernatant activity.

@Chitin synthase prepared and trypsinized as for chitinase; assayed as described in ref. 26 but at pH 6.0. Units are total activity from two 9 cm Petri dishes as assayed by t.l.c/autoradiography (c.f. Table 2): nmol GlcNAc converted to chitin/diacetylchitoboise. min^{-1}; and in parentheses (specific activity): nmol. min^{-1}. (g fresh wt)$^{-1}$.

Net chitin synthesis in vitro. Microsomal preparations of *Mucor mucedo* that had grown on agar differed from those of the same fungus grown in submerged culture (and from all other fungal preparations made by us) in that in a standard chitin synthase assay, i.e. incubation with radioactive UDP-*N*-acetylglucosamine and magnesium chloride in buffer followed by counting of insoluble radioactive product, virtually no activity was detected. On closer inspection however this was not due to lack of chitin synthase activity. Thin layer chromatography of the products (e.g. exp. 1, Table 2) revealed that the two enzymes were acting in consort, to give the product diacetylchitobiose.

$$2nUDPGlcNAc \longrightarrow (GlcNAc)_{2n} \longrightarrow n(GlcNAc)_2$$

Thus between pH values of 5.5 and 7.5, there was very little chitin, but appreciable amounts of diacetylchitobiose were formed (Table 2). In contrast, preparations from *Coprinus cinereus* gave virtually no diacetylchitobiose, but appreciable amounts of chitin (Table 2). (These results however must be interpreted in the light of the differing pH optima for these enzymes[17,26]: *Mucor mucedo* chitinase,5.55; *Mucor mucedo* chitin synthase, 5.5; *Coprinus cinereus* chitin synthase, 8.0).

Table 2. Percentage recoveries of radioactivity from thin layer chromatograms of assays of microsomal preparations from *Mucor mucedo* and *Coprinus cinereus* for chitin synthase*

Preparation	pH	Chitin	$(GlcNAc)_2$	UDPGlcNAc
Mucor exp.1	6.0	0.4	37.8	61.8
Mucor exp.2	5.5	1.8	28.3	69.9
	6.5	8.2	14.1	77.7
	7.5	2.6	15.9	81.5
Mucor exp.3	7.5	2.1	7.2	90.7
Mucor exp.4	7.5	0.8	13.6	85.6
Coprinus exp.2	5.5	40.5	0	59.5
	6.5	75.1	0.7	24.2
	7.5	85.3	1.2	13.5
Mucor & *Coprinus* exp.2	5.5	30.0	12.8	57.2
	6.5	83.0	5.7	11.3
	7.5	87.3	3.1	9.6

*Microsomal preparations were made as described before[17,27]. Assays contained 10 µl preparation, 5 µl in the case of the *Mucor* & *Coprinus* mixtures, together with 35 µl buffer (50 mM-MES, 10mM-$MgCl_2$ 1mM-EGTA, as final conc), and 5 µl substrate (0.6 mM-UDP[^{14}C]-GlcNAc, 13.75 nCi, 25 mM-GlcNAc, as final conc.). Incubation was 1 h at 30°C. Reaction was stopped at 100°C for 1 min. Assay mixture was chromatographed on Merck silica gel 60-F254 plates with propan-1-ol: water: ammonia (70:30:1), autoradiographed, and radioactive areas scraped and counted.

It is well known that nascent chitin (i.e. just synthesised, prior to its crystallisation) is more susceptible to chitinase degradation than α-chitin[14], so it may not be surprising that the *Mucor mucedo* chitin is rapidly degraded by the chitinase in the preparation. If it were just the fact that the two enzymes were both present in the preparation that could explain the results in Table 1, then it might be expected that the *Mucor mucedo* chitinase would degrade nascent chitin produced by *Coprinus cinereus* enzyme just as readily. To test this idea, mixtures of the two enzyme preparations were incubated (Table 2). It is seen however that chitin still accumulated in these mixtures. The amounts of the two products chitin and diacetylchitobiose that accumulate resembled those to be expected if the assays had been done separately, boiled and then mixed. This result is consistent with the idea that the chitinase in the *Mucor mucedo* microsomes is intimately associated with the chitin synthase. While "its own" enzyme is making chitin, other sources of chitin such as that from *Coprinus cinereus* microsomes are poor substrates.

Net chitin synthesis in situ. Chitin synthesis has been studied in fungi *in situ* by treating cells with toluene[28], toluene/ethanol/Triton X-100[29], and digitonin[30]. Here we have used the technique of freeze-thawing strips of agar-grown mycelium to permeabilize *Mucor mucedo* so that the tissue would incorporate radioactive UDP-GlcNAc. The insoluble radioactive product was equated with chitin, for although *Mucor mucedo* also makes chitosan[31] this will be via the action of a deacetylase on chitin [32], and so any chitosan will retain the radioactivity of the chitin. On incubating freeze-thawed rectangles of young mycelium with the chitin synthase assay mix, chitin accumulated for 20 minutes (Fig. 1a). After this time, however, radioactivity was lost from the tissue at a steady rate. Taking 20 minutes as an assay time, the rate of incorporation declined markedly as the mycelium aged (Fig.1b). That this incorporation

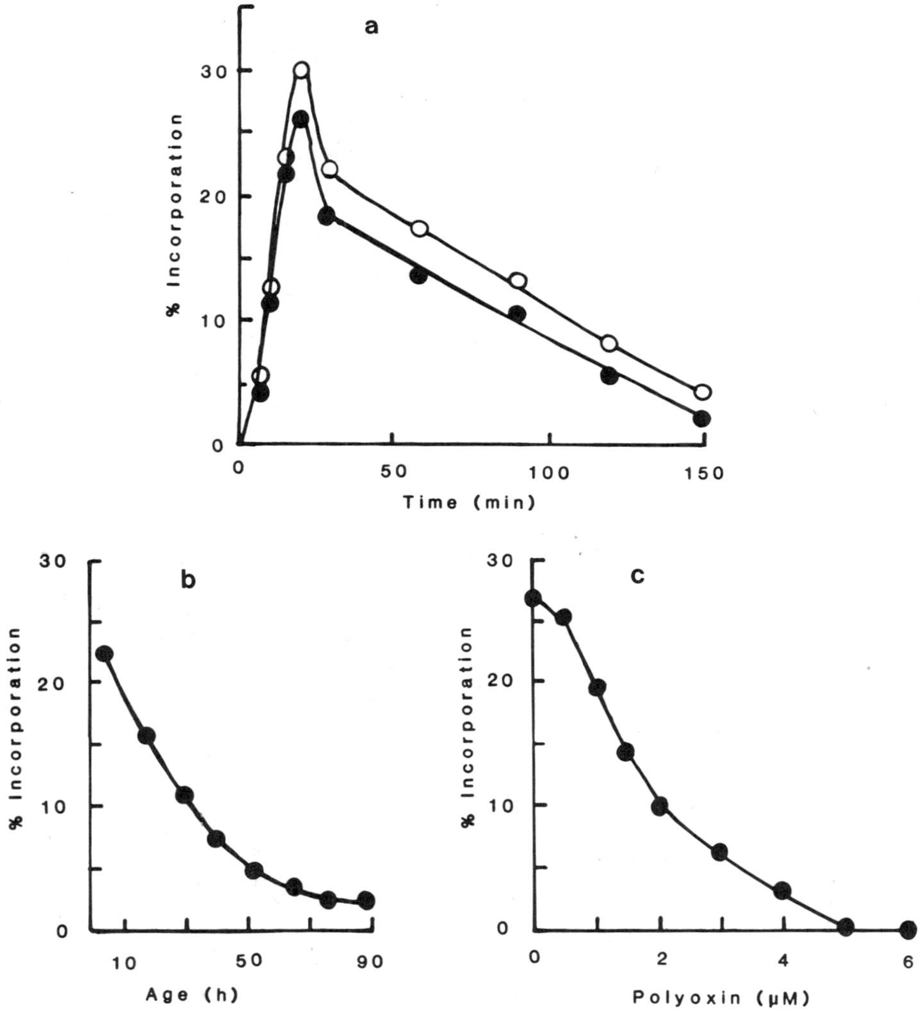

Fig.1. Chitin synthase assayed in situ in Mucor mucedo. Cultures were grown at 20°C in 100 mm square Petri dishes on Cellophane for 100 h on 40 ml 2% malt, 2% glucose, 2% agar, inoculated along one edge with spore suspensions in a strip of filter paper. Strips of mycelium (6 x 16 mm) were cut out, transferred to 1.2 ml centrifuge tubes, and subjected rapidly to six freeze-thaw cycles of immersion in liquid nitrogen and then a water bath at 20°C. Assay mixture was added (60 µl) containing a 50 mM potassium phosphate buffer pH 6.0,10 mM-$MgCl_2$, 1 mM-Na_2EDTA, 10mM-GlcNAc, 1mM-UDP-GlcNAc containing 12.5 nCi carbon-14, and incubated at 20°C. The reaction was stopped at 100°C for 2 min, the tubes were cooled, centrifuged and the supernatant removed for chromatography. The pellet of mycelium was washed with 0.5 ml each for 5 min of water (100°C), ethanol (20°C), acetone (20°C), 1M-NaOH: ethanol, 1:2 (100°C), 3 x ethanol (20°C), freeze-dried and dispersed in scintillant. Fig 1a; Time course for incorporation in youngest strips from colony; ● strainZ46 (+), ○ strain Z43 (-). Fig. 1b; Effect of age of strip on activity, 20 min assay. Fig 1c; Effect of polyoxin D on activity of youngest strips, 20 min assay. % Incorporation in each case represents the C14 of the insoluble residue expressed as % of the original total count.

was attributable to chitin synthase was shown by its abolition by 5 µM-polyoxin D (Fig.1c), a very specific inhibitor of chitin synthesis[32,33]. Properties of the chitin synthase in situ were comparable with those in vitro of microsomal preparations made from submerged shake-flask cultures, where the maximum activity was obtained after 15 hours growth from spores. The values for apparent Km for UDPGlcNAc were 0.77 and 1.85mM for assays in situ and in vitro respectively; pH optima were both 5.5, temperature optima were 20 and 15°C; polyoxin D concentrations for 50% inhibition were 1.7 and 1.25 µM.

The decline in incorporated radioactivity shown in Fig.1a can be attributed to the action of chitinase in situ. The evidence for this is that when the supernatants from these incubations were chromatographed (conditions as for Table 2) the major radioactive product accumulating with time was diacetylchitobiose. Thus again we have the situation of chitin synthase and chitinase acting in consort in the one preparation. The mycelium represented in Fig. 1 was young, from zero to 12 hours, so certainly this chitinase activity cannot be considered as a phenomenon of senescence.

DISCUSSION

From these results we can conclude that in our system of Mucor mucedo growing on an agar surface chitin synthase and chitinase activities are intimately associated, and we can suggest that they are regulated in consort. How might these two enzymes together act during growth and differentiation?

α-Chitin formation. The chitin in the fungal wall is in its most stable form, as α-chitin. In this form the chains are anti-parallel, i.e. adjacent chains run in opposite directions. Chitin formed in vitro from chitin synthase forms α-chitin after long incubations, presumably crystallising via the formation of the strong hydrogen bonding of the adjacent antiparallel chains. It is possible that a chitinase might regulate α-chitin formation in vivo via precise cleavage of a glycosidic bond. There is however no evidence for such a process, and there is circumstantial evidence against it as Kang et al[35] found the degree of polymerisation of chitin formed in vivo to be the same as that formed by the purified chitin synthase in vitro, at about 100; i.e. the chain length of its product appears to be an intrinsic property of the chitin synthase, and α-chitin formation can be envisaged as crystallisation of these relatively short nascent chains.

Chitin microfibril formation. Chitin microfibrils characteristically vary in length, width, height and orientation in different fungi and in different stages of morphogenesis[13,34]. Thus in hyphae of Mucor mucedo they were 18-27nm in diameter by up to about 1 µm long, and in Candida albicans they were 9-15 nm by up to about 0.1 µm[34]. As properties of chitin synthase are very similar in different fungi[33], these marked differences in its product suggest that there are important idiosyncratic differences in the regulation of chitin synthesis in other ways. Our finding of chitinase associated with chitin synthase suggests that controlled lysis of the chitin might play a role in the precise fashioning of the microfibril.

Spore germination. The initial swelling of the Zygomycete sporangiospore to form the germ sphere involves exponential synthesis of chitin/chitosan in the newly formed vegetative wall and distension and fracture of the spore wall[36]. Our results (Table 1) implicate chitin synthase and endogenously activated microsomal chitinase in this process. As there is no evidence for lysis of the original spore wall during swelling, the chitinase here could be participating in the formation of the vegetative wall which will eventually become the germ tube wall and then the hyphal wall, in the manner proposed by Bartnicki-Garcia[15].

Apical growth. The continual association of chitinase with chitin synthase in actively growing cells provides the circumstantial evidence for the role of chitinase in apical growth[15]. Recent years have brought the suggestion that the dynamic structure of the apex is supported by a cytoskeleton which plays a major role in the forward movement of the cell[14,37,38]. We may think of this as a tunnel-building machine, laying down the wall of the tunnel as it advances. The units laid down consist of cell membrane, being provided by membrane vesicles fusing with the apical cell membrane, and wall components, some such as chitin being synthesised through the cell membrane by intrinsic transmembrane proteins, others (e.g. by analogy with plants perhaps the polyglucuronide of Mucor) being synthesised in Golgi cisternae and released to the wall by fusion of the resulting vesicles. The chitin microfibrils, as they are formed by the chitin synthase enzyme aggregates in the apical membrane, may be sites of anchorage between the membrane and the wall. If so, the dynamic nature of the growing apex would require regular cleavage of the polysaccharide chains so that the synthetic system could move forward with the apex. This is an elaboration of the "unitary model" of Bartnicki-Garcia[15], invoking the extra feature of the attachment between wall and membrane via chitin microfibrils.

Branching. It is generally accepted that chitinase action will be required to form a branch. The cylindrical wall of a hypha unlike the apex is a rigid structure. Its chitin microfibrils are wider, more crystalline and in the case of higher fungi are cross-linked with glucans[13,14]. The site of a branch must be weakened by the breaking of covalent bonds by lytic enzymes, to enable a new apex to be formed, although there is yet to be direct cytological evidence for this. The membrane-bound zymogenic chitinase that we describe here, or a component of the supernatant chitinase, may be present in all of the cylindrical hyphal cell membrane awaiting localised activation at the site of branch formation.

Thus twelve years later, we still have to agree with Bartnicki-Garcia[15] that the evidence for the involvement of chitinase and other lytic enzymes in apical growth of fungal hyphae is circumstantial. Our results here with chitinases of Mucor mucedo merely add a fifth to the four bodies of evidence that he lists in support of this idea. One hope for more substantial evidence would be to identify contents of the apical vesicles, so clearly shown by freeze substitution electron microscopy[38]. Another would be to discover a convincing specific inhibitor of chitinase. The roles for chitinase suggested here are outside the cell membrane and so should be inhibited readily by a specific agent. The fact that none is known yet can be taken teleologically as evidence for a key role for chitinase, i.e. nature has it well protected.

REFERENCES

1. M.V.Tracey, Chitinase in some Basidiomycetes. Biochem J. 61: 579-586 (1955).
2. W.Iten and P. Matile, Role of chitinase and other lysomal enzymes of Coprinus lagopus in the autolysis of fruiting bodies. J. Gen.Microbiol. 61: 301-309 (1970).
3. J.Monreal and E.T.Reese, The chitinase of Serratia marcescens. Can. J. Microbiol. 15: 689-696 (1969).

4. H.Veldkamp, A study of the aerobic decomposition of chitin by microorganisms. Mededelingen van de Landbouwhogeschool te Wageningen, 55: 127-174 (1955).

5. O.M.H.de Vries and J.G.H.Wessels, Release of protoplasts of Schizophyllum commune by combined action of purified α 1-3 glucanase and chitinase. J. Gen. Microbiol.76:319-330 (1973).

6. K.Soderhall and T.Unestam, Properties of extracellular enzyme from Aphanomyces astaci and their relevance in the penetration process of crayfish cuticle. Physiol Plant. 35: 140-146 (1975).

7. T.A.Coudron, M.J. Kroha and C.M. Ignoffo, Levels of chitinolytic activity during development of three entomopathogenic fungi. Comp Biochem. Physiol. 79B: 339-348 (1984).

8. J.Huber, Untersuchungen zur Physiologie insektentotender Pilze. Arch. Mikrobiol. 29: 257-276 (1958).

9. R.F.Rosenberger, Endogenous lytic enzymes and wall metabolism, in "Fungal walls and Hyphal Growth", J.M.Burnett and A.P.J.Trinci, eds., pp.265-277. Cambridge University Press (1979).

10. A.P.J.Trinci and C.F.Thurston, Transition to the non-growing state in eukaryotic micro-organisms, in:"The Survival of Vegetative Microbes", T.R.G.Gray and J.R.Postgate, eds., pp.55-79. Cambridge University Press (1976).

11. F.Reyes, M.I.Perez-Leblic, M.J.Martinez and R.Lahoz, Protoplast production from filamentous fungi with their own autolytic enzymes, FEMS Microbiol. Lett. 24: 281-283 (1984).

12. S.Isaac and A.V.Gokhale, Autolysis: a tool for protoplast production from Aspergillus nidulans. Trans.Br.Mycol.Soc.78: 389-394 (1982).

13. J.H.Burnett, Aspects of the structure and growth of hyphal walls, in "Fungal Walls and Hyphal Growth", J.H.Burnett and A.P.J.Trinci, eds.,pp.1-25. Cambridge University Press (1979).

14. J.G.H.Wessels, Apical hyphal wall extension. Do lytic enzymes play a role? in "Microbial Cell Wall Synthesis and Autolysis", C.Nombela, ed.,pp.31-42, Elsevier, Amsterdam (1984).

15. S. Bartnicki-Garcia, Fundamental aspects of hyphal morphogenesis, in: "Microbial Differentiation", J.M.Ashworth and J.E.Smith., eds.,pp.245-268, Cambridge University Press (1973).

16. E.Lopez-Romero, J.Ruiz-Herrera and S.Bartnicki-Garcia. The inhibitory protein of chitin synthase from Mucor rouxii is a chitinase. Biochim.Biophys.Acta, 702: 233-236 (1982).

17. A.M.Humphreys and G.W.Gooday, Properties of chitinase activities from Mucor mucedo: evidence for a membrane-bound zymogenic form. J.Gen.Microbiol. 130: 1359-1366 (1984).

18. A.M.Humphreys and G.W.Gooday, Phospholipid requirement of microsomal chitinase from Mucor mucedo. Curr.Microbiol. 11: 187-190 (1984).

19. A.M.Humphreys and G.W.Gooday, Chitinase activities from Mucor mucedo, in "Cell Wall Synthesis and Autolysis", C.Nombela, ed., Elsevier Scientific Press, Amsterdam (1984).

20. R.J.Cohen, Some properties of chitinase from Phycomyces blakesleeanus Life Sciences, 15: 289-300 (1974).

21. P.R.Mahadevan and V.R.Mahadakar, Role of enzymes in growth and morphology of Neurospora crassa: cell-wall bound enzymes and their possible role in branching. J.Bact. 101: 941-947 (1970).

22. A.Zarain-Herzberg and A.Arroyo-Begovich, Chitinolytic activity from Neurospora crassa. J.Gen.Microbiol.129: 3319-3326 (1983).

23. Y.Polachek and R.F.Rosenberger, The distribution of autolysis in hyphae of Aspergillus nidulans: existence of a lipid mediated attachment to hyphal walls. J.Bact. 135: 741-754 (1978).

24. N.Elango, J.U.Correa and E.Cabib, Secretory nature of yeast chitinase. J.Biol.Chem. 257: 1398-1400 (1982).

25. K.Barrett-Bee and M.Hamilton. The detection and analysis of chitinase activity from the yeast form of Candida albicans. J.Gen.Microbiol. 130: 1857-1862 (1984).

26. G.W.Gooday and A.de Rousset-Hall, Properties of chitin synthetase from Coprinus cinereus. J.Gen.Microbiol. 89: 135-145 (1975).
27. G.W.Gooday, Activity of chitin synthetase during the development of fruit bodies of the toadstool Coprinus cinereus, Biochem.Soc. Trans. 1: 1105-1107 (1973).
28. A.Dominguez, M.V.Elorza, J.R.Villanueva and R.Sentandreu, Regulation of chitin synthase activity in Saccharomyces cerevisiae: effect of the inhibition of cell division and of synthesis of RNA and protein. Curr.Microbiol. 3: 263-266 (1980).
29. S.P.Ram, P.A.Sullivan and M.G.Shepherd, The in situ assay of Candida albicans enzymes during yeast growth and germ-tube formation. J.Gen.Microbiol. 129: 2367-2378 (1983).
30. M.P.Fernandez, J.U.Correa and E.Cabib, Activation of chitin synthase in permeabilized cells of a Saccharomyces cerevisiae mutant lacking proteinase B. J.Bact. 152: 1255-1264 (1982).
31. R.Datema, J.G.H.Wessels and H.van den Ende, The hyphal wall of Mucor mucedo 2. Hexosamine-containing polymers. Eur.J.Biochem. 80: 621-626 (1977).
32. L.L.Davis and S.Bartnicki-Garcia, Chitosan synthesis by tandem action of chitin synthase and chitin deacetylase from Mucor rouxii. Biochemistry. 23: 1065-1073 (1984).
33. G.W.Gooday, Biosynthesis of the fungal wall: mechanisms and implications. J.Gen.Microbiol. 99: 1-11 (1977).
34. N.A.R.Gow and G.W.Gooday, Ultrastructure of chitin in hyphae of Candida albicans and other dimorphic and mycelial fungi. Protoplasma, 115: 52-58 (1983).
35. M.S.Kang, N.Elango, E.Mattia, J. Au-Young, P.W.Robbins and E.Cabib, Isolation of chitin synthetase of Saccharomyces cerevisiae. J.Biol.Chem. 259: 14966-14972 (1984).
36. S.Bartnicki-Garcia, Cell wall construction during spore germination in Phycomycetes, in "The Fungal Spore: Morphogenetic Controls" G.Turian and H.R.Hohl, ed. pp. 533-556. Academic Press, London (1981).
37. G.W.Gooday, The hyphal tip, in: "Fungal Differentiation: A Contemporary Synthesis", J.E.Smith, ed., pp 315-356, Dekker, New York (1983).
38. R.J.Howard, Ultrastructural analysis of hyphal tip cell growth in fungi: Spitzenkörper, cytoskeleton and endomembranes after freeze-substition. J.Cell Sci. 48: 89-103 (1981).

Editor's Report

G. W. GOODAY

CHITIN ESTIMATION TO MEASURE FUNGAL BIOMASS

based on works by

M. T. ESQUERRE-TUGAYE, D. MOUSAIN, C. PLASSARD, D. ROBY, L. SALSAC, A. TOPPAN, and C. VIGNON

As chitin is a characteristic component of most fungi, and not of plants, its assay can give a quantitative estimate of the fungal biomass in plant/fungal relationships, such as pathogenesis and mycorrhizal associations. Here follow two investigations of the usefulness of this idea.

GLUCOSAMINE DETERMINATION AS A MEANS OF MEASURING CHITIN-CONTAINING FUNGAL PATHOGENS IN HOST PLANTS

(A. Toppan, D. Roby and M. T. Esquerré-Tugayé)

Measurement of fungal chitin can be used in order to quantify the amount of a chitin-containing fungal pathogen present in a host plant at any time after inoculation, and to correlate the spreading of pathogen to biochemical events occurring in the infected plant. The objective of this work was to develop a simple and accurate procedure for measuring the amount of fungal pathogens in diseased plants, using melon plants infected with *Colletotrichum lagenarium*.

Melon plants were grown in a growth chamber, inoculated at the two-leaf stage with a suspension of *Colletotrichum lagenarium* conidia, and harvested 1 to 7 days later. Healthy plants were simultaneously grown for controls. Extraction of *C. lagenarium* elicitors was performed by autoclaving the whole mycelium, and the crude preparation obtained was purified through a combination of solvent fractionation and chromatographic procedures (1). Elicitor treatment of melon seedlings was carried out by inoculation into a cotyledon (Fraction 1, 50 μg), four days before conidia spraying. Chitin measurement in infected plants was achieved by determination of the glucosamine content of the hydrolyzed cell wall fraction. Chitinase and protease activities were measured as already described (2, 3).

Assay of chitin. The procedure devised is illustrated in Figure 1. In order to ensure the accurate measurement of chitin, three steps were needed:

1 - Isolation of the cell wall fraction (4), so as to recover the whole fungal chitin and to eliminate cytoplasmic substances hindering glucosamine measurement,

2 - Selection of conditions for complete chitin hydrolysis and glucosamine recovery,

3 - Purification of the hydrolysate in order to discard compounds, notably sugars interfering with glucosamine assay (5).

Infected plants

↓ Homogeneisation

Cell wall extract

↓ 6 N HCl (4 ml/10 mg)

Acid hydrolysis

↓

Purification on Dowex 50 x 8

↓ H_2O washing, 0.3 N HCl elution

Glucosamine assay

Fig. 1. Chitin measurement in infected plants

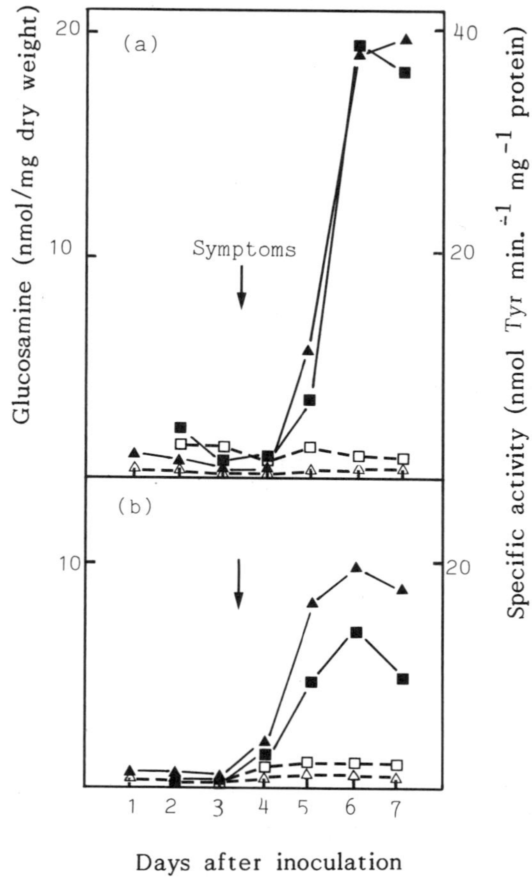

Fig. 2. Glucosamine (▲ , △) and proteolytic activity (■ , □) in melon stems (a) and leaves (b) harvested on different days after inoculation (closed symbols are used for infected plants, open symbols for controls).

Reliability of the method. Glucosamine recovery was strictly correlated to the amount of cell walls used for hydrolysis up to 100 mg of cell walls (corresponding to about 5 grams of infected tissues).

Glucosamine in infected melon plants. The method was applied to leaves and stems of infected melon seedlings (Fig. 2). Although the mycelium spread sooner in leaves than in stems, the chitin content in infected stems was higher 7 days after inoculation.

Suitability of the method. The pathogen growth strictly parallels the strong proteolytic activity that the fungus displayed at the same time in leaves and stems (3) (Fig. 2). A very close relationship exists between the two curves during the course of infection.

Treatment of melon plants with *Colletotrichum lagenarium* elicitors leads to an induced systemic resistance to this pathogen. This induction was correlated to an increase in chitinase activity, which was quicker and higher after infection in treated plants than in controls. Thus, an inverse relationship exists between host chitinase activity in melon plants, the amount of chitin and the extent of the disease.

In conclusion, we have developed an improved procedure (6) based on chitin measurement *via* glucosamine for the determination of chitin-containing fungal pathogens in host plants. This methods enables, for example, a correlation to be made between the amount of pathogen and the activity of the two different enzymes: a fungal protease and a plant chitinase.

REFERENCES

1. A. Toppan and M. T. Esquerré-Tugayé, Cell surfaces in plant-microorganism interactions. IV Fungal glycopeptides which elicit the synthesis of ethylene in plants, *Plant Physiol.*, 75:1133 (1984).
2. A. Toppan and D. Roby, Activité chitinasique de plantes de melon infectées par *Colletotrichum lagenarium* ou traitées par l'éthylène, *Agronomie*, 2:829 (1982).
3. M. T. Esquerré-Tugayé, Activité protéolytique et pathogénèse dans l'anthracnose du melon, *Ann. Phytopathol.*, 4:277 (1972).
4. M. T. Esquerré-Tugayé, C. Lafitte, D. Mazau, A. Toppan, and A. Touzé, Cell surfaces in plant-microorganism interactions. II Evidence for the accumulation of hydroxyproline rich glycoproteins in the cell wall of diseased plants as a defense mechanism, *Plant Physiol.*, 64:320 (1979).
5. W. D. Wagner, A more sensitive assay discriminating galactosamine and glucosamine in mixtures, *Anal. Biochem*, 94:394 (1979).
6. A. Toppan, M. T. Esquerré-Tugayé, and A. Touzé, An improved approach for the accurate determination of fungal pathogens in diseased plants, *Physiol. Plant Pathol.*, 9:241 (1976).

GLUCOSAMINE DETERMINATION AS A MEANS OF MEASURING FUNGAL MASS AND MYCORRHIZAL INFECTION

(C. Plassard, C. Vignon, D. Mousain and L. Salsac)

A colorimetric assay of fungal chitin, after acid hydrolysis of acetyl units and glycosidic bonds, or alkaline hydrolysis of acetyl units only, has been used to estimate the extent of mycorrhizal infection of maritime pines (*Pinus pinaster* Soland. *in* Ait.) associated with an ectomycorrhizal Basidiomycete, *Pisolithus tinctorius* (Pers.) Coker and Couch. This assay can be an alternative to the usual count of mycorrhizal roots, which is time consuming, subjective and sometimes inaccurate.

After acid hydrolysis, the tissues of the pine produced compounds which reacted during the colorimetric assay: the contaminating absorption had to be evaluated. For this, a mycelium-free root sample was often used, but this method was restraining and sometimes impossible. Thus, we propose another method, with the modified coloured reaction, avoiding the use of the control sample.

The colorimetric assay for glucosamine and its polymers involves two different stages: 1) Deamination of the glucosamine residues by nitrous

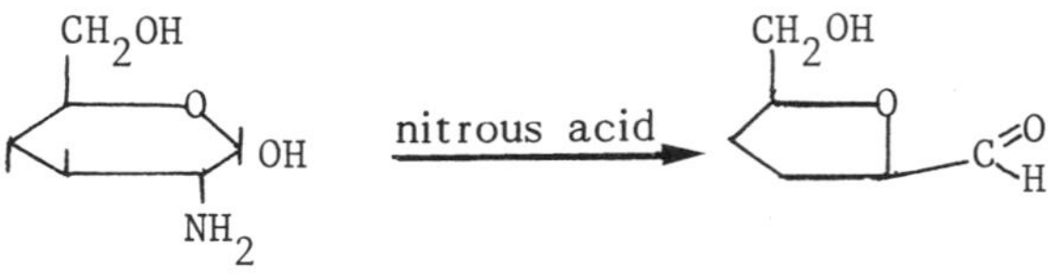

acid and formation of a free aldehyde function; 2) Reaction of these free aldehyde functions with 3-methyl-2-benzothiazolone hydrazone hydrochloride (MBTH) (a specific reagent for aldehyde functions).

Thus, the absence of nitrous acid during incubation would determine the content of aliphatic aldehyde functions present in the solution. Consequently, the content of amino sugars may be calculated by the difference between the absorbance values obtained with or without action of nitrous acid.

<u>Hydrolysis carried out on the two types of plant material: roots and mycelium</u>

After alkaline treatment, the glucosamine content of pine roots was negligible: the fungal chitosan could be directly and specifically assayed (Table I).

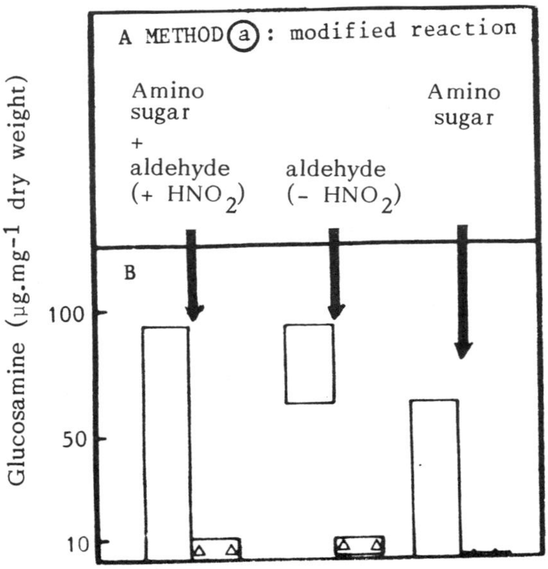

Figure 1. Colorimetric assay on plant material

(☐ : *P.tinctorius*, △ △ : *P. pinaster*)

Table I. Assay of fungal glucosamine on alkaline hydrolysis residues (mycelium, roots, mixtures of the two tissues; glucosamine content expressed in $\mu g.mg^{-1}$ fungal dry weight, with standard error at p = 0.05).

	Pine	Mycelium	Mycelium + Pine
Glucosamine	0.2 ± 0.13	46.8 ± 10	49.6 ± 2.5

Table II. Assay of fungal glucosamine on mixtures of roots and mycelium after acid hydrolysis (glucosamine in $\mu g.mg^{-1}$ fungal dry wt., with standard error at p = 0.05).

Pine roots (dry wt. in mg)	50.0	80.0
Fungal glucosamine	60.0 ± 2.4^a	61.1 ± 0.5^a
	59.7 ± 1.9^b	60.0 ± 1.6^b

a) Estimation by difference in absorbance with or without HNO_2
b) Estimation by the use of a control sample with Y = Y' - Y (Y = absorbance obtained with roots-mycelium mixtures; Y'= absorbance on roots alone).

After acid hydrolysis, the glucosamine content of the pine roots was negligible with the modified colorimetric assay (Fig. 1B). This acid hydrolysis method was more sensitive than the alkaline one: the glucosamine level of the mycelium was higher after acid hydrolysis than after alkaline hydrolysis (Fig. 1B and Table I).

On mixtures mycelium-pine roots the fungal glucosamine was estimated by two methods: a) modified colorimetric assay, and b) use of a control sample.

The results obtained were very similar with both methods (Table II).

Assay of fungal glucosamine on mycorrhizal pines

Roots of pine were experimentally inoculated (2), and subjected to the acid or alkaline hydrolysis. The extent of mycorrhizal infection was estimated by measuring the absorbance values at 432 nm of an acetone extract of the brown pigments of the fungus on the roots (3). At the same time the mycelial masses were estimated by means of the chitin assay.

After alkaline hydrolysis, the correlations obtained between the mycelial masses and the count of mycorrhizas (Fig. 2A) or the absorbance measured at 432 nm (Fig. 2B) were highly significant. The same was true after acid hydrolysis, and the curves obtained with mycelial masses estimated after method (a) or (b) were not significantly different regardless of the method employed (Fig. 3).

The choice of acid hydrolysis for the assay of fungal chitin of mycelium or mycorrhizal roots is justified by the facility and the sensitivity of the method as opposed to alkaline hydrolysis. On the other hand, acid hydrolysis is not specific for fungal glucosamine residues. But with the use of a nitrous acid-free reference (method a) the amino sugar content of roots is negligible compared to the fungal glucosamine, and the specificity of the acid hydrolysis is improved. The good correlation with the other methods usually employed suggests that the colorimetric chitin assay, after acid hydrolysis of plant tissues, constitutes a good means for estimating the intensity of mycorrhizal infection.

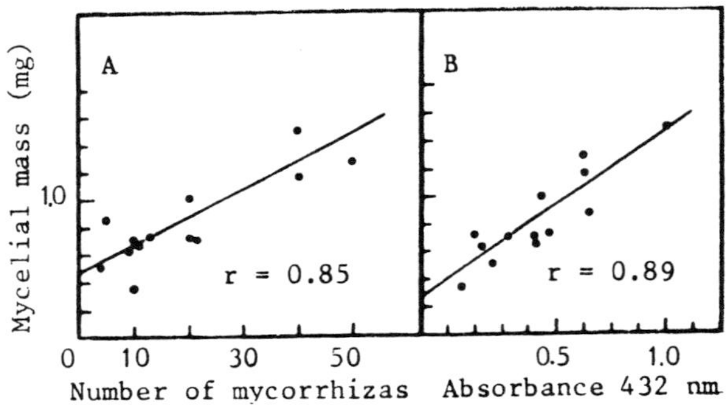

Figure 2. Correlation between the estimation of fungal mass after alkaline hydrolysis and the count of mycorrhizas (A) or the O.D. at 432 nm.

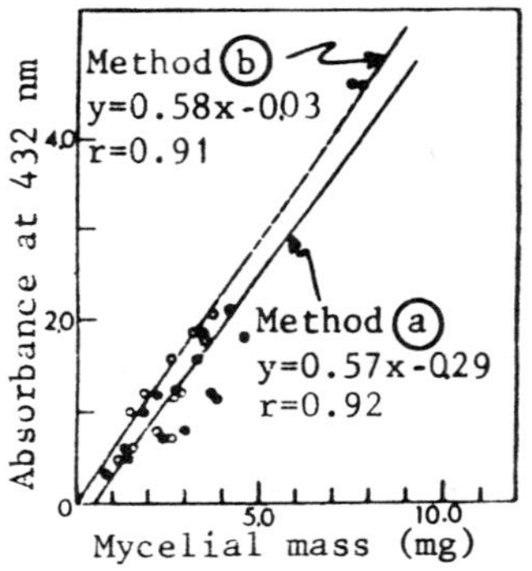

Figure 3. Acid hydrolysis of mycorrhizas: relation between fungal mass from (a) or (b) method and O.D. at 432 nm.

REFERENCES

1. A. Tsuji, M. Kinoshita, and M. Hoshino, Analytical chemical studies on amino sugar. II. Determination of hexosamine using 3-methyl-2-benzothiazolone hydrazone hydrochloride, Chem. Pharm. Bull., 17:1505 (1969).
2. J. A. Fortin, Y. Piché, and M. Lalonde, Technique for the observation of early morphological changes during ectomycorrhiza formation, Can. J. Bot., 58:361 (1980).
3. C. Plassard, D. Mousain, and L. Salsac, Dosage de la chitine sur des ectomycorhizes de pin maritime (Pinus pinaster) à Pisolithus tinctorius: évaluation de la masse mycéllienne et de la micorrhization, Can. J. Bot., 61:692 (1983).

Editor's Report

G. W. GOODAY

STUDIES OF AUTOLYTIC ENZYMES IN FUNGI

based on works by

J. CALATAYUD, F. LABORDA, R. LAHOZ,
M. J. MARTINEZ, M. I. PEREZ-LEBLIC, F. REYES,
and J. RODRIGUEZ

All fungi produce enzymes capable of autolysing their own walls. These enzymes tend to accumulate in mycelium and medium during long-term growth of the fungus. Although they break down the walls of the older mycelium, there is no clear evidence that the products are re-cycled as nutrients for the fungus (cf. Gooday et al., this Volume).

Here follow two studies on the production and action of autolytic chitinase and β-N-acetylglucosaminidase during submerged culture of an *Aspergillus* and a *Penicillium* species.

DEGRADATION OF CELL WALL CHITIN FROM THE FUNGUS ASPERGILLUS NIDULANS DURING AUTOLYSIS

(F. Reyes, M. J. Martinez, J. Calatayud and R. Lahoz)

INTRODUCTION

Autolysis in filamentous fungi is a natural degradative process which starts after the exhaustion of the external carbon supply. The fungus starts to metabolize the reserve substances and then a degradation of the cytoplasmic material and cell wall polysaccharides takes place (3, 10). Autolysis occurs by the lytic enzymes that the fungus has or synthesizes (5, 11) during the degradative process. These enzymes show increased activity in the culture fluid and decreased activity in the mycelium during autolysis (6). Each fungus is able to produce the necessary lytic enzymes to carry out its own cell wall degradation. The scope of this work was to investigate which enzymes were implicated in the degradation of cell wall chitin of *Aspergillus nidulans* (Eidam) Wint, and possible regulation among them.

MATERIALS AND METHODS

The experiments were performed with *A. nidulans* (CECT-2544, Coleccion Espanola de Cultivos Tipo, Valencia, Spain). The growth, obtention of enzymatic complex from autolyzed cultures and wall preparations were made as described before (9). Chitin was obtained from prawn shell and *A. nidulans* cell wall (1, 4). Oligosaccharides from chitin hydrolyzate with concentrated HCl (12) were purified by Bio-Gel P 2. Carboxymethylchitin (CMC) was obtained with a 43 % degree of substitution (14). β-N-Acetylglucosaminidase

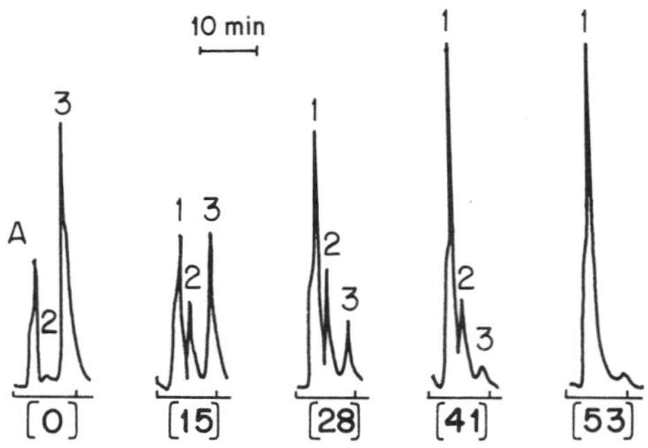

Fig. 1. HPLC analysis of hydrolysis of chitotriose (1.4 mM) by β-N-acetylglucosaminidase at 0, 15, 28, 41 and 53 min of incubation at 37° C.

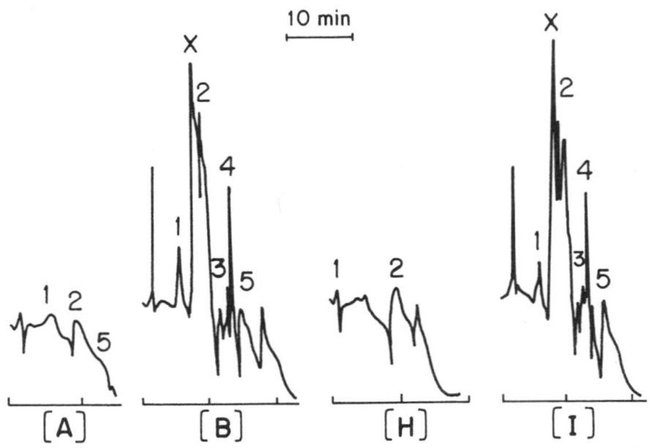

Fig. 2. HPLC analysis of degradation of colloidal prawn shell chitin (A, B) and A. nidulans cell wall chitin (H, I) by pure endochitinase (B, I) and endochitinase in presence of chitin deacetylase (A, H) after 72 h of incubation.

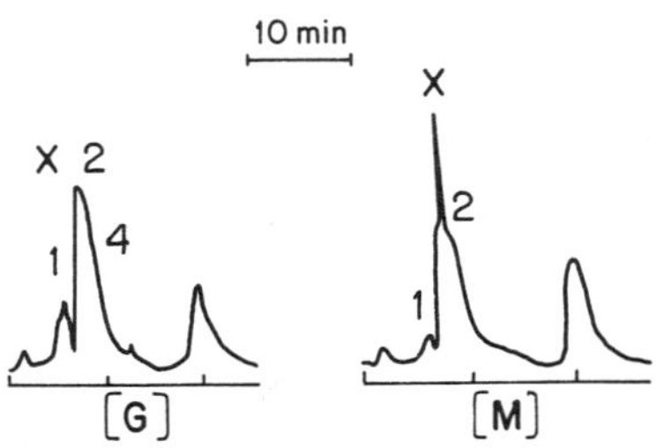

Fig. 3. HPLC analysis of joint action of both purified enzymes on colloidal prawn cell chitin (G) and A. nidulans chitin (M) after 72 h of incubation.

Peak identification: 1 = N-acetylglucosamine; 2 = chitobiose; 3 = chitotriose; 4 = chitotetrose; 5 = chitopentose; A = buffer, and X = dimer of N-acetylglucosamine and glucosamine.

(E.C. 3.2.1.30) was assayed by the hydrolysis of chitin oligomers (13) as substrates, endochitinase (E.C. 3.2.1.14) by viscosity reduction of a solution of 0.1 % CMC, and chitin deacetylase (E.C. 3.5.1.41) by the increase of the free amino groups with fluorescamine. β-N-Acetylglucosaminidase was purified to homogeneity by precipitation of autolyzed cultures with 1 % tannic acid. The dry precipitate was dissolved, filtered through a Millipore filter (0.22 μm) and purified through Sephadex G-200. The endochitinase, separated in the same way, was finally purified by electrophoresis in 7 % polyacrylamide gel. Products of hydrolysis were analysed with an HPLC system (2).

RESULTS AND DISCUSSION

In the enzyme system from autolyzed cultures of *A. nidulans*, three enzymes implicated in the chitin degradation were found. The β -N-acetylglucosaminidase was purified to homogeneity. This enzyme had a molecular weight of 190000 and isoelectric point of 4.3. It hydrolyzed chitobiose, chitotriose, chitotetrose, chitopentose and chitohexose, its affinity for these substrates being very similar, with Km values of 0.18, 0.25, 0.17, 0.46 and 0.14 mM, respectively. The Vmax were 0.37, 0.56, 0.36, 0.20 and 0.08 U/mg which decreased progressively with the length of the chain, giving N-acetylglucosamine as end product. The optimum pH was between 5 and 6. This enzyme did not hydrolyze chitin or chitosan. The hydrolysis of chitobiose was inhibited by N-acetylglucosamine and glucosamine in a non-competitive manner with Ki 0.22 and 0.40 mM, respectively, and was competitively inhibited by 2-acetoamido-2-deoxy-D-gluconolactone (Ki 0.11 mM). Similarly, it was inhibited in a mixed manner by the presence of some other oligomers. The degradation of these oligosaccharides by β-N-acetylglucosaminidase was studied by HPLC. Figure 1 shows the chitotriose degradation.

The endochitinase was purified to homogeneity. It had a mol. wt. of 27000, isoelectric point of 4.8, an optimum pH of 5.5, and it did not hydrolize chitosan. This enzyme produced a fast depolymerization of the chitin molecule and, after slowly degrading oligomers, produced N-N'-diacetylchitobiose as final product. This enzyme degraded the prawn shell chitin in the same way as the chitin from *A. nidulans* cell wall (Fig. 2).

The chitin deacetylase was present in the enzymic precipitate in small quantities and had a mol. wt. similar to the endochitinase. The chitin deacetylase did not affect the fast depolymerization of chitin by the endochitinase, but the slow action of this enzyme on the polymers giving chitobiose as final product was completely inhibited, as shown by HPLC (Fig. 2). The joint action of purified β -N-acetylglucosaminidase and endochitinase on chitin produced a slow degradation of this polymer giving N-acetylglucosamine (Fig. 3). The peak found between N-acetylglucosamine and chitobiose (X) was identified as a dimer formed by N-acetylglucosamine and glucosamine.

These results suggest than the degradation of cell wall chitin during the autolysis of *A. nidulans* is a slow process, probably regulated by the presence or absence of a chitin deacetylase.

REFERENCES

1. S. Bartnicki-Garcia and W. J. Nickerson, Isolation, composition and structure of cell walls of filamentous and yeast like form of *Mucor rouxii*, *Biochim. Biophys. Acta*, 58:102 (1962).
2. P. van Eikeren, and H. McLaughlin, Analysis of the lysozyme catalyzed hydrolysis and transglycosylation of N-acetyl-D-glucosamine oligomers by HPCL, *Anal. Biochem.*, 77:513 (1977).
3. J. W. Foster, Chemical nature of the mould mycelium, *in*: "Chemical Activities of Fungi", Academic Press, N. Y. (1949).
4. C. Jeuniaux, Chitinases, *in*: "Methods in Enzymology. Vol. III", E. F.

Newfeld and V. Ginsburg, eds., Academic Press, N. Y. (1966).
5. R. Lahoz, F. Reyes and M. I. Pérez-Leblic, Lytic enzymes in the autolysis of filamentous fungi, Mycopathol., 60:45 (1976).
6. M. I. Martinez, F. Reyes and R. Lahoz, Cell wall plant polysaccharides-degrading enzymes in autolysis of Botrytis cinerea, Trans. Br. Mycol. Soc., 78:395 (1982).
7. T. P. Mommsen, Chitinase and β-N-acetylglucosaminidase from the digestive fluid of the spider Cupiennius salei, Biochim. Biophys. Acta, 612:361 (1980).
8. A. Ohtakara, Studies on the chitinolytic enzymes of black-koji mould. Part VII. Degradation of glycol chitin and chitin by the chitinase system of Aspergillus niger, Agr. Biol. Chem., 28:811 (1964).
9. M. I. Pérez-Leblic, F. Reyes, R. Lahoz and S. Archer, Autolysis of Penicillium oxalicum with special reference to its cell walls, Can. J. Microbiol., 28:1289 (1982).
10. F. Reyes and R. Lahoz, Variation in lysis of walls of Sclerotinia fructigena with age of culture, J. Gen. Microbiol., 98:607 (1977).
11. F. Reyes, R. Lahoz and A. Val Moreno, Synthesis of 1,3- β-glucanase and β-N-acetylglucosaminidase during autolysis of Neurospora crassa, J. Gen. Microbiol., 126:347 (1981).
12. J. A. Rupley, The hydrolysis of chitin by concentrated hydrochloric acid, and the preparation of low molecular weight substrate for lysozymes Biochim. Biophys. Acta., 83:245 (1964).
13. M. W. Tracey, Chitin, in: "Modern Methods of Plant Analysis. Vol. 2", K. Paech and M. V. Tracey, eds., Springer-Verlag, Berlin (1955).
14. R. Trujillo, Preparation of carboxymethylchitin, Carbohydr. Res., 7:483 (1968).

EFFECT OF AMPHOTERICIN B ON THE SECRETION OF β-N-ACETYLGLUCOSAMINIDASE AND CHITINASE DURING THE AUTOLYSIS OF Penicillium oxalicum

(J. Rodriguez, M. I. Pérez-Leblic and F. Laborda)

INTRODUCTION

β-N-Acetylglucosaminidase and chitinase are lytic enzymes that have an active role in the autolysis of many fungi (1). Both enzymes were found in the culture filtrate and in the mycelial extract of Penicillium oxalicum (2).

Autolytic enzymes may be packed within cytoplasmic vesicles that could fuse with the plasma membrane for enzyme liberation to the culture filtrate or be trapped in the cell wall through a lipid-mediated attachment (3, 4). Synthesis and secretion of lytic enzymes implicated in mycelial degradation have been shown to happen during fungal autolysis (5).

Amphotericin B (Am B) is a polyenic antibiotic, with a high affinity for sterols, that causes severe alterations in the fungal plasma membrane (6).

In this work an effect of Am B on β-N-acetylglucosaminidase and chitinase production during P. oxalicum autolysis is shown.

MATERIALS AND METHODS

Penicillium oxalicum was grown in the absence and the presence of a sublethal doses of Am B and β-N-acetylglucosaminidase and chitinase activities were measured as previously described (2). The final concentration of Am B was 300 μg/ml.

RESULTS AND DISCUSSION

In the presence of a sublethal dose of Am B, no change in the general pattern of growth for P. oxalicum was found when compared with the fungal growth in the absence of antibiotic, autolysis began on the 4th day of growth in both cases. However, in the presence of Am B, the active growth of P. oxalicum was reduced to 30 % (Fig. 1), while a lower degree of autolysis was achieved (Fig. 2). This lower degree of autolysis could be due to a lower level of lytic enzymes produced in the presence of antibiotic (7).

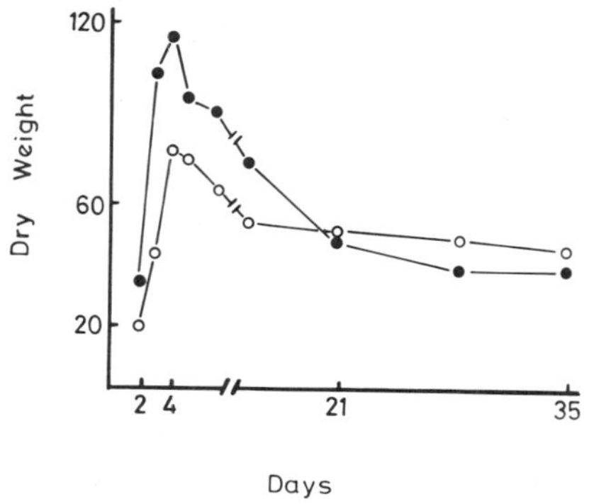

Fig. 1

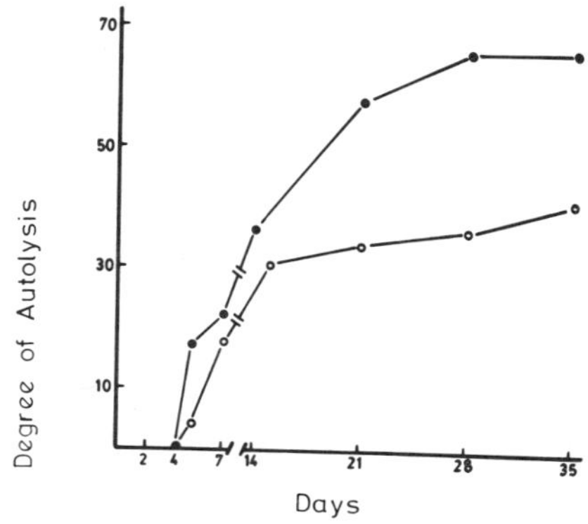

Fig. 2

Fig. 1. Growth of Penicillium oxalicum in the absence (●) and in the presence (o) of Amphotericin B.

Fig. 2. Degree of autolysis of Penicillium oxalicum in the absence (●) and in the presence (o) of Amphotericin B.

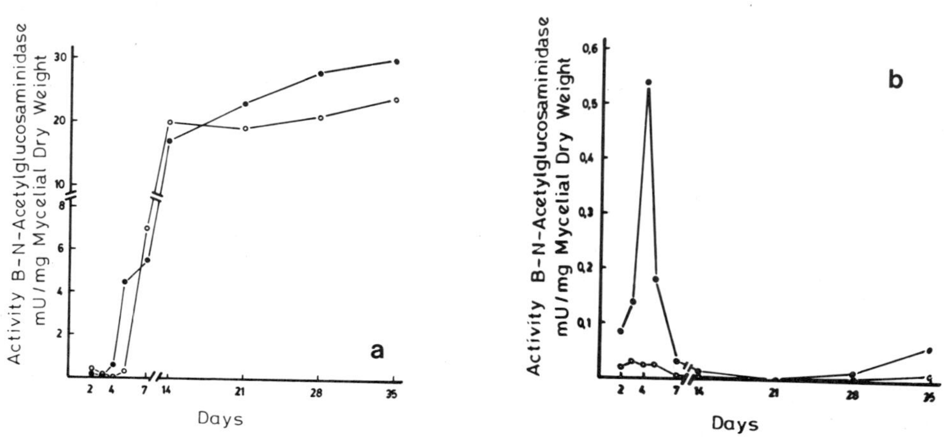

Fig. 3. β-N-Acetylglucosaminidase activity in the culture filtrate (a) and in the mycelial extract (b) when Penicillium oxalicum was grown in the absence (●) and in the presence (o) of Amphotericin B.

The level of β-N-acetylglucosaminidase detected in the culture filtrate was very low during the trophophase and it showed a great increase from the beginning of the autolysis until the 28th day of growth, when it became stable to the end of the period studied. This pattern for β-N-acetylglucosaminidase in the culture filtrate was not affected when P. oxalicum was grown in the presence of Am B (Fig. 3a). In the mycelial extract β-N-acetylglucosaminidase was detected during the trophophase, but a great reduction in the enzyme level was found once the autolysis began which was almost undetectable during this phase of growth. A reduction in the level of this enzyme was found when the fungus was grown in the presence of Am B (Fig. 3b).

The level of chitinase detected in the culture filtrate was similar to that described above for β-N-acetylglucosaminidase. However, when P. oxalicum was grown in the presence of Am B, a decrease is the enzyme level was found during the autolysis period (Fig. 4a). In the mycelial extract, chitinase activity was also found during the trophophase and its level progessively decreased during the autolysis period to the 28th day when no activity was detected. When the fungus was grown in the presence of Am B, a reduction in the chitinase activity was found in the mycelial extract during the trophophase but an increase of the enzyme level was found during the autolysis period (Fig. 4b).

Since it was found that neither β-N-acetylglucosaminidase nor chitinase activity was affected when measured in the presence of different amounts of Am B, the differences detected on the levels for either enzyme, when P. oxalicum was grown in the presence of Am B, could be due to an effect on the enzyme secretion pathway which may be different for each enzyme. Different localizations for the two enzymes in the mycelium of P. oxalicum have been described (2).

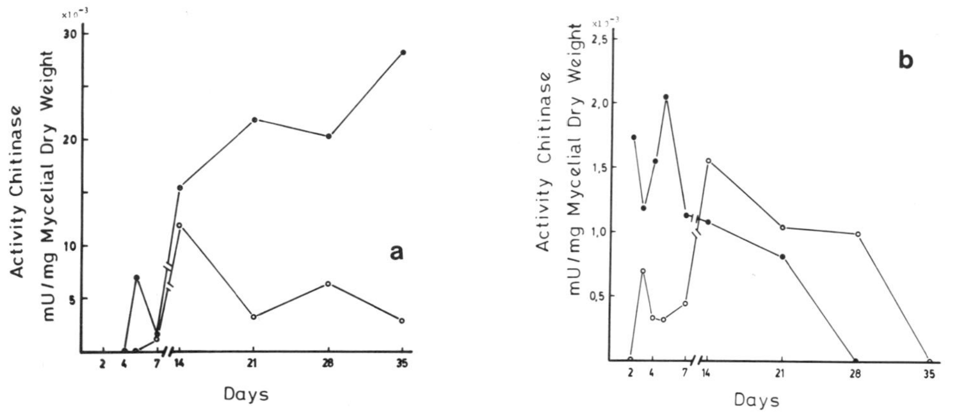

Fig. 4. Chitinase activity in the culture filtrate (a) and the mycelial extract (b) when Penicillium oxalicum was grown in the absence (●) and in the presence (o) of Amphotericin B.

REFERENCES

1. M. I. Pérez-Leblic, F. Reyes, M. J. Martinez and R. Lahoz, Cell wall degradation in the autolysis of filamentous fungi, Mycopathologia, 80:147 (1982).

2. M. I. Pérez-Leblic, F. Reyes, R. Lahoz and S. A. Archer, Autolysis of Penicillium oxalicum with special reference to its cell wall, Can. J. Microbiol., 28:1289 (1982).
3. I. Polacheck and R. F. Rosenberger, Distribution of autolysis in hyphae of Aspergillus nidulans: Evidence for a lipid-mediated attachment to hyphall walls, J. Bacteriol., 35:741 (1978).
4. P. Herrero, F. Moreno and S. Gascon, Role of vesicles on the transport and secretion of exocellular enzyme by yeast, Cell. Mol. Biol., 26:485 (1980).
5. F. Reyes, R. Lahoz and A. Val Moreno, Synthesis of 1,3-β-glucanase and β-N-acetylglucosaminidase during autolysis of Neurospora crassa, J. Gen. Microbiol., 126:347 (1981).
6. T. Sekiya, F. Yano and Y. Nozawa, Effects of amphotericin B and its methil ester on plasma membranes of Candida albicans and erythrocytes as examined by freeze-fracture electron microscopy, Sabouraudia, 20:303 (1982).
7. R. Lahoz, F. Reyes and M. I. Pérez-Leblic, Lytic enzyme in the autolysis of filamentous fungi, Mycopathologia, 60:45 (1976).

Affiliations

M. T. Esquerré-Tugayé, A. Toppan and D. Roby, Centre de Physiologie Végétale, UA CNRS 24, 118 route de Narbonne, F-31062 Toulouse cedex, France.

C. Plassard, C. Vignon, D. Moussain and L. Salsac, Laboratoire de Recherche sur les Symbiotes des Racines, INRA, 9 place Pierre Viala, F-34060 Montpellier cedex, France.

F. Reyes, M. J. Martinez, J. Calatayud and R. Lahoz, Institut Jaime Ferrain de Microbiologia CSIC, Joaquin Costa 32, S-28006 Madrid, Spain.

J. Rodriguez, M. J. Perez-Leblic and F. Laborda, Dept. Microbiologia, Universidad, Carretera Madrid-Barcelona Km 33, Alcalà de Henares (Madrid), Spain.

COMPARISON OF CHITIN FIBRIL STRUCTURE AND ASSEMBLY IN THREE UNICELLULAR ORGANISMS

W. HERTH, M. MULISCH and P. ZUGENMAIER

Zellenlehre, Universität Heidelberg, D-6900 Heidelberg

Institut für Allgemeine Zoologie, Freie Universität Berlin, D-1000 Berlin 33

Institut für Physikalische Chemie, Technische Universität Clausthal
D-3392 Clausthal-Zellerfeld, F.R. Germany

INTRODUCTION

The study of chitin fibril formation and deposition by multicellular systems involved in e.g. arthropod cuticle, polychaete bristle or gastropod radula development is complicated due to the interaction of several cell types (1-3).

Furthermore it is hard to distinguish cytologically between chitin formation and secretion of accompanying materials like acidic polysaccharides and proteins (3-7).

With unicellular organisms the involvement of certain cell organelles in chitin formation is easier to be traced. Using such unicellular systems, we expect to gain more insights into the cytological details of the polymerization, secretion and assembly steps of chitin microfibrils. We hope to find out some general principles of chitin fibril formation by comparison of different systems.

MATERIAL AND METHODS

The centric diatom alga *Thalassiosira fluviatilis*, the chrysoflagellate alga *Poterioochromonas stipitata*, and the ciliated protozoon *Eufolliculina uhligi* were cultivated as described previously (8-13). Procedures for light microscopy, inhibitor treatments, scanning electron microscopy, transmission electron microscopy, negative staining, freeze-fracture, chemical characterization, X-ray diffraction and electron diffraction have been also previously reported (14-18).

RESULTS

Thalassiosira. These centric diatom algae have a cylindrical shape and form numerous spine-like appendages which penetrate the valves at specific pores and extend like sunrays into the culture medium (Figs. 1a, b). These spines may be separated from the cells by air bubbling through the culture fluid and may then be collected for further analysis. The spines are band-like structures consisting of parallel subunits and showing kinking sites upon drying onto the electron microscopic support (Fig. 2a). These spines

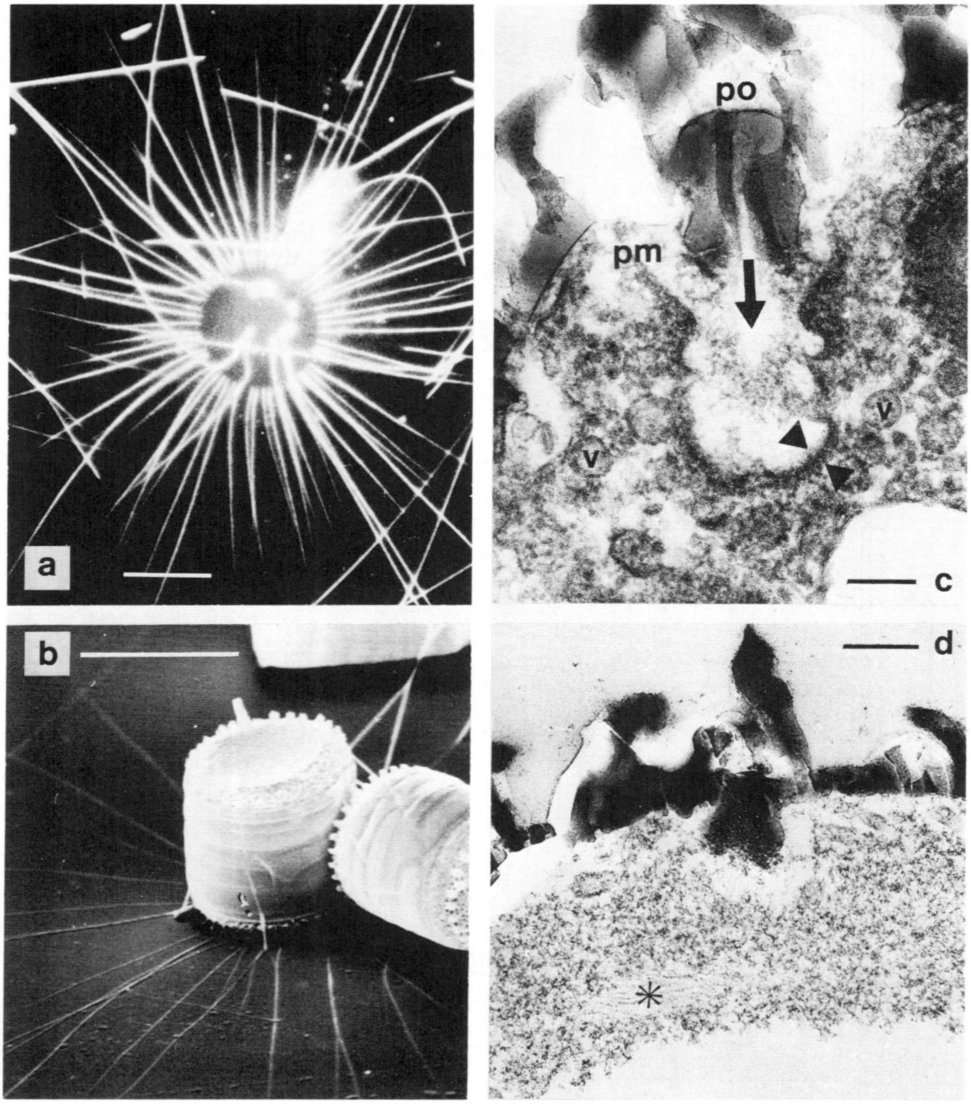

Fig. 1 Thalassiosira. a) Light micrograph, epifluorescence, of a CW-stained cell showing the numerous spines radiating from the cell surface, bar 10 µm. b) Scanning electron micrograph showing the marginal pore projections and the spines radiating from these pores, bar 10 µm. c) Electron micrograph, longitudinal section of the conical invagination of the plasma membrane (pm) underlying the pore (po) in the silica valve. The spine has a pointed basal end (arrow) and is assembled from fuzzy material projecting from the plasma membrane. The plasma membrane is coated on both external and cytoplasmic face (between arrowheads). Numerous small vesicles (v) and undulated surface of the pm indicate exocytosis in this region, bar 0.1 µm. d) As c), but after application of CW. Note the accumulation of fuzzy material and bundles of thin fibrils (asterisk), bar 0.1 µm.

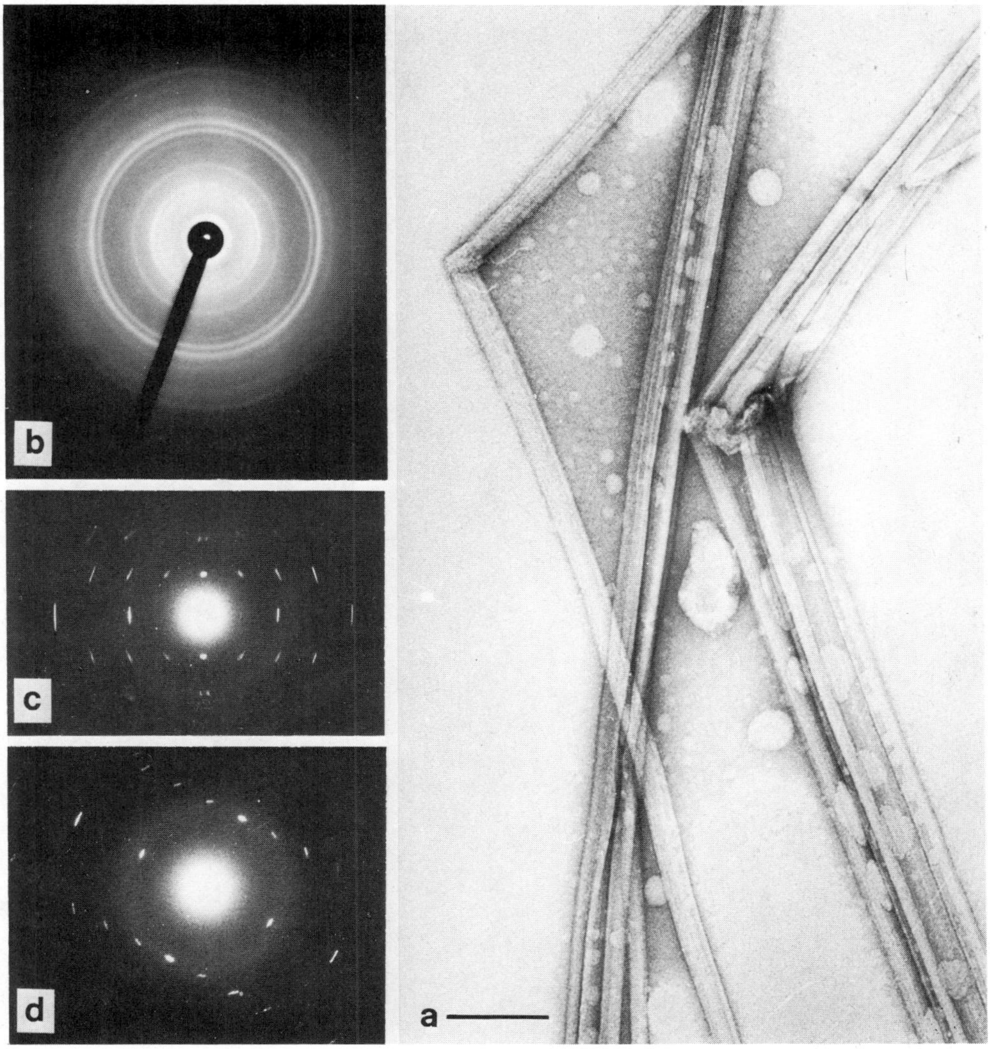

Fig. 2 Thalassiosira spine material. a) Ultrastructure of the ribbon-like spines revealing parallel subunits and kinking sites. Negative stain, bar 0.1 µm. b) X-ray diffraction pattern of T.-spines, typical ß-chitin pattern. c, d) Electron diffraction patterns of this ß-chitin, c) in b*, c*-projection, d) in a*, c*-projection.

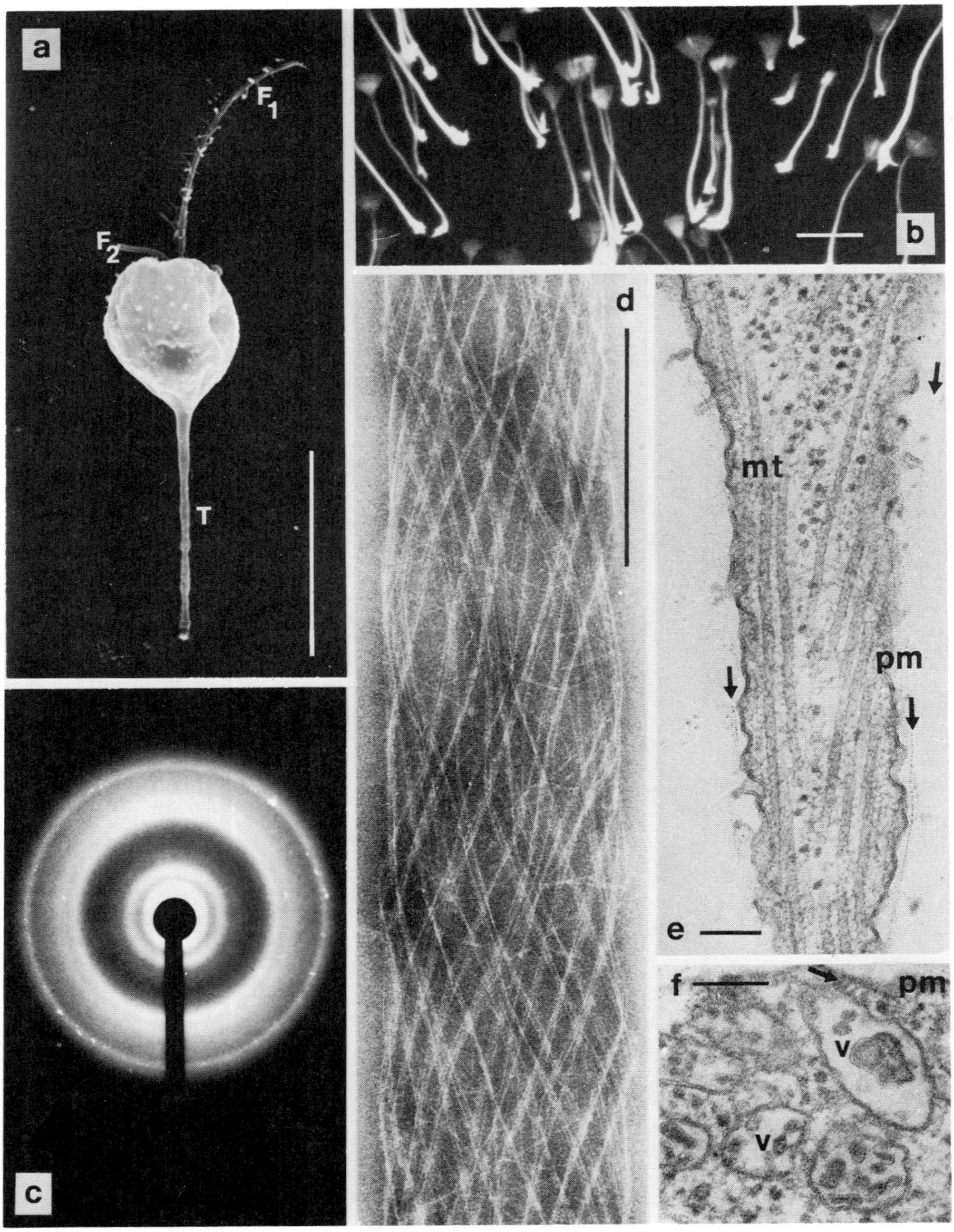

Fig. 3 Poterioochromonas. a) Scanning micrograph of a typical attached cell, with the long flagellum (F_1), the short flagellum (F_2) and the straight cytoplasmic tail (T), bar 10 µm. b) Light micrograph (epifluorescence) of CW-stained loricae attached to a coverslip, bar 10 µm. c) X-ray diffraction pattern of the extraction-resistant lorica material indicating its chitinous nature. d) Fibrillar arrangement in the stalk region of the lorica, negative stain, bar 0.5 µm. e) Longitudinal section of the tail region. Fibrils (arrows) are arranged parallel to the plasma membrane (pm) and parallel to the cortical microtubules (mt), which bear irregular projections toward the plasma membrane, bar 0.1 µm. f) Irregularly formed vesicles (v) with occassional cross-bridges (arrow) to the plasma membrane (pm) are found on the posterior flanks of the cell, bar 0.1 µm.

show an extremely sharp X-ray diffraction pattern characteric of ß-chitin, indicating their homogeneous composition and purity (Fig. 2b). In electron diffraction they reveal point diagrams (Figs. 2c, d) with a=4.28 Å, b=36.7 Å, c=10.33 Å, indicating a bigger unit cell than previously postulated for ß-chitin (19). These chitin spines are growing unilaterally at their base from specific regions of the cell underlying the pores in the silica valves. In thin section, these regions are seen to consist of conical invaginations of the plasma membrane, with specific coat layers on both faces of the plasma membrane (Fig. 1c) characterizing these differentiated plasma membrane regions. This suggests that chitin polymerization is a transmembrane process involving these specialized regions of the plasma membrane. Cytoplasmic vesicles (Fig. 1c) seem to contribute to the formation of this conical invagination of the plasma membrane.

The steps of polymerization and microfibril assembly may be experimentally separated by application of calcofluor white (CW) or congo red (CR). In the presence of these dyes, amorphous clouds plugging the pores are formed, instead of being assembled into the spines (Fig. 1d). Attempts to inhibit chitin microfibril assembly of Thalassiosira with the insecticides dimilin or diflubenzuron were unsuccessful in the tested range of 10^{-5} - 10^{-3} M inhibitor.

Poterioochromonas. The biflagellate motile alga Poterioochromonas (Fig. 3a) may attach to a substratum with its posterior tail region. It first secretes attachment material and then starts building a lorica, first the foot region, then the stalk region, and in the end the cup region housing the cell. The whole lorica looks like a wineglass (Fig.3b), and is completed within 1-2 h. The composition of the lorica is rather complex, with acidic matrix components (not yet fully characterized) and a fibrillar backbone (Fig. 3d). The native material does not show an identifiable X-ray diffraction pattern. After purification by extraction, the residual material yields an X-ray diffraction pattern showing the main reflections of chitin (Fig. 3c), and the hydrolysate of this fraction contains mainly glucosamine. Poterioochromonas deposits its chitin microfibrils parallel to the cell surface. The microfibrils are composed of parallel thinner subunits (Figs. 3d, 4a) and are deposited parallel to cortical microtubules in the cytoplasmic tail region (Fig. 3e). By treatment with CW or CR, chitin microfibril formation and orientation are inhibited, but amorphous to fuzzy material is continued to be formed instead of the microfibrils (compare Figs. 4a and b).

The rather extended plasma membrane region of the cytoplasmic tail of Poterioochromonas seems to be the site of chitin formation in this organism. The contribution of secretory vesicles (Fig. 3f) is still unresolved. With freeze-fracture, the tail plasma membrane region is characterized by the occurrence of densely packed particle aggregates consisting of small particles, and individual bigger particles (Figs. 4c, d). It is not yet clear which (if any) of these particles are chitin synthases, or part of the chitin synthase complex.

Attempts to inhibit the microfibril formation process in Poterioochromonas with diflubenzuron or dimilin did not show any effects in the concentration range from 10^{-5} M to 10^{-3} M.

Eufolliculina. After division of the mother animal, the swarmer of the ciliated protozoon Eufolliculina regularly leaves the mother lorica. After a while, it settles down on a substratum and builds a new bottle-like housing (Fig. 5b). Repetition of this process leads to the formation of colonies consisting of many housed animals attached to the substratum (Fig. 5a). The lorica formation process in this animal is characterized by secretion of

Fig. 4 _Poterioochromonas_. a) Higher magnification of untreated lorica microfibrils, which show parallel subunits (arrows). Negative stain, bar 0.1 µm. b) CW-treatment. Fuzzy material (arrow) and short rods (arrowheads) have assembled instead of microfibrils. Negative stain, bar 0.1 µm. c, d) Freeze-fracture, tail region plasma membrane. The extraplasmic fracture face (EF) shows imprints of particle aggregates (bigger arrows) and individual particles (small arrows). The protoplasmic fracture face (PF) shows particle aggregates (bigger arrows) and a higher number of indivudual particles (small arrows), bars in c, d) 0.1 µm.

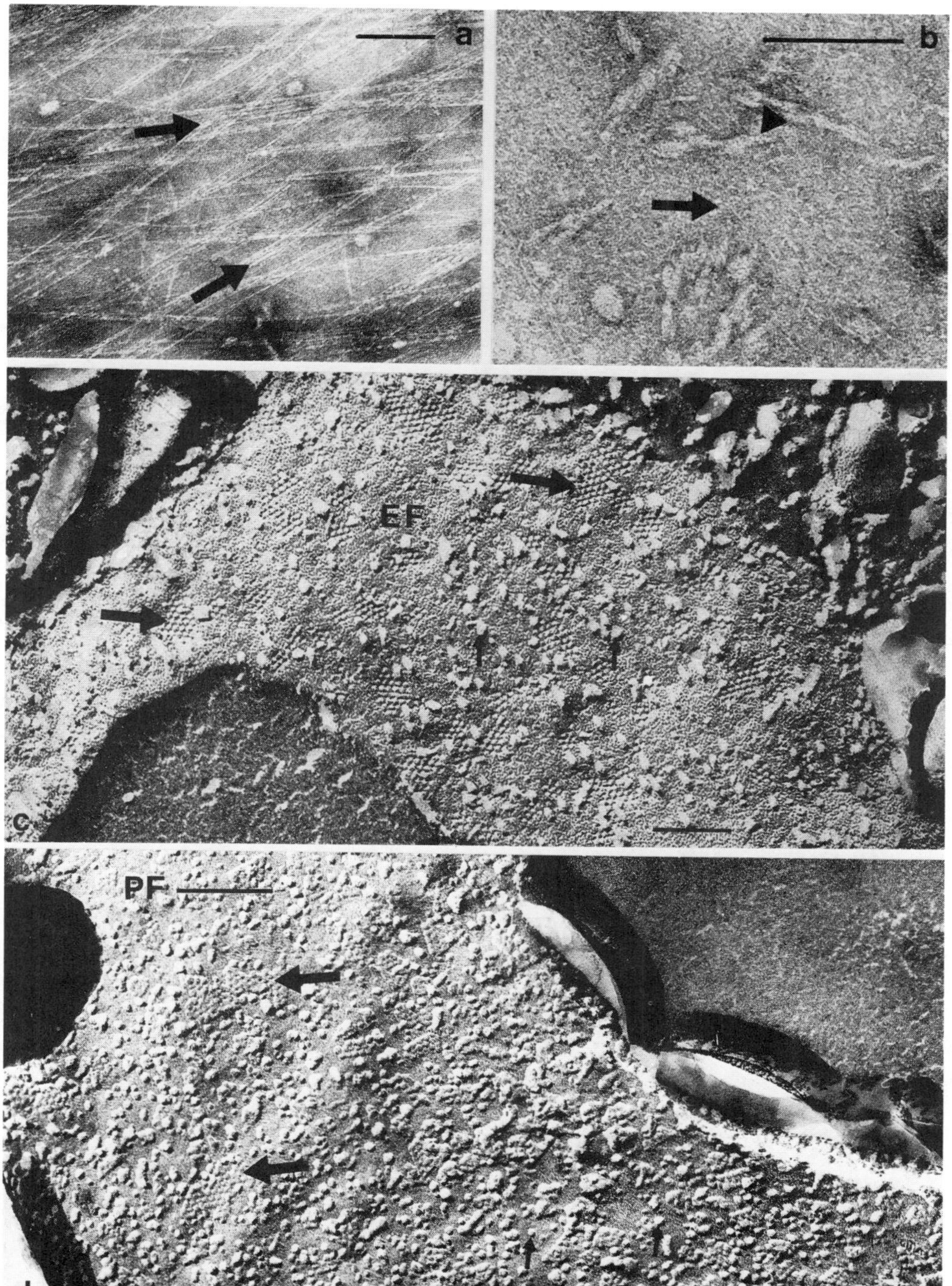
a
b
EF
c
PF
d

Fig. 5 Eufolliculina. a) Survey light micrograph of sessile cells in their loricae. The anterior wings are part of the oral apparatus, bar 100 µm. b) Scanning micrograph of loricae. The posterior part (ampulla) is attached to the substrate by fibrillar material (arrow), bar 100 µm. c) Tangential section of cortical ribs of the swarmer (TEM). The cortex contains a large number of vesicles (arrows) which include material of different electron density, bar 1 µm. d) For lorica secretion, the vesicles (v) fuse with the plasma membrane (pm) and release their contents to the outside, bar 0.2 µm. e) P-face (PF) of the plasma membrane with so called "fusion rosettes" (circles) above the cortical vesicles (v). Freeze-fracture, bar 0.5 µm. f) High magnification of such a rosette, bar 0.1 µm.

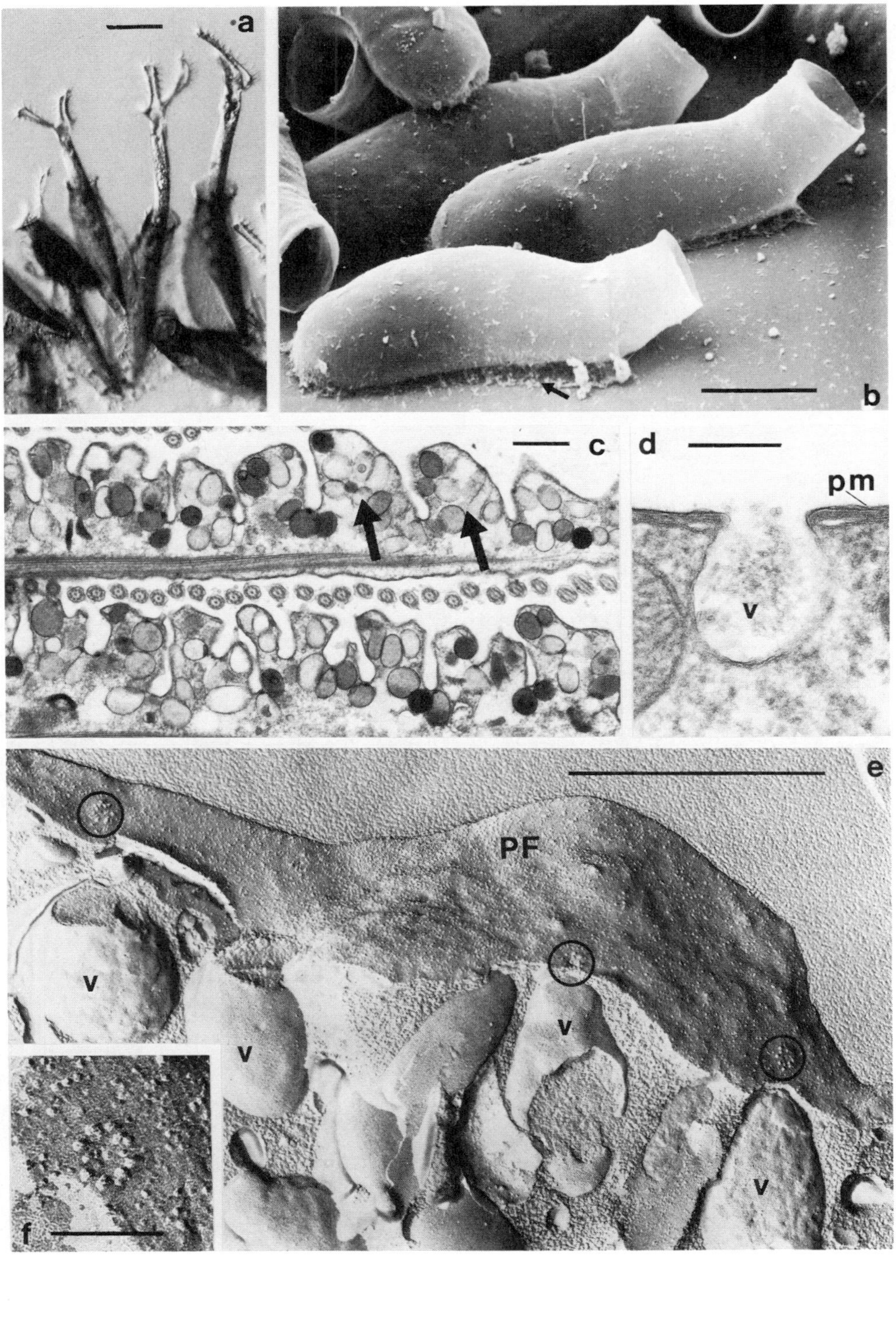
a
b
c
d
pm
v
e
PF
v
v
v
v
f

Fig. 6 Lorica of _Eufolliculina_. a) Extraction-resistant fibrillar material, showing a meshwork of ribbon-like fibrils, which are up to 20 nm broad. Note kinking sites (arrows). Negative staining, bar 0.1 µm. b) X-ray diffraction pattern of extracted lorica material with ß-chitin reflections. c-g) Ultrathin sections of the lorica wall of untreated control cells (c) and after application of CR (d, f) and CW (e, g). c) The lorica wall consists of irregularly arranged fibrils embedded and interconnected by electron-dense material (e.g., between arrows), bar 1 µm. d) Swarmers treated with CR during lorica construction form a thick-walled lorica made up of several fibrillar layers and apparently containing less electron-dense material, bar 1 µm. e) Loricae formed in the presence of CW have a fluffy appearance and consist of an accumulation of very fine fibrils (diameter ca. 1.5 nm), bar 1 µm. f) Higher magnification of the fibrils in the loricae of CR-treated cells which are thinner than normal (diameter ca. 8 nm), bar 0.2 µm. g) In the loricae of CW-treated cells, sometimes larger clumps of fuzzy material (asterisk), and, very rarely, thin fibrils (arrow) can be observed, bar 0.2 µm.

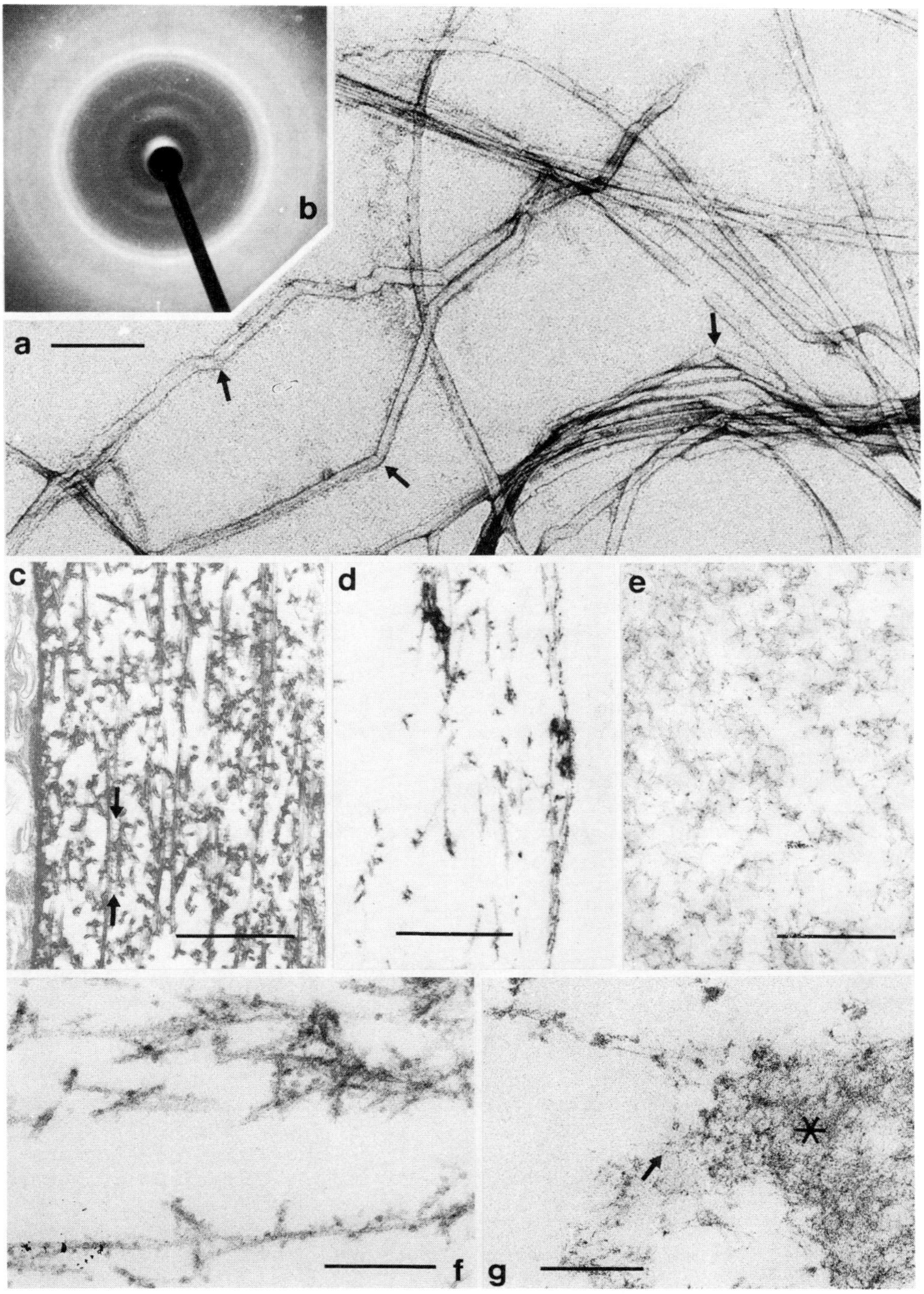
a
b
c
d
e
f
g

numerous cortical vesicles of different contents, which have been accumulated in the cortical cytoplasm before cell division (Figs. 5c, d). The secreted vesicular contents then assemble extracellularly and rather remote from their points of secretion to form fibrillar material embedded in an acidic matrix (Fig. 6c). The region of the plasma membrane where these vesicles are to be secreted is characterized by rather smooth appearance and occurrence of "fusion rosettes", that are ring-like particle aggregates consisting of 8-10 particles and one central particle (Figs. 5e, f). Particle arrangements on the vesicular membranes are irregular, different in the various vesicle types, and can not yet be attributed to chitin formation.

The fibrillar material assembled extracellularly consists of ribbon-like structures showing kinking sites and thinner parallel subunits (Fig. 6a). After removal of the encrusting substances by acid and alkaline extraction, the residual material yields an X-ray diffraction pattern similar to ß-chitin (Fig. 6b).

When CW or CR are applied during the process of lorica formation, secretion continues, but the secreted material stays fuzzy or almost amorphous. Only in places thin microfibrils are still assembled (Figs. 6d-g).

DISCUSSION

The comparison of the way of chitin microfibril assembly in the three organisms described above shows both, common principles and diversity. In all three cases the polymerization step may be separated from microfibril assembly by CW and CR, indicating that a certain gap may naturally be inserted in between these two steps. Corresponding results have also been obtained for chitin fibril formation in yeasts and for cellulose microfibril assembly in a variety of organisms (20-23).

It looks, however, as if this temporal gap were small in *Thalassiosira* and in *Poterioochromonas*, whereas in *Eufolliculina* the gap seems naturally to be greater, with polymer formation in the vesicles, secretion of the polymer perhaps protected from a precocious assembly by accompanying substances, and assembly into crystalline microfibrils outside the cell, presumably after changes in the ionic surroundings or removal of the protecting substances by enzymes.

In *Thalassiosira* and *Poterioochromonas*, specific regions in the plasma membrane (small in *Thalassiosira*, more extended in *Poterioochromonas*) seem to bear the chitin synthases. In both cases, however, the contribution of vesicles carrying the synthases to the plasma membrane, and perhaps already starting the polymerization reaction, seems also probable from the cytological pictures.

The membrane mediated control of microfibril deposition is rather direct in *Thalassiosira* and *Poterioochromonas*. In *Thalassiosira*, there are fixed sites exactly matching with the pores in the silica valves. In *Poterioochromonas*, the synthases probably are motile in the plasma membrane and must somehow interact with the cytoskeleton (for further discussion of the postulated orientation mechanisms se 11, 23, 24). Unfortunately, the freeze-fracture results do not yet allow a tentative identification of the chitin synthases as with the plasma membrane "rosettes" postulated to be involved in cellulose synthesis (e.g., 25-27).

There seems to be a relatively close similarity between the secretion/extracellular assembly process of chitin in *Eufolliculina*, and cuticle, bristle and radula assembly in multicellular systems (1-7). The way of locally restricted, membrane controlled assembly in the diatoms seems rather primitive, but also bears some overall resemblance to bristle formation in a mould in arthropods and polychaetes (3-7). The membrane mediated control of chitin deposition over a more extended region of the plasma membrane in Poterioochromonas compares well with the apical chitin deposition in fungal hypha and with bud scar deposition in yeasts (1, 28).

In all three cases described above, there are candidates for "chitosome"-like structures, which might be chitin synthase en route to its place of action (1, 29), but at least in Eufolliculina, the polymerization seems to be almost completed before exocytosis.

Taken together, the results on these unicellular organisms show considerable flexibility in the way of chitin fibrill assembly, which might reflect different regulatory mechanisms, and adaptations to the specific needs of the organism.

ACKNOWLEDGEMENTS

The authors thank Dr. G. Lieser for cooperation in the electron diffraction work, Prof. W. Barthlott, Dr. A. Beckett and Miss R. Benke for cooperation in scanning electron microscopy, and Mrs. B. Heck for skilful technical assistance. Valuable discussions with Prof. K. Hausmann and Prof. E. Schnepf are appreciated. This work was supported by Deutsche Forschungsgemeinschaft.

REFERENCES

1. R. A. A. Muzzarelli, Chitin. Pergamon Press, Oxford, New York (1977)
2. A. C. Neville, Biology of the arthropod cuticle. Springer-Verlag Berlin (1975)
3. J. Bereither-Hahn, A. G. Matoltsy, and K. S. Richards, eds., Biology of the integument 1, Invertebrates, Springer-Verlag Berlin (1984)
4. Y. Bouligand, Les soyes et les cellules associés chez deux annelides polychêtes. Etude en microscopie photonique à contraste de phase et en microscopie electronique. Z. Zellforsch. 79, 332-363 (1967)
5. R. M. Gustus, and R. A. Cloney, Ultrastructural similarities between setae of brachiopods and polychaetes. Acta Zoologica 53, 229-233 (1972)
6. R. M. O'Clair, and R. A. Cloney. Patterns of morphogenesis mediated by dynamic microvilli: chaetogenesis in Nereis vexillosa. Cell Tiss. Res. 151, 141-157 (1974)
7. R. F. Carter, and K. A. Wright, Formation of the odontostyle during molting of the nematode Xyphinema americanum (nematoda: Dorylaimoidea). J. Ultrastr. Res. 87, 221-241 (1984)
8. W. Herth, and P. Zugenmaier, Ultrastructure of the chitin fibrils of the centric diatom Cyclotella cryptica. J. Ultrastr. Res. 61, 230-239 (1978)
9. W. Herth, The site of ß-chitin fibril formation in centric diatoms. I. Pores and fibril formation. J. Ultrastr. Res. 68, 6-15 (1979)
10. W. Herth. The site of ß-chitin fibril formation in centric diatoms. II. The chitin-forming cytoplasmic structures. J. Ultrastr. Res. 68, 16-27 (1979)
11. W. Herth, A. Kuppel, and E. Schnepf. Chitinous fibrils in the lorica of the flagellate chrysophyte Poterioochromonas stipitata (syn. Ochromonas malhamensis). J. Cell Biol. 73, 311-321 (1977)
12. M. Mulisch, W. Barthlott, and K. Hausmann. Struktur und Ultrastruktur von Eufolliculina spec. Schwärmer und sessiles Stadium. Protistologica 17, 285-312 (1981)
13. M. Mulisch, and K. Hausmann. Lorica construction in Eufolliculina sp. (Ciliophora, Heterotrichida). J. Protozool. 30, 97-104 (1983)
14. W. Herth. Calcofluor white and congo red inhibit chitin microfibril assembly of Poterioochromonas: Evidence for a gap between polymerization and microfibril formation. J. Cell Biol. 87, 442-450 (1980)

15. P. Zugenmaier, W. Herth, and G. Lieser. Elektronenbeugungsuntersuchungen an Polymereinkristallen des ß-Chitins. Colloid and Polymer Science 259. 472-473 (1981)
16. W. Herth, W. W. Franke, H. Bittiger, A. Kuppel, and G. Keilich. Alkali-resistant fibrils of ß-1,3- and ß-1,4 glucans: structural polysaccharides in the pollen tube wall of Lilium longiflorum. Cytobiol. 9, 344-357 (1974)
17. W. Herth, A. Kuppel, W. W. Franke, and R. M. Brown. The ultrastructure of the scale cellulose from Pleurochrysis scherffelii under various experimental conditions. Cytobiol. 10, 268-284 (1975)
18. M. Mulisch, W. Herth, P. Zugenmaier, and K. Hausmann. Chitin fibrils in the lorica of the ciliate Eufolliculina uhligi: Ultrastructure, extracellular assembly and experimental inhibition. Biol. Cell 49, 169-178 (1983)
19. J. Blackwell. The macromolecular organization of cellulose and chitin. In: Brown, R. M., ed., Cellulose and other natural polymer systems. Plenum Press, New York, London, pp. 207-224 (1982)
20. G. L. Vannini, F. Poli, A. Donini, and S. Pancaldi. Effects of congo red on cell wall synthesis and morphogenesis in Saccharomyces cerevisiae. Plant Sci. Lett. 31, 9-17 (1983)
21. W. Herth, and I. Haußer. Chitin and cellulose fibrillogenesis in vivo and their experimental alteration. Proceed. of the seventh annual symposium in Bot., Structure, function and biosynthesis of plant cell walls, eds. Dugger, W. M., and Bartnicki-Garcia, S., Riverside, California, pp. 89-119 (1984)
22. C. Haigler, and M. Benziman. Biogenesis of cellulose I microfibrils occurs by self assembly in Acetobacter xylinum. In: Brown, R. M. ed., Cellulose and other natural polymer systems. Plenum Press, New York, London, pp. 273-298 (1982)
23. H. Quader. Morphology and movement of cellulose synthesizing (terminal) complexes in Oocystis solitaria: Evidence that microfibril assembly is the motive force. Eur. J. Cell Biol. 32, 174-177 (1983)
24. I. B. Heath, and R. W. Seagull. Oriented cellulose fibrils and the cytoskeleton: a critical comparison of models, In: Lloyd, C. W., ed., The cytoskeleton in plant growth and development. pp. 163-182, Academic Press, London (1982)
25. T. H. Giddings, D. C. Brower, and A. Staehelin. Visualization of particle complexes in the plasma membrane of Micrasterias denticulata associated with the formation of cellulose fibrils in primary and secondary cell walls. J. Cell Biol. 84, 327-338 (1980)
26. S. C. Mueller, and R. M. Brown. Evidence for an intramembrane component associated with a cellulose microfibril-synthesizing complex in higher plants. J. Cell Biol. 84, 315-326 (1980)
27. W. Herth. Plasma-membrane rosettes involved in localized wall thickening during xylem vessel formation of Lepidium sativum L. Planta 163, in press
28. J. G. H. Wessels, and J. H. Sietsma. Fungal cell walls: a survey. In: Tanner, W., and F. A. Loewus, ed., Plant Carbohydrates II. Encyclopedia of plant physiology, New series, vol. 13 B, pp. 352-394 (1981)
29. J. Ruiz-Herrera. Synthesis of chitin microfibrils in vitro. In: Brown, R. M., ed., Cellulose and other natural polymer systems. Plenum Press New York, London, pp. 207-224 (1982)

POLYGALACTOSAMINE PRODUCED BY A MICROORGANISM

H. Takagi and K. Kadowaki

Research Laboratory, Higeta Shoyu Co, LTD
2-8 Chuo-cho, Choshi-shi
Chiba, Japan

INTRODUCTION

There is a potential need for strong and safe flocculating agents in the fermentation and food industry. Since harvest or removal of microbial cells from cultured broth is a very important but energy consuming process in the fermentation industry, safety-proof and strong agents may be very useful for improving the process substituted for centrifugation or filtration.

Several workers (1 - 3) reported that some microorganisms produced flocculating substances. But these were not strong and safe enough to be used practically. We tried screening of flocculant producing fungi for practical use in industry and isolated a mould from soil, which produced a large amount of a strong microbial cell flocculant. In addition to its high productivity and strong flocculating activity, the substance has very interesting and unique properties as a flocculant.

The present paper reports studies on screening of flocculant producing fungi, microbiological properties of isolated mould, interesting characteristics of flocculating activity and purification and the chemical and physical properties of the flocculant produced by Paecilomyces sp. I-1. We found the flocculant to be a polymer of almost pure galactosamine, a rare material on earth, and discuss the possible mechanism of flocculation and its usefulness in industry.

RESULTS

Screening for flocculant producing fungi

For the screening of flocculant producing microorganisms, 25 strains of fungi from the culture collection of our laboratory and 50 strains of fungi isolated from soil sampled at several areas in Japan were examined.

As the purpose of flocculant in these studies is to remove and harvest the microbial cells from the fermentation broth efficiently, E. coli cells were chosen as a test microorganism for flocculating activity. 9 ml of washed E. coli cell suspension (OD 660 = 2.0) and 1 ml of culture filtrate or its dilution were mixed in the test tube and observed for whether flocculations occured immediately or not.

Table 1. Components of the culture medium

	GC1	GC2	GYe
Glucose	2 %	2 %	2 %
Yeast extract	-	-	0.3 %
Casamino acid	0.3 %	0.3 %	-
$CaCl_2$	-	0.5 %	-
pH	7.0	7.0	7.0

Culture filtrates of 75 strains in GYe medium were tested for their ability to flocculate E. coli cells. The ability was found in only one strain (named I-1) in this medium. In tests in GCl medium, the activity was detected in culture filtration of 5 strains, all of which were isolated from soil and whose colonies grown on Czapek's agar plate were whitish to pinkish. The strain which produced the strongest flocculant was the I-1 that was isolated from soil at Wakayama.

Taxonomical characteristics of strain I-1

Colonies grown at 25° C on Czapek's agar plates were whitish and velvety in appearance and reached a diameter of 4 - 5 cm in 14 days. The reverse was colorless at first and became yellow with age. The mould grew at a wide range of temperature, 0 - 30° C (optimum: 23 - 25° C) and of pH, 2 - 10.

Morphological observation was carried out with both optical and scanning electron microscopes. The mould had septa on hyphae and lacked conidiophores. We observed chains of 20 - 40 pieces of conidia which were produced from phialides borne directly on hyphae. The chains of conidia were easily broken by mild vibration. The conidia were cigar-shaped, smooth and 1.0 - 1.4 x 4 - 6 μm in size. Phialides were monophialidic, slender, 20 - 45 μm long and tapered with basal portion (0.5 - 1.5 μm diameter) and apex (0.5 - 1.0 μm diameter). The process of conidia formation was phialo-type (4) or enteroblastic-type (5).

These observations led us to conclude that the isolated mould is one of monophialidic species of Paecilomyces. Among species described in Onion and Barron's report (6), we could not determine the decisive species of the mould. Though it was closely related to P. bacillosporus in most morphological observations but little in physiological characteristics and in length of conidia, we called the strain Paecilomyces sp. I-1.

Characteristics of flocculating activity

The flocculant produced by I-1 was tested by culture filtrate in studies of cultural conditions for flocculant production and by water solution of ethanol precipitates of culture filtrate in cases of studies of characteristics of flocculating activity.

E. coli cells were chosen as a test microorganism for flocculating activity. A portion of culture filtrate was serially diluted on a microtiter plate or in a test tube, and mixed with an equivalent volume of washed E. coli cell suspensions (1 mg dry cells/ml). Flocculating activity was expressed by the maximum dilution rate of the flocculant. Flocculating activity = n (the n th power of 2). In case of flocculating reactions, except E. coli cells, 1 mg dry cells per ml of microbial cells and 20 mg per ml of cellulose, Sephadex, agar, charcoal, alminium, oxide, silica gel and kieselguhr were

Table II. Spectrum of flocculating activity of flocculant produced by *Paecilomyces sp.* I-1.

Microorganism	Flocculating activity	Miscellaneous	Flocculating activity
Bacillus subtilis 168	5	Blood cell A	4
B. subtilis K	3	Blood cell B	4
B. amylosacchariticus	3	Blood cell AB	4
B. natto 1212	5	Blood cell O	4
Escherichia coli K12	6	Activated sludge	5
E. coli B	6	Cellulose powder	8
Micrococcus sp.	6	CM-Cellulose	8
Torulopsis candida	8	DEAE-Cellulose	2
Debaryomyces hansenii	7	Sephadex G100	6
Rhodotorula marina	4	Agar powder	2
Candida utilis	6	Activated charcoal	6
C. mogii	7	Alminium oxide	6
Saccharomyces cervisiae	7	Silica gel	6
Cryptococcus kuetzingii	4	Kieselguhr	9

used. Judgement of positive flocculating reaction was done by whether the floc was formed and whether sedimentation occurred within a minute. Table II shows the spectrum of flocculating activity of the flocculant produced by *P. sp.* I-1. The substance could flocculate and sediment all kinds of microbial cells examined, though the intensity of flocculating activity depended on the species of microorganisms. The substance could flocculate almost all suspending solids in aqueous solution. The flocculating activity was not affected by temperature (0° - 100° C) and pH (4 - 8). It was not affected by most ions, except carbonate and iron, which strongly inhibited it.

Purification of the flocculant

To the culture filtrate 2, volumes of ethanol were added with stirring. Precipitates were collected by centrifugation, dissolved in water and dialyzed against water. The yield of the lyophilized material (named crude PF-101) was 0.5 g per liter of the culture filtrate.

Behavior of crude PF-101 for gel filtration materials was examined. When crude PF-101 solution was applied to a column of Sephadex G200 or Sepharose 6B and eluted with 0.04 M Tris-HCl buffer (pH 6.7), the flocculant did not appear in the eluate. However, when Bio-Gel P-300 was used, it was eluted at one void volume. The flocculant absorbed at Sepharose 6B could be recovered when elution was performed with buffers involving carbonate or carboxylate and with buffers over pH 7.5.

After crude PF-101 was dissolved in hot water and the insoluble materials were removed by centrifugation, the solution was applied to a column of Sepharose 6B previously equilibrated with 0.04 M Tris-HCl buffer (pH 6.7). After the column was washed with the same buffer, elution was performed with 0.04 M Tris-maleate buffer (pH 6.5). Gel filtration pattern of the flocculant is shown in Fig. 1. Active fractions were pooled and 2 volumes of ethanol were added, and washed with 95 % ethanol, and lyophilized (named PF-101).

Physicochemical properties

A 0.1 % solution of PF-101 in an aqueous solution gave 7 of flocculating activity and 1.08 of relative viscosity. Isoelectric focusing of PF-101 showed

Table III. Analytical data on PF-101.

Analysis item	Method	% Dry wt.	
Neutral sugar	Phenol sulfuric acid	2.8 %	
	Gas chromatography	not detected	a)
Amino sugar	Elson Morgan	72 %	a)
	Ninhydrin	88 %	a)
	Gas chromatography	86 %	b)
Amino acid	Liquid chromatography	trace	a)
Acetyl	Liquid chromatography	2.3 %	a)
Formyl	Liquid chromatography	5.7 %	a)

a) Determined after acid hydrolysis of PF-101 in 4N HCl at 100°C for 16h
b) Determined after acid hydrolysis of PF-101 in 6N HCl at 110°C for 16h

a single band at pH 8.5, which suggested homogeneity of PF-101. The molecular weight was estimated to be over 300,000 D by ultrafiltration method, which was supported by the result that the flocculant was eluted at a void volume by Bio-Gel P-300 column chromatography.

Component of PF-101

Elemental analysis showed that contents of carbon, hydrogen and nitrogen were 37.45 %, 6.30 %, 5.44 %, respectively, but neither sulfur nor phosphorus was detected.

A summary of the analytical data is presented in Table III. Fig. 2 shows the result of GLC analysis of reaction products as their TMS-ethers obtained from hydrolysate of PF-101. Only two peaks were obtained. Peak 2 (Fig. 2) was identified as galactosamine from its coincidence with the authentic galactosamine. Peak 1 was identified as tris-aminomethane which

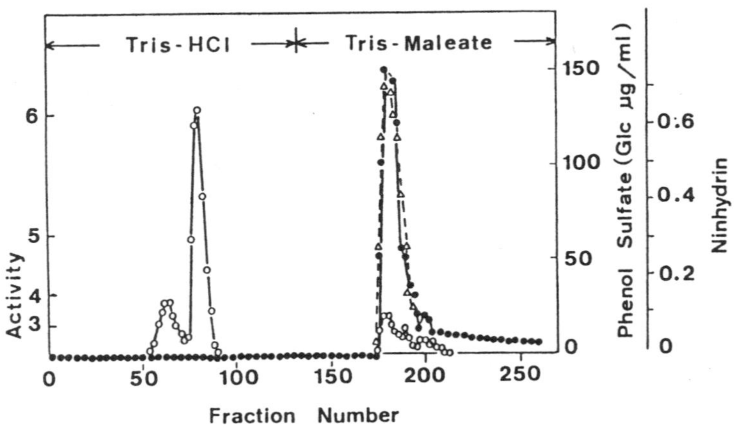

Fig. 1 Chromatography of partially purified PF-101 on a Sepharose 6B column.

The sample after ethanol precipitation was applied to a Sepharose 6B column (5 X 100cm) and eluted with 0.04M Tris-HCl buffer, pH 6.7 (135 fractions), followed by 0.04M Tris-Maleate buffer, pH 6.5. Elution was collected in 25ml fractions.
Elution profile is shown by flocculating activity (●), Phenol-H_2SO_4 assay (○) and Ninhydrin assay (△).

might be trapped by PF-101 during purification. LC analysis of the hydrolysate of PF-101 denoted the existence of galactosamine, ammonia and a very small amount of amino acids (under 3 %). These results suggest that PF-101 is characterized as a polysaccharide composed of pure galactosamine.

In the hydrolysate of PF-101, 2.3 % acetic acid and 5.7 % formic acid were detected (Table III). Since PF-101 contained no neutral sugar (Fig. 2), formic acid could not be a decomposition product of neutral sugar and might originate from N-formyl galactosamine. Assuming that the acid was derived from N-acetyl, only 8 % of the galactosamine units are N-acetylated.

Deaminative cleavage

An aminosugar that is not N-substituted is quantitatively cleaved with its concomitant conversion to a 2,5-anhydrohexose (7, 8). The effect of the nitrous acid on PF-101 was measured by quantitative determination of the 2,5-anhydrotalitol residues formed on deaminative cleavage. Direct deaminative cleavage of PF-101 yielded 33.6 % of the anhydrotalitol. In contrast, the amount of the anhydrotalitol from the acid hydrolysate of PF-101 before deamination was 67 %. PF-101 was treated with nitrous acid and the deamination products were analyzed by gel filtration on Bio-Gel P-2. A main peak was identified as 2,5-anhydrotalitol from the gel chromatogram of the deaminative cleavage of the authentic galactosamine. This sugests that many galactosamines with free amino residues (non-acetylated amines) are positioned consecutively.

Periodate oxidation and Smith degradation of PF-101

Periodate oxidation of PF-101 resulted in the uptake of 2.74 mol periodate per mol of monosaccharide residue, whereas, in control experiments, oxidation with glucose, galactosamine, dextran and starch consumed 3.43, 4.32, 2.01 and 0.88 mol per mol of monosaccharide residue, respectively. The formic acid formation from PF-101, glucose, galactosamine, dextran and starch afforded 0.64, 3.15, 4.01, 1.0 and zero mol per mol of monosaccharide residue, respectively. When Smith degradation was performed completely, the main product of acid hydrolysate of the degradation was identified as erythritol by gas chromatography. Glycerol, galactosamine and tris-aminomethane were found as minor products of acid hydrolysate of Smith degradation.

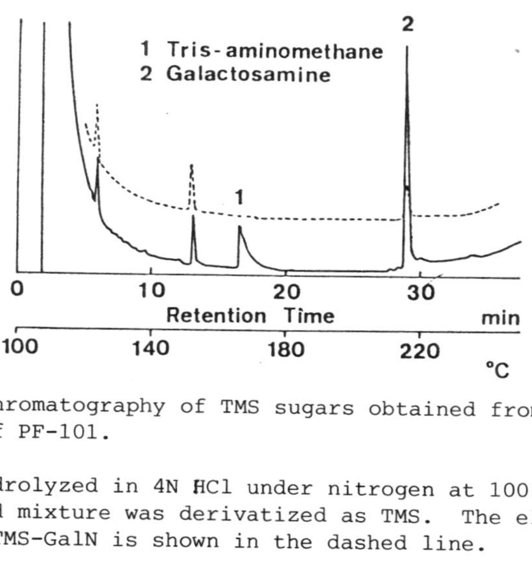

Fig. 2 Gas chromatography of TMS sugars obtained from the acid-hydrolysate of PF-101.

PF-101 was hydrolyzed in 4N HCl under nitrogen at 100°C for 16hr. The hydrolysed mixture was derivatized as TMS. The elution pattern of authentic TMS-GalN is shown in the dashed line.

Table IV. Effect of growth factors on PF-101 production by *Paecilomyces sp*. I-1.

Nitrogen source	Flocculating activity	Flocculant µg / ml	Mycelium wt. mg / 100ml
$NaNO_3$	2	59	249
NH_4Cl	3	71	199
Urea	2	43	249
Glutamic acid	5	154	309
Malt extract	None	15	160
Yeast extract	2	49	717
Meat extract	4	79	612
Polypeptone	9	564	490
Casamino acid	8	398	233

The mould was cultured at 25°C for 5 days with shaking. Growth factor was added at 0.3% to the basal medium (pH 7.0) containing 2% glucose, 0.1% KH_2PO_4, 0.01% K_2SO_4 and 0.5% $CaCl_2$.

Culture conditions for flocculant production by Paecilomyces sp. I-1

Table IV shows the effects of various kinds of growth factors on PF-101 production. PF-101 was accumulated in the media containing polypeptone and casamino acid. On the other hand, addition of yeast extract and meat extract stimulated only the growth of mycelium and little or no PF-101 production.

The mould efficiently produced PF-101 from all kinds of carbohydrates, such as glucose, fructose, mannose, galactose, sucrose, maltose, raffinose and starch, but never from lactose. Starch appeared especially favorable for both growth and PF-101 production.

Calcium ion drastically stimulated both growth and production. Sodium, magnesium and potassium ions affected neither of them, while iron and copper ions inhibited the growth. As for anions, phosphate, sulfate, carbonate and acetate inhibited both.

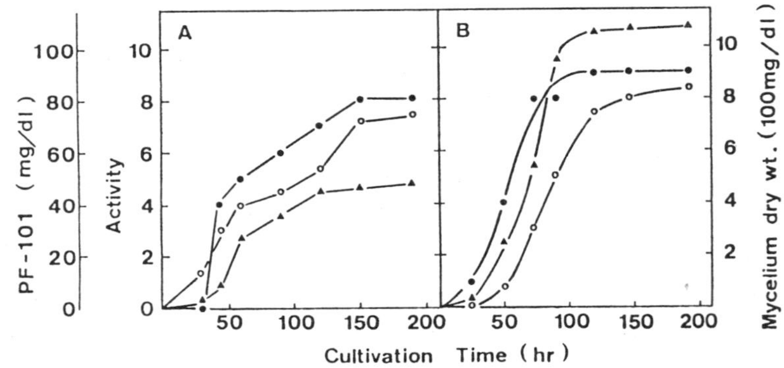

Fig. 3 Time course of flocculant production by *Paecilomyces* sp. I-1.

The mould was incubated in GC2 medium (A) and the medium composed of 2% starch, 0.3% polypeptone and 0.5% $CaCl_2$ (B). A portion of culture was sampled at the intervals and measured mycelium wt. (○), flocculating activity (●) and PF-101 (△).

Fig. 3 shows the time course of PF-101 production by P. sp. I-1. PF-101 accumulation and flocculating activity were found in the culture fluid with the growth of mycelium. At optimum conditions, the flocculating activity for E. coli cells resulted 2^9 for 1 g of PF-101 per liter accumulated.

DISCUSSION

We first reported in this paper that Paecilomyces sp. produced a new kind of strong flocculant named PF-101 in culture fluid. P. sp. I-1 produced the flocculant in a simple medium. However, some conditions were required for the efficient accumulation of PF-101. The best medium for producing PF-101 was composed of 2 % starch, 0.3 % polypeptone, 0.5 % $CaCl_2$ with pH adjusted to 7.0. When P. sp. I-1 was grown in 100 ml of this medium in a 500 ml flask in a rotary shaker at 25° C for 5 days, flocculation for E. coli cells was observed even when the culture filtrate was diluted by 2^9.

The substance could be purified simply by using the property to be absorbed to Sepharose. Its composition, determined by gas and liquid chromatography and colorimetric analysis of acid hydrolysate, is 85 % galactosamine, 2.3 % acetyl, 5.7 % formyl but no glucosamine, neutral sugar or amino acid was detected. These results suggest that PF-101 is characterized as a polysaccharide composed of galactosamine.

Polygalactosamine, in which about two thirds of galactosamine residues were N-unsubstituted, was first isolated from the culture fluid of Aspergillus parasiticus, a well-known mycotoxin producer, by Disther and Roseman (9) but they never described its flocculating activity in their report. Hayashi, Yamada and Miyazaki (10) found that contents of acetylated galactosamine in polygalactosamine, produced in the culture fluid of A. parasiticus, changed from 20 to 70 % by addition of ammonium tartrate. Polygalactosamine of A. parasiticus was characterized as a linear chain of $\alpha(1\rightarrow4)$-linked galactosamine residues (11).

In general, many amino sugar residues which compose the polysaccharides of biocomponents are N-acetylated, such as chitin, mucopolysaccharides, peptidoglycans from cell walls of bacteria and fungi and blood group substances. However, contents of acetylated galactosamine in PF-101 are 7 - 8 % of galactosamine residues from calculating the amount of acetyl. Thus, the fact that the proportion of N-acetyl of PF-101 is very low creates an interest in its synthetic pathway.

Acid hydrolysates of deaminative cleavage products of PF-101 contained 17 % galactosamine. This result suggests that about 80 % galactosamine residues are N-unsubstituted and that about 10 % galactosamine residues are N-substituted by other than acetyl.

PF-101 is a novel and original substance in that it is composed of only galactosamine and about 80 % of its galactosamine residues are N-unsubstituted.

The data of deaminative cleavage, periodate oxidation and Smith degradation suggest that N-unsubstituted galactosamine residues are mainly linked by $\alpha(1\rightarrow4)$-linkage, that the polymer may have a little $(1\rightarrow2)$-linkage and/or $(1\rightarrow6)$-linkage, and that the maximum continuous sequence of N-acetylated galactosamine is three, considering the gel filtration chromatography of deamination products of PF-101.

The mode of flocculation of PF-101, a cationic polyelectrolyte, is explained as follows. The polygalactosamine chain, which is adsorbing anionic particles and neutralizing their charges, may form intermolecular bridges that result in stable floc and precipitation.

Since PF-101 bears a chemical resemblance to chitosan, N-deacetylated chitin, it is very likely that PF-101 can favorably compare with chitosan in almost all functional aspects, such as polyelectrolyte behavior, chelating ability, gel and film forming ability, sterilizing power and so on.

PF-101 has several advantages over chitosan, for example, easy preparation of the pure material necessary for fine chemicals and stability in natural environment because of a rare occurrence of microorganisms or enzymes hydrolyzing polygalactosamine. No microorganism could be isolated nor detected in medium consisting of PF-101 as sole carbon and nitrogen source. On the other hand, chitin and chitosan are easily hydrolized by various kinds of microorganisms and enzymes.

It is also expected that PF-101 can be produced at a low cost in a large-scale culture for the following reasons; (a) a high growth rate in an inexpensive medium, (b) little if any anxiety of bacterial contamination, because PF-101 prevents bacterial growth, (c) a high productivity of PF-101, (d) a simple purification process of the product.

REFERENCES

1. J. Nakamura, S. Miyashiro and Y. Hirose, Agric. Biol. Chem., 40, 377 (1976).
2. P. Schenck, P. L. Foster, W. W. Walker Jr. and S. Fogel, Belg. Patent 822,947 (1975) C. A., 84, 3296 (1976)
3. R. Kurane, T. Suzuki and Y. Takahara, Agric. Biol. Chem., 43, 421 (1979).
4. G. C. Ainsworth, "Ainsworth and Bisby's Dictionary the Fungi", 6th ed., CMI, Kew (1971).
5. G. T. Cole and B. Kendrick, "Biology of Conidial Fungi", vol.1 and 2, Academic Press, New York (1981).
6. A. H. S. Onions and G. L. Barron, "Monophialidic Species of Paecilomyces", 14th ed., Mycological Papers, No. 107, CMI, Kew, England (1967).
7. P. C. Bardalaya and J. H. Nordin, J. Bacteriol., 125, 655 (1976).
8. Z. Dische and E. Borenfreund, J. Biol. Chem., 184, 517 (1950).
9. J. J. Distler and S. Roseman, J. Biol. Chem., 235, 2538 (1960).
10. O. Hayashi, H. Yamada and T. Miyazaki, Agric. Biol. Chem., 40, 1643 (1976).
11. H. Takada, Y. Araki and E. Ito, J. Biochem., 89, 1265 (1981).

Section III

CHITIN SYNTHESIS INHIBITION

BENZOYLARYL UREAS: INSECTICIDAL COMPOUNDS INTERFERING WITH CHITIN SYNTHESIS

A. G. M. WILLEMS, M. S. BROUWER and B. JONGSMA

Biochemistry and Synthesis Departments
Crop Protection Division DUPHAR B.V
's-Graveland, The Netherlands

INTRODUCTION

The development of selective crop protection products based on the interference with chitin biosynthesis in fungi and insects has been one of the aims in pesticide design for several decades. A major development in this field was the discovery of the mode of action of the fungicidal antibiotic polyoxin D by Japanese scientists[1] in the period 1968-1970. In the same period, Dutch scientists at Duphar B.V. discovered the insecticidal activity of the benzoylphenylureas.

In the course of investigations centered on the herbicide dichlobenil, the basic structure of that compound was combined with that of the urea herbicide diuron.

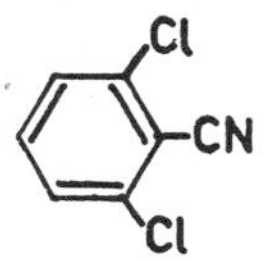

dichlobenil

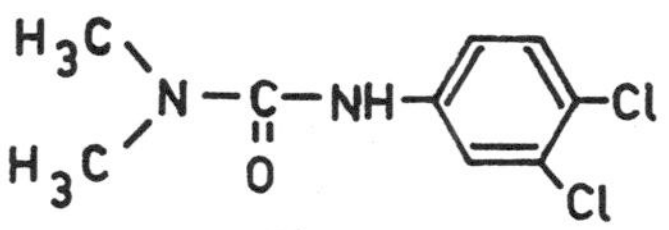

diuron

The resulting compound, 1-(2,6-dichlorobenzoyl)-3-(3,4-dichlorophenyl)urea, with code number DU 19111 was obtained and screened on its biological properties.

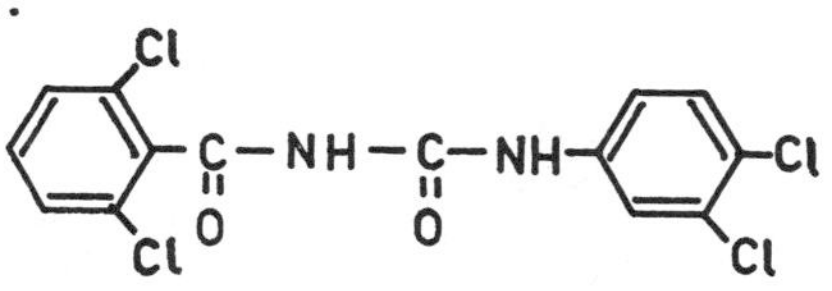

DU 19111

During this screening program no herbicidal or phytotoxic effects were found, but it was observed that larvae of several insect species showed abnormal symptoms after ingestion of the compound. At the time the moult should occur and the larvae had stopped feeding as usual, the treated larvae appeared incapable of shedding their exuviae. They turned black and died. Application of histological techniques revealed that affected larvae were not able to form a new cuticle[2]. Further studies, to be discussed later, soon revealed that interference of the chitin biosynthesis was the most probable mode of action.

The high insecticidal activity of DU 19111 against the larval stages of several insect species and its unique mode of action prompted Duphar chemists to synthesize several hundreds of benzoylarylureas and to evaluate their insecticidal potency in laboratory tests and small scale field trials[3,4,5]. From this research program diflubenzuron was selected for development. It is now fully available commercially under the trade name DIMILIN® and it is registered for various applications in 40 countries all over the world.

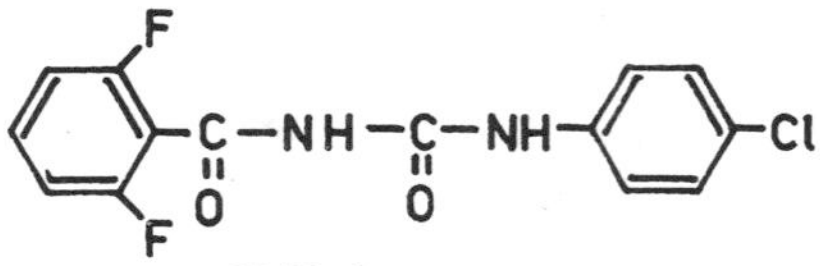

diflubenzuron

COMPOUNDS OF OTHER COMPANIES

After this discovery was published in the patent and scientific literature many other chemical companies started their own benzoylarylurea projects. Up to now about 30 companies in Germany, USA, Japan, Switzerland, and Great Britain filed patent applications for over 200 inventions of similar benzoylarylureas with insecticidal properties. At this time only four companies other than Duphar seem to be active in the field of the insecticidal benzoylarylureas.

Bayer was one of the first companies after Duphar that showed activity in this field. In the literature of 1978[6,7] two experimental compounds of Bayer have been mentioned: Bay SIR 6874 and Bay SIR 8514. Of these two compounds Bay SIR 8514 survived. It is now being cautiously introduced in a few selected countries with the trade name Alsystin. The common name of the active ingredient is triflumuron.

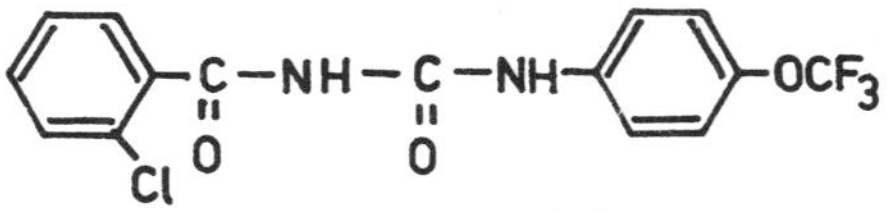

Bay SIR 8514

In 1979 Dow entered this field. This company introduced in that year in Egypt an experimental product called DC-806, which was a mixture of two active ingredients, the phosphor compound chlorpyrifos (Dursban) and a benzoylarylurea with the code name Dowco 439. This latter compound has never been referred to in literature until 1983. In that year Dow presented a new compound of this type, XRD-473, at the International Congress of Plant Pathology at Brighton and in that presentation[8] Dowco 439 was mentioned as a reference compound.

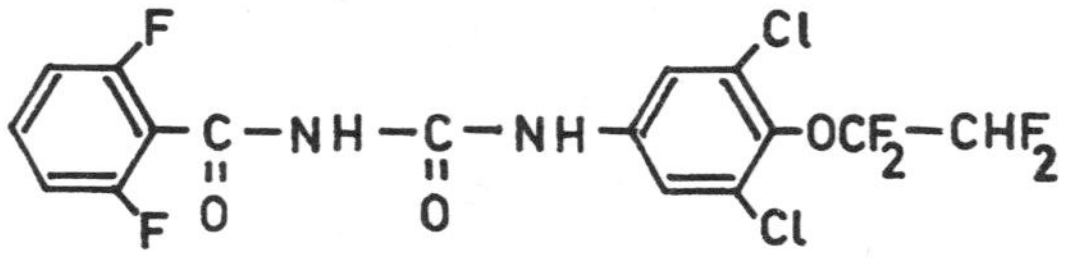

XRD-473

At the 5th International Congress of Pesticide Chemistry organised by IUPAC in 1982 in Kyoto the Japanese company Ishihara Sangyo Kaisha published in a poster the structure of a new benzoylarylurea, which contained a pyridyl group. This compound was named IKI-7899 and later on it became also known as CGA 112913. Its common name is chlorfluazuron. The first reference in open literature to this compound was by Granett et al.[9] in 1983.

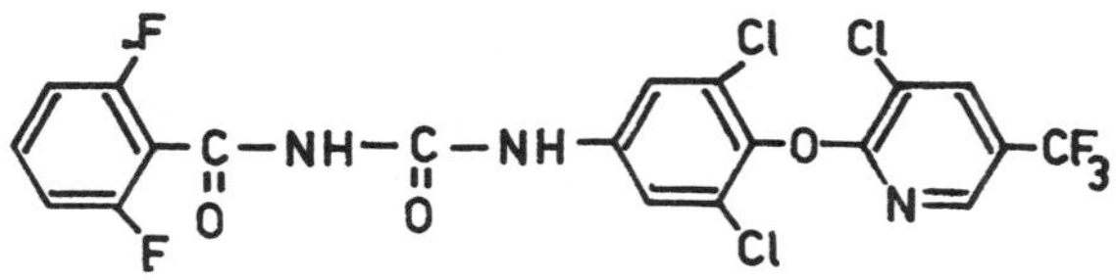

IKI 7899/CGA 112913

The German company Celamerck introduced its first insecticidal benzoylarylurea in 1983 at the Plant Pathology Conference at Brighton[10]. This compound was called CME-134.

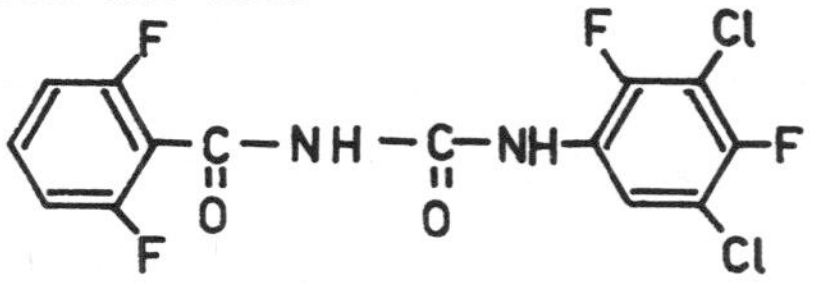

CME-134

With respect to the chemical structures it is remarkable that the first experimental benzoylarylureas of companies other than Duphar contained a 2-chlorobenzoyl moiety, whereas the more recent compounds all contain the 2,6-difluorobenzoyl group like Duphar's diflubenzuron. According to Hajjar and Cassida[11] and Nakagawa et al.[12] the insecticidal activity of benzoylarylureas depends on the substitution pattern in the benzoyl ring in the following order :

$$2,6\text{-}F_2 > 2\text{-}Cl > 2,6\text{-}Cl_2$$

Another difference between the more recently introduced compounds and the older ones is that the aniline moiety of all newer compounds is 3,5-dichloro substituted. This makes these compounds much more lipophilic, less water soluble, and presumably also more persistent

FIELDS OF APPLICATION

All compounds mentioned above show a considerable selectivity. In larvae they are mainly active as a stomach poison. They do not penetrate into leaves and as a consequence sucking insects like aphids, plant bugs, and spider mites can generally not be controlled by these compounds. The compounds act as a larvicide as well as an ovicide, which implies that in general adult insects are not affected.

As DIMILIN up till now is the only widely registered and practically used benzoylarylurea, we will use this chemical as example in the discussion of the biological properties of this class of compounds. DIMILIN is the brand name of the Duphar insecticide that contains diflubenzuron as active ingredient. It is successfully used in practice against a number of insect pests in forestry, in a number of agricultural crops, and in fruit growing. The dosages to be used are in general very low and vary from 25 to 100 g active ingredient per hectare, depending on the crop and type of insect. In dosages of 0.5 - 1.0 g per m^2 it is also very useful for controlling flies,

Forestry

In this field DIMILIN is for example very effective for the control of the gypsy moth (*Lymantria dispar*), the nun moth (*Lymantria monacha*), tent caterpillars (*Malacosoma spp.*), the browntail moth (*Euproctis chryssorhoea*), the processionary caterpillar (*Thaumetopoea pityocampa*), and the fall webworm (*Hyphantria cunea*).

Agricultural crops

The main field crops in which DIMILIN has found its use are cotton and soybean. In the USA and in most of the cotton producing countries of Central and South America DIMILIN is used to control the cotton boll weevil (*Anthonomus grandis*). In the Middle East area, the Egyptian cotton leafworm (*Spodoptera littoralis*) is the major insect problem in cotton and other crops. In Egypt combinations of diflubenzuron and a fast and relatively short acting conventional insecticide are used to control this leafworm.

In soybean cultures in the USA and Brasil DIMILIN is very effectively used against the velvetbean caterpillar (*Anticarsia gemmatalis*).

Fruit growing

DIMILIN provides excellent control of a number of important insect pests in apples and pears. Because of its selectivity and general characteristics it is a very useful tool in integrated control programs. The main target insects in apples and pears are leafminers (*Lithocolletis blancardella, Leucoptera scitella, Stigmella malella*) and the codling moth (*Laspeyresia pomonella*).

A very important application of DIMILIN will become the control of the citrus rust mite (*Phyllocoptruta oleivora*). Registration for this use is applied for in the USA.

Flies and mosquitoes

DIMILIN is an effective weapon in the control of various species of flies. A special application is the fly control in cattle-, poultry- and pig-farming by feed-through. In this technique the active ingredient is administered to the animals mixed in the feed or via a large pill (bolus). Also the fly problem in mushroom breeding farms can effectively be controlled by DIMILIN.

In an increasing number of tropical countries DIMILIN is used in mosquito control programs

TOXICOLOGICAL PROFILE

The insecticidal benzoylarylureas have a very specific mode of action. Therefore it is expected that these compounds will be much less toxic to mammals and human beings than conventional insecticides. In all toxicological studies performed with diflubenzuron, this expectation proved to be true.

The acute oral toxicity (LD-50) in rat and mouse of diflubenzuron and Bay SIR 8514 are very similar: >4640 mg/kg and >5000 mg/kg respectively.

Because of this very high LD-50 value the highest dosages in chronic toxicity studies with diflubenzuron had to be chosen at extremely high levels. In the 2-years rats and mice studies the animals were fed diets containing up to 10,000 mg/kg diflubenzuron, and up to 250 mg/kg bodyweight was administered daily to dogs in a 1-year study. In the chronic studies in mice and rats special attention was paid to the overall tumour incidence. From the data obtained it was concluded that diflubenzuron is not carcinogenic.

The main treatment related effect in all these long term studies was a marginal increase of the met- and sulphhaemoglobin levels in the blood. Although this increase is statistically significant, it was only a marginal effect and considered to be of no biological significance, because in blood of people with hereditary methaemoglobinemia much higher levels of met- and sulphhaemoglobin have been found without causing any adverse effect. Based on these effects the ADI (Acceptable Daily Intake) for men is calculated as 0.02 mg/kg body weight.

ENVIRONMENTAL FATE

About the environmental fate of benzoylarylureas other than diflubenzuron only very limited information is available and this Section will therefore deal almost completely with diflubenzuron.

Degradation in soil

The degradation of diflubenzuron in soil was extensively studied in laboratory experiments[13]. Five different agricultural soils and two hydrosoils were used in these studies. The chemical mixed through the soil at a concentration of 1 mg/kg, was degraded very fast: 50% in 0.5-1 week. The degradation rate was unrelated to soil type, but very much dependent on the particle size of the chemical. The mean particle size of the commercial formulations is 3 µm, which means that also in the practical situation a half life of less than a week can be expected. Results of field studies with diflubenzuron indicate that the degradation under practical conditions is very rapid indeed.

The primary degradation products in soil have been identified[13] as 4-chlorophenylurea and 2,6-difluorobenzoic acid. The urea contributes to soil bound residues and has a half life of 2-3 months[14]. The 2,6-difluorobenzoic acid is rapidly degraded to yield carbon dioxide and water. Its half life in soil is 1-2 weeks.

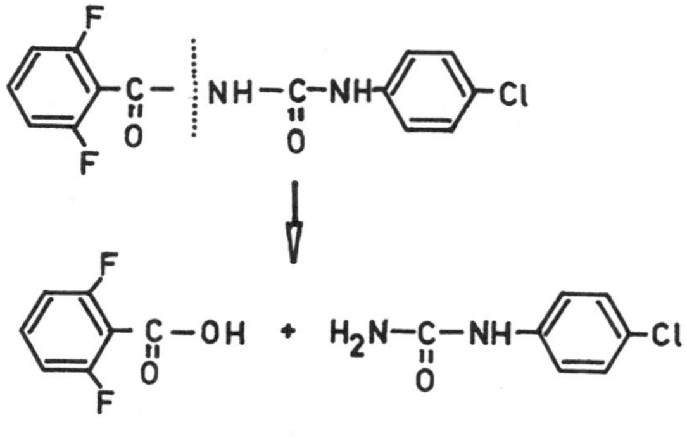

Primary degradation of diflubenzuron

Of the other benzoylarylureas mentioned in this contribution we can only present the results of an orientating experiment carried out in Duphar laboratories in which we compared the degradation rate of these compounds in a sandy clay with that of diflubenzuron. The degradation rates we found decreased in the following order:

diflubenzuron (Duphar)
Bay SIR 8514 (Bayer)
CME-134 (Celamerck)
XRD-473 (Dow)
IKI-7899 (Ishihara/Ciba-Geigy)

Stability in water

The persistence of diflubenzuron in water and in water-soil systems is related to the microbial activity in the system and to the particle size of the diflubenzuron applied. Ivie et al.[15] found that the degradation rate of diflubenzuron in aqueous solutions was related to pH and temperature. Schaefer and Dupras[16] found that Bay SIR 8514 behaved similarly. The stability in water of both compounds was greatly reduced as temperature and pH both become relatively high. Another important factor for the elimination of these compounds from water is adsorption to organic matter or sediment.

MODE OF ACTION

Insect cuticles mainly consist of chitin and protein. Consequently it is obvious that the disturbance of the formation of cuticular tissue by benzoylarylureas led to biochemical studies on the possible effects on these constituents.

Grosscurt[17], Grosscurt and Andersen[18], and Hunter and Vincent[19] studied the possible interference with protein synthesis and found no effect.

The studies of effects on chitin were initially aimed at the enzyme chitinase, which is involved in the degradation of chitin in the insect cuticle during moult. Yu and Terriere[20,21] thought of an indirect stimulation of this enzyme. They hypothesized that diflubenzuron caused an increase of β-ecdysone levels in the larvae which in its turn stimulated the activity of the enzyme chitinase. Ishaaya and Casida[22] assumed that benzoylarylureas had a direct stimulating effect on this enzyme. Both these approaches were refuted by experimental data and a new theory was developed based on work of Deul et al.[23], Ker[24], Mayer et al.[25] and Mitsui et al.[26]. These authors demonstrated that diflubenzuron inhibited the incorporation of radioactivity from precursors of chitin in insect tissues. Mayer's[25] experiments included also Bay SIR 8514 which gave similar results. The inhibition was a very fast process which occurred shortly after the treatment with the chemical. A possible explanation for this fast effect is inhibition of the enzyme chitin synthase. This explanation was, for many scientists, a reason to try to isolate this enzyme from various sources and to test the chitin synthesis inhibiting activity of benzoylarylureas in the isolated systems. An inhibitory effect was found in all systems that contained cells or fragments of cells which were not too small. Mayer et al.[27], Cohen and Casida[28,29], and Mitsui et al.[26] succeeded in isolating chitin synthase in cell free systems. Diflubenzuron and Bay SIR 8514 showed no inhibitory effect in these systems, whereas polyoxin D, a known inhibitor of the enzyme chitin synthase, was very active. A possible explanation for this difference is that benzoylarylureas, in contrast to polyoxin D, do not inhibit the enzyme chitin synthase, but act at some transport process needed for polymerisation of N-acetylglucosamine to chitin or for the deposition of the formed chitin. This action most likely takes place in the cell membrane. Cohen and Casida[30] in support of this idea suggest that diflubenzuron might act by alteration of a critical spatial organisation of the enzyme chitin synthase within the cell membrane. Also Mitsui et al.[31] assume that membranes are involved (see Mitsui, this Volume). They describe experiments which demonstrate that diflubenzuron might inhibit the transport of UDP-N-acetylglucosamine, a precursor of chitin, across the membranes of midgut cells which then results in an inhibition of the chitin synthesis in this peritrophic membrane.

CONCLUSION

Since the discovery of the insecticidal activity of benzoylarylureas by Duphar scientists in the late sixties a lot of derivatives have been synthesized, screened and described in patent applications by several companies. Only one compound, Duphar's diflubenzuron, is now fully commercially available and finds wide application in forestry, some agricultural crops, and fly and mosquito control. Due to this wide practical application diflubenzuron is the most extensively studied benzoylarylurea and as a result much more information is available on diflubenzuron than on any other compound of this type. Diflubenzuron is a powerful insecticide which is much less toxic than conventional insecticides. It degrades in soil faster than most other insecticidal benzoylarylureas and acts by interference with chitin biosynthesis, probably by interfering some transport process in a cell membrane.

REFERENCES

1. A. Endo, K. Kakiki, and T. Misato, Mechanism of action of the antifungal agent polyoxin D, J. Bacteriol. 104: 189 (1970).
2. R. Mulder and M. J. Gijswijt, The laboratory evaluation of two promising new insecticides which interfere with cuticle deposition, Pestic. Sci. 4: 737 (1973).
3. J. J. van Daalen, J. Meltzer, R. Mulder, and K. Wellinga, A selective insecticide with a novel mode of action, Naturwissenschaften 59: 312 (1972).
4. K. Wellinga, R. Mulder, and J. J. van Daalen, Synthesis and laboratory evaluation of 1-(2,6-disubsituted benzoyl)-3-phenylureas, A new class of insecticides I, 1-(2,6-dichlorobenzoyl)-3-phenylureas, J. Agric. Food Chem. 21: 348 (1973).
5. K. Wellinga, R. Mulder, and J. J. van Daalen, Synthesis and laboratory evaluation of 1-(2,6-disubstituted benzoyl)-3-phenylureas, A new class of insecticides II, Influence of the acyl moiety on insecticidal activity, J. Agric. Food Chem. 21: 993 (1973).
6. C. H. Schaefer, T. Miura, W. H. Wilder, and F. S. Mulligan III, New substituted benzamides with promising activity against mosquitoes, J. Econ. Entomol. 71: 427 (1978).
7. L. A. Lacey and M. S. Mulla, Biological activity of diflubenzuron and three new IGRs against Simulium vittatum (Diptera: Simuliidae), Mosquito News 38: 377 (1978).
8. R. J. Sbragia, B. Bisabri-Ershadi, R. H. Rigterink, D. P. Clifford, and R. Dutton, XRD-473, a new acylurea insecticide effective against Heliothis, Proc. 10th Int. Congr. Plant Prot. 1: 417 (1983).
9. J. Granett, B. Bisabri-Ershadi, and M. J. Hejazi, Some parameters of benzoylphenylurea toxicity to beet armyworms (Lepidoptera: Noctuidae), J. Econ. Entomol. 76: 399 (1983).
10. H. M. Becher, P. Becker, R. Prokic-Immel, and W. Wirtz, CME-134, a new chitin synthesis inhibiting insecticide, Proc. 10th Int. Congr. Plant Prot. 1: 408 (1985).
11. N. P. Hajjar and J. E. Casida, Insecticidal benzoylphenylureas: structure-activity relationships as chitin synthesis inhibitors, Science 200: 1499 (1978).
12. Y. Nakagawa, K. Kitahara, T. Nishioka, H. Iwamura, and T. Fujita, Quantitative structure-activity studies of benzoylphenylurea larvicides I. Effect of substituents at aniline moiety against Chilo suppressalis Walker, Pestic. Biochem. Physiol. 21: 309 (1984).

13. W. B. Nimmo, P. C. de Wilde, and A. Verloop, The degradation of diflubenzuron and its chief metabolites in soil. Part I: Hydrolytic cleavage of diflubenzuron, Pestic. Sci. 15: 574 (1984).
14. W. B. Nimmo. A. G. M. Willems, K. D. Joustra, and A. Verloop, The degradation of diflubenzuron and its chief metabolites in soil. Part II: Fate of 4-chlorophenylurea, Pest. Sci. 16: (submitted) (1985).
15. G. W. Ivie, D. L. Bull, and J. A. Veech, Fate of diflubenzuron in water, J. Agric. Food Chem. 28: 330 (1980).
16. C. H. Schaefer and E. F. Dupras, Factors affecting the stability of SIR-8514 (2-chloro-N-[[[4-(trifluoromethoxy)phenyl]amino]-carbonyl]benzamide) under laboratory and field conditions, J. Agric. Food Chem. 27: 1031 (1979)
17. A. C. Grosscurt, Effects of diflubenzuron on mechanical penetrability, chitin formation, and structure of the elytra of Leptinotarsa decemlineata, J. Insect Physiol. 24: 827 (1978).
18. A. C. Grosscurt and S. O. Andersen, Effects of diflubenzuron on some chemical and mechanical properties of the elytra of Leptinotarsa decemlineata, Proc. Kon. Ned. Akad. Wetensch. Amsterdam 83C: 143 (1980)
19. E. Hunter and J. F. Vincent, The effects of a novel insecticide on insect cuticle, Experientia 30: 1432 (1974).
20. S. J. Yu and L. C. Terriere, Activities of hormone metabolizing enzymes in house flies treated with some substituted urea growth regulators, Life Sciences 17: 619 (1975).
21. S. J. Yu and L. C. Terriere, Ecdysone metabolism by soluble enzymes from three species of Diptera and its inhibition by the insect growth regulator TH-6040, Pestic. Biochem. Physiol. 7: 48 (1977).
22. I. Ishaaya and J. E. Casida, Dietary TH 6040 alters composition and enzyme activity of housefly larval cuticle, Pestic. Biochem. Physiol. 4: 484 (1974).
23. D. H. Deul, B. J. de Jong, and J. A. M. Kortenbach, Inhibition of chitin synthesis by two 1-(2,6-disubstituted benzoyl)-3-phenylurea insecticides II, Pestic. Biochem. Physiol. 8: 98 (1978).
24. R. F. Ker, Investigations of locust cuticle using the insecticide diflubenzuron, J. Insect Physiol. 23: 39 (1977).
25. R. T. Mayer, S. M. Meola, D. L. Coppage, and J. R. DeLoach, Utilization of imaginal tissues from pupae of the stable fly for the study of chitin synthesis and screening of chitin synthesis inhibitors, J. Econ. Entomol. 73: 76 (1980).
26. T. Mitsui, C. Nobusawa, and J. Fukami, Inhibition of chitin synthesis by diflubenzuron in Mamestra brassicae L., J. Pestic. Sci. 6: 155 (1981).
27. R. T. Mayer, A. C. Chen, and J. R. DeLoach, Characterization of a chitin synthase from the stable fly, Stomoxys calcitrans (L.), Insect Biochem. 10: 549 (1980).
28. E. Cohen and J. E. Casida, Properties of Tribolium gut chitin synthetase, Pestic. Biochem. Physiol. 13: 121 (1980).
29. E. Cohen and J. E. Casida, Inhibition of gut chitin synthetase, Pestic. Bioch. Physiol. 13: 129 (1980).
30. E. Cohen and J. E. Casida, Insect chitin synthetase as a biochemical probe for insecticidal compounds, in: "Pesticide Chemistry: Human welfare and the environment", J. Miyamoto and P. C. Kearney eds, Pergamon Press, Oxford, (1983).
31. T. Mitsui, C. Nobusawa, and J. Fukami, Mode of inhibition of chitin synthesis by diflubenzuron in the cabbage armyworm, J. Pestic. Sci. 9: 1 (1984).

INHIBITION OF CHITIN SYNTHESIS IN INSECT SYSTEMS

E. COHEN

Department of Entomology, The Hebrew University of Jerusalem
Faculty of Agriculture
Rehovot, Israel

INTRODUCTION

Chitin is an amino sugar biopolymer which together with a variety of proteins form elaborate structures such as insect cuticles and peritrophic membranes. This polymer serves mainly as a supporting element and resembles in this respect cellulose and collagen in plants and vertebrates, respectively. It was shown using chitin synthesis inhibitors that disruption of the polysaccharide formation resulted in malformed cuticles, lacking their normal exoskeletal properties and unable to withstand increased internal pressure during molting[1]. Since the cuticle is obviously of critical importance to survival of insects, chitin as a major structural element has been considered as a desired target site for selective pesticides.

In vivo formation of chitin polymers and their subsequent crystallization to microfibrils are complex biochemical and physico-chemical events taking place at or some distance away from cell plasma membranes. Those events and their apparent regulation in insects are poorly understood. More information is available from intensive studies on chitin synthesis conducted with various fungal systems[2]. Chitin polymerization is the last cellular event which starts with trehalose or glucose via a cascade of biochemical transformations which include phosphorylation, amination, acetylation and formation of UDP-N-acetyl-D-glucosamine as the ultimate substrate. Since the above metabolites can be channeled into various metabolic pathways, the "commitment" to synthesize chitin resides in a particulate enzyme embedded in the plasma membranes of epidermal cells. This enzyme, chitin synthetase |DP-2-acetamido-2-deoxy-D-glucose: chitin 4-β-acetamidodeoxyglucosyltransferase (EC 2.4.1.16)| which catalyzes the polymerization of N-acetyl-D-glucosamine, has been recently extracted from several insect species, and characterized[3-5]. Similar in properties to fungal chitin synthetase[2], the insect enzymes are considerably stimulated by N-acetyl-D-glucosamine, require magnesium ions for their activity and are activated to a certain extent by trypsin.

The chitin synthetase associated with formation of the peritrophic membrane is continuously functional throughout the active life cycle of the insect. Yet, the epidermal polymerase is clearly under intricate regulation regarding the initiation, quantities and duration of chitin formation. Essentially, massive chitin biosynthesis is restricted to periods of molting although some activity can proceed into the intermolt phase. Since ecdysterone initiate molting, it has been suggested that chitin formation is under hormonal control. Indeed, various insect organ cultures require exogenous ecdysterone for initiating and maintaining chitin synthesis and deposition, and for cuticle formation[6-8]. Also an increase in chitin synthetase activity following injection of the molting hormone, was observed[4].

In theory there are many points of interference with chitin synthesis. They may include the polymerizing enzyme, availability of co-factors and substrate; regulatory elements related to de-novo-synthesis of the enzyme; its transport to the proper site in the plasma membrane or a proteolytic activation of a presumed zymogen. Other regulatory effects might be linked to the actual polymerization step or to extrusion of nascent polysaccharides through the plasma membrane. The process of crystallization and fibrillogenesis might be also disrupted, resulting in a severe interference in chitin formation. Currently the insect chitin synthetase systems are available as probes for reliable and rapid evaluation of compounds affecting chitin formation[3,4] or fibrillogenesis[9].

INHIBITION OF CHITIN SYNTHESIS

Research on chitin synthesis, dormant for many years, has been revived due to an unexpected discovery relative to the insecticidal properties of certain benzoylphenyl ureas [10-12] (cf. Willems et al., this Volume).

Histological observations revealed damaged cuticles, characterized by severe distortion of the endocuticular layers[13,14] where the chitin but not the protein component was affected[1]. Similar histological lesions were produced by polyoxin-D a nucleoside peptide antibiotic[15]. This compound and related polyoxins inhibit fungal growth[16] and act as powerful competitive inhibitors of the fungal chitin synthetase[2]. Table 1 summarizes data regarding the relative efficiency of diflubenzuron and polyoxin-D in inhibiting the incorporation of sugar precursors into chitin. Overall, diflubenzuron was highly effective and some organ cultures like the cockroach leg regenerate[6] are extremely sensitive to the insecticide (I_{50} = 6.1 x 10^{-11}M). Polyoxin-D is less effective and in most cases by more than two orders of magnitude compared with diflubenzuron. However, this difference might be attributed to slow penetration of the antibiotic or to accumulation of the benzoylphenyl urea from the medium, resulting in higher concentrations inside the organ culture[6]. The similarity in histological lesions and in symptoms of insect poisoning produced by diflubenzuron and polyoxin-D, as well as the reported accumulation of UDP-N-acetyl-D-glucosamine, led to the notion that insecticidal benzoylphenyl ureas act as chitin synthetase inhibitors[10].

INHIBITORS OF CHITIN SYNTHESIS

Inhibitors of insect chitin synthesis can be assayed with whole organisms and with isolated gut or integumental organ cultures. Inhibitory compounds can be also examined on the level of chitin polymerization and fibrillogenesis. Of all the currently available inhibitors, only the synthetic benzoylphenyl ureas and the nucleoside peptide antibiotics which are natural products, are of major importance. Both groups which effectively suppress chitin synthesis have commercial value and will be discussed in depth.

Benzoylphenyl Ureas

Following the discovery of the insecticidal properties of benzoylphenyl ureas, their great potential has been recognized. A good correlation between toxicity to the milkweed bug *Oncopeltus fasciatus* and the extent of chitin synthesis inhibition using isolated abdomens, for a large number of benzoylphenyl ureas, was reported[17]. The recently available cell-free systems capable of chitin polymerization were utilized to pinpoint the precise mode of action of benzoylphenyl ureas. It soon became apparent that chitin synthetases from the gut of *Tribolium castaneum*[3], integuments of *Trichoplusia ni*[4], *Hyalophora cecropia*[4] and *Chilo suppressalis*[20], and whole pupae of *Stomoxys calcitrans*[5], were insensitive to diflubenzuron. Thus, the urea compounds which in vivo block chitin synthesis do not act directly on the polymerizing enzyme. Moreover, chitin crystallization and microfibril formation proceed in vitro, in the presence of high concentrations of diflubenzuron[9].

Several possibilities have been raised to explain the mechanism by which chitin synthesis is inhibited by benzoylphenyl ureas. One theory invoked an active metabolite which interacts directly with the chitin synthetase[21]. This seems unlikely in light of the slow metabolism of the inhibitors in insects which preclude accumulation of such a metabolite in sufficient amounts at the polymerization site. Other suggestions implicate hypothetical regulatory elements such as an interference with uridine diphosphatase[22] or with proteolytic activation of a zymogenic chitin synthetase[23]. The first suggestion based on the inhibitory effect of uridine diphosphate on insect chitin synthetases[4,21], was adapted from the study on the fungus *Coprinus cinereus*, where uridine diphosphatase was associated with regulation of chitin polymerization[24]. The second postulation is problematic, since

Table 1. Inhibition of Chitin Synthesis in Insect Organ Cultures

Insect[a]	Inhibition[b], I_{50}(M) Polyoxin-D	Diflubenzuron	Reference
Leucophaea maderae	7.5×10^{-7}	6.1×10^{-11}	6
Oncopeltus fasciatus	1.5×10^{-5}	5.5×10^{-7}	17
Stomoxys calcitrans	1.3×10^{-5}	5.2×10^{-8}	18
Lucilia cuprina	6.0×10^{-7}	7.0×10^{-7}	19
Chilo suppressalis	1.0×10^{-5}	8.3×10^{-8}	20

a. Integumental organ cultures were utilized.

b. Inhibition of the incorporation of radioactive sugar precursors into chitin. I_{50} = 50% inhibition.

diflubenzuron drastically and rapidly inhibits already active chitin synthetases. Nevertheless, at a certain stage of the insect life cycle, some of the insect polymerases could be zymogenic, and interference at this level of regulation with either benzoylphenyl ureas or with other compounds is feasible. It has been postulated that benzoylphenyl ureas disrupt the access of substrate and co-factors to the site of chitin polymerization[22]. A recent attempt to examine the above possibility using *Mamestra brassicae* gut preparation failed to demonstrate any interference by diflubenzuron[25]. It appears that the precise biochemical lesion produced by insecticidal benzoylphenyl ureas is associated with sensitive control mechanisms at or near the chitin polymerization site. It has been proposed that insect cell-free systems are able to carry out chitin polymerization in the presence of diflubenzuron, because some unidentified sensitive regulatory components are lost upon extraction of the enzyme[21]. Alternatively, the insecticide might interact with membranal components, playing a crucial role in polymerization and in extrusion of nascent polymers at the plasma membrane site. Obviously more basic information with respect to chitin polymerization in-situ is necessary to elucidate the mode of action of the above inhibitory compounds.

Nucleoside Peptide Antibiotics

The nucleoside peptide antibiotics are natural products, synthesized by actinomycetes. Structurally they resemble UDP-N-acetyl-D-glucosamine and compete with this substrate for the active site of the polymerizing enzyme. Polyoxins (A-M) are one group of antibiotics isolated from *Streptomyces cacaoi* var. *asoensis* with a basic structure of pyrimidine, ribose and attached peptides[26]. A new group of nucleoside antibiotics, the nikkomycins, have been recently isolated from *S. tendae*[27]. Nikkomycins are closely related to polyoxins, the main difference is a pyridine ring attached to the last amino acid. Due to their polar nature, the antibiotics are inefficient as insecticides. Polyoxin-A for example was insecticidal only by injection into nymphs of *Melanoplus sanguinipes*[28]. Larvae of *Epilachna varivestis* fed on bean leaves sprayed with nikkomycin, developed defective cuticles[29]. The same compound inhibited molting in the mite *Tetranychus urticae* and an electron microscopy study revealed disrupted arrangement of chitin microfibrils[30]. The antibiotics were the most active compounds in suppressing the insect chitin synthetase (Table 2)[21]. The

Table 2. Inhibition of *Tribolium castaneum* chitin synthetase

Compound	Inhibition[a] I_{50} (μM)
Nikkomycin	0.02
Polyoxin-D	4
Captan	6
1-Geranyl-2 methyl benzimidazole	50
Barban	160
Diflubenzuron	NE[b]
BAY SIR 8514	NE[b]
IKI-7899	NE[b]

a. Inhibition of incorporation of tritiated UDP-N-acetyl-D-glucosamine into chitin. I_{50} = 50% inhibition.

b. NE - no effect at a concentration of 3×10^{-4}M.

Tribolium gut and *Trichoplusia* integumental chitin synthetases are extremely sensitive to nikkomycin, which is so far the most effective inhibitor of the insect polymerase.

Sulphenimides

The fungicide captan and its close analogs, captafol and folpet interact with sensitive thiol containing cellular components via thiophosgene, their decomposition product[31]. Captan inhibited formation of the peritrophic membrane in blowflies[32] and chitin synthesis in the cockroach leg regenerate system[33]. Sulphenimide fungicides like captan, captafol and dichlofluanid

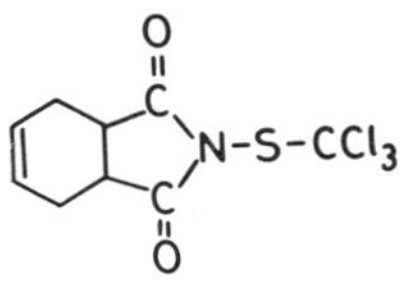

captan

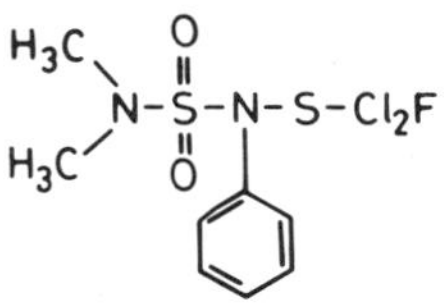

dichlofluanid

are among the most effective inhibitors of *Tribolium*[21] and *Trichoplusia*[4] chitin synthetase. The I_{50} value for captan (table 2) is similar to that of polyoxin-D. Apparently the mode of action involves sensitive thiol groups, crucial for the enzyme function, which readily interact with thiol-blocking reagents. Moreover, the insect chitin synthetase could be protected from the inhibitory effects of sulphenimide by sulfhydryl containing reagents such as β-mercaptoethanol or dithiothreitol[21].

Benzimidazoles, Phenyl Carbamates

Certain benzimidazoles were found to be toxic to houseflies, and one compound (table 2) blocked the molting of silkmoth larvae at a relatively low level of 0.5μg per insect[34]. Over 40 benzimidazoles and structurally related compounds with various substituents were assayed using the *Tribolium* chitin synthetase system, and their structure-activity relationship was analysed[35]. A terpenoyl moiety is required for inhibitory action and the best compounds in this series include 1-geranyl and 1-citronellyl 2-methyl benzimidazole. Both compounds are approximately 12 times less active than polyoxin-D. It is expected that further structure optimization might yield better chitin synthetase inhibitors.

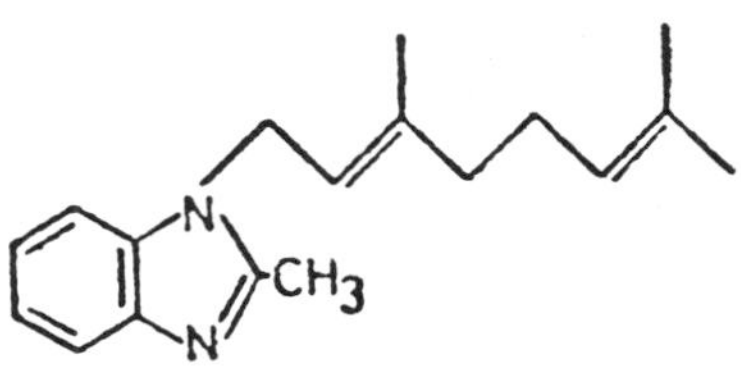

1-Geranyl-2-methyl benzimidazole

Of the phenyl carbamates which are weak or moderate *Tribolium* chitin synthetase inhibitors[21], only the herbicide barban displayed appreciable activity. Although benzimidazole and phenyl carbamates act on the polymerizing enzyme, their mode of action is yet to be resolved.

CONCLUSIONS

The insecticidal action of benzoylphenyl ureas has generated interest in insect chitin synthesis and eventually led to the successful extraction of cell-free systems capable of chitin polymerization and fibrillogenesis. These in vitro systems were utilized to correlate chemical structure and inhibitory activity. They will be helpful in establishing a basis for methodical and rational synthesis of improved and optimized bioactive molecules, for future use as safe and selective insecticides or fungicides of agricultural and clinical importance. Studies on chitin synthesis should contribute to the understanding of the functional organization of the chitin synthetase within the integument and its intricate regulation. Chemical interference with chitin formation at various levels such as catalysis, polymer extrusion, crystallization, as well as interaction with a variety of possible membranal and soluble component associated with the above levels, should assist in gaining a close insight into a highly complex system.

REFERENCES

1. R. Mulder, and M.J. Gijswijt, The laboratory evaluation of two promising new insecticides which interfere with cuticle deposition, Pestic. Sci. 4: 737 (1973).
2. G.W. Gooday, and A.P.J. Trinci, Wall structure and biosynthesis in fungi, in:"Eukaryotic Microbial Cell", Soc. Gen. Microbiol. Sym. 30: 202 (1980).
3. E. Cohen, and J.E. Casida, Properties of Tribolium gut chitin synthetase Pestic. Biochem. Physiol. 13: 121 (1980).
4. E. Cohen, and J.E. Casida, Properties and inhibition of insect integumental chitin synthetase, Pestic. Biochem. Physiol. 17: 301 (1982).
5. R.T. Mayer, A.C. Chen, and J.R. DeLoach, Characterization of a chitin synthase from the stable fly Stomoxys. calcitrans (L.), Insect Biochem. 10: 549 (1980).
6. B.A. Sowa, and E.P. Marks, An in vitro system for quantitative measurement of chitin synthesis in the cockroach: Inhibition by TH 6040 and polyoxin-D, Insect Biochem. 5: 855 (1975).
7. K. Kitahara, Y. Nakagawa, T. Nishioka, and T. Fujita, Cultured integument of Chilo suppressalis as a bioassay system of insect growth regulators, Agr. Biol. Chem. 47: 1583 (1983).
8. H. Oberlander, S.M. Ferkovich, F. Van Essen and C.F. Leach, Chitin biosynthesis in imaginal discs cultured in vitro, Wilhelm Roux's Archives, 185: 95 (1978).
9. E. Cohen, In vitro chitin synthesis in an insect: formation and structure of microfibrils, Europ. J. Cell Biol. 16: 284 (1982).
10. A. Vorlcop, and C.D. Ferrell, Benzoylphenyl ureas - A new group of larvicides interfering with chitin deposition, in:"Pesticide Chemistry in the 20th Century", J.R. Plimmer, ed., ACS Sym. Ser., No. 37. Amer. Chem. Soc., Washington, D.C., pp. 237-270 (1977).
11. G. Zoebelein, I. Hamman, and W. Sirrenberg, BAY SIR 8514 a new chitin synthesis inhibitor, Z. Ang. Ent. 89: 289 (1980).
12. T. Naga, T. Toki, T. Koyanagi, and R. Nishiyama, Structure-activity relationships of a series of benzoyl-pyridyloxyphenyl urea derivatives, Paper II d-7 of the Fifth International Congress of Pesticide Chemistry. Abstracts, Kyoto, Japan, August 29 - September 4 (1982).
13. A.C. Grosscurt, Diflubenzuron: Some aspects of its ovicidal and larvicidal mode of action and an evaluation of its practical possibilities, Pestic. Sci. 9: 373 (1978).
14. S.J. Lim, and S.S. Lee, The toxicity of diflubenzuron on Oxya japonica (Willemse) and its effect on molting, Pestic. Sci. 13: 537 (1982).
15. M.J. Gijswijt, D.H. Deul, and B.J. DeJong, Inhibition of chitin synthesis by benzoyl-phenyl urea insecticides, III. Similarity in action in Pieris brassicae (L.) with polyoxin-D, Pestic. Biochem. Physiol. 12: 87 (1979).

16. K. Isono, J. Nagatsu, K. Kobinata, K. Sasaki, and S. Suzuki, Studies on polyoxins, antifungal antibiotics. V: Isolation and characterization of polyoxins C,D,E,F,G, H and I, Agr. Biol. Chem. 31: 190 (1967).
17. N.P. Hajjar, and J.E. Casida, Structure-activity relationships of a series of benzoylphenyl ureas as toxicants and chitin synthesis inhibitors in Oncopeltus fasciatus, Pestic. Biochem. Physiol. 11: 33 (1979).
18. R.T. Mayer, S.M. Meola, D.L. Coppage, and J.R. DeLoach, Utilization of imaginal tissues from pupae of the stable fly for the study of chitin synthesis and screening of chitin synthesis inhibitors, J. Econ. Entomol. 73: 76 (1980).
19. I.F. Turnbull, and A.J. Howells, Effect of several larvicidal compounds on chitin biosynthesis by isolated larval integuments of the sheep blowfly Lucilia cuprina, Aust. J. Biol. Sci. 35: 491 (1982).
20. K. Kitahara, Y. Nakagawa, T. Nishioka, and T. Fujita, Cultured integument of Chilo suppressalis as a bioassay system if insect growth regulators, Agr. Biol. Chem. 47: 1583 (1983).
21. E. Cohen, and J.E. Casida, Inhibition of Tribolium gut chitin synthetase, Pestic. Biochem. Physiol. 13: 129 (1980).
22. E. Cohen, and J.E. Casida, Insect chitin synthetase as a biochemical probe for insecticidal compounds, in:"Pesticide Chemistry. Human Welfare and the Environment", J. Miyamoto, and P.C. Kearney, Eds., Pergamon Press, Oxford, pp. 25-32(1983).
23. T. Leighton, E. Marks, and F. Leighton, Pesticides, insecticides and fungicides as chitin synthesis inhibitors, Science, 213: 905 (1981).
24. A. de Rousset-Hall, and G.W. Gooday, A kinetic study of a solubilized chitin synthetase preparation from Coprinus cinereus, J. Gen. Microbiol. 89: 146 (1975).
25. T. Mitsui,C. Nobusawa, and J.-I. Fukami, Mode of inhibition of chitin synthesis by diflubenzuron in the cabbage armyworm Mamestra brassicae L., J. Pestic. Sci. 9: 19 (1984).
26. M. Hori, K. Kakiki, S. Suzuki, and T. Misato, Studies on the mode of action of polyoxins. Part III. Relation of polyoxin structure to chitin synthetase inhibition, Agr. Biol. Chem. 35: 1280 (1971).
27. U. Dähn, H. Hagenmäier, H. Höhne, W.A. König, W.A. Wolf, and H. Zähner, Stoffwechselprodukte von Mikroorganismen. 154. Mitteilung. Nikkomycin, ein neuer Hemmstoff der Chitinsynthese bei Pilzen, Arch. Microbiol. 107: 143 (1976).
28. A. Vardanis, Polyoxin fungicides: Demonstration of insecticidal activity due to inhibition of chitin synthesis, Experientia, 34: 228 (1978).
29. U. Schluter, Ultrastructural evidence for inhibition of chitin synthesis by nikkomycin, Wilhelm Roux's Archives, 191: 205 (1982).
30. U. Mothes, and K.-A. Seitz, Action of a microbial metabolite chitin synthesis inhibitor nikkomycin on the mite Tetranychus urticae; an electron microscope study, Pestic. Sci. 13: 426 (1982).
31. M.R. Siegel, Reaction of certain trichloromethyl sulfenyl fungicides with low molecular thiols. In vitro studies with glutathione, J. Agr. Food Chem. 18: 819 (1970).
32. B. Becker, Effect of 20-hydroxy-ecdysone, juvenile hormone, Dimilin, and captan on in vitro synthesis of peritrophic membranes in Calliphora erythrocephala, J. Insect Physiol. 24: 699 (1978).
33. E.P. Marks, and B.A. Sowa, Cuticle formation in vitro, in:"The Insect Integument", H.R. Hepburn, Ed., Elsevier Scientific Publishing Co., Amsterdam, pp. 339-357 (1976).
34. E. Kuwano, N. Sato, and M. Eto, Insecticidal benzimidazoles with a terpenoid moiety, Agric. Biol. Chem. 44: 1709 (1982).
35. E. Cohen, E. Kuwano, and M. Eto, The use of Tribolium chitin synthetase assay in studying effects of benzimidazoles with a terpene moiety and related compounds, Agric. Biol. Chem. 48: 1617 (1984).

CHITIN BIOSYNTHESIS IN INSECTS AND ITS DISRUPTION AS A MEANS OF PEST CONTROL

A. RETNAKARAN

Forest Pest Management Institute
Canadian Forestry Service
Sault Ste. Marie, Ont. Canada

INTRODUCTION

The epidermis is one of the most versatile tissues seen in insects, and is capable of performing a spectacular array of functions related to growth and development.[1,2] It plays a pivotal role in the moulting cycle, the sequential polymorphic changes seen in holometabolous insects and the formation of the distinct surface architecture including those of the sense organs and the epidermal glands.[3,4] It has been proposed that genes responsible for the larval, pupal and adult characteristics are successively derepressed during metamorphosis to permit the manifestation of these three distinct morphologies.[5] At the same time, the two hormones, 20-hydroxy ecdysone and juvenile hormone, have been shown to act at the genetic level in the epidermis regulating morphogenesis.[6,7] The different functions of the epidermis are thus genetically predetermined but hormonally sustained.

One of the major functions of the epidermis that is intimately associated with growth and development is the formation of the chitinous exoskeleton. The importance of this material to the well being of insects has been dramatically brought into focus by the inhibition of chitin biosynthesis by benzoylphenyl ureas that are currently being developed as environmentally acceptable pesticides.[8,9] The biosynthesis and degradation of chitin in insects has been extensively studied but still certain critical steps of its formation as well as the exact nature of inhibition by benzoylphenyl ureas are poorly understood.[10,1,11] In this chapter, the current status of chitin biosynthesis in insects and the inhibition of chitin formation by benzoylphenyl ureas as an ecologically safe method for controlling insect pests will be critically examined.

INTEGUMENT OF INSECTS

The body wall or integument in insects is composed of the epidermis and the extracellular layer that it secretes called the cuticle. This cuticle is multilayered and consists of (i) an outermost non-chitinous epicuticle which may be tanned, untanned or membranous, (ii) followed by an exocuticle which contains chitin with associated proteins and is usually heavily tanned, (iii) and finally the chitinous endocuticle which shows the characteristic growth layers or lamellae and is immediately next to the epidermis (Fig. 1). The exo and endocuticular layers are together known as the procuticle. Pore canals that are cytoplasmic extensions of the epidermis are present in the procuticle.[12-19] (See also previous chapters, this Volume).

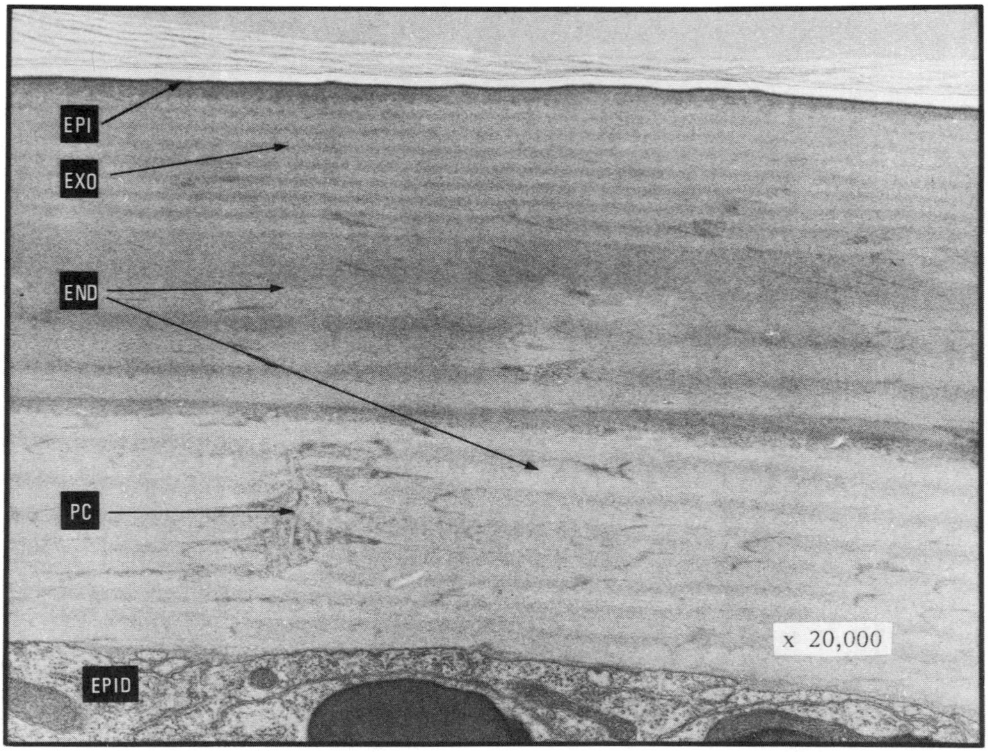

Fig. 1. Electronmicrograph of the sternite of the housefly in cross section illustrating the ultrastructure of the insect integument. EPI - epicuticle, EXO - exocuticle, END - endocuticle, PC - pore canal, EPID - epidermis (Courtesy of Dr. J. Percy).

CHITIN BIOSYNTHESIS

The synthesis of chitin in fungi is relatively well understood [20,16,21], and will serve as a useful backdrop to understand the scheme in insects. By a series of steps glucose (Glc) is changed to N-acetylglucosamine (GlcNAc) which is then converted to the energy-rich, Uridine diphospho-N-acetylglucosamine (UDP-GlcNAc). UDP-GlcNAc serves as the basic building block for chitin biosynthesis and the reaction is catalyzed by a polymerase, chitin synthase or synthetase (UDP-2-acetamido-2-deoxy-D-glucose:chitin 4-β-acetamidodeoxy-D-glucosyltransferase, EC2.4.1.16) (Fig. 2). This enzyme is membrane bound and occurs in an inactive form as a zymogen. It is activated by a protease which probably removes an inhibitor that is attached to the enzyme. The enzyme occurs as granules called chitosomes. Divalent cations such as Mg^{2+} and Mn^{2+} and free GlcNAc stimulate the activity of this enzyme whereas polyoxins produced by *Streptomyces cacaoi* act as powerful competitive inhibitors. There is no evidence of the involvement of a lipid intermediate, dolichol phosphate or inhibition by tunicamycin which specifically blocks synthesis of GlcNAc-dolichol phosphate.[21,22]

Chitin synthesis in insects is essentially similar to that of fungi excepting for the ultimate step of polymerization by the enzyme, chitin synthase. The non-reducing disaccharide, trehalose made up of two Glc units is the most common sugar in insects. It is hydrolyzed by trehalase and the glucose units are converted to UDP-GlcNAc similar to fungi.[23,24,25,10] The terminal polymerization step has been the contentious issue in chitin biosynthesis in insects for which several explanations and speculations have been given.[26] *In vitro* assays for chitin synthase with integument preparations have consistently shown an extremely low turnover rate compared to the *in vivo* system.[27,28,29,30] The chitin synthase from gut preparations may not reflect the situation in the integument.[31] The low turnover rates indicate a sub-optimal assay system

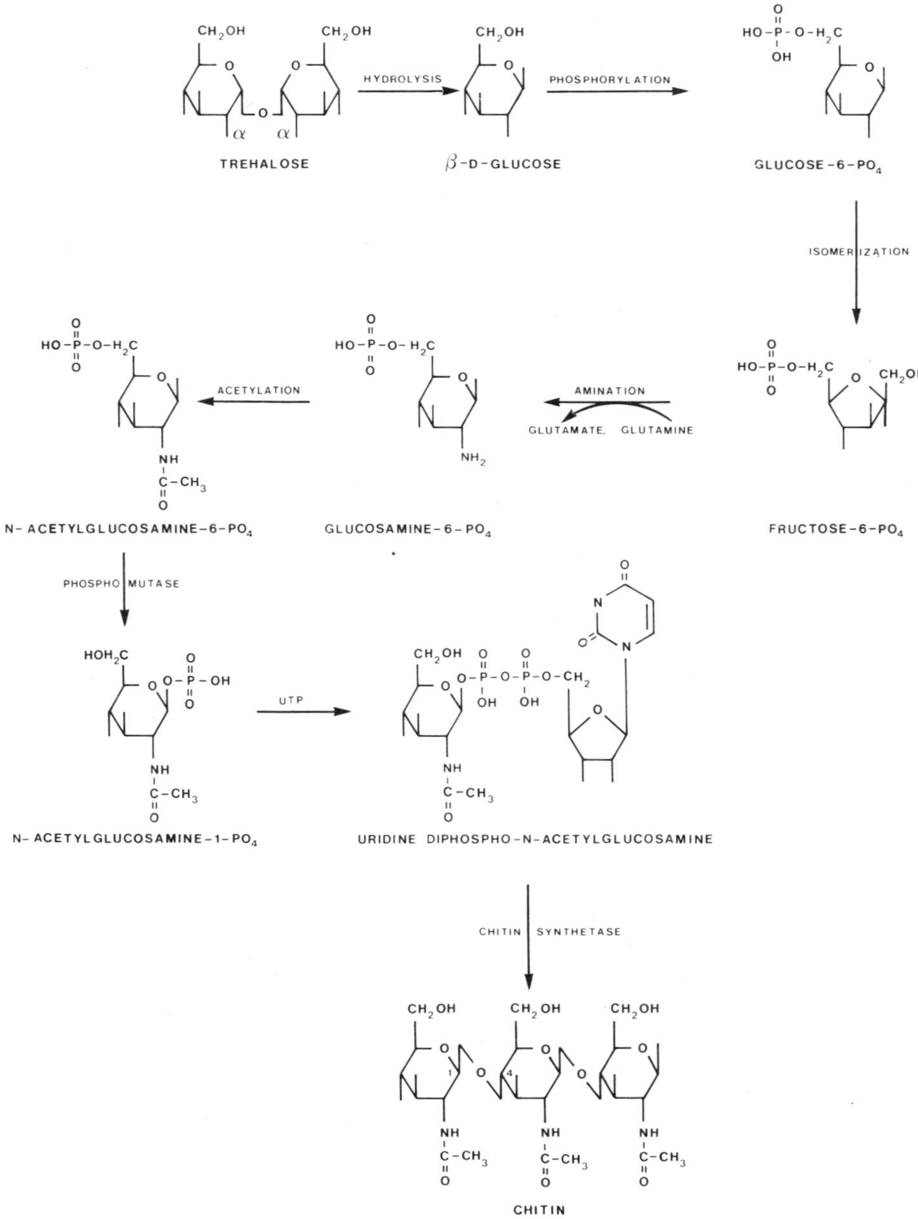

Fig. 2. A generalized scheme of chitin biosynthesis.

for the integument chitin synthase and any conclusions based on such results are at best equivocal.

There are two factors in integumental chitin synthesis that are fundamental and need to be considered to explain the biosynthetic scheme. (i) Chitin is a highly insoluble material that occurs in the non-cellular, cuticle region of the epidermis. How is the insoluble material transported to this region? (ii) The chitin architecture is discrete and varies from one part of the integument to the other and suggests that its synthesis and arrangement are precisely regulated. If chitin is synthesized and merely extruded out of the epidermal cell, how is the qualitative and quantitative distribution of this polymer controlled?

Chitin is synthesized predominantly at the beginning of a new instar soon after moulting. At this stage there are no chitinases present in the integument.[32,33,18,34,29] The presence of oligosaccharides in reactions catalyzed with enzyme preparations from such early instar integuments suggests that they were formed by a biosynthetic and not a biodegradative process.[29] These oligosaccharides varying from 2 to 6 GlcNAc units in length are soluble.[35]

In several vertebrate tissues and cell lines, oligosaccharides are initially linked with a polyprenol such as dolichol phosphate prior to forming an asparagine-linked oligosaccharide enroute to becomming glycoproteins.[36] Such polyprenol linked intermediates have been isolated in brine shrimps.[37] The linking of UDP-GlcNAc with dolichol phosphate to form GlcNAc-PP-Dol and UMP is selectively inhibited by the antibiotic, tunicamycin and is used as a probe to detect oligosaccharides linking with polyprenols.[22] In the blood sucking bug, *Triatoma infestans* it has been shown that a material like $(GlcNAc)_n$-PP-Dol was formed during chitin synthesis and also tunicamycin inhibited chitin formation.[38,39]

In certain vertebrate systems that contain glycoproteins, it has been shown that the oligosaccharide-dolichol phosphate is linked to the N-asparagine residues of an acceptor peptide or protein with the subsequent release of the dolichol moiety.[40] Aspartic acid is one amino acid always detected with chitin samples from insects.[41,13] It is conceivable that during extraction the asparagine residue is hydrolysed to aspartic acid. If specific receptor proteins are to be laid first to which oligosaccharides are linked and polymerised to form microfibrils, then the inhibition of synthesis of these specific receptor proteins would result in shutting off chitin formation. The backing-up of the system would ultimately result in feedback inhibition of the chitin biosynthetic pathway. This is indeed what happens in the imaginal discs of the Indian meal moth, *Plodia interpunctella* and in the crayfish.[42,43] The obvious advantage of such a mechanism is that the epidermal cell can regulate chitin deposition by controlling the synthesis of the receptor protein.

Based on the available evidence, it is proposed that the epidermis in insects using UDP-GlcNAc as the precursor, synthesizes soluble oligosaccharides that are linked to polyprenol lipids such as dolichol phosphate, transported outside the cell and attached to specific residues, possibly asparagine, on a receptor protein which is concomitantly synthesized. The oligosaccharides are joined together to form the chitin microfibrils. In the words of Hackman,[44] "It should be remembered that speculations stimulate further research but they become counterproductive if they reach the stage of 'proof by repeated affirmation' and steady progress demands an uncompromising methodology."

INHIBITION OF CHITIN BIOSYNTHESIS

Many compounds ranging from natural products such as plumbagin[45] to antibiotics such as polyoxins and nikkomycins[46], insecticides such as benzimidazoles[47] and fungicides such as captan[48] have been shown to block chitin synthesis in insects. But one group, benzoylphenyl ureas, by far has become the newest and most actively pursued class of insecticides at the present time. (cf. Willems et al., this Volume).

The mode of action of benzoylphenyl ureas such as diflubenzuron have not been elucidated at the molecular level. It has been shown that larvae that have ingested this material suffer from severe inhibition of chitin deposition in the cuticle (Fig. 3,4). The larvae usually die attempting to moult. This *in vivo* inhibition however, has not been unequivocally shown in an *in vitro* system primarily due to the unavailability of an optimal assay system. The following are some of the theories that have been put forward to explain the mode of action of benzoylphenyl ureas:

(1) Inhibits chitin synthase. This is the most widely held theory and is based on the accumulation of UDP-GlcNAc in diflubenzuron treated insects.[49,50]

(2) Inhibits the transport of UDP-GlcNAc across biomembranes.[51]

(3) Inhibits the protease responsible for activating the chitin synthase zymogen.[52]

(4) Inhibits ecdysone metabolizing enzymes resulting in the accumulation of ecdysone which stimulates chitinase production resulting in moult disruption.[53] But O'Neill et al.[54] have shown that diflubenzuron treatment did not result in ecdysone accumulation.

(5) Blocks conversion of glucose to fructose-6-phosphate resulting in inhibition of chitin synthesis.[55]

(6) An active metabolite of the benzoylphenyl urea is responsible for the inhibition of chitin syntheis.[56]

(7) *In vivo* regulatory system sensitive to benzoylphenyl urea is disrupted upon preparation of an *in vitro* system. Therefore there is no longer any inhibition of chitin synthase by the benzoylphenyl urea.[57]

(8) Chitin is synthesized in two steps. First, soluble oligosaccharides are formed and transported outside the cell. Second, these oligosaccharides are covalently bound to specific receptor proteins and linked together to form chitin microfibrils. It is proposed that this second polymerisation step is inhibited by benzoylphenyl ureas.[11,29]

Once an optimal assay for chitin synthase is developed, it will be possible to obtain definite results on benzoylphenyl urea inhibition of this enzyme. At present the explanations put forward are all but hypotheses requiring further supportive evidence.

Benzoylphenyl ureas besides inhibition of chitin synthesis have been shown to inhibit the pentose cycle in some lepidopterans,[58] increase cuticular lipids,[59] increase microsomal oxidase levels,[60] depress carbo-

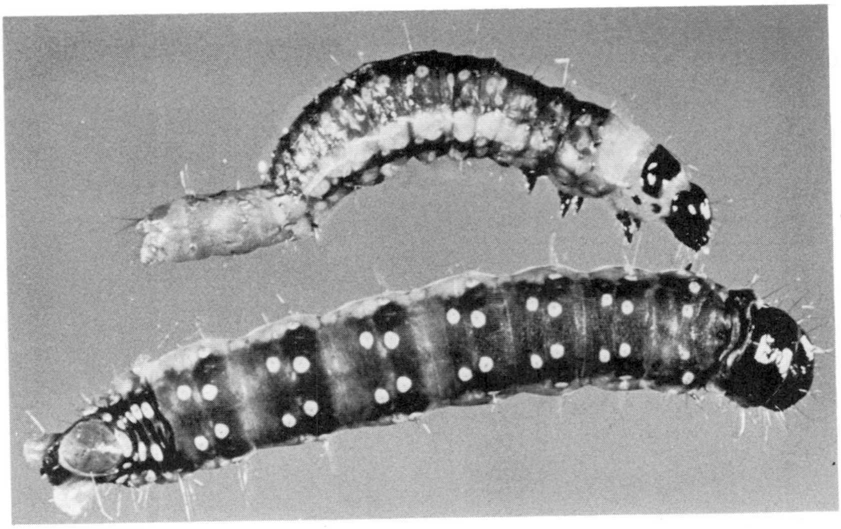

Fig. 3. Effect of chlorfluazuron on 5th instar spruce budworm, *Choristoneura fumiferana* (top) as compared to a normal larva (bottom) that has successfully moulted into the 6th instar (Retnakaran[89]).

hydrate hydrolases such as trehalase, invertase and amylase,[61] increase phenoloxidase activity [62] and also inhibit DNA synthesis in at least 3 species.[63]

Benzoylphenyl ureas have ovicidal (or embryocidal), larvicidal and other minor effects. Upon ingestion, most insect larvae show various degrees of malformation of the cuticle depending on the dosage, species and the benzoylphenyl urea analog tested (Fig. 5.)[64] A few species show contact effects.[65] The effects are manifested at the time of the subsequent moult but sometimes delayed to the pupal stage or in rare cases, to the adult stage. Sometimes there are no visible effects but the larvae fail to feed and this has been traced to displaced mandibles.[9] Ovicidal effects are caused by direct contact of the analog with the eggs or congenitally transferred to the eggs. Adult females can induce the ovicidal effect either through contact or by ingestion of the benzoylphenyl urea. Sometimes males that come in contact with the material contaminate the female during mating which subsequently results in having a lethal effect on the eggs.[66] The embryos are fully developed inside the egg and die while attempting to emerge or in some cases the larvae die soon after emergence. Some of the minor effects are weakening of the elytra in some beetles,[67] inhibition of peritrophic membrane formation[68] and absence of polysaccharides in the oenocytes.[69] In general, benzoylphenyl ureas interfere with the normal growth and development in insects and for this reason they have been referred to as Insect Growth Regulators or IGRs.

Bioassays to determine the control potential of benzoylphenyl ureas are conducted over a longer period than for contact, neurotoxic pesticides. Knock-down effects are seldom observed instead the effects are observed primarily at the moult succeeding treatment (Fig. 6). Younger instars are in general more sensitive than older instars but there are instances where the reverse has been shown to be true.[70,71] The differential sensitivity of several benzoylphenyl ureas to selected lepidopterans is summarized in Table 1.

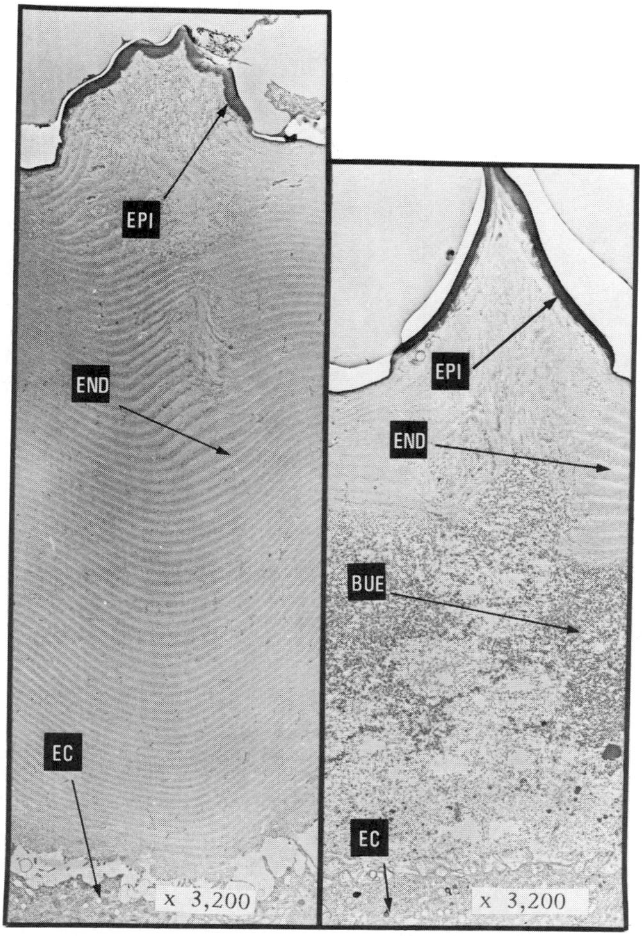

Fig. 4. Ultrastructural effect of chlorfluazuron on the spruce budworm, Choristoneura fumiferana. The benzoylphenyl urea treated 6th instar larva (right) shows the typical effect of chitin inhibition with no lamellae in the endocuticular region as compared to the untreated insect (left). EPI = epicuticle, END = endocuticle, EC = epidermal cell, BUE = benzoylphenyl urea effect (Courtesy of Dr. J. Percy).

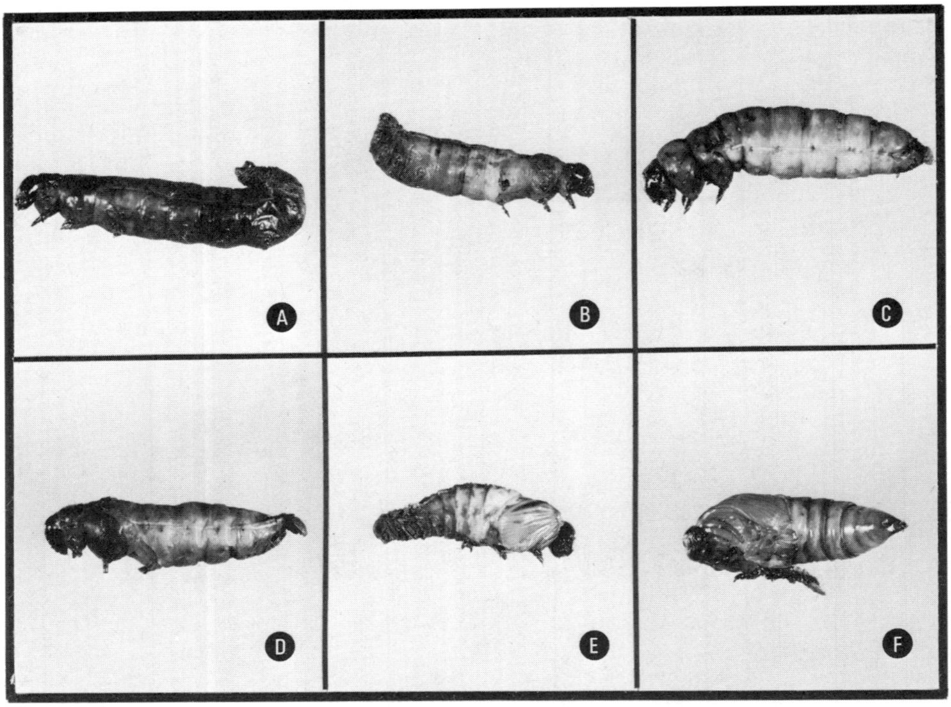

Fig. 5. Effect of diflubenzuron on 6th instar spruce budworm, *Choristoneura fumiferana*. A - most severe effect, completely larval; B - a part of abdomen pupating; C - abdomen is pupal; D - abdomen and part of thorax pupal; E - mostly pupal; F - least affected; completely pupal but with the larval exuvium still attached.

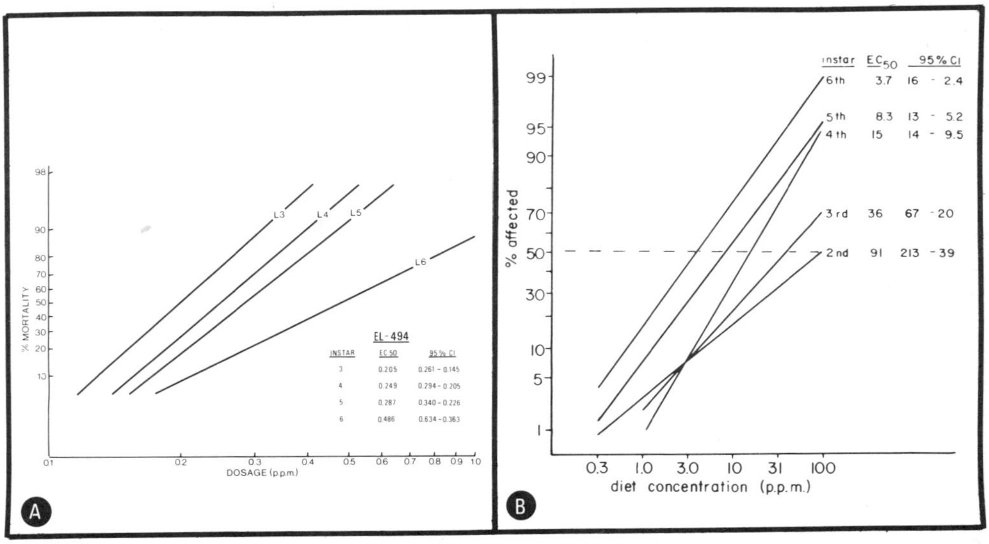

Fig. 6. Effect of 2 benzoylphenyl ureas on the different larval stages of the spruce budworm, *Choristoneura fumiferana*. A - EL-494 has the typical effect i.e. the younger stages are more susceptible. B - diflubenzuron has the reverse effect i.e. older stages are more susceptible.[70,71]

Table 1. Differential susceptibility of several lepidoptera to selected benzoylphenyl ureas. Chlorfluazuron is fast-acting compared to the others (Retnakaran et al [9]).

Insect	Toxicity of benzoylphenyl urea (LC_{50} in ppm)				
	Diflubenzuron	Alsystin	EL-494	Chlorfluazuron	Penfluron
Spruce budworm - 3rd instar (*Choristoneura fumiferana*)	36	1.65	0.205	0.27	0.31
Western spruce budworm - 3rd instar (*Choristoneura occidentalis*)	192	6.95	1.15	0.26	1.94
Omnivorous leafroller - 1st instar (*Platynota stultana*)	6938	173	-	1.36	-
Beet armyworm - 1st instar (*Spodoptera exigua*)	1.36	8.06	-	0.23	0.27
Douglas fir tussock moth - 3rd instar (*Orgyia pseudotsugata*)	0.058	0.059	2.56	-	0.029
Gypsy moth - 3rd instar (*Lymantria dispar*)	0.013	-	0.30	-	-

Table 2. Experimental aerial application of chlorfluazuron (25% WP in 7 N oil) on a spruce budworm, *Choristoneura fumiferana* infestation in the French River area, Ontario, Canada in 1980 (Retnakaran[89]).

Plot size (ha)	Treatment (g in 4.7 l/ha)		Pre-spray density (no./46 cm branch tip) in		Post-spray density (no./46 cm branch tip) in		% population* reduction in		% defoliation in	
			Balsam fir	White spruce	Balsam fir	White spruce	Balsam fir	White spruce	Balsam fir	White spruce
20	Chlorfluazuron	(35)	44.3	41.3	1.4	4.4	86	61	23	36
	Control	(-)	21.8	48.3	4.8	13.3			89	91
20	Chlorfluazuron	(70)	34.4	70.0	0.5	1.6	93	88	50	60
	Control	(-)	21.8	51.9	4.8	10.1			89	94
90	Matacil	(70)	16.7	26.0	0.1	0.8	96	80	2	4
	Control	(-)	16.5	26.8	2.5	3.7			84	94
90	Matacil	(70+70)	16.7	37.5	1.2	7.3	52	0	33	37
	Control	(-)	16.5	35.2	2.5	4.4			100	98
85	Dipel-88	(19.8 BIU/ha)	11.0	35.1	0.4	3.4	75	23	21	51
	Control	(-)	11.8	35.2	1.7	4.4			40	98

$$* \ \% \text{ population reduction} = \left[1 - \left(\frac{\text{post-spray density in treatment}}{\text{pre-spray density in treatment}} \times \frac{\text{pre-spray density in control}}{\text{post-spray density in control}}\right)\right] \times 100.$$

Table 3. Control of some insect pests of agriculture, stored products, forests and public health by benzoylyphenyl ureas.

Insect species	Host	Analog	Activity	Reference
Anthonomus grandis (Cotton boll weevil) Fam: Curculionidae Ord: Coleoptera	Cotton	Diflubenzuron; Penfluron	Females lay eggs that fail to hatch. Excellent control potential.	Wright and Villavaso[66]
Spodoptera littoralis (Egyptian cotton leafworm) Fam: Noctuidae Ord: Lepidoptera	Cotton, Beet, Tomato, Alfalfa	Diflubenzuron; Alsystin, Penfluron	Contact toxicity for eggs, larvae and pupae. Excellent control.	Abbassy *et al.* [72]; Ascher and Eliyahu[73]; El-Guindy *et al.* [74]
Laspeyresia pomonella (Codling moth) Fam: Olethreutidae Ord: Lepidoptera	Apple, Pear	Diflubenzuron; Alsystin, Penfluron	Excellent control in the field by aricidal action	Burts[75]; Moffitt *et al.* [76]
Anticarsia gemmatalis (Velvet bean caterpillar) Fam: Noctuidae Ord: Lepidoptera	Soybean	Diflubenzuron	Excellent control in the field	Greene [77]; Turnipseed *et al.* [78]
Leptinotarsa decemlineata (Colorado potato beetle) Fam: Noctuidae Ord: Lepidoptera	Potato	Diflubenzuron; Alsystin	Excellent control by both ovicidal and larvicidal effects.	Cooper *et al.* [79]; Tamaki *et al.* [80]
Lycoriella mali (Mushroom fly) Fam: Sciaridae Ord: Diptera	Mushroom	Diflubenzuron; Alsystin	Excellent control	Cantelo[81]; Kalberer and Vogel[82]

Table 3. Control of some insect pests of agriculture, stored products, forests and public health by benzoylyphenyl ureas. (con't)

Insect species	Host	Analog	Activity	Reference
Sitophilus species (Various weevils) Fam: Curculionidae Ord: Coleoptera	Various stored grains	Diflubenzuron; Alsystin	Excellent control at egg and young larval stages at very low concentrations.	McGregor and Kramer[83]; Mian and Mulla[84]
Thaumetopoea pityocampa (Pine processionary moth) Fam: Notodontidae Ord: Lepidoptera	Pine and other conifers	Diflubenzuron	Excellent, long-lasting control.	Robredo[85]
Lymantria dispar (Gypsy moth) Fam: Lymantriidae Ord: Lepidoptera	Mostly hardwoods	Diflubenzuron; Alsystin	Excellent control at very low dosages.	Abdel-Monem and Mumma[86]; Granett and Dunbar[87]
Choristoneura fumiferana (Spruce budworm) Fam: Tortricidae Ord: Lepidoptera	White spruce and balsam fir	Diflubenzuron; Alsystin and Chlorfluazuron	Excellent control with chlorfluazuron but refractory to others.	Retnakaran[88,89]
Croesia semipurpurana (Oak-leaf tier) Fam: Tortricidae Ord: Lepidoptera	Oak	Diflubenzuron; Alsystin	Excellent control in the field with diflubenzuron	Retnakaran and Grant[90]; Retnakaran and Tomkins[91]
Malacosoma disstria (Forest tent caterpillar) Fam: Lasiocampidae Ord: Lepidoptera	Hardwoods especially aspen	Diflubenzuron	Excellent control in the field	Retnakaran and Smith[92]; Retnakaran et al.[93]

Insect species	Host	Analog	Activity	Reference
Haematobia irritans (Horn fly) Fam: Muscidae Ord: Diptera	Range Cattle	Diflubenzuron	Excellent control by ovicidal effect. Congenital through both sexes.	Schmidt and Kunz[94]; Wright and Harris[95]
Stomoxys calcitrans (Stable fly) Fam: Muscidae Ord: Diptera	Feed lot of live stock	Diflubenzuron	Excellent control by surface treatment of animals or incorporating the material in the feed.	Schmidt and Kunz[94]; Wright and Harris[95]
Aedes aegypti (Yellow fever mosquito) Fam: Culicidae Ord: Diptera	Vector for yellow fever virus	Diflubenzuron; Alsystin; Penfluron	Ovicidal and larvicidal effects; excellent control at levels as low as 52 ppb	Saleh *et al.* [96]
Aedes triseriatus Fam: Culicidae Ord: Diptera	Vector for California encephalitis virus	Diflubenzuron	Excellent control	Hsieh and Steelman[97]
Anopheles albimanus Fam: Culicidae Ord: Diptera	Vector for malaria in Central America	Diflubenzuron	Excellent control	Mulla *et al.* [98]
Culex pipiens fatigans (House mosquito) Fam: Culicidae Ord: Diptera	Vector for the filanal parasite	Diflubenzuron; Penfluron	Excellent control	Miura *et al.* [99]
Culex tarsalis Fam: Culicidae Ord: Diptera	Vector for St. Louis encephalitis virus	Diflubenzuron	Excellent control	Schaefer *et al.* [100]

Benzoylphenyl ureas are generally effective on insect larvae when they ingest the material. Open-feeders are more susceptible than internal feeders as shown in Table 2. Some insects are able to detoxify and eliminate certain benzoylphenyl ureas and hence are refractory.[34] Contact effect is the exception rather than the rule and is seen primarily among dipterans, coleopterans and possibly a few lepidopterans. A representative selection of insect pests of agriculture, stored products, forests and public health that can be effectively controlled by benzoylphenyl ureas is shown in Table 3.

Resistance to benzoylphenyl ureas have been observed in several insects but because the mode of action is different from that of cholinesterase inhibitors (carbamates and organophosphates), cross resistance is less likely.[101] Inhibitors of mixed function oxidases were not synergistic[102] but esterase inhibitors showed strong synergism.[103]

Benzoylphenyl ureas have very little adverse effect on parasites, predators, pollinators and most non-target species. Adult parasites, predators, pollinators and non-targets are seldom affected. The external parasites that are in the larval phase at the time of treatment are susceptible.[104,105,106] Diflubenzuron is non-toxic to predatory mites.[107] Juvenile stages of freshwater and marine crustacea are very sensitive to benzoylphenyl ureas but since adults are unaffected, they soon repopulate the area as soon as the diflubenzuron is degraded by the microorganisms in the water, the net effect being a temporary lowering of the population density.[108]

Benzoylphenyl ureas are non-toxic to vertebrates and are degraded generally within 24 h by microorganisms including fungi.[109,110] Fungi are unaffected by these materials. In fact there is a strong suspicion that most *in vitro* preparations of insect tissue where chitin synthesis has been demonstrated, were contaminated with fungi, probably yeast. This is probably the reason why diflubenzuron has never been shown to inhibit chitin synthesis in *in vitro* preparations whereas in every *in vivo* test on insects inhibition of chitin synthesis has been unequivocally established.[111]

Benzoylphenyl ureas combine the environmental safety of biological control agents such as microbials with the efficacy of broad-spectrum pesticides such as carbamates and organophosphates.[11]

ACKNOWLEDGMENTS

I thank my colleague, Dr. Jean Percy for providing me with the electronmicrographs of the cuticle, Mr. Bill Tomkins for the art work and the librarian, Mrs. Sandy Burt and her assistant Mrs. Nancy Jean Dukes for obtaining the reference material especially the ones that are hard to get.

REFERENCES

1. A. C. Neville, "Biology of the Arthropod Cuticle", Springer, Berlin (1975).
2. V. B. Wigglesworth, "The Control of Growth and Form: the Study of the Epidermal Cell in an Insect", Cornell Univ. Press, Ithaca (1959).
3. S. Canevey, J. Insect Physiol., 16:1087 (1970).
4. G. Richards, Biol. Rev., 56:501 (1981).
5. C. M. Williams, and F. C. Kafatos, Mitt. Schweiz. Ent. Ges., 44:151 (1971).

6. C. A. D. De Kort, and N. A. Granger, Ann. Rev. Entomol., 26:1 (1981).
7. J. Koolman, K.-D. Spindler, in: "Endocrinology of Insects", R. G. H. Downer and H. Laufer, ed., Alan R. Liss. Inc., New York (1983).
8. W. Maas, R. Van Hes, A. C. Grosscurt, and D. H. Deul, in: "Chemie der Planzenschutz und Schadlingsbekampfungsmittel", R. Wegler, ed., Vol. 6, Springer-Verlag, Heidelberg (1981).
9. A. Retnakaran, J. Granett, and T. Ennis, in: "Comprehensive Insect Physiology Biochemistry and Pharmacology", G. A. Kerkut, and L. I. Gilbert, ed., Vol. 12, Pergamon Press, Oxford (1985).
10. R. H. Hackman, in: "Biology of the Integument", J. Bereiter-Hahn, A. G. Matolsty, and K. S. Richards, ed., Vol. 1, Springer, Berlin (1984).
11. A. Retnakaran, in: Chitin and Benzoylphenyl Ureas", J. E. Wright, and A. Retnakaran, ed., Dr. W. Junk, The Hague (1985).
12. B. K. Filshie, in: "Insect Ultrastructure", R. C. King, and H. Akai, ed., Vol. 1, Plenum Press, New York (1982).
13. R. H. Hackman, Aust. J. Biol. Sci., 13:568 (1960).
14. R. H. Hackman, Comp. Biochem. Physiol., 49:457 (1974).
15. R. H. Hackman, in: "Chitin and Benzoylphenyl Ureas", J. E. Wright, and A. Retnakaran, ed., Dr. W. Junk, the Hague (1985).
16. R. A. A. Muzzarelli, "Chitin", Pergamon Press, Oxford (1977).
17. F. A. Rutherford, and P. R. Austin, in: "Proceedings of the First International Conference on Chitin/Chitosan", R. A. A. Muzzarelli and E. R. Pariser, eds., M.I.T., Cambridge (1978).
18. D. Koga, J. Jilka, and K. J. Kramer, Insect Biochem., 13:295 (1983).
19. K. M. Rudall, and W. Kenchigton, Biol. Rev., 49:597 (1973).
20. S. Bartnicki-Garcia, J. Ruiz-Herrera, and C. E. Bracker, in: "Fungal Walls and Hyphal Growth", J. H. Burnett and A. P. J. Trinci, ed., Cambridge Univ. Press, Cambridge (1979).
21. E. Cabib, and R. Roberts, Ann. Rev. Biochem., 51:763 (1982).
22. R. T. Schwarz, and R. Datema, Meth. Enzymol., 83:432 (1982).
23. S. O. Anderson, Ann. Rev. Entomol., 24:24 (1979).
24. D. J. Candy, and B. A. Kilby, J. Exp. Biol., 39:129 (1962).
25. J. F. Gwinn, J. R. Stevenson, Comp. Biochem. Physiol., 45B:777 (1973).
26. E. Cohen, in: "Chitin and Benzoylphenyl Ureas", J. E. Wright and A. Retnakaran, ed., Dr. W. Junk, the Hague (1985).
27. E. Jaworski, L. Wang, and G. Marco, Nature, 198:790 (1963).
28. R. T. Mayer, A. C. Chen, and J. K. DeLoach, Insect Biochem., 10:549 (1980).
29. A. Retnakaran, and R. H. Hackman, Arch. Insect. Biochem. Physiol., (1985) (in press).
30. I. F. Turnbull, and J. Howells, Aust. J. Biol. Sci., 36:251 (1983).
31. E. Cohen, and J. E. Casida, Pestic. Biochem. Physiol., 13:121 (1980).
32. T. Fukamizo, and K. J. Kramer, Insect Biochem., 15:1 (1985).
33. C. A. Porter, and E. G. Jaworski, J. Insect Physiol., 11:1151 (1965).
34. A. Retnakaran, J. Granett, and J. Robertson, J. Insect Physiol., 26:385 (1980).
35. R. F. Powning, and H. Irzykiewicz, J. Chromatogr., 29:115 (1967).
36. M. R. Rosner, S. C. Hubbard, R. J. Ivatt, and P. W. Robbins, Meth. Enzymol., 83:399 (1982).
37. M. N. Horst, Arch. Biochem. Biophys., 223:254 (1983).
38. L. A. Quesada-Allué, Biochem. Biophys. Res. Commun., 105:312 (1982).
39. L. A. Quesada-Allué, Acta Physiol. Lat. Am., 26:349 (1976).
40. J. J. Elting, and W. J. Lennarz, Meth. Enzymol., 83:408 (1982).
41. C. J. Brine, and P. R. Austin, Comp. Biochem. Physiol., 70:173 (1981).
42. H. Oberlander, S. Ferkovich, E. Leach, and F. V. Essen, Wilhelm Roux's Arch., 188:81 (1980).
43. J. R. Stevenson, and D. A. Tung, Comp. Biochem. Physiol., 39B:559 (1971).
44. R. H. Hackman, Ann. Rev. Entomol., 27:75 (1982).
45. I. Kubo, M. Uchida, and J. A. Klocke, Agric. Biol. Chem., 47:911 (1983).
46. M. Hori, K. Kakiki, S. Suzuki, and T. Misato, Agric. Biol. Chem., 35:1280 (1971).

47. E. Kuvano, N. Sato, and M. Eto, Agric. Biol. Chem., 46:1715 (1982).
48. B. Becker, J. Insect Physiol., 24:699 (1978).
49. L. C. Post, B. J. DeJong, and W. R. Vincent, Pestic. Biochem. Physiol., 4:473 (1974).
50. W. H. Van Eck, Insect Biochem., 9:295 (1979).
51. T. Mitsui, C. Nobusawa, and J. Fukami, J. Pesticide Sci., 9:19 (1984).
52. T. Leighton, E. Marks, and E. Leighton, Science, 213:905 (1981).
53. S. J. Yu, and L. C. Terriere, Pestic. Biochem. Physiol., 7:48 (1977).
54. M. P. O'Neill, G. M. Holman, and J. E. Wright, J. Insect Physiol., 23: 1243 (1977).
55. S. C. Saxena, and V. Kumar, Ind. J. Exp. Biol., 10:1199 (1981).
56. E. Cohen, and J. E. Casida, Pestic. Biochem. Physiol., 13:129 (1980).
57. E. Cohen, and J. E. Casida, in: "International Congress of Pesticide Chemistry, 5th, Proceedings", J. Miyamoto, ed., Vol. 3 (1983).
58. R. Moreau, C. Castex, and M. Lamy, Ann. Zool.-Ecol. Anim., 7:161 (1975).
59. H.' S. Salama, Z. A. Motgally, and U. Skatulla, Z. Ang. Ent., 80:396 (1976).
60. S. J. Yu, and L. C. Terriere, Life Sci., 17:619 (1975).
61. I. Ishaaya, and K. R. S. Ascher, Phytoparasitica, 5:149 (1977).
62. D. H. Deul, G. J. DeJong, and J. A. M. Kortenbach, Pestic. Biochem. Physiol., 8:98 (1978).
63. N. Soltani, M. T. Besson, and J. Delachambre, Pestic. Biochem. Physiol., 21:256 (1984).
64. A. Retnakaran, and L. Smith, Can. Ent., 107:883 (1975).
65. K. R. S. Ascher, and N. E. Nemny, Pestic. Sci., 7:447 (1976).
66. J. E. Wright, and E. J. Villavaso, in: "Cotton Insect Management with Special References to the Boll Weevil", E. P. Lloyd, R. L. Ridway, and W. C. Cross, eds., Plenum Press, New York, (1983).
67. A. C. Grosscurt, and S. O. Anderson, Entomol. Proc. C., 83:143 (1980).
68. L. Clark, G. H. R. Temple, and J. F. V. Vincent, J. Insect Physiol., 23:241 (1977).
69. J.-Cl. Denneulin, and M. Lamy, Ann. d'Endocrin. (Paris), 38:405 (1977).
70. J. Granett, and A. Retnakaran, Can. Ent., 109:893 (1977).
71. A. Retnakaran, Can. Ent., 111:847 (1979).
72. M. A. Abbassy, M. Ashry, and M. A. Salama, Med. Fac. Landbouw Rijkuniv. Gent, 45/3:721 (1980).
73. K. R. S. Ascher, and M. Eliahu, Phytoparasitica, 9:133 (1981).
74. M. A. El-Guindy, M. M. Abdel-Sattar, and A. R. M. El-Refai, Pestic. Sci., 14:253 (1983).
75. E. C. Burts, J. Econ. Entomol., 76:936 (1983).
76. H. R. Moffitt, K. D. Mantey, and G. Tamaki, Can. Ent., 116:1057 (1984).
77. G. L. Greene, Proc. World Soybean Res. Conf., 602 (1975).
78. S. G. Turnipseed, E. A. Heinrichs, R. F. P. Da Silva, and J. W. Todd, J. Econ. Entomol., 67:760 (1974).
79. R. M. Cooper, R. K. Lindquist, and D. E. Simeonet, J. Econ. Entomol., 76:563 (1983).
80. G. Tamaki, R. L. Chauvin, H. R. Moffitt, and K. D. Mantey, Can. Ent., 116:197 (1984).
81. W. W. Cantelo, Mushroom Sci., 11:255 (1981).
82. P. Kalberer, and E. Vogel, Zeit. Pflanzenkrankheiten Pflanzenschutz, 85:328 (1978).
83. H. E. McGregor, and K. J. Kramer, J. Econ. Entomol., 69:479 (1976).
84. L. S. Mian, and M. S. Mulla, J. Econ. Entomol., 75:80 (1982).
85. F. J. Robredo, Agricultura, 49:269 (1980).
86. A. H. Abdel-Monem, and R. O. Mumma, J. Econ. Entomol., 74:176 (1981).
87. J. Granett, and D. M. Dunbar, J. Econ. Entomol., 68:99 (1975).
88. A. Retnakaran, Inf. Rep. FPM-X-45, (1981).
89. A. Retnakaran, Can. Ent., 114:523 (1982).
90. A. Retnakaran, and G. G. Grant, Can. Ent., 117:363 (1985).

91. A. Retnakaran, and W. Tomkins, C.F.S. Res. Notes, 2:5 (1982).
92. A. Retnakaran, and L. Smith, C. F. S. Bi-mon. Res. Notes, 32:2 (1976).
93. A. Retnakaran, L. Smith, B. Tomkins, and J. Granett, Can. Ent., 111:841 (1979).
94. C. D. Smith, and S. E. Kunz, Econ. Entomol., 73:702 (1980).
95. J. E. Wright, and R. L. Harris, J. Econ. Entomol., 69:728 (1976).
96. M. S. Saleh, I. A. Gaaboub, and S. M. I. Kassem, J. Agric. Sci., Camb., 97:87 (1981).
97. M. Y. G. Hsieh, and C. D. Steelman, Mosquito News, 34:278 (1974).
98. M. S. Mulla, H. A. Darwazeh, and R. L. Norland, J. Econ. Entomol., 67:329 (1974).
99. T. Miura, C. H. Schaefer, R. M. Takahashi, and F. S. Mulligan III, J. Econ. Entomol., 69:655 (1976).
100. C. H. Schaefer, W. H. Wilder, and F. S. Mulligan III, J. Econ. Entomol., 68:183 (1975).
101. G. D. Pimprikar, and C. P. Geroghiou, Pestic. Biochem. Physiol., 12:10 (1979).
102. A. Retnakaran, C.F.S. Res. Notes, 2:1 (1982).
103. J. Granett, and M. Hejazi, J. Econ. Entomol., 76:399 (1983).
104. A. B. Broadbent, and D. J. Pree, Environ. Entomol., 13:133 (1984).
105. B. J. Emmett, and B. M. Archer, Pl. Pathol., 29:177 (1980).
106. J. Granett, and R. M. Weseloh, J. Econ. Entomol., 68:577 (1975).
107. D. W. Anderson, and R. H. Elliott, Can. Ent., 114:733 (1982).
108. M. S. Mulla, G. Majori, and H. A. Darwazeh, Mosquito News, 35:211 (1975).
109. S. L. Seuferer, H. D. Braymer, and N. J. Quinn, Pestic. Biochem. Physiol., 10:174 (1979).
110. H. Wilcox III, and T. Coffey Jr., Environmental impacts of diflubenzuron (Dimilin) insecticide, Forest Insect and Disease Management, Forest Service, U.S.D.A., Broomall, Pa. U.S.A. (1978).
111. M. L. Bade, J. Appl. Polymer Sci., 37:165 (1983).

Editor's Report

G.W. GOODAY

COMPARATIVE EFFECTS OF BENZOYLARYLUREAS ON INSECTS

based on works by

K. R. S. ASCHER, I. ISHAAYA, V. MELAMED-MADJAR, and S. YATHOM

The range of benzoylarylureas has been reviewed by Willems et al. (this Volume), and compared with other inhibitors of chitin synthetis by Cohen (this Volume). Here follow two investigations of the comparative effects of different compounds on several insects.

EFFECT OF SOME BENZOYLPHENYLUREAS ON LARVAE AND EGGS OF DIFFERENT INSECT SPECIES

(K. R. S. Ascher, S. Yathom and V. Melamed-Madjar)

INTRODUCTION

In recent years we have witnessed the development of a series of "second generation" benzoylphenylurea chitin synthesis inhibitors. If we consider diflubenzuron, triflumuron and penfluron as the most prominent representatives of the "first generation" benzoylphenylureas, the "second generation" benzoylphenylureas comprise, inter alia, CME 134 (teflubenzuron) (1, 2), XRD-473 (3) and IKI 7899 (chlorfluazuron) (4-7). Structures are given by Willems et al. (this volume). In the present paper we report the effects of benzoylphenylureas on Bemisia tabaci, Ostrinia nubilalis, Spodoptera littoralis, Carpophilus hemipterus and Liriomyza trifolii.

Bemisia tabaci

The toxic effects of the three new benzoylphenylureas teflubenzuron, XRD-473 and chlorfluazuron were assayed against preimaginal stages of Bemisia tabaci. Cotton seedlings were sprayed and females were allowed to lay eggs for two days on treated leaves after various periods of aging of the residues. Alternatively, seedlings on which females had been allowed to oviposit were sprayed immediately after the elapse of the two-day oviposition period or at about weekly intervals thereafter, thus treating different preimaginal stages. With the first mode of treatment, it was shown that the benzoylphenylurea residues caused no mortality in the ovipositing females; but teflubenzuron residues gave 100% kill of preimaginal stages for 38 days at a spray concentration (all concentrations expressed as % a.i.) of 0.005%, were still highly active at 0.001% and 0.0005%, and their activity deteriorated at 0.0001%; XRD-473 residues gave 100% kill until 49 days at 0.001% and broke down at 0.0005%; and chlorfluazuron residues gave 100% kill until 60 days at 0.005% but broke down at 0.001%. With all three compounds, mortality occurred in first-instar nymphs.

When teflubenzuron was sprayed on different preimaginal stages (second mode of treatment), it was highly active if sprayed directly on eggs, and mortality set in in the egg and in one subsequent nymphal stage at 0.1%, and at lower concentrations (0.05, 0.01 and 0.001%), also in two subsequent nymphal stages. The treatment of first-instar larvae produced analogous protracted effects, but treatment of third-instar nymphs and pupae produced no mortality.

Ostrinia nubilalis

Tests against *Ostrinia nubilalis* were conducted by spraying maize plants with the benzoylphenylureas and subsequently infesting them with 4- to 6-day-old larvae of the European corn borer. After 3 days, the larvae were transferred to a semisynthetic diet and observed for mortality until pupation and adult emergence. In those assays teflubenzuron was slightly better than XRD-473, which, in turn, was much better than chlorfluazuron.

Spodoptera littoralis

The ovicidal effects of benzoylphenylureas in eggs of *Spodoptera littoralis* provide a good example of the influence of structural modification of the parent compound, diflubenzuron, on toxicity. In *S. littoralis*, the order of efficacy of the benzoylphenylureas as ovicides (2, 8-11) is completely different from that as larvicides (2, 7, 9, 12-14) in which the "second generation" compounds have shown the best activity. When the compounds were compared as ECs (or liquid formulations), the order of activity as *S. littoralis* ovicides was as follows: diflubenzuron $\simeq$ triflumuron > teflubenzuron > XRD-473 >> PH 60-38 and chlorfluazuron (inactive). When WP formulations of diflubenzuron and some of its analogs were compared, the order of activity was penfluron (PH 60-44) $\geq$ PH 60-43 = diflubenzuron >> PH 60-45. The latter was nearly ineffective.

Carpophilus hemipterus

When adults of *Carpophilus hemipterus* were exposed for 24 h to benzoylphenylurea residues on a solid semisynthetic diet and then transferred to untreated diets, the females laid sterile eggs during the 14-day observation period. Chlorfluazuron, XRD-473 and diflubenzuron (in that order) were highly active in this sense.

Liriomyza trifolii

Several benzoylphenylureas were assayed against *L. trifolii* by spraying bean plants, which, after different periods of residue aging, were exposed for either 4 or 24 h to ovipositing females (preventive treatment). Also, bean plants previously infested and exhibiting mines were sprayed (curative treatment). The development of mines, the number of larvae and especially of pupae and finally of adults emerged served as criteria of effectiveness. XRD-473 was by far the most active among the benzoylphenylureas tested, though it was not as effective as the standard treatment with cyromazine (Trigard; Larvadex); it was followed by the EC of chlorfluazuron, which was much less effective; chlorfluazuron formulated as WP was inactive. The EC of triflumuron and the liquid formulation of diflubenzuron had very slight activity and their WPs were inactive.

REFERENCES

1. H.-M. Becher, P. Becker, R. Prokic-Immel, and W. Wirtz, CME 134, a new chitin synthesis inhibiting insecticide, *Proc. 10th Int. Congr. Pl. Prot.*, Brighton, pp. 408-415 (1983).

2. K.R.S. Ascher and N.E. Nemny, The effect of CME 134 on Spodoptera littoralis eggs and larvae, Phytoparasitica 12:13-27 (1984).
3. R.J. Sbragia, B. Bisabri-Ershadi, R.H. Rigterink, D.P. Clifford, and R. Dutton, XRD-473, a new acylurea insecticide effective against Heliothis, Proc. 10th Int. Congr. Pl. Prot., Brighton, pp. 417-421 (1983).
4. T. Haga, T. Toki, T. Koyanagi, and R. Nishiyama, Structure-activity relationships of a series of benzoyl-pyridyloxyphenly-urea derivatives, Abstr. 5th Int. Congr. Pestic. Chem., Tokyo, IId-7 (1982).
5. R. Neumann and W. Guyer, A new chitin synthesis inhibitor, CGA 112'913: its biochemical mode of action as compared to diflubenzuron, Proc. 10th Int. Congr. Pl. Prot., Brighton, pp. 445-451 (1983).
6. R. Scheurer, Th. Schläpfer, M.A. Ruzette, and F. Buholzer, CGA-112'913 (IKI 7899), a new insecticide against cotton pests, Mitt. dtsch. Ges. allg. angew. Ent. 4:127-129 (1983).
7. I. Ishaaya, N.E. Nemny, and K.R.S. Ascher, The effect of IKI 7899, a new chitin synthesis inhibitor, on larvae of Tribolium castaneum and Spodoptera littoralis, Phytoparasitica 13:193-197 (1984).
8. K.R.S. Ascher and N.E. Nemny, The ovicidal effect of PH 60-40 [1-(4-chlorophenyl)-3-(2,6-difluorobenzoyl)-urea] in Spodoptera littoralis Boisd., Phytoparasitica 2:131-133 (1974).
9. K.R.S. Ascher, N.E. Nemny, M. Eliyahu, and I. Ishaaya, The effect of BAY SIR 8514 on Spodoptera littoralis (Boisduval) eggs and larvae, Phytoparasitica 7:177-184 (1979).
10. K.R.S. Ascher, N.E. Nemny, and I. Ishaaya, The toxic effect of diflubenzuron on Spodoptera littoralis eggs and on their respiration, Pestic. Sci. 11:90-94 (1980).
11. K.R.S. Ascher, M. Eliyahu, N.E. Nemny, and I. Ishaaya, The toxicity of some novel pesticides - synthetic pyrethroids and benzoyl phenylurea chitin synthesis inhibitors - for eggs of Spodoptera littoralis (Boisd.), Z. angew. Ent. 94:504-509 (1982).
12. K.R.S. Ascher and N.E. Nemny, Toxicity of the chitin synthesis inhibitors, diflubenzuron and its dichloro-analogue, to Spodoptera littoralis larvae, Pestic. Sci. 7:1-9 (1976).
13. K.R.S. Ascher and N.E. Nemny, Contact toxicity of diflubenzuron against Spodoptera littoralis larvae, Pestic. Sci. 7:447-452 (1976).
14. K.R.S. Ascher and M. Eliyahu, The residual contact toxicity of BAY SIR 8514 to Spodoptera littoralis larvae, Phytoparasitica 9:133-138 (1981).

TOXICITY AND BIOCHEMICAL ASPECTS OF CHITIN SYNTHESIS INHIBITORS ON MALATHION-SUSCEPTIBLE AND -RESISTANT STRAINS OF TRIBOLIUM CASTANEUM

(I. Ishaaya and K. R. S. Ascher)

INTRODUCTION

Benzoylphenylureas are selective insecticides acting on insects of various orders by inhibiting chitin formation (1, 2). Diflubenzuron (DFB), the most investigated compound of this series, affects the postapolytic stage of larvae and exhibits in some species suppression of fecundity (3), ovicidal activity (4), and contact toxicity (5). Recently, potent acylureas have been developed by Ishihara Sangyo Kaisha Ltd. (IKI-7988), Celamerck GmbH (CME-134) and Dow Chemical Co. Ltd. (XRD-473). They are generally an order of magnitude more potent than DFB on several agricultural pests (6-11). The present study evaluates the toxicity and biochemical aspects of IKI-7899, CME-134 and XRD-473 in comparison with those of DFB on malathion-resistant and -susceptible strains of *Tribolium castaneum*. For structures and references, see Willems et al. (this Volume).

MATERIALS AND METHODS

The rearing procedure, using wheat flour containing 5 % dried yeast as the basic diet, and the bioassay method were the same as described previously (12, 13). Each treatment was carried out with 15-20 replicates of 15 first-instar larvae and held at 28° C for determination of larval mortality, pupation and emergence. For comparison, log dosage-percentage mortality curves on a probit scale were used to determine LC-50 and LC-95 (concentrations needed for 50 and 95 % cumulative larval and pupal mortality) or I-50 and I-95 (concentrations needed for 50 and 95 % larval weight gain inhibition at day 14) values for the various chitin synthesis inhibitors. Delayed effect, as expressed by pupal-adult intermediates and abnormal adults, was determined (14).

A malathion-resistant (CTC-12) and a malathion-susceptible (bb) strain of *T. castaneum* were used. Biochemical and genetic determinations (15, 16) indicate that the CTC-12 strain exhibits about four-fold more activity of cytochrome P-450 and of epoxide hydrase than does the bb strain.

RESULT AND DISCUSSION

According to LC-50 and LC-95 values, XRD-473, CME-134 and IKI-7899 were considerably more toxic than DFB and exhibited similar toxicity on both the bb and CTC-12 strains (Table I). On the other hand, DFB was much less toxic on the CTC-12 than on the bb strain, which seems to have been due to DFB's susceptibility to oxidative and hydrolytic activities present in the CTC-12 strain (16). Accordingly, the potency of IKI-7899, CME-134 and XRD-473 relative to diflubenzuron (DFB) increased considerably in the CTC-12 strain, reaching ratios of 14-23-fold at the LC-95 value. This phenomenon was even more conspicuous when larval weight gain inhibition was determined. The potency of the new acylureas on the CTC-12 strain was

Table 1. LC-50 and LC-95 values of four chitin synthesis inhibitors obtained with malathion-susceptible (bb) and -resistant (CTC-12) strains of *Tribolium castaneum*. Data are means of 15-20 replicates of 15 first-instar larvae each.

Compound	strain bb	strain CTC-12
	LC-50, $mgkg^{-1}$	
DFB	0.480a	0.720a
IKI-7899	0.106b (4.5)*	0.108b (6.5)
CME-134	0.092b (5.2)	0.104b (6.9)
XRD-473	0.068c (7.1)	0.070c (10.2)
	LC-95, $mgkg^{-1}$	
DFB	1.060a	2.630a
IKI-7899	0.156b (6.8)	0.176b (14.9)
CME-134	0.148b (7.2)	0.188b (14.0)
XRD-473	0.108c (9.8)	0.116c (22.6)

[a,b,c]Data differ significantly from each other at *P*=0.05.

*In parentheses, potency relative to DFB.

over 20-fold relative to DFB at I-50 and over 100-fold at I-95. With all the new acylureas, i.e. IKI-7899, CME-134 and XRD-473, a considerable number of pupal-adult intermediates and abnormal adults were observed. This effect may result from the stability of these compounds at the larval stage and from their persistence during pupal and adult formation.

Apparently all the new acylureas are not appreciably affected by metabolizing enzymes present in the malathion-resistant strain. These findings concur with those obtained by Neumann and Guyer (7), who showed that DFB is more susceptible than IKI-7899 to detoxification in *Spodoptera littoralis* larvae.

The high potency of IKI-7899, CME-134 and XRD-473 in malathion-susceptible and -resistant strains of *T. castaneum* and in other agricultural pests (6-12), along with their low mammalian toxicity, renders these compounds potential insecticides for controlling important agricultural pests, and especially those resistant to organophosphorus compounds.

Acknowledgments. The authors thank Mrs. Sara Yablonski and Mrs. Zmira Mendelson for their technical assistance and Dr. D. Wool of Tel-Aviv University for providing the CTC-12 and bb strains of *Tribolium castaneum*. This paper constitutes contribution No. 1383-E, 1985 series, from the Agricultural Research Organization, The Volcani Center, Bet Dagan, Israel.

REFERENCES

1. I. Ishaaya and J. E. Casida, Dietary TH6040 alters composition and enzyme activity of housefly larval cuticle, *Pestic. Biochem. Physiol.*, 4:484 (1974).
2. L. C. Post, B. J. de Jong and W. R. Vincent, 1-(2,6-Disubstituted benzoyl)-3-phenylurea insecticides: inhibitors of chitin synthesis, *Pestic. Biochem. Physiol.*, 4:473 (1974).
3. M. J. Sarasua and C. Santiago-Alvarez, Effect of diflubenzuron on the fecundity of *Ceratitis capitata*, *Entomologia exp. appl.*, 33:223 (1983).
4. K. R. S. Ascher and N. E. Nemny, The ovicidal effect of PH 60-40 [1-(4-chlorophenyl)-3-(2,6-difluorobenzoyl)-urea] in *Spodoptera littoralis* Boisd. *Phytoparasitica*, 2:131 (1974).
5. K. R. S. Ascher and N. E. Nemny, Contact activity of diflubenzuron against *Spodoptera littoralis* larvae, *Pestic. Sci.*, 7:447 (1976).
6. H.-M. Becher, P. Becker, R. Prokic-Immel and W. Wirtz, CME-134, a new chitin synthetis inhibiting insecticide, *10th Int. Congr. Pl. Protection*, Brighton, 408 (1983).
7. R. Neumann and W. Guyer, A new chitin synthesis inhibitor CGA 112'913: its biochemical mode of action as compared to diflubenzuron, *10th Int. Congr. Pl. Protection*, Brighton, 445 (1983).
8. R. J. Sbragia, B. Bisabri-Ershadi, R. H. Rigterink, D. P. Clifford and R. Dutton, XRD-473, a new acylurea insecticide effective against *Heliothis*, *10th Int. Congr. Pl. Protection*, Brighton, 417 (1983).
9. K. R. S. Ascher and N. E. Nemny, The effect of CME-134 on *Spodoptera littoralis* eggs and larvae, *Phytoparasitica*, 12:13 (1984.
10. I. Ishaaya, N. E. Nemny and K. R. S. Ascher, The effect of IKI-7899, a new chitin synthesis inhibitor, on larvae of *Tribolium castaneum* and *Spodoptera littoralis*, *Phytoparasitica*, 12:193 (1984).
11. I. Ishaaya and S. Yablonski, Toxicity and biochemical aspects of IKI-7899 and diflubenzuron on malathion-susceptible and -resistant strains of *Tribolium*, *in*: "Chitin and Benzoylphenyl Ureas", J. W. Wright and A. Retnakaran, eds., Dr. W. Junk Publishers, The Hague, in press (1985).
12. I. Ishaaya, K. R. S. Ascher and S. Yablonski, The effect of BAY SIR 8514, diflubenzuron, and Hercules 24108 on growth and development of *Tribolium confusum*, *Phytoparasitica*, 9:207 (1981).

13. I. Ishaaya and K. R. S. Ascher, Effect of diflubenzuron on growth and carbohydrate hydrolases of Tribolium castaneum, Phytoparasitica, 5:149 (1977).
14. I. Ishaaya, Biological and biochemical aspects of the disruption of adult formation in Tribolium by a novel bis (thiocarbamate) R-31026 and the juvenoid R-20458, Pestic. Sci., 13:204 (1982).
15. E. Cohen, Epoxide hydrase activity in the flour beetle Tribolium castaneum (Coleoptera, Tenebrionidae), Comp. Biochem. Physiol., 69B:29 (1981).
16. D. Wool, S. Noiman, D. Manheim and E. Cohen, Malathion resistance in Tribolium strains and their hybrids: inheritance patterns and possible enzymatic mechanisms (Coleoptera, Tenebrionidae), Biochem. Genet., 20:621 (1982).

Affiliations

K. R. S. Ascher, S. Yathom, M. Melamed-Madjart and I. Ishaaya, Department of Toxicology and Entomology, Agricultural Research Organization, Volcani Center, Bet Dagan, Israel.

BIOSYNTHESIS OF CHITIN BY INSECT EPIDERMAL CELLS: EFFECT OF BENZOYLPHENYLUREAS

B. MAUCHAMP and E. CAMINADE

Laboratoire de Phytopharmacie INRA
Route de Saint Cyr
78000 Versailles F.

INTRODUCTION

Insect epidermal cells derived from ektoderm, are involved in the main physiological functions. They are covered by a continuous secretion : the cuticle. The epidermal cell layer and the cuticle constitute the integument, the function of which is much more important than that of the skin of other animals. In fact the integument has to be considered as a skin, a skeleton and also a food reserve. Epidermal cells have a cyclic activity related to the molt/intermolt cycle. During this cycle many changes occur which affects for instance the shape of the cells, the cohesion between the cuticle and epidermal cells beneath it, or enzymatic activities Since very good reviews[1-4] were recently published on these aspects of epidermal cell activity, we focus our study on the relation between epidermal cell activity and chitin synthesis. High concentrations of chitin are located in insect cuticle ; also, several authors suggested that chitin synthesis could be considered as a very specific target for potential insecticides. Such compounds as benzoylphenylureas have been reported to be chitin synthesis inhibitors. We developed in our laboratory the available methods of chitin detection to evaluate ability of epidermal cells to synthesize chitin.[5,6]

IN VIVO OBSERVATION OF STRUCTURES CONTAINING CHITIN

Several techniques were developed to detect chitin directly on thin cross sections of integument. We demonstrated[7] that chitin can be located by incubation with FITC-WGA and observation by fluorescence microscopy. To prevent interferences with GlcNAc-rich-glycoproteins, integuments were digested by hot concentrated alkali that induce a partial deacetylation, then washed with buffer and incubated with FITC-WGA. The affinity of WGA with N-acetylglucosamine had also been used to demonstrate the presence of chitin by electron microscopy.[8] Simultaneously electron microscopy reveals the specific pattern of fibrillar structure resulting from chitin-protein association (Fig.1). However tissue as imaginal wing discs produce material that has not the usual oriented arrangement of chitin rods (Fig.2). Absence of such arrangement was also observed in scales (Fig.3). Entire tissues (integument, imaginal wing discs, entire adult wings covered with scales) were heated in saturated aqueous potassium hydroxide at 120°C for 6 h. After cooling the remaining structure was washed until

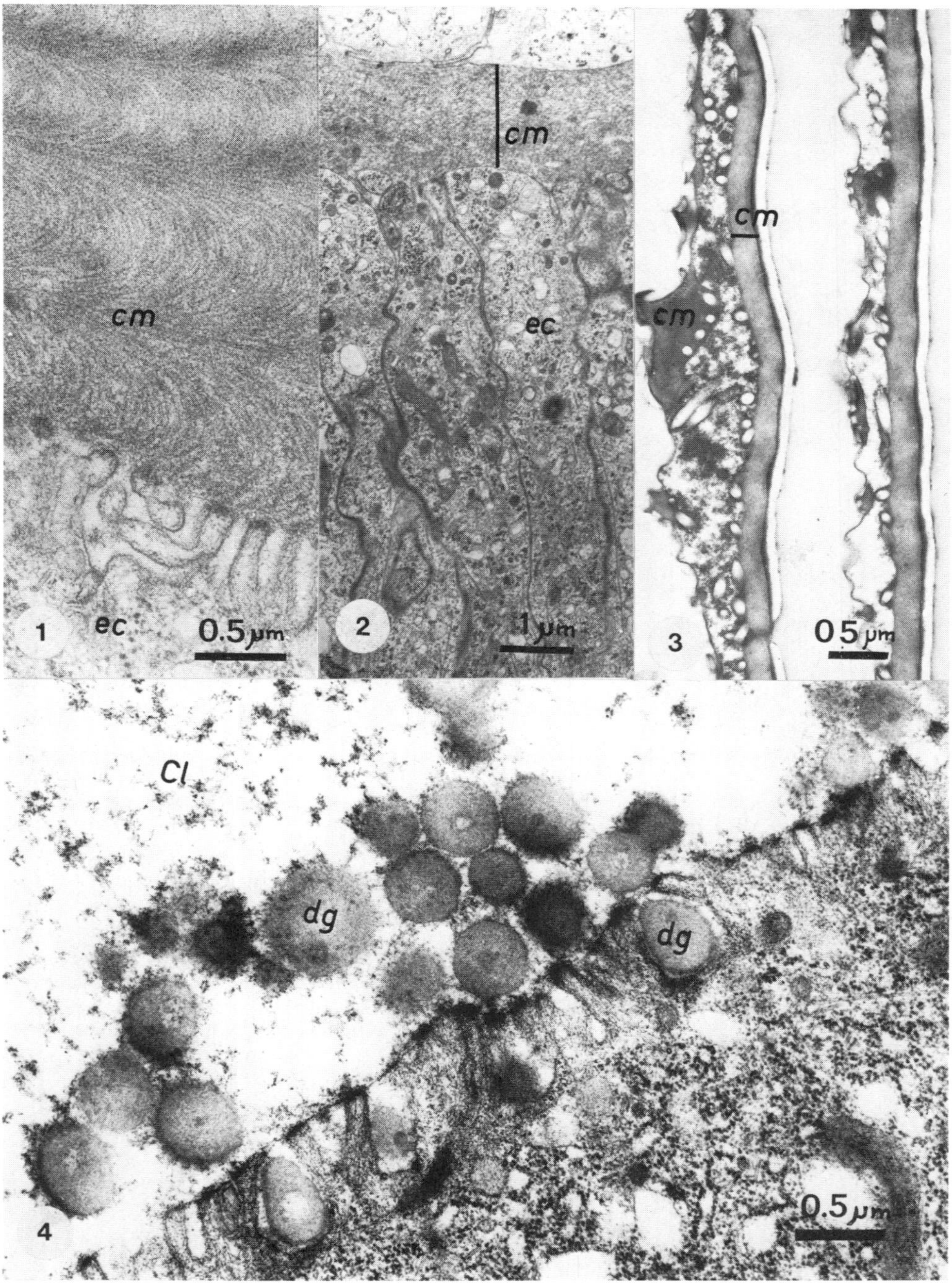

Fig.1-3. Electron micrographs of section of epidermal cell secretions in which chitin material was detected.
1-cuticle deposition by integument epidermal cells.
2-Apicale face of imaginal wing disc cells.
3-Section of flattened scale.
ec:epidermal cells;cm:chitin detected material.

Fig.4. Electron micrograph of section of integument after treatment with Diflubenzuron. Cl:Chitin like material,dg:dense granules.

neutralization. This material can be examined either after chitin/chitosan reaction by flooding with 0.2% aqueous solution of iodine in potassium iodine this solution being replaced with 1% aqueous sulfuric acid, or after incubation with FITC-WGA. We demonstrated by physico-chemical techniques the presence of chitin in the material giving a positive reaction. After treatment with hot alkali, residue was hydrolyzed. Acetic acid obtained was detected by GLC when glucosamine was analyzed by spectrophotometry or by coupled HPLC-mass spectrometry. Other methods such as IR, X-ray diffraction and more recently laser Raman spectrometry[9] were considered for the detection of chitin. In spite of the high sensitivity of these methods quantitative estimation of synthesized chitin cannot be obtained with accuracy and changes in cuticle formation during various postembryonic stages cannot be directly correlated with chitin synthesis activity. Correlations were established with biochemical parameters. Electron microscopic observations or spectrocolorimetric titration suggested that glycogen could be used to construct chitin[10]. Zaluska[11] demonstrated a reciprocal relationship between glycogen content and chitin content. Activity of epidermal cells can be modified by drugs, mainly those that interfere with cuticle deposit. After treatment with benzoylphenyl urea integument showed very dramatic changes. Changes occurred both in the cuticle structure and inside the cells. More precise data were obtained by electron microscopy (Fig.4). In cuticle fibrous material appeared between dense granules. Regular architecture of the cuticle was completly broken down. By the different techniques that we developed we demonstrated that the material localized between the dense granules was chitin-like. This material is similar to the material deposited by imaginal wing discs. Numerous vesicles appeared inside the cells ; some of them were filled with dense material, the others displayed coating filaments on inner side of vesicle membranes.

IN VITRO EVALUATION OF CHITIN SYNTHESIS

Epidermal cell activity can be evaluated *in vitro* either by tissue culture[12,13] or by obtaining chitin synthesis in cell-free systems. Results obtained by tissue culture were very close to those described in *in vivo* studies. Benzoylphenylureas prevented cuticle formation and induced modification of vesicle distribution in the cells. Initiation of filament formation appeared at the tip of the microvilli but it seems that size and nature of the microfibrils was not in appropriate configuration enough to produce the filaments giving the parabolic pattern observed in controls. Cell-free systems had been developed in our laboratory to study the mode of action of potential chitin synthesis inhibitors. Several insect cell-free systems to evaluate chitin polymerization were developed from different Arthropod species[14-16] or from yeast and fungi[17] . Our results and those of other authors revealed differences between the enzymes isolated from different Arthropods species or yeast and fungi. The major difference was revealed by enzyme prepared from a microsomal preparation of the brine shrimp *Artemia salina* larvae ; the enzyme was inhibited by UDP and Benzoylphenylurea but not by polyoxin D. Enzyme prepared from insects was demonstrated to be inhibited by Polyoxin D but not by benzoylphenylureas. In our experiments we found the latter result. These contradictory observations suggested that chitin synthetase from insects, crustaceans or fungi are not the same although the same product was synthesized.

CONCLUDING REMARKS

Chitin biosynthesis studies were intensified due to the consideration of chitin as a specific target of potential insecticides. *In vivo* and *in vitro* studies are necessary to detect active compounds and to explain how chitin synthesis was affected. The results so far are not sufficient to demonstrate the mode of action of all the compounds that revealed inhibitory activity. It seems clear that inaccurate definition of the cell-free systems accounts for differences in chitin synthetase activity between the different species. Our observations suggest that several step are involved in chitin synthesis and confirm the fact that chitin synthetase could be present as a zymogen.

REFERENCES

1. M. Locke, Epidermal cells, *in* "Biology of the Integument : 1. Invertebrates" J. Bereither-Hahn, A.G. Matoltsy and K.S. Richards, eds, Springer-Verlag, Berlin, Heidelberg, New-York, Tokyo (1984)
2. R.H. Hackman, Cuticle : Biochemistry, *in* "Biology of the Integument : 1. Invertebrates" See Ref. 1.
3. A.C. Neville, Cuticle : Organization, *in* "Biology of the Integument : 1. Invertebrates" See Ref. 1.
4. W. Gnatzy and F. Romer, Cuticle : Formation, Moulting and Control. *in* "Biology of the Integument : 1. Invertebrates". See Ref. 1.
5. R.H. Hackman, Chemistry of the insect cuticle. *In* "The Physiology of Insecta". M. Rockstein ed., Academic Press, New-York (1974)
6. R.H. Hackman, The estimation of chitin in arthropod cuticles. *In* "Chitin and Chitosan" S. Hirano and S. Tokura eds, Jap. Soc. Chitin and Chitosan (1982)
7. B. Mauchamp and J. Schrevel. Observation en microscopie à Fluorescence de la cuticule des Insectes : une méthode faisant appel aux propriétés spécifiques de la WGA vis-à-vis des glycoconjugués de la Chitine. *C.R. Acad. Sci. Paris* 285 1107 (1977)
8. M. Horrisberger and M. Vonlanthen, Localization of mannan and chitin on thin sections of budding yeasts with gold markers. *Arch. Microbiol.* 115 1. (1977)
9. M. Truchet and B. Mauchamp. Characterization of the insect cuticle layers in histological sections by laser Raman spectrometry. In this volume
10. B. Surholt. Studies *in vivo* and *in vitro* on chitin synthesis during the larval adult moulting cycle of the migratory *Locusta migratoria. J. Comp. Physiol.* 102, 135 (1975)
11. H. Zaluska. Glycogen and chitin metabolism during development of the silkworm (*Bombyx mori* L.) *Acta Biol. exp. Wars.* 19, 339 (1959)
12. H. Oberlander, C.E. Leach, and C. Tomblin. Cuticle deposition in imaginal disks of three species of Lepidoptera : effects of ecdysones *in vitro*. *J. Insect Physiol.* 19, 192 (1973)
13. E.P. Marks, B.A. Sowa. Cuticle Formation *in vitro*. *In* "The Insect Integument". H.R. Hepburn ed. Elsevier. Amsterdam (1976)
14. E. Cohen and J.E. Casida, Properties of *Tribolium* gut chitin synthetase. *Pest. Biochem. Physiol.* 13, 121 (1980)
15. R.T. Mayer, A.C. Chen and J.R. DeLoach, Characterization of a chitin synthase from the stable fly. *Insect Biochem.* 10, 549. (1980)
16. M.N. Horst, The biosynthesis of crustacean chitin by a microsomal enzyme from larval brine shrimp. *J. Biol. Chem.* 256, 1412 (1981)
17. J. Ruiz-Herrera, E. Lopez and S. Bartnicki-Garcia. Properties of chitin synthetase in isolated chitosomes from yeast cell of *Mucor rouxii*. *J. Biol. Chem.* 252, 3338 (1977)

EFFECT OF DIFLUBENZURON ON THE CUTICLE OF CRAB LARVAE

M. E. CHRISTIANSEN

Zoological Museum, University of Oslo
N-0562 Oslo 5, Norway

INTRODUCTION

During the last few years the chitin synthesis inhibitor diflubenzuron (Dimilin[R], TH 6040) [1-(4-chlorophenyl)-3-(2,6-difluorobenzoyl)-urea] has been used for the control of a few insect pests. Marx (1) indicated that inhibitors of chitin synthesis might not be totally specific for insect pests, but could also affect other arthropods including such economically important crustaceans as crayfish, crab, lobster and shrimp.

It has been shown that diflubenzuron reduces populations of mayflies and water fleas in field tests (2), kills crustaceans, especially water fleas, clam shrimps, and tadpole shrimps, in laboratory tests (3), and reduces reproduction in brine shrimps (4). This has led Nimmo et al. (5) to test the effects of diflubenzuron on the small estuarine crustacean, _Mysidopsis bahia_, a vital link between primary producers and consumers in estuarine and marine food webs. Their results show that diflubenzuron is acutely toxic to the mysid in concentrations of a few ppb (µg/l).

Diflubenzuron has also been shown to be effective on marine crustaceans such as the Mediterranean shore crab (6, 7), barnacles (8), and copepods (9) and Kasyanov and Costlow (10) have observed that this insecticide disturbs larval moulting in horseshoe crabs.

DIFLUBENZURON IS TOXIC TO CRAB LARVAE

In recent years laboratory experiments studying the effects of diflubenzuron on crab larvae have been carried out by Christiansen et al. (11), Forward and Costlow (12), Costlow (13), Christiansen and Costlow (14, 15) and Christiansen et al. (16). The results show that crab larvae are severely affected by the insecticide.

It appears that concentrations of 10 ppb diflubenzuron or less - levels which are lethal to several insect pests (e.g. salt marsh mosquitoes) - are also lethal to the larvae of four different crab species investigated. The larvae of all four species live, at least during some part of their development, in estuarine and coastal waters along the east coast of the United States, often adjacent to the salt marsh areas where salt marsh mosquitoes occur (for _Rhithropanopeus harrisii_ (Gould) and _Sesarma reticulatum_ (Say) see Christiansen et al. (11), for the commercial blue crab Callinectes sapidus Rathbun and the commercial stone crab Menippe mercenaria (Say) see Costlow (13)).

Since larvae of the estuarine crab R. harrisii are easy to culture, and survival from hatching to metamorphosis is high when larvae are kept in optimal conditions, this species has been used in various experiments to determine the effects of diflubenzuron on crustacean larvae. When larvae of R. harrisii were exposed to 0.5, 1, 3, 5, 7 and 10 ppb diflubenzuron from hatching, mortality increased in tact with increasing concentrations of the insecticide (Fig. 1). In another experiment, all larvae treated with 10 ppb diflubenzuron during the entire intermoult period of each of the 4 zoeal stages (each stage lasts for ca. 3 days at 25°C, 20‰ salinity) died during moulting to the next stage (Table 1). A high percentage of the larvae died during ecdysis, or did not survive through the remaining zoeal stages, when first zoeal larvae were exposed to 10 ppb diflubenzuron on the same day as they were expected to moult. This is when preecdysial cuticle is deposited.

The mode of dying resembled that described by many authors for insect larvae. The crab larvae did not seem to be affected by the insecticide before moulting to the succeeding stage had started. They were, however, unable to complete moulting, and most of them died with their exuviae partially shed. Many larvae in the lethal and sublethal concentrations which

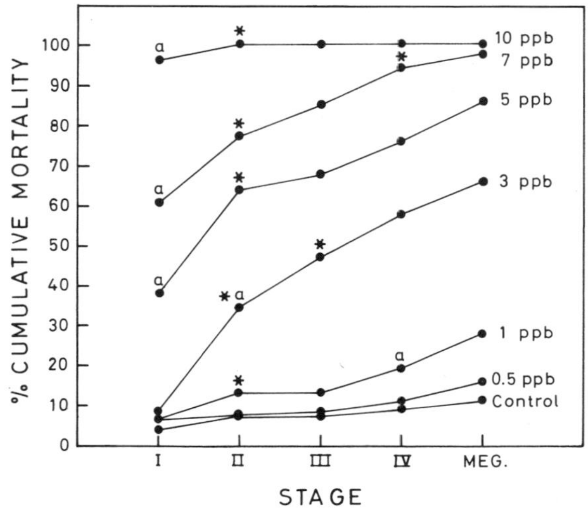

Fig. 1. Rhithropanopeus harrisii. Cumulative mortality during stages of development in various concentrations of diflubenzuron. a: significantly higher than control, includes the marked and also all the succeeding stages within the same concentration; *: significant increase from previous stage; I-IV = zoeal stages; MEG. = megalopa stage. (From Christiansen et al. (11).)

Table 1. Rhithropanopeus harrisii. Survival (%) to succeeding stage of larvae exposed to 10 ppb diflubenzuron during 1st, 2nd, 3rd or 4th zoeal stage. (From Christiansen et al. (11).)

Larvae exposed during	No. of repli- cates	Control		Exposed to 10 ppb	
		Initial no.	Survival (%)	Initial no.	Survival (%)
1st zoea	3	120	97.5	120	0
2nd zoea	3	146	98.6	150	0
3rd zoea	3	144	99.3	139	0
4th zoea	4	167	98.8	170	0

managed to moult were abnormal with deformed swimming setae and deformed rostral and dorsal spines. They were not able to swim normally, and most never reached the last larval or the first crab stage.

DIFLUBENZURON INTERFERES WITH CUTICLE FORMATION IN CRAB LARVAE

Ultrastructural Studies

The ultrastructure of crab larval cuticle during the moult cycle has been studied by Christiansen and Costlow (15). First zoeal larvae of Rhithropanopeus harrisii were used as test organisms, and the cuticle of control larvae and larvae exposed to 10 ppb diflubenzuron were compared. Normally, the epicuticle, which probably consists of six different regions, and the exocuticle are deposited before ecdysis (in first zoea before hatching). The endocuticle is laid down after ecdysis, and the secretion of this layer seems to start during or just after ecdysis.

Distinct differences in the structure of both endocuticle and exocuticle were observed between control larvae and diflubenzuron-treated larvae. Endocuticular secretion was inhibited when larvae were exposed to the insecticide (Fig. 2). Some endocuticular lamellae were laid down after the exposure had started, but this part of the cuticle was deformed (Fig. 3). Formation of the epicuticle was not disturbed by diflubenzuron, whereas the structure of the new exocuticle in exposed larvae was different from that in the control larvae (Fig. 4). This indicates that diflubenzuron inhibits chitin synthesis somehow or another, and that the epicuticle of crab larvae probably does not contain chitin, similar to what has been suggested by a number of authors for many arthropods. The disorganized appearance of the exocuticle may be due to lack of chitin filaments.

These results are similar to those found for insect larvae. Both Mulder and Gijswijt (17) and Turnball et al. (18) observed abnormalities in the endocuticle layer of insect larvae exposed to diflubenzuron, and Deul et al. (19) demonstrated complete inhibition of chitin synthesis in the cuticle of insect larvae within 15 min of application of diflubenzuron.

Incorporation Studies

The rate of chitin synthesis at different stages of the moult cycle has been studied in in vivo experiments on first zoeal larvae of the crab Rhithropanopeus harrisii (16). Incorporation of ^{3}H-labelled glucose and

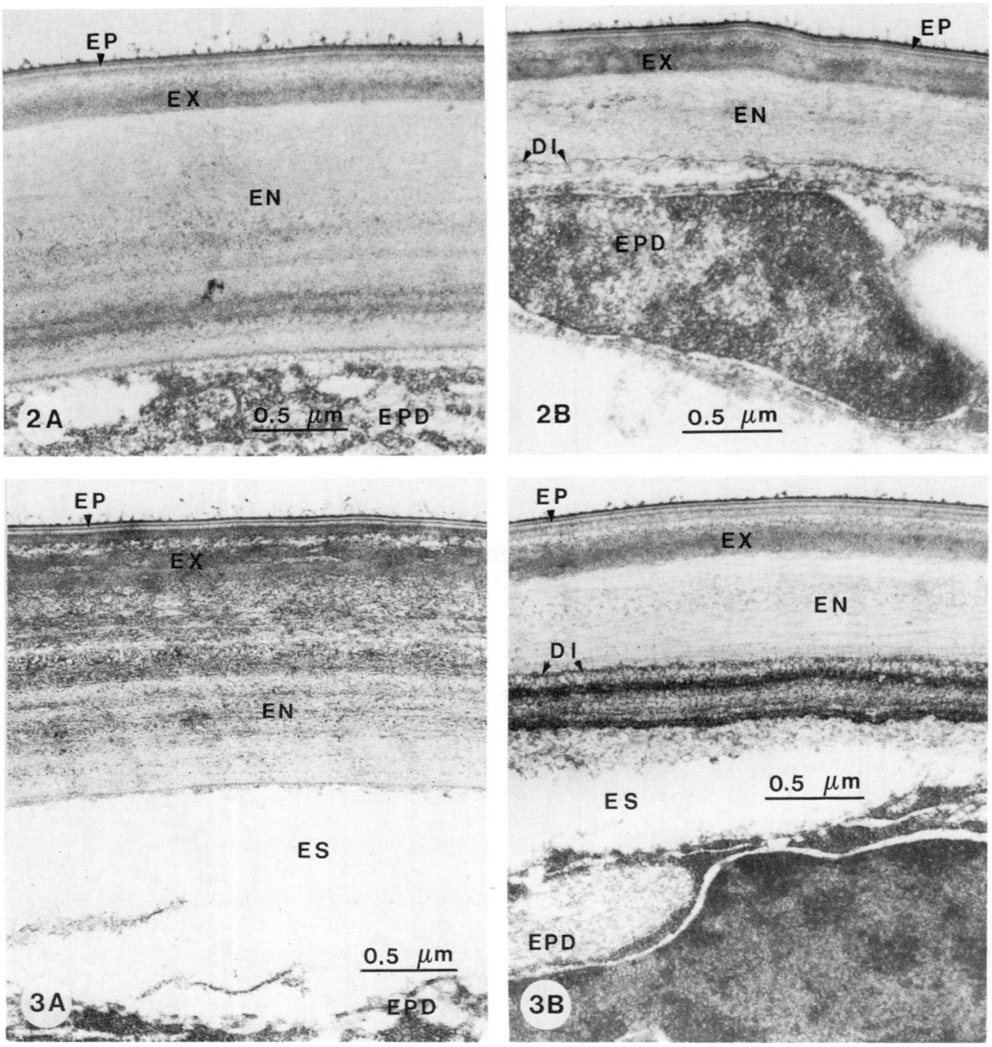

Figs. 2-3. Rhithropanopeus harrisii. Electron micrographs of transverse sections through the proximal part of the dorsal spine close to carapace. - 2A. Control larva between 26 and 38 h old. - 2B. Larva exposed to diflubenzuron for 24 h, same age as the larva in 2A; addition of endocuticular lamellae has barely taken place since exposure to the insecticide was started. - 3A. Control larva between 2 and 2.5 days old, in the apolysis stage. - 3B. Larva exposed to diflubenzuron for 48 h, same age as the larva in 3A; endocuticular secretion has taken place since exposure was started, but the structure of the inner part of the cuticle is different from that of the control. DI: probably when endocuticle formation was disturbed by diflubenzuron; EN: endocuticle; EP: epicuticle; EPD: epidermis; ES: exuvial space; EX: exocuticle. (From Christiansen and Costlow (15).)

^{3}H-labelled N-acetylglucosamine, GlcNac, both precursors to chitin, into the cuticle has been examined both in control larvae and larvae treated with 10 ppb diflubenzuron.

In the control larvae, the incorporation of both precursors was greatly dependent on the moult cycle. It was high during secretion of endocuticle, and low during the intermoult stage after secretion of the cuticle is completed. In larvae treated with diflubenzuron, the incorporation rate of glucose into chitin was particularly low when endocuticle was produced. GlcNac incorporation was also reduced in this stage, but not as much as for glucose (Table 2).

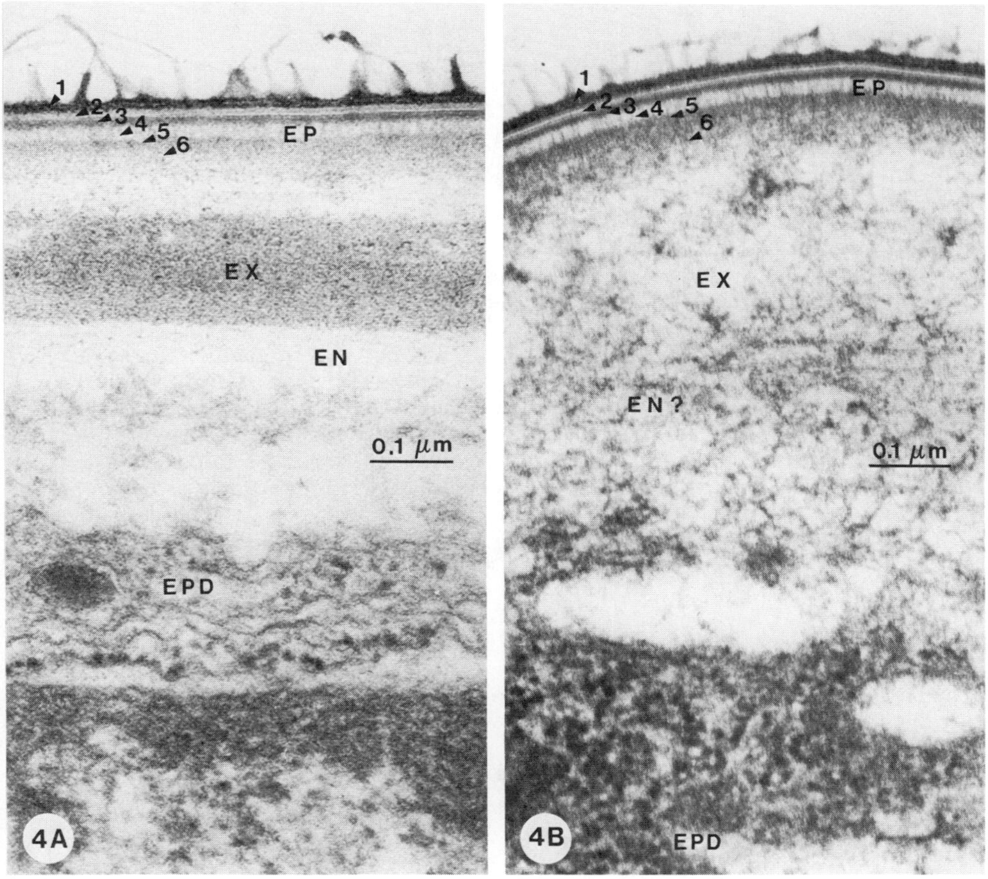

Fig. 4. Rhithropanopeus harrisii. Electron micrographs of transverse sections through the proximal part of the dorsal spine close to carapace. - 4A. Control larva about 3 days old and recently moulted to 2nd zoeal stage; epi- and exocuticle are complete and endocuticle secretion has begun. - 4B. Larva exposed to diflubenzuron for about 60 h, same age as the larva in 4A and fixed in the process of trying to emerge from its exuvium; structure of the epicuticle is similar to the control, whereas structure of the remaining part of the new cuticle, which may be exo- and endocuticle or only swollen exocuticle, is different from the control. 1-6 indicate the six different regions of the epicuticle. Abbreviations: see Figs. 2-3. (From Christiansen and Costlow (15).)

Table 2. *Rhithropanopeus harrisii*. Incorporation of tritiated glucose and tritiated GlcNac into the cuticle of control larvae and diflubenzuron-treated larvae in postmoult and intermoult stages. DPM/larva: decays per min per larva. (Data from Christiansen et al. (16).)

Stage	Control DPM/larva		Diflubenzuron DPM/larva	
	Glucose	GlcNac	Glucose	GlcNac
Postmoult	1720	1573	103	850
Intermoult	393	407	113	110

Autoradiographic studies (16) carried out in the premoult stage when exocuticle is produced, showed that diflubenzuron treatment markedly affected the incorporation of both GlcNac and glucose into the cuticle of crab larvae. As expected, the autoradiographs also showed that in the control larvae, GlcNac was a more specific precursor of cuticle than glucose in the premoult stage.

Marks et al. (20) have summarized hypotheses proposed in later years for the mode of action of benzoylphenylureas. In their opinion, diflubenzuron appears to inhibit a number of enzyme systems in different insects, and its effects in some cases on chitin synthesis may be secondary. The results reported on crab larvae indicate that diflubenzuron affects crabs in a way similar to that suggested by Marks et al. for insects.

CONCLUSIONS

It is obvious from studies referred to in this article that diflubenzuron affects a number of non-target arthropods. Whatever its mode of action, the effects of diflubenzuron appear to be similar in both insects and crustaceans. Therefore, extreme care must be taken when diflubenzuron or other chitin synthesis inhibitors are used for insect control in areas where crustaceans also occur. Otherwise, as Muzzarelli has concluded (21) a generalized use of insecticides that inhibit chitin synthesis could possibly lead to enormous ecological disequilibria with consequences for the feeding, metabolism, growth, reproduction and survival of countless organisms whose existence should not be indiscriminately jeopardized.

REFERENCES

1. J. L. Marx, Chitin synthesis inhibitors: new class of insecticides, *Science* 197:1170 (1977).
2. T. Miura and R. M. Takahashi, Effects of the IGR, TH 6040, on nontarget organisms when utilized as a mosquito control agent, *Mosquito News* 35:154 (1975).
3. T. Miura and R. M. Takahashi, Insect developmental inhibitors. Effects of candidate mosquito control agents on nontarget aquatic organisms, *Envir. Ent.* 3:631 (1974).

4. P. A. Cunningham, Effects of Dimilin (TH 6040) on reproduction in brine shrimp, Artemia salina, Envir. Ent. 5:701 (1976).
5. D. R. Nimmo, T. L. Hamaker, J. C. Moore, and C. A. Sommers, Effect of diflubenzuron on an estuarine crustacean, Bull. envir. Contam. Toxicol. 22:767 (1979).
6. H. Cardinal, G. Vernet, and G. Sinegre, Quelques effets d'un inhibiteur de croissance: le diflubenzuron sur un crabe Carcinus mediterraneus (Czerniavsky), C. R. Soc. Biol. 173:1105 (1979).
7. H. Cardinal and G. Vernet, Perturbations physiologiques du crabe Carcinus mediterraneus en présence de diflubenzuron, C. R. Soc. Biol. 176:210 (1982).
8. G. Gulka, C. M. Doscher, and N. Watabe, Toxicity and molt-accelerating effects of diflubenzuron on the barnacle, Balanus eburneus, Bull. envir. Contam. Toxicol. 25:477 (1980).
9. P. A. Tester and J. D. Costlow, Jr., Effect of insect growth regulator Dimilin^{R} (TH 6040) on fecundity and egg viability of the marine copepod Acartia tonsa, Mar. Ecol. Prog. Ser. 5:297 (1981).
10. V. L. Kasyanov and J. D. Costlow, Effect of insecticide Dimilin on larval molting in the horse-shoe crab, Limulus polyphemus, Biol. Morya 1:46 (1984).
11. M. E. Christiansen, J. D. Costlow, Jr., and R. J. Monroe, Effects of the insect growth regulator Dimilin^{R} (TH 6040) on larval development of two estuarine crabs, Mar. Biol. 50:29 (1978).
12. R. B. Forward, Jr. and J. D. Costlow, Jr., Sublethal effects of insect growth regulators upon crab larval behavior, Wat. Air Soil. Pollut. 9:227 (1978).
13. J. D. Costlow, Effect of Dimilin^{R} on development of larvae of the stone crab Menippe mercenaria, and the blue crab, Callinectes sapidus, in: Marine pollution: functional responses, W. B. Vernberg, A. Calabrese, F. P. Thurberg, and F. J. Vernberg, eds., Academic Press, New York (1979).
14. M. E. Christiansen and J. D. Costlow, Jr., Persistence of the insect growth regulator Dimilin^{R} in brackish water: a laboratory evaluation using larvae of an estuarine crab as indicator, Helgoländer Meeres-unter. 33:327 (1980).
15. M. E. Christiansen and J. D. Costlow, Jr., Ultrastructural study of the exoskeleton of the estuarine crab Rhithropanopeus harrisii: effect of the insect growth regulator Dimilin^{R} (diflubenzuron) on the formation of the larval cuticle, Mar. Biol. 66:217 (1982).
16. M. E. Christiansen, E. Gosling, and M. A. Williams, Effect of the insect growth regulator diflubenzuron (Dimilin^{R}) on the uptake of glucose and N-acetylglucosamine into the cuticle of crab larvae, Mar. Biol. 83:225 (1984).
17. R. Mulder and M. J. Gijswijt, The laboratory evaluation of two promising new insecticides which interfere with cuticle deposition, Pestic. Sci. 4:737 (1973).
18. I. F. Turnbull, N. A. Pyliotis, and A. J. Howells, The effects of DOPA decarboxylase inhibitors on the permeability and ultrastructure of the larval cuticle of the Australian sheep blowfly, Lucilia cuprina, J. Insect Physiol. 26:526 (1980).
19. D. H. Deul, B. J. de Jong, and J. A. M. Kortenbach, Inhibition of chitin synthesis by two 1-(2,6-disubstituted benzoyl)-3-phenylurea insecticides. II, Pestic. Biochem. Physiol. 8:98 (1978).
20. E. P. Marks, T. Leighton, and F. Leighton, Modes of action of chitin synthesis inhibitors, in: Insecticide mode of action, J. R. Coats ed., Academic Press, New York (1982).
21. R. A. A. Muzzarelli, Biochemical modifications of chitin, in: The insect integument, H. R. Hepburn ed., Elsevier, Amsterdam (1976).

TOXICITY OF DIFLUBENZURON TO NON-TARGET ORGANISMS

R. A. A. MUZZARELLI

Faculty of Medicine, University of Ancona
I-60100 Ancona, Italy

Farm animals, birds and arthropods might suffer from exposure to chitin synthesis inhibitors, in terms of biochemical damage, chemical intoxication and, for arthropods, chitin synthesis inhibition. Thus, the aim of a number of studies has been to assess the toxicity of diflubenzuron to non-target insects and animals (1, 2).

This paper attempts to survey these studies. The effects of diflubenzuron on the honey bee Apis mellifera have been studied by Stoner and Wilson (3) and by Baulm (4). For 10 days they feld the insecticide at rates of 0.59, 5.9 and 59 μg/g in sucrose syrup to nurse bees in laboratory cages. A reduction of sealed brood occurred only when the bees were fed 59 μg/g diflubenzuron (5, 6). Barker and Waller (7) fed diflubenzuron at rates of 100 μg/g in water and 60 μg/g in sucrose syrup to small colonies of bees for 40 days. In both cases, the treated colonies has less brood. Egger (8) observed 1.5 % brood damage to two honey bee colonies 25 days after an Austrian forest was sprayed with diflubenzuron.

On another occasion, diflubenzuron (75 g in 40 l water), was sprayed on spruce stands in Bavaria to control the nun moth Lymantria monacha. Three apiaries within the area were compared with three apiaries outside and the results showed that diflubenzuron caused no damage to bee larvae and that the weight of the bee population was not influenced. Feeding experiments in the field using 1 and 10 μg/g of active agent did not result in larval mortality. Robinson (9) also observed no effects on small 5-frame bee colonies when they were exposed to diflubenzuron sprayed on the test hives and on cotton plots weekly for 8 weeks at various rates.

On the contrary, risks may exist in the case of crops, such as cotton, on which the honey bees forage. For crops treated with diflubenzuron at levels of 10 μg/g, bees could contaminate the broodnest at levels of 1 μg/g or more and brood damage would occur.

Crustaceans and arthropods appear to be potential victims of diflubenzuron, due to their biochemical vulnerability in the early development stages. Histopathological studies on the cirripede crustacean Balanus eburneus indicated that diflubenzuron caused a disruption in exoskeleton development similar to that observed in insects. Barnacles exposed to diflubenzuron (0.75 and 1.00 μg/g) for ten days or longer were delayed in the premolt phase of cuticle secretion (10).

In *Carcinus mediterraneus* a disturbance in the formation of exocuticle during *proecdysis was observed* after having used diflubenzuron. Cardinal and Vernet (11) have demonstrated that while the level of hemolymphatic ecdysteroids is normal among the contaminated animals with regard to the control animals (analogous results being obtained on the level of epidermal chitinases during proecdysis), the number of hemocytes significantly decreased in diflubenzuron-contaminated animals. As hemocytes are the storage organs of glycogen and polysaccharides, the decrease in the number of hemocytes could be responsible for the cuticle perturbation.

Diflubenzuron levels which are toxic to insects (e.g., saltwater mosquitoes) are also toxic to zoeal larvae of crabs (12, 13). *Artemia sp.* exposed to diflubenzuron concentrations greater than 10 ng/g did *not survive* beyond 3 days (14). Diflubenzuron was acutely and chronically toxic to the estuarine mysid *Mysidopsis bahia* at such low levels as 2.1 g/l and 1.24 g/l, respectively (*15, 16). Effects* of diflubenzuron on fecundity and egg viability of the marine copepod *Acartia tonsa* have also been described (17). Ultrastructure of larval cuticle *during the m*olt cycle of the estuarine crab *Rhitropanopeus harrisii* was studied by Christiansen and Costlow (18) in *control larvae as well as* in larvae exposed to 10 ng/g of diflubenzuron. When first zoeal larvae were exposed to ng/g diflubenzuron mortality resulted. Furthermore, distinct differences in structure of the cuticle were found between the two groups of larvae. Both endocuticle and exocuticle deformation appeared in treated larvae, whereas epicuticle formation did not seem to be affected. Several other authors for insect larvae have also shown that diflubenzuron inhibits chitin synthesis in crab larvae (see Christiansen, this Volume).

The effects on birds have also been studied. When De Reede (19) examined the possibility that great tits *Parus major* and tree sparrows *Passer montanus* might be poisoned by the *application* of diflubenzuron to *control woodland* insects, he found no significant effects on the breeding results or on the growth of nestlings. The maximum possible daily intake of diflubenzuron by wild nestlings was estimated at 15 mg/kg bodweight per day for the great tit, 10 mg/kg bodyweight per day for the tree sparrow, and 4 mg/kg bodyweight per day for the red-backed shrike. From these data, and in view of the toxicological dietary feeding tests on other species, it was concluded that the poisoning of birds, after spraying the recommended amounts of diflubenzuron in orchards, is highly improbable.

In other studies, diflubenzuron feeding at levels up to 250 μg/g did not affect the hyaluronic acid content of combs of male broiler or layer-breed chickens (20). Feeding diglubenzuron to growing chickens at the same levels did not affect body, testes, liver, comb or foot weights either. When two breeds of laying hens were fed 10 μg/g diflubenzuron for 15 weeks, no significant effects were observed on body weight gain, egg production, fertility or hatchability (21, 22). These laying hens were placed on the diflubenzuron diet after the onset of production and their reproductive characteristics were measured 11 weeks later. Kubena (23) has concluded that egg production, egg weight, eggshell weight, fertility, hatchability and effects on the progeny were not affected by diflubenzuron at levels up to 250 g/g.

It seems, therefore, that the direct toxic effects of diflubenzuron are limited to species within the phylum Arthropoda. This specificity makes diflubenzuron and related compounds more promising as insect pest control agents than the more broadly acting insecticides now in use. However, its toxicity is not specific to pest populations and the ramification of damage to non-target species must be considered in assessing the benefits gained from its use. For instance, results by Hensen and Garton (24) indicate that the control of the dipteran family Chironomidae would require a 10 μg/l treatment. This concentration would not only control the target population, but would severely reduce and/or eliminate the more sensitive mayflies and stoneflies as well.

Due to differences in their generation times, the dipterans could recover in a couple of months, where as the mayflies and stoneflies would require a one- to two-year recovery period. Therefore, the dipteran populations would regain pest density, while the other more sensitive non-target species would continue to suffer long-term reductions.

While several aquatic field studies have already elucidated the specificity of diflubenzuron (25 - 32), further assessment in natural communities is necessary. These studies should include complex systems which contain many interacting species, representing several trophic levels. Only in this way, can an informed decision be made on the general applicability of chitin synthesis inhibitors. The effects of diflubenzuron and other chitin synthesis inhibitors on fungi (mainly *Streptomyces* and *Mucor*) and yeasts was studied by Lyr and Seyd (33) and by Fischer (34).

REFERENCES

1. R. A. A. Muzzarelli, *in*: "The Insect Integument", H. R. Hepburn, ed., Elsevier, Amsterdam, p. 63 (1976).
2. R. A. A. Muzzarelli, *Inquinamento 16*, 29 (1974).
3. A. Stoner, and W. T. Wilson, *Am. Bee J. 122*, 579:582 (1982).
4. W. Bäuml, *Anz. Schädlingsk. Pflanzens. Umwelts. 55*, 97:101 (1982).
5. J. L. Marx, *Science*, 197:1070 (1977).
6. J. R. Barker, and S. Taber, *Environ. Entromol.*, 6:167 (1977).
7. R. J. Barker, and G. D. Waller, *Environ. Entomol.*, 7:534 (1978).
8. A. Egger, *Centralblatt Gesamte Fortwesen*, 94:65 (1977).
9. F. A. Robinson, *Am. Bee. J.*, 119:193 (1979).
10. G. Gulka, C. M. Gulka, and N. Watabe, *Arch. Environ. Contam. Toxicol.*, 11:16 (1982).
11. H. Cardinal, and G. Vernet, *Comp. Rend. Soc. Biol.*, 176:210 (1982).
12. M. E. Christiansen, J. D. Costlow, and R. J. Monroe, *Mar. Biol.*, 50:29 (1978).
13. J. D. Costlow, *in*: "Marine Pollution: Functional Responses", W. B. Vernberg, A. Calabrese, F. P. Thurberg, and F. J. Vernberg, eds., Academic Press, New York, 355 (1979).
14. P. A. Cunningham, *Environ. Entomol.*, 5:701 (1976).
15. D. R. Nimmo, T. L. Hamaker, J. C. Moore, C. A. Sommers, *Contam. Toxicol.*, 22:767 (1979).
16. G. Gulka, C. M. Doscher, and N. Watabe, *Bull. Environ. Contam. Toxicol.*, 25:477 (1980).
17. P. A. Tester, and J. D. Costlow, *Mar. Ecol. Progr. Ser.*, 5:297 (1981).
18. M. E. Christiansen, and J. D. Costlow, *Mar. Biol.*, 66:217 (1982).
19. R. H. DeRede, *Agro-Ecosystems*, 7:327 (1982).
20. H. R. Crookshank, B. A. Sowa, L. F. Kubena, G. M. Holman, H. E. Smalley, and R. Morison, *Poultry Sci.*, 57:804 (1978).
21. L. F. Kubena, *Poultry Sci.*, 60:1175 (1981).
22. R. W. Miller, C. Corley, and S. R. Shafelt, *J. Econ. Entomol.*, 69:741 (1976).
23. L. F. Kubena, *Poultry Sci.*, 61:268 (1982).
24. S. R. Hansen, and R. R. Garton, *Arch. Environ. Contam. Toxicol.*, 11:1 (1982).
25. T. Miura, and R. M. Takahashi, *Mosquito News*, 35:154 (1975).
26. M. S. Mulla, W. L. Kramer, and D. R. Barnard, *J. Econ. Entomol.*, 69:285 (1976).
27. C. D. Steelman, J. E. Farlow, T. P. Breand, P. E. Schilling, *Mosquito News*, 35:67 (1975).
28. A. Ali, and M. S. Mulla, *J. Econ. Entomol.*, 70:191 (1977).
29. A. Ali, and M. S. Mulla, *J. Econ. Entomol.*, 70:571 (1977).
30. A. Ali, and M. S. Mulla, *Environ. Entomol.*, 7:21 (1978).

31. A. Ali, and M. S. Mulla, Arch. Environ. Contam. Toxicol., 7:483 (1978).
32. A. E. Colwell, and C. H. Schaeffer, Can. J. Fish Aquat. Sci., 37:632 (1980).
33. H. Lyr and W. Seyd, Zeits. Allgemeine Mikrobiol., 18:721 (1978).
34. P. Fischer, Umschau, 79:218 (1979).

CHITIN INHIBITORS IN PEST MANAGEMENT IN INDIA

M. BALASUBRAMANIAN

Tamil Nadu Agricultural University
Coimbatore - 641 003, India

INTRODUCTION

Increase in the use of conventional pesticides in India has resulted in problems like pesticide resistance, pest resurgence, secondary pest outbreak and hazards to non-target species. This has led to the evaluation of chitin inhibitors as one of the alternate components in integrated pest management.

Among them, diflubenzuron (Dimilin^R) and BAY SIR 8514 have been extensively tested in the laboratory against various agricultural crop pests, and a few studies have been made to evaluate the efficacy of these compounds under field conditions. Results from extensive laboratory studies have shown that the effects of diflubenzuron and BAY SIR 8514 were mainly organismal which included disruption of ecdysis, delayed mortality, ovicidal effect, larvicidal action on different instars and action on pupae and adults. The moult inhibitors had varying biological activities on eggs, larvae and pupae of many of the pests tried. The action on larvae differed depending on stages, concentration and method of feeding. Feeding behaviour was also found to be altered. There was reduction in food consumed, weight gained and percentage of adults emerged.

FIELD EVALUATION OF CHITIN INHIBITORS

The performance of diflubenzuron (Dimilin^R) and BAY SIR 8514 in the control of selected pests on certain crops in the field are summarized in Table I.

Diflubenzuron has been found to control most of the pests satisfactorily at concentrations ranging from 75-300 g a.i/ha or 0.01 % to 0.1 %. The efficacy of BAY SIR 8514 was more or less similar at 100 -300 g a.i/ha. Almost all of the field tests involved the testing of efficacy against lepidopterous pests in the larval stage. There is only one report of a field study with diflubenzuron on the control of a coleopteran pest (15). Successful cases of control involved mostly leaf feeding caterpillars like *Spodoptera litura* F. on cotton (1), groundnut (6, 7), tobacco (12 - 14) and cauliflower (17) and the red hairy caterpillar of groundnut, *Amsacta albistriga* Wlk (8). For a leaf miner pest like *Aproarema modicella* Dev. on groundnut, only 50 - 63.7 % field mortality was reported. Diflubenzuron gave 100 % control of the coconut black headed caterpillar *Opisina arenosella* Walker in 14 days (9). However, a subsequent study (10) revealed that diflubenzuron, when

Table 1: Field efficacy of diflubenzuron and BAY SIR 8514

Crop	Pest	Chemical	Dose	Result	Reference No.
Cotton	Spodoptera litura F.	Diflubenzuron	300 g ai/ha	84.8% control	1
	Pectinophora gossypiella Saund	,,	75 g ai/ha	As effective as carbaryl	2
	Bollworm complex	,,	0.02%	As effective as carbaryl	3
	,,	,,	75 g ai/ha	Control ranged from 40-75%. Efficacy inferior to synthetic pyrethroids	4
	,,	SIR 8514	300 g ai/ha	More effective than diflubenzuron on Earias spp. For Pectinophora control, both were equal in efficacy.	5
Ground-nut	Spodoptera litura F.	Diflubenzuron	0.02%	77.14% reduction in larval population after 144 hours	6
	S.litura	SIR 8514 and diflubenzuron	100-300 g ai/ha	46.14 to 58.1% mortality	7
	Aproaerema modicella Dev.	,,	,,	50 to 63.7% mortality	7
	Amsacta albistriga Wlk.	Diflubenzuron	20 g ai/ha	86.67% larval mortality in 144 hours	8
Coconut	Opisina arenosella Wlk.	,,	20 g ai/palm	100% control in 14-18 days	9
	Parasierola nephantidis Mues. on Opisina arenosella Walker	,,	10-20 g ai/palm	Directly or indirectly affected the development. Adult emergence ranged from 17.35-48.04% only.	10
Sun-flower	Diacrisia obliqua Wlk.	,,	0.02%	As good as 0.05% endosulfan	11
Tobacco	Spodoptera litura F.	,,	100 g ai/ha	Control though as good as chlorpyriphos, less cost-effective than	12
		Triflumuron	125 g ai/ha	chlorpyriphos	
	S.litura	Diflubenzuron	0.02%	0.02% more effective than 0.01%	13
	S.litura	,,	0.01%	90.66% control after 7 days of spraying. More effective than nuclear polyhedrosis virus.	14

Crop	Pest	Chemical	Dose	Result	Reference No.
Egg Plant	*Epilachna vigintioctopunctata* F.	Diflubenzuron ,,	0.05%	Better than BAY SIR 8514 0.13% and monocrotophos.	15
Ladies Finger	*Earias* spp.	,,	0.015%	Recorded 8.87% incidence as against 20.35% in control. Not as effective as synthetic pyrethroids.	16
Cauli-flower	*Spodoptera litura* F.	,,	250 g ai/ha	100% mortality after 4 days in 2 experiments	17
Lab lab	*Heliothis armigera* Hb.	,,	1000 ppm	Recorded 3% pod damage as against 28.3% in control. As effective as endosulfan 0.07%.	18
Tea	*Buzura suppressaria* Guenee	,,	0.0625%	99.2% mortality after 7 days As effective as 0.125%.	19

applied at 20 g a.i/palm for the control of O. arenosella, affected the development of the larval parasite Parasierola nephantidis Mues.

Several investigators have tried diflubenzuron as well as BAY SIR 8514 for the control of bollworms on cotton. Diflubenzuron has been reported to be as effective as carbaryl in the control of bollworms (2, 3). BAY SIR 8514 was more effective on Earias spp. than diflubenzuron (5). However, diflubenzuron was not comparable to the superior efficacy of the synthetic pyrethroids (4).

PROBLEMS AND PROSPECTS

One of the major problems faced with the field use of chitin inhibitors is the absence of a quick knock-down effect. In most of the cases studied, appreciable mortality was recorded only four or six days after application. However, better results could be achieved by properly timing the application. The chitin inhibitors have been reported to produce ovicidal effects (20, 21) and, by synchronizing the application of the chemical with peak brood emergence and oviposition periods, better results could be achieved. When the chemical is directed against the larval population, the timing of application of the chemical should be such that it coincides with the presence of a susceptible stage of the target insect, i. e. in active feeding. Another important factor to be borne in mind is the cost-effectiveness. On tobacco for instance, control of S. litura with diflubenzuron at 100 g a.i/ha was less cost-effective than chlorpyriphos (12). In view of the adverse effect of diflubenzuron on the parasites (10), the use of such compounds in an integrated pest management system, where parasites play a prominent part, would be very critical. In such cases, a knowledge on the host-parasite population ecology would help to identify an appropriate time for the application of the chitin inhibitors to control the pest without adversely affecting the parasite population.

The prospect of integrating the use of chitin inhibitors with other non-chemical methods is bright. For S. litura control, for instance, the nuclear polyhedrosis virus (NPV) was found to be promising (14). Studies on the compatibility of diflubenzuron or other moult inhibitors with the NPV, as well as its combined effect on the pest, would yield valuable information.

Further studies on the efficacy of different formulations and the effect of synergists and adjuvants on the toxicity of the moult inhibitors are necessary. It would be desirable to have a compound with the spectrum of activity limited to target pest species so that the harmful effects on parasitic insects could be avoided. The use of moult-inhibitors should be such that they would fit into a cost-effective integrated pest management system.

REFERENCES

1. Anonymous, Final report on biological testing of the proprietory product dimilin 25 % WP, Department of agriculture, Directorate of agriculture, Madras (1981).
2. Anonymous, Annual report of all India coordinated cotton improvement project, Hissar, Haryana agricultural University, Hissar, India (1980).
3. Anonymous, Studies on the efficacy of synthetic pyrethroids and other insecticides against bollworms on cotton, Report of the department of arl. entomology, Mahatma phule agrl. university, Rahuri, (1980).
4. M. W. Bhamburkar and T. V. Kathane, Role of synthetic pyrethroids in the control of bollworms in relation to yield of H4 cotton under rainfed condition. Pestology, 8: 9 (1984).
5. N. Natarajan, T. Kumarasamy and M. Balasubramanian, Efficacy of moult inhibitor BAY SIR 8514 on cotton pests under field conditions. Abstract of papers: Seminar on behavioural and physiological approaches in the management of crop pests, Tamil nadu agricultural University, Coimbatore, June 21 - 23, p. 41 (1984).

6. R. Natesan and M. Balasubramanian, Efficacy or diflubenzuron (inhibitor of chitin deposition) in the control of tobacco caterpillar Spodoptera litura, Indian J. Plant Prot. 7:141 (1979).
7. N. Natarajan, T. Kumarasamy and M. Balasubramanian, Field efficacy of BAY SIR 8514 on groundnut pests. Abstract of papers: Seminar on behavioural and physiological approaches in the management of crop pests, Tamil nadu agricultural University, Coimbatore June 21 - 23 p. 42 (1984).
8. R. J. Rabindra and M. Balasubramanian, Diflubenzuron and some new insecticides for the control of groundnut red hairy caterpillar Amsacta albistriga Walker. Abstract of papers: Seminar on behavioural and physiological approaches in the management of crop pests, Tamil nadu agricultural University, Coimbatore, June 21 - 23, p. 43 (1984).
9. V. T. Sundaramurthy, Effect of diflubenzuron on the field population of coconut black headed caterpillar Nephantis serinopa Meyrick (Lepidoptera: Gelechidae) and its parasite Parasierola nephantidis (Muesebeck) (Hymenoptera: Bethylidae) in India. Bull. Ent. Res., 70:25 (1980).
10. V. T. Sundaramurthy and S. Jayaraj, Effect of moult inhibitor on the parasitizing behaviour of Parasierola nephantidis. Abstract of papers: Seminar on behavioural and physiological approaches in the management of crop pests, Tamil nadu agricultural University, Coimbatore, June 21 - 23, p. 43 (1984).
11. Anonymous, Studies on the comparative efficacy of some insecticides against Bihar hairy caterpillar Diacrisia obliqua on sunflower. Report of dept. of entomology, Mahatma phule agrl. University, Rahuri, (1980).
12. M. S. Chari, T. M. Bharpoda and S. N. Patel. Studies on integrated management of Spodoptera litura Fb. in tobacco nursery. Paper presented at the Vth National symposium on tobacco, Rajamundry, Feb. 15 - 17 (1984).
13. R. Natesan and M. Balasubramanian, Effect of diflubenzuron against tobacco caterpillar, Spodoptera litura on tobacco, Nicotiana tabacum L. Madras Agric. J., 66:799 (1979).
14. G. Santharam and M. Balasubramanian, Note on the control of Spodoptera litura on tobacco with nuclear polyhedral virus and diflubenzuron. Indian J. Agric. Sci., 50:726 (1980).
15. B. Rajendra Kumar, Studies on the efficacy of certain chitin inhibitors in the control of brinjal pests. M. Sc. (Ag) thesis, Andhra pradesh agrl. University, Bapatla (1982).
16. Anonymous, Control of pest complex of okra by spraying modern insecticides. Report of Dept. of agrl. entomology, Mahatma phule agrl. University, Rahuri, (1981).
17. M. Ramzan and D. Singh, Chemical control of the tobacco caterpillar, Spodoptera litura Fabricius on cauliflower. J. Res. Punjab Agric. Univ., 17:236 (1980).
18. K. Mohanvel, Studies on the effects of diflubenzuron (1-(4-chlorophenyl) -3(2-6-Diflubenzoyl) urea) on gram pod borer, Heliothis armigera Hubner (Lepidoptera: Noctuidae). M. Sc. (Ag) thesis, Tamil nadu agrl. University, Coimbatore, (1981).
19. Anonymous, Annual scientific report (1982-83), Tocklai experimental station, Tea research association, Jorhat, p. 51, (1983).
20. M. Balasubramanian, R. Natesan and K. Mohanvel, The effect of diflubenzuron - a chitin inhibitor on Spodptera litura and Heliothis armigera. Andhra Agric. J., 27:74, (1980).
21. S. Chockalingam, and A. Noorjahan, Ovicidal effect of diflubenzuron on hemipterous bugs, Dysdercus cingulatus and Chrysocoris purpureus. Curr. Sci., 53:112, (1984).

MODE OF INHIBITION OF CHITIN SYNTHESIS BY DIFLUBENZURON

T. MITSUI

The Institute of Physical and Chemical Research Hirosawa, Wako Saitama 351-01, Japan

INTRODUCTION

Diflubenzuron and related benzoylphenyl ureas have been found to inhibit chitin synthesis in insect integuments and midguts but the mode of action of the insecticides still remains unknown. Diflubenzuron was reported to inhibit chitin synthetase from brine shrimp (1), Lucilia cuprina and two other species of Dipteran insects (2). However, the insecticide was neither an in vitro inhibitor of the integumental chitin synthetase in Trichoplusiani and Hyalophora cecropia (3) nor the whole-body or midgut enzyme preparations in Tribolium castaneum (4), Stomoxys calcitrans (5, 6) and Mamestra brassicae (7, 8). Other hypotheses have been proposed including an inhibition of a proteolytic enzyme required for activation of the enzyme from a zymogen form (9) and an inhibition via active metabolites or as a blocking agent by direct binding to a non-chitin synthetase site important in chitin polymerization and fibrillogenesis (4). We proposed another mechanism (10, 11), that the insecticide may inhibit UDP-N-acetylglucosamine transport across the microvilli membranes in the midgut epithelial cells.

Inhibition of chitin synthesis by diflubenzuron and polyoxin D in the peritropic membranes of the insect midguts

The precursors of chitin (acetylglucosamine and UDP-N-acetylglucosamine) and the inhibitors (diflubenzuron and polyoxin D) were applied in vitro to the midguts of the newly molted last instar larvae of the cabbage armyworm, Mamestra brassicae, following the three application methods shown in Fig. 1.

The inhibition of chitin synthesis by both inhibitors in the peritrophic membranes is summarized in Table I. Chitin synthesis from ^{14}C-acetylglucosamine (^{14}C-GlcNAc) in the peritrophic membranes was inhibited by both inhibitors, when they were applied either inside or outside the midguts. When ^{14}C-UDP-GlcNAc was applied outside the midguts, diflubenzuron inhibited chitin synthesis, although the total amount of chitin formed was about 1/20 as much as the amount formed from ^{14}C-GlcNAc. When ^{14}C-UDP-GlcNAc was treated, however, inside the midguts, diflubenzuron did not block chitin synthesis whereas polyoxin D inhibited it. To confirm the result, the midguts were turned inside out and incubated in Grace's medium containing 0.05 μCi/ml of ^{14}C-UDP-GlcNAc with or without inhibitors. Chitin synthesis was inhibited by polyoxin D but not by diflubenzuron.

Table I. Inhibition of chitin synthesis by diflubenzuron and polyoxin D in the peritrophic membrane of the midguts from Mamestra brassicae larvae in vitro, by the application methods shown in Fig. 1.

Inhibitor (μg/ml)	Inhibition (%) Inside GlcNAc	Inside UDP-GlcNAc	Outside GlcNAc	Outside GlcNAc	Inside-outside GlcNAc
Diflubenzuron (1 μg/ml)	78	0	87	71	82
Polyoxin D (5 g/ml)	80	74	83	--	74

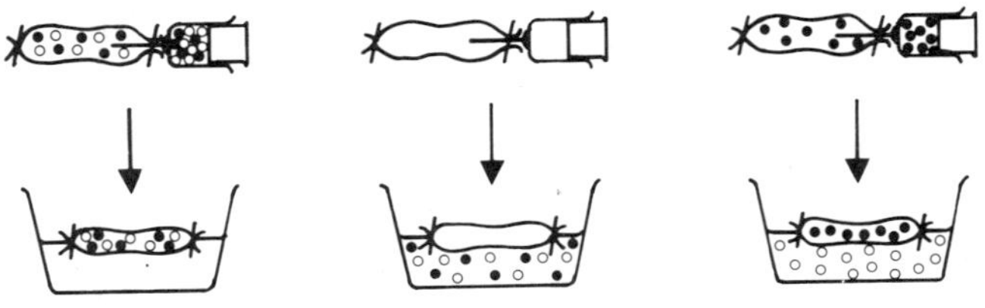

Fig. 1. Application methods of the inhibitors and the precursors to the ligated midguts. (a) inside; (b) outside; (c) inside-outside; : inhibitor; o: ^{14}C precursor.

Accumulation of ^{14}C-UDP-GlcNAc in the midgut tissues

When the midguts were cultured for 16 h in the Grace's medium containing 0.1 μCi/ml of ^{14}C-GlcNAc in the presence or absence of 1 μg/ml of diflubenzuron, ^{14}C-UDP-GlcNAc was found to accumulate in the diflubenzuron-treated tissues 1.7 times as much as in the untreated tissues (Table II).

Effect of diflubenzuron on the synthesis of phospholipids

The midguts were cultured in vitro for 16 h in Grace's medium in the presence of 0.1 μCi/ml of ^{14}C-acetic acid with or without diflubenzuron. Phospholipids were extracted by the method of Bling and Dyers (12) and separated by thin layer chromatography. Neither qualitative nor quantitative differences were observed between the diflubenzuron-treated and the untreated tissues.

Effect of diflubenzuron on the trasport of UDP-GlcNAc across the microvilli membranes of midgut epithelial cells

The midguts were turned inside out by using a pin and a forceps. One end was ligated with cotton thread, 0.075 μCi ^{14}C-GlcNAc was injected into it and the other end was also ligated. The sample midguts thus turned were incubated for 2 h in the Grace's medium in the presence or absence of inhibitors. Chitin and UDP-GlcNAc in the medium and the tissues were analy-

Table II. Accumulation of ^{14}C-UDP-GlcNAc in the midguts of final instar Mamestra brassicae larvae.

Inhibitor	Concentration (μg/ml)	^{14}C-UDP-GlcNAc formed (pmol)	^{14}C-Chitin formed (pmol)
Control	--	222	158
Diflubenzuron	1.0	367	4

zed separately. The amount of ^{14}C-GlcNAc was about 3 times higher in the control and the polyoxin D-treated medium than in the diflubenzurontreated medium. The reason why the amount of ^{14}C-UDP-GlcNAc in the control medium is small in comparison with the polyoxin D-treated medium may be that the isolated ^{14}C-UDP-GlcNAc was caught by chitin synthetase, producing chitin as a result. The amount of radioactivity incorporated into chitin in the polyoxin D and diflubenzuron-treated tissues was about 1/3 as much as in the control tissues (Table III). Thus, the total amount of radio-active compound(s) transported across the microvilli membranes in the diflubenzuron-treated tissues was about 1/4 of the control tissues. On the other hand, ^{14}C-UDP-GlcNAc in the tissues was found again to accumulate in the diflubenzuron-treated tissues 1.6 times as much as in the control tissues.

The binding of UDP-GlcNAc to the midgut brush-border membrane vesicles

The brush-border membranes were collected by scraping the microvilli off the surface of the turned-out midguts, followed by sucrose gradient centrifugation. The 50 % sucrose fraction was rich in alkali phosphatase and bound well to UDP-GlcNAc *in vitro* at pH 7.0. Almost 90 % of the binding of UDP-GlcNAc to the vesicles in this fraction was inhibited by polyoxin D. This suggests that most of the binding might be UDP-GlcNAc and chitin synthetase binding. The other binding of UDP-GlcNAc to the polyoxin D-treated vesicles, optimum pH 7.5, was partially inhibited by diflubenzuron, 50 % inhibition, at $4.2 \cdot 10^{-5}$ M.

DISCUSSION

The midgut is better than the integument as a target to study the mode of inhibition of chitin synthesis by diflubenzuron, because the precursor(s) and inhibitor(s) could be applied either inside (chitin is synthesized in this side) or outside (blood circulation). As shown in Table I, chitin synthesis was not inhibited only when diflubenzuron and UDP-GlcNAc were applied inside the midguts. The result suggests the following 3 possibilities to explain

Table III. Analysis of ^{14}C-chitin and ^{14}C-UDP-GlcNAc in the medium and the midgut tissue after application of ^{14}C-GlcNAc to the turned-out midguts.

Inhibitor (μg/ml)	UDP-GlcNAc in medium (pmol) (a)	Chitin (pmol eq.) (b)	Total (a+b)	UDP-GlcNAc in tissue (pmol)
Control	50.5	52.8	103.3	52.7
Polyoxin D (10)	53.5	11.5	65.0	62.1
Diflubenzuron (2.5)	18.0	8.2	26.2	85.7
Polyoxin D (10)+ Diflubenzuron (2.5)	24.2	8.2	32.4	85.7

the inhibitory effect of diflubenzuron on chitin synthesis. (1) Diflubenzuron blocks UDP-GlcNAc formation from GlcNAc in the epithelial cells. (2) Diflubenzuron blocks synthesis of phosphatidylcholine and other phospholipids which are important components of biomembranes. As a result, diflubenzuron disrupts the membrane function in the same way as IBP, an organo phosphorus fungicide (13). (3) Diflubenzuron blocks the transport of UDP-GlcNAc, formed from GlcNAc in the epithelial cells, across the microvilli membrane of the cells. Therefore, diflubenzuron does not block chitin synthesis when UDP-GlcNAc is directly applied to the outer face of the microvilli. On the other hand, polyoxin D blocks chitin synthesis even when UDP-GlcNAc is applied to the outer face of microvilli, because it inhibits chitin synthetase which seems to be located on the outer face of the biomembranes.

The first two hypotheses could not account for the results observed and were not considered further. To examine the third possibility, a new technique was devised. The midguts were turned inside out, ^{14}C-GlcNAc was injected into the turned-out midguts, incubated for 2 h in the presence or absence of inhibitor(s) and the radio-labelled compounds which appeared across microvilli membranes were analyzed. As shown in Table III, diflubenzuron reduced the amount of the radio-labelled compounds transported across the microvilli to about 1/4 as much as in the control medium. The results lead to the conclusion that the mode of action of diflubenzuron seems to be the inhibition of UDP-GlcNAc transport across the microvilli membranes. The binding experiments using the brush-border membrane vesicles showed that there might be two substances bound to UDP-GlcNAc, one of which might be chitin synthetase. The binding of UDP-GlcNAc to the other substance was partially inhibited by diflubenzuron. However, the amount of UDP-GlcNAc bound to the substance was too small in this experimental condition to conclude that there exists a UDP-GlcNAc transporter in the microvilli membranes.

REFERENCES

1. M. N. Horst, J. Biol. Chem., 256:1412 (1981).
2. I. F. Turnbull and A. J. Howells, Aust. J. Biol. Sci., 36:25 (1983).
3. E. Cohen and J. E. Casida, Pestic. Biochem. Physiol., 17:301 (1982).
4. E. Cohen and J. E. Casida, Pestic. Biochem. Physiol., 13:129 (1980).
5. R. T. Mayer, A. C. Chen and J. R. DeLoach, Insect Biochem., 10:549 (1980).
6. R. T. Mayer, A. C. Chen and J. R. DeLoach, Experientia, 37:337 (1981).
7. T. Mitsui, J. Fukami and L. M. Riddiford in: "Regulation of Insect Development and Behaviour", F. Sehnal, A. Zabza, J. J. Menn, and B. Cymborowski, eds., Sci. Papers Inst. Org. Phys. Chem. Wroclaw Technical University, 22:1131 (1981).
8. T. Mitsui, C. Nobusawa and J. Fukami, J. Pesticide Sci., 6:155 (1981).
9. L. Leighton and E. P. Marks, Science, 213:905 (1981).
10. T. Mitsui, C. Nobusawa, J. Fukami and I. Yamaguchi, J. Pesticide Sci., 9:19 (1984).
11. T. Mitsui, M. Tada, C. Nobusawa and I. Yamaguchi, J. Pesticide Sci., 10:55 (1985).
12. E. G. Bligh and W. J. Dyer, Can. J. Biochem. Physiol., 37:911 (1959).
13. O. Kodama, K. Yamashita and T. Akasuka, Agri. Biol. Chem., 44:1045 (1980).

NEW SYNTHETIC POLYOXIN ANALOGS FOR CHITIN SYNTHESIS INHIBITION

H. A. SMITH, P. SHENBAGAMURTHI, F. NAIDER and J. M. BECKER

Microbiology Department, University of Tennessee
Knoxville, TN, 37996 USA

Chemistry Department, College of Staten Island
Staten Island, NY, 10301 USA

The natural polyoxins are a class of peptidyl nucleoside antibiotics produced by *Streptomyces* and characterized by their ability to compete with UDP-GlcNAc for active site binding to chitin synthetase (Hori, *et al*., 1971 and 1974). As chitin synthetase inhibitors they offer the promise of a new source for clinical antifungal agents. Previous reports have demonstrated that the polyoxins are strong inhibitors of chitin synthetase *in vitro*, effective at μM concentrations against chitin synthetase of *C. albicans* (Becker, *et al*., 1983). However, when assayed for toxicity to zoopathogenic fungi such as *Candida albicans* or *Cryptococcus neoformans* concentrations in the mM range were required to observe toxic effects (Becker, *et al*., 1983).

It is generally believed that the lack of toxicity of the polyoxins to certain fungi is due to the inability of the polyoxins to effectively cross the cellular membrane and accumulate intracellularly to concentrations high enough to inhibit chitin synthetase within the cell. Previous reports from our laboratory have attempted to address the problems associated with poor entry of polyoxin into the fungal cell (Naider, *et al*., 1983 and Shenbagamurthi, *et al*., 1983). To this end we synthesized a series of dipeptidyl and tripeptidyl analogues of uracil polyoxin C. The dipeptidyl analogues were strongly inhibitory to chitin synthetase and at millimolar concentrations were effective in killing the yeast *C. albicans* (Shenbagamurthi, *et al*., 1983). We also reported on the synthesis of a polyoxin prodrug leucyl-norleucyl-uracil polyoxin C (Naider, *et al*., 1983). This compound was a competitive inhibitor of trimethionine uptake and although inactive to chitin synthetase in a membrane preparation was toxic to yeast cells of *C. albicans*.

Our studies indicated that a major difference between active and inactive chitin synthetase inhibitors was the stability of the synthetic compounds to hydrolysis by cellular peptidases. Thus we have chosen chemical modifications designed to prevent the breakdown of polyoxins by the target cell. The analogs synthesized include the following (see Fig. 1): the utilization of D-amino acids within the peptide portion of the molecule (compound 7), N-alkylation of the terminal amine with methyl, propyl or lipophilic groups (compounds 5, 6, and 10, respectively), pyridinylation of amino acid side chains (compound 3),

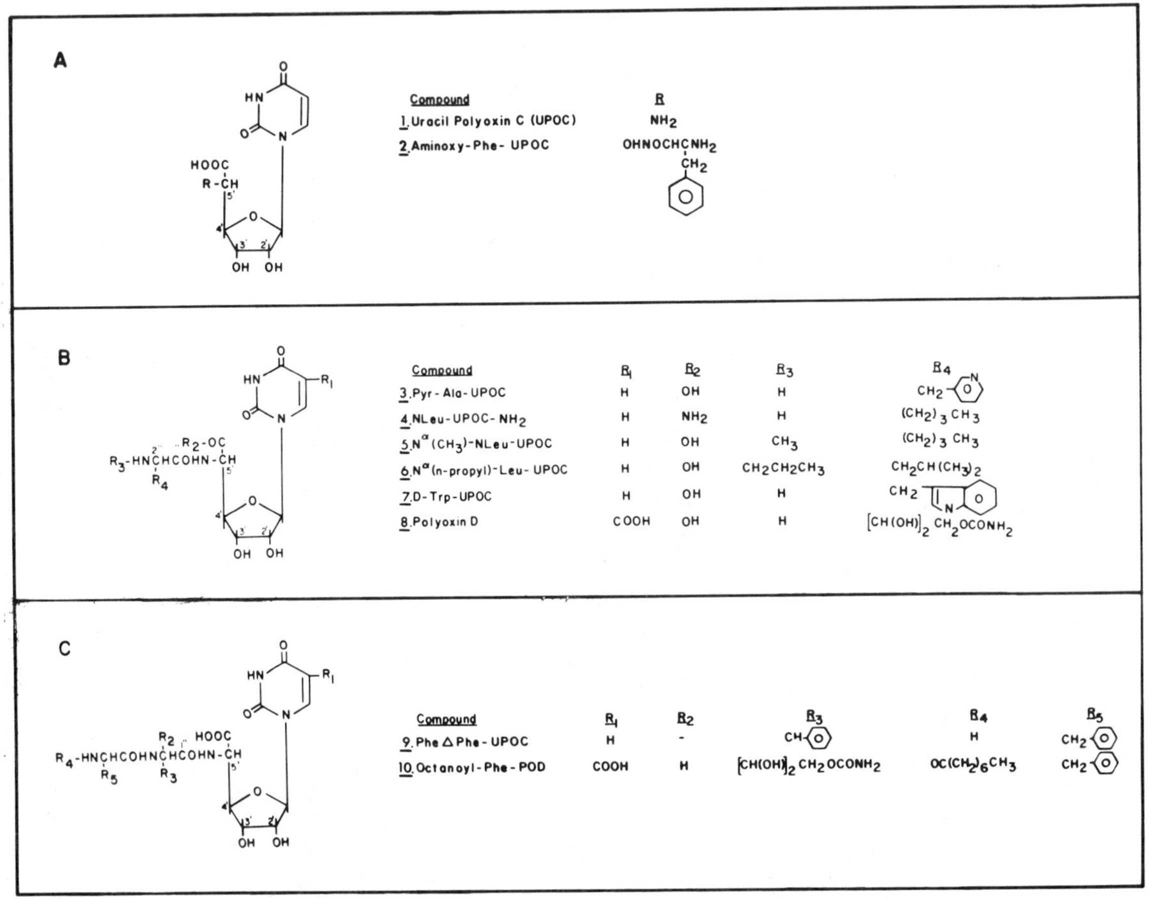

Fig. 1. Structural formulae of polyoxin analogs

inclusion of a terminal amide in place of the unprotected carboxyl group (compound 4), utilization of dehydro amino acids in the peptide portion of the polyoxin (compound 9), and the use of the aminoxy type peptide linkage (compound 2). Included also in Fig. 1 are structural formulas of uracil polyoxin C (UPOC, compound 1), which is the parent compound for all the synthetic derivatives except compound 10, and polyoxin D (compound 8), which is one of the naturally-occurring polyoxins.

The synthetic procedure involved coupling of aminoacid active esters to uracil polyoxin C, subsequent removal of the protecting group and purification by preparative HPLC. A new method for introducing the n-propyl group into the α-amino group of peptides was developed. n-C_7H_{15}CO-Phe-Polyoxin D was synthesised by the stepwise coupling of active esters to Polyoxin D. Synthesis of aminoacyl derivatives of polyoxin C and L have been reported previously (Azuma, _et al_., 1977).

The vast majority of biological studies on polyoxins have utilized polyoxin D. Polyoxin L differs from polyoxin D only in that the 5-carboxyl is replaced with a hydrogen atom (Isono, _et al_., 1969). The compounds that we synthesized were prepared from uridine and are, thus, analogues of polyoxin L. Our first objective was to prepare 1-(5'-amino-5'deoxy-β-D-allofuranosyluronyl)uracil, which we called uracil polyoxin C (UPOC). This intermediate corresponds to the product that would be obtained by alkaline hydrolysis of the dipeptidyl bond of polyoxin L. Starting with 2',3'-O-cyclohexylideneuridine, UPOC was synthesized by the procedure of Damodaran, _et al_. (1971). This intermediate was then used to prepare a variety of analogs by selectively reacting the 5'-amino group of UPOC with the suitable activated form of the amino acids (active ester or azlactone) in DMF at room temperature. The crude deblocked polyoxins were purified on preparative HPLC using CH_3OH:H_2O:TFA as the eluent and the homogeneity of final products were checked by analytical HPLC in two different solvent systems (CH_3OH:H_2O:TFA and CH_3CN:H_2O:TFA). Their K' values were in conformity with their expected hydrophobicities. The yields ranged from 23-30% based on uracil polyoxin C (UPOC).

Utilizing two separate assays for chitin synthetase activity, one employing a mixed membrane preparation (Chiew, _et al_., 1980) and the other digitonin permeabilized cells (Fernandez, _et al_., 1982) we were able to examine the inhibitory effects of polyoxin compounds upon _C. albicans_ chitin synthetase activity in the presence and absence of cellular peptidases. A comparison of the ID_{50} values for both assays reveals that Pyr-Ala-UPOC is active in the mixed membrane assay but without activity in the permeabilized cell (Table 1). This compound was also hydrolyzed by a cell extract at C_1'' peptide bond (Fig. 1, Panel B). The results indicate that pyridinylation of the amine terminal residue side chain does not result in a compound stable to peptidase hydrolysis. Most of the other polyoxin compounds were found to have similar ID_{50} values in both chitin synthetase assays and were insensitive to hydrolysis by cell extract peptidases (Table 1). We were therefore successful in the synthesis of polyoxin derivatives which were resistant to peptidase hydrolysis and which in certain cases exhibited chitin synthetase inhibitory activity.

These results suggest that the requirements of chitin synthetase will not allow the inclusion of D-amino acids in the peptide portion of the polyoxin, although we have not inserted D-amino acids at positions other than C_2'' (Table 1; Chart 1, Panel B). The substitution of D-residue in place of an L-amino acid reduced the chitin synthetase inhibitory activity by over 1500 fold as the K_I values for L-Trp-UPOC and D-Trp-UPOC were 6.8×10^{-7} M and 1×10^{-3} M, respectively.

TABLE 1. Biological activities of synthetic polyoxin derivatives

	Polyoxin Compounds	mmf ID_{50} (M)[a]	Perm. Cell ID_{50} (M)[b]	Hydrolysis[c]	MIC[d]
1.	Uracil Polyoxin C	$> 10^{-4}$	$> 10^{-4}$	-	> 4
2.	Aminoxy-Phe-UPOC	$> 10^{-4}$	$> 10^{-4}$	-	> 3
3.	Pyr-Ala-UPOC	3.3×10^{-5}	$> 10^{-4}$	+	> 4
4.	NLeu-UPOC-NH_2	2.0×10^{-4}	$> 10^{-4}$	+	> 4
5.	$N(CH_3)$NLeu-UPOC	1.2×10^{-4}	9.8×10^{-5}	-	> 4
6.	N(npropyl)Leu-UPOC	$> 10^{-4}$	$> 10^{-4}$	-	> 2
7.	D-Trp-UPOC	$> 10^{-4}$	$> 10^{-4}$	-	> 4
8.	Polyoxin D	1.8×10^{-6}	1.0×10^{-6}	-	0.06
9.	Phe-dehydro-Phe-UPOC	4.0×10^{-5}	8.0×10^{-5}	-	> 2
10.	Octanoyl-Phe-POD	4.75×10^{-5}	5.0×10^{-5}	-	> 1

[a]Molar concentration at which the enzyme in a mixed membrane fraction (mmf) is inhibited by 50%.

[b]Molar concentration at which chitin synthetase is inhibited by 50% in the permeabilized cell assay.

[c]A plus (+) sign indicates compound was hydrolyzed by *C. albicans* cell extract. A minus sign (-) indicates no hydrolysis by cell extract.

[d]Minimum inhibitory concentration (MIC), the lowest concentration (mM) of compound which inhibited growth.

The N-alkylation of NLeu-UPOC by methylation or propylation at R_4, or amidation of the unprotected carboxyl group at R_2 reduced the activity of the parent compound by approximately 100-fold. However, such modifications do confer resistance to peptidase hydrolysis. Replacement of the normal peptide bond structure with (-CONHO-) in the Aminoxy-Phe-UPOC compound resulted in 1000-fold reduction of inhibitory activity to chitin synthetase in comparison to the related compound HomoPhe-UPOC. The Phe-dehydro-Phe-UPOC compound was moderately active against chitin synthetase and also resistant to hydrolysis by cellular peptidases (Table 1). The utilization of dehydro amino acids may therefore be a feasible modification in future synthetic attempts. The Octanoyl-Phe-POD derivative was an active chitin synthetase inhibitor. This compound is also resistant to peptidase hydrolysis. These data suggest that the addition of a lipid soluble group at R_4 (Fig. 1, Panel C) is a synthetic option to consider to confer peptidase resistance and increase the permeability of the compound to the cell. However, Octanoyl-Phe-POD was not an effective inhibitor of growth of *C. albicans* (Table 1).

In a previous study we reported on the morphological alterations and inhibition of growth produced by novel dipeptidyl and tripeptidyl polyoxins synthesized in our laboratory (Naider, *et al.*, 1983 and Shenbagamurthi, *et al.*, 1983). As noted previously Polyoxin D causes severe alterations in the morphology of *C. albicans* (Becker, *et al.*, 1983). For the analogs the minimum inhibitory concentration (MIC) was determined as the lowest concentration of drug which inhibited clearly visible growth. In comparison to Polyoxin D with an MIC equal to 0.06 mM the novel polyoxins had little inhibitory activity with MIC values greater than 1 to 4 mM (Table 1).

A goal of our research has been to synthesize polyoxins which utilize the peptide transport system for cell entry. We were also interested in determining whether the polyoxin compounds were able to utilize the uridine transport to effect cell entry as the polyoxins we synthesized are either dipeptidyl or tripeptidyl uridine nucleoside derivatives. To determine the ability of the polyoxins to enter the cell the initial rate of uptake of either radiolabeled trimethionine or uridine was measured in the presence of a 10-fold concentration of the uptake substrate (Table 2). The results of these experiments indicate that the polyoxin compounds do not utilize either transport system to enter into the cell. As shown in Table 2, non-labeled uridine decreased uridine uptake by 86% and Gly-Met-Gly decreased trimethionine uptake 81% when these competitive inhibitors were present at a 10-fold M excess.

TABLE 2. Effect of polyoxin compounds on uridine and trimethionine transport

	Polyoxin Compounds	Uridine Transport Inhibition (%)[a]	Trimethionine Transport Inhibition (%)[b]
1.	Gly-Met-Gly	0	81
2.	AminoxyPhe-UPOC	-19	15
3.	Pyr-Ala-UPOC	1	3
4.	NLeu-UPOC-NH_2	5	-5
5.	$N(CH_3)$NLeu-UPOC	19	9
6.	N(npropyl)Leu-UPOC	-27	9
7.	D-Trp-UPOC	17	-3
8.	Phe-dehydro-Phe-UPOC	-3	6
9.	Octanoyl-Phe-POD	-12	16
10.	Uridine	86	0
11.	Uracil	-14	0

[a]The value shown is the percent inhibition of the initial rate of transport when the compound was present at a 10-fold M excess with respect to labeled uridine

[b]The value shown is the percent inhibition of the initial rate of transport when the compound was present at a 10-fold M excess with respect to labeled trimethionine.

In conclusion, we have synthesized eight novel polyoxins some of which incorporate peptidase resistant chemical modifications in the peptide portion of the polyoxin molecule. The results indicate that effective chemical modification might incorporate dehydroamino acids or N-acylation with lipophilic groups. Future studies should be directed at retaining this intracellular stability while improving the permeability of the polyoxins. Synthetic efforts toward this goal are presently underway in our laboratory.

ACKNOWLEDGEMENTS

We gratefully acknowledge support for this work from grant A1-14387 and contract NO1-A1-42651 from the National Institute of Allergy and Infectious Diseases, and grant CH-290 from the American Cancer Society.

REFERENCES

1. T. Azuma, T. Saito and K. Isono, Polyoxin analogs. III. Chem. Pharm. Bull. 25:1740 (1977).
2. J.M. Becker, N.L. Covert, P. Shembagamurthy, A.S. Steinfeld and F. Naider, Polyoxin D inhibits the growth of zoopathogenic fungi. Antimicrobial Agents Chemotherapy, 23:926 (1983).
3. Y.Y. Chiew, M.G. Shepherd and P.A. Sullivan, Regulation of chitin synthesis during germ tube formation in C. albicans. Arch. Microbiol. 125:97 (1980).
4. N.P. Damodaran, G.H. Jones and J.G. Moffat, Synthesis of the basic nucleotide skeleton of the polyoxin complex. J. Am. Chem. Soc. 93: 3812 (1971).
5. P. Fernandez, J.U. Correa and E. Cabib, Activation of chitin synthetase in permeabilized cells of Saccharomyces cerevisiae mutant lacking protein-aseB. J. Bacteriol. 152:1255 (1982).
6. M. Hori, K. Kakiki, S. Suzuki and T. misato, Relation of polyoxin structure to chitin synthetase inhibition. Agr. Biol. Chem. 35:1280 (1971).
7. M. Hori, K. Kakiki and T. Misato, Interaction between polyoxin and active center of chitin synthetase. Agr. Biol. Chem. 38:699 (1974).
8. K. Isono, T. Azuma and S. Suzuki, Polyoxin analogs. I. Chem. Pharm. Bull. 19:505 (1971).
9. F. Naider, P. Shen bagamurthy, A.S. Steinfeld, H.A. Smith, C. Boney and J.M. Becker, Synthesis and biological activity of tripeptidyl polyoxins as antifungal agents. Antimicrobial Agents Chemotherapy, 24:787 (1983).
10. P. Shenbagamurthy, H.A. Smith, J.M. Becker, A. Steinfeld and F. Naider, Design of anticandidal agents: synthesis and biological properties of analogues of polyoxin L. J. Med. Chem. 26:1518 (1983).

NIKKOMYCIN Z AS INHIBITOR OF CANDIDA ALBICANS GROWTH

M. GONNEAU, J. C. YADAN, P. SARTHOU and F. LE GOFFIC

C.N.R.S.-C.E.R.C.O.A.
2 rue Henri Dunant
94320 Thiais, France

INTRODUCTION

As the pathogenic yeast *Candida albicans* can take up oligopeptides through active transport systems, *Lichliter et al.* (1) first proposed to use these systems to deliver toxic compounds into the cell.

Among antifungal agents of natural origin, polyoxins and especially nikkomycin is an interesting substance since it possesses: (i) a pyridine moiety with phenolic and benzylic hydroxyl groups leading possibly to a so called "suicide substrate" of chitin synthetase, a fundamental enzyme in yeast; (ii) a peptide backbone susceptible to be actively taken up through peptide transport system. As a negative consequence, one could expect peptides to compete with this drug and protect the cell against its toxic effect. Indeed it has been demonstrated that classical polyoxins are almost completely inactive in such media containing peptides whereas nikkomycin Z is fully active in rich medium.

We therefore synthesized ^{3}H-nikkomycin to study: (i) its interactions with chitin synthetase and report here our results; (ii) its uptake by *Candida albicans* and report here that it does utilize a peptide transport system. We also describe a nikkomycin resistant mutant deficient in dipeptide uptake which simultaneously increased its capacity to take up tripeptides. The material and methods used in the experiments reported herein have been previously described (2, 4).

RESULTS AND DISCUSSION

Activity of nikkomycin on yeast chitin synthetase

Nikkomycin was assayed as an inhibitor of *S. cerevisiae* chitin synthetase by the use of permeabilized cells. The drug behaves indeed as a competitive inhibitor of the enzyme. As it appears chemically as a potent "suicide substrate" of the enzyme, we decided to look at this possibility and synthesized tritiated nikkomycin through a transient iodonikkomycin easily available by iodination of nikkomycin itself. ^{3}H-nikkomycin was incubated under different experimental conditions with permeabilized *S. cerevisiae* cells and the residual chitin synthetase activity determined. There was found no retention of radioactive materials within the permeabilized cells after extensive dialysis and the chitin synthetase activity was fully recovered. Hence it was concluded that nikkomycin behaves as a simple competitive inhibitor of chitin synthetase.

Inhibition of C. albicans by nikkomycin and polyoxin D

Both nikkomycin (5 μl/ml) and polyoxin D (10 μl/ml) caused growth inhibition of C. albicans ATCC 26278 in a liquid medium devoid of peptones; when grown in YEPG liquid medium C. albicans was resistant to polyoxin D whereas 95 % of the population remained fully sensitive to nikkomycin.

Isolation of a peptide transport deficient mutant of C. albicans

A peptide transport deficient mutant of C. albicans on the basis of its resistance to nikkomycin (>500 μg/ml) on YEPG medium was isolated. The strain was simultaneously resistant to bacilysin, a toxic dipeptide. This simultaneous resistance was presumably due to the lack of a common transport system most probably a general peptide transport system.

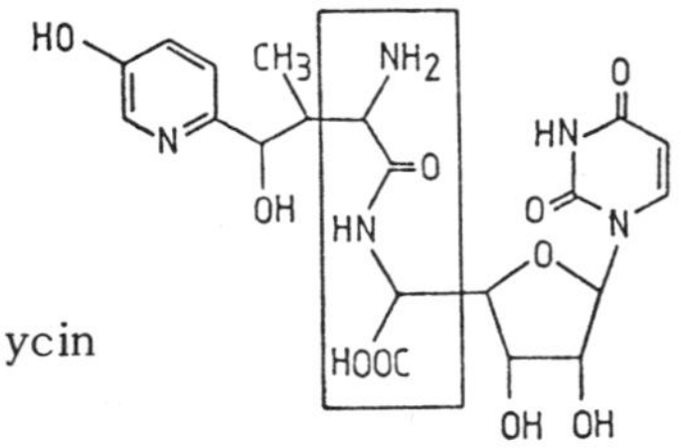

Nikkomycin

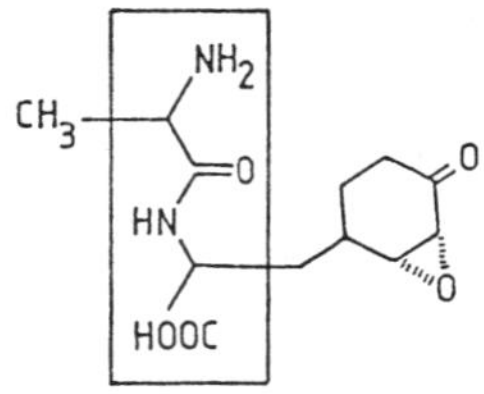

Bacilysin

Transport of ^{3}H-nikkomycin and methionyl peptides in C. albicans

The rates of ^{3}H-nikkomycin uptake were found to be identical in sensitive cells grown from rich and minimal media (similar Km and Vmax). As nikkomycin uptake was expected to be mediated through a peptide transport system in rich medium in wild type but not in nikkomycin mutant, we compared their ability to transport peptides and therefore studied ^{14}C-di-and trimethionine as compared to ^{3}H-nikkomycin uptake in the parent and nikkomycin mutant. The mutant lost almost all its capacity to take up both nikkomycin and dimethionine as compared with the wild strain. The Km values for dipeptides, dimethionine and nikkomycin were similar in wild type strain but maximum rates differed by a factor of 25 fold in favor of dimethionine. A somewhat surprising observation was the considerable enhancement of ^{14}C-trimethionine uptake in the nikkomycin mutant as compared with the wild strain.

Analysis of trimethionine and nikkomycin uptake in C. albicans ATCC 26278 by competition experiments

Experiments were conducted in which nikkomycin competed with ^{14}C-trimethionine and vice versa. The results show that the Ki obtained for nikkomycin (20 μM) is fivefold greater than its Km for its uptake. This and the deviation of the response obtained at high nikkomycin concentrations may suggest the occurrence of two (or more) distinct transport systems for trimethionine.

CONCLUSION

Even though nikkomycin is a powerful inhibitor of chitin synthetase, it does not behave as a so called "suicide substrate". Its uptake in C. albicans cells is handled by a peptide transport system. Taking in account the observation presented here, we suggest that tripeptidyl derivatives of nikkomycin should prove useful for inhibition of both wild type and dipeptide transport deficient mutants of C. albicans.

REFERENCES

1. W. D. Lichliter, F. Naider and J. M. Becker, Basis for the design of anticandidal agents from studies of peptide utilization in Candida albicans, Antimicrob. Agents Chemother, 10:483 (1976).
2. V. Guillermet, J. M. Masson and F. Le Goffic, Inhibition de la chitinesynthétase par la polyoxyne: comparaison des levures perméabilisées avec les extraits acellulaires, C. R. Acad. Sc. Paris, 295:7 (1982).
3. P. Sarthou, M. Gonneau and F. Le Goffic, Photoaffinity inhibition of peptide transport in yeast, Biochem. Biophys. Res. Commun., 110:884 (1983).
4. J. C. Yadan, M. Gonneau, P. Sarthou and F. Le Goffic, Sensitivity to nikkomycin Z in Candida albicans. Role of the peptide permeases, J. Bact.,160:884 (1984).

Section IV

MOLECULAR RECOGNITION AND BIODEGRADATION

CHITOSAN BOTH ACTIVATES GENES IN PLANTS AND INHIBITS RNA SYNTHESIS IN FUNGI

L. A. HADWIGER, D. F. KENDRA, B. W. FRISTENSKY and W. WAGONER

Department of Plant Pathology
Washington State University
Pullman WA 99164-6430, U.S.A.

INTRODUCTION

The regulatory role of chitosan in eucaryotic organisms may have far reaching consequences because chitosan and its acetylated form, chitin, are prevalent in nature. We have found two different regulatory functions for chitosan in a simple fungal-plant interaction[1]. Chitosan which is a normal component of the cell walls of some fungi (Fig. 1) can activate specific genes in plants and at similar concentrations can completely inhibit all RNA synthesis in some fungal organisms and thus suppress gene activity. In this paper we review data which partially explains this paradoxical action.

CHITOSAN-DNA INTERACTION

Our explanations of chitosan's action are based on the interaction between chitosan's positively charged amino groups and the negatively charged phosphate groups of nucleic acids. We have shown the chitosan polymer must be 7 or more sugar units in length, both to optimally induce plant genes and inhibit fungal growth[2]. This length requirement suggests that a series of positive charges match up with phosphate negative charges in the grooves of the DNA helix in the B form. Recently a positively charged compound, "octamer X", has been synthesized which upon binding to the DNA sequence poly (dA-dT) · poly (dA-dT), can convert this DNA which normally exists in the B form, to the Z-form[3]. Chitosan could well possess similar properties. Also, as expected, chitosan complexes very securely with DNA in solution and alters its melting temperature curve[1]. Equivalent concentrations of chitosan cause DNA to be precipitated from solution. The action of chitosan in peas in inducing certain genes is mimicked by compounds or physical processes whose specificity for DNA is well characterized[4]. These plant genes are induced by short wave U.V. light, actinomycin D (a DNA intercalating agent) or by trimethylpsoralen activated by long wave ultraviolet light (which covalently bonds the psoralen with the pyrimidine bases of DNA).

Radio-labelled chitosan, when applied to plant cells, localizes in the nucleus[5]. When chitosan is applied to fungal cells in concentrations sufficient to inhibit growth the efficiency for isolating and solubilizing DNA from this tissue drops precipitously (Table 1). Finally, the glucosamine residues of radio-labelled chitosan can be recovered following

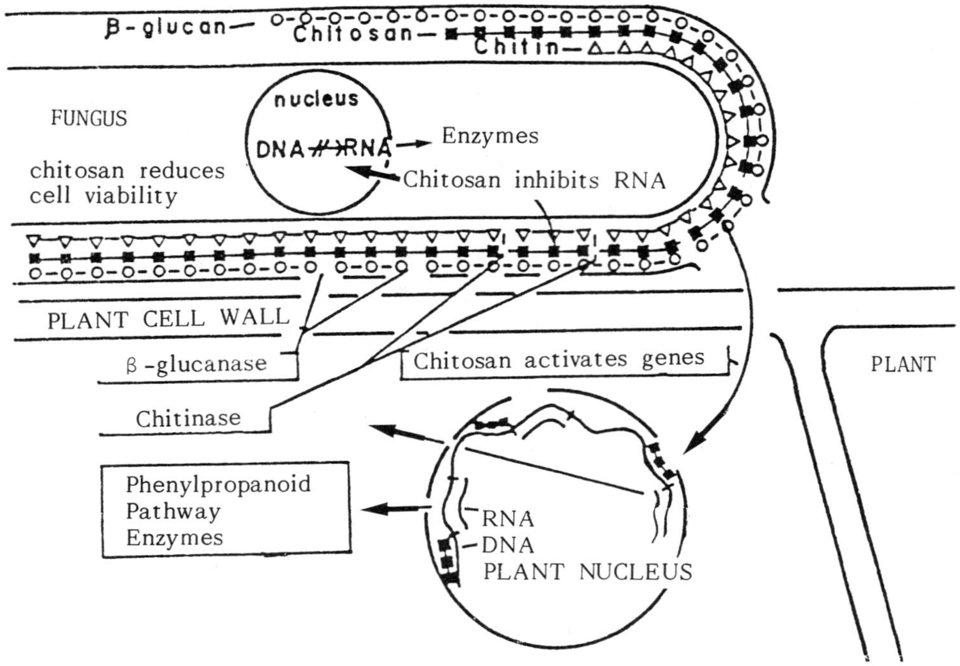

Fig. 1. A proposed mode of action of chitosan endogenous to the Fusarium solani pea interaction. Chitosan is synthesized as a normal, but minor component of the fungal cell wall along with chitin and β-glucan. When the fungal cell contacts the plant host cell, endo-chitinase and endo-β-1,3-glucanase are present and progressively release fungal wall fragments. Some of the fragments contain chitosan segments. Since, we have found hexosamine oligomers seven units or longer to be biologically active, we hypothesize that such oligomers may both induce host resistance responses and inhibit fungal growth. Chitosan fragments entering the host cell may influence chromatin structure via their ability to complex with DNA. The mRNAs from some of the activated genes produce proteins which enhance the activity of the phenylpropanoid pathway which in turn produce phenolic compounds potentially adverse to the pathogen such as phytoalexins or lignin-like compounds. The resistance response also increases the activity of hydrolytic enzymes, endo-β-glucanase and endo-chitinase, thus providing an enhanced host potential to degrade fungal walls. Finally, chitosan-rich fragments also accumulate in the fungal cell. Chitosan blocks the synthesis of RNA and reduces cell viability so that fungal growth is stopped.

the isolation and hydrolysis of nuclei or chromatin from plant cells (Kendra and Hadwiger, unpublished). These findings suggest that chitosan actually interacts with cellular DNA.

Contrasts in the effects of chitosan on plant and fungal cells may be related to the relative genome sizes. Fusarium species have a genomic size of 2.4×10^7 bases compared with peas with 4.7×10^9 bases[6,7]. Table 1 indicates the effect of chitosan on the fungal cell. The high level chitosan treatment which completely stopped fungal growth, progressively reduced isotope uptake by the cell and prevented the uridine-^{3}H which entered from being incorporated into RNA (Table 1). Chitosan also lowered the rate at which DNA was labelled but since this reduction relates

Table 1. Effect of chitosan on the synthesis of total DNA and RNA in *Fusarium solani* f. sp. *phaseoli*.

Treatment[a] to culture medium	Period	Uridine-^{3}H into RNA	RNA extracted	Uridine-^{3}H into DNA	DNA extracted	Isotope Uptake
	h	cpm/mg	μg	cpm/mg	μg	% x 10^{-2}
No treatment	0	41,727	68	11,312	44	9.2
	2.6	80,437	68	17,636	56	15.3
Chitosan 460 μg/ml	0	0	38	3,381	20	3.7
	2.6	0	7	606	16	1.4
Chitosan 180 μg/ml	0	980	61	3,403	38	2.6
	2.6	354	20	900	21	2.5
Chitosan 90 μg/ml	0	3,436	61	1,566	38	2.2
	2.6	925	44	3,130	23	2.9

[a] *Fusarium solani* f. sp. *phaseoli* was grown in Vogel's medium with casein hydrolysate for 3 days. Following transfer to fresh medium 10 ml aliquots of suspended fungal mycelium were added to 250 ml flasks containing the treatments. RNA was labelled for a 1 h pulse with 10 μC of Uridine-5-^{3}H. The mycelia were washed on a 1.2 micron filter and the RNA extracted as described[8]. An aliquot of ethanol-precipitated total nucleic acid was analyzed separately for labelled DNA after the RNA was enzymatically removed.

closely with availability of isotope entering the cell the actual influence of chitosan on DNA synthesis is probably minimal. Thus it appears that chitosan reduces isotope uptake and stops or reduces RNA synthesis. Since the inhibitory effect on RNA synthesisis is complete within 1 h this may be the primary site of chitosan's fungicidal action.

CHITOSAN AND FUNGAL REGULATION

The growth of some fungi, especially those which contain chitosan as a major cell wall component, is not inhibited by chitosan. These fungi presumably alleviate chitosan's potential destruction through exclusion, rapid degradation, etc. *Fusarium* species may also have some potential to regulate chitosan levels. For example, *Fusarium* macroconidia placed in a chitosan concentration which is just adequate to prevent germination or growth for 24 h can resume growth when the chitosan medium is replaced with fresh medium without chitosan (Kendra and Hadwiger, unpublished). We observed that chitosan also accumulates in dormant chlamydospores, but the level is much lower in mycelia following germination of these spores[1]. Thus chitosan may be utilized by the fungus as a natural regulator molecule to manipulate the dormant state. The chitosan concentration is also high in the inhibited growing tip of *Fusarium* strains which have been resisted by pea tissue[1]. Have pea plants acquired and fostered disease resistance through a selected ability to prematurely induce dormancy in an invading fungus?

CHITOSAN AND THE RESISTANCE OF PLANTS TO INFECTION

Chitosan treatments induce a response in plants very similar to those induced by challenges with *Fusarium solani* f. sp. *phaseoli*, a normally virulent pathogen of beans which cannot successfully infect peas. When chitosan is applied to peas prior to the normally virulent pea pathogen, *F. solani* f. sp. *pisi*, the tissue can resist this pathogen[1]. The extensively-researched resistance response incited by chitosan results in the enhanced synthesis of plant RNA species homologous with several cloned and partially characterized pea genes[8] (See example in Fig. 2). There are at least 20 proteins selectively synthesized in this resistance response[9]. Finally, in this chitosan-induced resistance response there is a resultant increase in the activity of enzymes in major secondary pathways for the production of the antifungal phytoalexin, pisatin, as well as hydrolytic enzymes such as chitinase and β-glucanase[10,11], with the potential to attack and release carbohydrate oligomers from the fungal cell walls (See Figure 1).

HYPOTHESIZED EXPLANATION FOR THE DIFFERENTIAL EFFECT OF CHITOSAN ON PLANT AND FUNGAL CELLS

In contrast to its potential to inhibit RNA synthesis in fungi, chitosan clearly activates some plant genes. The enhanced transcription of genes is associated with changes in conformation of nuclear chromatin, and has been well established in eucaryotic organisms[12]. It is likely that a small amount of chitosan entering a plant nucleus, containing a large quantity of chromatin, may more easily alter the chromatin conformation in selected regions thus causing transcriptional changes, rather than completely suppressing RNA synthesis as is possible in the fungus which contains much less DNA.

EFFECTS OF CHITOSAN ON OTHER EUCARYOTIC ORGANISMS

Chitosan can inhibit an array of fungi from Epidermophyton floccosum an athlete foot-inciting fungus[13] to Ceratocystis ulmi the causal agent of Dutch Elm disease (Hadwiger, unpublished). Chitosan's gene activation effect appears to stimulate genes in a broad range of plant genera. For

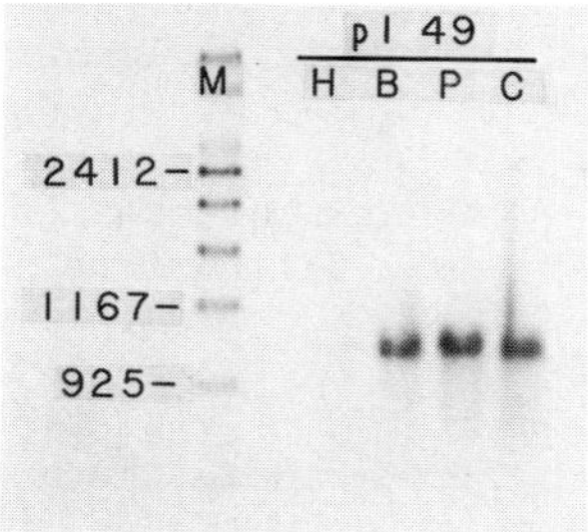

Fig. 2. Northern blot analysis of RNA from pea tissue treated with Fusarium solani or chitosan. The cross hybridization of total RNA extracted from pea endocarp tissue, which had been treated 20 h with H_2O (lane H), with a DNA-P^{32} probe from the cDNA pea clone pI 49 is light but detectable. The corresponding bands are much darker in lanes containing RNA from tissue treated with F. solani f. sp. phaseoli (B), f. sp. pisi (P) or chitosan (C), indicating a greater synthesis and/or accumulation of this RNA species in response to these treatments. Lane M contains size markers.

example, chitosan can induce increases in proteinase inhibitor production in tomato plants[14]. More recently chitosan has been shown to stimulate monoterpene production in Lodgepole pine[15]. Both chitin and chitosan can induce lignification in wheat leaves[16].

Without information on the specific effects of chitosan on crops other than peas, we embarked on a series of tests to determine if field-grow crops could benefit from chitosan treatments. The results indicate mixed benefits, however chitosan treatments of wheat seeds prior to planting can afford the wheat plant protection from some of the detrimental effects of root rotting pathogens which cause lodging (fallen wheat) and yield loss[17].

REFERENCES

1. L. A. Hadwiger and J. M. Beckman, Chitosan as a component of pea-Fusarium solani interactions, Plant Physiol. 66:205 (1980).
2. D. F. Kendra and L. A. Hadwiger, Characterization of the smallest chitosan oligomer that is maximally antifungal to Fusarium

solani and elicits pisatin formation in Pisum sativum, Exp. Mycol. 8:276 (1984).

3. D. S. Zarlenga, H. B. Halsall and R. A. Day. A polycationic amine that induces unique conformational changes in poly (dA-dT) in low salt, Nucl. Acid Res. 12:6325 (1984).
4. D. C. Loschke, L. A. Hadwiger and W. Wagoner, Comparison of mRNA populations coding for phenylalanine ammonia lyase and other peptides from pea tissue treated with biotic and abiotic phytoalexin inducers, Physiol. Plant Pathol. 23:163 (1983).
5. L. A. Hadwiger, J. M. Beckman and M. J. Adams, Localization of fungal components in the pea-Fusarium interaction detected immunochemically with anti-chitosan and anti-fungal cell wall antisera, Plant Physiol. 67:170 (1981).
6. A. Szecsi, Kinetic complexity and repetitiveness of Fusarium graminearum nuclear DNA, Exp. Mycol. 5:323 (1981).
7. M. G. Murray, R. E. Cuellar and W. F. Thompson, DNA sequence organization in the pea genome, Biochem. 17:5781 (1978).
8. B. Fristensky, R. C. Riggleman, W. Wagoner and L. A. Hadwiger, Gene expression in susceptible and disease resistant interactions of peas induced with Fusarium solani pathogens and chitosan, Physiol. Plant Pathol. (In press).
9. W. Wagoner, D. C. Loschke and L. A. Hadwiger, Two-dimensional electrophoretic analysis of in vivo and in vitro synthesis of proteins in peas inoculated with compatible and incompatible Fusarium solani, Physiol. Plant Pathol. 20:99 (1982).
10. E. J. Nichols, J. M. Beckman and L. A. Hadwiger, Glycosidic enzyme activity in pea tissue and pea-Fusarium solani interactions. Plant Physiol. 66:199 (1980).
11. F. Mauch, L. A. Hadwiger and T. Boller, Ethylene: symptom, not signal for induction of chitinase and β-1,3-glucanase in pea pods by pathogens and elicitors, Plant Physiol. 76:607 (1984).
12. R. Reeves, Transcriptionally active chromatin, Biochim. Biophys. Acta 782:343 (1984).
13. C. R. Alan and L. A. Hadwiger, The fungicidal effect of chitosan on fungi of varying cell wall composition, Exp. Mycol. 3:285 (1979).
14. M. Walker-Simmons, D. Jin, C. A. West, L. Hadwiger and C. A. Ryan, Comparison of proteinase inhibitor-inducing activities and phytoalexin elicitor activities of a pure fungal endopolygalacturonase, pectic fragments, and chitosans, Plant Physiol. 76:833 (1984).
15. R. H. Miller, A. A. Berryman and C. A. Ryan, Biological elicitors of defense reactions in Lodgepole pine. Forest Science (In press).
16. R. B. Pearce and J. P. Ride, Chitin and related compounds as elicitors of the lignification response in wounded wheat leaves. Physiol. Plant Pathol. 20:119 (1982).
17. L. A. Hadwiger, B. Fristensky and R. C. Riggleman, Chitosan a natural regulator in plant-fungal interactions increases crop yields, in: "Chitin, Chitosan and Related Enzymes," J. P. Zikakis, ed., Academic Press, New York (1984).

CHITOSAN AND OTHER POLYAMINES: ANTIFUNGAL ACTIVITY AND INTERACTION WITH BIOLOGICAL MEMBRANES

J. L. LEUBA and P. STOSSEL

Nestlè Research Department, Nestec Ltd
CH-1800 Vevey, Switzerland

INTRODUCTION

The antifungal effect of chitin has been demonstrated in field experiments. A marked reduction of root-rot in beans and of vascular wilt in radishes, both caused by *Fusarium* spp., were observed subsequent to addition of chitin to soil (1). Chitin amendment suppressed total fungal population and stimulated lytic and antibiotic-producing microorganisms such as actinomycetes (2-4). The antifungal effect of chitin thus was indirect, via antagonistic soil microorganisms.

In vitro, the deacetylated form of chitin, chitosan, inhibited fungal growth significantly more than chitin (5,6). Although some fungal strains or species were less sensitive, growth of all species except those containing chitosan as wall constituent was affected (5).

Chitosan does not only interact specifically with fungi, but it also affects plants: it inhibits growth of plant cells in suspension (7), induces the accumulation of phytoalexins (low molecular weight antimicrobial agents) (7,8) and of proteinase inhibitors (9,10), increases wall-associated phenolic compounds, induces the deposition of callose (7), and changes membrane permeability (11,12). Chitosan also agglutinates a variety of bacteria and yeasts as well as cells of mammalian origin (13). Chitosan has been thought to play a key role in the host-parasite interaction between *Pisum sativum* and *Fusarium solani* f. sp. *pisi*. According to the proposed model (14), chitosan molecules - liberated from the pathogen's wall by host hydrolytic enzymes - penetrate host and pathogen cells and nuclei (15) and interfere with mRNA and protein synthesis. The pathogen thus may be impaired by both its own chitosan and host phytoalexin also induced by the liberated chitosan. A recent investigation on the antifungal and phytoalexin-inducing effect of chitosan showed maximum activity for molecules containing seven or more glucosamine residues (16).

Apart from this model which demands direct (specific) DNA-polymer interaction, the available literature appears to attribute the biological activity of chitosan to its polycationic properties interfering with negatively charged residues of macromolecules at the cell surface. The present study was thus undertaken to establish the relationship between the effect of chitosan and other synthetic polyamines on growth of fungal plant pathogens and their effect on the plasmalemma, i.e. the change in cell permeability and the interaction with cell membranes (erythrocytes).

MATERIALS AND METHODS

Chitin (pract. grade) from Fluka was bleached (19), lyophilized and milled. Crab shell chitosan was purchased from Sigma, and shrimp chitosan (medium-high viscosity) was from Hercules Inc., Wilmington, DE (USA). The powder form was prepared as previously described (6), and deacetylation was determined by potentiometric titration (20). The diameter of chitin and chitosan particles was $\leqslant 50$ µm.

Solutions of the polycations (suspension of chitin and chitosan) in slightly modified (6) Vogel's growth medium (21), solidified with 1.5 % agar, were used for bioassays with the following phytopathogenic fungi: the Oomycetes *Aphanomyces euteiches* Drechs. CMI 170483, *Phytophtora megasperma* Drechs. f. sp. *glycinea* Kuan et Erwin race 6, and *Pythium paroecandrum* Drechs. CMI 92552; the Ascomycetes/Deuteromycetes *Fusarium oxysporum* f. sp. *phaseoli* Kendrick and Snyder CMI 141119, *Fusarium solani* (Mart.) Sacc. f. sp *pisi* (Jones) Snyder and Hansen CMI 167691, *Rhizoctonia solani* Kühn ATCC e 18618, and *Thanatephorus cucumeris* (Frank) Donk (= perfect stage of *R. solani*) CMI 190141 a. The media, 5 ml per Petri dish (Ø 50 mm), were inoculated with a 5 mm mycelial plug from a culture on Vogel's medium, subsequently incubated in the dark at 25°C, and growth was recorded when the colony radiu of controls reached 15 ± 3 mm. Growth is expressed in % inhibition of radial growth recorded on the control medium.

To study leakage of proteinaceous and UV-absorbing material induced by chitosan, 20 ml of maltose-asparagine medium (22) were inoculated with a 5 mm mycelial plug from a culture of *P. paroecandrum*. Following incubation at 25°C for 6 days, mycelial mats were washed carefully in 5 mM Sørensen's phosphate buffer, pH 5.8, and transferred to a suspension of either chitin or chitosan, respectively, in the same buffer (pH 5.8 or 7.5), containing 1 % maltose. The flasks were capped with Parafilm and incubated on a shaker at 100 rpm, 25°C for 30 min to 2 h. After centrifugation of mycelium and polymer particles, protein in the supernatant was determined (11,23). Leakage of UV-absorbing material into the incubation medium was recorded at 255 nm. Data indicate means of 4 replicates.

Agglutination of sheep erythrocytes, purchased from Behringwerke, Marburg, FRG, by polyamines was done in freshly prepared 50 mM acetic acid solution, pH 6. Buffer conductance was adjusted to 14.1 mS/cm by addition of NaCl. The preserved sheep blood was washed three times in the buffer, diluted to 6 % with buffer, and 150 µl samples were dispensed into microtiter plates. To each sample, 50 µl of freshly prepared, suitably diluted polyamine solution in the same buffer were added and mixed with suspended erythrocytes. Agglutination was recorded the following day.

RESULTS AND DISCUSSION

Antifungal activity

The assay system used to determine the antifungal effect of polyamines is demonstrated with PEI and three different fungal species (Fig. 1). It is shown that the growth of *A. euteiches* and *F. solani* was fairly well-controlled by PEI whereas *R. solani* was rather less sensitive.

Fusarium oxysporum, a fungus sensitive to chitosan (6), was then used to compare the effect of L-lysine oligomers, PEI and chitosan on fungal growth (Table 1). With L-lysine oligomers, growth inhibitory activity markedly increased with increasing degree of polymerization (DP), except with the monomer whose activity can be disregarded. Apparently poly-L-lysine DP ca. 19 represented a critical value of DP from which the extent of growth inhibition (%) vs. DP values was strongly modified. This point must be confirmed using intermediate poly-L-lysine chain lengths. Data on the induction of phenylalanine ammonia-lyase (PAL) and of the phytoalexin, pisatin, in pea suggest poly-L-lysine with MW ca. 7,200 (DP 50) the most active pisatin inducer whereas data on PAL induction are inconclusive (24).

PEI was more active than poly-L-lysine of comparable molecular weight (Table 1), but the difference was less for comparable DP's, suggesting that chemical structure (e.g. secondary or primary amino groups) plays only a minor role.

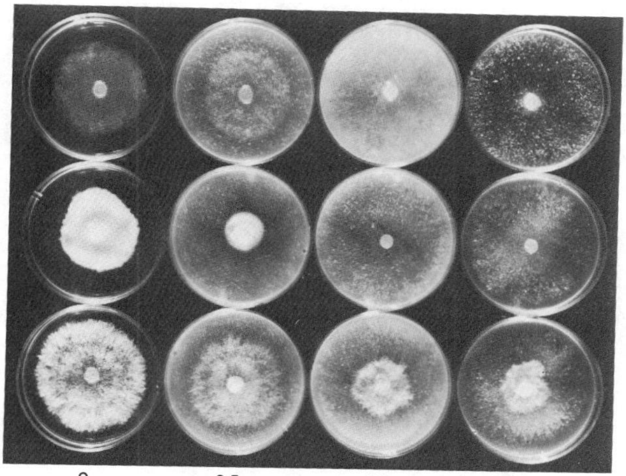

Fig. 1. Growth of *Aphanomyces euteiches*, *Fusarium solani* f. sp. *pisi* and *Rhizoctonia solani* on Vogel's medium containing 0, 0.5, 2.5, and 5 mg PEI per ml, respectively

Chitosan was also able to reduce fungal growth (Table 1). However, the glucosamine polymer was less effective than the synthetic polyamines. Despite of its high MW and DP, chitosan was slightly less effective than poly-L-lysine MW 4,000/DP 19. This relatively low effect may be explained by incomplete ionization and partial solubilization of chitosan (pKa 6.2) in the growth medium (pH 6) (25), whereas poly-L-lysine (pKa ε-NH^3: 10.5) and PEI were fully charged at this pH.

The polyanion, poly-L-glutamic acid, showed as little activity as L-lysine (Table 1).

Polyethyleneimine inhibited radial growth of *P. megasperma*, *P. paroecandrum*, *F. solani* and *I. cucumeris* more effectively than chitosan (results not shown).

Table 1 Effect of poly-L-lysine, polyethyleneimine, chitosan and poly-L-glutamic acid on Fusarium oxysporum at pH 6.0

POLYMER	MW	DP	GROWTH INHIBITION %	
			2.5 mg/ml	5 mg/ml
L-lysine	146	1	2	7
Poly-L-Lysine	730	5	21	37
,,	4,000	19	64	73
,,	30,000-50,000	200-340	70	86
,,	240,000	1,150	72	92
Polyethyleneimine	30,000-40,000	750-950	98	100
Chitosan	>300,000	>1,500	33	60
Poly-L-glutamic ac.	15,000-50,000	100-340	3	6

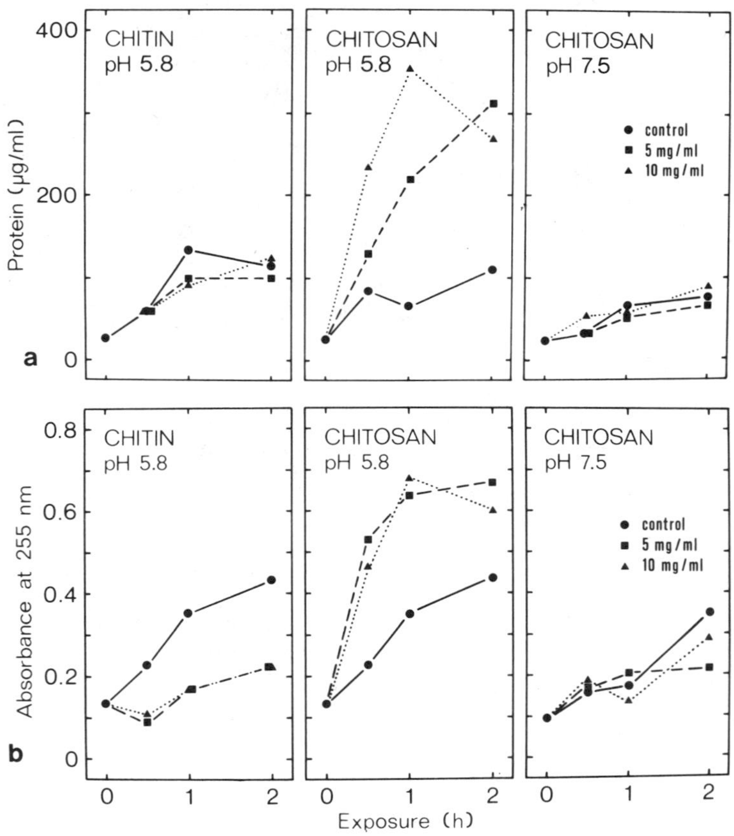

Fig. 2. Leakage of proteinaceous (a) and UV-absorbing (b) material induced by either chitin or chitosan in Pythium paroecandrum.

Membrane permeability

The disturbance of membrane function of a filamentous fungus by chitosan was shown with *P. paroecandrum*. Chitosan induced considerable leakage of proteinaceous (Fig. 2a) and UV-absorbing (Fig. 2b) material at pH 5.8, i.e. below its pKa. In contrast, chitosan at pH 7.5 - when most amino groups are thus in the free base form - as well as chitin at pH 5.8 did not induce significant leakage (Fig. 2a, b). Therefore loss of cell material into the medium by *P. paroecandrum* correlates well with inhibition of its hyphal growth.

Hemagglutination

The agglutination of sheep erythrocytes was chosen as a further model to demonstrate the effectiveness of the interaction between the polyamines and biological membranes. This physico-chemical model, however, cannot be directly compared with fungal growth inhibition which involves a multitude of biological reactions. Nevertheless, agglutination tests provide information on molecular properties essential for the interaction of polyamines and cell membranes.

The hemagglutination titer i.e. the lowest polymer concentration causing hemagglutination, varied considerably (Fig. 3). It decreased with increasing DP of poly-L-lysine. Even at the highest concentration tested, 1000 µg/ml, poly-L-lysine with DP 19 (MW 4,000) was inactive. PEI, DP 750-950, had a considerably (Fig. 3) lower hemmaglutination titer than poly-L-lysine of comparable molecular weight but a DP of only 200-340. The hemagglutination titer of PEI, however, was only slightly below that of poly-L-lysine with a comparable DP of 1,500 but a molecular weight of 240,000. These results clearly show the correlation between DP of the two polyamines and their hemagglutination activity which is inversely related to titer.

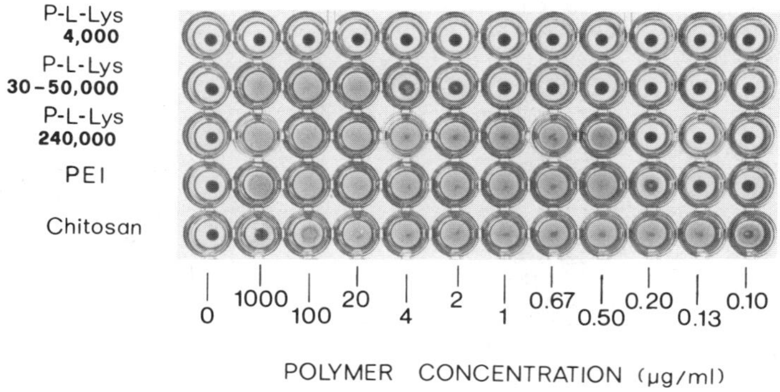

Fig. 3. Agglutination activity of poly-L-lysine oligomers, PEI and chitosan (MW and DP as indicated on Table 1) on sheep erythrocytes. Positive interaction results in wide dispersion of agglutinated cells which contrasts with the spotted sediment from reference. Concentrations (µg/ml) are those of the polymer solutions and not the final concentration in the assay.

Agglutination of erythrocytes occurred even at the lowest titre of chitosan (Fig. 3). No agglutination however was observed at 1000 μg/ml, suggesting repulsive effects as reported for the agglutination of bacteria with high concentrations of poly-L-lysine (26)

Though chitosan has the highest agglutination activity of the polyamines tested, it was the least active in inhibiting fungal growth. This apparent variation in the biological activities of the hexosamine may be due (a) to underestimation of the antifungal activity because of the low solubility in growth medium, as discussed above, (b) to the presence or absence of the cell wall which may limit the access to target site(s), and (c) to the chain length and distribution of ionic charge favouring hemagglutination, whereas shorter but fully ionized polyamines, e.g. PEI, favour inhibition of fungal growth.

CONCLUSION

According to Katchalski and co-workers (26), the antibacterial activity of basic polyamino acids may rely on the interaction with nucleic acids or certain enzymes, but most likely with electronegative cell surface. Moreover, Siegel and Daly (27) suggested competitive displacement of Ca^{2+} from anionic membrane sites by poly-L-lysine, rendering the membrane leaky. The release of Ca^{2+} (11) and subsequent efflux of proteinaceous and UV-absorbing material from suspension-cultured soybean cells (12) as well as the data presented in this paper confirm the nonspecific action of polyamines on membrane integrity. Moreover, Ca^{2+} appears to be important for glucan and chitin synthesis (28), and displacement of Ca^{2+} may thus limit growth.

The responses of plant cells to chitosan, i.e. accumulation of phytoalexins and proteinase inhibitors, lignin-like compounds and callose, are general stress responses triggered by pathogens lacking chitosan as a wall constituent or induced by wounding (29-32). Therefore, these responses as well as cessation of growth in plants and fungi may be caused by changes in membrane function and impairment of cell wall-synthesizing enzymes.

ACKNOWLEDGMENTS

The authors are grateful to Mrs. M. Weber and Mr. B. Pugin for photographic work.

REFERENCES

1. R. Mitchell and M. Alexander, Chitin and the biological control of Fusarium diseases, Plant Disease Rep. 45:487 (1961).
2. R. Mitchell and M. Alexander, Microbiological processes associated with the use of chitin for biological control, Soil. Sci. Soc. Proc. 26:556 (1962).
3. R. Mitchell, Addition of fungal cell wall components to soil for biological disease control, Phytopathology 53:1063 (1963).
4. Y. Henis, B. Sneh, and H. Katan, Effect of organic amendments on Rhyzoctonia and accompanying microflora in soil, Can. J. Microbiol. 13:643 (1967).
5. C. R. Allan and L. A. Hadwiger, The fungicidal effect of chitosan on fungi of varying cell wall composition, Exp. Mycol. 3:285 (1979).
6. P. Stössel and J. L. Leuba, Effect of chitosan, chitin and some aminosugars on growth of various soilborne phytopathogenic fungi, Phytopath. Z. 111:82 (1984)

7. H. Köhle, D. H. Young, and H. Kauss, Physiological changes in suspension-cultured soybean cells elicited by treatment with chitosan, Plant Sci. Lett. 33:221 (1984).
8. L. A. Hadwiger and J. M. Beckman, Chitosan as a component of pea - Fusarium solani interactions, Plant Physiol. 66:205 (1980).
9. M. Walker-Simmons, L. A. Hadwiger, and C. A. Ryan, Chitosan and pectic polysaccharides both induce accumulation of the antifungal phytoalexin pisatin in pea pods and antinutrient proteinase inhibitors in tomato leaves, Biochem. Biophys. Res. Commun. 110:194 (1983).
10. M. Walker-Simmons and C. A. Ryan, Proteinase inhibitor synthesis in tomato leaves. Induction by chitosan oligomers and chemically modified chitosan and chitin, Plant Physiol. 76:787 (1984).
11. D. H. Young, H. Köhle, and H. Kauss, Effect of chitosan on membrane permeability of suspension-cultured Glycine max and Phaseolus vulgaris cells, Plant Physiol. 70:1449 (1982).
12. D. H. Young and H. Kauss, Release of calcium from suspension-cultured Glyine max cells by chitosan, other polycations, and polyamines in relation to effects on membrane permeability, Plant Physiol. 73:698 (1983).
13. E. E. Evans and S. P. Kent, The use of basic polysaccharides in histochemistry and cytochemistry: IV. Precipitation and agglutination of biological materials by Aspergillus polysaccharide and deacetylated chitin, J. Histochem. Cytochem. 10:24 (1962).
14. L. A. Hadwiger and D. C. Loschke, Molecular communication in host - parasite interactions: hexosamine polymers (chitosan) as regulator compounds in race-specific and other interactions, Phytopathology 71:756 (1981).
15. L. A. Hadwiger, J. M. Beckman, and M. J. Adams, Localization of fungal components in the pea-Fusarium interaction detected immunochemically with anti-chitosan and anti-fungal cell wall antisera, Plant Physiol. 67:170 (1981).
16. D. F. Kendra and L. A. Hadwiger, Characterization of the smallest chitosan oligomer that is maximally antifungal to Fusarium solani and elicits pisatin formation in Pisum sativum, Exp. Mycol, 8:276 (1984).
17. J. R. Rubini, M. A. Stahmann, and A. F. Rasmussen, Agglutination of red cells by synthetic lysine polypeptides, Proc. Soc. Exp. Biol. Med. 76:659 (1951).
18. A. Katchalsky, D. Danon, A. Nevo, and A. de Vries, Interaction of basic polyelectrolytes with the red blood cell. II. Agglutination of red blood cells by polymeric bases, Biochim. Biophys. Acta 33:120 (1959).
19. Y. Lingappa and J. L. Lockwood, Chitin media for selective isolation and culture of actinomycetes, Phytopathology 52:317 (1962).
20. R. A. A. Muzzarelli, Chitin, Pergamon Press, Oxford (1977).
21. H. J. Vogel, A convenient growth medium for Neurospora, Microb. Genet. Bull. 13:42 (1956).
22. G. Défago, K. F. Memmen, and H. Kern, Influence de cholestérol et des saponines sur le pouvoir pathogène de Pythium paroecandrum, Phytopath. Z. 83:167 (1975).
23. O. H. Lowry, N. J. Rosebrough, A. L. Farr, and R. J. Randall, Protein measurement with the folin phenol reagent, J. Biol. Chem. 193:265 (1951).
24. L. A. Hadwiger, A. Jafri, S. von Broembsen, and R. Eddy, Mode of pisatin induction. Increased template activity and dye-binding capacity of chromatin isolated from polypeptide-treated pea pods, Plant Physiol. 53:52 (1974).

25. J. W. Park, K. H. Choi, and K. K. Park, Acid-base equilibria and related properties of chitosan, Bull. Korean Chem. Soc. 4:68 (1983).
26. E. Katchalski, L. Bichowski-Slomnitzki, and B. E. Volcani, The action of some water-soluble poly-α-amino acids on bacteria, Biochem. J. 55:671 (1953).
27. S. M. Siegel and O. Daly, Regulation of betacyanin efflux from beet root by poly-L-lysine, Ca-ion and other substances, Plant Physiol. 41:1429 (1966).
28. H. Kauss, H. Köhle, and W. Jeblick, Proteolytic activation and stimulation by Ca^{2+} of glucan synthase from soybean cells, FEBS Lett. 158:84 (1983).
29. E. W. B. Ward, P. Stössel, and G. Lazarovits, Similarities between age-related and race-specific resistance of soybean hypocotyls to Phytophtora megasperma var. sojae, Phytopathology 71:504 (1981).
30. T. R. Green and C. A. Ryan, Wound-induced proteinase inhibitor in plant leaves: a possible defense mechanism against insects, Science 175:776 (1972).
31. R. B. Pearce and J. P. Ride, Specificity of induction of the lignification response in wounded wheat leaves, Physiol. Plant Pathol. 16:197 (1980).
32. R. C. Nims, R. S. Halliwell, and D. W. Rosberg, Wound healing in cultured tobacco cells following microinjection, Protoplasma 64:305 (1967).

CHITINASE: A DEFENSE OF HIGHER PLANTS AGAINST PATHOGENS

T. BOLLER

Botanishes Institut, Universität Basel
CH-4056 Basel, Switzerland

INTRODUCTION

Higher plants often contain large amounts of chitinase (EC 3.2.1.14) activities. This is surprising in view of the fact that chitin, the substrate for these enzymes, occurs in lower animals, in fungi, and in certain algae but has never been found in higher plants. Apparently, chitinases have no function in the plant's own metabolism and may be considered dispensable enzymes. Why, then, do they occur in higher plants?

Many fungi and, more specifically, many fungal plant pathogens have cell walls rich in chitin. It has been suggested, therefore, that plant chitinases might be directed against the cell walls of fungi and thereby might function as an important defense against potential pathogens (1,2). Here, the available evidence supporting this hypothesis is summarized, along with a physical and biochemical characterization of plant chitinase.

OCCURRENCE AND PROPERTIES OF PLANT CHITINASES

In higher plants, chitinolytic activity has first been observed in almond emulsin (3). Even earlier, an enzyme related to chitinase, lysozyme (EC 3.2.1.17) was detected in higher plants, specifically, in turnip roots (4). The turnip lysozyme was later purified and shown to be a chitinase (5). Therefore, for the purpose of this review, plant lysozymes are equated with chitinases.

Ubiquity of Chitinase in Higher Plants

Powning and Irzykiewicz (6) screened a number of seeds, representing eleven families, for chitinase and found activity in all seeds tested. Most contained small but detectable levels of chitinase. Some seeds, e.g. from bean and wheat, contained very high chitinolytic activities.

Similarly, we assayed leaves from a variety of plants representing six different families and found chitinase activity in all of them; the specific activity differed by more than two orders of magnitude between the leaves with the smallest and largest chitinolytic power, cucumber and tomato, respectively (7).

Chitinase was also found in stems (8) and roots (9) of various plants. Some latices (10-12) contained particularly high activities of chitinases (analyzed and purified as lysozymes). Taken together, these results indicate that chitinases are ubiquitous in higher plants but present in widely differing quantities.

Physical Properties of Plant Chitinases

Chitinases can be purified easily from many sources by affinity chromatography on the insoluble substrate, colloidal chitin (13). Table 1 summarizes the physical properties of the plant chitinases, including those analyzed and purified as lysozymes. All chitinases examined have molecular weights in the order of 25,000 to 35,000, as determined by sodium dodecyl sulfate electrophoresis. Similar molecular weights were found for the native enzymes (5, 7, 10-13), indicating that they occur as monomers in solution. Most chitinases have a basic isoelectric point. However, chitinases with acidic isoelectric points exist as well, e.g. in yam roots (15). In some cases, several different isoenzymes are present, e.g. two in pea pods (Mauch, unpublished) or three in yam roots (15).

Table 1: Physical Properties of Purified Plant Chitinases

Plant, tissue	Activities measured	Mol.Wt.	IEP	Ref.
Wheat germ	Chitinase	30,000[a]	7.5 - 9.2	(13)
Tomato stem	Chitinase	31,000[a]	8.5	(8)
Bean leaf	Chitinase, lysozyme	30,000[a]	9.0	(7)
Soybean seed	Chitinase	31,600[a]	-	(14)
Pea pod	Chitinase	32,000[a]	9.1	-[d]
	Chitinase	36,000[a]	9.0	-[d]
Yam root	Chitinase	33,000[a]	3.8 - 4.1	(15)
Papaya latex	Lysozyme, chitinase	28,000[b]	10.5	(10)
Fig latex	Lysozyme, chitinase	29,000[b]	9.0	(11)
Rubber tree latex	Lysozyme, chitinase	26,000[a]	9.0	(12)
Turnip root	Lysozyme, chitinase	25,000[c]	10.0	(5)

[a] by sodium dodecyl sulfate gel electrophoresis
[b] by analysis of sedimentation velocity
[c] by gel filtration
[d] F. Mauch, unpublished

Preliminary experiments with cucumbers indicated that acidic and basic chitinases co-existed in the same plant. Interestingly, cucumber fruits contained primarily basic chitinases while virus-infected cucumber leaves contained primarily acidic chitinases (Boller, unpublished).

Enzymatic Properties of Plant Chitinases

Chitinases can be classified as exo- or endochitinases on the basis of the initial reaction products formed. Exochitinases yield N-acetylglucosamine only while endochitinases produce chitooligosaccharides from chitin. Initially, plant chitinase was claimed to exhibit exochitinase activity on the basis of results with ethylene-treated bean leaves (16). However, reanalysis of the reaction products formed by bean chitinase showed that this enzyme was an endochitinase forming very little N-acetylglucosamine (7). Similarly, the chitinases from wheat germ (13) and from tobacco and cucumbe leaves (Boller, unpublished) were endochitinases.

For measurements of plant chitinases, it is important to use an endochitinase assay. The assay of Abeles et al. (16) which has often been employed in plant work is useful only for exochitinases. It often yields erratic results with plant chitinases and may underestimate the activity of chitinolytic enzymes from plants by orders of magnitude (7). In quantitative analyses of plant chitinases, attention should be paid also to the non-linear relationship between the amount of enzyme and the amount of reaction product formed (7, 13).

Plant chitinases appear to be highly specific with regard to the polysaccharide accepted as substrate. Wheat germ chitinase hydrolyzed chitin and glycol chitin but not cellulose (β-1,4-glucan), β-1,3-glucan or β-1,6-glucan; it did not degrade the polysaccharide side chain of ovalbumin or ovomucoid (13). Bean chitinase had a similarly restricted specificity (7). The only natural substrate for bean chitinase other than chitin was bacterial cell wall peptidoglycan (7). Plant lysozymes showed a similar substrate specificity, chitin being the preferred substrate (5, 10-12).

Generally, plant chitinases show broad pH optima with regard to chitin, with more than 50 % of the maximal activity in the range between pH 3 and pH 9 (7, 13). However, the pH range for lysozyme activity is often much more restricted, with a sharp pH optimum in the range of pH 4.5 and almost no activity left at pH 6.2, the pH value where animal lysozyme is conventionally measured (2, 10-12). This might be one reason why lysozyme activity has not been detected with some of the purified plant chitinases (8, 13-15).

Localization of chitinase

In ethylene-treated bean leaves, most of the chitinase activity was found to be located in the central vacuole of the cells (17). It will be interesting to investigate how the enzyme is transported into the vacuole. In this context, we found that in vitro translation of polyadenylated RNA from bean leaves in a rabbit reticulocyte lysate and immunoprecipitation with antiserum against chitinase yielded a single product which was 2,000 dalton larger than native chitinase. This indicates that chitinase is formed as a pro-enzyme and is processed proteolytically (Vögeli, unpublished). The chitinase (studied as a lysozyme) in rubber tree latex was localized in the lutoid fraction, a fraction of small membrane-bound organelles in the latex which are considered homologous with the vacuole (12).

The vacuolar location of chitinase is intriguing with regard to its postulated function in defense. Since fungi attack plant cells from outside, the functioning of plant chitinase in defense requires either a specific transport from the vacuole into the cell wall - a possibility without precedent in higher plants - or the disruption of the vacuole with the subsequent release of chitinase (17). The latter possibility is realized in the hypersensitive reaction. There, a few cells in the immediate vicinity of an invading fungus break open and release their vacuolar contents.

INDUCTION OF CHITINASE IN RESPONSE TO HORMONES

The basal level of chitinase varies much between different plants and between different organs of the same plants. In specific tissues, e.g. in laticifers or in certain seeds, chitinase is constitutively present in large amounts. In many other tissues, chitinase is present at a low basal level and becomes elevated only upon hormone application or stress (2).

Induction by Ethylene

Abeles et al. (16) first reported the induction of chitinase by ethylene in bean leaves. We have studied this effect in some detail (7). Chitinase began to increase about 6 h after the onset of the ethylene treatment and was induced thirtyfold within 24 h (7). In vitro translation of RNA and immunoprecipitation of the translation products by an antiserum against bean chitinase indicated that the induction by ethylene was due to a dramatic increase in the level of translatable mRNA for chitinase (Vögeli, unpublished). Chitinase induction appeared to be linked closely to the presence of ethylene since it could be halted again, even after 12 h, by the withdrawal of ethylene. About 1 nl ml^{-1} ethylene was required for half-maximal chitinase induction (7).

Ethylene also caused an increase in chitinase activity in melon seedlings (18), in pea pods (19) and in many other plants (7). In bean leaves, a number of hydrolytic enzymes were assayed in ethylene-treated and control leaves; only chitinase and β-1,3-glucanase were induced by ethylene (7).

Induction by Auxin

In pea pods, auxin increased chitinase activity. This effect appeared to be mediated by ethylene. Auxin stimulated ethylene formation; inhibition of ethylene biosynthesis prevented the induction of chitinase (19).

INDUCTION OF CHITINASE IN RESPONSE TO INFECTIONS

Induction in Response to Fungal Infections

If plant chitinase has a function in defense against fungi, its activity might be expected to increase in response to fungal infections. This was indeed observed in several studies. In tomato plants (20) and in pea pods (19) β-1,3-glucanase and chitinase increased in parallel after infection. Earlier Abeles et al. (16) had oberved parallel induction of chitinase and β-1,3-glucanase in response to ethylene. On this basis, they proposed that the two enzymes functioned in defence since β-1,3-glucanase and chitinase were already known to co-operate in the degradation of fungal cell walls (16).

The activities of chitinase and β-1,3-glucanase in infected plant tissues might be due to enzymes of both plant and fungal origin. In the study of tomato plants infected by *Verticillium albo-atrum*, plant and fungal chitinases were separated and characterized; it became evident that the increase in activity upon infection was due to the induction of plant chitinase (20).

Induction in Response to Fungal Elicitors

The complication of a possible contribution of fungal enzymes can be avoided if plants are treated with heat-killed fungal spores or fungal cell wall components. Treatment of tomato stems with autoclaved spores of *Verticillium albo-atrum* increased chitinase and β-1,3-glucanase activities (20). In pea pods, it was found that autoclaved spores or chitosan, an elicitor of phytoalexin production in this tissue, increased chitinase and β-1,3-glucanase in a similar way as did infection (19). This indicated that the increases in enzyme activities observed in infected plants were also due to the induction of the plant enzymes, and not to fungal enzyme activities.

In cultured parsley cells, an elicitor from cell walls of *Phytophthora megasperma* caused a tenfold increase in chitinase activity within 12 h (21)

Induction in Response to Viruses

Plants infected with incompatible viruses often react hypersensitively by the formation of local necrotic lesions. Chitinase (Gehri, unpublished) and β-1,3-glucanase (22) were strongly induced in tobacco leaves reacting hypersensitively to TMV. When the first leaves of young cucumber plants were infected with TNV, chitinase increased up to 600fold in the infected areas (Métraux and Boller, in preparation). Interestingly, chitinase was also induced up to 100-fold in the uninfected second leaves of the plants. Thus, in cucumber, chitinase was induced both locally and systemically in response to a localized infection.

It is unlikely that virus-induced chitinase has a function in the prevention of virus spread. Rather, the induction of chitinase and of β-1,3-glucanase during the hypersensitive reaction may be one element of a general rearmament of the plant's defense arsenal against fungal pathogens.

Is Stress Ethylene the Signal for the Induction of Chitinase in Response to Infection or Elicitors?

In immature pea pods, fungal infection or elicitors caused a rapid, transient burst of stress ethylene within 3 h. Later, after a lag of 4 - 8 h, chitinase and β-1,3-glucanase started to increase (19). Exogenous ethylene similarly induced chitinase and β-1,3-glucanase activities. However, when the endogenous ethylene formation was reduced below control levels in infected of elicitor-treated plants, chitinase and β-1,3-glucanase remained strongly induced. Thus, elicitors appear to induce chitinase and β-1,3-glucanase directly; ethylene production is a symptom, but not a necessary signal for the observed induction of hydrolases in infected pea pods (19).

POSSIBLE FUNCTIONS OF CHITINASE IN DEFENSE AGAINST PATHOGENS

In certain plant-pathogen interactions, the fungal hyphae first invade the plant tissue but later disappear in the course of the resistance response of the plant (1, 23). It has been suggested that this reflects the action of plant chitinases and other hydrolases degrading the hyphal cell walls. However, it might also simply be a case of fungal autolysis. Up to now, there is no direct evidence for the importance of plant chitinase in resistance against fungi. Attempts to correlate susceptibility or resistance with a lower or higher level of chitinase activity in the plants remained equivocal (20).

Nevertheless, as discussed below, in vitro studies have clearly established the ability of plant chitinase to attack and to degrade fungal cell walls. In addition, the enzyme products formed during this process may act as elicitors of phytoalexin formation and thereby contribute to the plant's defense.

Lysis of Pathogen Cell Walls

Crude plant extracts have been found to enzymatically digest isolated cell walls of potential fungal pathogens (9, 23, 24). Purified plant chitinase by itself was found to be very active in releasing soluble fragments from the cell walls of fungi (7, 24). Purified β-1,3-glucanase further increased the ability of chitinase to hydrolyze the fungal chitin (24).

Models of fungal cell walls indicate that chitin and β-1,3-glucan represent the innermost wall layer which is masked and protected by an outer layer of various different polysaccharides, mannoproteins, or even of lipids (25, 26). Thus, chitin might not be exposed to the action of chitinase in the walls of intact fungi. It will be interesting to find out whether plants possess additional enzymes which can unmask the chitin in plant-pathogen interactions.

However, complete lysis may not be necessary to stop a growing fungus. Growth of the hyphal tip appears to represent a sensitive equilibrium between synthetic and hydrolytic processes; addition of an exogenous hydrolase may disturb the balance sufficiently to cause tip rupture and lysis of the hypha (25). In this context, it is interesting that plant chitinase degrades "nascent chitin" formed by chitin synthetase much more efficiently than preformed chitin (13).

Release of Elicitors

Cell wall fragments isolated by biochemical techniques from various fungi can act as elicitors of defense reactions in the plant (27). In plant-pathogen interactions, elicitors might be released from the fungal cell wall by the action of plant hydrolases like chitinase and β-1,3-glucanase. In support of this concept, crude enzyme preparations from pea (28) or soybean (29) released elicitor-active fragments from the cell walls of *Fusarium solani* and *Phytophthora megasperma* cell walls. More recently, a β-1,3-glucanase was purified from soybean; this enzyme released elicitors from the *Phytophthora* cell walls (30).

In the case of chitinase, it is presently unknown whether soluble chitin fragments, generated by chitinase from fungal cell walls, have elicitor activity. In wheat leaves, insoluble chitin was found to act as an elicitor of lignification, a process thought to be important in resistance against pathogens (31). It was proposed that wheat enzymes produced soluble fragments from the chitin which then acted as the natural elicitors. With regard to this hypothesis, it is interesting to note that crude plant homogenates containing chitinase readily hydrolyze chitin to oligosaccharides but usually are almost completely devoid of the exochitinase and chitobiase activities necessary for the further degradation of the oligosaccharides (7, 13). Thus, fungal cell walls might be broken down only partially, and chitooligosaccharides can be expected to persist for a considerable time.

CONCLUSION

In recent years, chitinase has received considerable attention in the context of the resistance of plants against fungal pathogens. The proposal has been put forward to introduce chitinase genes from bacteria into the genome of higher plants; it has been argued that expression of these genes would enhance disease resistance (32). However, as summarized here, plants already possess highly active chitinases and seem to regulate them according to their neeeds. Chitinase activities are generally low in healthy tissues but may be very high in specific cells, e. .g. in laticifers, or may be induced strongly in response to exogenous ethylene, pathogen attack, or elicitors. The enzyme protein responsible for chitinase activity may accumulate in considerable amounts in these cases. In papaya latex, chitinase (assayed and purified as lysozyme) accounted for up to 30 % of the total soluble protein (33). In ethylene-treated bean leaves, chitinase acccumuated to more than 1 % of the total leaf protein within 24 h (7).

The mere quantity of chitinase present there appears to indicate an important function. Since chitinase generally accumulates in parallel with β-1,3-glucanase, and since chitin and β-1,3-glucans represent the backbone of many fungal cell walls, the hypothesis of a function in defense against fungi suggests itself very easily.

Support for this hypothesis comes from indirect evidence only. It has clearly been shown in vitro that chitinase has the ability to attack fungal cell walls. Furthermore, it is likely that chitinase or related enzymes release elicitor-active material from fungal cell walls. However, it remains to be shown to what extent the defense barrier drawn up in the form of chitinase contributes to resistance in individual plant-pathogen interactions, and by what means truly virulent pathogens overcome the barrier.

REFERENCES

1. G. F. Pegg, Glucanohydrolases of higher plants: a possible defence mechanism against parasitic fungi, _in_: "Cell Wall Biochemistry Related to Specificity in Host-Plant Pathogen Interactions," B. Solheim and J. Raa, eds., Universitetsforlaget, Tromsø: 305 (1977).
2. T. Boller, Induction of hydrolases as a defense reaction against pathogens, _in_: "Cellular and Molecular Biology of Plant Stress," J. L. Key and T. Kosuge, eds., A. R. Liss, New York: 247 (1985).
3. L. Zechmeister and G. Toth, Chromatographie der in der Chitinreihe wirksamen Enzyme des Emulsins, _Enzymologia_ 7: 165 (1939).
4. A. Fleming, On a remarkable bacteriolytic element found in tissues and secretions, _Proc. Roy. Soc. Lond._ 93: 306 (1922).
5. J. Bernier, E. Van Leemputten, M. Horisberger, D. A. Bush, and P. Jollès, The turnip lysozyme, _FEBS Lett._ 14: 100 (1971).
6. R. F. Powning and H. Irzykiewicz, Studies on the chitinase system in bean and other seeds, _Comp. Biochem. Physiol._ 14: 127 (1965).
7. T. Boller, A. Gehri, F. Mauch, and U. Vögeli, Chitinase in bean leaves: induction by ethylene, purification, properties, and possible function, _Planta_ 157: 22 (1983).
8. G. F. Pegg and D. H. Young, Purification and characterization of chitinase enzymes from healthy and _Verticillium albo-atrum_-infected tomato plants, and from _V. albo-atrum_, _Physiol. Plant Pathol._ 21: 389 (1982).
9. P. M. Wargo, Lysis of the cell wall of _Armillaria mellea_ by enzymes from forest trees, _Physiol. Plant Pathol._ 5: 99 (1975).
10. J. B. Howard and A. N. Glazer, Papaya lysozyme. Terminal sequences and enzymatic properties, _J. Biol. Chem._ 244: 1399 (1969).
11. A. N. Glazer, A. O. Barel, J. B. Howard, and D. M. Brown, Isolation and characterization of fig lysozyme, _J. Biol. Chem._ 244: 3583 (1969).
12. S. J. Tata, J. J. Beintema, and S. Balabaskaran, The lysozyme of _Hevea brasiliensis_ latex: Isolation, purification, enzyme kinetics and a partial amino acid sequence. _J. Rubb. Res. Inst. Malaysia_ 31: 35 (1983).
13. J. Molano, I. Polacheck, A. Duran, and E. Cabib, An endochitinase from wheat germ, _J. Biol. Chem._ 254: 4901 (1979).
14. S. A. Wadsworth and J. P. Zikakis, Chitinase from soybean seeds: Purification and some properties of the enzyme system. _J. Agric. Food Chem._ 32: 1284 (1984).
15. T. Tsukamoto, D. Koga, A. Ide, T. Ishibashi, M. Horino-Matsushige, K. Yasishita, T. Imoto, Purification and some properties of chitinases from yam, _Dioscorea opposita_ Thumb, _Agric. Biol. Chem._ 48: 931 (1984).
16. F. B. Abeles, R. P. Bosshart, L. E. Forrence, W. H. Habig, Preparation and purification of glucanase and chitinase from bean leaves, _Plant Physiol._ 47: 129 (1971).
17. T. Boller and U. Vögeli, Vacuolar localization of ethylene-induced chitinase in bean leaves, _Plant Physiol._ 74: 442 (1984).

18. A. Toppan and D. Roby, Activité chitinasique de plantes de melon infectées par Colletotrichunm lagenarium ou traitées par l'ethylène, Agronomie 2: 829 (1982).
19. F. Mauch, L. A. Hadwiger, and T.Boller, Ethylene: symptom, not signal for the induction of chitinase and β-1,3-glucanase in pea pods by pathogens and elicitors, Plant Physiol. 76: 607 (1984).
20. G. F. Pegg and D. H. Young, Changes in glycosidase activity and their relationship to fungal colonization during infection of tomato by Verticillium albo-atrum. Physiol. Plant Pathol. 19: 371 (1981).
21. J. Chappell, K. Hahlbrock, and T. Boller, Rapid induction of ethylene biosynthesis in cultured parsley cells by fungal elicitors and relation to the induction of phenylalanine ammonia-lyase, Planta 161: 475 (1984).
22. A. E. Moore, B. A. Stone, Effect of infection with TMV and other viruses on the level of β-1,3-glucan hydrolase in leaves of Nicotiana glutinosa, Virology 50: 791 (1972).
23. G. F. Pegg and J. C. Vessey, Chitinase activity in Lycopersicon esculentum and its relationship to the in vivo lysis of Verticillium albo-atrum mycelium, Physiol. Plant Pathol. 3: 207 (1973).
24. D. H. Young and G. F. Pegg, The action of tomato and Verticillium albo-atrum glycosidases on the hyphal wall of V. albo-atrum, Physiol. Plant Pathol. 21: 411 (1982).
25. V. Farkas, Biosynthesis of cell walls of fungi, Microbiol. Rev. 43: 117 (1979).
26. J. G. H. Wessels and J. H. Sietsma, Fungal cell walls: a survey, in: Encyclopedia Plant Physiol., New Series, Vol. 13B, W. Tanner and F. A. Loewus, eds., Springer, New York: 352 (1981).
27. A. G. Darvill and P. Albersheim, Phytoalexins and their elicitors - a defense against microbial infection in plants, Annu. Rev. Plant Physiol. 35: 243 (1984).
28. L. A. Hadwiger and J. M. Beckman, Chitosan as a component of pea-Fusarium solani interactions, Plant Physiol. 66: 205 (1980).
29. M. Yoshikawa, M. Matama, H. Masago, Release of a soluble phytoalexin elicitor from mycelial walls of Phytophthora megasperma var. sojae by soybean tissues, Plant Physiol. 67: 1032 (1981).
30. N. T. Keen, M. Yoshikawa, β-1,3-endoglucanase from soybean releases elicitor-active carbohydrates from fungus cell walls, Plant Physiol. 71: 460 (1983).
31. R. B. Pearce and J. P. Ride, Chitin and related compounds as elicitors of the lignification response in wounded wheat leaves, Physiol. Plant Pathol. 20: 119 (1982).
32. W. Nitzsche, Chitinase as a possible resistance factor for higher plants, Theor. Appl. Genet. 65: 171 (1983).
33. E. L. Smith, J. R. Kimmel, D. M. Brown, and E. O. P. Thompson, Isolation and properties of a crystalline mercury derivative of a lysozyme from papaya latex, J. Biol. Chem. 215: 67 (1955).

Editor's Report

G. W. GOODAY

PLANT CHITINASES AND LYSOZYMES

based on works by

P. BERNASCONI, M. T. ESQUERRE-TUGAYE, P. JOLLES,
P. E. PILET, D. ROBY, and A. TOPPAN

As shown in detail by Boller (this Volume), chitinase activities are of importance in the defence of plants against pathogens. Here are two further accounts of the occurrence of chitinase and the closely related enzyme lysozyme in plants.

CHITINASES: PLANT DEFENSE PROTEINS RELATED TO RESISTANCE AGAINST FUNGAL PATHOGENS

(D. Roby, A. Toppan and M. T. Esquerré-Tugayé)

Chitinase is thought to be a defense protein of plants against various pests (fungi, insects)[1,2,3], since chitin is not present in higher plants, but is an important component of the fungal cell-wall and insect cuticle. The aim of this work was to check this hypothesis by studying :

- the extent of induction, number and origin of chitinase in melon seedlings infected with *Colletotrichum lagenarium*, a fungal pathogen of melons,

- the regulation of chitinase by elicitors, fungal glycopeptides isolated from the cell-surface of the pathogen, and by the hormone ethylene, which was produced in large quantities by infected plants,

- the relation of chitinase with resistance of plants against the fungal pathogen.

MATERIALS AND METHODS

Melon seedlings were grown in a growth chamber, inoculated at the two-leaf stage with a suspension of *Colletotrichum lagenarium* conidia and harvested 1 to 7 days later. The disease symptoms appeared 3 to 4 days after inoculation, leading to death of the seedlings after 7 days. Healthy plants were simultaneously grown for controls.

Elicitor treatment was performed using the purified elicitor fraction 1[4] ; excised melon leaves were enclosed in test tubes stoppered with serum caps, with the bottom of the petioles immersed in the elicitor solution. Ethylene was measured by gas chromatography.

Induction of resistance against C. lagenarium was obtained by inoculating the elicitor (50 µg Glc eq., fraction 1) on a cotyledon of the melon seedlings, prior to (4 days) inoculation with the pathogen.

Chitinase assays : Exochitinase activity was measured using chitin as a substrate, by the colorimetric assay of Reissig. Endochitinase activity was determined on radiolabelled chitin.

RESULTS

Chitinase activities in healthy and infected melon plants. Exo- and endo-chitinase activities were highly stimulated in infected plants, where they started to increase 3 days after inoculation. Six days after inoculation, there was a 10-fold increase in exo- and a 50-fold increase in endo-chitinase activity, the level of activity being very low in healthy seedlings. Upon purification, chitinase appeared as a multienzymatic chitinolytic system comprising 3 peaks of exo- and 2 major peaks of endo-chitinase activity (figure a), which were also detected, but in low amounts, in healthy plants (figure b). It was checked with antisera prepared against purified endo 1, and purified exo 3, that these two enzymes are present in low amount in healthy plants, by the Ouchterlony double-diffusion procedure.

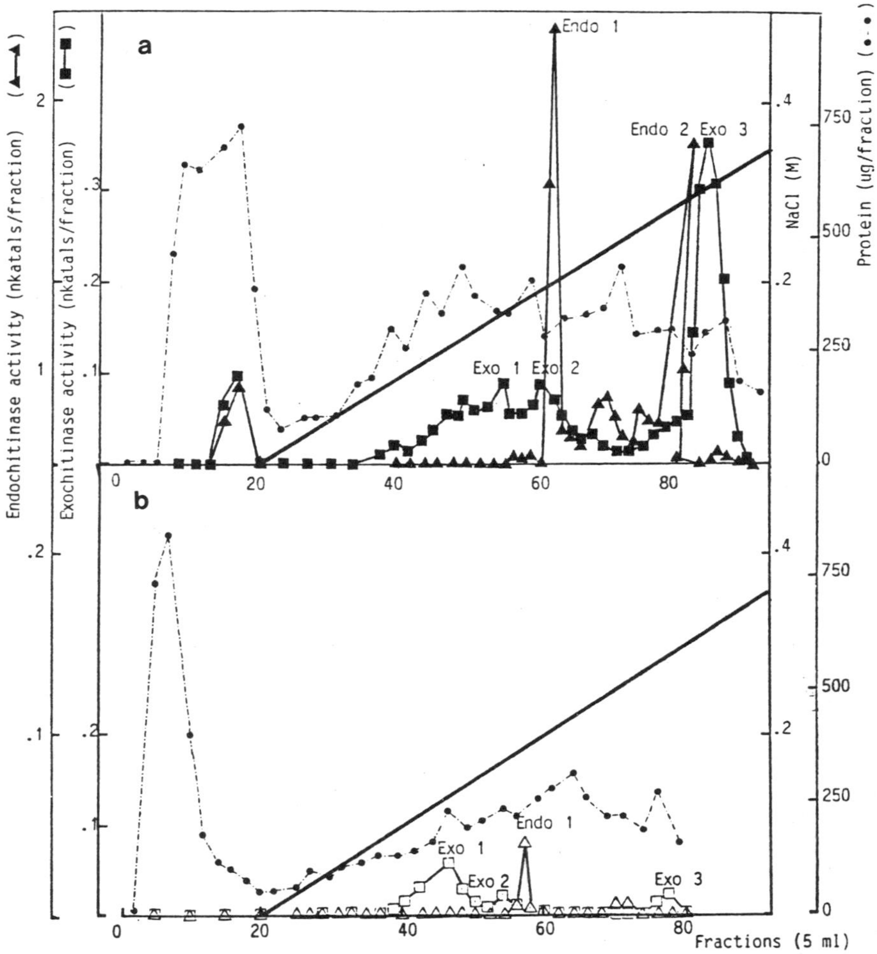

Figure 1. Ion exchange chromatography of partially purified extract of infected melon plants (a) and of healthy plants (b).

Elicitation of chitinase activity and ethylene. Chitinase activity was enhanced within the first 6 hours in elicitor-treated leaves, and became 3 times (exo) and 8 times (endo) higher than in control leaves, after 24 hours. Ethylene was also increased, and played a role in chitinase elicitation[5], as demonstrated by using AVG, a specific inhibitor of ethylene biosynthesis ; AVG inhibited partially elicitor-induced ethylene synthesis, and this inhibition was accompanied by a decrease of elicitor-induced chitinase. The nature of the role of ethylene is not yet understood, but the effect of AVG was significant in as much as it was found repeatedly.

Chitinase activities, chitin content, and elicitor-induced resistance. The elicitor treatment induced a systemic increase in exo- and endo-chitinase activities ; subsequent inoculation by C. lagenarium also enhanced the enzymes, but to a quicker and higher extent in elicitor-treated plants than in control plants. So chitin content was lower in treated plants than in controls.

CONCLUSION

The reported data show that several chitinases, forming a chitin-degrading system, are induced upon infection with a fungal pathogen, or treatment with an elicitor isolated from the cell-surface of the fungus. Ethylene is produced in both situations, and plays a role in the elicitation of chitinase : when partially inhibited by AVG, there is also a partial decrease in chitinase activity. Similar data have been obtained by Chappell et al.[6], about elicitation of phenylalanine ammonia lyase in cultured parsley cells.

This work also shows a positive correlation between the systemic elicitation of chitinase and the systemic protection of melons against C. lagenarium, which strongly supports the role of chitinase as an antifungal enzyme.

REFERENCES

1. P.M. Wargo, Lysis of the cell-wall of Armillaria mellea by enzymes from forest trees, Physiol. Plant Pathol. 5:99-105 (1975).
2. A. Toppan and D. Roby, Activité chitinasique de plantes de Melon infectées par Colletotrichum lagenarium ou traitées par l'éthylène, Agronomie 2:829-834 (1982).
3. L.A. Hadwiger AND D.C. Loschke, Molecular communication in host-parasite interactions : hexosamine polymers (chitosan) as regulator compounds in race-specific and other interactions, Phytopathol. 71(7): 756-862 (1981).
4. A. Toppan and M.T. Esquerré-Tugayé, Cell-surfaces in plant-microorganism interactions. IV. Fungal glycopeptides which elicit the synthesis of ethylene in plants, Plant Physiol. 75:1133-1138 (1984).
5. M.T. Esquerré-Tugayé, D. Mazau, B. Pélissier, D. Roby and A. Toppan, Elicitors and ethylene trigger defense responses in plants, in Ethylene : biochemical, physiological and applied aspects, Fuchs and E. Chalutz eds., Abstract p 217-218 (1984).
6. J. Chappell, K. Hahlbrock and T. Boller, Rapid induction of ethylene biosynthesis in cultured parsley cells by fungal elicitor and its relationship to the induction of phenylalanine ammonia lyase, Planta, 161:475-480 (1984).

PURIFICATION OF LARGE AMOUNTS OF LYSOZYME WITH CHITINASE ACTIVITY FROM RUBUS HISPIDUS CULTURED IN VITRO

(P. Bernasconi, P. Jollès and P. E. Pilet)

Lysozyme (E.C.3.2.1.17) and chitinase (E.C.3.2.1.14) activities in plants have first been characterized by Fleming (1) in several organs of different plants. Studies of papaya and fig latex (2,3), turnip roots (4), wheat germs (5), bean leaves (6) and soja seeds (7) showed that plant lysozymes and chitinases present similar enzymatic and, in part, molecular properties.

RUBUS HISPIDUS IN VITRO CULTURE

In vitro culture of Rubus was first initiated by Gautheret (8), using Heller's medium solidified with 0.8 % agar. Several investigations were then reported (9,10), including lysozyme studies (11,12) in Rubus hispidus callus obtained from cambium cells of stem explants.

RUBUS HISPIDUS SUSPENSION CULTURE

Suspension cultures of Rubus hispidus cells were obtained from previous callus cultures. The growth medium contained 5 % glucose as a carbon source. Heller's mineral medium supplied oligo- and macroelements. No addition of plant hormones or growth substances was needed. 300 ml flasks containing 100 ml of suspension were incubated on an orbital shaker (150 rpm) at 24°C in the dark. The growth (increase in fresh weight and in protein content) started after a two days lag phase. The stationary phase occurred after 15 days of growth. After 17 days, the biomass of the starting material increased 20 times.

CHITINASE, LYSOZYME AND α-MANNOSIDASE ACTIVITIES MEASUREMENTS

Endochitinase activity was determined by the release of soluble fragments from tritiated colloïdal chitin (13). Exochitinase activity was measured by the colorimetric determination of free N-acetylglucosamine monomers released from colloïdal chitin (14,15). Lysozyme activity was followed by the clearence of a suspension of Micrococcus luteus dried cells (12,16). Determination of α-mannosidase was performed using PNP-α-D-mannoside.

CHITINASE AND LYSOZYME ACTIVITIES IN SUSPENSION CULTURES

The cells of Rubus hispidus, cultured in suspension, showed high chitinase and lysozyme activities (17). As described for other plant materials, these activities were localized, in part, in the vacuoles (18,19); a very large amount was found in the growth medium. It was then of interest to compare the subcellular localization of these two activities with that of α-mannosidase, a typical intravacuolar enzyme (18,19). Table I and II indicate the results obtained with a 17 days old culture. As shown, chitinase and lysozyme activities were preferentially exported in the growth medium.

PURIFICATION OF AN ENZYME SHOWING BOTH CHITINASE AND LYSOZYME ACTIVITIES

Investigations performed with the growth medium showed that two enzymes were responsible for the chitinase and lysozyme activities. One of them was comparable to enzymes previously obtained from turnip (4) or bean

Table 1. α-Mannosidase and chitinase activities in vacuoles, whole cells and growth medium of 17 days old suspension culture.

SAMPLE	A α-MANNOSIDASE ΔA/min/0.1 ml	B CHITINASE cpm/min/0.1 ml	B / A
VACUOLES	0.0034 ± 0.0002	1.6 ± 0.3	470 ± 80
CELLS	0.103 ± 0.001	16.2 ± 0.1	157 ± 2
MEDIUM	0.0026 ± 0.0001	7.1 ± 0.1	2700 ± 143

Table 2. Lysozyme activity in vacuoles, whole cells and growth medium of 17 days old suspension culture and lysozyme activity compared to α-mannosidase (see Table 1).

SAMPLE	C LYSOZYME % lysis/min/0.1 ml	C / A
VACUOLES	0.0260 ± 0.0007	7.6 ± 0.5
CELLS	1.502 ± 0.002	14.6 ± 0.2
MEDIUM	1.890 ± 0.005	727 ± 30

leaves (6). This enzyme possessed a high pI value allowing its purification by a one-step procedure (17). Starting from the filtered growth medium it was performed using a DEAE A-50 Sephadex column and a 20 mM TRIS-HCl buffer, pH 9. The enzyme was not retained on the column. By SDS-PAGE, only one band was characterized (17) after purification. A batch procedure, using the same gel and buffer conditions, was also developed. It gave the same results. When compared to the activity characterized in the initial growth medium, the purified enzyme behaved as a lysozyme with a certain chitinase activity. Another protein, responsible for a part of the chitinase activity and retained on the gel, must be present in the growth medium.
Very little exochitinase activity was detected with the purified lysozyme sample. As incubation of the pure form with colloïdal chitin released only a very small amount of free N-acetylglucosamine, this enzyme seemed to hydrolyse high molecular weight chitin polymers but not smaller chitin oligomers. After incubation of these hydrolysis products with Helix pomatia β-glucuronidase, specific for low molecular weight chitin polymers, free monomers were released.

CONCLUSION

The preparation and maintenance of Rubus hispidus suspension cultures was relatively easy. The cells exported chitinase and lysozyme activities in the growth medium. An enzyme showing both activities was purified in a one-step procedure from the filtered growth medium. The purified enzyme was, in part, similar to other plant chitinases or lysozymes.

REFERENCES

1. A. Fleming, Proc. Roy. Soc. 26 (1932) 71-84.
2. J.B. Howard and A.N. Glazer, J. Biol. Chem. 244 (1969) 1399-1409.
3. J.B. Glazer, A.O. Barel, J.B. Howard and D.M. Brown, J. Biol. Chem. 244 (1969) 3583-3589.
4. I. Bernier, E. Van Leemputten, M. Horisberger, D.A. Bush and P. Jollès, FEBS Lett. 14 (1971) 100-104.
5. J. Molano, I. Polacheck, A. Duran and E. Cabib, J. Biol. Chem. 254 (1979) 4901-4907.
6. T. Boller, A. Gehri, F. Mauch and U. Voegeli, Planta 157 (1983) 22-31.
7. S.A. Wadsworth and J.P. Zikakis, J. Agric. Food Chem. 32 (1984) 1284-1288.
8. R.G. Gautheret, La culture des tissus végétaux, (1959) Masson, p. 179 and 238.
9. A. Benbadis, Rev. Cytol. Biol. Vég. 27 (1964) 225-235.
10. P.E. Pilet and J.C. Roland, Cytobiol. 4 (1971) 41-61.
11. P.E. Pilet, J. Jollès and P. Jollès, Z. Pflanzenphysiol. 111 (1983) 91-95.
12. P.E. Pilet and P. Bernasconi, Plant Sc. Lett. 35 (1984) 147-151.
13. J. Molano, A. Duran and E. Cabib, Anal. Biochem. 83 (1977) 648-656.
14. C. Jeuniaux, Methods Enzymol. 8 (1966) 644-650.
15. J.A. Rupley, Bioch. Biophys. Acta 83 (1964) 245-255.
16. P. Jollès, Methods Enzymol. 5 (1961) 137-140.
17. P. Bernasconi, P. Jollès and P.E. Pilet, FEBS Lett. (submitted)
18. T. Boller and H. Kunde, Plant Physiol. 63 (1979) 1123-1132.
19. U. Voegeli and T. Boller, Plant Physiol. 74 (1984) 442-444.

Affiliations

D. Roby, A. Toppan, M. T. Esquerré-Tugayé, Centre de Physiologie Végétale, U.A. CNRS 241, 118 route de Narbonne, F-31062 Toulouse, France.

P. Bernasconi, P. Jollès and P. E. Pilet, Institut del Physiologie et Biologie Végétales, Université, Bâtiment Biologie, CH-1015 Lausanne, Switzerland.

THE ROLE OF CHITINASE AND CHITIN SYNTHETASE INHIBITORS IN CONTROLLING PLANT PATHOGENIC FUNGI

I. CHET, E. COHEN and I. ELSTER

The Hebrew University of Jerusalem, Faculty of Agriculture
Rehovot 76100, Israel

INTRODUCTION

The cell walls of *Sclerotium rolfsii* are composed of β-1,3-glucan and chitin[1,5], whereas members of the Oomycetes e.g. *Pythium* sp. also contain cellulose[1].

Since these components are the most important ones any biodegradation or inhibition of their synthesis will significantly affect the viability of the pathogens.

In recent years the increasing use of potentially hazardous fungicides in agriculture has been the subject of growing concern of both environmentalists and public health authorities. Application of biocontrol agents such as *Trichoderma harzianum* capable of degrading cell walls of *Sclerotium rolfsii*, *Rhizoctonia solani* and *Fusarium* spp. or using chitin-synthetase inhibitors against *S. rolfsii* are two potential means for controlling soil-borne plant pathogenic fungi without using common fungicides.

Use of Chitinase and β-1,3-Glucanase-Producing *Trichoderma* for Biological Control

When *Trichoderma harzianum* was grown in liquid culture chitinase excretion into the growth medium was enhanced by concentrations of chitin up to 5 CU within 44 h of incubation. Only slight enzyme activity was detected when N-acetylglucosamine or glucose was the carbon source in the medium[9].

Optimal growth conditions of *T. harzianum* for chitinase excretion when grown on chitin were 28°C and pH 5. It was found that 45 and 64% of chitinase activity was lost after incubation at 40 and 60°C, respectively, and that activity was totally lost after incubation for 1 h at 70°C or higher. At pH 9 the enzyme activity was minimally damaged. Optimal activity of chitinase in cell-free extract was found at 35°C and pH 5-6 in the reaction mixture. Similarly the fungus excreted β-1,3-glucanase when grown on laminarin[9].

In many experiments showing a successful biological control the antagonistic Trichoderma was found to be a mycoparasite[2,8].

From recent work it appears that mycoparasitism is a complex process including several successive steps. The initial detectable interaction shows that the hyphae of the mycoparasite grow directly towards its host[3]. This phenomenon appears as chemotropic growth of the Trichoderma to some stimuli in the host's hypha. When the mycoparasite reaches the host its hyphae often coil around it (Fig. 1) or are attached to it. This attachment is apparently rather specific and depends on lectin-carbohydrate binding between the antagonist and its host.

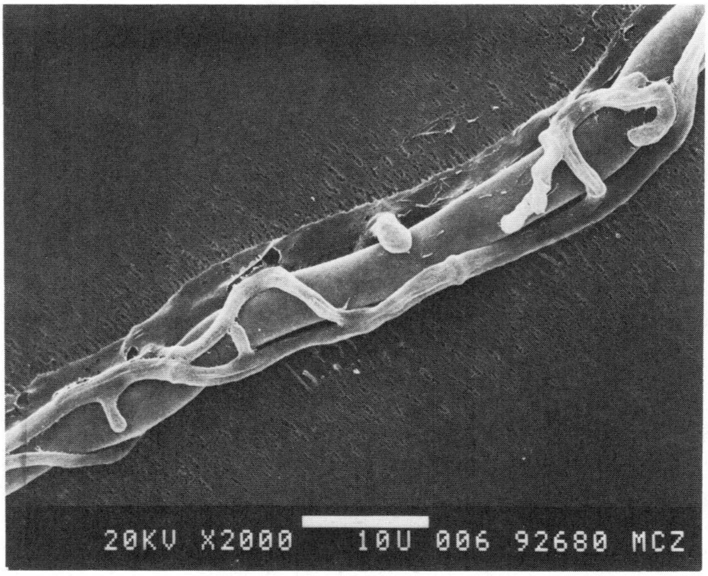

Fig. 1. Trichoderma harzianum coils around Sclerotium rolfsii.

Following these interactions, the mycoparasite sometimes penetrates the host mycelium apparently by partially degrading its cell wall by the two key lytic enzymes chitinase and β-1,3-glucanase. Both enzymes must act simultaneously to decompose hyphal walls because of the linkage between the chitin and β-glucan. Sclerotia contain melanin in the rind, which apparently increases cell wall resistance to biodegradation[4]. Trichoderma harzianum enzymes even degrade sclerotia of S. rolfsii[7].

Control of Plant Diseases Caused by Sclerotium rolfsii Using Polyoxin D

A chitin synthetase preparation from mycelium of the plant pathogenic fungus Sclerotium rolfsii was partially characterized. Enzyme activity was stimulated by N-acetyl-D-glucosamine and Mg^{++}. Optimal temperature for the enzyme is 30°C and optimum pH about 6.4. The extracted chitin synthetase requires proteolytic activation[6].

The fungal enzyme was significantly inhibited by the antibiotic polyoxin D. Hyphae of S. rolfsii treated with Polyoxin-D became swollen and showed many depressions. Moreover, the chemical also inhibited sclerotial formation[6].

In greenhouse experiments we applied Polyoxin to soil and found about 85% reduction in disease incidence of musk melons at 10 mg/Kg soil of the antibiotics (Fig. 2). Similar results were also obtained in bean, as can be observed in Table 1.

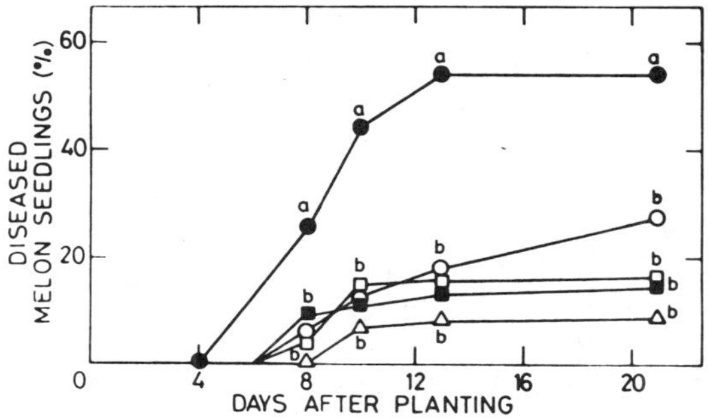

Fig. 2. The effect of Polyoxin-D in soil on disease incidence in musk melon caused by Sclerotium rolfsii.

Amount of Polyoxin applied: (●) - 0; (O) - 5 (□) - 7.5; (Δ) - 10 mg/Kg soil; ■ - 25 μg as a seed coating.

Table 1. Control of disease caused by Sclerotium rolfsii in beans by Polyoxin D

Polyoxin D (mg/Kg soil)	Pre-emergence disease incidence (%)	Dampiness disease incidence (%)	Total Disease incidence (%)	Disease reduction (%)
0	29	49	78	-
2.5	13	19	32	59
5	15	2	17	79
7.5	4	3	7	91
10	0	0	0	100

Polyoxin-D was mixed in 20% water solution of the polysaccharide "Pelgel" (Nitreagin, Wisconsin USA) which served as an adhesive material. Each 3 ml of solution was used to coat 150 gr of chick-pea seeds. Comparison between soil application and seed coating of Polyoxin-D can be seen in Fig. 3.

Seed coating was calculated as mg Polyoxin-D per Kg soil (calculated by multiplication of coating level per one seed x 20 seeds).

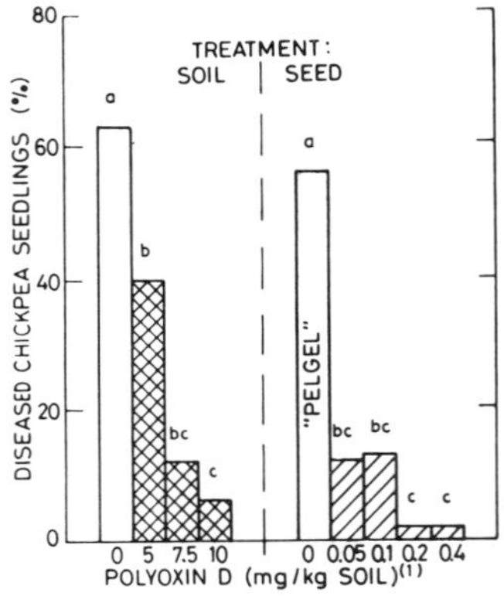

Fig. 3. The effect of Polyoxin-D application technique on chick-pea disease incidence caused by S. rolfsii.

We have shown that Polyoxin-D can serve as an efficient fungicide against Sclerotium rolfsii in three different crops. Seed coating enables the use of very small amounts. This antibiotic has the potential to replace some hazardous toxic fungicides against soil borne plant pathogenic fungi.

Acknowledgement. This work was partially supported by a grant from the National Council for Research and Development, Israel, and the G.S.F., München, F.R. Germany.

REFERENCES

1. S. Bartnicki-Garcia, Fungal cell wall composition, in: "Handbook of Microbiology", Vol. 2, pp. 201-214, Chemical Rubber Co., Cleveland, Ohio (1973).
2. I. Chet and R. Baker, Isolation and biocontrol potential of Trichoderma hamatum from soil naturally suppressive of Rhizoctonia solani, Phytopathology 71:286 (1981).
3. I. Chet, G.E. Harman, and R. Baker, Trichoderma hamatum: its hyphal interactions with Rhizoctonia solani and Pythium spp., Microb. Ecol. 7:29 (1981).
4. I. Chet and Y. Henis, Effect of catechol and disodium EDTA on melanin content of hyphal and sclerotial walls of Sclerotium rolfsii Sacc. and the role of melanin in the susceptibility of these walls to β-1,3-glucanase and chitinase. Soil Biol. Biochem. 1:131 (1969).
5. I. Chet, Y. Henis, and R. Mitchell, Chemical composition of hyphal and sclerotial walls of Sclerotium rolfsii Sacc., Can. J. Microbiol. 13:137 (1967).
6. E. Cohen, I. Elster, and I. Chet, Properties and inhibition of Sclerotium rolfsii chitin synthetase (in press) 1985.
7. Y. Elad, R. Barak, and I. Chet, Parasitism of sclerotia of Sclerotium rolfsii by Trichoderma harzianum, Soil Biol. Biochem. 16:381 (1984).
8. Y. Elad, R. Barak, I. Chet, and Y. Henis, Ultrastructural studies of the interaction between Trichoderma spp. and plant pathogenic fungi, Phytopath. Z. 107:168 (1983).
9. Y. Elad, I. Chet, and Y. Henis, Degradation of plant pathogenic fungi by Trichoderma harzianum. Can. J. Microbiol. 28:719 (1982).

Editor's Report
G. W. GOODAY

CHITINASE ACTIVITIES IN ANIMALS, FUNGI AND BACTERIA

based on works by

T. AMATATSU, A. ANDO, E. DANULAT, Z. ER-EL, T. FUJII, W. P. KAMMEL, D. KLEIN, C.P. KUBICEK, A. B. SAADA, M. SHIMADA, R. A. SMUCKER, E. M. KUBICEK-PRANZ, K. MIZUSHINA, F. STREICHSBIER, A. TODAR, J. TOTH, D. A. WRIGHT, M. YABUKI, M. YAMASHITA, and E. ZOMER

Digestive chitinase activities are widely distributed among animals, fungi and bacteria. Here follow a series of accounts of specific occurrences and properties of these and related enzymes.

ORIGIN OF CHITINASE ACTIVITY IN COD, GADUS MORHUA (L.) (E. Danulat)

Previous studies have demonstrated the occurence of high levels of chitinolytic activity in the stomach and the intestine of *Gadus morhua*[1]. However, in common with the situation in other fish[2,3], it remains unclear whether the activity was primarily produced by the fish or was of bacterial origin. The present paper describes experiments designed to resolve this problem monitoring levels of chitinolytic activity in the gut of cod following the selective inhibition of bacterial growth and metabolism by means of antibiotic treatment.

MATERIALS AND METHODS

These experiments were carried out in the Biologische Anstalt Helgoland where 6 *G. morhua* (23-34 cm) were kept in aerated 330 l-tanks, supplied with filtered sea water (17±1°C; 31‰ S). Fish food consisted of live shrimps (*Crangon* sp.). Either Ampicillin and Sisomicin (A-S)+, or Chloramphenicol, were mixed with the tank water of the fish to give final concentrations of 50 ppm each. Control fish were maintained in similiar conditions except that no antibiotics were added. Every 24 h, the water was changed and the addition of antibiotics was repeated. After 24, 48 and 72 h, each two treated and two control fish were killed.

Chitinase activity was assayed in enzyme extracts of stomach contents, stomach tissue, pyloric caeca, intestinal contents and intestine tissue[1,4]. The number of chitinolytic bacteria was determined on chitin agar plates after incubation at 18°C for 4 weeks.

RESULTS AND DISCUSSION

Both Ampicillin/Sisomicin and Chloramphenicol proved to be very effective in eliminating bacteria from the digestive tract of cod: having subjected the fish to the antibiotics for 48 or 72 h, all of the test plates with sample homogenates diluted from 10^{-3} to 10^{-6} were free of bacterial colonies. TABLE 1 presents a survey of the results obtained for the stomach contents and the intestinal contents. Similiar reduction of bacterial numbers was also obtained in samples of stomach and intestinal tissues and the pyloric caeca. In contrast, chitinoclasts in numbers of up to 10^9 were determined for samples of the control fish in both experiments.

The Ampicillin/Sisomicin-experiment resulted in high levels of chitinolytic activity in samples of the stomach, the pyloric caeca and the intestinal contents. TABLE 2 shows average chitinolytic activities, which in case of medicated fish were calculated exclusively if no bacterial colonies were detected on the test plates. Comparing the results of antibiotically treated cod to the ones obtained for control fish, no significant differences in enzyme activity can be found for any of the samples of G. morhua.

Table I. Maximum numbers of chitinoclasts (CB/g ww) in stomach contents (st.c.) and intestinal contents (int.c) of medicated cod, compared to control fish.

Treat-ment	Chitinolytic bacteria per gram wet weight (CB/g ww) 24 h		48 h		72 h	
	st.c.	int.c.	st.c.	int.c.	st.c.	int.c.
+ A-S	9.3×10^8	2.2×10^9	$< 10^4$	$< 10^4$	$< 10^4$	$< 10^4$
- A-S	7.2×10^8	3.0×10^9	9.0×10^7	2.6×10^8	5.4×10^7	9.0×10^8
+ CP	(-)		$< 10^4$	$< 10^4$	$< 10^4$	$< 10^4$
- CP	(-)		1.2×10^8	1.8×10^8	(-)	7.5×10^9

+/- A-S : with/without Ampicillin-Sisomicin-treatment via the tank-water
+/- CP : with/without Chloramphenicol-treatment via the tank-water

Table II. Chitinolytic activities (GlcNAc, $\mu g \times h^{-1} \times g^{-1}$) in samples of cod treated with antibiotics, compared to control fish.

Sample		Chitinolytic activity (GlcNAc, $\mu g \times h^{-1} \times g^{-1}$ wet weight) + A-S	- A-S	+ CP	- CP
Stomach contents	$\bar{x}$	406.24	334.31	281.74	229.11
	$\pm$ S.E.	227.82	116.65	220.11	92.51
	n	4	6	2	2
Stomach tissue	$\bar{x}$	557.77	695.57	553.85	692.36
	$\pm$ S.E.	331.10	284.34	170.93	432.10
	n	5	6	4	4
Pyloric caeca	$\bar{x}$	242.47	276.10	90.29	139.10
	$\pm$ S.E.	113.90	78.79	32.30	31.06
	n	5	6	4	4
Intestinal contents	$\bar{x}$	369.95	486.70	130.14	142.04
	$\pm$ S.E.	140.95	210.73	65.67	8.11
	n	4	6	4	4
Intestinal tissue	$\bar{x}$	35.99	51.89	15.76	43.67
	$\pm$ S.E.	32.39	29.89	7.68	30.26

In general, levels of chitinolytic activity determined in the Chloramphenicol experiment are lower than the ones obtained in the Ampicillin/Sisomicin study (TABLE 2). In previous studies, it was found that chitinase secretion is greatly stimulated by feeding crustaceans to the cod repeatedly. Since the Chloramphenicol experiment was started 4 days after the fish were captured and the cod were fed only once at the beginning of the 72 h experimental period, the comparatively low results may be explained by a lack of enzyme stimulation.

In the pyloric caeca and the intestinal tissue of Chloramphenicol treated fish, average chitinolytic activities (48 h; 72 h) are significantly lower than those obtained in control fish. A more detailed look at the data reveals that enzyme activities decreased considerably also in samples of the stomach contents and the stomach tissue after the cod were exposed to Chloramphenicol for 72 h. The prolonged medication with this antibiotic may have toxic effects not only on the bacteria but also on the fish. As described by Oppenheimer[5], Chloramphenicol in a concentration of 50 ppm had lethal effects on small fish within 12 h. Goodrich and Morita[3] treated juvenile *Enophrys bison* with Chloramphenicol (50 ppm; for 24, 72, 144h) before assaying chitinase activity and bacterial numbers in the stomach. Having eliminated the bacteria in the stomach contents, chitinolytic activity had decreased by 90 % after 72/144 h[3]. The authors concluded that there is only bacterial chitinase in the digestive tract of *Enophrys bison*, not mentioning possible toxic effects of Chloramphenicol on the fish.

As to cod, the results obtained from experiments with Ampicillin/Sisomicin and with Chloramphenicol clearly indicate that chitinolytic activities in the digestive tract of this fish primarily originate in tissues. In addition to fish chitinases, some bacterial chitinase may occur, especially in the pyloric caeca and the intestine where the pH conditions are much more favorable for bacteria than in the stomach.

Acknowledgement. This project has been funded by the Deutsche Forschungsgemeinschaft (DFG). Thanks to Bayer AG, Germany, for generous donations of the antibiotics α-Amoniobenzyl-Penicillin (Ampicillin) and Sisomicinsulfate. The results of this paper are part of the author's Ph.D. dissertation. Authorization for pre-publication was given by the Graduation Committee of the Biology Department, University of Hamburg.

REFERENCES

1. E. Danulat & H. Kausch, Chitinase activity in the digestive tract of the cod, Gadus morhua (L.), *J. Fish Biol.* 24: 125-133 (1984)
2. K. Okutani, Studies of chitinolytic systems in the digestive tracts of Lateolabrax japonicus, *Bull. Misaki Mar. Biol. Inst.* 10: 1-47 (1966)
3. T.D. Goodrich & R.Y. Morita, Bacterial chitinase in the stomachs of marine fishes from Yaquina Bay, Oregon, USA, *Mar. Biol.* 41: 355-360 (1977)
4. C. Jeuniaux, Chitinases, *Meth. Enzym.* 8: 644-650 (1966)
5. C.H. Oppenheimer, The effect of marine bacteria on the development and hatching of pelagic fish eggs, and the control of such bacteria by antibiotics, *Copeia* 1: 43-49 (1955)

N-ACETYL-β-D-GLUCOSAMINIDASE ENZYMES FROM TRICHODERMA REESEI

(E. M. Kubicek-Pranz, W. P. Kammel, F. Streichsbier, J. Toth and C. P. Kubicek)

The filamentous fungus *Trichoderma reesei* is a highly efficient protein producing organism and yields of up to 20 mg/ml have been reported (1).

T. reesei is generally used for the production of cellulolytic enzymes (2), but other enzymes can also be produced in a high yield (3 - 6). The application of *Trichoderma* sp. to produce enzymes capable of lysing fungal cell walls and their application for protoplast formation is also well established (7); however, the chitinolytic enzymes have so far not been investigated. In the present paper we document some results on N-acetyl-β-D-glucosaminidase enzymes from *T. reesei*.

MATERIALS AND METHODS

The organism used was *T. reesei* QM 9414 obtained from ATCC. The fungus was cultivated in shake flasks on a rotary shaker at 30° C in a medium containing 1 % (w/v) of the carbon source indicated in "Results" and the other nutrients as described elsewhere (8). The pH was controlled by the aid of citrate-phosphate buffer (9). During growth, at appropriate intervals portions of the culture broth were aseptically taken off, centrifuged (34000 x g, 4° C, 20 min) and the supernatants filtered through a sintered-glass filter (G 4) filled with a layer of glass wool. Samples could then be stored at -20° C for at least 4 weeks. Controls showed that different times of storage did not affect the enzyme activities nor the multiple band pattern of N-acetyl-β-D-glucosaminidase from *T. reesei*.

N-Acetyl-β-D-glucosaminidase was assayed by the hydrolysis of p-nitrophenyl-N-acetyl-β-D-glucosaminide. Substrate was used at a final concentration of 1 mM; 50 mM citrate buffer pH 4.8 was used as a buffering agent. Incubation was carried out at 50° C for 10 min, after which the reaction was terminated by the addition of 2 ml 1 M Na_2CO_3. The yellow colour was diluted appropriately with distilled water, and the extinction read at 405 nm. A molar extinction coefficient of 18.5 x 10^6 cm^2/mol was used for the calculations. All other enzymes are assayed as described previously (10, 11). When N,N'-diactelychitobiose and N,N',N"-triacetylchitotriose were used as the substrate, the activity waş determined (12). All activities are expressed as µmoles of product formed per min under the conditions of the assay. Specific activities are given as µmoles product formed per min (= 1 U) per mg of protein, which was determined by the dye binding procedure (13).

Cell walls from *T. reesei* were prepared as described (10). Incubation of cell walls with agents in the attempt to release N-acetyl-β-D-glucosaminidase was carried out (14).

Separation of active proteins was carried out by chromatofocusing using a fast protein liquid chromatography system (15). SDS-PAGE of individual peak fractions was carried out (16), and proteins were stained by the Silver procedure (17).

RESULTS

N-Acetyl-β-D-glucosaminidase was formed by *T. reesei* during growth on various carbon sources (Table I). Although the presence of an "inducing" carbon source like chitin or chitobiose or also isolated *T. reesei* cell walls lead to the highest activities, high activities were also found during growth on slowly metabolizable carbon sources (e.g. lactose). Hence the formation of N-acetyl-β-D-glucosaminidase appears to be controlled by induction and (de)repression.

During these experiments it was found that part of N-acetyl-β-D-glucosaminidase always remained associated with the cellular debris after centrifugation or filtration of the culture broth. No intracellular activity was found (Table II). Therefore all values reported above (Table I) are given as total activities (i.e. cell wall bound plus truely "extracellular" activity).

Table I. Formation of β-N-acetyl-glucosaminidase and β-N-acetyl-galactosaminidase by T. reesei QM 9114 during growth on different carbon sources.

Carbon source (1 %, w/v)	β-GlcNAc-ase 72 h	β-GlcNAc-ase 168 h	β-GalNAc-ase 72 h	β-GalNAc-ase 168 h	Ratio
Glucose	18.3	24.6	6.4	12.0	2.8/2.0
Cellulose	39.2	53.6	24.0	29.0	1.6/1.8
Starch	44.5	58.0	11.3	49.1	3.9/1.2
Lactose	201.7	175.2	17.1	31.3	11.8/10.3
Maltose	57.2	115.8	10.0	28.6	5.7/4.0
Cellobiose	46.2	81.1	9.3	74.6	5.0/1.1
Chitobiose	234.4	268.2	56.3	68.5	3.4/3.9
Chitin	267.7	380.5	47.3	77.9	5.6/4.9
Cell walls	50.5	199.1	21.8	44.9	2.3/4.4

Activities were assayed using PNP-N-acetyl-glucosaminide and PNP-N-acetyl-galactosaminide as substrates and are expressed as mU/ml of total culture (culture supernatant plus homogenized cell debris). Except for the carbon source used, the cultivation medium of Labudova and Farkas (9), was used. Cell walls were isolated from T. reesei QM 9414 as described by Kubicek (10).

The term "Ratio" is defined as $\frac{\beta\text{-GlcNac-ase (mU/ml)}}{\beta\text{-GalNAc-ase (mU/ml)}}$;

Abbreviations: β-GlcNAc-ase; β-N-acetyl-glucosaminidase; β-GalNAc-ase, β-N acetyl-galactosaminidase.

Table II. Localization of β-N-acetyl glucosaminidase from T. reesei QM 9414 (values as mU/ml).

Carbon source	Glucose	Chitobiose
Culture supernatan	8.2	122.4
Cell free extract	ND	ND
Homogenized mycelia	10.1	112.0

Carbon sources were used at a final concentration of 1 % (w/v). Activities were determined on cultures grown for 72 h. At this time the degree of autolysis- determined by measuring malate dehydrogenase distribution (23)- was below 18 %. All values are related to volume of the original culture. ND below the detection limit.

Some studies were carried out in order to release the bound activity from isolated cell walls. However, neither distilled water, tap water and 8 M LiCl nor various detergents (ETDA, SDS, Triton X-100, Tween 80, β-octyl-glucoside) released more than 7 % of the bound activity into the supernatant. In this respect N-acetyl-β-D-glucosaminidase closely resembles β-glucosidase from the same organism (14).

Initial attempts to purify the enzyme by ion exchange chromatography revealed the presence of several, poorly separated peaks of activity. Consequently a more potent separation technique (chromatofocusing by means of a fast protein liquid chromatography system) was applied. By the aid of this technique the N-acetyl-β-D-glucosaminidase pool from the T. reesei culture fluid could be resolved into 7 peaks of activity. This profile (Fig. 1) could reproducibly be obtained with different enzyme samples (obtained under the same controlled culture conditions) and is thus no artefact. Isopropanol preci-

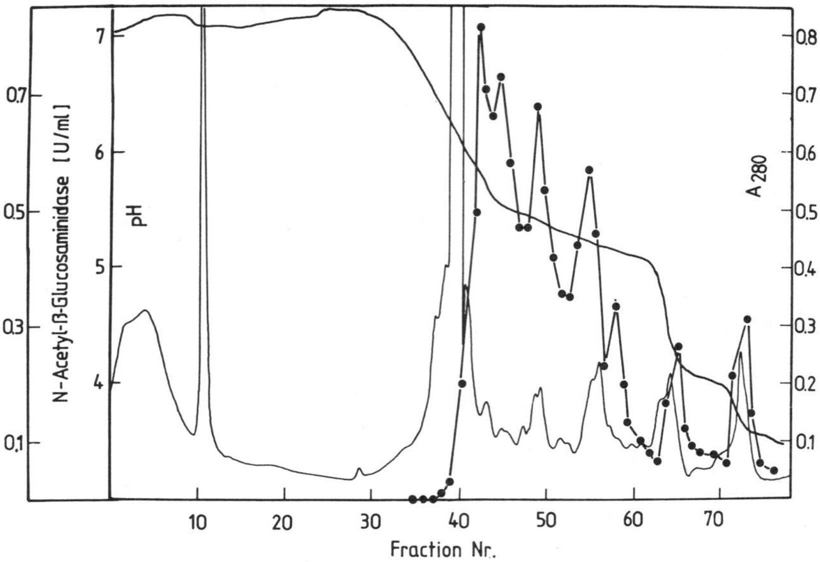

Fig. 1. Separation of β-N-acetyl-glucosaminidase enzymes from a freshly harvested culture filtrate of T. reesei QM 9414 by chromatofocusing on Mono P. Conditions for the experiment were: flow rate 1 ml/min; starting buffer 0.025 M imidazole HC1 pH 7.2; the gradient was developed with polybuffer 74, diluted 1:12. pH and protein (A_{280}) were monitored automatically. Fractions of 1.5 ml were collected and assayed for activity by means of PNP-GlcNAc.

pitated samples or samples concentrated by lyophilization showed less peaks of activity, but since these procedures were always accompanied by losses in total activity this might be due to denaturation of some of the multiple forms.

Fig. 2 shows silver stained SDS-PAGE of the proteins present in the 7 N-acetyl-β-D-glucosaminidase peaks: only fraction 1 exhibited only one major protein peak corresponding to 56 kD; all other fractions were composed of several proteins of different molecular weight. Peaks 4 and 5 had a too low protein content to be even detected by means of the silver staining technique. This is surprising in view of the high A_{280} peak in peak 4, but has not been investigated further.

The presence of several additional proteins in all of the 7 peaks was also found by assaying for various additional hydrolytic enzymes: as can be seen from Table III, all peaks also contained activity of acid phosphatase, and all but two (3 and 7) also contained β-1,6-glucanase. Coincident with the SDS-PAGE (Fig. 2) profile, peak 7 contained the most hydrolytic enzymes.

Although we had not obtained purified entities of the various N-acetyl-β-glucosaminidase peaks, some preliminary investigation were carried out on their substrate specificity (Table IV). All fractions were shown to hydrolyze p-nitrophenil-N-acetyl-β-D-glucosaminide, p-nitrophenyl-N-acetyl-β-Dglucosaminide, and p-nitrophenyl-diacetyl-chitobiose, although the proportions of activities obtained were not always identical. All but one peak (IV), hydrolysed diacetyl-chitobiose. Hydrolysis of triacetylchitotriose was generally close to or below the detection limit.

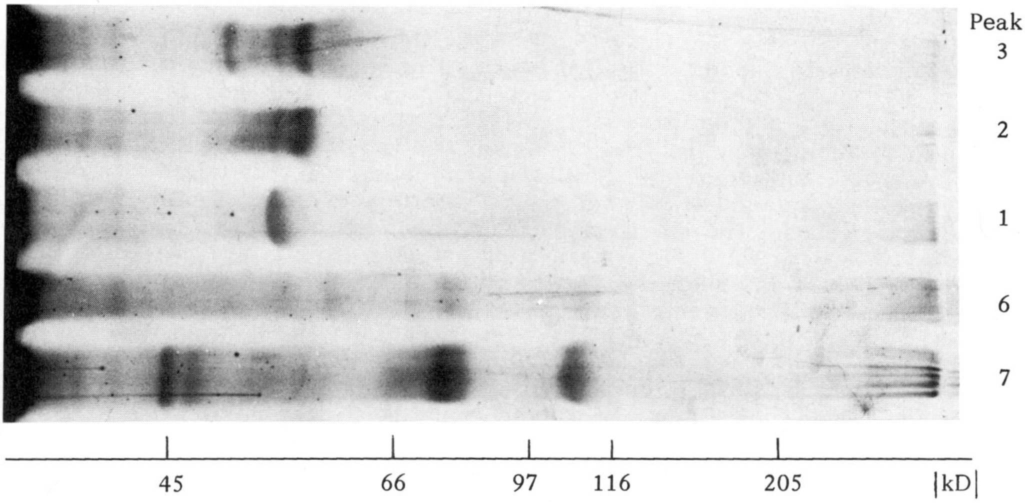

Fig. 2. SDS-PAGE 4 (16) of fractions I - VII. Proteins were stained by the silver procedure (17).

Table III. Presence of various hydrolytic enzyme activities in peak fractions containing β-N-acetyl-glucosaminidase activity.

Enzyme	Substrate	Activities (mU/ml)						
		I	II	III	IV	V	VI	VII
β-1,3-glucanase	Yeast laminarin	-	-	-	-	-	-	-
β-1,6-glucanase	Pustulan	14.6	37.5	-	27.6	28.3	11.9	-
Chitinase	Chitin	-	-	149	187	117	95	-
Xylanase	Larch wood xylan	-	-	-	-	-	-	-
Mannanase	Yeast mannan	-	-	-	-	-	-	-
β-galactosidase	PNP-galactoside	-	-	-	-	-	-	28
β-xylosidase	PNP-xyloside	-	-	-	-	-	-	230
β-glucuronidase	PNP-glucuronide	-	-	-	-	-	-	-
Acid phosphatase	PNP-phosphate	451	243	121	47	61	43	172
β-glucosidase	PNP-glucoside	-	-	-	-	-	-	27

Abbreviation: PNP= p-nitro-phenyl

Table IV. Substrate specificity of N-acetyl-β-D-glucosaminidase from *T. reesei*.

Substrate	Peak fractions						
	I	II	III	IV	V	VI	VII
PNP*-β-D-N-acetyl-GlcN	100	100	100	100	100	100	100
PNP-β-D-N-acetyl-GalN	31	83	53	50	42	42	34
PNP-β-D-diacetyl-chitobioside	32	39	31	18	19	11	44
Diacetyl-chitobiose	9	11	13	21	8	BD	11
Triacetyl-chitotriose	4	4	BD	BD	3	BD	5

* PNP= para-nitrophenyl-; activities are presented as percentage of activity with PNP-N-acetyl-β-D-glucosaminide to enable a better comparison. Peak fractions refer to the various peaks shown in the chromatofocusing experiment from Fig. 1.

BD: below detection limit (3 %).

DISCUSSION

The present report is to the best of our knowledge the first on N-acetyl-β-D-glucosaminidase from *T. reesei*; although the enzyme has previously been characterized from other fungi (12, 18, 19). In none of these cases, however, have so many multiple forms of this enzyme been detected as in the present case. Therefore all possible precautions were undertaken to ascertain that their appearance was not due to an artefact, e.g. unspecific retardation on the column. The high reproducibility with which they were repeatedly obtained, however, would argue against an artefact. This is also supported by the findings that the substrate specificity of the individual protein fractions was quantitatively differing; hence the proteins in the individual peaks appeared to be slightly different. The appearance of multiple forms has been documented with a number of other extracellular enzymes, although the reason for their origin is still unclear. In the case of *T. reesei* and cellulases proteolytic attack in the culture filtrate might be responsible for this phenomenon (20, 21), but recent evidence shows that this is not the case (9, 15). We have shown that the cellulases which are secreted by T. reesei protoplasts induced *de novo* already appear in multiple forms (22, 23); thus it is likely that the event responsible must be looked for intracellularity. From the results of the present paper it appears that N-acetyl-β-D-glucosaminidase should be an attractive candidate for investigating the reason for multiplicity of extracellular enzymes in *T. reesei*.

As mentioned above, the individual peaks of activity showed some variability in the efficiency to hydrolyse various substrates, which was also true for p-nitrophenyl-N-acetyl-β-D-glucosaminidase. This appears to be important in view of the findings that the ratio of activity (p-nitrophenyl-N-acetyl-glucosaminidase/p-nitrophenyl-N-acetyl-galactosaminidase) varied with regard to the carbon source used (Table I). This is most probably due to partial inactivation of some of the multiple forms during cultivation, which has also been shown for carboxymethyl cellulase and β-glucosidase from *T. reesei* (15). The different susceptibility of different peaks of activity to inactivation is also reflected by the finding that lyophilization of the crude culture filtrate leads to some loss of activity and also of some peaks, although the protein profile during isoelectric focusing remains constant (unpublished experiments). We have thus no reason for assuming that the activity of N-acetyl-β-D-galactosaminidase is due to a separate enzyme. A variation in the ratio (see above) appears rather to be due to inactivation of individual multiple forms of the enzyme which however will only be observed by using better protein separation techniques.

Acknowledgement. This paper was supported by Osterreichischer Founds zur Förderung wissenschaftlicher Forschung, project P 5423.

REFERENCES

1. M. Mandels, *Ann. Rep. Ferm. Proc.*, 5:35 (1982).
2. Y. H. Lee, and L. H. Fan, *Adv. Biochem. Engin.*, 17:101 (1980).
3. D. M. Alabran, D. H. Ball, and E. T. Reese, *Carbohyd. Res.*, 123:179 (1983).
4. T. Haltmeyer, M. Leissola, D. Ulmer, R. Waldner, and A. Fiechter, *Biotechnol. Bioengin.*, 25:1685 (1983).
5. S. Hasegawa, J. Nordin, and S. Kirkwood, *J. Biol. Chem.*, 244:5460 (1969).
6. S. Bielicki, and E. Galas, *Proc. Bioconversion Symp.*, IIT Delhi, 203 (1977).
7. O. M. H. De Vries, and J. G. H. Wessels, *J. Gen. Microbiol.*, 73:13 (1972).
8. M. Mandels, and R. E. Andreotti, *Proc. Biochem.*, 13:6 (1978).
9. I. Labudova, and V. Farkas, *Biochim. Biophys. Acta*, 744:135 (1983).

10. C. P. Kubicek, Arch. Microbiol., 132:349 (1982).
11. W. Harrer, J. Wurth, C. P. Kubicek, M. Röhr, and J. Marihart, Eur. J. Appl. Microbiol. Biotechnol., 17:339 (1983).
12. F. Reyes, and R. J. W.-Byrde, Biochem. J., 131:381 (1973).
13. M. M. Bradford, Anal. Biochem., 72:248 (1976).
14. C. P. Kubicek, Eur. J. Appl. Microbiol. Biotechnol., 13:226 (1981).
15. W. P. Kammel, and C. P. Kubicek, J. Appl. Biochem., (1985) (in press).
16. U. K. Laemmli, Nature, 227:680 (1970).
17. C. R. Merril, D. Goldman, and M. Ebert, Science, 211:1437 (1981).
18. T. Mega, T. Ikenaka, and Y. Matsushima, J. Biochem., 68:109 (1970).
19 C. S. Jones, and D. J. Kosman, J. Biol. Chem., 255:11861 (1980).
20 M. Nakayama, Y. Tomita, H. Suzuki, and K. Nishizawa, J. Biochem., 79:955 (1976).
21. C. S. Gong, M. Ladisch, and G. T. Tsao, Adv. Chem. Ser., 105:261 (1979).
22. H. Kolar, H. Mischak, W. P. Kammel, and C. P. Kubicek, J. Gen. Microbiol., (1985) (in press).
23. H. Mischak, C. P. Kubicek, and M. Röhr, Appl. Microbiol. Biotechnol., 21:27 (1985).

ENZYMATIC DEGRADATION OF FUNGAL CELL WALLS WITH RESPECT TO POLYSACCHARIDE COMPOSITION

(E. Zomer, A. B. Saada, D. Klein, A. Todar and Z. Er-el)

Lytic enzymes effective on fungal cell walls have been reported extensively in the literature. The wide range of lytic organisms is not surprising considering the diversity in wall composition of fungi.

The current study was to develop an enzymatic model for a) evaluation of fungal cell wall polysaccharide composition as a screening method of fungi with high content of chitin; b) mild degradation of fungal cell wall for the extraction of valuable intracellular components, proteins and nucleic acids; c) preparation of fungal protoplasts for genetic engineering.

Available commercial enzymes were found inadequate for this study. The first objective of the project was to develop glucanase and chitinase with purity and activity suitable for this study.

β-1-3-Glucose glycanohydrolase (glucanase) was prepared and purified from *Oerskovia xanthineolytica*. The method was modified after Scott and Schekman (1) using Baker's yeast and/or *Aspergillus niger* ghosts as inducers. The process was optimized and scaled up to a 500 1 fermentor. The enzyme was purified by ultrafiltration using a Romicon Hollow fiber system, precipitated with cold acetone and finally dried by vacuum filtration. The preparation was low in chitinase, protease and nuclease activity.

β-1-4 N-Acetyl glucosamine-glycanohydrolase (chitinase) was prepared from a selected strain of *Streptomyces griseus* grown on crude crab chitin and *Calliphora erythrocephala* puparium. The enzyme was purified again by the ultrafiltration system, precipitated with cold acetone and vacuum dried. No glucanase activity could be detected in this preparation. Chitinase activity was 10-fold higher than the Sigma preparation from an identical strain.

Five strains of fungi were chosen to test the enzymatic model: 1. *Rhizopus arrizus*; 2. *Aspergillus niger*; 3. *Morchella crassipes*;

4. Candida utilis; 5. Pachysolen tannophilus. They represent various groups of fungi and are known to contain different compositions of cell wall polysaccharides. The degradation products from each enzyme were examined by total increase in reducing ends and HPLC or TLC for product composition analysis. According to the results, a mixture of chitinase and glucanase was prepared. The fungi were treated at 1-5% concentration (DW/DW) at 4°C to 45°C. Lytic activity was measured by the decrease in optical density at 650 nm, or by the formation of spheroplast (microscopically). Release of preinduced intracellular enzymes was also recorded.

The results were very encouraging. Intracellular enzymes were successfully extracted with high yield and higher specific activity than the conventional way of physical disruption.

Uricase was produced from Aspergillus niger. Extraction done by combined treatment with glucanase and chitinase gave the highest yield (Table 1).

Table 1. Release of uricase from A. niger: combined action of glucanase and chitinase

TREATMENT	URICASE ACTIVITY (u/g)		
	INCUBATION TIME, h		
	1.5	3	5
Control	0.9	0.6	1.4
Glucanase (2%)	1.9	3.7	5.7
Chitinase (2%)	6.0	11.0	6.9
Glucanase (2%) + Chitinase (2%)	10.5	11.7	9.7

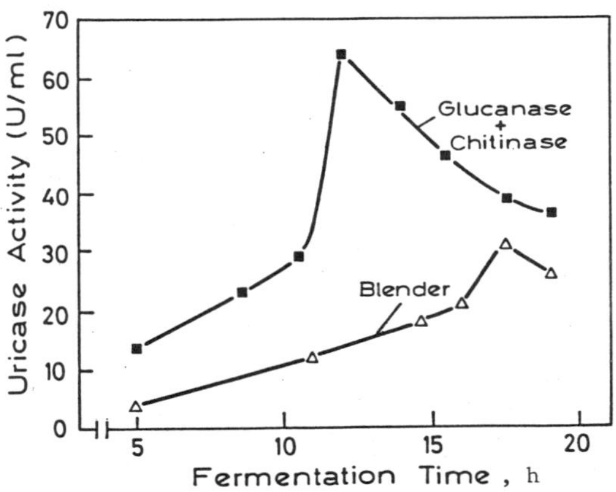

Fig. 1. Production of uricase from Aspergillus niger after induction with uric acid. Comparison of two methods of extraction: enzymatic vs. mechanical.

Enzyme yield with the enzymatic method gave twice the amount than the conventional method using a blender or homogenizer (Fig. 1). The enzyme was obtained at the end of the exponential growth with shorter fermentation time, while mechanical disruption required an additional 5 hrs for maximum yield

Another advantage for the enzymatic extraction of intracellular enzymes is the purified form of the crude extract. Formate dehydrogenase (FDH) was extracted from Torulopsis glabrata using glucanase. In less than an hour over 90% of the FDH was obtained in highly purified form (Fig. 2).

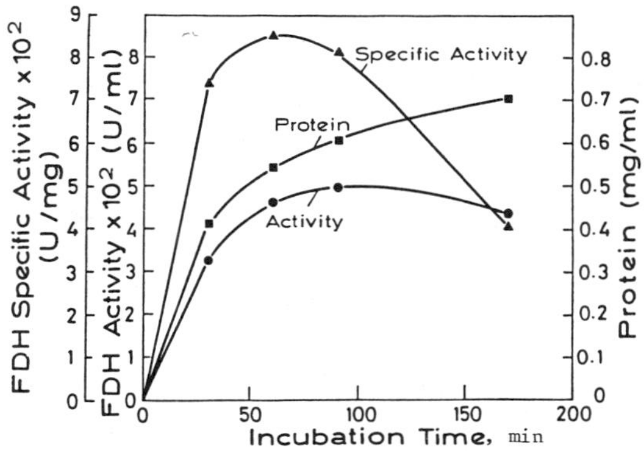

Fig. 2. Extraction of formate dehydrogenase from Torulopsis glabrata by glucanase.

The use of the chitinase/glucanase complex simplified the method and better control of the solubilization was achieved. Scale up from mgs to kgs amounts was simple and gave a good yield even at a high concentration of cells (10% DW/V).

Intracellular enzymes produced by this method include: Glycerol kinase from Candida utilis, alcohol dehydrogenase and xylose dehydrogenase from Pachysolen tannophilus, formate dehydrogenase from Torulopsis glabrata, uricase from Aspergillus niger, mushroom flavor components from Morchella crassipes; spheroplast formation was observed with Candida utilis, Pachysolen tannophilus, Asperigillus niger and Rhizopus arrizus.

REFERENCE

J.H. Scott and R. Schekman, J. Bacteriol. 148:414 (1980).

OYSTER CHITINASE ACTIVITY AS A FUNCTION OF IONIC CONCENTRATION OF THE MEDIUM

(R. A. Smucker and D. A. Wright)

Chitin is known as a potent chelating natural polymer (1, 2) which has the property of effectively binding transition metal ions. Various metal ions are also known to inhibit enzymatic activities of the chitinase complex, although some may actually stimulate chitinase activity at low levels (3, 4). In this work we describe investigations of nutrient ion effects on chitinase complexes isolated from the crystalline style of the American oyster, Crassostrea virginica (5).

MATERIALS AND METHODS

Oysters were collected from the Patuxent estuary and stored in the through flow seawater system of Chesapeake Biological Laboratory (15 °/°°) until use. Crystalline styles used for chitinase determination were removed by making a lateral incision mid-way along the style sac. Styles were either used immediately or stored in liquid nitrogen before use.

Styles were pooled from 10 - 15 animals and homogenized in a Teflon-glass homogenizer. The extract was filtered through a Gelman A/E glass fiber filter before use. Enzyme extract for use in ionic and trace metal studies was dialyzed over 3 days at 4° C. This was followed by a 2 day dialysis against EDTA buffered to pH 4.4 using HEPES. Final dialysis was against 3 volumes of distilled water over 24 h. Major cations and trace metals in extracts were checked using atomic absorption spectroscopy, and were found to be negligible.

The method of Hirano (6), modified by Molano (7), was used to produce highly acetylated chitin. This was washed with distilled water of a medium porosity glass filter and was analysed for major ions (Na^+, Ca^{++}, Mg^{++}) and trace metals as before.

Experimental media were prepared as follows: temperature, 10 °/°° seawater, buffered to pH 5.0 with citrate-phosphate buffer; pH, natural sea-water was diluted to 10 °/°° salinity with distilled water and buffered to a 3 - 7 range using a 20 mM citrate-phosphate buffer system; salinity, natural seawater diluted with distilled water; major cations, magnesium, calcium and sodium salts (as chloride) were added singly or in combination to distilled water buffered to pH 5.0 with 10 mM sodium acetate buffer.

RESULTS

Chitinase activity occurred over a broad temperature range (0 - 58° C), but with an optimum of 33 - 34° C. All subsequent experiments were conducted at 34° C.

Figure 1 and 2 illustrate pH and salinity optima, respectively, for <u>C. virginica</u> cystalline style chitinase. The pH was not controlled in the salinity experiment and varied from 7.1 in 5 °/°° salinity. De-ionized water "controls" had a pH of 6. Although Figure 2 shows marked pH effects over the range 6

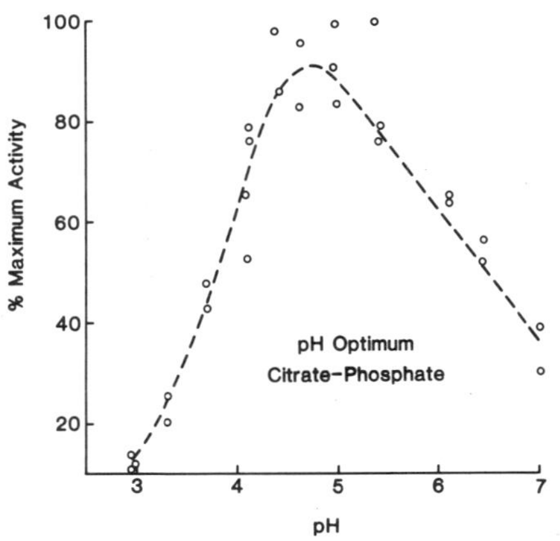

Fig. 1. Effect of pH on oyster chitinase activity.

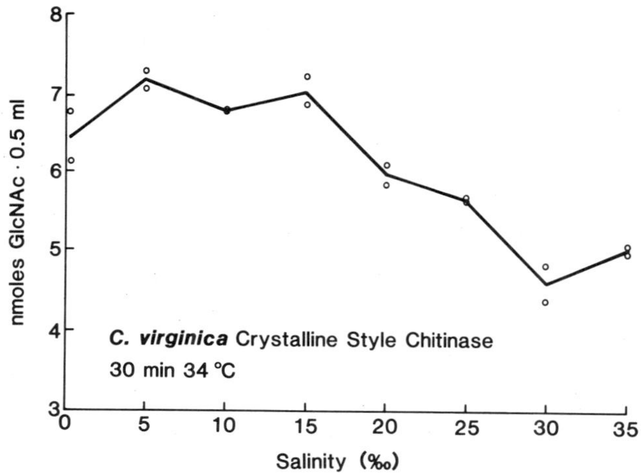

Fig. 2. Effect of salinity on oyster chitinase activity.

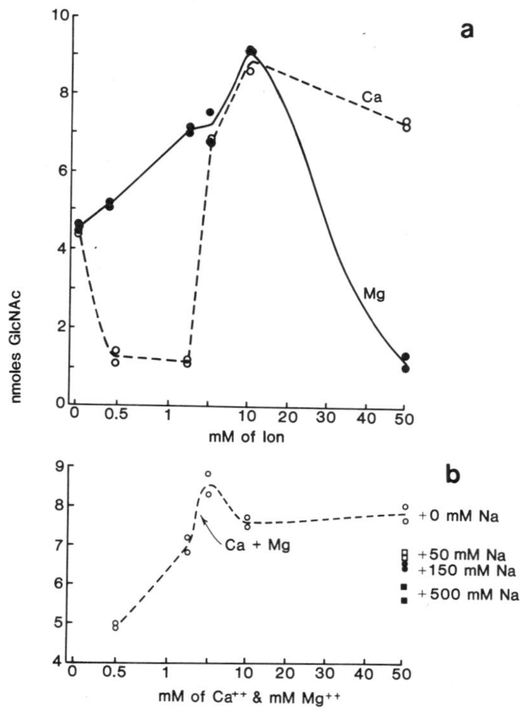

Fig. 3. Effect of various cations on oyster chitinase activity.

- 7, the stability of chitinase activity over the 0 - 15 °/°° salinity range indicates that total ionic strength overrides pH in this respect. In an experiment designed to test the effect of the individual major physiological cations on chitinase activity, enzyme substrate mixtures were incubated in a series of media containing various cation mixtures. Calcium and magnesium concentrations ranged from 0.5 to 50 mM. These ions were tested either singly (Fig. 3a) or in combination (Fig. 3b) and were compared with distilled water controls. Sodium was added to a Ca/Mg mixture (50 mM of each) at concentrations of 50 mM, 150 mM and 500 mM. Results are included in Figure 3b.

DISCUSSION

The current investigation was designed to characterize the relative influence of major ions on chitinase activity in the crystalline style of the American oyster. When considering the current data, it is useful to bear in mind the concentrations in which these metals appear in seawater: Ca, 10 mM; Mg, 50 mM; Na, 470 mM.

It is apparent from the current work that calcium and magnesium play important roles as co-factors in the collective chitinase activity, and that they tend to be antagonistic to each other. In this regard, it is particularly interesting that when these ions are considered singly, they each have an inhibitory action over part of their normal concentration range. For example from 0.5 to 2.0 mM, Ca on its own inhibits chitinase activity. Magnesium on the other hand stimulates enzyme activity over this concentration range and tends to override the inhibitory effect of calcium when both ions are present in equimolar concentrations. The reverse was true when these ions were present in concentrations of 50 mM, which is close to the natural magnesium concentration in 100 % seawater. In normal (100 %) seawater it is clear that this inhibitory effect is offset by the presence of calcium. Sodium, too, when considered singly tends to be more inhibitory at increasingly higher concentrations. In summary, chitinase activity in full strength seawater may be seen as the resultant of the strongly inhibitory effect of magnesium and the slightly inhibitory effect of sodium on one hand, and the stimulatory effect of calcium on the other.

REFERENCES

1. R. A. A. Muzzarelli, "Chitin", Pergamon Press, New York (1977).
2. R. Maruca, B. J. Suder, and J. P. Wightman, Interaction of heavy metals with chitin and chitosan III chromium, J. Appl. Polymer Sci., 27:4827 (1982).
3. A. Ohtakara, M. Mitsutomi, and Y. Uenida, Purification and some properties of chitinase from Vibrio sp., J. Ferm. Tech., 57(3):169 (1979).
4. J. Skujins, A. Pukite, and A. D. McLaren, Enzymologia, 39(6):353 (1970).
5. R. A. Smucker, and D. A. Wright, Chitinase activity in the crystalline style of the American oyster Crassostrea virginica, Comp. Biochem. Physiol, 77A:239 (1984).
6. S. Hirano, Y. Ohe, and H. Ono, Selective N-Acylation of chitosan, Carbohidrate Res., 47:315 (1976).
7. J. Molano, A. Duran, and E. Cabib, A rapid and sensitive assay for chitinase using tritiated chitin, Anal. Biochem., 83:648 (1977).

INTERPRETATIVE MODEL FOR ENZYMATIC CHITIN HYDROLYSIS

(R. A. Smucker)

Kinetic studies of insoluble substrates such as cellulose and chitin present the researcher with problems that classic solution kinetics does not satisfactorily explain. Lee and Lee (1) report, as invalid, any comparisons of initial rates for cellulose hydrolysis to data from a prolonged hydrolysis period. Molan, Duran and Cabib (2) attributed the non-linearity of enzymatic chitin hydrolysis to structural heterogeneity of the chitin substrate but not to instability of enzyme preparations or to methodology.

During our course of study regarding extant cell-free chitinase in natural water, sediment, and soil, there was an unexpected plateau in activity as a function of time (Fig. 1). There was apparent end-product inhibition with only a small amount of end product present. The present work provides kinetic and direct microscopic evidence supporting a heterogeneous substrate hypothesis for colloidal chitin. An interpretative kinetic model is presented for enzymatic hydrolysis of colloidal chitin.

MATERIALS AND METHODS

Tritiated (C-1)-colloidal chitin was prepared as described previously (2, 3). Chitosan (Sigma) (in 10 % acetic acid) was diluted 1:4 with methanol and reacetylated with excess acetic anhydride containing ^{3}H-(C-1)-acetic anhydride. Details of preparation and dialysis for low background were as described previously (2).

Because of tritium lability being a problem with environmental samples, ^{14}C-(C-1)-acetic anhydride was used for labeling colloidal chitin at the C-1 position of acetate in the N-acetyl-D-glucosamine moiety.

Chitinase from <u>Streptomyces griseus</u> (Sigma), <u>Crassostrea virginica</u> crystalline style (Smucker and Wright, this Volume), and in natural sediment were separately assayed for chitin hydrolysis. Reactions were intitiated by addition of 30 - 50 μl radiolabeled substrate to 500 μl enzyme solution. In most cases toluene was added to the reaction mixture prior to addition of substrate. Controls were enzyme solutions heated at 100° C for 20 minutes or treated with 0.2 μl 10 % trichloracetic acid prior to addition of substrate. At appropriate time intervals, reactions were terminated with 0.2 ml of 10 % TCA per 0.5 μl liquid scintillation counting. Pattern of end product formation was determined for <u>S. griseus</u> and oyster style by HPLC (4).

Regenerated chitin (2, 5) and <u>Thalassiosira fluviatilis</u> chitin were prepared by freeze-etch replication for transmission electron microscopy (6). Samples were examined and photographed in an RCA-EM-U4.

RESULTS AND DISCUSSION

Chitin hydrolysis data from both sediment (Fig. 1) and from oyster crystalline style homogenate (Fig. 2) reflect multiple product release rates. The image in Fig. 3 is typical of the fibril arrangement in regenerated chitin colloidal preparations. Water background and the colloidal bulk phase are readily distinguished by the colloidal boundary. Small fibrils at the boundary are available to the enzyme populations. In contrast, most of the remaining fibril populations are in large interconnected groups or are on the "inside" of the colloid, not available to enzymes because of diffusion constraints and crystallinity regions. The surface microfibrils should be readily available to chitinase whereas the bulk phase becomes functional after a short lag period. Figure 4 also demonstrates fibril ends in native diatom chitin fibers which are generally considered highly crystalline.

Colloidal chitin clearly has multiple substrate types which should be considered when studying chitin-chitinase interactions, especially when evaluating low levels of chitinase activity. Barrett-Bee and Hamilton (7) demonstrated a corroborative enzyme-substrate phenomenon when using low levels of <u>Candida</u> chitinase and regenerated chitin in their binding studies.

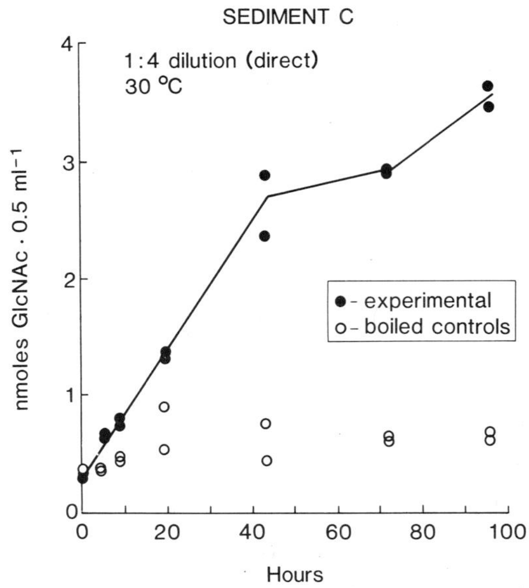

Fig. 1. Chitinase activity of clay sediments.

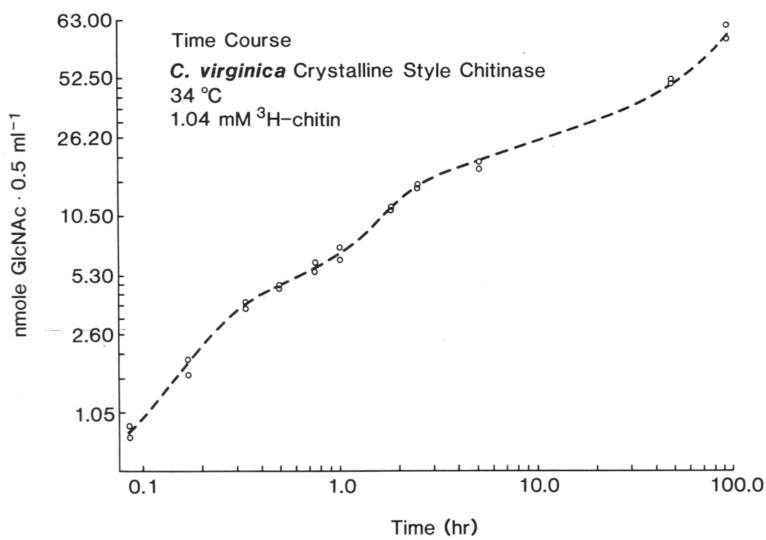

Fig. 2. Chitinase activity from cell-free oyster styles.

Figure 5 summarizes the major points of a kinetic curve using Streptomyces griseus chitinase and regenerated chitin as a model. The lag period (phase II) can be explained by a projected low-turnover of chitinase while the more crystalline regions of the colloid surface are opened, exposing new fibril ends. New fibril ends then are formed continuously and steady state is achieved in phase III.

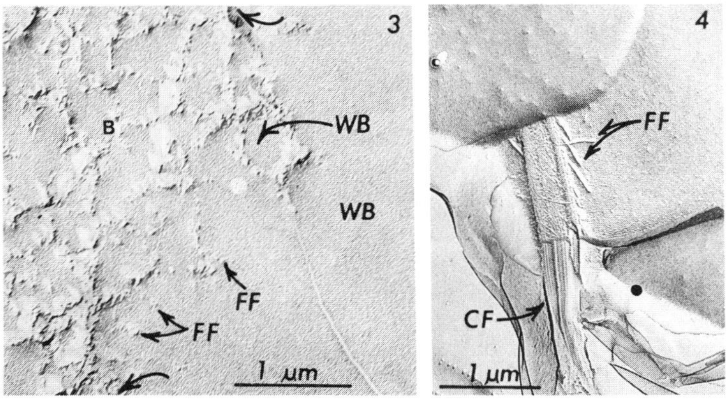

Fig. 3 - 4. Freeze-etch replicas of colloidal chitin (3) and Thalassiosira chitin (4). Free fibrils (FF); bulkphase (B); water background (WB); crystalline fibrils (CF); arrows indicate nominal colloid-water interface.

The present study provides kinetic and structural data supporting a multiple phase kinetic model for regenerated chitin and plausibly for other "particulate" chitin preparations. These phases must be taken into account when using data from assays with low chitinase:chitin ratios or from short term assays.

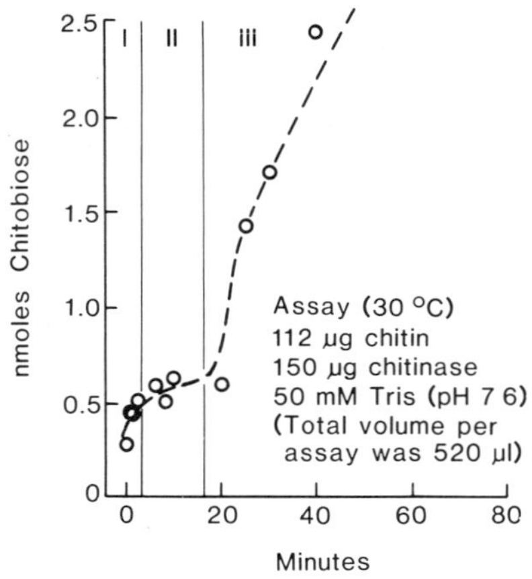

Fig. 5. Streptomyces griseus chitinase activity. Model phases: I) early linear phase; II) fibril opening phase; III) second linear phase.

REFERENCES

1. L. T. Lee, and Y. H. Lee, Kinetic studies of enzymatic hydrolysis of insoluble cellulose: Analysis of the initial rates, Biotechnol. Bioengin., 24:2383 (1982).
2. J. Molano, A. Duran, and E. Cabib, A rapid and sensitive assay for chitinase using tritiated chitin, Anal. Biochem., 83:648 (1977).
3. R. A. Smucker, and C. K. Kim, Effects of phosphate on Streptomyces griseus chitinase production, in: "Chitin, Chitosan, and Related Enzymes", J. P. Zikakis, ed., Academic Press, 423 (1984).
4. R. A. Smucker, C. E. Warnes, and C. J. Haviland, Chitinase production by a freshwater pseudomonas, in: "Biodeterioration VI.", G. C. Llewellyn, and E. C. O'Rear, eds., (in press).
5. S. Hirano, Y. Ohe, and H. Ono, Selective N-acetylation of chitosan, Carbohydr. Res., 47:315 (1976).
6. R. L. Steere, R. A. Smucker, and E. F. Erbe, Inexpensive modification of Balzer's free-etch units for production of high resolution specimens, in: "Proc. of 41st Ann Mtg., EMSA", G. W. Bailey, ed., San Francisco Press, Inc., San Francisco, CA, (1983).
7. K. Barrett-Bee, and M. Hamilton, The detection and analysis of chitinase activity from the yeast of Candida albicans, J. Gen. Microbiol., 130:857 (1984).

PURIFICATION AND CHARACTERIZATION OF CHITINASE AND CHITOBIASE PRODUCED BY AEROMONAS HYDROPHILA SUBSP. ANAEROGENES A52

(M. Yabuki, K. Mizushina, T. Amatatsu, A. Ando, T. Fujii, M. Shimada, and M. Yamashita)

In making fungal protoplasts it is necessary to lyse cells with an enzyme mixture generally composed of glucanase(s) and chitinase(1). For this purpose, we isolated a bacterium which was a strong producer of chitinase (EC 3.2.1.14) and chitobiase (EC 3.2.1.29). Characterization of the enzymes and application of them in formation of fungal protoplasts were studied.

EXPERIMENTS AND RESULTS

(1) Isolation of chitinase producing bacterium and taxonomic study

Isolation of the bacterium was performed on chitin-agar plates containing 1% colloidal chitin, 0.1% peptone, 0.1% KH_2PO_4, 0.05% $MgSO_4 7H_2O$, 2% agar and pH was adjusted to 7.0. A loopful suspension of soil sample was inoculated on the plates, and after appropriate incubation at 30°C colonies surrounded by clear zones resulting from lysis of chitin were picked up. The most powerful producer of chitinase was selected on the basis of size of a clear zone formed by the organism.

The isolated bacterium was characterized by the systematic methods(2,3). Major taxonomic features are listed in Table 1. The organism resembled much *Aeromonas hydrophila* subsp. *anaerogenes* biotype 2, but it assimilates glucosamine or N-acetylglucosamine as sole source of carbon and nitrogen and can produce large amount of chitinase extracellularly. We designated the isolate as *A. hydrophila* subsp. *anaerogenes* A52.

(2) Formation of the enzymes by the isolate

Standard composition of culture medium for production of the enzyme was established as follows: 1.0% shrimp shell chitin, 0.5% peptone, 0.5% yeast extract, 0.1% K_2HPO_4, 0.3% NaCl and pH was adjusted to 7.0. In shaking culture, 500 ml-conical flasks containing 75 ml of the medium were cultivated at 30°C under rotary shaking of 210 rpm. In jar fermenter culture, 200-l jar containing 100 l of the medium was cultivated at 28°C with agitation of 150 rpm under aeration of 100 l/min, but chitin concentration was reduced to 0.5%.

In flask culture, Fig. 1, chitinase production reached its maximum after 24 h incubation, however, appearance of chitobiase was observed after 48 h and reached its maximum after 96 h. In jar fermenter, Fig 2, formation of chitinase reached its maximum at 18 h faster than in flask culture and the amount of the enzyme was also 50% larger.

Table 1. Major taxonomic features of the isolate

Cells straight, rod shaped with rounded ends, 1 X 2.4 µm	Not sensitive to vibrio-static agent (0/129)
Motile by polar flagellum, monotrichous	Acetoin was formed from 2,3-butanediol but not from glucose
Gram negative	Gas was not produced from both glucose and glycerol
Growth not dependent on NaCl, but max growth at 1% NaCl	Grown in KCN broth
Did not produce water soluble brown pigment	Chitin hydrolyzed
Cytochrome oxidase and catalase positive	Glucosamine and N-acetylglucosamine were used as sole source of carbon and nitrogen
Gluconate oxidase negative	

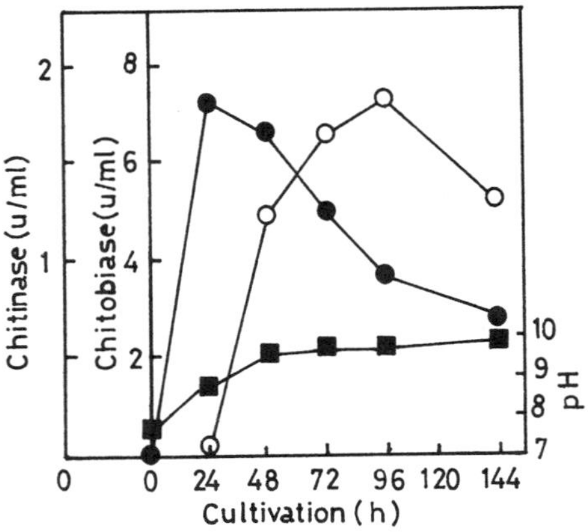

Fig. 1. Production of chitinase and chitobiase by flask culture. Chitinase(●), chitobiase(○), and pH(■). Activity measurements were performed as previously reported(4).

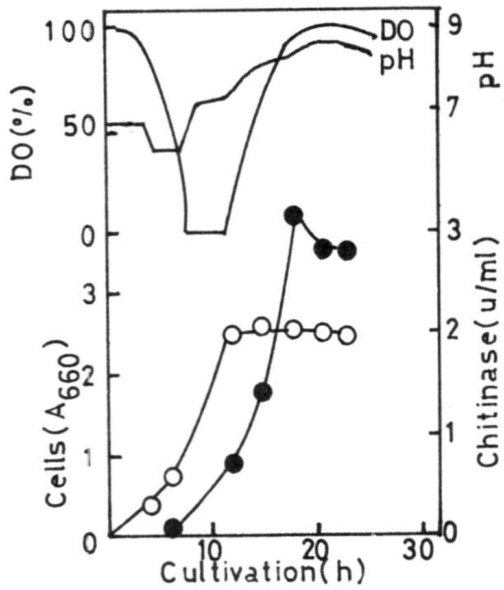

Fig. 2. Production of chitinase by jar fermenter culture. Chitinase(●), cells(○), dissolved oxygen(DO) and pH are presented.

Table 2. Characteristics of chitinase and chitobiase

	Chitinase	Chitobiase
Inhibitor	Hg^{2+}, AsO_2^-, PO_4^{3-}, monoiodoacetate, N-ethylmaleimide, p-chloromercuri-benzoate	Hg^{2+}, monoiodoacetate, N-ethylmaleimide, p-chloromercuribenzoate
Molecular weight		
by gel-filtration	115,000	106,000
by SDS-PAGE	110,000	109,000
Michaelis constant (Km)	1.35 mg/ml (chitin)	1.7×10^{-4} M (p-nitrophenyl- β-D-N-acetyl glucosamine
Opt pH for activity	7.0	7.0
Stable range of pH	6.0-8.5	7.0-9.0
Opt temp for activity	45°C	50°C
Stable range of temp	-45°C	-45°C

(3) Purification and properties of chitinase and chitobiase

Culture filtrate was concentrated by ultrafiltration followed by affinity adsorption on chitin. Chitobiase remained in liquid fraction. Chitin-bound chitinase was released by autodigestion and further purified by CM-Sephadex C50 column chromatography. Chitobiase fraction was purified by successive column chromatography with CM-cellulose and DEAE-Sephadex A50. The purified enzyme preparations showed homogeneity on polyacrylamide gel disc electrophoresis. Characteristics of the enzymes are listed in Table 2.

Molecular weight of the enzymes were approximately 110,000 for chitinase and 100,000 for chitobiase. The values are two to three fold higher than those reported previously (5,6,7). In addition, pI's of the enzymes were 4.60 for chitinase and 5.35 for chitobiase differing from those reported earlier (7,8). The enzymes showed a glycoprotein nature.

(4) Formation of fungal protoplasts with chitinase preparation

Partially purified chitinase through affinity adsorption was lyophilyzed and used for preparing protoplasts from fungal cells. The enzyme was combined with Zymolyase(β-glucanases) and lytic system was composed as indicated in Table 3.

Table 3. Composition of protoplasting system

Component	Initial concn	Final concn	volume(ml)
Mycelia	40mg/ml	20 mg/ml	5.0
KCl	4 M	0.5 M	1.25
Zymolyase(β-glucanases)	2.0 mg/ml	0.12 mg/ml	0.6
Chitinase	11.1 mg/ml	1.11 mg/ml	1.0
Distilled water			2.15
Total volume			10.0

The 16-h cultured mycelia of *Aspergillus oryzae* M-13 was incubated in the lytic system. After 60 min incubation at 37°C, 10^6/ml protoplasts were released and were collected. Protoplasts were washed and purified then incubated on surface of agar plate. Observed regeneration ratio was over 99%.

REFERENCES

1. J.F. Peberdy, Fungal protoplasts: isolation, reversion and fusion. Ann. Rev. Microbiol., 33:21 (1979).
2. R.E. Buchanan and N.E. Gibbons, "Bergey's Manual for Determinative Bacteriology", Williams and Wilkins, Baltimore (1974).
3. P. Gerhardt, "Manual of Methods for General Bacteriology". Am. Soc. Microbiol., Washington, (1981).
4. M. Yabuki, Tech. Bull. Fac. Hort. Chiba Univ., 34:21 (1983).
5. S. Hara, Y. Yamamura, Y. Fujii and T. Ikenaka, in "Proceedings Second International Conference on Chitin/Chitosan", S. Tokura and S. Hirano, eds., Japanese Society Chitin, Tottori, Japan, 1982.
6. R.L. Roberts, Anal. Biochem.-127:402 (1982).
7. T. Tsukamoto, Agric. Biol. Chem., 48:931 (1984).
8. J. Molano, J. Biol. Chem., 254:4901 (1979).

Affiliations

E. Danulat, Institut Hydrobiologie & Fischereiwissenschaft, Zeiseweg 9, D-2000 Hamburg 50, F. R. Germany.

E. M. Kubicek-Pranz, W. P. Kammel, F. Streichsbier, J. Toth and C. P. Kubicek, Institut Biochemie Tech. Mikrobiol., Tech. Universitat, Getreidemarkt 9/172 A-1060 Wien, Austria.

E. Zomer, A. B. Saada, D. Klein, A. Todar and Z. Er-el, Unikoor, P. O. Box 1172, IL-91010 Jerusalem, Israel.

R. A. Smucker and D. A. Wright, Chesapeake Biological Lab., Center Environ. Estuarine Studies, University Maryland, Solomons MD 20688, USA.

M. Yabuki, K. Mizushina, T. Amatatsu, A. Ando, T. Fujii, Dept. Agricultural Chemistry, Fac. Horticulture, Univ. Chiba, Matsudoshi, Chiba 271, Japan.

M. Shimada and M. Yamashita, Central Research Lab., Godo Shusei Co., Matsudoshi, Chiba 271, Japan.

Editor's Report

G. W. GOODAY

CHITIN DEACETYLASES IN INVERTEBRATES

based on works by

M. ARUCHAMI, N. GOWRI, and G. SUNDARA-RAJULU

Chitin deacetylase activities have been characterised from fungi in those cases where the resulting chitosan forms a structural component of the wall, and from chitinoclastic bacteria. Here are two accounts of their activity in invertebrates, being apparently associated here in softening of chitinous tissues to allow their expansion.

DISTRIBUTION OF DEACETYLASE IN ARTHROPODA

(M. Aruchami, G. Sundara Rajulu and N. Gowri)

A chitin deacetylase has been described from the queens of the termite *Macrotermes estherae* and its properties have been investigated (1). The enzyme was involved in the deacetylation of the cuticle chitin to chitosan in the physogastric phase to facilitate extensive enlargement of the abdomen to accommodate the proliferating huge ovary. It was felt to be of interest to find out if deacetylase occurs in other arthropods. For this purpose a survey was made on a few species from different orders of the phylum and the investigation disclosed that the enzyme is widely distributed.

MATERIALS AND METHODS

The materials investigated were 4 species of insects, 3 of crustaceans, 1 arachnid and 1 myriapod (Table I). The presence of chitosan was detected by the van Wisselingh test (2). The enzyme was isolated in the fresh moult stage from the sub-epidermal tissues underlying the cuticular regions having chitosan. It was unified by zone electrophoresis on starch gel as outlined elsewhere (3). The enzyme in the tissue homogenate as well as in a purified state was characterized for its molecular weight, pH and temperature optima (4).

RESULTS AND DISCUSSION

Frozen sections of the cuticle covering the various parts of the body in the individuals selected for investigation reacted differently to 5 % glacial acetic acid in the intermoult condition. The cuticles covering the general parts of the body were unaffected by the organic acid. The cuticle covering some of the specialized parts (Table II) showed swelling reactions on the addition of the acid.

The cuticular parts which showed a swelling reaction to the dilute acetic acid solution yielded a reddish violet colour reaction to the iodine-sulphuric acid reagent employed in van Wisselingh's test. But the cuticle covering the general parts of the body turned faintly brown with such treatment. The latter type of cuticle gave reddish or typical violet colour reaction to the iodine-sulphuric acid reagent only after treatment with hot saturated alkali at 160° C for 15 to 20 minutes. The above observations suggest that the cuticle covering the general parts of the body has chitin while some specialized regions are overlaid by cuticle containing chitosan.

Purified chitosan dissolves completely in dilute organic acids (5). The failure of the cuticle covering the specialized regions to dissolve in acetic acid could be explained by the intimate association of the chitosan with the protein moiety. Support for this suggestion came with the following observation.

The specialized cuticles dissolved completely in the organic acid after removal of the protein by treatment at 60° C in 5 % KOH solution for 15 minutes or until the material was no longer positive to tests for protein. In the light of these observations, it can be inferred that chitosan does occur in patches in same specialized regions of the cuticle even in intermoult condition in the species of arthropods studied.

Table I. Distribution of chitosan and deacetylase in the different classes of Arthropoda.

Class	Species	Specialized parts
Insecta:	Oecophila longinoda	Abdominal cuticle of physogastric queen
	Anax immaculifrons	Eye-lens cuticle
	Lethocerus indicus	Eye-lens cuticle
	Apis cerana indica	Eye-lens cuticle
Crustacea:	Lepas sp.	Stalk cuticle
	Sacculina rotundata	Cuticle of the externa
	Neptunes sanguinolentus	Eye-lens cuticle
Arachnida:	Palamnaeus swammerdami	Cuticle of lung-books leaves
Myriapoda:	Scutigera longicornis	Eye-lens cuticle

Table II. Molecular weights, optimum temperatures and pH for the deacetylase of Arthropoda.

Species	Homogenate		Purified enzyme		
	° C	pH	M.W.	° C	pH
O. longinoda	21-23	8.4	35,000	20-22	8.2
A. Immaculifrons	23-26	8.4	42,000	21-24	8.2
L. indicus	22-25	8.3	42,000	20-23	8.1
A. cerna indica	23-25	8.4	41,000	21-23	8.2
Lepas sp.	17-19	8.4	32,000	15-17	8.2
S. rotundata	24-27	7.9	42,000	22-24	8.8
N. sanguinolentus	20-22	8.4	42,000	18-20	8.2
P. swammerdami	24-26	8.2	35,000	22-24	8.4
S. longicornis	23-25	8.4	42,000	21-23	8.2

Table I also presents the list of the different specialized cuticular parts that contain chitosan. It was found in the cuticle covering the compound eyes which forms the transparent cornea (lens) in the insects, crustaceans and the myriapod, in the cuticle of the abdomen of the physogastric queen ant, stalk of the barnacle, externa cuticle of the rhizocephalan and that of the leaves of the book-lungs of the scorpion investigated. The question is whether these specialized cuticular regions ever contain the fully formed chitin. In order to answer this, the cuticle in these regions was studied in the different stages of the moult cycle for dissolution in the dilute acetic acid and for its reaction to van Wisselingh's test. It was observed that in the fresh-moult condition the cuticle covering the body of the entire organism contained fully formed chitin in all the species studied. This situation remained unchanged subsequently in all the generalized parts of the body, but in the case of the specialized regions mentioned in Table I, the chitin was later deacetylated to chitosan. Corresponding to the occurrence of chitosan the epidermis and the subepidermal tissues underlying such specialized regions had the enzyme deacetylase in the freshmoult and postmoult stages.

The enzyme was prepared and studied in the homogenate of the tissues and in a purified condition for its molecular weight, pH optima and temperature optima (Table II). The molecular weight of the enzyme involved in the deacetylation of the transparent lens cuticle in the different species was similar. In the other cases, the enzyme had varying molecular weights and different temperature optima. The pH optima for the enzyme, however, from all the different species were similar.

The role of deacetylase and the significance of natural deacetylation have been discussed in respect of the physogastric queens of termites (1). The deacetylation enables the chitin to stretch so that the cuticle can expand. This kind of stretching mechanism can account for the facilitation of the abdomen in O. longinoda in the physogastric phase (6), softening of the stalk cuticle in Lepas sp. (7), and expansion of the externa cuticle in S. rotundata (8). The corneal cuticle has the same number of lamellae of similar thickness as the adjacent general cuticle with which it is continuous in the freshmoult condition. Subsequently the lamellae of the cornea swell-up in the middle line so that the eye-lens protrudes out as a dome (9). Deacetylation of chitin presumably facilitates this swelling. The significance of the natural deacetylation of the chitin in the leaves of book-lungs deserves further investigation.

REFERENCES

1. G. Sundara Rajulu, M. Aruchami, and N. Gowry, Natural deacetylation of chitin to chitosan in the abdominal cuticle of the physogastric queen of Macrotermes estherae, in: "Proceeding Second International Conference Chitin/Chitosan", S. Tokura and S. Hirano, eds., Japanese Soc. Chitin, Tottori, Japan, 1982.
2. E. van Wisselingh, Detection of chitin, Jahrb. Wiss. Bot., 31:619 (1898).
3. G. Sundara Rajulu, A comparative study of the haemolymph of Arthropoda, J. Anim. Morphol. Physiol., 17:86 (1970).
4. G. Sundara Rajulu, Cuticle of queen termites, Ind. Zool., 5:103 (1981).
5. A. G. Richards, The integument of Arthropoda, University of Minnesota Press, Minneapolids.
6. A. Sannasi, Studies on the integument of social insects, Ph. D. Thesis, Madras University, Madras (1979).
7. H. G. Cannon, On the anatomy of the pedunculate barnacle Lithotrya. Phil. Trans. Roy. Soc. London B., 233:89 (1947).
8. H. Boschma, Three successive layers of external cuticle is Sacculina septodia, Kon. Ned. Akad. Wetenschab. Proc., 50:3 (1947).
9. E. Link, Uber die Stirnaugen einiger Lepidopteren und Neuropteren, Zool. Anz., 33:445 (1908).

NATURAL DEACETYLATION OF THE CUTICLE IN SACCULINA ROTUNDATA

(N. Gowri, M. Aruchami and G. Sundara Rajulu)

Rhizocephalan parasites are the most successful among the crustacean parasites (1, 2). Following an active nauplius period the larvae attach to the abdomen of the host crab by means of an antenna and metamorphose into a "Kentrogen". All the embryonic cells in this stage conglomerate into a ball of cells and pass into the host's body to form a button-like "interna" which attaches itself to the gut of the host. The parasite then gives out roots which ramify through all the organ systems. The body of the "interna" grows in size and projects out from the ventral side of the crab's abdomen. Later it enlarges about 100 times in size and becomes "externa" (3, 4).

Cannon and Travis (5, 6) observed that crustaceans, like other arthropods, undergo periodical moulting in order to facilitate the enlargements of body. Sacculinids seem to be exceptions to this generalization that there is no moulting during the enormous enlargement of the "externa" and the roots (7). This phenomenon could not be explained in known terms.

MATERIALS AND METHODS

Sacculina rotundata, investigated in the present study, are parasites on the crabs *Neptunes sanguinolentus* commonly found in the Bay of Bengal. Methods for the preparation of the materials, purification of chitin, estimation of acetyl groups, molecular weight and temperature as well as pH optima have been given elsewhere (8).

RESULTS AND DISCUSSION

The "externa" of *Sacculina rotundata* projecting out of the host's abdomen 9 to 13 mm broad, about 6 mm wide and about 4 mm thick. After complete enlargement, the corresponding dimensions increased to 625 x 1100 x 38 mm. "Externa" of 8 different widths were studied (Table I) to follow the sequence of changes in the nature of the chitin during enlargement. Similar studies were also performed on the different roots.

The cuticle chitin of "interna" contains 7.48 % of acetyl groups. As the parasite becomes "externa" the acetyl content fell to 6.06 %. With the gradual increase in the size of the "externa", the acetyl content also slowly declines to reach the least value of 0.77 % which is equivalent to 89.7 % deacetylation.

Table I. Acetyl content of the purified "chitin" of the cuticle of *Sacculina rotundata.*

Material	Size, mm	Acetyl content (%)		Degree of deacetylation (%)
Interna:		7.48	0.19	0
Externa:	100	6.81	0.21	8.9
	200	6.06	0.22	18.9
	300	5.33	0.18	28.7
	500	4.12	0.23	44.9
	600	2.90	0.23	61.2
	800	0.95	0.07	87.2
	900	0.86	0.06	88.5
	1100	0.77	0.02	89.7
Chitosan of prawn		0.47	0.03	93.7

Table II. Properties of the deacetylase of Sacculina rotundata compared to those of queen termite.

Feature	Enzyme in the homogenate	Purified enzyme	Queen termite
Molecular weight	NA	42,000	35,000
Temperature optimum, ° C	24-27	22-24	20-22
pH optimum	7.9-8.1	7.9-8.1 8.8-9.0	8.2-8.4

Table III. Acetyl content of the purified "chitin" of the root cuticle of Sacculina rotundata.

Material		Acetyl content (%)	
Interna roots		3.35	0.57
Externa:	a. Primary root	4.06	0.98
	b. Secondary root	3.78	0.94
	c. Tertiary root	1.42	0.33
	d. Ampulla (root tips)	0.56	0.21

When an homogenate of the epidermal tissues from a 600 mm "externa" was incubated with samples of purified chitin, they were deacetylated to various extents. Properties of the enzyme are shown in Table II, in comparison to those of the deacetylase from Sacculina rotundata and from the queen termite Macrotermes estherae. It can be seen that the deacetylases from the two sources have different properties. The enzyme of the crustacean has higher molecular weight than that of the queen termite; further there are two pH optima for the enzyme of Sacculina.

Estimation of the acetyl content of the "interna" showed that it had only 0.57 - 3.35 % acetyl content while the cuticle covering the main body contained 7.48 % (Tables I - III). The primary and branch roots of the "interna" also had a lower acetyl content ranging from 0.56 to 4.06 %. These results indicate that the cuticle of root systems of both the "interna" and "externa" have deacetylated chitin.

The significance of the occurrence of deacetylase could be the same as that proposed for the queen termite. It can be therefore suggested that whenever an expansion of the cuticle is needed without moulting, a deacetylation involving a deacetylase may occur.

REFERENCES

1. Y. Delage, Evolution de la Sacculine, Arach. Zool. Esp. Gen., 2:2 (1884).
2. J. H. Day, The life history of Sacculina, Quart. Jl. Microsc. Sci., 77:549 (1935).
3. G. E. H. Foxon, Notes on the life history of Sacculina carcini, J. Mar. Biol. Assn., 24:253 (1940).
4. M. J. George, Stages of development of Sacculina sp., Proc. Indian Acad. Sci., 30:207 (1949).
5. H. G. Cannon, On the anatomy of the pedunculate barnacle Lithotrya, Phil. Trans. Roy. Soc. Lon. B., 233:89 (1947).
6. D. F. Travis, The moulting cycle of the spiny lobster Panulirus argus, Acta Histochem., 120:192 (1965).

7. H. Boschma, Three successive layers of external cuticle in Sacculina septodia, Kon. Ned. Akad. Wetenschab. Proc., 50:3 (1947).

8. G. Sundararajulu, M. Aruchami and N. Gowri. 1982. Natural deacylation of chitin to chitosan in the abdominal cuticle of the physogastric queen of Macrotermes estherae, Proc. II Int. Conf. Chitin and Chitosan, Sapporo, Japan, 140.

Affiliations

M. Aruchami, N. Gowri and G. Sundara-Rajulu, Dept. Zoology, Kongunadu Arts & Science College, Coimbatore, 641029, India.

RECOGNITION OF HUMAN ERYTHROCYTES BY A CHITIN-BINDING LECTIN OF A FUNGUS, CONIDIOBOLUS LAMPRAUGES

K. OISHI and F. ISHIKAWA

Institute of Applied Microbiology, The University of Tokyo
1-1-1 Yayoi, Bunkyo-ku
Tokyo 113, Japan

INTRODUCTION

A number of chitin- or chitosaccharide-binding lectins have been found in plant kingdom.[1] They are listed in Table 1. Some of them were purified from plant extracts by affinity chromatography technique using chitin as affinity adsorbent.[2 - 4]

Some of these lectins are biologically active. For example, wheat germ agglutinin (WGA) is known to block chitin biosynthesis in molds and inhibits their growth.[5] In most cases, however, these lectins are used for the detection and determination of special chitosaccharide-containing sugars in the fields of biochemistry and cell physiology. Specificity of these lectins for the cell surface antigens might not be so high as that of monoclonal antibodies, but the lectins are still excellent probes for the comparative analysis of chitosaccharide-containing antigens on animal and plant cell surfaces. Transformed cell lines of murine are much more sensitive to

Table 1. Chitosaccharide-Binding Lectins from Plants

Lectin	Origin	Reference
WGA	*Triticum vulgaris* (wheat) germ	6
STA	*Solanum tuberosum* (potato) tuber	7
BS II	*Bandeiraea simplicifolia* seed	4
PWM	*Phytolacca americana* (pokeweed) root	8
RBM	*Oryza sativa* (rice) bran	9
DSA	*Datura stramonium* seed	10
Tomato	*Lycopersicon esculentum* fruit juice	11
VML	*Cucurbita pepo* (vegetable marrow) fruit juice	12
Melon	*Cucumis melo* fruit juice	12
Cucumber	*Cucumis sativus* fruit juice	12
CSA	*Cytisus sessilifolius* seed	13
LAA	*Laburnum alpinum* seed	14
CFA	*Cerastrium fomentosus* seed	–
UEA II	*Ulex europeus* seed	16

WGA than the normal line is.[17] Tunicamycin, a UDP-GlcNAc analog which specifically inhibits GlcNAc transferase in chitosaccharide-containing sugar synthesis, shows no fatal effect on normal murine cells, but it is toxic for the transformed cells.[18] Chitosaccharide-binding lectins would further be used for studying the structure and function of such cell surface antigens.

This paper deals with the diagnostic properties of a chitosaccharide-binding lectin, CLA, from a fungus *Conidiobolus lamprauges*.

MATERIALS AND METHODS

Preparation of CLA

Conidiobolus lamprauges No. 454 was isolated from Japanese soil in 1978.[19] Its morphological properties are shown in Fig. 1. This organism was cultured at 27°C for 3 to 5 days with shaking in 0.05M phosphate buffer (pH 6.0) containing 0.5% polypepton, 0.3% yeast extract, 0.3% malt extract and 1% lactose. CLA produced in the culture broth was purified by affinity chromatography using *p*-aminobenzyl-1-thio-β-N-acetyl-D-glucosamine-C_6-Sepharose 4B as an affinity ligand.[20] Electrophoretically pure preparation was used throughout the study.

Hemagglutinating and Sugar-Binding Activities of CLA

Hemagglutinating activity and sugar-binding specificity were assayed by the conventional procedures as described previously.[19]

Receptor of CLA on Human Erythrocytes

Purified CLA preparation (5.2 mg), CNBr-Sepharose 4B (1.7 g), and 0.5M GlcNAc-0.05M $NaHCO_3$ (10 ml) were mixed and incubated at 20°C for 5 hr. After the incubation, 1 g of glucosamine was added to this mixture and kept at 4°C overnight. The CLA-Sepharose 4B gel thus obtained was washed well, packed into a column and equilibrated with 0.5% Triton X-100-0.05M phosphate buffered saline (PBS, pH 6.0). Stroma of human O erythrocytes was prepared by the method of Dodge et al,[21] solubilized by Triton X-100 and Applied to CLA-Sepharose 4B column. Elution was made by 0.5M GlcNAc. CLA receptor activity was determined by hemagglutination inhibition test. Active fractions were electrophoresed in the presence of SDS and stained

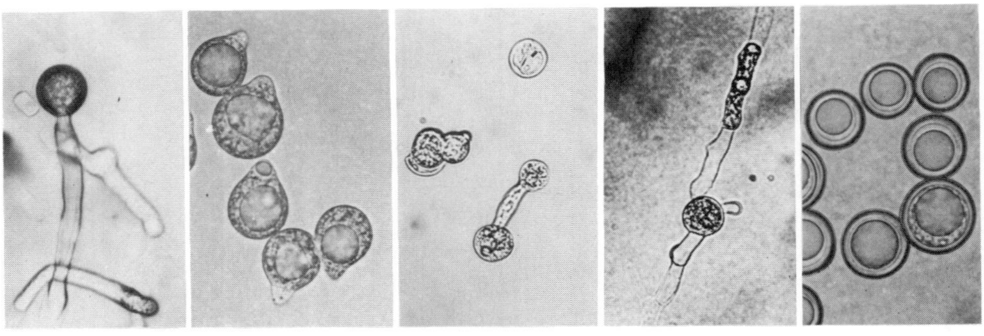

(a) (b) (c) (d) (e)

Fig. 1. Morphological properties of *Conidiobolus lamprauges*. (a) Conidiophore; (b) primary conidia; (c) secondary conidia; (d) zygospore formation, (e) matured zygospores.

with Coomassie Brilliant Blue and periodic acid-Schiff reagent (PAS).

Chitosaccharide-Binding Lectins from Plants

WGA and STA were prepared in a similar manner to the case of CLA from HCl-extract of ground wheat germ and PBS-extract of potato tuber, respectively. UEA II and CSA were obtained from Maruzen Oil Biochemical, Tokyo.

RESULTS

Sugar-Binding Specificity of CLA

Sugar-binding specificity of CLA and typical chitosaccharide-binding plant lectins was investigated by hemagglutination inhibition test. WGA and STA are nonspecific for human blood group, whereas UEA II and CSA are blood group O(H) specific. The results are shown in Table 2. All the lectins examined were not inhibited by neutral sugars such as Glc, Gal, and Man. Fru, Fuc, Rha, Xyl, Ara, Rib, Suc, Lac, Mal, Cel, and Raf were also noninhibitory. On the other hand, strong inhibition was observed in the presence of chitooligosaccharides. $GlcNAc_3$ and $GlcNAc_4$ were the most potent inhibitors against all the lectins examined. These results indicate that the sugar-binding specificity of CLA is common to those of plant lectins, though CLA was rather sensitive to GlcN, GlcNAc, and ManN. WGA is considered to bind $GlcNAc_3$ with the corresponding 3 subsites,[22] and this must be the case in CLA.

Erythrocyte-Binding Specificity of CLA

Contrary to the sugar-binding specificity, erythrocyte-binding specificity of CLA is quite different from those of plant lectins. As shown in Table 3, CLA agglutinated 250 specimens of human O erythrocytes without exception. Minimal concentration of CLA for the agglutination of 2% human O erythrocyte suspension was as low as 25μg protein/ml. Erythrocytes from

Table 2. Inhibition of Chitosaccharide-Binding Lectins by Sugars

	Minimal concn (mM) for complete inhibition				
	CLA	WGA	STA	UEA II	CSA
Glc	>45	>40	>40	>50	>100
GlcN	20	>40	>40		>100
GlcNAc	13	>226	>450	>50	>100
$GlcNAc_2$	0.74	2.9	5.9	2.5	3
$GlcNAc_3$	0.25	0.5	0.25	1.25	
$GlcNAc_4$	0.19	0.38	0.05	2.5	
Ethylene glycol chitin* (μg/ml)	15.6	>2000	31.3		
p-NP-α-GlcNAc	>20	>20	>20		
p-NP-β-GlcNAc	0.62	>10	>10	10	2.5
3-GlcNAc	>45	>45	>45		
6-GlcNAc	>45	>45	>45		
Gal	>40	>40	>40	>50	>100
GalN	>40	>40	>40		
GalNAc	>45	>45	>45	>50	>100
Man	>40	>40	>40	>50	>100
ManN	20	>40	>40		
ManNAc	>45	>45	>45		

Table 3. Agglutination of Vertebrate Erythrocytes by Chitosaccharide-Binding Lectins

Erythrocyte	No. of specimen	Relative agglutinating activity*				
		CLA	WGA	STA	UEA II	CSA
Mammalia						
Homo sapiens	250	1	1	1	1	1
Pan troglodytes	5	64** >10000**	8	64	4	>1000
Macaca fascicularis	50	>10000	8	128	4	16
Cercopithecus aethiops	10	>10000	32	32	64	256
Saimiri sciurea	10	>10000	8	32	>1000	32
Lepus cuniculus v. *domesticus*	1	>10000	0.25	0.5	4	16
Mesocricetus auratus	2	>10000	16	16		+
Rattus rattus	2	>10000	2	8		64
Mus musculus	5	>10000	4	8		
Canis familiaris	14	>10000	2	4		
Felis catus	5	>10000	8	32		
Sus scrofa v. *domesticus*	5	>10000	1	2		+
Ovis aries	5	>10000	4	32		
Capra hircus	5	>10000	8	16		
Bos taurus domesticus	5	>10000	8	64		+
Equus caballus	5	>10000	32	16		
Aves						
Gallus gallus v. *domesticus*	5	>10000	0.25	8	256	>1000
Anas platyrhnchos domesticus	2	>10000				
Columba livia	1	>10000				
Reptilia						
Geoclemys reevesii	5	>10000	16	128	64	64
Eumeces latiscutatus	2	>10000				
Takydromus tachydromoides	4	>10000	16	16		
Elaphe quadrivirgata	3	>10000	4	64	>1000	>1000
Amphibia						
Triturus pyrrhogaster	5	>10000				
Rana nigromaculata	3	>10000	+	+	>1000	>1000
Rhacophorus schlegelli	5	>10000	+	+	+	>1000
Pisces						
Cyprinus carpio	5	>10000	128	128	>1000	>1000
Girella punctata	2	>10000				
Sillago sihama	2	>10000				
Mugil cephalus	2	>10000				
Sardinops melanosticta	2	>10000				

* : Relative concentration of lectins required for hemagglutination.
**: Two specimens were 64 and 3 specimens were >10000.

2 individuals of chimpanzee were agglutinated by CLA at the concentration of 1,600μg/ml, while those from other 3 individuals were not agglutinated even when 250,000μg/ml CLA was added. None of the erythrocytes from other mammals including 3 species of monkeys, fowls, reptiles, amphibians, and fishes were agglutinated by CLA at the concentration of 250,000μg/ml. These results mean that CLA is highly human erythrocyte-specific. In other words, human erythrocytes have highly specific surface antigen(s) with chitosaccharide nature.

Trypsinized human erythrocytes were agglutinated by CLA 4 times more than nontreated cells, whereas trypsinized rabbit, sheep, cattle, and

Table 4. Agglutination of Human A, B, and O(H) Erythrocytes by CLA

HAU of CLA*	No. of specimen agglutinated		
	A	B	O(H)
2			
4	1	1	
8	2		
16	3	1	
32	13	5	
64	29	13	
128	6	22	6
256	2	9	18
512		1	1
1024			
Total	56	52	25

* Dilution of original CLA solution.

chicken erythrocytes were still not agglutinated by CLA, suggesting that insensitiveness of these erythrocytes to CLA comes from the lack of CLA receptor but not from the masking of the receptor by non receptor sugar chains.

WGA and STA, which are nonspecific for human blood group, agglutinated all the erythrocytes so far tested, and UEA II and CSA, which are specific for human O(H) group, agglutinated the erythrocytes from several animals.

ABO blood group on human erythrocytes related to the sensitivity of these cells to CLA. As shown in Table 4, average agglutinating activity of CLA on A and B cells was 1/4 and 1/2 of that on O cells, respectively. Diversion of the activity was also seen in the cases of A and B cells. These results suggest the presence of interference by the GalNAc and Gal at the nonreducing terminal of ABO-active sugar chains, which is often seen at the hemagglutination by O(H)-specific lectins and antibodies.

Receptor of CLA on Human Erythrocytes

When solubilized erythrocyte stroma was applied to a column of CLA-Sepharose 4B, most of the proteinaceous materials and no CLA receptor activity were observed in the pass-through fractions. The activity was adsorbed to the gel and eluted as a single band by addition of 0.5M GlcNAc. The eluate was about 8 times active on protein basis than the original solubilized stroma. The active eluate electrophoresed was scarcely stained with PAS, but it gave with CBB stain a band corresponding to Band 3 glycoprotein.[23]

ABO active substances on human erythrocytes are trace components with a very high antigenicity. Most of them are glycolipid nature[24] and some are glycoprotein.[15] ABO active glycoproteins are stained with PAS. CLA receptor on human erythrocytes, therefore, differs from the known ABO active sugars, though it is under the influence of A and B antigens.

DISCUSSION

Homopoly-GlcNAc, chitin, is widely distributed in animal kingdom,[3] but is not found in Echinoderma, Protochordata, and Vertebrata. In these animals, GlcNAc is present as a component of heteroglycoconjugates. Some of these sugars are known in these animals as molecular markers of physiological and pathological changes and as fatal substances to regulate their biological activity.

In the studies on the structure and function of such GlcNAc-containing sugars, chitosaccharide-binding lectins have often been used as conventional probes. Monoclonal antibodies are not always the best probes for such studies because of their hyperspecificity for the antigens and, on the contrary, limited specificity for a special sugar structure. Chitosaccharide-binding lectins with novel properties are still required in this field.

By using CLA, we have found that a highly human-specific sugar antigen is present on erythrocyte surface. Structure and function of this antigen are not yet clear, but CLA is considered to be applicable to the diagnosis of human erythrocytes in police science and to a detector of a molecular marker of some groups of apes in breeding of experimental animals.

CONCLUSION

A chitosaccharide-binding lectin, CLA, was obtained from a fungus *Conidiobolus lamprauges*. Sugar-binding specificity of this lectin was similar to those of known chitosaccharide-binding plant lectins, whereas the erythrocyte-binding specificity was quite different. CLA agglutinated human erythrocytes and, though very weakly, cells of some chimpanzees, but not agglutinated erythrocytes from other mammals, fowls, reptiles, amphibians, and fishes, indicating the occurrence of a novel chitosaccharide-relating, highly human-specific antigen on human erythrocytes.

REFERENCES

1. F. Ishikawa and K. Oishi, Chitin-binding lectins from fungi, in: "Chitin, Chitosan, and Related Enzymes," J. P. Zikakis, ed., Academic Press, Inc., Tokyo (1984).
2. R. Bloch and M. M. Burger, Purification of wheat germ agglutinin using affinity chromatography on chitin, Biochem. Biophys. Res. Commun., 58:13 (1974).
3. R. A. A. Muzzarelli, "Chitin," Pergamon Press, Oxford, (1977).
4. P. N. S. Iyer, K. D. Wilkinson, and I. J. Goldstein, An N-acetyl-D-glucosamine binding lectin from *Bandeiraea simplicifolia* seed, Arch. Biochem. Biophys., 177:330 (1976).
5. D. Nirelman, E. Galun, N. Sharon, and R. Lotan, Inhibition of fungal growth by wheat germ agglutinin, Nature, 256:414 (1975).
6. I. J. Goldstein, S. Hammarström, and G. Sundblad, Precipitation and carbohydrate-binding specificity studies on wheat germ agglutinin, Biochim. Biophys. Acta, 405:53 (1975).
7. A. K. Allen and A. Neuberger, The purification and properties of the lectin from potato tubers, a hydroxyproline-containing glycoprotein, Biochem. J., 135:307 (1973).
8. K. Yokoyama, T. Terao, and T. Osawa, Carbohydrate-binding specificity of pokeweed mitogens, Biochim. Biophys. Acta, 538:384 (1978).
9. M. Tsuda, Purification and characterization of a lectin from rice bran, J. Biochem., 86:1451 (1979).
10. D. C. Kilpatrick and M. M. Yeoman, Purification of the lectin from *Datura stramonium*, Biochem. J., 175:1151 (1978).
11. D. C. Kilpatrick, Purification and some properties of a lectin from the fruit juice of the tomato, Biochem. J., 185:269 (1980).
12. A. K. Allen, A lectin from the exudate of the fruit of the vegetable marrow that has a specificity for β-1,4-linked N-acetylglucosamine

oligosaccharides, Biochem. J., 183:133 (1979).
13. W. M. Watkins and W. T. J. Morgan, Further observations on the inhibition of blood group specific serological reactions by simple sugars of known structure, Vox Sang., 7:129 (1962).
14. T. Osawa, Inhibition tests of plant agglutinins from *Laburnum alpinum* and *Cytisus sessilifolius* with various sugars and sugar derivatives. Indication of the type of linkage in certain disaccharides, Biochim. Biophys. Acta, 115:507 (1966).
15. N. B. Whittemore, N. C. Trabold, C. F. Reed, and R. I. Weed, Solubilized glycoprotein from human erythrocyte membranes possessing blood group A, B, and H activity, Vox Sang., 17:289 (1969).
16. I. Matsumoto and T. Osawa, Purification and characterization of an antiH(O) phytohemagglutinin of *Ulex europeus*, Biochim. Biophys. Acta, 194:180 (1969).
17. R. E. Pollack and M. M. Burger, Surface-specific characteristics of a contact-inhibited cell line containing the SV40 viral genome, Proc. Natl. Acad. Sci., USA, 62:1074 (1969).
18. K. Kohno, A. Hiragun, H. Mitsui, A. Takatsuki, and G. Tamura, Effect of tunicamycin on cell growth and morphology of nontransformed and transformed cell lines, Agric. Biol. Chem., 43:1553 (1979).
19. F. Ishikawa, K. Oishi, and K. Aida, Chitin-binding hemagglutinin produced by *Conidiobolus* strains, Appl. Environ. Microbiol., 37: 1110 (1979).
20. F. Ishikawa, K. Oishi, and K. Aida, Purification and immunological properties of chitin-binding hemagglutinin and β-N-acetylglucosaminidase produced by *Conidiobolus lamprauges*, Agric. Biol. Chem., 47:149 (1983).
21. J. T. Dodge, C. Mitchell, and D. J. Hanahan, The preparation and chemical characteristics of hemoglobin-free ghosts of human erythrocytes, Arch. Biochem. Biophys., 100:119 (1963).
22. A. K. Allen, A. Neuberger, and N. Sharon, The purification, composition and specificity of wheat-germ agglutinin, Biochem. J., 131: 155 (1973).
23. G. Fairbanks, T. L. Steck, and D. F. H. Wallach, Electrophoretic analysis of the major polypeptides of the human erythrocyte membrane, Biochemistry, 10:2606 (1971).
24. W. M. Watkins, Blood group substance, Science, 152:172 (1966).

Section V

CHEMISTRY OF CHITIN

DRY CHITOSAN SALTS AND COMPLEXES OF ALIPHATIC CARBOXYLIC ACIDS

P. R. AUSTIN and S. SENNETT

College of Marine Studies, University of Delaware
Newark, DE 19716, USA

INTRODUCTION

Chitosan is the major outlet for chitinous products, largely because of its utility in the treatment of a broad range of potable and waste water systems for purification, clarification and solids concentration. Its safety and efficacy in reducing biological oxygen demand (BOD) are especially valuable. Currently, the practical use of chitosan requires the transport and storage of the formic or acetic acid solubilizing component and some expertise in control of solution preparation. These limitations might be avoided in many instances if dry, free-flowing carboxylic acid salts of chitosan were available. They could then simply be dissolved in water for their application.

Chitosan is soluble in a long list of aqueous acid systems[1,2,3], but only a few dry chitosan salts have been described, such as the hydrochloride and sulfamate salts[4,5,6]. Cupery[6] indicates that unlike the sulfamate salt, addition of a solution of chitosan in aqueous acetic acid to methanol leads to a gel-like mass. Our experience with other organic acids and coagulants has been similar. In another approach, a water soluble chitin-chitosan of about 50% deacetylation has been prepared from alkali-chitin by low temperature hydrolysis[7], but the adsorptive and other properties of this product may be different from normal chitosan of 80±% deacetylation.

We report here methods for the preparation of dry chitosan salts and complexes of the lower aliphatic carboxylic acids in a disperse system of chitosan, acid and an inert organic medium.

Heterogeneous reactions of polymers are notoriously difficult to control as reaction with the disperse polymer phase depends not only on particle size, but on its crystallinity, intra- and inter-molecular hydrogen bonding, rate of diffusion or penetration, and accessibility of reactive centers, in this case, the free amino groups of chitosan. This polymer itself is prepared in a heterogeneous system from chitin by reaction with hot 40-50% sodium hydroxide. Thus, chitosan is commonly deacetylated only to the extent of 75-90%, adequate even though non-uniform, to induce its solubility in 1-4% formic or acetic acid. However, with the strong attractive forces of the acid-base reaction, chitosan salt formation apparently proceeds effectively even in a heterogeneous system.

MATERIALS AND METHODS

Low viscosity chitosan, Kytex from brown shrimp (*Penaeus aztecus*), was provided by Hercules Inc. (no longer suppliers); a low viscosity chitosan from Alaskan pink shrimp (*Pandalus borealis*) was obtained from the former Food Chemical and Research Laboratories (FCRL), now Protan, Redmond, Washington, and high viscosity crab shell chitosan, pharmaceutical grade, was supplied by Bioshell, Inc., Albany, Oregon. The materials varied in color; some had been bleached. All had specific gravities in the range of 1.44-1.45. Acids and reaction media were obtained from chemical supply firms and used as received.

In general, chitosan (0.5-25g, 8-40 mesh) was suspended in a selected organic medium, the appropriate acid added and the mixture allowed to stand for 1 to 3 hours. Temperature of reaction and drying varied between 12-35°C, but was usually at 20-25°C. For small sample reactions, occasional shaking appeared satisfactory; larger batches were stirred magnetically. The chitosan salts so formed were separated by filtration, washed with fresh medium, and the residual reaction media evaporated.

Viscosities were determined of a 1% aqueous chitosan salt solution with a Brookfield Synchro-Lectric Viscometer, Model LVF; measurements were made after solutions stood at ambient temperature for about two hours. Solution pH was determined with an Orion Research Digital Ionalyser 501. Specific gravities were approximated by the sink and float method.

Specifications indicated that the Kytex and Bioshell chitosans were deacetylated to the extent of 77-83%; hence, the figure of 80% was taken for unit equivalent weight calculations, i.e., a calculated unit molecular weight of 169. All of the chitosan samples were assumed to have this same value. Mole ratio of acid reacted or complexed per mole of chitosan was calculated from the gain in weight of the chitosan sample reacted.

RESULTS AND DISCUSSION

Exploratory studies indicated that finely-divided chitosan can absorb a substantial amount of a low molecular weight carboxylic acid when the polymer is dispersed in organic media that are solvents for the carboxylic acids and non-solvents for chitosan. In general, most of the lower fatty acids up through four carbon atoms formed soluble chitosan salts[9], which are dry, free-flowing solids. Reactions were somewhat specialized as to the media employed, but chlorohydrocarbons, ketones, esters, ethers, nitroaliphatics and acetonitrile were all applicable (Table 1).

Particle size of the chitosan appeared important in rate and extent of reaction; the finer sizes (20-40 mesh) are preferred. The acids vary in reactivity; formic and pyruvic acids appear most effective and their chitosan salts are the most stable. Temperature of reaction was more important than first thought (to be discussed later); ambient conditions were frequently satisfactory, but the 20-25°C range was consistently favored. Similarly, ambient temperatures were preferred for drying, as even at 40°C, some discoloration was encountered.

Initially, we chose reaction media on the basis of their solubility parameters, following the leads developed in solvent studies of the parent chitin[8]. However, it was found that several other factors may also be involved such as the dissociation and dielectric constants of the acids and media, complexities contributed by the presence of water and the molecular size of the acids.

Table 1. Chitosan Salts Formed in Various Media[a]

Acid	mL	Medium	mL	Yield g	Moles Acid Per Mole Chitosan	Sol. Rate[b]
Formic, 88%	1.5	Tetrahydrofuran	10	0.63	0.94	++
Formic, 95-97	1.0	Ethyl Acetate	10	0.65	1.09	+++
Formic, 95-97	1.0	Heptane	10	0.63	0.94	+
Formic, 88	NA	Acetone[c,d]	1	NA	-	+
Acetic (glacial)	1.0	1,2-Dichloroethane	5	0.69	1.06	++
Acetic (glacial)	1.0	Acetonitrile	5	0.65	0.83	++
Acetic (glacial)	1.0	Ethyl Acetate	5	0.57	0.39	±
Propionic	1.0	5% Water in Acetonitrile	5	0.55	0.23	+
Isobutyric	2.0	Acetonitrile[c]	4	0.58	0.31	+
Chloroacetic	1.0	1,2-Dichloroethane	5	0.68	0.62	++
Dichloroacetic	1.0	1,2-Dichloroethane	5	0.75	0.63	+++

[a]In most cases, 0.5 g, 40 mesh, Kytex low viscosity chitosan was used with an excess of a molar equivalent of acid.
[b]Rate of solution: +++ = rapid; ± = viscous, not all soluble; pH of 1% solutions ranged from 5.3-5.8.
[c]High viscosity chitosan.
[d]Qualitative; NA = Not Available.

Another factor in this type of heterogeneous system is that of a distribution or partition coefficient of a third component that may be involved, that is, the distribution of a third component between two immiscible phases. In this case, the lower molecular weight carboxylic acid may distribute itself between the immiscible chitosan and the organic medium.

In a few cases, as with chitosan formate and acetate, the salt may be formed by reaction with an excess of the neat acid, i.e., without an inert medium, in which case the excess acid acts as its own medium. However, an additional inert reaction medium is usually preferred.

In these experiments, the physical recovery of a product is nearly quantitative. Yields lower than theoretical indicate that less than a stoichiometric amount of the acid reacted with chitosan and formed only a partial salt, or possibly that some of the salt dissolved in the medium. Such partial salts are water soluble, but in some cases may contain insoluble particles. Solubility of the partial salts may result from their heterogeneity. Perhaps they parallel the behavior of the heterogeneous but soluble 50% hydrolyzed chitin-chitosan of Sannan et al.[7] mentioned earlier.

Water content of the reaction system must be considered, as chitosan itself absorbs a substantial amount of moisture, particularly at higher humidities. In certain instances, it may be advantageous to add small amounts of water to act as a carrier or acid-dissociation medium and increase the reactivity of the system. The reaction of propionic acid and chitosan in acetonitrile containing 5% of water is a case in point (Table 1).

Table 2. Hydroxy Acid Salts of Chitosan

Acid		mL	Medium	mL	Yield g	Moles Acid Per Mole Chitosan	Sol. Rate[a]	pH
Glycolic,	70%	3	THF[b]	40	2.2[c]	0.22	+++	5.3
Lactic,	85%	1	Ethyl Acetate	5	.73	.85	++	-
Lactic, ex.	85%	d	DCE	18	.76	.96	+++	6.4

[a]Rate of Solution: +++ = Most rapid.
[b]THF = Tetrahydrofuran; DCE = 1,2-Dichloroethane.
[c]From 2g high viscosity chitosan; low mole ratio may result from partial solubility of salt in reaction medium.
[d]1,2-Dichloroethane extraction from 85% lactic acid; reacted with 0.5g low viscosity chitosan.

Commercial formic, glycolic and lactic acids normally contain up to 30% of water for stabilization or avoidance of their decomposition on distillation. Such acids may require special media or techniques for best results. Thus, 95-97% formic acid may be used with 1,2-dichloroethane as the reaction medium, but with 70% glycolic acid or 85% lactic acid, tetrahydrofuran, ethyl acetate or other media that dissolve both water and acid are preferred for carrying out the reaction with the insoluble chitosan (Table 2). In some cases, the existence of azeotropes aids in the evaporative removal of water and media; the water dielectric may promote the reaction itself.

As an alternate procedure for acids containing significant amounts of water, e.g., 85% lactic acid, one may first extract the acid into the medium, remove an aqueous fraction and utilize the medium-acid portion for reaction with chitosan (Table 2). The chitosan lactate may be one of the most useful compositions because of its excellent solubility, stability and the higher pH of its aqueous solutions.

The pH of the different chitosan salt solutions varies with the acid component and perhaps the amount of acid absorbed by chitosan, as may be seen from Tables 1 and 2. However, they fall in the range of 5.3-6.4.

CHITOSAN SALT ACID SOLVATES OR COMPLEXES

The unusually slow drying of the chitosan formate and acetate led to the discovery that these salts were, in fact, carrying a surplus of the acid as a solvate or complex (Fig. 1). The residual reaction medium evaporated rapidly; slow but measurable release of the solvated acid then ensued. The solvated salts themselves are dry, free-flowing powders, relatively stable if kept cool in a closed container. Under ambient conditions the volatile acid is lost slowly, even beyond the simple stoichiometric chitosan salt composition. The chitosan formate is nearly odorless and remains soluble for up to a year, whereas the chitosan acetate continues to have a sharp odor and gradually loses acid until the salt becomes only partially soluble.

Examples of the preparation of the acid solvates are given in Table 3, including that of chitosan pyruvate. Significantly, only these smaller

Table 3. Acid Solvates or Complexes of Chitosan Salts

Chitosan Salt Solvate	Mole Ratio Acid to Chitosan: Reaction[a]	Product	pH[b]	Vis.,[b] cp
Formate	4.5	2.47	4.2	8.3
Acetate	5.8	1.93	5.1	7.6
Pyruvate	4.8	2.67	3.5	5.1

[a]Kytex; low viscosity from brown shrimp; 1,2-dichloroethane reaction medium; 20-22°C for reaction and drying.
[b]One percent solution in water.

acid molecules appear to participate in solvate or complex formation involving more than a stoichiometric amount of acid.

The development of the chitosan salt acid solvates or complexes apparently is importantly a function of the molar excess of acid during salt formation. In the case of acetic acid (Table 4), a 5.6 molar ratio of acid to chitosan led to a substantial acid solvate, whereas a 2-3 molar ratio gave merely the simple soluble chitosan acetate partial salts. This series of chitosan acetates also showed a gradual decrease in specific gravity with increasing acetic acid content.

Pyruvic acid reacts rapidly with chitosan to yield both the normal salt and its acid solvate or complex, depending upon medium, the ratio of the reactants and temperature (Table 5). As a reaction medium, 1,2-dichloroethane was found most satisfactory. Since pyruvic acid is non-

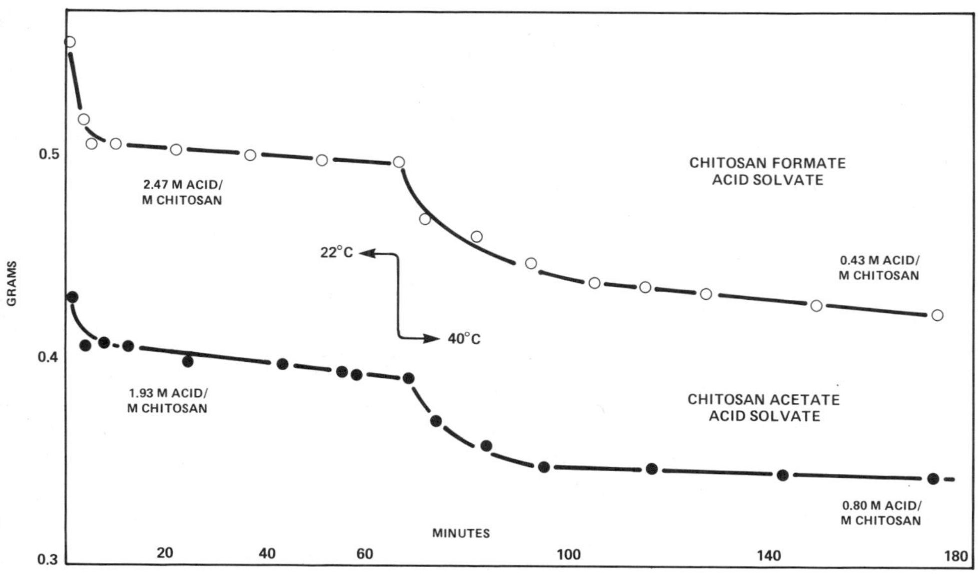

Fig. 1. Loss Rate of Acid Solvates of Chitosan Salts.

Table 4. Acid Ratios in Chitosan Acetate-Acid Solvate or Complex Preparation

Chitosan[a]		Acetic Acid		Moles Acid: Chitosan		Sol. Rate[b]	Sp.Gr.
g	Moles	mL	Moles	React	Product		
5	0.03	10	0.17	5.6	1.56	+++	1.37
10	0.06	12	0.20	3.3	0.83	++	1.39-1.41
5	0.03	3	0.05	1.7	0.55	+	1.44

[a]FCRL pink shrimp; 1,2-dichloroethane medium; reacted at 14-20°C; 1-2.5h.
[b]+++ = most rapid.

volatile, a range of compositions appears quite stable. The pyruvic acid reactions also illustrate the effect of raising the temperature over even a modest range in increasing the uptake of acid by chitosan. There appeared to be no regular effect on specific gravity of the salt by the amount of acid combined with chitosan.

A possible explanation for the formation of the acid-solvated or complexed chitosan salts, as well as perhaps some portion of the regular chitosan salts, may be that they fall into the class of intercalated biological systems[10], that is, guest molecules inserted in a lamellar host. Several aspects of the preparation, physical properties and solution behavior of the solvates contribute to this postulate: (a) chitosan, like chitin, has a multiplanar structure[11,12], a prerequisite for intercalation; (b) solvates or complexes are formed best with small, low molecular weight acids; the acid molecules, too, are flat; (c) the acid components of simpler salts as well as the complexes are present in variable ratios; (d) there is an apparent volume increase during the salt formation of certain solvates or complexes which is manifest in the lower specific gravity of the dry products; and (e) there may be a parallel in the behavior of chitin, which forms solvates with a number of low molecular weight alcohols and ketones[13]. Resolution of the conformation of the chitosan salts and their acid solvates or complexes, including their relation to the chirality of chitosan[14], remains an intriguing problem.

CONCLUSION

Dry chitosan salts or complexes can be obtained by the direct action of a lower aliphatic carboxylic acid on chitosan dispersed in an organic medium. These dry, free-flowing products dissolve readily in water and yield viscous solutions like the dissolution of chitosan itself in dilute aqueous acids. The chitosan salts or complexes of formic, pyruvic and lactic acid are favored for their ease of preparation, solubility and stability.

The heterogeneous reactions are carried out under ambient conditions in one to three hours; the product is filtered, washed with fresh medium and dried. Aliphatic monocarboxylic acids with up through four carbon atoms operate well. An excess of a molar equivalent of the acid is usually employed with an inert medium, such as chlorohydrocarbon, ester, ether, ketone or acetonitrile. Some water can be tolerated in the

Table 5. Effect of Temperature on Chitosan Pyruvate Preparation

React,[a] moles Acid : Chitosan	Temp. °C	Product, moles Acid : Chitosan	Sol. Rate[b]	Sp.Gr.
1 : 1	30-35	0.31	-	1.45
3 : 1	30-35	1.59	+++	1.44-1.45
5 : 1	30-35	2.32	+++	1.45-1.46
4.7 : 1	18	1.21	+++	1.42-1.44
3 : 1	13	0.57	±	1.44
5 : 1	13	0.88	++	1.42-1.44

[a]0.34g Kytex low viscosity chitosan, Sp.Gr., 1.45; 5 mL 1,2-dichloroethane.
[b]+++ = most rapid, ± = viscous solution but not all soluble.

reaction, as both 88% formic acid and 70% glycolic acid can be used in an appropriate medium.

A novel feature of this method is the development of chitosan salt acid solvates or complexes with more than a molar equivalent of acid when a substantial excess of the acid component is employed in the reaction. It is possible that these solvates or complexes may be intercalation products. Chitosan complexes with formic and acetic acids lose acid slowly unless confined, but complexes of less volatile acids such as pyruvic acid are quite stable.

These solid chitosan salts or complexes, soluble in water, offer advantages of convenience, ease of control and simplicity in handling. The complexes with more than a molar equivalent of acid provide a reservoir of acid for good solubility, for adjusting pH and for solubilizing basic components in certain formulations.

ACKNOWLEDGMENT

This report is based on research sponsored by the National Sea Grant Program, National Oceanic and Atmospheric Administration, U.S. Department of Commerce, Grant No. NA83AA-D-00017, and funds from the University of Delaware, U.S.A.

REFERENCES

1. G. W. Rigby, Deacetylated Chitin and Processes for Same, U.S. 2, 040,879 and U.S. 2,040,880 (1936); R. A. A. Muzzarelli, "Chitin," Pergamon Press, London (1977).
2. E. R. Hayes, D. H. Davies and V. G. Munroe, Organic solvent systems for chitosan, in: "Proceedings of the First International Conference on Chitin/Chitosan," R. A. A. Muzzarelli and E. R. Pariser, eds., M.I.T. Sea Grant Report MITSG 78-7, Cambridge (1978).

3. P. Gross, E. Konrad and H. Mager, Chitosan as a natural film forming ingredient, *in*: "Chitin and Chitosan, Proceedings of the Second International Conference on Chitin and Chitosan," S. Hirano and S. Tokura, eds., The Japanese Society of Chitin and Chitosan (1982), Tottori, Japan.
4. E. R. Hayes, Characterization of chitosan, *in*: "Proceedings of the First International Conference on Chitin/Chitosan," R. A. A. Muzzarelli and E. R. Pariser, eds, M.I.T. Sea Grant Report MITSG 78-7, Cambridge (1978).
5. M. Sugano, T. Fujikawa, Y. Hiratsuji, K. Fukuda and Y. Hasegawa, A novel use of chitosan as a hypocholesterolemic agent in rats, *Am. J. Clin. Nut.* 33:787 (1980).
6. M. E. Cupery, Sulfamic Acid Salts of Polymeric Amino Compounds, U.S. 2,201,762 (1940).
7. T. S. Sannan, K. Kurita, and Y. Iwakura, Studies on chitin, 2. Effect of deacetylation on solubility, *Makromol. Chem.* 177: 3589-3600 (1976).
8. P. R. Austin, Chitin solvents and solubility parameters, *in*: "Chitin, Chitosan and Related Enzymes," J. P. Zikakis, ed., Academic Press, New York (1984).
9. P. R. Austin, Dry Free-Flowing Water-Soluble Complexes of Chitosan, U.S. Patent Application, Serial No. 703,070 (1985).
10. W. D. Wilson and R. L. Jones, Intercalation in biological systems, *in*: "Intercalation Chemistry," M. S. Whittingham and A. J. Jacobsen, eds., Academic Press, New York (1982).
11. K. Ogawa, S. Hirano, T. Myanishi, T. Yui, and T. Watanabe, A new polymorph of chitosan, *Macromolecules* 17:973-975 (1984).
12. J. Blackwell and M. A. Weih, The structure of chitin-protein complexes, *in*: "Chitin, Chitosan and Related Enzymes," J. P. Zikakis, ed., Academic Press, New York (1984).
13. P. R. Austin, Chitin Complexes with Alcohols and Carbonyl Compounds, U.S. 4,063,016 (1977); F. A. Rutherford, III and P. R. Austin, "Chitin Solvents, Complexes and Physical Properties," Delaware Sea Grant Technical Report DEL-13-78 (1978).
14. P. R. Austin, Renatured Chitosan and Process of Making Same, U.S. 4,309,534 (1982).

CHEMICAL MODIFICATIONS OF CHITIN AND CHITOSAN

K. KURITA

Department of Industrial Chemistry
Faculty of Engineering, Seikei University
Musashino-shi
Tokyo, Japan

INTRODUCTION

Although chitin may be less advanced than cellulose with regard to research and utilization, it has recently begun to attract much attention in both basic and applied research fields, including not only biology and biochemistry but also organic and polymer chemistry, pharmacology, and medicine. The number of reports and patents is increasing at a remarkable rate. Besides basic research, many attempts have been made to find new applications for chitin. Further basic studies, however, seem to be necessary to realize its full potential. One possible breakthrough in chitin chemistry and technology is the development of chemical modifications of chitin which are being studied more and more actively to explore highly sophisticated functions. Chitin can be regarded as a new type of polymeric material having greater possibilities than cellulose in many respects. Since it is an amino polysaccharide, chitin is capable of undergoing many additional modification reactions. It is anticipated that widespread applications will be found in the near future.

This review deals with the chemical modification of chitin and chitosan, focusing on the recent developments, to see the possibility of exploring highly useful functions in perspective. The following modification reactions will be discussed along with the resulting properties: deacetylation, acylation, aldimination, carboxymethylation, sulfation, complexation with metal cations, and some miscellaneous reactions.

Deacetylation

On treatment of chitin with base, the acetamido groups are hydrolyzed to give free amino groups. This hydrolysis to chitosan is usually carried out with sodium hydroxide or potassium hydroxide at elevated temperatures under heterogeneous conditions using chitin flakes or powders. In spite of the alkaline hydrolysis, the polysaccharide main chain is also degraded to some extent. The reduction of molecular weight becomes pronounced with increasing reaction time (1). An attempt was recently made to reduce the extent of depolymerization as much as possible in preparing fully deacetylated chitosan (2). The deacetylation was carried out similarly but with added thiophenol which traps oxygen thus preventing degradation and exerting a catalytic effect.

For instance commercial chitosan with 80 % deacetylation becomes fully deacetylated by repeating three times the treatment with sodium hydroxide in water/DMSO in the presence of thiophenol at 100° C for 1 h. The molecular weight decreases with increasing number of treatments, but the reduction is less extensive than with conventional treatment.

The hydrolysis can be conducted under homogeneous conditions as well as the conventional heterogeneous conditions. The conventional hydrolysis at gentle boiling depends on the alkali concentration, and at low concentrations, the degree of deacetylation levels off at relatively low values. With 40 % sodium hydroxide, deacetylation proceeds to a high extent. The samples with 45 to 55 % deacetylation were found to be water-soluble, whereas those with above 60 % or below 40 % deacetylation are naturally insoluble in water (3-5). This remarkable water-solubility is probably brought about by random deacetylation along the chains, since the sample with 50 % deacetylation prepared by heterogeneous deacetylation is not water-soluble.

N-Acylation

The N-acetylation of chitosan leads to fully N-acetylated chitin. In studying the influence of the reaction medium on the N-acetylation of chitosan, it was found (6) that maximum reaction rate is achieved using binary mixtures of ethanol and methanol or methanol and formamide. Complete N-acetylation may be achieved in three minutes at room temperature using a highly swollen chitosan in organic aprotic solvents (7). Alternatively, the preparation of partially acylated chitosans was made by treating solutions of chitosan in acetic acid-methanol with acid anhydrides at room temperature. Rapid N-acylation of chitosan can be achieved, in fact, under heterogeneous conditions if highly swollen gels are prepared or the sample has been steeped in solvent (7-9). Homogeneous acylation is also possible in trichloroacetic acid/dichloroethane (10). Although methanesulfonic acid is a strong acid, it is used as a solvent (7). Chitin is insoluble in the solvent, but the acylated products go into solution. The structure of the acyl residue has considerable influence on the ease of O-acetylation, and with bulky acyl groups, steric hindrance may prevent substitution beyond 0.5. Under these conditions, the formation of highly acetylated products requires prolonged reaction times.

Long-chain di-O-acyl chitins may be prepared upon reaction with acyl chlorides or anhydrides in methanesulfonic acid at low temperature (11). Chitosans boiled with a large excess of hexanoyl, decanoyl or dodecanoyl chlorides in dry pyridine-chloroform afford fully acylated derivatives (12). Highly benzoylated chitin shows good solubility in several organic solvents (13). An aspirin carrier was prepared by reaction of chitosan with 2-acetoxybenzoic anhydride (14).

Carboxyacyl chitosans have been prepared from reactions with anhydrides of alkyl and aryl dicarboxylic acids (15-17). The acylation with cyclic anhydrides such as trimellitic anhydride and pyromellitic anhydride on water-soluble chitin gives amic acid-chitins, which are transformed into imide-chitins. These derivatives have newly introduced carboxyl and acid anhydride groups, respectively, available for further modification (15).

Partially succinylated chitosans and O-hydroxyethyl chitosan containing both amino and carboxyl groups show varying solubility in water, dilute acid and dilute alkali, depending on the degree of substitution (16). By further reaction with carbodiimides, gels with a low degree of cross-linking have been prepared (16-18). Acetylated chitosan is obtained as porous particles when an aqueous chitosan solution is treated with a toluene solution of acetic anhydride (19). These particles are useful as molecular sieves, and dextrans with different molecular weights were demonstrated to be effectively separated. Wetting characteristics of acylated chitin film surfaces and blood clotting on the surfaces have been studied to evaluate these derivatives as biocompatible materials (20). Table I summarizes some examples of blood clotting time. In this series, 2.0 acetylchitin shows the best result. All acylated chitins behave similarly to siliconized glass.

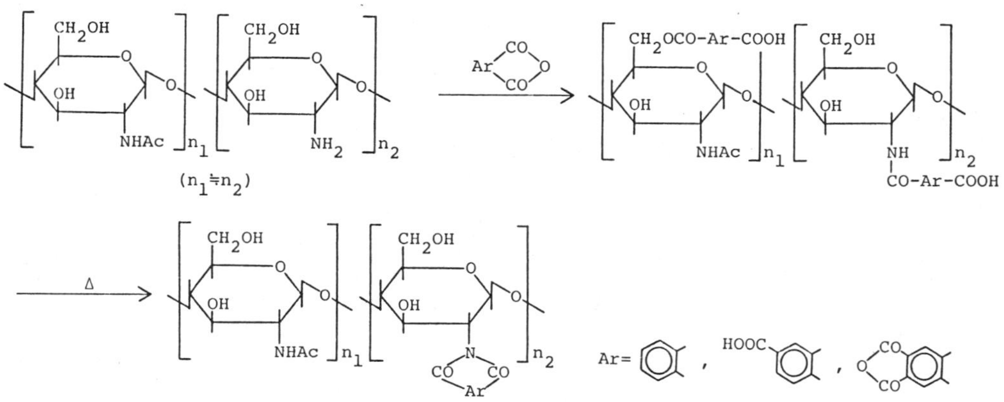

Table I   Relative clotting time index

Sample	Clotting time ratio	
	Sample/Glass	Sample/Siliconized glass
Chitin	1.3	0.7
1.4 Formylchitin	1.6	1.0
2.0 Acetylchitin	3.8	1.9
1.9 Propionylchitin	2.4	0.9
1.8 Butyrylchitin	2.0	1.2
2.0 Caproylchitin	2.5	1.1

Some of these derivatives show solubilities analogous to proteins, in being soluble in dilute sodium hydroxide, precipitating at pH 4 - 5 and dissolving at low pH values.

Although the maleylation of amino groups in proteins may be carried out in water at pH 7.1 - 9.5, the heterogeneous reaction with chitosan was better performed in formamide (21). The copolymers of N-maleylchitosan and acrylamide are stable at all pH values; they swell in water and give gels with good mechanical properties. Immobilized antibodies on these materials are effective in reducing hepatitis virus antigen values in blood plasma (22).

Schiff bases and their reduction products

The Schiff reaction between chitosan and aldehydes or ketones gives the corresponding aldimines and ketimines, which can be hydrogenated to products less susceptible to hydrolysis (23). N-Alkyl chitosans from simple aldehydes and ketones have been isolated as white powders with degrees of substitution 23 - 33 % (24). Intramolecular hydrogen bonds are apparently weakened by the presence of the bulky substituents and thus, despite their hydrophobicity, the N-alkyl chitosans swell enormously in water; they retain the film-forming ability of chitosan and membranes can be cast from their acetic acid solutions.

The reductive amination reaction of chitosan with polyfunctional aldehydes and ketones provides access to polyampholytes, such as the one obtained from salicylaldehyde.

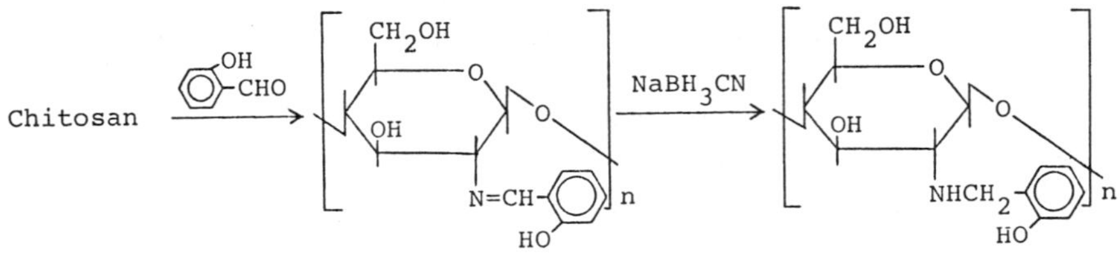

The addition of phthalaldehydic acid or glyoxylic acid to aqueous suspensions of chitosan resulted in immediate dissolution, with accompanying gel formation at suitable pH values (25-27). The Schiff bases, upon hydrogenation, afforded N-(o-carboxybenzyl) chitosan (NCBC) and N-(carboxymethyl) chitosan (NCMC), soluble in both acidic and alkaline media (see Figure 1). In this compound, each carboxymethylamino residue is simultaneously a secondary amine and a carboxylic acid.

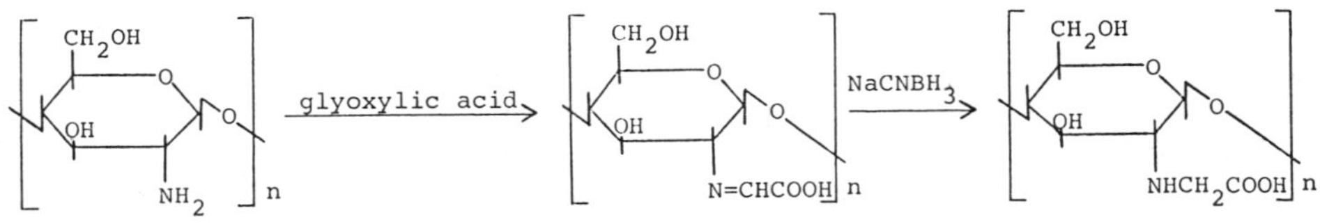

The n.m.r. spectrum for a typical NCMC shows signals attributable to the N-carboxymethyl substituent, at 168.7 and 47.7 ppm, for COO^- and $N\text{-}CH_2$, respectively. A downfield shift (~6 ppm) is observed for the carbon bearing the N-carboxymethyl group, C-2', with respect to the corresponding

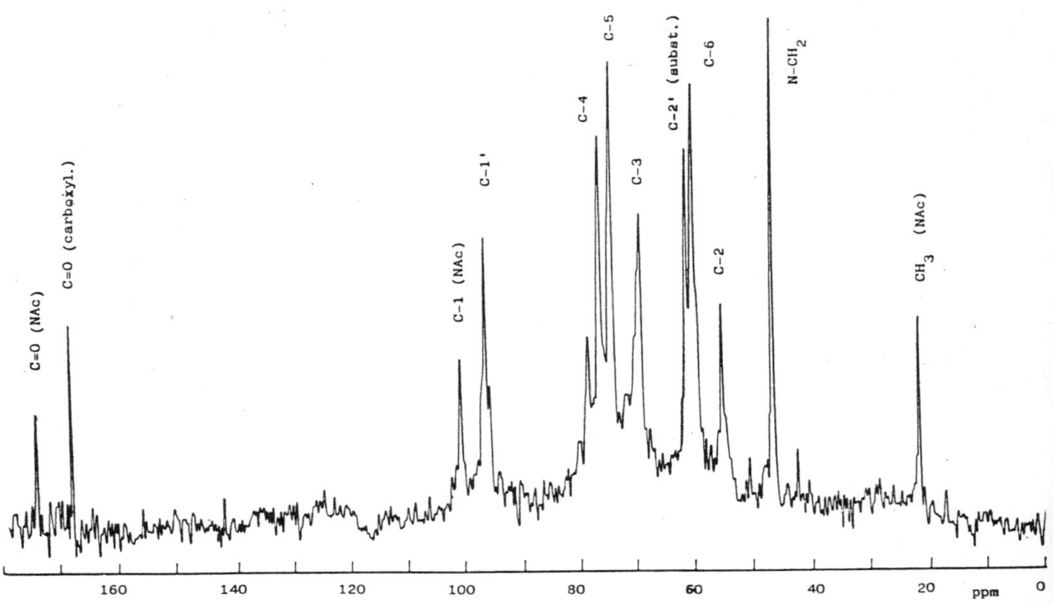

Fig. 1. ^{13}C-n.m.r. spectrum of a typical N-carboxymethyl chitosan with degree of acetylation 40 %, degree of N-carboxymethylation 50 % and free amine 10 %. Courtesy of R.A.A. Muzzarelli (29).

carbon of unmodified residues, C-2 at 57.8 ppm. From the ratio of the areas of signals C-2 and C-2' (as well as the corresponding anomeric carbons C-1 and C-1'), N-acetylation is about 40 % and N-carboxymethylation more than 50 %.

Some similarities exist between O-carboxymethyl chitosan and N-carboxymethyl chitosan in physical properties (28) (moisture content, fall in viscosity during the first few days after preparation, coagulation by solvents and neutral salts), but the primary alcohol groups of NCMC are available for further reactions such as cross-linking (29).

The attachment of reducing carbohydrates as side-chains to the 2-amino functions of chitosan transforms it into branched-chain water-soluble derivatives. Easy conversion can be achieved by reductive alkylation using sodium cyanoborohydride (23, 30). Further, specific chemical modifications can be effected with the introduction, via D-galactose oxidase treatment, of aldehyde functions into the pendant D-galactose residue. Reductive alkylation of chitosan with lactose affords 1-deoxylactit-1-yl chitosan whose solution has unusual rheological properties.

O-Carboxymethylation

Carboxymethylation is achieved with monochloroacetic acid and sodium hydroxide (31). The reaction takes place preferentially at C-6 hydroxyl groups (32). In this reaction, however, some part, sometimes over 50 %, of acetamido groups are inevitably hydrolyzed on account of the strongly basic conditions. In a modified method, DMSO is used to swell chitin (33).

Crosslinked carboxymethylchitins or chitosan show high capability of separating bovine serum fibrinogen and albumin. The adsorption behavior of bovine blood proteins on carboxymethylchitins with various substitution degrees was also studied by Tokura and coworkers (34).

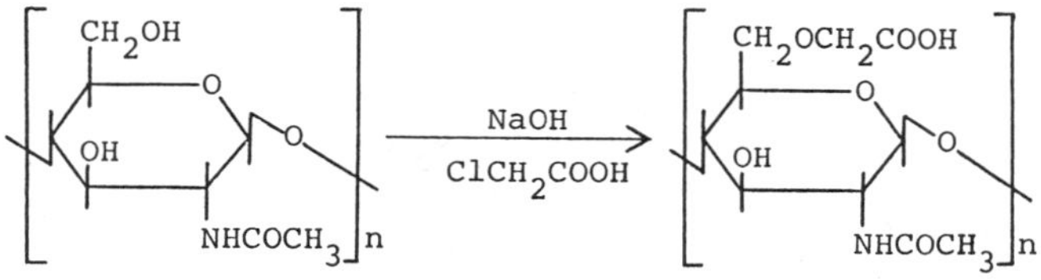

An interesting application of carboxymethylchitin was reported recently in the preparation of artificial red blood cells (35). A W/O type emulsion obtained from sheep hemolysate and lecithin in dichloromethane is dispersed in an aqueous carboxymethylchitin solution to form a W/O/W type emulsion. Removal of the organic solvent results in an aqueous suspension of artificial red blood cells.

Sulfation

Sulfation of chitin and chitosan has been one of the most attractive modification fields owing to the possibility of preparing anticoagulant polysaccharides in view of the structural similarity to heparin. For sulfation, various reagents have been used including concentrated sulfuric acid, oleum, sulfur trioxide/pyridine, sulfur trioxide/sulfur dioxide, and chlorosulfonic acid, but the last one is most commonly used. The present Volume includes Chapters devoted to the preparation and uses of sulfated chitosans.

Miscellaneous Reactions

Cyanoethylation with acrylonitrile is carried out in a manner similar to

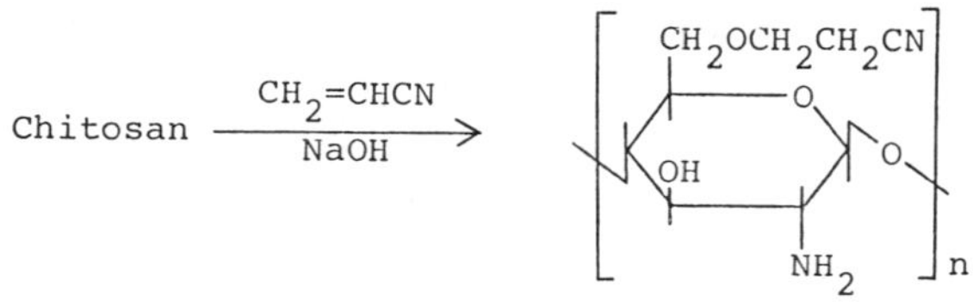

the preparation of cyanoethylcellulose. Cyanoethylchitosan, on blending with cellulose nitrate, gives membranes which are good for microfiltration (36). An advantage of these membranes is that almost no shrinkage occurs on autoclave sterilization.

Graft polymerization onto chitin is another interesting way to modify its properties, but only a few reports have been published. The grafting of methyl methacrylate onto chitin takes place with tributylborane as the initiator by a radical mechanism (37). Styrene is also grafted with γ-ray irradiation (38). In both cases, water plays an important role in promoting the graft polymerization. Further reactions are described in the following Chapters.

REFERENCES

1. A. C. M. Wu and W. A. Bough, in: "Proc. 1st Int. Conf. Chitin/Chitosan", R. A. A. Muzzarelli and E. R. Pariser (eds.), M.I.T. Cambridge USA (1978).
2. A. Domard and M. Rinaudo, Int. J. Biol. Macromol., 5:49 (1983).
3. T. Sannan, K. Kurita, and Y. Iwakura, Makromol. Chem., 176:1191 (1975).
4. T. Sannan, K. Kurita, and Y. Iwakura, Makromol. Chem., 177:3589 (1976).
5. K. Kurita, T. Sannan, and Y. Iwakura, Makromol. Chem., 178:3197 (1977).
6. G. K. Moore and G. A. F. Roberts, Intl. J. Biol. Macromol., 3:292 (1981).
7. N. Nishi, J. Noguchi, S. Tokura, and H. Shiota, Polymer J., 11:27 (1979).
8. S. Hirano, and Y. Yagi, Carbohydr.Res., 83:103 (1980).
9. G. K. Moore, and G. A. F. Roberts, Intl. J. Biol. Macromol., 4:246 (1982).
10. T. Ando and S. Kataoka, Kobunshi Ronbunshu, 37:1 (1980).
11. K. Kaifu, N. Nishi, and T. Komai, J. Polym. Sci., 19:2361 (1981).
12 S. Fujii, H. Kumagai, and M. Noda, Carbohydr. Res., 83:389 (1980).
13. O. Somorin, N. Nishi, S. Tokura, and J. Noguchi, Polymer J., 11:391 (1979).
14. S. Hirano, and Y. Ohe, Carbohydr. Polymers, 4:15 (1984).
15. K. Kurita, H. Ichikawa, S. Ishizeki, H. Fujisaki, and Y. Iwakura, Macromol. Chem., 183:1161 (1982).
16. R. Yamaguchi, Y. Arai, T. Itoh, and S. Hirano, Carbohydr. Res., 88:172 (1981).
17. S. Hirano, and T. Moriyasu, Carbohydr. Res., 92:323 (1981).
18. R. Yamaguchi, Y. Arai, T. Kaneko, and T. Itoh, Biotechnol. Bioengin., 24:1081 (1982).
19. J. Koshugi, Eur. Pat. Appl., 21:750; 23:789 and 26:618 (1981).
20. K. Kaifu, and T. Komai, J. Biomed. Mater. Res., 14:477 (1980).
21. L. A. Berkovich, M. P. Tsyurupa, and V. A. Kavankov, J. Polymer Sci., 21:1281 (1983).
22. K. Kifune, K. Inoue, I. Sakamoto, and T. Unigame, Japan Kokai Tokkyo Koho, 59:105 (1984).
23. L. D. Hall and M. Yalpani, JCS Chem. Comm., 1153 (1980).
24. R. A. A. Muzzarelli, F. Tanfani, and M. Emanuelli, J. Membrane Sci., 16:295 (1983).
25. R. A. A. Muzzarelli, F. Tanfani, M. Emanuelli, and S. Mariotti, Carbohydr. Res., 107:199 (1982).

26. R. A. A. Muzzarelli, F. Tanfani, M. Emanuelli, and S. Mariotti, Carbohydr. Polymers, 2:145 (1982).
27. R. A. A. Muzzarelli, and F. Tanfani, Pure Appl. Chem., 54:2141 (1982).
28. J. Koshugi, Eur. Pat. Appl., 0018131 (1980).
29. R. A. A. Muzzarelli, F. Tanfani, M. Emanuelli and L. Bolognini, Biotechnol. Bioengin., (in press).
30. M. Yalpani, L. D. Hall, M. A. Tung, and D. E. Brooks, Nature, 302:812 (1983).
31. A. Zimmermann, Swed., 136:717 (1952).
32. T. Mivazaki, and Y. Matsushima, Bull. Chem. Soc. Japan, 41:2723 (1968).
33. R. Trujillo, Carbohydr. Res., 7:483 (1968).
34. N. Nishimura, Y. Ikeuchi, and S. Tokura, Carbohydr. Res., 134:305 (1984).
35. A. Kato, M. Arakawa, and T. Kondo, J. Microencapsulation, 1:105 (1984).
36. R. E. Kesting, U.S., 4:220 (1980).
37. K. Kojima, M. Yoshikuni, and T. Suzuki, J. Appl. Polym. Sci., 24:1587 (1979).
38. Y. Shigeno, K. Kondo, and K. Takemoto, J. Macromol. Sci.-Chem., A17:571 (1982).

Editor's Report

R. A. A. MUZZARELLI

CHEMICALLY MODIFIED CHITOSANS

based on works by

O.P. AGRAWAL, B. CASU, M. COLOMBO, T. COMPAGNONI, A. DOMARD, J.C. DOMSZY, A. EBINA, B. FOCHER, K. FUJI-I, A. GADELLE, A. GERVASINI, S. HIRANO, Y. KONDO, E. KONRAD, G. LANG, Y. MAEKITA, A. MASSOLI, F. MORAZZONI, A. NAGGI, A. NISHI, S. NISHIMURA, E. PIVARI, M. RINAUDO, G.A.F. ROBERTS, K. SCHELLER, S. TOKURA, S. TORRI, A. TSUTSUMI, H.WENDEL, and F.A. WOOD

PREPARATION AND CHARACTERIZATION OF N-ACETYL GLUCOSAMINE OLIGOMERS BY HYDROFLUORIC ACID HYDROLYSIS; CONFORMATIONAL STUDIES

(A. Domard and A. Gadelle)

Oligomers have been primarily obtained by chemical or enzymatic hydrolysis of chitin. As opposed to previous processes, the use of hydrogen fluoride for the preparation of these compounds offers an alternative and convenient approach. Whereas the use of other mineral acids may give N-deacetylated products, depolymerisation using HF preserves this linkage. Oligomers produced from the HF hydrolysis of chitin can be separated by gel permeation chromatography (GPC) on Biogel P-4 equilibrated with ammonium acetate buffer (0.05 M, pH 4.5). An example of a chromatogram is given in Figure 1.

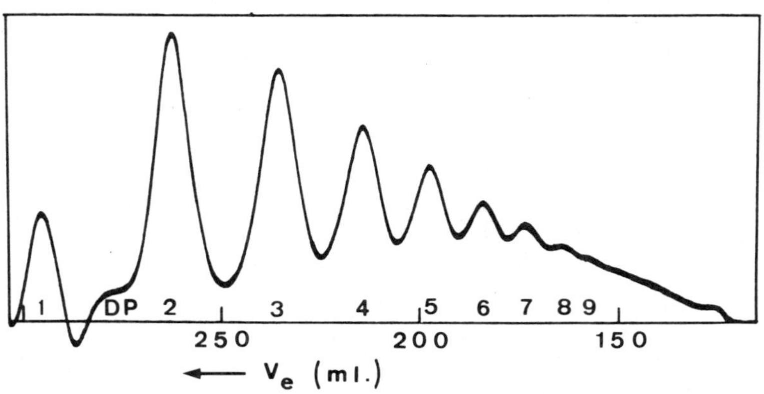

Fig. I Chromatogram of a sample of N-acetyl glucosamine oligomers (Chitin treated for 2 hours with HF).

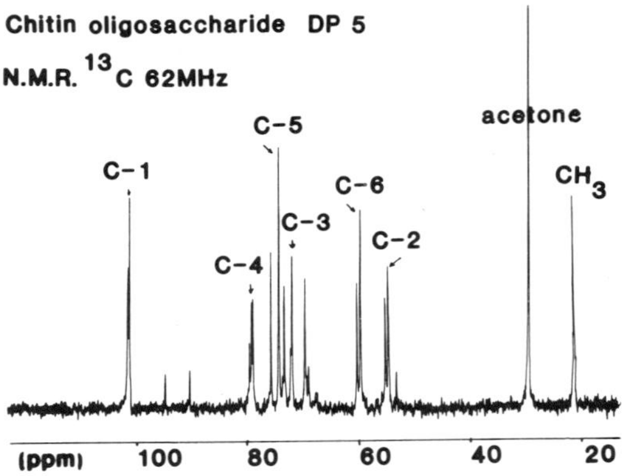

Figure 2. ^{13}C-N.m.r. spectrum of the pentasaccharide.

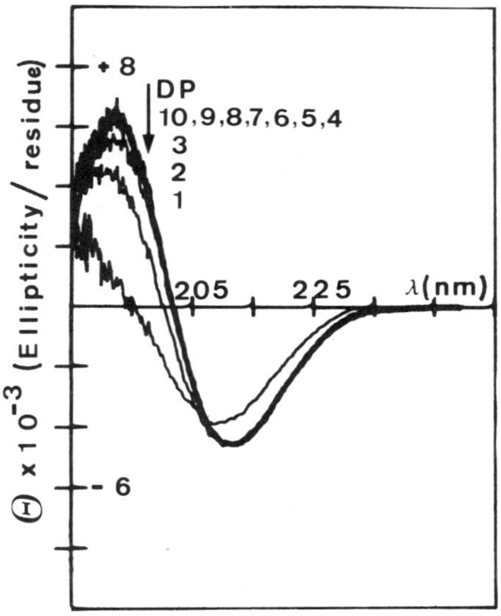

Figure 3. Circular dichroism spectra of a series of N-acetylglucosamine oligomers in water.

We readily obtained on preparative scale chromatography all the products having a DP range from 1 to 10. GPC of each oligomer, i.r., ^{13}C-n.m.r. and mass spectra all confirmed the high purity of the products, and the absence of any N-deacetylation. An example of a ^{13}C-n.m.r. spectrum is given in Figure 2.

Each oligomer was studied by ^{13}C- and ^{1}H-n.m.r. and by circular dichroism (CD) spectroscopy. According to these techniques, the conformations of oligomers having degree of polymerization larger than 4 are all equivalent as shown in Figure 3.

PREPARATION AND CHARACTERIZATION OF PHOSPHORYLATED CHITIN AND CHITOSAN

(N. Nishi, S. Nishimura, Y. Maekita, A. Ebina, A. Tsutsumi and S. Tokura)

To the mixture of chitin or chitosan (degree of deacetylation: 45 % and 97 %) in methanesulfonic acid was added phosphorus pentoxide (0.5-4.0 mol. equiv. to the residue) and the mixture was stirred at 0-5° C for 2 h . The products were precipitated with ether and centrifuged. The residue was washed with ether (x5), acetone (x3), methanol (x3) and ether (x1), and then dried. Chitin or chitosan phosphates thus obtained was insolubilized by the crosslinking reaction with adipoyl dichloride as follows. After the reaction with phosphorus pentoxide (2.0 mol. equiv.) for two h in methanesulfonic acid, adipoyl dichloride (0.5-2.0 mol. equiv.) was added in three portions and stirred for another 2 h at 0-5° C. The products were worked up by precipitation and washing with ether, followed by treatment with aqueous EDTA and exhaustive washing with deionized water.

Adsorption of metal ions on the insolubilized chitin phosphate and insolubilized chitosan phosphates was determined as follows, and the results were compared with those for chitin or chitosans. Samples (50 mg) were suspended in the aqueous solution containing 0.04 M metal ions ($MgCl_2$, $CaCl_2$, $SrCl_2$,

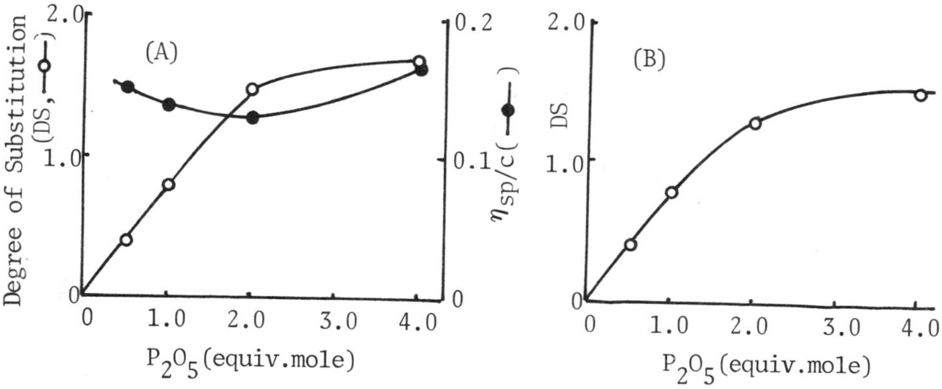

Fig.1. Amount of P_2O_5 - Degree of Substitution Relationships(—o—) for (A)Chitin Phosphate and (B)Chitosan Phosphate. Reaction Time: (A) 3 h , (B) 2 h. Chitin or Chitosan Conc. : g/7 ml CH_3SO_3H. —●— : η_{sp}/c(c=0.5 g/dl in 0.05 M NaCl at 25°C).

$BaCl_2$, $MnCl_2$, $Cu(NO_3)_2$, $Zn(NO_3)_2$, $CdCl_2$ or $NiCl_2$ in 5 ml of 50 mM tris-HCl buffer or 41 mM NaCl solution) and stirred at 30° C. The mixtures were centrifuged, and amounts of the metal ions in the supernatants were determined by titration with EDTA.

The phosphorylation reactions of chitin and chitosans in. phosphorus pentoxide-methanesulfonic acid system were found to be very efficient as in the cases of other chemical modifications in methanesulfonic acid (1-3). The DS of the products estimated from i.r. spectra, ^{13}C-n.m.r. spectra and elemental analyses (C, H, N and P) increases with increase in the amount of phosphorus pentoxide, and the products become nearly fully phosphorylated when 4.0 mol. equiv. of phosphorus pentoxide per residue are employed. The relations between the amount of phosphorus pentoxide employed and the degree of substitution are shown in Fig. 1 together with the reduced viscosity of the products of various DS in aqueous solution containing 0.05 M NaCl. Anyhow, it is clear that chitin or chitosan phosphate (4,5) of any DS can be prepared by regulating the amount of reagent. Among them, chitin phosphates are easily soluble in water, independent of DS. However, chitosan phosphates of high DS are insoluble in water while those of low DS are soluble in it. The water-soluble products were found to behave as typical polyelectrolytes viscometrically. The i.r. spectra of these derivatives show that as the degree of phosphorylation increases there is a corresponding decrease in the absorption at 1310 cm^{-1}, which is due to the hydroxyl group, and new absorption frequencies appear at 1240 cm^{-1} and 920 cm^{-1}, which are characteristic of the phosphoryl group. Further, the ^{13}C-n.m.r. study suggested that the hydroxyl groups of C-3 and C-6 positions are equally substituted with the phosphoryl group.

Chitin or chitosan phosphates thus obtained can easily be insolubilized by the cross-linking reaction with adipoyl dichloride as described in the initial part above. The i.r. spectra of the insolubilized products have strong absorption at 1730 cm^{-1} which is due to the di-ester linkage of the adipoyl residue.

The results on the adsorption experiments for various metal ions by the prepared resins, insolubilized chitin and chitosan phosphates, are shown in Table I compared with those for chitin or chitosans. Table I indicates that alkaline earth metals and Mn^{2+} were adsorbed more strongly to these deri-

Table I Adsorption of Metal Ions to Insolubilized Chitin Phosphate, Insolubilized Chitosan Phosphates, Chitin and Chitosans[a] (meq./g)

	Chitin	Chitosan(I)[b]	Chitosan(II)[c]	Insol-P-Chitin[d]	Insol-P-Chitosan(I)[e]	Insol-P-Chitosan(II)[f]
Mg^{2+}	0.4	0.3	0.5	1.1	1.5	1.3
Ca^{2+}	0.4	0.8	0.4	1.4	1.9	1.5
Sr^{2+}	0.4	1.5	0.6	1.4	2.4	1.7
Ba^{2+}	0.3	1.1	0.8	1.3	1.9	2.0
Mn^{2+}	0.5	1.1	0.5	1.1	2.2	1.8
Ni^{2+}	0.3	3.5	2.3	1.6	0.4	2.3
Cu^{2+}	0.4	5.3	4.8	1.0	2.2	2.6
Zn^{2+}	0.4	5.5	3.2	2.1	1.7	2.7
Cd^{2+}	0.6	6.5	4.9	0.9	2.0	3.0

[a] Metal ion: 0.04 M, buffer: tris-HCl buffer pH 7.4, Incubation condition : 20 h at 30°C except Insol-P-Chitin(pH 7.0, 1 h at 30°C)
[b] 45% deacetylated [c] 97% deacetylated [d] degree of phosphorylation: 1.0
[e] 45% deacetylated, degree of phosphorylation: 1.3
[f] 97% deacetylated, degree of phosphorylation: 1.5

vatives than to chitin or chitosans. However, it was also found that the transition elements, except Mn^{2+}, were adsorbed more strongly to chitosans than to these derivatives in our present reaction condition. The pH dependence for the Ca^{2+} adsorption to these materials showed that the amount of adsorbed Ca^{2+} increased with a rise in pH, suggesting the large contribution of phosphoryl group. However, the differential i.r. spectra of these materials between presence and absence of Ca^{2+} suggested that not only the phosphoryl group but also other groups such as the hydroxyl group and also the acetylamino group contributed cooperatively to the metal binding.

REFERENCES

1. N. Nishi, J. Noguchi, S. Tokura, and H. Shiota, Studies on Chitin I. Acetylation of Chitin, Polym. J., 11:27 (1979).
2. O. Somorin, N. Nishi, S. Tokura, and J. Noguchi, Studies on Chitin II. Preparation of Benzyl and Benzoylchitin, Polym. J., 11:391 (1979).
3. N. Nishi, H. Ohnuma, S. Nishimura, O. Somorin, and S. Tokura, Studies on Chitin VII. Preparation of p-Substituted Benzoylchitins, Polym. J., 14:919 (1982).
4. T. Sakaguchi, T. Horikoshi, and A. Nakajima, Adsorption of Heavy Metal Ions by Chitin Phosphate and Chitosan Phosphate, Nippon Nogei Kagak Kaishi, 53:149 (1979).
5. T. Sakaguchi, T. Horikoshi, and A. Nakajima, Adsorption of Uranium by Chitin Phosphate and Chitosan Phosphate, Agric. Biol. Chem., 45:2191 (1981).

PERACETYLATED, MODIFIED CHITOSACCHARIDES PRODUCED BY THE NITROUS ACID DEPOLYMERIZATION OF PARTIALLY N-ACETYLATED CHITOSAN

(S. Hirano, Y. Kondo and K. Fuji-i)

Several attempts have been made to prepare some depolymerized compounds of chitin and chitosan by partial hydrolysis with conc. HCl (1), phosphoric acid (2), trichloroacetic acid (3), formic acid (4), and by acetolysis (5). However, yields of the chitosaccharides produced by these methods are generally low (<16 %) because of random cleavage.

The nitrous acid-deaminative cleavage (6, 7, 8, 9, 10) of partially N-acetylated chitosan is considered to be more specific.

Chitosan was dissolved in 2 % aqueous acetic acid, and the solution was diluted with two volumes of methanol. Acetic anhydride (0.52 mol/GlcN) was added with vigorous stirring at room temperature (11). The corresponding product (d.s. for N-Ac) was isolated in a quantitative yield, $[\alpha]_D^{29}$ $-16°$ (c 0.8, 10 % aqueous acetic acid). I.r. data: ν_{max}^{KBr} 1660 and 1560 cm^{-1} (C=O and NH of N-Ac). Anal. Calc. for $C_6H_{10}O_4N(C_2H_3O)_{0.48}(H)_{0.52}\cdot 0.42H_2O$ n: C, 44.25; H, 6.83; N, 7.42. Found: C, 44.24; H, 6.82; N, 7.40.

The partially N-acetylated chitosan (8.8 g) obtained above was dissolved in 20 % aqueous acetic acid (300 ml), and solid $NaNO_2$ (9.5 g) was added. The mixture was stirred at room temperature for 30 min, and allowed to stand at room temperature for 18 h. A small amount of insoluble material was removed by filtration. The filtrate was neutralized with 6 M NaOH solution, and concentrated to dryness at below 45° in vacuo. The residue was peracetylated with acetic anhydride-pyridine (1:1, 80 ml) by stirring at room temperature for three days. The reaction mixture was poured into ice-water (~300 ml). The peracetylated product was extracted with chloroform (3 x

Table 1. Chromatographic Fractionation of Peracetylated, Modified Chitosaccharides

Fraction Number	Yield, g(%)[a]	R_{AcGlcN}[b]	Compound
37-80	0.98 (14)	1.8-1.4	1
81-89	0.07 (1)	1.0	1, 2
90-109	1.03 (15)	0.96	2
110-139	0.22 (3)	0.91, 0.62	2,3
140-180	0.66 (10)	0.53	3
181-209	0.36 (5)	0.53, 0.30	3, 4
210-264	0.73 (11)	0.30	4
265-289	0.66 (10)	0.64, 0.34	4, 5
290-340	0.45 (7)	0.34	5
340-400	1.04 (15)	0.00	6

[a]The peracetylated product(6.9 g) was applied to a column (3.0x63.0 cm) of silica gel(see text for experimental details).
[b]T.l.c. analysis was performed with acetone-benzene(1:1) for fraction number 37-264, and with acetone-benzene(3:1) for fraction number 265-400 as solvents.

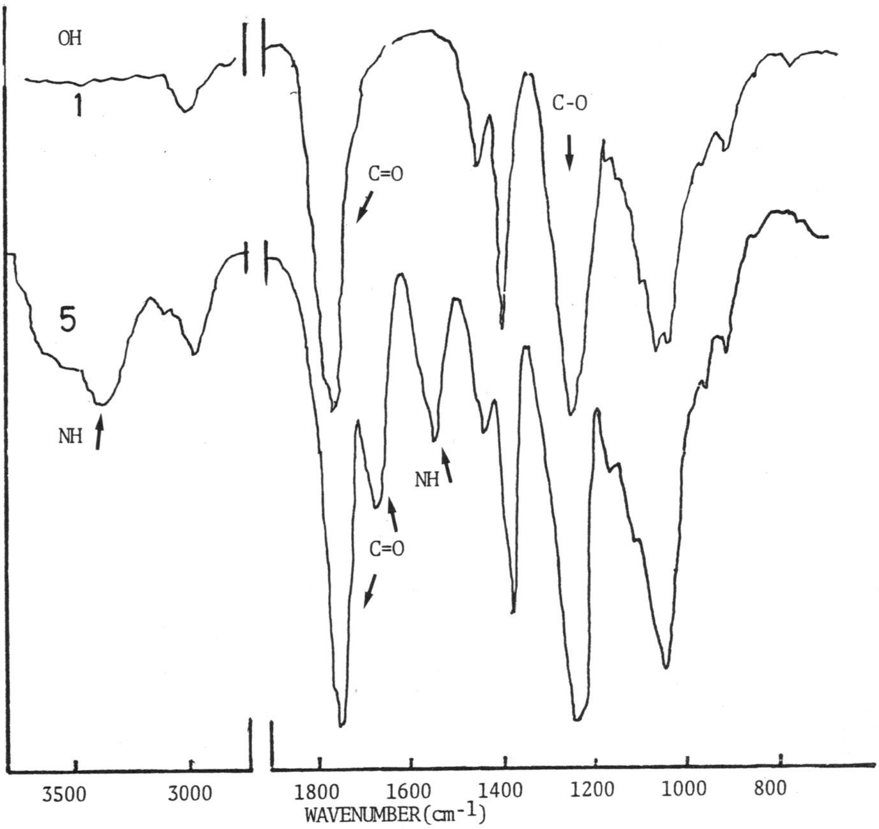

Fig. 1. A comparison in the i.r. absorptions of 1 and 5. Compounds 2-6 had an identical i.r. absorptions, which differed from that of 1 in N- and O-Ac absorptions, indicating that 1 had O-Ac and 2-6 had both N- and O-Ac in the molecule.

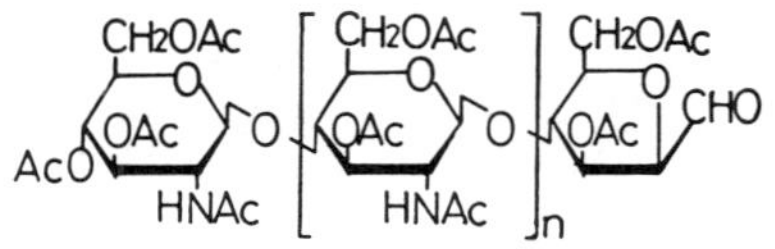

100 ml), and concentrated in vacuo to afford an amorphous mixture of peracetylated products (6.9 g).

The peracetylated product (6.9 g) obtained above was dissolved in chloroform (∿30 ml), and the solution was applied to a column (3.0 x 63.0 cm) of silica gel. The column was eluted with benzene-ethylacetate (3:1) at a flow rate of 17 ml/h, and 10 ml fractions were collected. Each of the fractions was monitored by tlc (Table I). A total of six compounds were isolated, and rechromatographed on a column of silica gel.

A portion (36-102 mg) of samples 2, 3, 4, 5 were O-deacetylated by treatment with 0.1 M aqueous NaOH solution at room temperature for 18 h to afford a syrupy, hydroscopic compound, which was soluble in water. The compound had i.r. absorptions at 1650 and 1540 cm^{-1} (C=O and NH of N-Ac) but no i.r. absorptions at 1750 and 1240 cm^{-1}(C=O and C-O of O-Ac). Each of the O-deacetylated products gave a single spot in tlc. The molecular weights of the O-deacetylated compounds of 2-5 were estimated by gel chromatography on a column (1.4 x 142 cm) of Sephadex G-25. A calibration curve was obtained by analysis of the relationship between Ve/Vo and the log of the molecular weight of standard N-acetylchito-oligosaccharides, mol. wt. 830, 627 and 424 (Seikagaku Kogyo Co., Tokyo). The eluates were monitored by u.v. absorption at 194 nm (Fig. 3).

Table 2. Specific Rotations of Peracetylated, Modified Chitosaccharides

Compound	$[\alpha]_D$ degree (c, chloroform, temperature)
1	+36 (0.9, 17°)
2	+16 (1.3, 16°)
3	+ 1 (1.0, 24°)
4	- 8 (1.0, 24°)
5	-13 (1.0, 24°)
6	-12 (1.0, 27°)

Table 3. Elemental Analysis Data of Peracetylated Modified Chitosaccharides

Compound	Formula	Calc.(%)			Found(%)		
		C	H	N	C	H	N
2	$C_{24}H_{33}O_{15}N \cdot 0.82H_2O$	48.83	5.92	2.37	48.73	5.89	2.38
3	$C_{36}H_{50}O_{22}N_2 \cdot 0.96H_2O$	49.13	5.95	3.18	48.98	5.95	3.09
4	$C_{48}H_{67}O_{29}N_3 \cdot 1.26H_2O$	49.16	5.98	3.58	49.21	6.03	3.49
5	$C_{60}H_{84}O_{36}N_4 \cdot 1.03H_2O$	49.50	5.96	3.85	49.50	5.92	3.80

Partially N-acetylated chitosan (d.s. for N-Ac 0.48) was prepared in quantitative yield (11). The nitrous acid-deaminative cleavage was performed in 20 % aqueous acetic acid solution, and the reaction product was peracetylated to afford an amorphous solid (83 %). The peracetylated mixture was fractionated into six fractions by chromatography on a column of silica gel (Table I). Compounds 1 - 5 had a single spot as shown by t.l.c., but 6 had a tailing spot. The yield (62 %) of 2 - 6 was much higher than those previoysly reported (<16 %). Compound 1 had i.r. absorptions at 1750 and 1240 cm^{-1} (C=O and C-O of O-Ac), but no i.r. absorption at 1650 and 1540 cm^{-1} (C=O and NH of N-Ac), and 1 was identified as 2,5-anhydro-D-mannose by reference to an authentic sample (9) by g.l.c. Compounds 2 - 5 had both i.r. absorptions (Fig. 1), and consisted of D-glucosamine and 2,5-anhydro-D-mannose as revealed by g.l.c. and t.l.c.. Only 2,5-anhydro-D-mannitol was detected by g.l.c., but D-glucosaminitol was only detected in the acid hydrolyzate of $NaBH_4$-treated 2 - 6, indicating that the reducing end-group is 2,5-anhydro-D-mannose. An aldehyde proton signal at δ 9.50 was detected in the ^{1}H-n.m.r. spectra. Table II shows the specific rotations of the peracetylated, modified chitosaccharides containing 2,5-anhydro-D-mannose at the reducing end-group whose structure was supported by the elemental analysis (Table III).

The DP of these compounds were estimated by two methods: 1) The molar ratios of D-glucosamine per 2,5-anhydro-D-mannose were 1.0, 2.1, 3.0 and 4.2 for 2 - 5, respectively. 2) Gel filtration on a column of Sephadex G-25 after O-deacetylation (Fig. 2) indicated that 2 - 5 were di-, tri-, tetra- and penta-saccharides, and 6 was a mixture of saccharides (DP 6).

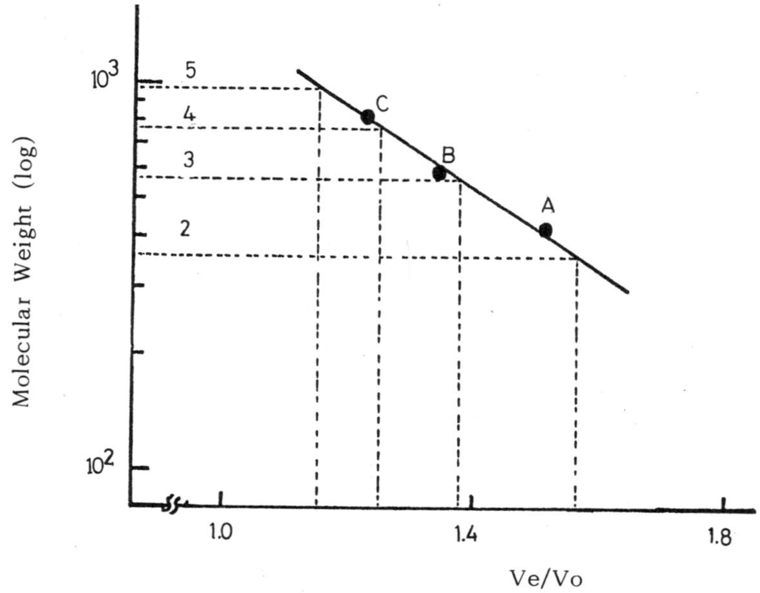

Fig. 2. Analysis of molecular weights of 2-5 by gel filtration on a column(1.4 x 142 cm) of Sephadex G-25 by elution with distilled water. The calibration curve was obtained by plotting Ve/Vo against the log of the molecular weight of standard N-acetylchito-oligosaccharides(A, 424; B, 627; C, 830).

REFERENCES

1. S. A. Barker, A. B. Foster, M. Stacey, and J. M. Webber, Isolation and properties of oligosaccharides obtained by controlled fragmentation of chitin, J. Chem. Soc., 2218 (1958).
2. P. R. Austin and C. J. Brine, Chitin powder, U. S. pat. No. 4:286 (1981).
3. T. Ando, and S. Kataoka, Acylation of chitin with acid anhydrides in trichloroacetic acid system, Kobunshi Ronbunshu, 37:1 (1980).
4. R. Yamaguchi, Y. Arai, and Itoh, A microfibril formation from depolymerized chitosan by N-acetylation, Agric. Biol. Chem., 46:2379 (1982).
5. T. Osawa, Lysozyme substrates. Synthesis of p-nitrophenyl 2-acetamido-4-O-(2-acetamido-2-deoxy-(β-D-glucopyranosyl)-2-deoxy-β-D-glucopyranoside and its β -D-(1→6) isomer, Carbohydr. Res., 1:435 (1966).
6. Z. Yosizawa, Deamination of the partially N-deacetylated mucopolysaccharides, Nature, 201:926 (1964).
7. D. Horton, Monosaccharide amino sugars, in: "The amino sugars", R. W. Jeanloz ed., Academic Press, New York, 1A:1 (1969).
8. M. Isemura, and K. Schmith, Studies on the carbohydrate moiety of acid glycoprotein by using alkaline hydrolysis and deamination by nitrous acid, Biochem. J., 124:591 (1971).
9. D. Horton and Philips, The nitous acid deamination of glycosides and acetates of 2-amino-2-deoxy-D-glucose, Carbohydr. Res., 30:367 (1973).
10. J. M. Williams, Deamination of carbohydrate amines and related compounds, in "Advances in carbohydrate chemistry and biochemistry", R. S. Tipson and D. Horton, eds., Academic Press, New York, 31:9 (1975).
11. S. Hirano, Y. Ohe, and H. Ono, Selective N-acylation of chitosan, Carbohydr. Res., 47:315 (1976).

CHITOSAN DERIVATIVES: WATER-SOLUBLE PRODUCTS BY REACTION WITH EPOXIDES

(G. Lang, E. Konrad and H. Wendel)

About ten years ago Konrad et al. found chitosan salts to be very well suited for applications in the field of cosmetics (1, 2). Part of their investigations was presented at the 2nd Int. Conf. on Chitin and Chitosan (3). Indeed one of the few commercial applications of chitosan until today is in setting lotions as a film forming and hair fixing component. However a disadvantage of chitosan salts concerning their use in cosmetics is their insolubility in most of the commercially used anionic detergents. The aim of our investigations was to find chitosan derivatives being clearly soluble in water and compatible with anionic detergents like sodium dodecylsulfate (SDS) or sodium laurethsulfate (SLES). (4 - 6).

Low and high M.W. chitosans were purchased from Kyowa Oil & Fat Co., with degrees of acetylation 14 and 24 % and limited viscosity numbers 140 ml/g-1600 l/g respectively. Glycidol and glycidyltrimethylammoniumchloride were purchased from Degussa AG, Hanau, both technical grades. Limiting viscosity numbers were determined in a DIN Ubbelohde viscosimeter in 0.2 M acetic acid/0.1 M sodium acetate. Potentiometric titrations were carried out in pure acetic acid using 0.1 M perchloric acid in the same solvent as titrant. Colloid titration (7) was carried out in double distilled water using 0.002 M PVSK (Polyvinylsulfate, K salt) as titrant under conductivity detection. All reactions were carried out in 2 l double-jacked reaction vessels with suitable stirring mechanism, adding epoxides or mixtures of epoxides to the suspensions of chitosan in water. During the reaction, viscosity increases, giving finally

(6 h, 80° C) clear, highly viscous solutions. For analytical procedures the products were dialysed 3 days against distilled water. The crude reaction products were filtered through gauze, precipitated by injection in acetone, collected by suction and dried.

N-(Dihydroxypropyl) chitosan (DHPC)

Primary amines are strong nucleophiles and react easily with epoxides (8). One of the most easily processable epoxides is glycidol (2, 3-epoxypropanol, GCD). Compared to other commercially used epoxides, GCD has the advantages of having a high boiling point (161° C) and of introducing two hydrophilic OH-groups in one step. Indeed reaction of three moles GCD with one mole of chitosan suspended in water leads to quantitative amounts of N, N-bis-(dihydroxypropyl) chitosan (DHPC). Unlike chitosan salt, DHPC is clearly soluble in water without neutralization and compatible in solutions with anionic detergents like SLES, but has nevertheless good filmforming properties, being suitable for application in hairsetting lotions.

N-(Glycidyltrimethylammonium) chitosan chloride (GTCC)

One of the most important advantages of cationic polymers in hair care is improving wet combability. Because DHPC possesses no real cationic group, we tried to overcome this by substituting glycidol by glycidyltrimethylammoniumchloride (GTMAC). GTMAC reacts with chitosan under the same conditions yielding N-substituted cationic products clearly soluble in water and having degree of substitution 1.0-1.5. Unfortunately GTCC is insoluble in solutions of anionic surfactants.

Mixed substituted products (GGTCC)

Chitosan can also be reacted with mixtures of glycidol and glycidyltrimethylammoniumchloride under the same reaction conditions as above. The sum of degrees of substitution of the mixed products is about 1.5-2,

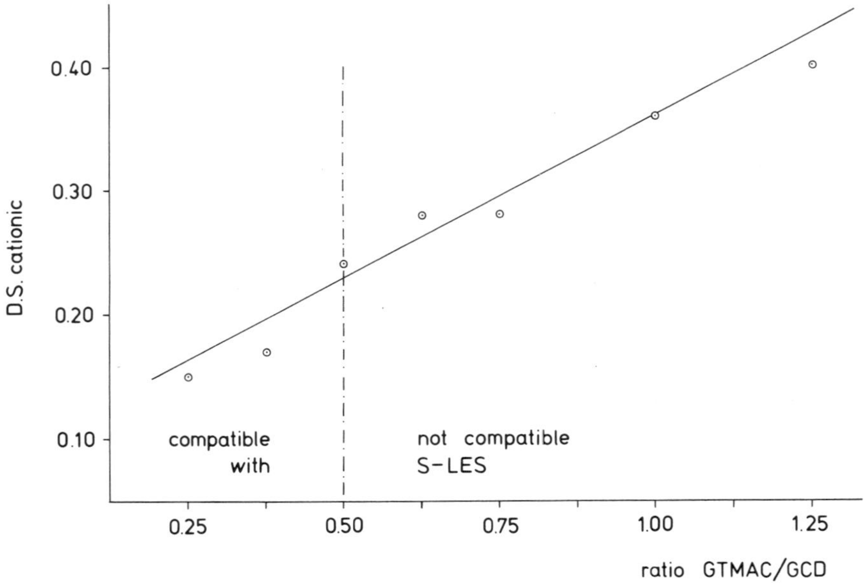

Fig.1. Degree of cationic substitution of mixed products *versus* ratio GTMAC / GCD.

Table I

Reaction conditions, analytical values and properties of products

Chitosan Mol Weight	Chitosan D.S. Acetyl	GCD (moles/mol Chitosan)	GTMAC (moles/mol Chitosan)	Solubility Water	Solubility 12%SLES	Limiting viscosity ml/g	Potent. Titr. mmol/g	Coll. Titr. mmol/g	D.S. GTMAC	D.S. GCD
L.M.	0,14	3,0	-	+	+	65	2,93	-	-	1,9
L.M.	0,14	-	3,0	+	-	68	-	3,2	1,00	-
H.M.	0,24	2,4	0,6	+	+	752	3,17	0,47	0,15	1,76
H.M.	0,24	2,18	0,82	+	+	683	3,50	0,58	0,17	1,33
H.M.	0,24	2,00	1,0	+	+-	713	3,39	0,80	0,24	1,31
H.M.	0,24	1,85	1,15	+	-	722	3,38	0,95	0,28	1,23
H.M.	0,24	1,7	1,3	+	-	753	3,33	0,92	0,28	1,30
H.M.	0,24	1,5	1,5	+	-	788	3,30	1,18	0,36	1,17
H.M.	0,24	1,33	1,67	+	-	792	3,34	1,33	0,40	1,04
L.M.	0,14	2,0	1,0	+	+	59	3,01	1,20	0,40	1,5

and the degree of cationic substitution is determined by the chosen ratio of GTMAC to GCD (Fig. 1). Products with cationic substitution degrees up to 0.25 are clearly soluble in SLES, low molecular weight products giving small, high molecular weight products giving large improvements in wet combability.

Degrees of substitution

The degrees of substitution have been calculated from the results of potentiometric titration, colloid titration of cationic groups, and elementary analysis. Reaction conditions, analytical values and properties of all products are summarized in Table I.

REFERENCES

1. Wella AG, Patent DE 26 27 419
2. Wella AG, Patent EP 0 002 506
3. P. Gross, E. Konrad, H. Mager in "Proc. 2nd Int. Conf. of Chitin and Chitosan", S. Tokura and S. Hirano (Eds.), Tottori, 205 (1982).
4. R. A. A. Muzzarelli: "Chitin," Pergamon Press, Oxford (1977)
5. R. A. A. Muzzarelli and E. R. Pariser, (Eds.) "Proc. 1st Int. Conf. of Chitin and Chitosan", M. I. T. Cambridge USA (1982).
6. R. A. A. Muzzarelli, F. Tanfani and S. Mariotti, J. Membrane Sci., 16:285 (1983).
7. H. Terayama, J. Polym. Sci., 8:243 (1951).
8. H. Kleemann, R. Wagner, "Glycidol", Huthing Verlag, Heidelberg, (1981).

6-O SULFATION OF CHITOSAN IN THE FORM OF COPPER(II) COMPLEX BY SULFUR TRIOXIDE-PYRIDINE

(B. Focher, A. Massoli, G. Torri, A. Gervasini and F. Morazzoni)

The selective sulfation of chitosan has been performed by protecting the amino groups with cupric ions. Cu(II)-chitosan complexes were obtained as a precipitate, by adding dropwise a 1.0 M copper sulfate solution into formic acid solutions of chitosan in the 2.0 - 7.0 pH range and allowing them to stand for 16 h under stirring.

The extent of Cu(II) adsorbed increases with increasing pH and Cu(II)/chitosan ratio. The quantity of Cu(II) complexed by chitosan rises rapidly at lower Cu(II) concentrations, whilst it levels off to a saturation value at higher concentrations (1).

E.s.r. spectra of Cu(II), obtained from a Cu(II)/chitosan molar ratio of 1.0, were recorded at -150°C with 9.095 GHz microwave frequency (2). The spectra of the precipitates from 2.0 - 4.5 pH solutions show a good resolution and the values of the magnetic tensor are well detectable. Two different paramagnetic species are present, and their relative quantities are changed by changing the pH of the starting chitosan solution. Due to spin-spin interaction of copper centers, the spectra broaden at higher pH, but remain quite similar to those at lower pH, as verified on the complex obtained from a Cu(II)/chitosan ratio of 0.01. The values of the g and A tensor components (species 1: $g_{1\parallel}$ = 2.28, $g_{1\perp}$ = 2.06, $A_{1\parallel}$ = 160 G; species

2: $g_{2\parallel}$ = 2.24, $g_{2\perp}$ 2.06, $A_{2\parallel}$ = 160 G) seem to indicate that the copper ion lies in a ligand field of two nitrogen and two oxygen atoms, arranged in square planar symmetry. The difference between the two species is in metal-ligand bond strength, species 2 revealing a higher bond strength. The two paramagnetic species could belong to two isomeric forms, cis and trans, of the same CuN_2O_2 chromophore rather than to two different chromophores (3).

The e.s.r. results are in accordance with ^{1}H-n.m.r. titration experiments, which indicate that H-2 protons are involved in the Cu(II) complex formation. The precipitation of chitosan by sulfate ions allows easier diffusion of Cu(II) into the chitosan matrix (4) and induces a decrystallization of the parent chitosan as confirmed by x-ray diffraction. The more amorphous structure, obtained independently of the pH of the starting solution, has raised the chitosan susceptibility to the sulfation reaction which has been performed in anhydrous DMF at different temperatures and with different quantities of sulfating agent (Fig. 1).

As shown by the ^{13}C-n.m.r. spectra (Figure 2), under optimized reaction conditions and independently of the pH of the starting solution, a selective 6-O sulfation similar to that obtained with sulfuric-chlorosulfonic acid has been achieved (5).

Comparison of the ^{13}C-n.m.r. spectra of chitosan 6-O sulfate and the corresponding Cu(II) chitosan 6-O sulfate (Fig. 2) shows that complexation with Cu(II) induces significant upfield shifts of C-1 and C-4, and a small downfield shift of C-2. In addition, a new signal shows up in the region of C-3/C-5 signals. Whereas shifts of C-1, C-4, and C-2 correspond

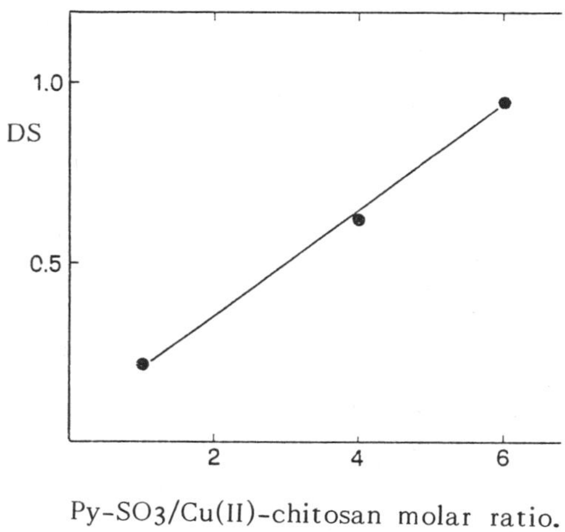

Fig. 1. Degree of substitution (DS) of chitosan 6-O sulfate as a function of the pyridine-SO_3/Cu(II) molar ratio. Reaction conditions: Temperature 20°C; time 24 h.

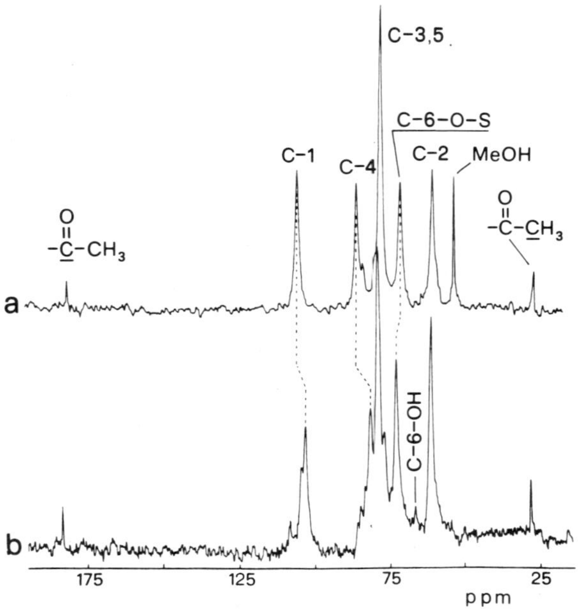

Fig. 2. ^{13}C-n.m.r. of chitosan 6-O sulfate (a), and Cu(II)-chitosan 6-O sulfate (b).

to those observed on decreasing pH in the absence of Cu(II) ions, the new C-3/C-5 signal is typically observed only in the presence of Cu(II). These results are in accordance with the e.s.r. spectra of Cu(II) chitosan 6-O sulfate, indicating that the sulfation reaction does not interfere with the coordination field.

The product shows i.r. bands characteristic of O-sulfate esters at 805 cm^{-1} (ν_{as} C-O-S) and 1230 cm^{-1} (ν_{as} S=O) (6). Finally, the limiting viscosity numbers ($|\eta|$) of the chitosan 6-O sulfate solutions in 0.1 M acetic acid, 0.2 M NaCl (pH = 3.5), and 0.2 M $NaHCO_3/Na_2CO_3$-0.2 M NaCl (pH 9.0) were found to be 0.8 and 1.05 (dlg^{-1}), respectively.

Acknowledgement. This work was supported by the Italian National Research Council, Progetto Finalizzato Chimica Fine e Secondaria, Roma.

REFERENCES

1. T. Koshijima, R. Tanaka, E. Muraki, A. Yamada, and F. Yaku, Cellulose Chem. Technol. 7:197 (1973).
2. R.A.A. Muzzarelli, F. Tanfani, M. Emanuelli, and S. Gentile, J. Appl. Biochem. 2:380 (1980).
3. J. Peisach and W.E. Blumberg, Arch. Biochem. Biphys. 165:691 (1974).
4. S. Hirano and S. Tokura, "Proc. Second Intl. Conference on Chitin /Chitosan" Japanese Soc. Chitin, Tottori, Japan, (1982).
5. A. Naggi, G. Torri, T. Compagnoni, and B. Casu, this volume.
6. F. Cabassi, B. Casu, and A.S. Perlin, Carbohydr. Res. 63:1 (1978).

BRANCHED CHITOSANS WITH N-ACETYL-GLUCOSAMINE OLIGOSACCHARIDE SIDE-CHAINS

(B. Casu, M. Colombo, T. Compagnoni, A. Naggi, E. Pivari and G. Torri)

Accessibility of linear polysaccharides to solvents and reagents is usually increased by branching, which prevents intermolecular association of the chains (1). Carbohydrate side-chains, introduced into chitosan by reacting the free amino group of the polysaccharide with the reducing end-groups of mono- and disaccharides (2), bring about dramatic changes in the rheological properties of chitosan (3).

When side-chains are made up of D-glucosamine or N-acetyl-glucosamine oligosaccharides, it is expected that these oligosaccharide chains are more accessible than similar segments in the original linear polymers.

In principle, GlcAm or GlcNAc oligosaccharide chains can be attached to chitosan by reacting (under conditions of reductive alkylation) the "aldehydo" groups of their terminal reducing residues, with the amino groups of the polysaccharide (2). However, as shown for heparin oligosaccharides, the reaction is more efficient when the aldehydo group is that of 2,5-anhydromannose (4).

In this work, chitosan oligosaccharides with terminal anhydromannose residues were obtained by partial depolymerization of chitosan (degree of N-acetylation 0.25) with nitrous acid (5). The oligosaccharides were N-acetylated (6) to prevent self-condensation of fragments, and reacted with chitosan (or chitosan oligosaccharides) in the presence of sodium cyanoborohydride, as illustrated in Fig. 1.

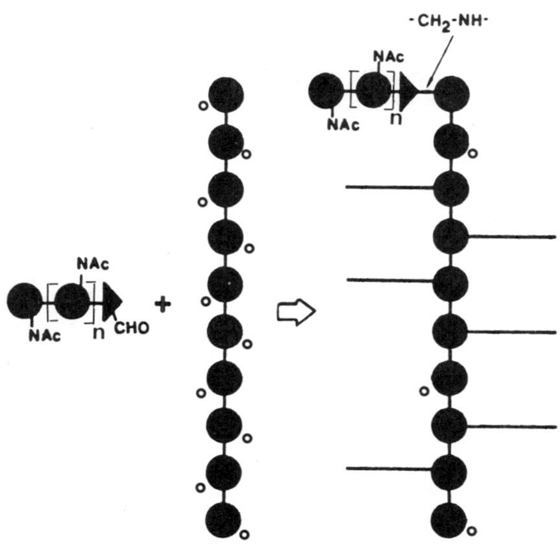

Fig. 1. Idealized scheme of a reductive alkylation reaction between GlcNAc oligosaccharides terminating with a 2,5-anhydromannose residue, and chitosan.

Average values of n (neocuproin end-group analysis) were 2 and 13. The monomer anhydromannose was also attached to chitosan. Degrees of substitution varied for different products, from 0.2 to 0.8 (by titration of free amino groups).

Products have different solubilities in water, depending on molecular weight, type and degree of branching. Solubility data for three typical products is given in Table I.

Table I. Degrees of substitution (d.s.) and solubility in water (g/l) of typical products.

		d.s.	solubility
1)	AM ≻ CHITOSAN	0.2	10
2)	$(GlcNAc)_3$AM ≻ CHITOSAN	0.15	0.5
3)	$(GlcNAc)_3$AM ≻ $(GlcAm)_{14}$AMol	0.7	400

Accessibility to complexing cations was evaluated with copper(II) ions, using a potentiometric method (7), and taking pH values (i.e., the difference in pH values measured for different additions of standard NaOH, in the absence and in the presence of copper(II) ions) as a measure of the extent of complexation. These values were qualitatively as expected, i.e., inversely proportional to the number of available free amino groups along the main chain. This number is small (chitin-like) for product 3, which is the most extensively substituted with N-acetylated oligosaccharides.

Acknowledgements. This work was supported by the Italian National Research Council, Progetto Finalizzato Chimica Fine e Secondaria, Roma.

REFERENCES

1. D. A. Rees, E. R. Morris, D. Thom and J. K. Madden, in "The Polysaccarides", G. O. Aspinall, ed., Academic Press, New York (1982).
2. L. D. Hall and M. Yalpani, JCS Chem. Comm., 1153, (1980).
3. M. Yalpani, L. D. Hall, M. A. Tung and D. E. Brooks, Nature, 302:812, (1983).
4. J. Hoffman, O. Larm and E. Scholander, Carbohydr. Res., 117:328 (1983).
5. F. Yaku, E. Muraki, K. Tsuchiya, Y. Shibata and T. Koshijima, Cell. Chem. Techn., 11:421, (1977).
6. I. Danishefsky, Methods Carbohydr. Chem., R. L. Whistler and J. N. Bemiller, eds., Academic Press, New York, 5:407, (1965).
7. Z. Tamura, M. Miyazaki and T. Suzuki, Chem. Pharm. Bull., 13:330, (1965).

THE HOMOGENEOUS REACTION OF CHITOSAN WITH SALICYLALDEHYDE

(J.G. Domszy, G.A.F. Roberts and F.A. Wood)

INTRODUCTION

The heterogeneous reaction of chitosan with salicylaldehyde has been reported previously[1] and has formed the basis of a method for determining the amine group content of chitosans[2], but no study of the homogeneous reaction has been reported. Since one method for determining the concentration of primary amines involves homogeneous reaction with salicylaldehyde in the presence of acetic acid followed by measurement of the absorbance[3], and since dilute acetic acid is an excellent solvent for chitosan, a study of the formation of N-salicylidenechitosan under homogeneous conditions has been carried out to evaluate its usefulness as an analytical procedure for determining the concentration of chitosan in solution.

EXPERIMENTAL

Materials

The chitosan used was a sample of Kytex L (Hercules Inc.) having a molecular weight of 3.82×10^5 (from LVN measurements[4]) and 84% free amine groups as determined by infrared spectroscopy[5]. This gives an equivalent weight of 199.7 for the amino groups.

Formation of N-salicylidenechitosan

Chitosan dissolved in dilute acetic acid was diluted with 4 volumes of methanol containing the required amount of salicylaldehyde. The reaction mixture was kept at 25°C and the absorbance measured as a function of time, using an equivalent salicylaldehyde solution as reference solvent.

RESULTS AND DISCUSSION

The uv-visible spectrum of N-salicylidenechitosan shows an intense absorption band at 320nm and a much less intense band at 410nm. The spectrum is the same as that of N-salicylidene-D-glucosamine produced under similar conditions. Formation of N-salicylidenechitosan can be readily followed by monitoring the increase in absorbance at λmax (410nm).

Effect of Variation in Salicylaldehyde Concentration

Solutions were prepared containing chitosan in aqueous acetic acid, at an acetic acid:amine group molar ratio of 3:2, diluted with 4 volumes of methanol containing sufficient salicylaldehyde to give molar ratios of salicylaldehyde:amino group of 3:1, 5:1, 10:1 and 20:1. The resultant solutions were maintained at 25°C and the rate of formation of N-salicylidenechitosan measured. The results are given in Figure 1 and show that the rate of reaction increases with increase in salicylaldehyde concentration, but that the extent of reaction of the chitosan is independent of salicylaldehyde concentration above a 3:1 molar ratio. This suggests that under these conditions the equilibrium position in the reaction

$$\text{Chit-NH}_2 + \text{Ar-CHO} \rightleftharpoons \text{Chit-N=CH-Ar} + H_2O$$

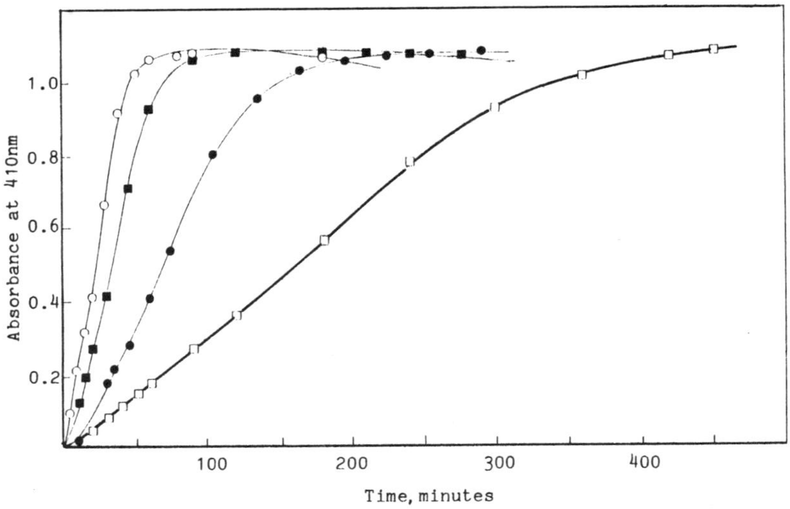

Fig. 1. Rates of formation of N-salicylidenechitosan at 25°C for a constant chitosan concentration of 1.1g dm^{-3} and different salicylaldehyde:amine group ratios: □ = 3:1; ● = 5:1; ■ = 10:1; ○ = 20:1. Absorbance measurements carried out using a 4cm cell.

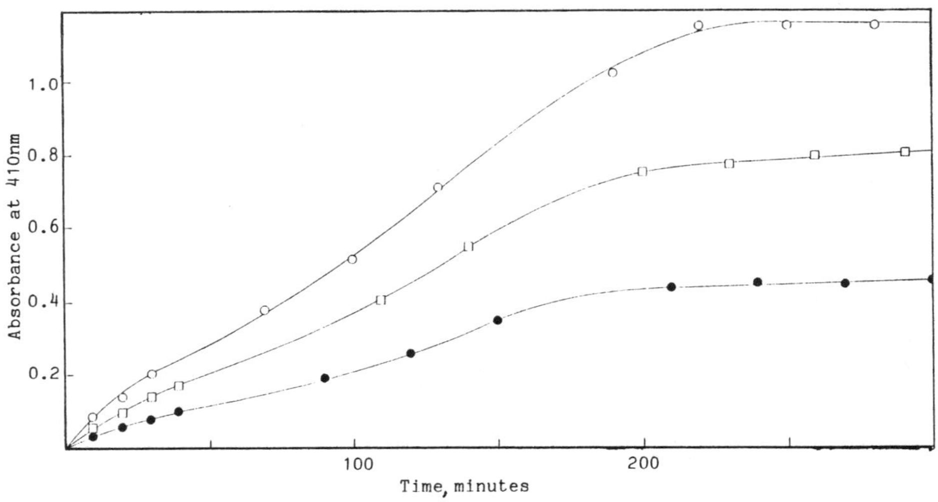

Fig. 2. Rates of formation of N-salicylidenechitosan at 25°C for a constant salicylaldehyde concentration of 5.36×10^{-3} mol dm^{-3} and different chitosan concentrations: ● = 8.7×10^{-2} g dm^{-3}; □ = 15.3×10^{-2} g dm^{-3}; ○ = 21.9×10^{-2} g dm^{-3}. Absorbance measurements carried out using a 4cm cell.

lies completely over on the right hand side. On this assumption the extinction coefficient for the 410nm absorption band is $\varepsilon = 255$.

Effect of Variation in Chitosan Concentration

The effect on the rate of reaction of variation in chitosan concentration was studied using a constant concentration of salicylaldehyde, the concentration chosen being that which gave a 5:1 molar ratio of salicylaldehyde:amine group for the highest concentration of chitosan used. The results are given in Figure 2 and show that the rate of reaction increases with increase in chitosan concentration despite the relatively large excess of salicylaldehyde present. In each case the formation of N-salicylidenechitosan is complete, as calculated using the extinction coefficient derived above.

The curves for all three concentrations have the same shape with an apparent decrease in the rate of formation in the first half of the curve, followed by an increase in the rate of formation before finally levelling off. The same behaviour can be seen in Figure 1 for the lowest salicylaldehyde:amine group ratio, but it is not found in the plot for the rate of formation of N-salicylidene-D-glucosamine under similar conditions (Figure 3). This difference in behaviour is more clearly seen in Figure 4 which is a plot of $\ln(OD_\infty - OD_t)$ _versus_ Time. The reaction between salicylaldehyde and D-glucosamine shows a single rate whilst that between salicylaldehyde and chitosan shows two distinct rates with the second one greater than the first.

A possible explanation for the latter behaviour is that a change in the accessibility of the amine groups on the chitosan occurs during the course of the reaction. The polymer chains will become progressively less hydrophilic and more compatible with the methanol, which is the major component of the mixed solvent, as the reaction proceeds. This causes coil expansion, which has the effect of making the remaining amine groups more accessible to the salicylaldehyde, leading to an increase in the rate

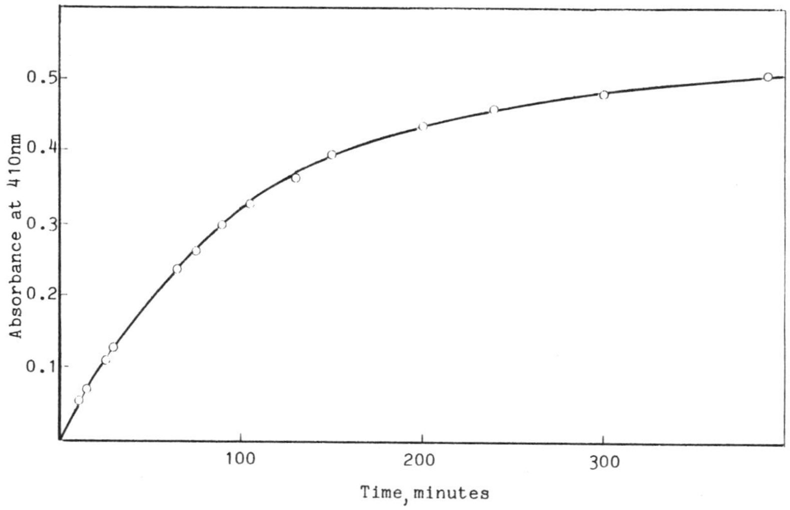

Fig. 3. Rate of formation of N-salicylidene-D-glucosamine at 25°C. Concentration of D-glucosamine = 1.07×10^{-3} mol dm^{-3}: concentration of salicylaldehyde = 7.49×10^{-3} mol dm^{-3}: molar ratio of acetic acid:D-glucosamine = 2:1.

of reaction. Similar behaviour has been reported for the acid hydrolysis of poly[vinylphenoxy(trimethyl)silane] by aqueous HCl in dioxan[6].

Initial analysis of the curves in Figure 3 showed the reaction to be a second order reaction, being first order with respect to the concentration both of chitosan and salicylaldehyde. Analysis of Figure 4 gives a rate constant of 1.85 x 10^{-2} $dm^3mol^{-1}s^{-1}$ for the lower part of the plot for N-salicylidenechitosan formation and 3.52 x 10^{-2} $dm^3mol^{-1}s^{-1}$ for the upper part. This compares with a value of 2.97 x 10^{-2} $dm^3mol^{-1}s^{-1}$ for the reaction between D-glucosamine and salicylaldehyde.

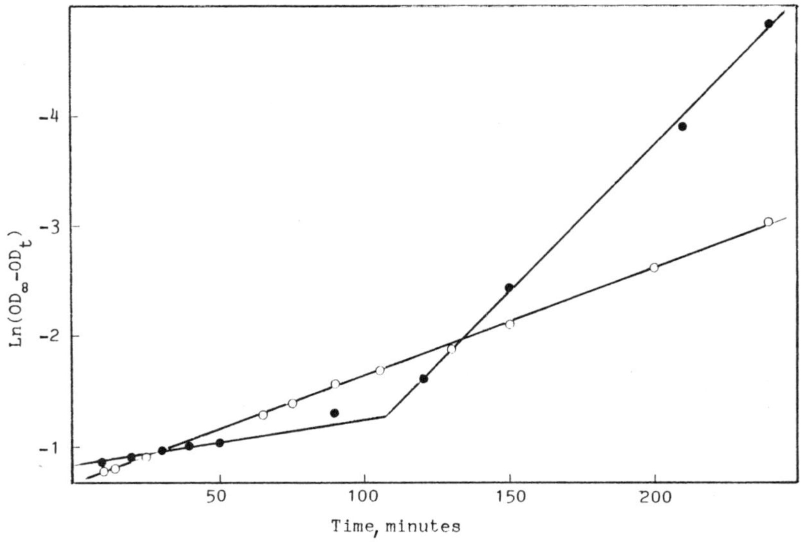

Fig. 4. Optical density plot at 410nm. Ln(OD_∞ - OD_t) versus Time for the formation of N-salicylidenechitosan (●) and N-salicylidene-D-glucosamine (○).

CONCLUSIONS

The uv/visible spectrum of N-salicylidenechitosan has λmax = 410nm and this absorption band can be used to follow formation of the polymeric Schiff's base. The reaction is second order, being first order with respect to the concentration of both chitosan and salicylaldehyde, and shows an increase in the rate constant from 1.85 x 10^{-2} $dm^{-3}mol^{-1}s^{-1}$ to 3.52 x 10^{-2} $dm^3mol^{-1}s^{-1}$ during the course of the reaction. This change in rate constant is not observed in the reaction between D-glucosamine and salicylaldehyde and has been attributed to polymer chain expansion, because of increased compatibility with the methanol component of the solvent (80% aqueous methanol), on introduction of salicylaldehyde residues.

Acknowledgements. The authors wish to thank Hercules Inc. for the sample of chitosan used in this work. One of the authors (JGD) wishes to thank the Science and Engineering Research Council and the Paper Industry Research Association for the award of a CASE research studentship.

REFERENCES

1. G.K. Moore and G.A.F. Roberts, Reactions of chitosan: 3.

Preparation and reactivity of Schiff's base derivatives of chitosan. Int. J. Biol. Macromol.. 3:337 (1981).
2. J.G. Domszy and G.A.F. Roberts. Evaluation of infrared spectroscopic techniques for analysing chitosan. Makromol. Chemie. 186:1671 (1985).
3. A.J. Milun. Colorimetric determination of primary amine in fatty amine acetates and fatty amines. Anal. Chem.. 29:1502 (1957).
4. J.G. Domszy and G.A.F. Roberts. Determination of the viscometric constants for chitosan. Int. J. Biol. Macromol.. 4:374 (1982).
5. G.K. Moore and G.A.F. Roberts. Determination of the degree of N-acetylation of chitosan. Int. J. Biol. Macromol.. 2:115 (1980).
6. A.D. Jenkins. K. Petrak. G.A.F. Roberts and D.R.M. Walton. Monomers and polymers with silicon-containing functional groups. Eur. Polymer J.. 11:653 (1975).

FULLY DEACETYLATED CHITOSAN

(A. Domard and M. Rinaudo)

In a recent paper (1), we have proposed to adapt to chitosan the procedure described by Kenne and Lindberg (2, 3) for lipopolysaccharides. The main difference from previous methods is the use of thiophenol which traps oxygen thus preventing degradation and exerting a catalytic effect. This results in better deacetylation with less degradation. Several experimental parameters have been studied in order to find the optimal conditions for the preparation of a 100 % deacetylated chitosan with relatively high molecular weight.

The acetyl content can be determined with more or less accuracy by

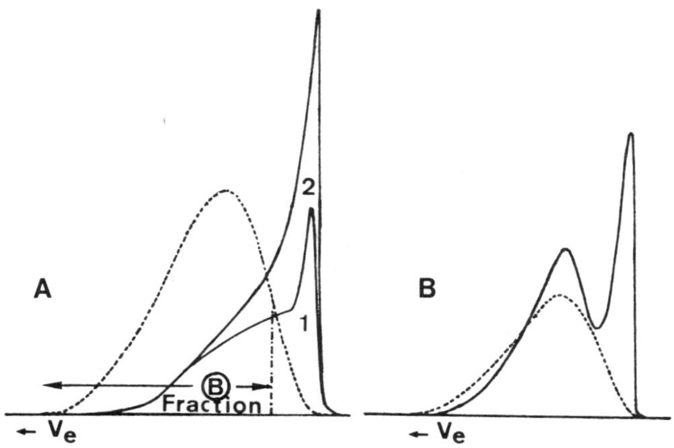

Fig. 1. Evidence of aggregates of chitosan in aqueous solution (0.05 M ammonium acetate) by gel permeation chromatography. (A), 1 and 2 are direct, after preheating and two days after injections, respectively. (B) Injections of fraction B (not excluded) isolated and redissolved in the eluent. ---RI and ——DDL

n.m.r., i.r., u.v.-vis. and mass spectroscopy, potentiometry and chemical titrations (See following Chapters; especially Muzzarelli and Rocchetti).

Molecular weight determination

Various methods have been tested to characterize this fully deacetylated sample by its average molecular weights. Membrane osmometry and light scattering measurements gave $DP_n \simeq 1100$ and $DP_w \simeq 3360$ for number average and weight average polymerization degrees, respectively. Gel permeation chromatography on cationic porous silica gels (4) gave $DP_n \simeq 960$, $DP_w \simeq 3370$ in good agreement with the above results.

Aqueous solutions of chitosan exhibit ageing, easily tested by light scattering or viscosimetric measurements. This behavior studied by gel permeation chromatography (Fig. 1) with low-angle laser light scattering detector on line indicates that aggregates are always present in aqueous solutions of chitosan, explaining partially the relatively large distribution of molecular weights.

REFERENCES

1. A. Domard, and M. Rinaudo, Preparation and characterization of fully deacetylated chitosan, Int. J. Biol. Macromol. 5:49 (1983).
2. C. Erbing, K. Granath, L. Kenne and B. Lindberg, A new method for the N-deacetylation of carbohydrates, Carbohydr. Res. 47:C5 (1976).
3. L. Kenne and B. Lindberg, N-Deacetylation of polysaccharides. Methods Carbohydr. Chem. 3:295 (1980).
4. A. Domard and M. Rinaudo, Gel permeation chromatography of cationic polymers on cationic porous silica gels. Polymer Commun. 25:55 (1984).

THE FORMATION OF THE CHITIN-ARYLPHORIN COMPLEX IN VITRO

(O.P. Agrawal and K. Scheller).

INTRODUCTION

Arthropod cuticle appears to be a polymeric network in which chitin occurs as microfibrils surrounded by and bound to protein by covalent and non-covalent bonds. This complex is immersed in a matrix of loosely bound proteins (1, 2). It is generally accepted that N-acetyldopamine is the immediate precursor for the compounds affecting sclerotization. The amino groups of proteins are added or condensed to an o-quinone of it (3). Little information is available on the characteristics of cuticular proteins and their interaction with other components during sclerotization. There are few reports that arylphorins are components of insect cuticle and may participate in the process of sclerotization (4-10). *In vitro* reactions of cuticular proteins with chitin (11, 12) and also of quinones with arylphorins (8) have been studied. Here we demonstrate arylphorin-chitin interaction with the idea that such *in vitro* studies may increase the understanding of molecular mechanism of sclerotization.

MATERIAL AND METHODS

The integuments of 7 days old *Calliphora vicina* larvae were extracted in Hackman's series of solvents (13). The chitin so obtained and a sample of crustacean chitin (a generous gift of Dr. K. D. Spindler, Düsseldorf, Germany) were purified by washing in several changes of 1M NaOH and 0.1M phosphate buffer (pH 7) at room temperature until no proteins could be detected in the supernatants. The pellets of pure chitin were lyophilized. The haemolymph, fat body and integument from 7 day larvae and 7 day puparia were homogenized and the clear supernatants were diluted to give protein concentration of 4-6 mg/ml in phosphate buffer. ^{3}H-tyrosine and ^{3}H-leucine were selected to radiolabel arylphorin _in vivo_ in 5 day _Calliphora_ larvae. Arylphorin was purified from the haemolymph.

The protein-chitin binding was done in 0.05M phosphate buffer at pH 5.8, since binding of cuticular proteins is maximum at electrofocussing point of the protein and low salt concentration (12). The amount of unbound protein was determined spectrophotometrically at 280 nm after centrifugation. The pellet was resuspended in fresh protein solution (1 mg/ml of chitin); a total of 10 incubations, at 3 h intervals were made at 25° C. The binding with radioactive arylphorin included 4 incubations only and an aliquot was taken before and after incubation for counting.

RESULTS

Similar results were obtained with crustacean and blowfly chitin. Therefore, we present here the results of only insect chitin (Table 1). It is clear that cuticular proteins, both larval and puparial, as well as pure arylphorin, have a higher affinity to chitin than haemolymph and fat body proteins, and BSA has no affinity at all. The chitin-protein complexes formed were further processed to elucidate the nature of binding. On treating with phosphate buffer only a minor quantity of unbound proteins was liberated. The pellet was shaken twice in 7M urea at 4° C for 12 h each. A substantial amount of protein was released, which is considered to be non-covalently bound protein. Further treatment of the pellet in urea gave no more protein. Another fraction was extracted into 2 changes of 1M NaOH at 4° C within 48 h to release covalently bound protein. Even after this treatment a considerable amount of proteins was still bound to the chitin pellet (Table 1). Analysis of chitin complex after 1-4 incubations showed that proteins firstly bind at covalent level. The proteins extracted were resolved on SDS-PAGE after dialysis and lyophilization. Arylphorin band can be seen in all the urea extracted samples. On the other hand, in NaOH fractions instead of clear protein bands a smear in low molecular weight region appears (Fig. 1). This indicates the possibility of degradation of protein including arylphorin during extraction by strong alkaline medium. Incubation of pure arylphorin in 1M NaOH under similar conditions confirmed the degradation of protein, whereas incubation in urea had no effect (Fig. 2).

Table 1. _In vitro_ binding of chitin to blowfly proteins

Protein source	% of bound protein	mg of bound protein per mg chitin				
		Total	Unbound	Non-covalently bound	Covalently bound	
					NaOH extracted	still bound
Haemolymph	22	1.52	0.08	0.66	0.35	0.43
Fat body	58	3.43	0.05	1.41	0.86	1.11
Integument	76	6.05	0.01	2.85	1.91	1.28
Puparia	71	4.41	0.01	0.88	2.95	0.57
Arylphorin	75	6.11	0.01	2.12	2.54	1.44
BSA	0	0	0	0	0	0

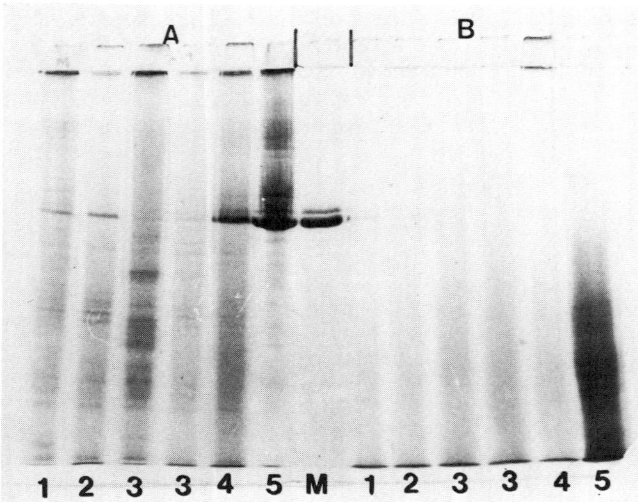

Fig.1. SDS-PAGE of proteins extracted from the chitin complexes formed with the following proteins:
(1) haemolymph (2) fat body
(3) integument (4) puparium
(5) arylphorin (M) arylphorin-marker

A: urea extracted fractions
B: NaOH extracted fractions

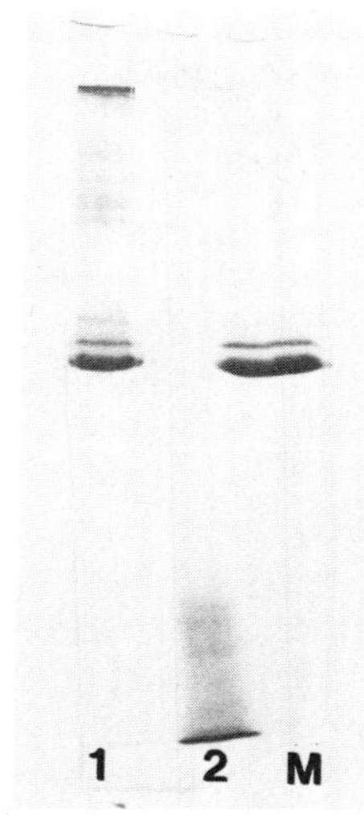

Fig.2. SDS-PAGE of arylphorin incubated in:
(1) 7M urea
(2) 1M NaOH
(M) arylphorin-marker

The binding of ^{3}H-arylphorin is shown (Table 2). The difference in the amount of protein and radio-counts seems to be due to the presence of non-labelled impurities which did not bind. The binding is mainly due to covalent bonds, non-covalent fraction is small. SDS-PAGE-fluorography of the samples showed that arylphorin without chitin gives a double band at 80,000 Dalton, typical of arylphorin, but it disappears from chitin-arylphorin incubation. Obviously, it was bound by chitin. Urea extracted fraction again resolves in a double band, whereas in NaOH fraction a labelled smear is observed.

Table 2. *In vitro* binding of chitin with ^{3}H-arylphorin

% incorporation of arylphorin (OD 28o nm)	% incorporation of radioactivity	% radioactivity liberated from chitin-arylphorin complex		
		Unbound	Non-covalent	Covalent
72	87	1	7	69

Total counts available for incubation: 8.4×10^5 dpm

DISCUSSION

Arylphorin is the generic name of larval serum proteins of holometabolous insects. They are synthesized in the fat body of feeding last larval instars and serve the functions of storage, ecdysteroid carriers, and in cuticle formation (review 14). There are reports that arylphorin is an essential component of insect cuticle. The protein has been thought to contribute tyrosine and phenylalanine for sclerotization and melanization. Application of radioactively labelled arylphorin coupled with immunological techniques provide evidence for the incorporation of this protein into the cuticles of Calliphora and Manduca (4-10). Unpublished observations in our laboratory show that arylphorin is present in other tissue having a chitinous component, such as tracheae, imaginal discs, peritrophic membranes, foregut, hindgut, and Malpighian tubules. Furthermore, it has been shown that 20-OH-ecdyson affects the incorporation of arylphorin in the cuticle of Calliphora. On the basis of solubility extraction it appears that blowfly arylphorin forms covalent and non-covalent links within the cuticle (10).

Due to inherent difficulties in isolating cuticle derivatives by non-destructive methods, in vitro models are invaluable for clarifying the nature and functional role of proteins in cuticle formation and sclerotization. It has been reported than nascent quinones will precipitate unspecified proteins from cuticle extracts of a lepidopteran species (15). Calliphora larval cuticular proteins also bind in vitro with tannin (16). Chitin can selectively bind with cuticular proteins including those of Calliphora (11, 12). These findings deal with cuticular proteins, in general, without any discussion of arylphorin. Grün and Peter (8) described arylphorins of Calliphora, Drosophila, and Manduca, but not BSA and ferritin, crosslink with sclerotizing agent quinones. The present results demonstrate that arylphorin, like cuticular proteins, bind with chitin and forms covalent and non-covalent bonds. Chitin is able to selectively precipitate arylphorin and other proteins from haemolymph, fat body, and cuticle extracts. The unsaturated chitin firstly forms covalent bonds, perhaps to form a glycoprotein and then proteins are included also in non-covalent bonding. This supports the involvement of arylphorin in the Calliphora cuticle as a structural component.

Acknowledgements. The award of a fellowship to O.P.A. by Deutscher Akademischer Austaushdienst is gratefully acknowledged. This work was supported by a grant of the Deutsche Forschungsgemeinschaft.

REFERENCES

1. R. H. Hackman, The Interactions of cuticular proteins and some comments on their adaptation to function, in: Insect Integument, H. R. Hepbrun, ed. Elsevier, Amsterdam (1976).
2. A. G. Richards, The chemistry of insect cuticle, in: Biochemistry of Insects, M. Rockstein, ed., Academic Press, New York (1978).
3. H. Lipke, M. Sugumaran, and W. Henzel, Mechanism of sclerotization in dipterans, in: Advances in Insect Physiology, vol. 17, M. J. Berridge, J. E. Treherne and V.B. Wigglesworth, ed., Academic Press, New York (1983).
4. K. Scheller, H. P. Zimmermann and C. E. Sekeris, Calliphorin, a protein involved in cuticle formation in the blowfly, Calliphora vicina, Z. Naturforsch. 35c:387 (1980).
5. H. Schenkel, C. Myller, M. König, P. Hausberg, and K. Scheller, Calliphorin: Studies on its biosynthesis and function, in: The Larval Serum Proteins of Insects, K Scheller, ed., Gg. Thieme Verlag, Stuttgart, New York (1983).
6. L. Levenbook, The structure and function of calliphorin, in: The Larval Serum Proteins of Insects, K. Scheller, ed., Gg. Thieme Verlag, Stuttgart, New York (1983).
7. L. Levenbook and A. C. Bauer, The fate of the larval storage protein calliphorin during adult development of Calliphora vicina, Insect Biochem. 14:77, (1984).

8. L. Grün and M. G. Peter, Selective crosslinking of tyrosinerich larval serum proteins and of soluble Manduca sexta cuticle proteins by nascent N-acetyldopamine-quinone and N- β -alanyldopamine-quinone, in: The Larval Serum Proteins of Insects, K. Scheller, ed., Gg. Thieme Verlag, Stuttgart (1983).
9. L. Grün and M. G. Peter, Incorporation of radiolabelled tyrosine, N-acetyldopamine, N- β -alanyldopamine, and the arylphorin manducin into the sclerotized cuticle of tobacco hornworm (Manduca sexta) pupae, Z. Naturforsch. 39c:1066 (1984).
10. O. P. Agrawal and K. Scheller, Arylphorin, a constituent of the sclerotized insect cuticle, XVIIth International Congress of Entomology, Hamburg, Abstract No. P4.1.-1. p. 207 (1984).
11. R. H. Hackman, Studies on chitin III adsorption of proteins to chitin, Aust. J. Biol. Sci. 8:530 (1955).
12. R. H. Hackman and M. Goldberg, The non-convalent binding of two insect cuticular proteins by a chitin, Insect Biochem. 8:353 (1978).
13. R. H. Hackman and M. Goldberg, Proteins of the larval cuticle of Agrianome spinicollis (Coleoptera), J. Insect Physiol. 2:221 (1958).
14. K. Scheller, The Larval Serum Protein of Insect: Function-Biosynthesis-Genetic, Gg. Thieme Verlag, Stuttgart (1983).
15. R. H. Hackman and A. R. Todd, Some observations on the reaction of cetechol derivatives with amines and amino acids in presence of oxidizing agents, Biochem. J., 55:631 (1953).
16. R. H. Hackman and M. Goldberg, Molecular crosslinks in cuticles, Insect Biochem., 7:175 (1977).

Affiliations

A. Naggi, G. Torri, T. Compagnoni and B. Casu, Istituto di Chimica e Biochimica "G. Ronzoni", V. G. Colombo, 81, I-20133 Milano, Italy.

B. Focher and A. Massoli, Stazione Sperimentale Cellulosa, P. Leonardo, I-20133 Milano, Italy.

A. Gervasini and F. Morazzoni, Dipartimento di Chimica Inorganica e Metallorganica, Università di Milano, Milano, Italy.

A. Tsutsumi, S. Nishimura, Y. Maekita, A. Ebina, S. Tokura and A. Nishi, Faculty of Sciences, Hokkaido University, Kita-ku, Sapporo 060, Japan

A. Domard, A. Gadelle and M. Rinaudo, Centre Recherches Macromolecules Végétales, C.NR.S. Grenoble, B.P. 68, F-38402 St. Martin d'Hères, France.

J.C. Domszy, G.A.F. Roberts and F.A. Wood, Physical Sciences Department, Trent Polytechnic, Burton st., Nottingham, England NG1 4BU.

H. Wendel, E. Konrad and G. Lang, Wella A.G., Berliner allée 65, D-6100 Darmstadt, F.R. Germany.

O.P. Agrawal and K. Scheller, Zoologisches Institut der Universität, Röntgenring 10, D-8700 Würzburg, F.R. Germany.

S. Hirano, Y. Kondo and K. Fuji-i, Department of Agricultural Biochemistry, University of Tottori, Tottori 680, Japan.

THE CHELATING ABILITY OF AMINO ACID GLUCANS AND SUGAR ACID CHITOSANS

R. A. A. MUZZARELLI

Faculty of Medicine, University of Ancona
I-60100 Ancona, Italy

INTRODUCTION

The industrial and laboratory chitosans, currently obtained from chitin by deacetylation, carry both acetamido (about 40 %) and amino (about 60 %) groups on carbon 2, and their pK values are between 6 and 7. Fully deacetylated chitosan (polyglucosamine) can be obtained by extended and repeated deacetylation of chitin (1 - 3).

The characterization of fully deacetylated chitosan | (1-4)-2-amino-2-deoxy-β-D-glucan | has so far been the object of very limited study (4, 5, see also this Volume). It posesses higher structural regularity than chitosan, because each carbon 2 carries a primary amino group, and therefore is better suited than chitosan as a starting material for the production of chelating media.

Enhanced performances for the chelation of metal ions can be obtained from β (1→4) polysaccharides when the acetamido groups are replaced with aminoacid groups. The acetamido groups are in fact poor chelators, have scarce specificity of interaction, establish hydrogen bonds and hinder the crystallization of the polysaccharide. A convenient way of replacing the acetamido groups with aminoacid groups, which lend themselves to metal ion chelation, is to react polyglucosamine with aldehydoacids, ketoacids and sugar acids (Schiff reaction) to produce aldimines and ketimines which, upon hydrogenation, yield the desired aminoacid- or sugar acid-substituted glucans (6 - 9). These products possess a β (1→4) glucan backbone carrying aminoacid moieties at C-2, or secondary amines derived from sugar acids. While a number of chelating resins can be used for the removal of transition metal ions from waters, the chitosan-derived chelating agents may find use in certain fields where their hydrophilicity, high capacity at low metal concentrations, and their selectivity or indifference to high salt concentrations represent advantages over synthetic chelating resins.

One such field is that of nuclear chemistry. This Chapter presents data on the removal of ^{60}Co from mineral acid solution as a way to reduce the volume of radioactivity-bearing wastes before final disposal, and on the recovery of uranium from dilute solutions or natural waters, including sea water.

EXPERIMENTAL

The derivatives of fully deacetylated chitosan so far studied from the standpoint of the chelating ability are those listed in Table I. They include 9 aminoacid glucans (10) and two sugar acid glucans (11 - 13) obtained by reacting chitosan from Euphausia superba (Antarctic krill) having a residual acetylation degree of less than 3 % (4).

Metal ion chelation

Metal sulfate solutions of the indicated concentrations were contacted with the polysaccharide gels both in batch and in column operations. The samples for batch measurements were kept on a shaking machine at 80 r.p.m. for 16 h, and than filtered on Whatman filter paper; standard columns were 6.0 x 0.6 cm, filled with sieved gel particles in the range 100 - 200 mesh in the dry state. The columns were fed with Gilson peristaltic pumps. The gels were dried at 60° C overnight, milled, sieved, weighed (200 mg) and slurried with water.

Preparation of the aminoacid glucans

Polyglucosamine (10 g) with pK value 6.2 and acetylation degree 3 % was suspended in water (1.5 l) and an excess (1.5 mol per mol glucosamine) of ketoacid was added (Table I). At this stage, green and black colors were observed for β-hydroxypyruvic acid and for p-hydroxyphenylpyruvic acid, respectively. The resulting pH value (around 3) was slowly adjusted to 4.5 with 0.2 M NaOH. The insoluble product was then reduced with sodium cyanoborohydride dissolved in water (5.85 g in 50 ml) and the pH value was finally adjusted to about 7 with 1 M NaOH. After 48 h the polysaccharides in the aminoacid form were washed, dialyzed against water and isolated. They were extracted in Soxhlet with ethanol, acetone and ether.

Table I. Aminoacid and sugar acid derivatives of chitosan obtained by Schiff reaction and subsequent hydrogenation.

Reagent	Product
Aminoacid glucans	
$CHO.COO^-$, glyoxylate	Glycine glucan
$CH_3CO.COO^-$, pyruvate	Alanine glucan
$CH_2OH.CO.COO^-$, β-hydroxypyruvate	Serine glucan
$C_6H_4OH.CO.COO^-$, p-hydroxyphenyl pyruvate	Tyrosine glucan
$C_6H_5CH_2CO.COO^-$, β-phenylpyruvate	Phenylalanine glucan
$COOH.(CH_2)_2CO.COO^-$, α-ketoglutarate	Glutamate glucan
$CH_3S.(CH_2)_2CO.COO^-$, α - ketomethyolbutyrate	Methionine glucan
$(CH_3)_2CH.CH_2CO.COO^-$, α-ketoisocaproate	Leucine glucan
$COOH.CH_2CO.COO^-$, oxalacetate	Aspartate glucan
Sugar acid glucans	
$CH_2OH.CO.(CHOH)_3.COO^-$, 5-ketogluconate	Gluconate chitosan
$CH_2OH.CHOH.CH.CO.CO.CO$, dehydroascorbate (CH and terminal CO linked through O)	N-(Dihydroxyethyl) tetrahydrofuryl chitosan

Preparation of the sugar acid glucans

Ascorbic acid besides being a suitable acid for the formation of a water-soluble chitosan salt (chitosan ascorbate) can also react with chitosan to form chitosan derivatives *via* Schiff reaction. Mild oxidising agents such as iodine and air easily convert ascorbic acid to dehydroascorbic acid, whose carbonyl groups are able to react with amines; for example, one of the most widely accepted methods for the quantitative determination of ascorbic acid involves oxidation followed by a reaction with 1,2-phenylenediamine to produce a fluorophor (11 - 13).

The reaction of chitosan with dehydroascorbic acid was followed by infrared spectrometry (bands at 1710 and 1760 cm^{-1}, due to α-β-unsaturated cyclic ketones), by alkalimetry (typical shape of the alkalimetric curve), by viscometry (viscosity decrease from 108 to 77 cP within 6 h and to 25 cP within 24 h, for a 1 % solution) and by circular dichroism spectropolarimetry (negative band shifted from 194 to 213 nm, positive band at 244 nm, new negative band at 310 - 320 nm and positive at 272 nm). The water soluble ketimines obtained from dehydroascorbate and from 2- or 5-ketogluconate were reduced to insoluble, fibrous and white products.

RESULTS AND DISCUSSION

Semi-quantitative description of the chelating ability of glutamate glucan and gluconate chitosan

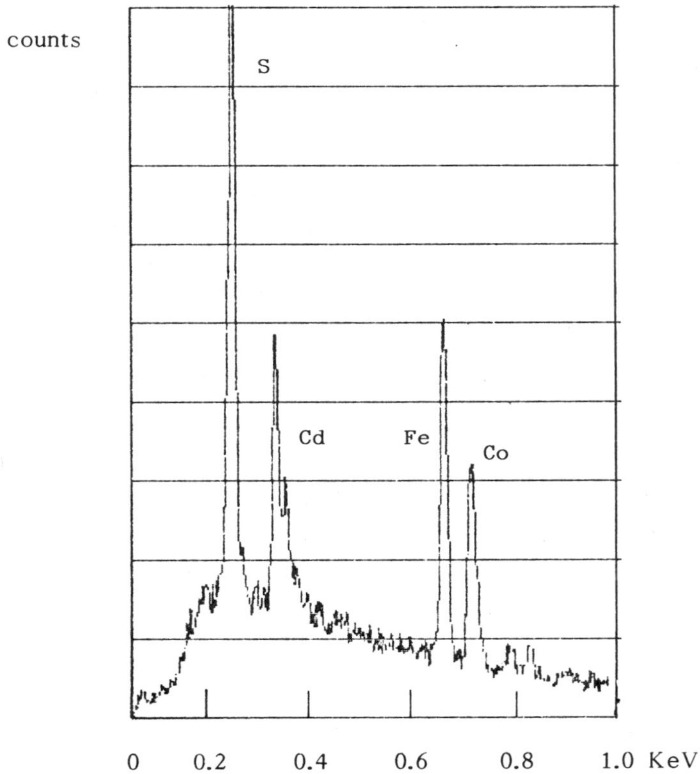

Figure 1. Energy dispersion analysis of x-rays (EDAX) spectrum for glutamate glucan (100 mg) contacted with a mixture of Fe, Co and Cd (sulfates) at the initial concentrations of 100 mg/l each (pH ca. 5, 25 ml, 1 h).

While not all of the modified polysaccharides listed in Table I possess the same level of chelating ability, some of them are powerful chelating polymers. The EDAX spectra offer evidence of the chelating ability for keto acid and sugar acid derivatives of chitosan.

Fig. 1 reports a representative EDAX spectrum obtained with glutamate glucan after stirring it (100 mg) for 1 h with a solution (25 ml, pH 5) containing Fe, Co and Cd (100 mg/l each). The molar ratio obtained was 1.5:1.0:1.0, thus confirming that glutamate glucan is active on a variety of metal ions which can be simultaneously chelated. Additionally the spectrum shows an important peak for sulfur, from the sulfate counter ion.

Under similar conditions, gluconate chitosan collects Cr, Cu, Zn, Co and Ni in the molar ratios 14:13:10:4:2, sulfur being present at the level of 55.

Characterization of glutamate, aspartate, serine and glycine glucan as chelating polymers

The serine glucan and the cross-linked glycine glucan were contacted with cobalt, nickel and copper solutions at pH values between 5 and 6 in the batch mode. The initial concentration ranges were: Co and Cu, 25 - 250 mg/l, and Ni, 50 - 1000 mg/l. After 16 h the capacities were measured and plotted against the equilibrium concentrations. The most satisfactory results (Figure 2) were obtained with Asp-glucan, Gly-glucan and Glu-glucan; the capacity of Ser-glucan for Co is between 1 and 20 % by weight for 6 mg/l and 850 mg/l equilibrium concentrations.

Adsorption of metal ions on glycine glucan

The cross-linked glycine glucan was contacted with cobalt, nickel, copper and cadmium solutions at pH values between 2.9 and 3.5, adjusted with sulfuric acid in the batch mode. The initial concentrations were 0.2 and

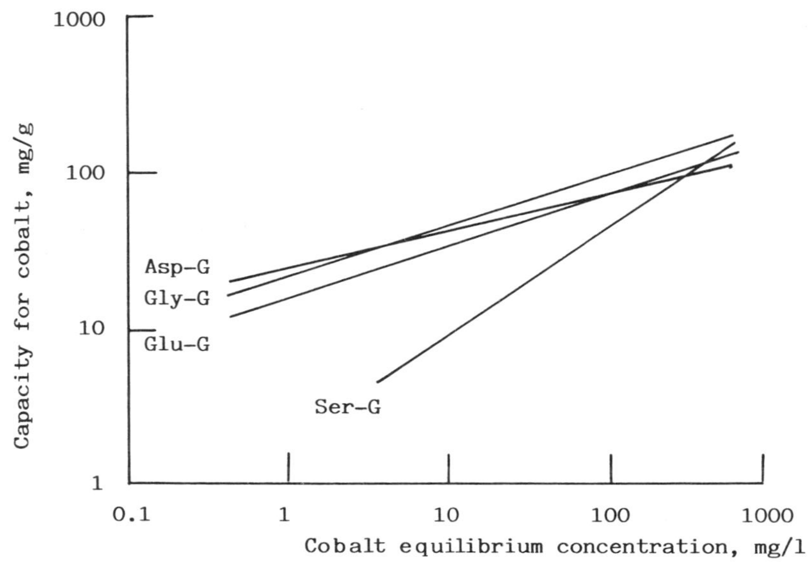

Figure 2. Isotherms for the collection of cobalt on aspartate glucan (Asp-G), glycine glucan (Gly-G), glutamate glucan (Glu-G) and serine glucan (Ser-G), at pH 5 - 6.

0.5 mM for each metal ion. The collection percentages after 16 h contact were: Co, 84.5 - 85.5; Ni, 98.6 - 97.6; Cu, 98.8 - 99.3; Cd, 99.86 - 98.88. The polymers exhibited typical colors, pink, green, blue and yellow, for Co, Ni, Cu and Cd, respectively. Cross-linked glycine glucan was therefore very effective in collecting transition metal ions even from acidic solutions.

Recovery of traces of cobalt and copper from a water by means of cross-linked glycine glucan

Two chromatographic columns of cross-linked glycine glucan gel were fed with solutions of cobalt and copper, respectively at the concentration of 10 mg/l and at the flow-rate of 0.5 ml/min. The breakthrough points were after the passage of 1.8 l of solution. The columns became shorter (4.5 cm) due to the shrinkage of the gel upon chelation, and red and blue colored for Co and Cu, respectively. The capacity was found to be about 12 % by weight. A similar experiment was carried out with 1.0 mg/l solutions: the breakthrough points were at 4.0 l and at 12.0 l for Co and Cu, respectively, with capacities of 2.0 % and 6.0 %, respectively.

Removal of trace metals from brines

A saturated solution of sodium fluoride (300 ml) containing traces of cobalt and copper (10 and 70 micrograms, respectively), and a 13 % solution of sodium chloride (200 ml) containing traces of Co and Cu (11 and 24 micrograms, respectively) were assayed in batch conditions for the definition of the optimum pH values for the recovery of cobalt. The results, presented in Figs. 3 and 4 indicated that quantitative removal of cobalt could be achieved from fluoride-nitrate solutions at pH 7.0 while recovery of Co from chloride solutions could be as high as 92 % at pH 5.0. The same solutions were percolated through columns of cross-linked glycine glucan. After washing with demineralized water (5 ml), the metals were eluted with 0.25 M sulfuric acid and determined by hot graphite atomic absorption spectrometry.

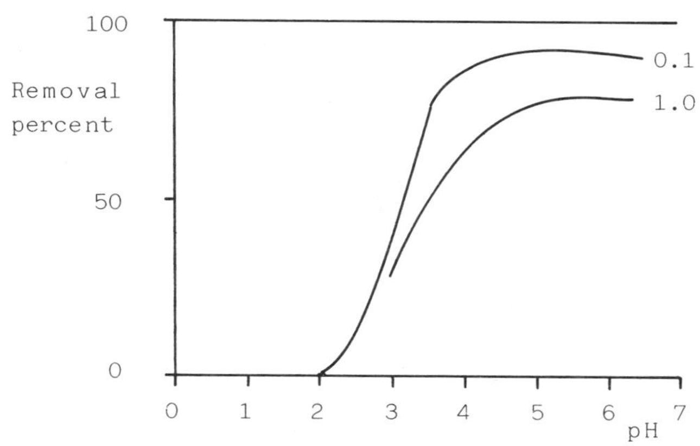

Figure 3. pH dependence of the removal of cobalt from sodium chloride solutions, by glycine glucan. Initial concentrations of cobalt: 1.0 and 0.1 mg/l.

Cobalt and copper were separated, their peaks appearing in the second and in the eighth 2-ml fractions, respectively. Chromium was also sharply separated, its peak appearing in the 15th fraction. These results refer to a

13 % NaCl solution, from which the said amounts of transition metal ions were recovered with recovery yields of 103 %, 90 % and 98 %, for Cr, Co and Cu, respectively. Chromium gave a violet chromatographic band. The recovery yields relevant to the fluoride solution were 115 % and 87 % for Co and Cu, respectively. The chromatographic profile after elution under the same conditions showed the sequence: Co, Cr and Cu.

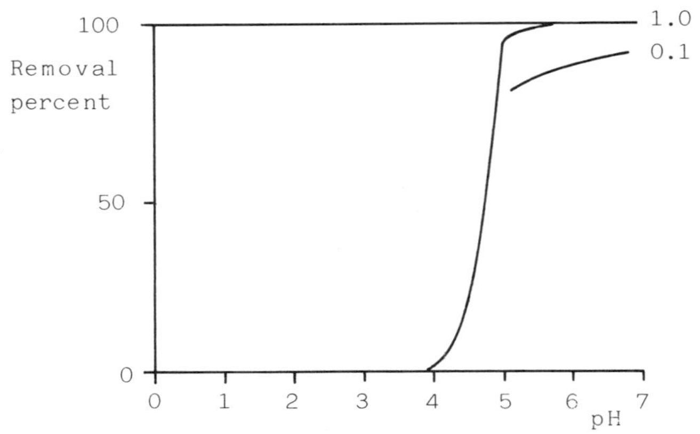

Figure 4. pH dependence of the removal of cobalt from sodium fluoride and nitrate solutions, by glycine glucan. Initial concentrations of cobalt: 1.0 and 0.1 mg/l.

Removal of ^{60}Co from nuclear plant waste waters

Non-core parts of nuclear reactors may become contaminated by radioactive nuclides as a consequence of the deposition of corrosion products. These metallic parts (mainly pipes) can be totally decontaminated by contac-

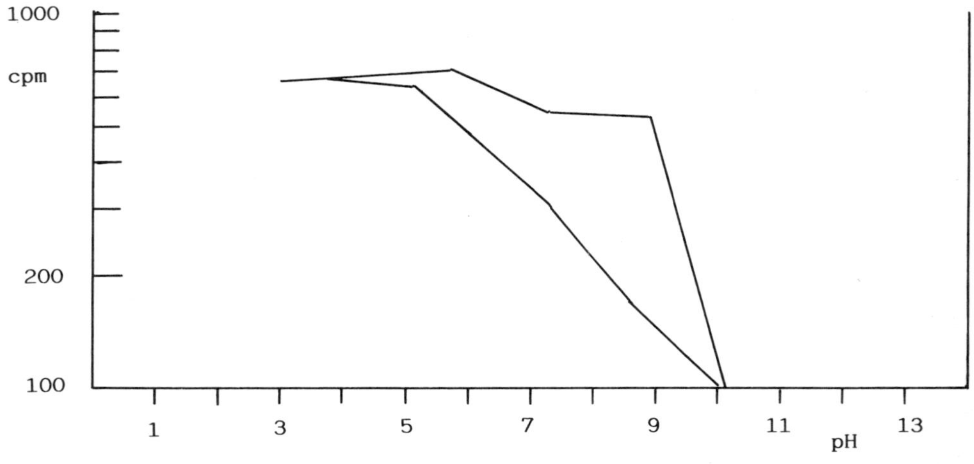

Figure 5. Effect of glycine glucan on the removal of 60-cobalt (carrier-free) from sodium fluoride and nitrate solutions at various pH values (ca. 13 % salt concentration). Upper curve, abatement of radioactivity by iron hydroxide precipitation; lower curve, abatement obtained with glycine glucan.

ting them with hydrochloric acid solution or with a mixture of hydroflucric and nitric acids. These operations give rise to relatively large volumes of strong acid solutions, containing radioactive nuclides, mainly ^{60}Co, in amounts below ponderal detection limits.

The radioactivity has to be concentrated to much smaller volumes for final disposal. For this purpose, solutions from nuclear plants have been treated with minor amounts of polysaccharide (ca. 0.5 g/l) to afford a volume reduction factor of 2000.

They have been brought to the optimum pH value with the consequence of a partial precipitation of iron hydroxide. This precipitate reduced appreciably the amount of radioactivity, depending on the pH value, as described by the upper curve in Fig. 5. In batch operations, the presence of glycine glucan abated the ^{60}Co concentration more effectively. For instance, at pH 7.6, the radioactivity level dropped from 550 to 250 cpm (Fig. 5, gap between the two curves at pH 7.6). As a consequence, the percolation of the radioactive solutions through a column of glycine glucan after pH adjustment,permits the discharge of a neutral salt solution from which the ^{60}Co has been completely removed.

Removal of uranium from solutions and brines

Although collection of uranium on polyaminosaccharides has been known and documented for more than 15 years (14 - 17) thirty N-substituted chitosans have been unsuccessfully tested for their uranium collection ability (18). Our recent discovery of the ability of the ascorbic acid derivative of chitosan to collect uranyl ions (12, 13) permits us to reach capacity values which are orders of magnitude higher than with other biopolymers. Uranium is collected to such an extent in terms of capacity that a compound results whose inorganic part is predominant by weight over the organic one.

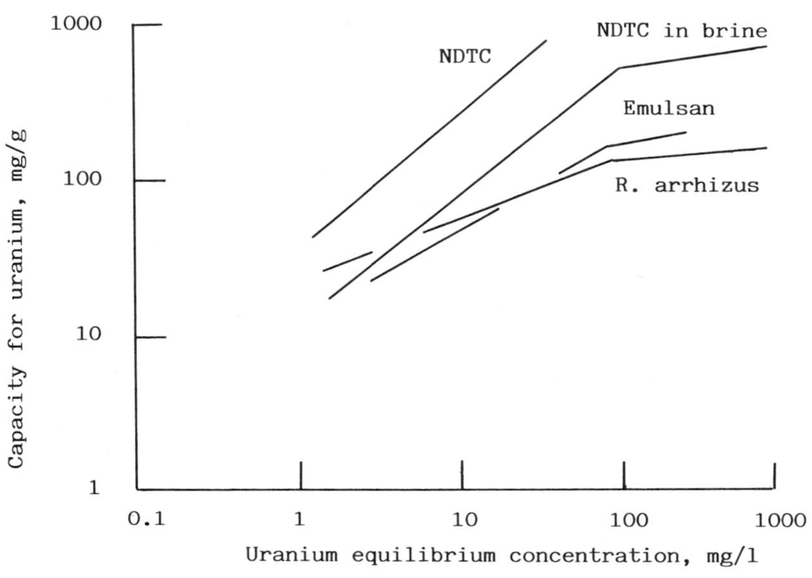

Figure 6. Isotherms for the collection of uranium on the chitosan derivative obtained from dehydroascorbic acid (NDTC) and on other polyaminosaccharides described in references 23 and 26.

Our early study on the chelating ability of Aspergillus niger micelia as well as Mucor rouxii, Choanephora cucurbitarum, Phycomyces blakesleeanus and Streptomyces mycelia showed that the fungal chitosan-glucan complexes are more effective than chitosan itself in collecting metal ions (19 - 22). Further studies on Aspergillus terreus, Streptomyces niveus, Penicillium crysogenum, Pseudomonas fluorescens and Rhizopus arrhizus showed that the latter has a capacity for uranium of 180 mg/g (23, 24), while the data reported by Galun et. al (25) on Penicillium digitatum are limited to capacity values as low as 3 - 9 mg/g for equilibrium concentrations of 51 - 58 mg/l.

N-(Dihydoxyethyl) tetrahydrofuryl chitosan (NDTC) possesses extremely high capacity for uranium because it is a polysaccharide carrying aliphatic chains with dihydroxyethyl and secondary amino groups.

For the batch measurements, initial uranium concentration ranged from 10 to 1000 mg U/l, all below the solubility limits set by the solution pH values. Most of the measurements were carried out at pH 4.5; it was noticed, however, that initial higher pH values were lowered to 4.7 - 5.1 after the period of time allowed for contact. The amounts of NDTC contacted with uranyl acetate solution were weighed quantities between 10 and 100 mg. The solution volume was 25 ml. Following the 16 h contact period, the NDTC mass was separated from the solution by vacuum filtration on 0.45 micron Sartorius membrane filters. The equilibrium uranium concentration of the filtrate (C_{eq}) was determined by spectrophotometry on 8-hydroxyquinoline extracts.

The collection percentages from 5 mM uranium solutions on NDTC (100 mg) for various initial pH values were: 87 % at pH 4.5; 97 % at pH 5.5; 100 % at pH 6.8 and 7.4, a few minutes after contact. After reacting with uranyl ions, NDTC had a bright yellow orange color. In the concentration interval 0.5 - 5.0 mM, uranium was collected in the batch mode to extents between 90 and 98 %.

Figure 6 shows the double logarithmic plot of the uranium adsorption isotherms obtained in this study and from data relevant to Rhizopus arrhizus cells, under the same conditions (the only difference being the counter ion, acetate instead of nitrate). Notice that a major portion of the curve relevant to NDTC is an order of magnitude higher than that of R. arrhizus cells. The capacity of NDTC is so high that equilibrium concentrations higher than 20 mg/l were not recorded, the slope of the NDTC curve being steeper than that of R. arrhizus cells. In practice, when R. arrhizus collects 90 mg/g, NDTC collects about 700 mg/g at the same equilibrium concentration (20 mg/l).

All the experimental points relevant to uranium collection from sodium chloride solutions on NDTC lie above those for R. arrhizus. While no data have been published for R. arrhizus cells in brines, it is interesting to note that NDTC in brine performs better than R. arrhizus cells in no-salt solutions. For instance, the NDTC capacity is 200 mg/g at equilibrium concentration of 20 mg/l, against 90 for R. arrhizus, and it is 800 mg/g for NDTC against 200 mg/g for R. arrhizus, both at 3C0 mg/l equilibrium concentration. This means that NDTC is superior to R. arrhizus cells not only under comparable conditions, but even when NDTC is under tha adverse effect of high sodium chloride concentration.

Under comparable conditions, emulsan, the extracellular bioemulsifier of Acinetobacter calcoaceticus has capacity for uranium (about 200 mg/g, based on Figure 2 of Ref. 26) similar to the R. arrhizus treated mycelia capacity which was determined as 180 mg/g (23). As Figure 6 shows, it is evident that the novel chitosan derivative possesses much higher (about 1000 mg/g) capacity for uranium, and that it is more effective than the other products even at equilibrium concentration 2 mg/l. Incidentally, the emulsan backbone is composed of N-acetyl galactosamine, N-acetyl amino uronic acid and another unidentified amino sugar (27). It is interesting to note that, among the other chemical derivatives of chitosan studied for uranium collection (18), the one obtained from dehydroascorbic acid is unique, because none was found to collect more uranium than chitosan itself.

Comparison of chitosan derivatives with cellulose derivatives

Recently, cellulose derivatives containing amino acid residues were tested for their chelating capacity; among them, glycine cellulose, obtained from the isocyanato cellulose with glycine methyl ester in triethylamine and DMSO, can be taken for comparison with glycine glucan obtained from chitosan and glyoxylic acid. From the data in Table III of Ref. 28, it can be calculated that glycine cellulose has the capacity of abcut 1.6 mg Cu/g at equilibrium concentration 4.4 mg Cu/l, and about 0.7 mg/g at equilibrium concentration 2.8 mg Cu/l. If we compare these data with those of Ref. 10, we can see that glycine glucan is about 31 and 60 times more powerful than glycine cellulose, under the conditions indicated; this also means that the slope of the isotherm curve for glycine cellulose is greater than for glycine glucan, i.e. the latter works much better at low metal ion concentrations. Thus, the straightforward reaction of chitosan with glyoxylic acid provides the most effective chelating polysaccharide so far available.

CONCLUSIONS

The aminoacid substituted glucans are stable amphoteric substances which can be used as chromatographic media for the chelation of metal ions. These modified polysaccharides can be easily prepared and can be dried at 60° C for the purpose of milling, sieving and storing; in the cross-linked form, they lend themselves to excellent column packing and high flow-rates. They selectively chelate transition metal ions and collect them from neutral and acidic solutions, as well as from highly concentrated brines (up to 15 % salt content), and therefore are particularly suitable for the removal of metals, including carrier-free radioisotopes, from acidic solutions originated in a number of industrial operations, and for the recovery of uranium from natural and industrial waters. These gels are superior to other chelating polymers in terms of capacity and their open and hydrophilic structure permits extremely fast fixation of transition metal ions.

ACKNOWLEDGEMENT

The original results presented in this Chapter have been partially obtained with the financial contribution of the Italian National Research Council, Progetto Finalizzato Chimica Fine e Secondaria, Roma, Italy. The EDAX spectra have been recorded in the Electron Microscopy Laboratory of this Faculty. The radiochemical measurements were done in cooperation with ENEL, Milano, Italy.

REFERENCES

1. R. A. A. Muzzarelli, "Chitin", Pergamon Press, Oxford (1977).
2. R. A. A. Muzzarelli and E. R. Pariser, eds., "Proceedings of the First International Conference on Chitin / Chitosan", M.I.T., Cambridge USA (1978).
3. S. Hirano and S. Tokura, eds., "Proceedings of the Second International Conference on Chitin / Chitosan", Japanese Soc. Chitin, Tottori Japan (1982).

4. S. Mima, M. Miya, R. Iwamoto and S. Yoshikawa, J. Appl. Polymer Sci. 28:1909 (1983).
5. E. Domard and M. Rinaudo, Intl. J. Biol. Macromol. 5:45 (1983).
6. R. A. A. Muzzarelli, F. Tanfani, M. Emanuelli and S. Mariotti, Carbohydr. Res. 107:199 (1982).
7. R. A. A. Muzzarelli and F. Tanfani, Pure Appl. Chem. 54:2141 (1982).
8. R. A. A. Muzzarelli, F. Tanfani, M. Emanuelli and S. Mariotti, Carbohydr. Polymers 2:145 (1982).
9. R. A. A. Muzzarelli, Italian Patent No 24179 A/83, 15 DEC 1983.
10. R. A. A. Muzzarelli, F. Tanfani, M. Emanuelli and L. Bolognini, Biotechnol. Bioengin. (in press).
11. R. A. A. Muzzarelli, F. Tanfani and M. Emanuelli, Carbohydr. Polymers 4:137 (1984).
12. R. A. A. Muzzarelli, F. Tanfani and M. Emanuelli, Enhanced uptake of uranium on a new derivative of chitosan and ascorbic acid. in: "Recovery of uranium from sea water", K. Kaikan and T. Kaikan, eds., Atomic Energy Soc. Japan, Tokyo (1984).
13. R. A. A. Muzzarelli, Carbohydr. Polymers 5:85 (1985).
14. S. M. Manskaya and T. V. Drozdova, "Geochemistry of organic substances", Pergamon Press, Oxford (1968).
15. R. A. A. Muzzarelli, U. S. Patent No 3,635,818 (1972).
16. R. A. A. Muzzarelli and O. Tubertini, Talanta 16:1571 (1969).
17. R. A. A. Muzzarelli, G. Raith and O. Tubertini, J. Chromatogr. 47:414 (1970).
18. S. Hirano, Carbohydr. Res. 100:431 (1982).
19. R. A. A. Muzzarelli, German Patent No 3,923,820 (1979).
20. R. A. A. Muzzarelli, New derivatives of chitin and chitosan, in: "New developments in industrial polysaccharides", V. Crescenzi and I. C. M. Dea, eds., Gordon & Breach, London (1985).
21. R. A. A. Muzzarelli, F. Tanfani and M. Emanuelli, J. Appl. Biochem. 3:322 (1981).
22. R. A. A. Muzzarelli, F. Tanfani and G. Scarpini, Biotechnol. Bioengin. 24:385 (1980).
23. M. Tzesos and B. Volesky, Biotechnol. Bioengin. 23:583 (1981).
24. M. Tzesos and B. Volesky, Biotechnol. Bioengin. 24:385 (1982).
25. M. Galun, P. Keller, D. Malki, H. Feldstein, E. Galun, S. M. Siegel and B. Z. Siegel, Science 219:285 (1983).
26. Z. Zosim, D. Gutnick and E. Rosenberg, Biotechnol. Bioengin. 25:1725 (1983).
27. Z. Zosim, D. Gutnick and E. Rosenberg, Biotechnol. Bioengin. 24:281 (1982).
28. T. Sato, K. Karatsu, H. Kitamura and Y. Ohno, Sen-I Gakkaishi 39:519 (1983).

IONIC INTERACTIONS BETWEEN CHITOSAN AND OXIDISED CELLULOSE

J. G. DOMSZY and G. A. F. ROBERTS

Physical Sciences Department, Trent Polytechnic
Burton street
Nottingham, NG1 4BU, England

INTRODUCTION

Although there are a number of patents claiming the use of chitosan and chitosan derivatives in paper-making[1-4], very little fundamental work on the adsorption of chitosan on cellulose has been reported. An extensive study of the effectiveness of chitosan as a strength additive, using various methods of application, has been reported[5-8]. The results were interpreted in terms of the effect of chitosan on interfibre bonding but again no fundamental studies on the adsorption process itself were carried out. A water-soluble chitosan derivative, glycol chitosan, was one of nine cationic polymers, or potentially cationic polymers, used in the study of the pH dependence of the adsorption of cationic polyelectrolytes on bleached kraft pulp[9]. The interest within the paper industry in the use of chitosan as an additive, and the lack of any detailed studies of the adsorption process, have prompted the authors to investigate the effects of a number of variables on the adsorption of chitosan on cellulose. The effects of temperature, mechanical pretreatment of the cellulose, molecular weight of chitosan and its degree of *N*-acetylation have been reported previously[10].

EXPERIMENTAL

Materials

a) Chitosan: The chitosan used in most of the adsorption studies was a sample of Kytex L (Hercules Inc.) which had $\bar{M}v = 3.82 \times 10^5$, calculated from the LVN value measured in 0.1M acetic acid/0.2M sodium chloride[11]. The extent of residual *N*-acetylation was found to be 16% by infrared spectroscopy[12].

b) Cellulose substrates: The standard cellulose used a a reference substrate, and for preparation of oxidised and dyed celluloses, was a sample of purified cotton. Oxidised cellulose samples having varying carboxylic acid group contents were prepared and analysed as described previously[10].

Cellulose samples containing sulphonic acid groups were prepared by dyeing with C.I. Reactive Red 13. Purified cotton (10g) was suspended in distilled water ($400cm^3$) containing the dye and sodium chloride (32g). The suspension was stirred vigorously for 20 minutes to ensure uniform mixing, then left for 72 hours with occasional stirring. The temperature was raised to 80°C and sodium carbonate (10g) added. After heating at 80°C for 5 hours the cotton was filtered off and washed on the filter until the filtrate was colourless. It was then soxhlet-extracted with water to leach out any remaining unreactive dye before drying. Three 10g samples were treated by this process using 0.1, 0.5 and 1.0g C.I. Reactive Red 13 respectively.

The amount of dye reacted was determined by spectroscopic analysis of a sample of the dyed cotton dissolved in 70% sulphuric acid. The absorbance at 573nm was measured using a solution of purified cotton in 70% sulphuric acid as the reference solvent. The molar extinction coefficient of purified C.I. Reactive Red 13 in 70% sulphuric acid was found to be 29,960 at 573nm.

A sample of purified cotton (10g) was dyed with commercial C.I. Direct Yellow 12 (0.5g) at a liquor ratio of 40:1 and in the presence of sodium chloride (2g). Dyeing was carried out for 40 minutes at 100°C, after which the cotton was filtered off and washed several times with distilled water.

Adsorption Measurements

Except where stated otherwise the adosrption measurements were carried out using solutions containing 1g dm^{-3} chitosan at a temperature of 20°C and a liquor ratio of 30:1. The concentration of chitosan remaining in solution at equilibrium was determined by metachromatic titration[13]. As reported previously[10] the adsorption process requires less than 48 hours to reach equilibrium under these conditions and a minimum adsorption time of 48 hours was used in all the experiments.

RESULTS AND DISCUSSION

Effect of Added Electrolyte on the Adsorption of Chitosan

The effect of the addition of varying concentrations of sodium chloride on the adsorption of chitosan was studied using an oxidised cotton sample (carboxylic acid group content = 162 m.eq.kg^{-1}) as substrate. The concentration of chitosan in solution was 0.5g dm^{-3}, the solutions being prepared by dilution of a stock 1g dm^{-3} solution with appropriate volumes of distilled water and 2M sodium chloride solution.

The equilibrium adsorption, when plotted against sodium chloride concentration, shows an initial rapid increase to a maximum, followed by a gradual decrease (Figure 1). Assuming that electrostatic interaction between anionic groups on the cellulose and protonated amine groups on the chitosan are important in the adsorption process then the gradual decrease in the equilibrium adsorption can be explained as arising from two effects:

a) suppression of the negative surface charge on the cellulose by the sodium chloride, thereby reducing the electrostatic interaction with chitosan;
b) increasing competition between the cationic groups on the chitosan and the sodium ions from the added electrolyte for the anionic sites on the cellulose substrate.

Because of the sodium chloride concentration at which the maximum occurs, the second of these two effects is considered to be the more important.

The initial rise in equilibrium adsorption with increase in added sodium chloride is due to the polyelectrolyte nature of chitosan in acid solution. In the absence of added electrolyte the polymer chains will be highly expanded due to both electrostatic repulsion effects between like charged chain segments and to osmotic effects due to the high counter ion concentration within the electrostatic domain of the polyelectrolyte chains. Addition of an electrolyte reduces these effects thereby causing a reduction in the hydrodynamic volume of the chains. The authors have recently shown[10] that the equilibrium adsorption of chitosan on cellulose increases with decrease in the molecular weight of the chitosan and with decrease in the amine group content, both of which represent increasing adsorption with decreasing hydrodynamic volume. Thus the increase in adsorption of chitosan with increase in added electrolyte is most readily explained as due to the decreasing hydrodynamic volume.

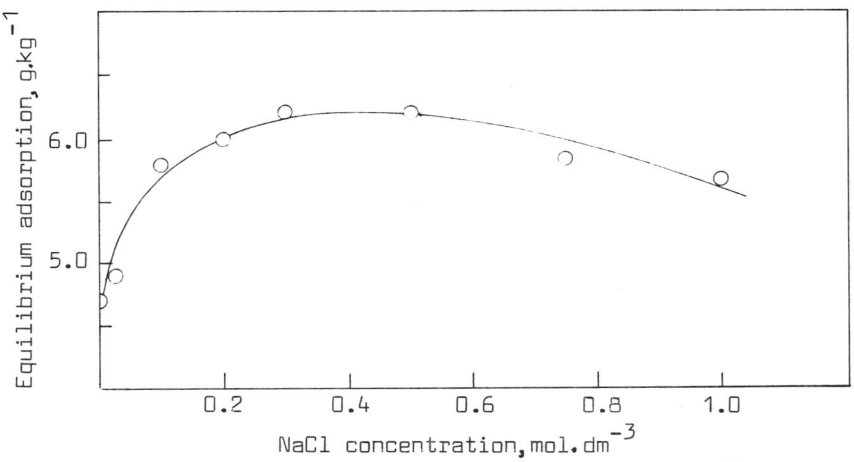

Fig. 1. Equilibrium adsorption of chitosan on oxidised cellulose *versus* Sodium chloride concentration.

The Influence of the Anionic Group Content of Modified Cellulose Substrates on the Adsorption of Chitosan

The equilibrium adsorptions of chitosan on oxidised cellulose substrates were determined and these are plotted as a function of carboxylic acid group content in Figure 2. The increase in adsorption with increase in carboxyl group content may arise from two effects:

a) Increased electrostatic interaction between chitosan chains and the modified cellulose:

b) Increased substrate surface area due to increased swelling on immersion in water because of increase in the hydrophilic character of the substrate with increase in the extent of modification. An increase in the substrate surface area through mechanical treatment has been shown to lead to an increase in chitosan adsorption[10].

In an attempt to separate surface charge and surface area effects further adsorption studies were carried out using modified cotton substrates in which sulphonic acid groups had been introduced by treatment with the reactive dye C.I. Reactive Red 13 (Figure 3a). It was considered that, because of the hydrophobic nature of the rest of the molecule, there would be little or no increase in the extent of swelling of the cellulose through introduction of anionic groups in this way. However the concentration of sulphonic acid groups introduced was small, the maximum of 34 milliequivalents kg^{-1} being well below the carboxylic acid group content of even the least extensively oxidised cotton sample. The results are given in Figure 2 and show, surprisingly, that there is a decrease in the equilibrium adsorption of chitosan with increase in the sulphonic acid group content.

This result may be explained in terms of van der Waals forces. These are greater the greater the area of contact between the two bonding molecules and the smaller the distance separating them. Thus planar systems show increased interaction compared with non-planar systems.

In the case of the oxidised cotton substrates the chitosan chain, which has a flat ribbon-like shape similar to cellulose, can approach close to the surface of the cellulose chains and hence maximise the contribution of the van der Waals forces to the forces promoting adsorption. However in the case of the reactive-dyed substrates the dye molecules protrude from the polymer chain surface, due to the steric constraints imposed by

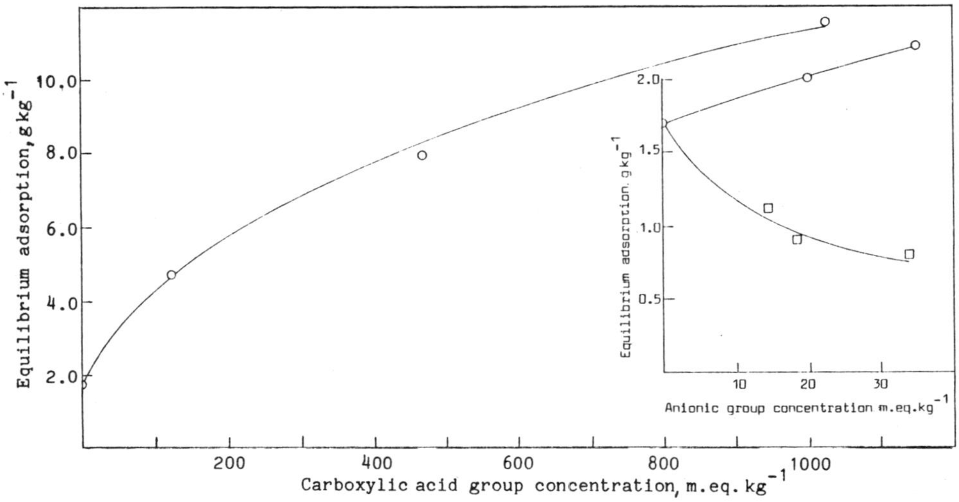

Fig. 2. Equilibrium adsorption of chitosan on oxidised cellulose *versus* Carboxylic acid group content. Inset: Equilibrium adsorption *versus* Anionic group concentration: ○ = carboxylic acid groups; □ = sulphonic acid groups.

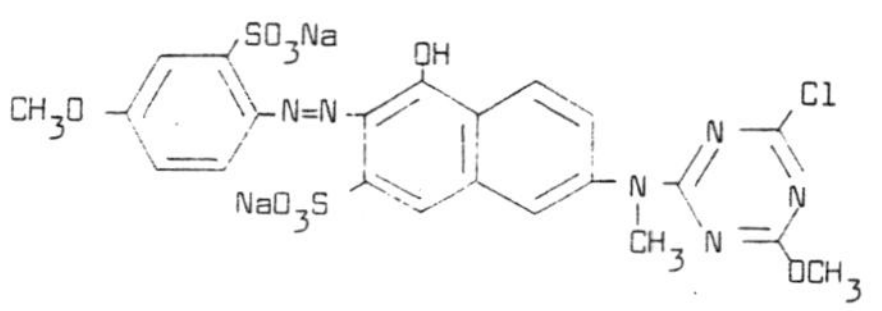

Fig. 3a. Structure of C.I. Reactive Red 13

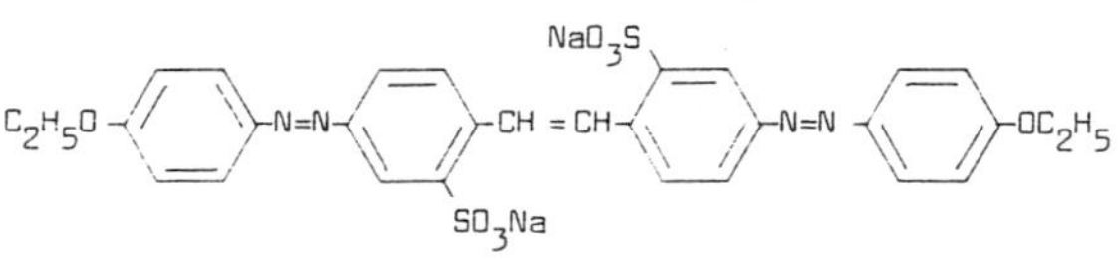

Fig. 3b. Structure of C.I. Direct Yellow 12

the covalent bond between dye and polymer. This disrupts the planarity of the surface presented to the chitosan chains, thereby reducing the potential area of contact. Thus although the electrostatic interaction is increased by introduction of the sulphonic acid groups on the reactive dye, the van der Waals forces between cellulose and chitosan are reduced and the results obtained indicate that these latter forces are the more important.

To test this a further sample of purified cotton was dyed with a direct dye, C.I. Direct Yellow 12 (Figure 3b). Direct dyes are planar and are adsorbed onto cellulose chains in a planar conformation[14]. Hence the chitosan chains can still approach close to the surface of the cellulose chains and maximise the van der Waals forces between themselves and the cellulose/direct dye substrate, in addition to the increased electrostatic attraction. Because of the poor wash fastness properties of direct dyes on cellulose, measurement of the extent of adsorption was carried out using a chitosan solution containing 0.5 mol dm^{-3} sodium chloride to prevent desorption of the dye. Under these conditions a small increase in adsorption was found with the dyed cotton substrate compared to the undyed reference cotton substrate (also from a solution containing 0.5 mol dm^{-3} sodium chloride).

Extent of Ionic Interaction Between Chitosan and Oxidised Cellulose

The above results suggest that, although important, ionic interactions between the cationic polyelectrolyte and anionic groups on cellulose are not the sole force of attraction and indeed are not likely to be dominant. In order to determine the extent of ionic bonding between chitosan and anionic sites on cellulose a sample of oxidised cotton (carboxylic acid group content = 162 m.eq. kg^{-1}) was dyed with the cationic dye Methylene Blue at pH 8 to bring about an exchange of dye ions for the counter ions of the carboxylic acid groups in the oxidised cotton. After careful washing to remove excess dye the sample was immersed in a solution of chitosan using standard conditions. After the adsorption process had reached equilibrium, but prior to determination of the residual chitosan concentration, the concentration of Methylene Blue in solution was determined. The results showed that 1.98×10^{-8} mols of dye were displaced for 1.63×10^{-5} mols of amine group adsorbed, a ratio of $\sim$1:800. Hence only 1 amine group in every 800, approximately, is involved in electrostatic interactions with the carboxylic acid groups of the cellulose substrate, which corresponds to an average of 2-3 ionic linkages per chitosan chain adsorbed. This has implications for the conformation of the adsorbed chitosan chains and is being investigated further.

CONCLUSIONS

The addition of sodium chloride has two opposing effects on the adsorption of chitosan on oxidised cellulose. At low sodium chloride concentrations the dominant effect is reduction of the hydrodynamic volume of the chitosan molecules leading to an increase in the equilibrium adsorption. At higher sodium chloride concentrations competition between sodium ions and cationic groups on the chitosan for the carboxylic acid groups on the cellulose predominates, leading to a gradual decrease in the equilibrium adsorption of chitosan.

Introduction of carboxylic acid groups into the cellulose substrate through oxidation causes an increase in the equilibrium adsorption of chitosan, the increase being greater the higher the concentration of carboxylic acid groups. However introduction of sulphonic acid groups into the cellulose substrate, through treatment with a reactive dye, has the reverse effect with the equilibrium adsorption of chitosan decreasing with increasing sulphonic acid group content. This latter effect has been tentatively ascribed to disruption of the planarity of the absorbent surface due to steric constraints imposed on the dye molecules by the dye/cellulose bond.

Determination of the concentration of counter ions (Methylene Blue ions) displaced during adsorption of chitosan on oxidised cellulose indicates that only 1 out of every 800 (approximately) charged groups on the chitosan interacts electrostatically with carboxylic acid groups on the cellulose. This corresponds to an average of 2-3 ionic bonds between cellulose and chitosan for each chitosan chain adsorbed.

ACKNOWLEDGEMENTS

The authors wish to thank Hercules Inc. for supplying the chitosan used in this work, and Mrs F.A. Wood for assistance in preparing the manuscript. One of the authors (JGD) wishes to thank the Science and Engineering Research Council and the Paper Industry Research Association for the award of a CASE research studentship.

REFERENCES

1. Oxford Paper Co., Improvement in coating compositions and planographic printing plates, Brit. Pat. 930,765 (1963).
2. R. Slagel and G. Sinkovitz, Paper products of improved dry strength, U.S. Pat. 3.709.780 (1973).
3. R. Slagel and G. Sinkovitz, Chitosan graft copolymers for making paper products of improved dry strength, U.S. Pat. 3,770,673 (1973).
4. E.A. Plisko, V.N. Baranova and L.A. Nud'ga, Electrical insulating paper, U.S.S.R. Pat. 428,053 (1974).
5. G.G. Allen, G.D. Crosby, J.H. Lee, M.L. Miller and W.M. Reif, New bonding systems for paper, Proc. Symp. Man-made Polym. Paper-making, 85 (1973).
6. G.G. Allen, J.F. Friedhoff, M. Korpela, J.E. Laine and J.C. Powell, Marine polymers. V: Modification of paper with partially deacetylated chitin, A.C.S. Symp. Ser., 10:172 (1975).
7. G.G. Allen, J.R. Fox, G.D. Crosby and K.V. Sarkanen, Chitosan, a mediator for fibre-water-interactions in paper, Fibre-water-Interact. Paper-making, Trans. Symp., 2:265 (1978).
8. V.N. Baranova, E.A. Plisko, R.E. Tarasova, S.N. Stepanov and V.A. Kondraklov, New auxiliary substance, chitosan, used in the manufacture of printing paper, Poligr. Prom-st., (Ser.) Ref. Inf., 12 (1980): Through Chem. Abs., 94:141476n (1981).
9. H. Tanaka, K. Tachiki and M. Sumimota, Adsorption of cationic polymers onto bleached kraft pulp, Tappi, 62:41 (1979).
10. J.G. Domszy, G.K. Moore and G.A.F. Roberts, The adsorption of chitosan on cellulose, Proc. Cellucon '84, in press.
11. G.A.F. Roberts and J.G. Domszy, Determination of the viscometric constants for chitosan, Int. J. Biol. Macromol., 4:374 (1982).
12. G.K. Moore and G.A.F. Roberts, Determination of the degree of N-acetylation of chitosan, Int. J. Biol. Macromol., 2:115 (1980).
13. B.D. Gummow and G.A.F. Roberts, Metachromatic behaviour of sodium 2'-hydroxy-1,1'-azonaphthalene-4-sulphonate in the presence of chitosan, Makromol. Chemie, (1985) in press.
14. I.D. Rattee and M.M. Breuer, The adsorption of dyes by cellulosic substrates, in: "The Physical Chemistry of Dye Adsorption", Academic Press, London (1974).

Editor's Report

R. A. A. MUZZARELLI

CHITIN AND CHITOSAN SOLUTIONS

based on works by

A. COSANI, A. FORNASA,
C.A. KIENZLE-STERZER, C.K. RHA, D. RODRIGUEZ-SANCHEZ,
M. SCANDOLA, M. TERBOJEVICH,
and M. VINCENDON

INTRODUCTION

Polyelectrolytes ionize forming polyions when dissolved in water or other polar solvents. The rheological behavior of polyions in solutions differs substantially from that of non-charged polymers because of the presence of electrostatic interactions between neighboring groups and between polyions-counterions in the medium. Current literature provides partial descriptions of the behavior of selected polyelectrolytes as a phenomenological view of causes-effects. The lack of knowledge of these causes-effects makes it difficult and uncertain to interpret and predict the behavior of polyelectrolytes in solution, at different solution conditions.

The rheological behavior of chitosan, a typical random coil polyelectrolyte, is an important subject of study because most of the uses of chitin are made possible by the preparation of chitosan solutions. Only recently, some analysis of the solution behavior of chitosan has begun.

The flow behavior of polyelectrolyte polysaccharides is dictated by their overall molecular conformation and the degree of hydrogen bonding or electrostatic repulsion between neighboring chain segments. The intrinsic viscosity determined for chitosan in solution indicates that chitosan adopts conformations which range from a random coil to a more compact, quasi-globular shape. The intrinsic viscosity, $|\eta|$, dl/g, of a linear chain polymer, follows the Mark-Houwink equation: $|\eta| = k.M^{a}$. In the case of random coil polymers, the exponent "a" lies between 0.5 and 0.8, and the overall shape of the hydrodynamic volume is spherical. Exponents larger than 0.8 reflect a highly expanded conformation, which may be rodlike in shape. Low values indicate globular conformation that can be cause by local ordering due to attractive hydrogen bonding forces or to electrostatic repulsive forces. Studies on the relationship between intrinsic viscosity of chitosan and the molecular weight demonstrated the significance of hydrogen bonding on the molecular conformation of chitosan in solution. For comparison, alginate and hyaluronate have Mark-Houwink exponents approaching 1.0 because they have a high degree of chain extension in solution.

As for all the polyelectrolyte polysaccharides, the Mark-Houwink exponent is pH- and ionic-strength-dependent: therefore, the hydrodynamic volume of chitosan can be manipulated by altering the intramolecular interactions between charged neighboring ions by simple changes in pH and ionic strength.

SOLUTION BEHAVIOR OF CHITOSAN IN DILUTED SOLUTIONS: EFFECT OF THE INTRAMOLECULAR INTERACTIONS BETWEEN NEIGHBORING SEGMENTS

(C. A. Kienzle-Sterzer, D. Rodriguez-Sanchez and C. K. Rha)

The overall objective of this paper is to elucidate the mechanisms (causes-effects) responsible for the solution behavior of a model cationic biopolymer:chitosan. The solution behavior of chitosan is complex because of the electrical charge on its backbone For instance, an increase in the reduced viscosity is observed with dilution at low ionic strength in the medium although an increase in the ionic strength of the medium suppresses such behavior [1 - 5].

The hypothesis in this study is that the electrostatic charge on the chitosan backbone affects its solution behavior via two major mechanisms: a) through intramolecular interactions which determine both the size and shape of the hydrodynamic domain of the isolated molecule, and b) through electrostatic interactions between the polyion and the solvent-counterions. To evaluate this hypothesis, the effect of pH, ionic strength and molecular weight upon the intrinsic viscosity of chitosan solutions reported elsewhere will be used[3,5,6,7].

DISCUSSION

One of the most simple and effective sources of experimental information regarding chain dimension and chain conformation is the evaluation of the rheological behavior of dilute polymer solutions. Dilute chitosan solutions have been shown to demonstrate a characteristic polyelectrolyte behavior at both low and high ionic strength of the medium[4,5]. These studies demonstrated the effect of solution conditions on the rheological behavior of chitosan solutions. The values of $[\eta]$ for chitosan solutions as affected by the degree of ionization and counterion concentration are shown in Table 1[3].

The Manning theoretical approach is commonly used to explain the poly electrolyte phenomena in solution[8]. Recently, an extension of this approach has been made to account for the polyelectrolyte coil expansion[9]. In this approach the polyelectrolyte backbone is considered to behave like a rigid rod with elastic bending capability. Specifically the electrostatic contribution to the persistence length (P_1) is represented by:

$$P_1=P_o+\xi^{-1}(\ell^2/12b)[2\xi-1+Ln(I_1/I_z)+Ln\ \kappa b]$$

where P_0 is the non-electrostatic component, ℓ is the segment length, b is the axial charge spacing, ξ is the charge density, κ is the Debye-Huckel screening parameter and I_1 and I_z are respectively the local and bulk counterion concentration. Therefore, if this term is introduced into the Yamakawa-Fujii expression of the intrinsic viscosity ($[\eta]$) for a wormlike chain in the random-coil limit[10]:

$$[\eta]=\Phi P_1^{3/2}/M$$

being Φ a constant, and M the polymer molecular weight, a linear relationship would be established between $[\eta]^{2/3}$ and the logarithm of the reciprocal of the counterion concentration. Chitosan in solution displays such linearity for degrees of ionization between 0.90 and 0.60 as reported elsewhere[3]. This finding suggests that chitosan molecules in dilute solution can be modelled as wormlike chains with some degree of flexibility. This suggested model is supported by : a) the stiff nature of the glucosamine ring, b) the limited rotation of the β(1-4) glucosidic link, c) the values of the characteristic ratio (C_∞=15)[11], and d) the "stiffness parameter" (B=0.08)[3].

In order to evaluate the effect of the degree of ionization and the counterion concentration upon the neighboring interactions, the electrostatic contribution to the excluded volume (Be,exp) is calculated from the slope of the linear relationship between $[\eta]$ and the reciprocal of the square root of the counterion concentration[3,5,11] as:

$$B_{e,exp} = A_2 C_s^{-1/2}/(0.51\Phi_0 M)$$

where A_2 is the slope of $[\eta]$ versus $C_s^{-1/2}$, and Φ_0 is the Flory's constant. The values of the electrostatic contribution to the excluded volume thus calculated are compared with the values of the theoretical electrostatic contribution to the excluded volume (Be,th) estimated by the expression[12]:

$$B_{e,th} = \left(\frac{\ell^2}{\kappa m_0}\right) R(w)$$

where ℓ = 0.533 nm for the glucosamine monomer, m_0 is the monomer molecular weight,

$$R(w) = \int_0^{\pi/2} d\theta \, \sin^2\theta \int_0^{w/\sin\theta} x^{-1}(1-e^{-x})dx$$

and

$$w = \frac{136.4\alpha^2}{\ell^2 C_s^{-1/2}}$$

The expression for Be,th assumes that the polyelectrolyte molecule can be modelled as a random-flight chain consisting of many rod-like segments with negligible near neighboring interactions and that the Debye-Huckel approximation is valid to estimate the excluded volume.

Table 1 presents the values of both Be,exp and Be,th for chitosan solutions as a function of the degree of ionization and counterion concentration. Both Be,th and Be,exp decrease as the degree of ionization decreases and the counterion concentration increases. This decrease in the electrostatic contribution to the excluded volume of chitosan molecules in solution is due to a decrease in the electrostatic repulsive forces between charged points on the chitosan backbone. However, it must be pointed out that, for all α values, Be,exp values are higher than Be,th for counterion concentration above 0.100 mol/L. This outcome is related to the collapse of the chitosan molecule with increasing the counterion concentration. As the collapsing proceeds, near neighboring interactions will appear which

Table 1. Intrinsic viscosity, experimental-, and theoretical values of the electrostatic contribution to the excluded volume as a function of the degree of ionization (α), and the counterion concentration

		$[Cl^-]$ / mol/L					
α		0.500	0.300	0.200	0.100	0.075	0.050
0.90	$[\eta]/dLg^{-1}$	6.29	6.87	7.44	8.71	9.50	10.43
	10^{26} $B_{e,exp}/cm^3$	1.16	1.50	1.84	2.60	3.00	3.68
	10^{26} $B_{e,th}/cm^3$	0.83	1.12	1.54	2.42	2.91	3.77
0.80	$[\eta]/dLg^{-1}$	6.22	6.74	7.31	8.43	9.18	10.14
	10^{26} $B_{e,exp}/cm^3$	1.07	1.39	1.70	2.40	2.77	3.40
	10^{26} $B_{e,th}/cm^3$	0.75	1.04	1.35	2.26	2.72	3.54
0.70	$[\eta]/dLg^{-1}$	5.94	6.50	6.96	7.98	8.67	9.69
	10^{26} $B_{e,exp}/cm^3$	0.97	1.25	1.53	2.17	2.50	3.07
	10^{26} $B_{e,th}/cm^3$	0.69	0.97	1.27	1.99	2.37	3.09
0.60	$[\eta]/dLg^{-1}$	5.71	6.21	6.73	7.55	8.34	9.20
	10^{26} $B_{e,exp}/cm^3$	0.89	1.14	1.40	2.00	2.29	2.80
	10^{26} $B_{e,th}/cm^3$	0.60	0.89	1.18	1.85	2.18	2.96

in turn will increase the value of $B_{e,exp}$ with respect to that of $B_{e,th}$ for a given counterion concentration.

Therefore, from the relationship between $[\eta]$ and the degree of ionization and the counterion concentration, it can be concluded that chitosan molecules in dilute solutions can be modelled as wormlike chains, with some local flexibility, consisting of many rod-like segments, and that the Debye-Huckel approximation can be used to estimate their excluded volume.

It is a well known fact that the intrinsic viscosity of a linear polymer in solution is related with the molecular weight by the Mark-Houwink equation[13]:

$$[\eta] = KM^{a}$$

where "K" and "a" are characteristic constants of the polymer-solvent system. In general, for a random-coil polymer "a" is between 0.5 and 0.8 such as in the case of pullulan-water[14], and sodium carboxymethylamylose-water[15]; for a highly expanded conformation "a" is higher than 0.8 such as in the case of sodium alginate-water[16]; whereas "a" values lower than 0.5, as for levan hydrolysates-water[17], reflect a globular compact conformation. Therefore, from the values reported in the literature [6,7], it is clear that chitosan conformation changes from a compact spherical structure (a=0.147) when dissolved in acetic acid - NaCl (1%-2.8% respectively) to a more random coil conformation (a=0.71) when urea is present. The increase in the value of "a" in the presence of urea shows the significant role of intramolecular hydrogen bonding upon the conformation of chitosan molecules in dilute solutions. In order to represent the effect of the intramolecular hydrogen bonding on the conformation of chitosan molecules in solution, the Kratky and Parod length and the Kuhn statistical segment were calculated by the equation[18]:

$$[\eta]^{-1}=-A_{2}+A_{1}M^{-1/2}$$

where A_1 is approximately equal to K_0^{-1}, and the Kuhn statistical segment K_1 can be calculated as:

$$K_{1}=\left(\frac{K_{\theta}}{\Phi}\right)^{2/3}M_{L}$$

where M_L is defined as the ratio of the molecular weight to the contour length of the polymer. The values of the unperturbed Kratky-Parod persistence length (K) on the Kuhn statistical segment are:

a) For the extended coils (chitosan in the presence of urea):

$K = 6.08$ nm $\qquad K_1 = 12.16$ nm

b) For compact structure (chitosan without urea):

$K = 2.64$ nm $\qquad K_1 = 7.28$ nm

These results indicate the role of the intramolecular hydrogen bonding on the structure-properties of chitosan in dilute solutions.

However, it is worth mentioning that a recent article by Domard and Rinaudo[19] dealing with the relationship between intrinsic viscosity and molecular weight have questioned the validity of the viscometric law. Similarly, Roberts and Domszy[20], have reported values for "a" considerably different to those reported by Berkovich and coworkers[6]. These two papers clearly indicate the need of a critical evaluation of the effect of charge density, polydispersivity, draining effect, and degree of deacetylation upon the Mark-Houwink relationship for chitosan in solution. It is the opinion of the authors that the evaluation of all these variables upon the relationship between [η] and chitosan molecular weight will help to understand these apparently contradictory findings.

Acknowledgements. This work was partially supported by MIT Sea Grants 92391 and 90877 from the NOAA Department of Commerce. The authors want to thank the Agricoltural Experiment Station, O.S.U. for its support.

REFERENCES

1. R. A. Cox, The influence of ionic strength on the viscosity of ribonucleic acid and other polyelectrolytes, J. Polym. Sci. 47:441 (1960).
2. H. W. Chien, C. H. Isihara, and A. Isihara, Intrinsic viscosity of polyelectrolytes in salt solutions, Polym. J. 3:288 (1976).
3. C. A. Keinzle-Sterzer, "Transport Properties of a Cationic Polyelectrolyte in Dilute and Concentrated Solutions: Chitosan," Ph.D. Thesis, M.I.T., Cambridge, MA USA (1984).
4. C. A. Kienzle-Sterzer, D. Rodriguez-Sanchez, and C. K. Rha, Dilute solution behavior of a cationic polyelectrolyte. J. Appl. Polym. Sci. 27:4467 (1982).
5. D. Rodgriguez-Sanchez, C. A. Kienzle-Sterzer, and C. K. Rha, Intrinsic viscosity of chitosan solutions as affected by ionic strength, in "Proc. 2nd Int. Conf. on Chitin-Chitosan," S. Hirano, S. Tokura, Eds. The Japanese Society of Chitin and Chitosan, Tottori (1982).
6. L. A. Berkovich, G. I. Timofeyera, M. P. Tsyurupa, and V. A. Davankov, Hydrodynamic and conformational parameters of chitosane, Vysokomol. Soedin., Ser. A. 22:1834 (1980).
7. R. A. A. Muzzarelli, "Chitin" Pergamon Press, Oxford (1977).
8. G. S. Manning, Limiting laws and counterion condensation on polyelectrolyte solutions. I. Colligative properties, J. Chem. Phys. 51:924 (1969).
9. G. S. Manning, The molecular theory of polyelectrolyte solutions with applications to the electrostatic properties of polynucleotides, Q. Rev. Biophys. 2:179 (1978).
10. H. Yamakawa, and M. Fujii, Intrinsic viscosity of wormlike chains. Determination of the shift factor, Macromolecules 7:128 (1974).
11. C. A. Kienzle-Sterzer, D. Rodriguez-Sanchez, and C. K. Rha, Solution properties of chitosan: chain conformation, in "Chitin, Chitosan and Related Enzymes," Pergamon Press, New York (1984).
12. M. Fixman, and J. Skolnick, Polyelectrolyte excluded volume paradox, Macromolecules 11:863 (1978).
13. P. J. Flory, "Principles of Plymer Chemistry," Cornell University Press, Ithaca (1953).
14. T. Kato, T. Okamoto, T. Tokuya, and A. Takahashi, Solution properties and chain flexibility of pullulan in aqueous solutions, Biopolymers 21:1623 (1982).
15. K. D. Goebel, and D. A. Brant, The configuration of amylose and its derivatives in aqueous solutions. Experimental results, Macromolecules 3:634 (1980).
16. O. Smidsrod, Solution properties of alginate, Carbohydr. Res. 13:359 (1970).
17. S. S. Stivala, J. E. Zweig, and J. Ehrlich, Dilute solution properties of Streptococcus salivaricus levan and its hydrolysates in " A.C.S. Symposium Series. Solution Properties of Polysaccharides," Washington, (1981).
18. G. Staikos, and A. Dondos, Determination of the statistical segment of wormlike polymers using viscosimetric data, Eur. Polym. J. 19:555 (1983).
19. A. Domard, and M. Rinaudo, Preparation and characterization of fully

deacetylated chitosan, Int. J. Biol. Macromol. 5:49 (1983).
20. G. A. F. Roberts and J. C. Domszy, Determination of the viscometric constants for chitosan, Int. J. Biol. Macromol. 4:374 (1982).

N.M.R. CONFORMATIONAL ANALYSIS OF CHITIN IN LITHIUM CHLORIDE SOLUTION

(M. Vincendon)

The dissolution of chitin presents an important problem inherent to the operating difficulties. These are mainly due to the existence of both intra and intermolecular hydrogen bonds, for this natural polymer in its solid state. There have been important efforts concerning the study of chitin dissolution which matters both from a practical point of view (eg. film and fiber spinning) and from a fundamental aspect (the conformation of macromolecules in solution). NMR spectroscopy could yield some information about the nature and the conformation of macromolecules in solution.

A 5 % lithium chloride solution in N,N-dimethylacetamide (DMA) has been used, as solvent for chitin. This solvent retains the aprotic character of pure DMA, and thus allows the observation of labile protons by proton NMR (1, 2).

RESULTS AND DISCUSSION

The solvent system obtained by the dissolution of 5 parts of LiCl in N,N-dimethylacetamide is known to have solvating properties towards highly non-soluble polymers.

As it has been shown for cellulose oligomers, the dissolution in the above mentioned solvent of chitin models, such as N-acetyl-D-glucosamine and chitobiose takes place via the interaction of one LiCl molecule per labile proton group (OH or NH). This was shown by the individual observation of each hydroxyl (or amino) proton signal by means of proton NMR spectroscopy. The interaction of LiCl with hydroxyl or amino groups induces a large downfield shift ($\Delta\delta$ = + 1 ppm) for these protons (3, 4).

Figure 1 shows the lowfield part of the ^{1}H NMR spectrum of methyl-β-D-chitobioside in a DMA-LiCl solution. Several remarks can be made about this spectrum :

1. all the hydroxyl and amino protons are visible on the spectrum.
2. all the signals (except these of non hydroxylated anomeric protons H-1 and H-1') have anomalous high δ values, which means that they have undergone a downfield shift.
3. the OH-3 hydroxyl group has a relatively low δ value (δ = 4.8 ppm) as compared to that of other secondary hydroxyl protons (OH-3', δ = 5.9 ppm ; OH-4', δ = 5.96 Ppm).
4. the ^{3}J(H-C-O-H) coupling constant for OH-3 (J = 0.9 Hz) is lower than those of other hydroxyls (J = 2.75 Hz).

The low δ value for the OH-3 hydroxyl proton can be explained by a deficiency in the accessibility to the solvating LiCl molecules. The specificity of OH-3 has also been noticed for the cellulose oligomers series, starting at DP = 2 (cellobiose) (3). We postulated an intramolecular hydrogen bond between this hydroxyl proton and the O-5' heterocyclic oxygen atom of the neighbouring D-glucopyranosyl unit. The same configuration of the D-glucopyranosyl units is observed in the cellulose and chitin chains, allowing a possibility of a similar hydrogen bond. The OH-3 methyl-β-D-chitobioside involvement in an intramolecular hydrogen bond - as shown on figure 1 - will restrict its accessibility to LiCl molecules and then explain its rather low δ value.

An other NMR parameter, i.e. the temperature chemical shift dependence, can eventually be used to check the presence of hydrogen bonds in aprotic solvents. In all proton signals, there is a high field shift as the temperature increases ; however, labile protons (OH and NH) are by far the more affected. Yet, the labile protons involved in an intramolecular hydrogen bond are less sensitive to temperature variation than those usually involved in solvent-solute intermolecular associations.

The temperature dependence of labile protons chemical shift $(\Delta\delta/\Delta T)$ has been observed for N-acetyl-D-glucosamine, and methyl-β-D-chitobioside in three typical aprotic solvents, e.g. dimethylsulfoxide, pure N,N-dimethylacetamide and DMA-LiCl.

The results are to be found in Table I. The main feature is low sensitivity to the temperature related variation of the methyl-β-D-

Table I. Hydroxyl and acetamido proton chemical shift increment with temperature $\Delta\delta/\Delta T$ (10^{-3} ppm/°C) of chitin and chitin models in various aprotic solvents.

		H-1	NH-2	OH-3	OH-4	OH-6	H-1'	NH-2'	OH-3'	OH-4'	OH-6'
N-acetyl	DMSO	1.0	3.4	5.4	4.8	6.1					
D-glucosa-	DMA	0.3	5.1	6.9	6.9	8.6					
mine	DMA-LiCl[4]	0.4	5.8	4.65	4.5	9.25					
Methyl- -	DMSO	0.1	5.2*	3.0		5.9	0	5.8*	6.8	6.9	6.8
D-chito-	DMA	-0.6	5.0*	2.5		7.7	-0.7	6.0*	7.8	7.9	8.0
bioside	DMA-LiCl[4]	0.9	4.25*	1.8		3.9	0.1	8.7*	3.3	6.75	5.8
Chitin	DMA-LiCl[4]			1.25		4.5	0.05	5.25			

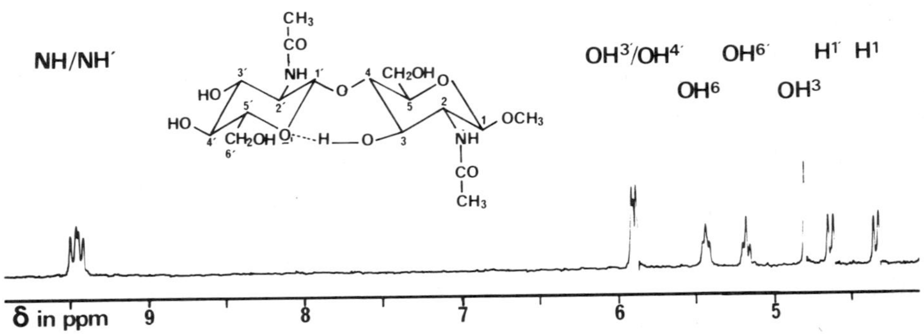

Figure 1 : 250 MHz ^{1}H NMR spectrum of methyl-β-D-chitobioside in N,N-dimethylacetamide-LiCl solution.

chitobioside OH-3 signal, for which an intramolecular hydrogen bond is possible, as compared to other hydroxyl signals which are to be found only in intermolecular solvent-solute associations. This factor also implies the existence of an intramolecular hydrogen bond for the chitin dimer in aprotic solvents solutions.

The observation of labile protons on the ^{1}H NMR spectrum of chitin in this solvent (4) allows to study their dependence on temperature variation. The OH-3 hydroxyl proton is found to be less sensitive to temperature increases, as for methyl-β-D-chitobioside, as indicated by its increment value $\Delta\delta / \Delta T = 1.25\ 10^{-3}$ ppm/°C (Table I). The chitin OH-3 proton signal has the same peculiar features as those found for its dimer ; these are low δ value, a low coupling constant value, and a low sensitivity to temperature increases. These observations have led us to postulate an OH-3 ... O-5' interunit intramolecular hydrogen bond.

The existence of such a bond has been observed for a cellulose solution in the same solvent. This intramolecular hydrogen bond has also been noticed for chitin in its solid state by means of X-ray diffraction. The dissolution of chitin in DMA-LiCl aprotic solvent occurs via the interaction of LiCl molecules with each labile proton of the chain, as shown by their high δ value. In this aprotic medium, the OH-3 ... O-5' intramolecular hydrogen bond is not destroyed, giving a rigid linear backbone to chitin chains in solution (3, 5).

REFERENCES

1. P. R. Austin, U.S. Patent 4 050 457 (1977).
2. J. Saint-Germain and M. Vincendon, ^{1}H, ^{13}C and ^{7}Li nuclear magnetic resonance study of the lithium chloride-N,N-dimethylacetamide system, Org. Magnetic Reson. 21:371 (1983).
3. D. Gagnaire, J. Saint-Germain and M. Vincendon, NMR evidence of hydrogen bonds in cellulose solutions, J. Appl. Polym. Sci. Polymer Symposia, 37:261 (1983).
4. M. Vincendon, Proton NMR study of chitin dissolution mechanism, Makromol. Chem., 186:1671(1985).
5. J. Blackwell, R. Hinke, and K. H. Gardner, Determination of the structures of α and β chitin by X-ray diffraction, "Proceedings of the First International Conference on chitin-chitosan", R. A. A. Muzzarelli, ed., 108, (1978).

DIFFUSION OF SMALL MOLECULES IN CHITOSAN SOLUTIONS: MICROVISCOSITY

(C. Kienzle-Sterzer and C.K. Rha)

The diffusion of low molecular weight compounds through both dilute and non-dilute polyelectrolyte solutions has received a great deal of attention in the last twenty five years. The importance of electrostatic interactions upon the diffusion of small particles even under conditions with no net electrical flow has been shown. This effect is caused by the influence of the counterions cloud surrounding the polyion, which produces a cooperative effect, on the self-diffusion coefficient (Florence and Johnson, 1975; Yoshida, 1978). Therefore, self-diffusion experiments can be used to study the effect of charge density, and environmental conditions on the microenvironment surrounding the polysaccharide in solution. It is the objective of this study to quantify the effect of pH, ionic strength, and intermolecular interactions upon the transport properties of polysaccharide solutions, using chitosan as the model polyelectrolyte.

MATERIALS AND METHODS

Chitosan, Madera chitosan (Bio-Shell Inc., Albany, OR) obtained from Tanner (snow) crab, *Chionoecetes bairdi*, with less than 1% of protein and ash, and with 80% deacetilation was selected. D-6-^{3}H-glucose aqueous solution (Smershan/Searle Corp., Illinois) was used without further purification.

Self-diffusion experiments were performed in a constant temperature both thermostated at 25.0±0.1°C. The open-end capillary method was employed. Capillaries of fixed diameter and length were filled with a known volume of chitosan solution containing 1% D-6-^{3}H-glucose. The capillaries were then immersed in a bath of radioactive-inert solution of identical composition to that of the capillaries. Self-diffusion coefficients were calculated by the expression (Crank, 1979):

$$D = \frac{\pi}{4}\left(1-\frac{C}{C_o}\right)\frac{L^2}{t}$$

where t is the diffusion time; C_o and C are the radioactive content at time zero and at time "t"; and L is the solution height inside the capillary. The radioactive content inside each capillary was monitored with a Beckman, Liquid Scintillation System LS-230.

RESULTS AND DISCUSSION

The self-diffusion coefficient of glucose in semiconcentrated chitosan solutions (chitosan concentrations between 0.05g/dL and 0.20g/dL) as affected by both the pH (pH values between 3.0 and 6.0) and counterion concentration (Cl^- concentration ranging from 0.10M to 1.00M) has been evaluated. Figure 1 shows the effect of chitosan and counterion concentrations on the relative self-diffusion coefficient of glucose (D/D_o) at pH 5.0. Similar relationships between D/D_o and chitosan concentration were observed for all other pH's selected in this study. From Figure 1 we observed: 1) a linear relationship between D/D_o and the chitosan concentration, 2) the ratio D/D_o decreased with increasing concentration, 3) for a given chitosan concentration, the ratio D/D_o increases as the counterion concentration increases, 4) for a given chitosan concentration and counterion concentration, D/D_o increases with increasing the pH of the solution.

These results can be explained in the content of the obstruction theory developed by Wang (1954), where D/D_o is related to the polymer concentration by the expression:

$$\frac{D}{D_o} = 1 - f(\bar{v})c$$

where c is the polymer concentration, and $f(\bar{v})$ is a function of the polymer specific volume. Then: 1) the insensitivity of $f(\bar{v})$ toward chitosan concentration indicates a compensating effect between the increase in the overlapping of the macromolecular domains and the reorganization of the surrounding solvation layers, as intermolecular interactions increases; 2) the effect of pH and counterion concentration upon D/D_o indicates the primary role of the hydrodynamic volume on the transport properties in chitosan solutions; 3) the decrease of D/D_o as the chitosan concentration increases suggests the formation of a "loose network" due to an increase in the physical entanglements; and 4) in agreement with the obstruction theory, semiconcentrated chitosan solutions can be envisioned as consisting of two distinct regions: a polymer rich region (polymer-solvation layer), and a free solvent region. Similar results have been reported for other polysaccharides such as hydroxypropylcellulose dextran and its derivatives.

The self-diffusion coefficients of small molecules through chitosan solution have been used to calculate the values of the microviscosity coefficient, with the assumption that in pure solvent both micro- and microviscosity values are the same, with the Stokes-Einsteing equation

$$D = \frac{RT}{6r\eta^*}$$

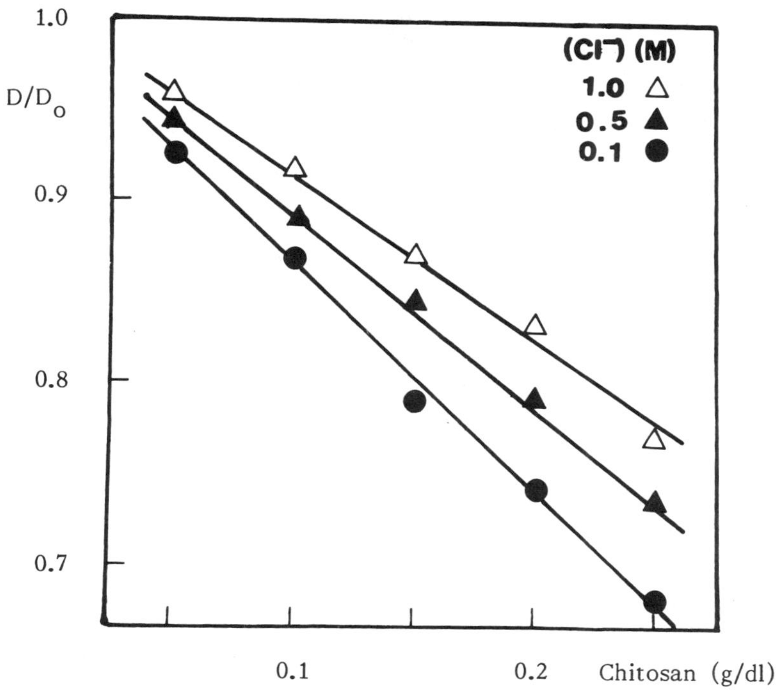

Figure 1. Relative self-diffusion coefficient of glucose as affected by chitosan concentration, and counterion concentration. pH = 5.0

where r is the radious of the diffusant molecule, and η^* is the microviscosity of the solution. Figure 2 shows the contribution of the microviscosity on the macroviscosity of semiconcentrated chitosan solutions (η_r) as the overlapping parameter ($C[\eta]$) increases. The values of η_r of chitosan solutions for all the conditions used in this study have been reported elsewhere (Kienzle-Sterzer et al., 1985).

Figure 2 shows that in sufficiently dilute chitosan solutions the molecules act as individual units and the Brownian motion, represented by the microviscosity, governs their rheological response. On the other hand, as the chitosan concentration increases, the intermolecular interactions became increasingly more important leading to the formation of inter-

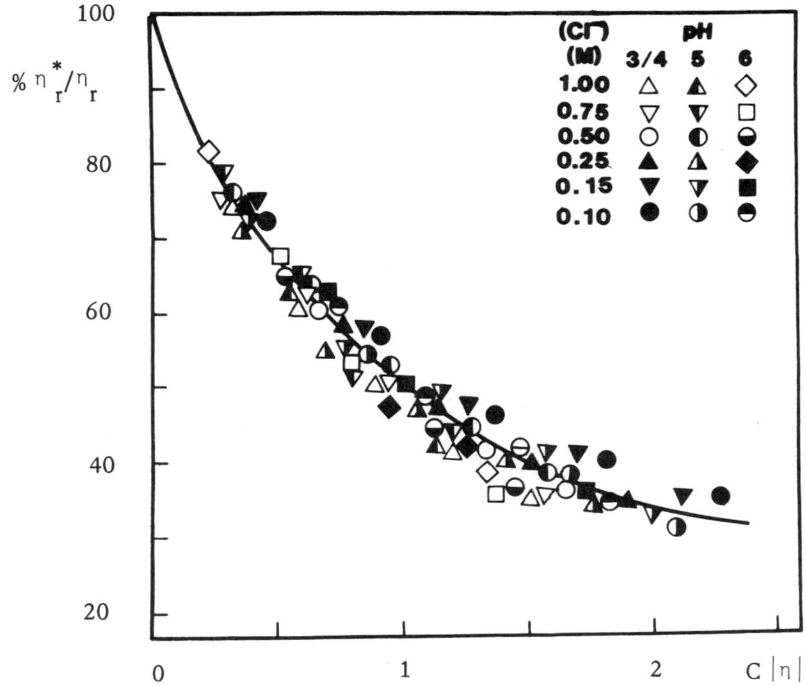

Figure 2. Contribution of microviscosity on the overall macroviscosity as a function of the overlapping parameter, pH and counterion concentration.

molecular entanglements. At this point the contribution of the microviscosity on the overall macroviscosity is small, and we can conclude that their rheological response is primarily controlled by the intermolecular interactions.

In conclusion, the evaluation of the self-diffusion coefficient of glucose through semiconcentrated chitosan solution clearly establishes the importance of pH, ionic strength, and intermolecular interactions upon the contribution of the microviscosity on the macroviscosity of polyelectrolyte solutions.

Acknowledgement. This work was partially supported by MIT Sea Grants 92391 and 90877 from the NOAA Department of Commerce. The authors want to thank the Agricultural Experiment Station, O.S.U. for its support.

REFERENCES

1. J. Crank, "The mathematics of diffusion", Clarendon Press, Oxford (1979).
2. A. T. Florence, and J. R. Johnson, Transport of anions in aqueous polymer solutions, J. Pharmacol, 27 Suppl. 70 P. (1975).
3. C. A. Kienzle-Stenzer, D. Rodriguez-Sanchez, and C. K. Rha, Flow behavior of cationic biopolymer: Chitosan, Polym. Bull. (Berlin), 13:1 (1985).
4. R. Namikawa, H. Ozazaki, K. Nakanishi, R. Matsuno, and T. Kamikubo, Diffusion of amino acids and saccharides in solutions of dextran and its derivatives, Agric. Biol. Chem., 41:1003 (1977).
5. R. Rymden, and J. Calfors, Diffusion of glucose derivatives in aqueous solutions of hydroxypropylcellulose, Polymer, 23:325 (1982).
6. N. Yoshida, Self-diffusion of small ions in polyelectrolyte solutions, J. Chem. Phys., 69:4867 (1978).
7. J. H. Wang, Theory of the self-diffusion of water in protein solutions. A new method for studying the hydration and shape of protein molecules, J. Am. Chem. Soc., 76:4755 (1954).

SOLUTION PROPERTIES AND MESOPHASE FORMATION OF CHITOSAN

(M. Terbojevich, A. Cosani, M. Scandola and A. Fornasa)

The rigidity of polysaccharide chains is usually ascribed both to steric effects and to local interactions, which result in a decreased number of relative orientations for adjacent residues. Lack of flexibility accounts for a number of properties in this class of compounds, including high intrinsic viscosity, short- and long-range ordering, mesophase- and gel-formation. These features are all relevant also from a biological point of view.

In the present paper, chitosan samples at varying degrees of N-acetylation are examined by intrinsic viscosity measurements at different ionic strengths to evaluate the effects of deacetylation on chain rigidity. Smidsrod's parameter B (1) is directly related to polyelectrolyte chain flexibility in the unperturbed state.

The rigidity of chitosan chains is well comparable to that found in similar anionic polysaccharides (see B values in Table I). The B parameter exhibits a tendency toward decreased values on increasing degree of acetylation. The observed behaviour might be explained in terms of hydrogen bond formation between an acetyl function and the $-CH_2OH$ group belonging to the next residue. As a result extended conformers of the glycosidic linkage are further stabilized.

To compare B with other accepted stiffness parameters (K_θ, b_θ, σ, C_∞, and A_m), a series of chitosans (degree of N-acetylation $\simeq$ 40 %) of various molecular weights were prepared by dissolving chitosan (8 g) in acetic acid ($\simeq$1 % v/v; 800 ml) and heating the solution under reflux. Aliquots of the solution were removed at intervals and the polymer was recovered by precipitation with NaOH, followed by washing with water to neutral pH and lyophilization.

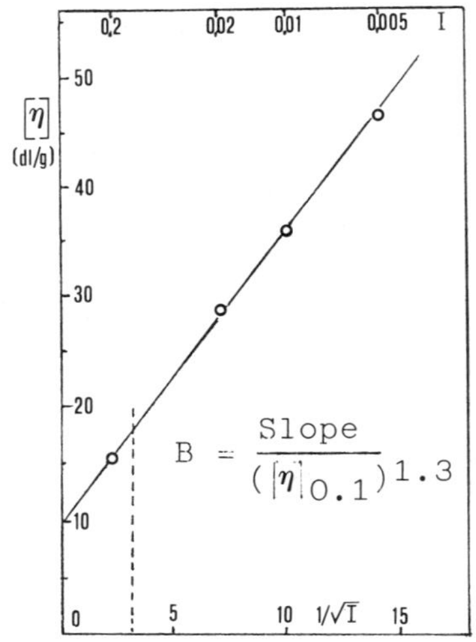

Fig. 1. Plot $[\eta]$ vs. $1/\sqrt{I}$ for chitosan-42

Table 1. B values for chitosan samples and selected polyelectrolytes.

Sample	Origin	% of acetylation(a)	B
Chitosan-52	Alkaline deacetylation of Chitin (b)	52.2	0.043
Chitosan-42	Rybex Krill	42.1	0.061
Chitosan-20	Anic	20.1	0.060
Chitosan 12	Senn Chemical	12.1	0.091
Polyacrylate[1]	---	---	0.23
Carboxymethyl-amylose[1]	---	---	0.20
Carboxymethyl-cellulose[1]	---	---	0.065
Hyaluronate[4]	Fidia S.p.A.	---	0.080
DNA [1]	---	---	0.0055

(a) First derivative ultraviolet spectrophotometry[2]
(b) NaOH 50 %; 60 °C; 75'[3]

The present light scattering and viscosity results on chitosan samples (degree of acetylation ≃ 40 %) obtained by mild acid hydrolysis, do not allow a safe evaluation of the Mark-Houwink parameters. The rigidity parameters in the unperturbed state were also difficult to determine as the second virial coefficient exhibits an anomalous pattern. Further experiments are in progress to shed light on the above results.

Mesophase formation was also investigated (5). Preliminary results suggest the onset of this process for chitosan (degree of acetylation ≃ 42 %) at a polymer concentration 6 - 7 % in water/acetic acid (88/12 v/v) mixtures

Table 2. Characteristics of samples investigated.[a]

Sample	$[\eta]$ dl/g	$\overline{M}_w \times 10^{-3}$ (b) (dn/dc) = 0.18	R_G (Å)(b)	$A_2 \times 10^3$ mol ml g^{-2}
t_o	15.11	2500	2400	2.0
z_1	8.12	1200	1700	0.3
t_1	5.38	680	1330	0.5
t_2	3.71	208	720	0.5
t_3	3.24	170	612	0.5
t_4	2.53	Association occurs		< 0
z_2	2.06	Association occurs		< 0

a) Solvent: 0.1 M CH_3COOH, 0.2 M NaCl. A degree of acetylation of ≃40 % was found for all samples.
b) Estimated error ± 10 %.

with no added salt. Isotropic solutions yield cholesteric-like films on controlled evaporation of the solvent.

Acknowledgement. This work was carried out with the financial support of the Italian National Research, Council, Rome, Progetto Finalizzato Chimica Fine e Secondaria.

REFERENCES

1. O. Smidsrod and A. Haug, Estimation of the relative stiffness of the molecular chain in polyelectrolyte from measurements of viscosity at different ionic strengths, Biopolymers, 10:1213 (1971).
2. R. A. A. Muzzarelli and R. Rocchetti, this Volume.
3. S. Mima, M. Miya, R. Iwamoto, and S. Yoshikawa, Highly deacetylated chitosan and its properties, J. Appl. Polym. Sci., 28:1909 (1983).
4. A. Fornasa, Conformational rigidity of polysaccharide chains, Graduate thesis, University of Padua, Italy (1984).
5. E. Bianchi, A. Ciferri, and G. Conio (unpublished).

Affiliations

D. Rodriguez-Sanchez and C.K. Rha, Department of Nutrition and Food Science, M.I.T., Cambridge, MA 02139, USA.

C.A. Kienzle-Sterzer, Department of Food Science & Technology, Oregon State University, Corvallis, OR 97331, USA.

M. Vincendon, Département de Recherche Fondamentale, Centre Etudes Nucléaires de Grenoble, 85-X, F-38041 Grenoble cédex, France.

M. Terbojevich, A. Cosani, M. Scandola and A. Fornasa, Dipartimento di Chimica Organica, Università di Padova, Via Marzolo 1, I-35100 Padova, Italy.

SYNTHESIS OF OLIGOSACCHARIDES REPRESENTING THE CHITOBIOSYL LINK IN GLYCOPROTEINS

E. KALLIN, S. SVENSSON,
◊G. GRONBERG and ◊O. MANSSON

Department of Carbohydrate Chemistry, Chemical Center
University of Lund, P. O. Box 124, S-221 00 Lund, Sweden
and ◊BioCarb AB, P. O. Box 739, S-220 07 Lund, Sweden

INTRODUCTION

The carbohydrate portion of the link region of N-glycosidically linked oligosaccharide chains in glycoproteins is composed of a chitobiosyl residue which sometimes has a 6-linked α-L-fucopyranosyl unit at the innermost 2-acetamido-2-deoxy-D-glucose residue.

The N-glycosidically linked oligosaccharide chains can be released by hydrazinolysis or by trifluoroacetolysis, and in order to provide models for these oligosaccharides, the oligosaccharides representing the linkage region have been synthesized. The synthetic models have been subjected to spectral and conformational studies.

RESULTS AND DISCUSSION

6-O-α-L-Fucopyranosyl-2-acetamido-2-deoxy-D-glucose was synthesized by the reaction sequence shown in Figure 1.

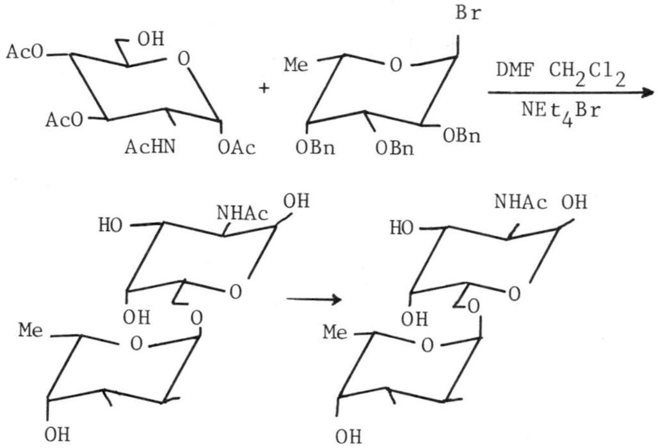

Figure 1. Reaction scheme.

The corresponding N-trifluoroacetyl derivative was obtained by a brief trifluoroacetolysis. The 6-O-α-L-fucopyranosyl-2-acetamido-2-deoxy-D-glucose and 6-O- α -L-fucopyranosyl-2-deoxy-2-trifluoroacetamido-D-glucose were characterized by their ^{1}H-n.m.r. spectra (500 MHz), and, for the latter, also by its ^{19}F-n.m.r. spectrum (CF_3 α : -78.75 ppm, β : -77.70 ppm).

4-O-2-Acetamido-2-deoxy-β-D-glucopyranosyl-2-acetamido-2-deoxy-D-glucose (chitobiose) was converted to 4-O-2-acetamido-2-deoxy-β-D-glucopyranosyl-2-deoxy-2-trifluoracetamido-D-glucose and 4-O-2-deoxy-2-trifluoracetamido- β -D-glucopyranosyl-2-deoxy-2-trifluoracetamido-D-glucose by using two different trifluoroacetolysis conditions as follows:

GNAc-GNAc	TFAA / TFA 100:1, 1 h, 100°C	GNAc-GNAc, 8 % GNAc-GNTF, 70 % GNTF-GNTF, 20 %
	6 h, 100°C	GNTF-GNTF, 100 %

Separation of mixtures was achieved by HPLC on a C 18-column using acetonitrile/water as eluent. A remarkable separation of the different components and their anomers was obtained (Figure 2).

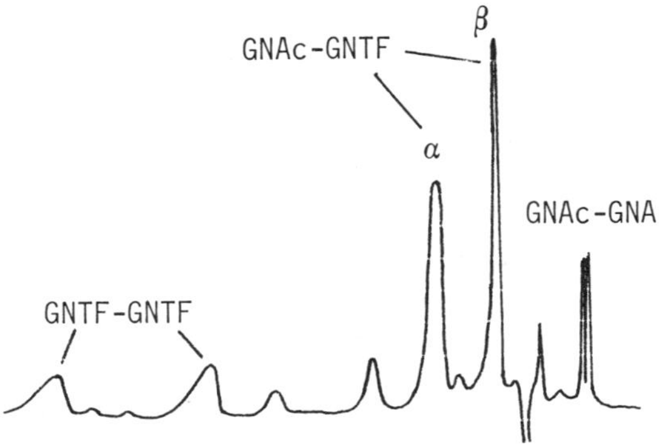

Figure 2. The separation of chitobiosyl derivatives by HPLC.

The different chitobiosyl derivatives were characterized by their ^{1}H- and ^{19}F-n.m.r. spectra (Table I).

4-O-2-Acetamido-2-deoxy- β-D-glucopyranosyl-(6-O-α-L-fucopyranosyl-)-2-acetamido-2-deoxy-D-glucose was synthesized and easily purified by silica gel chromatography and HPLC (C 18-column, water as eluent). This compound was characterized by its ^{1}H- and ^{19}F-n.m.r. spectra ($-CF_3$ α -78.5, β -77.5 ppm).

The conformation of 4-O-2-acetamido-2-deoxy- β-D-glucopyranosyl-(6-O-α -L-fucopyranosyl)-2-acetamido-2-deoxy-D-glucose and the corresponding N-trifluoroacetylated reducing sugar was calculated using the HSEA-method. The results showed indications of more than one allowed conformation, as shown in Figure 3.

Preliminary studies show that mass spectral analysis of the N-trifluoroacetylated derivatives by either the FAB-technique or by E.I. on these derivatives gives information on sequence and linkage types.

Table I. Characterization of chitobiosyl derivatives by n.m.r.

H^1NMR ppm*

Structure	HI1 α	HI1 β	HII1 α	HII1 β	HI-CH_3α	HII-CH_3β
II I GNAc $^{1}\underset{\beta}{-}^{4}$ GNAc	5.188 J 3.1,d	4.695 J 8.2,d	4.591 J 8.2,d	4.582 J 8.2,d	2.037 s	2.068 s
GNAc $^{1}\underset{\beta}{-}^{4}$ GNTF	5.269 J 3.0,d	4.804 J~8.1,m (virt.)	4.561 J 8.6,d	4.593 J 8.2, d		2.072 s
GNTF $^{1}\underset{\beta}{-}^{4}$ GNTF	5.263 J 2.6,d	4.796 J~8.2,m (virt.)	4.723 J 8.6,d	4.713 J 8.6,d		

$F^{19}NMR$ ppm**

Structure	I-CF_3α	I-CF_3 β	II-CF_3α	II-CF_3β
II I GNAc $^{1}\underset{\beta}{-}^{4}$ GNTF	-77.624 s	-78.667 s		
GNTF $^{1}\underset{\beta}{-}^{4}$ GNTF	-77.640	-78.665	-78.463 s	-78.517 s

* Shift rel. intern. Acetone (=2.225 ppm (20°C)).
** Shift rel. intern. Trifluoroacetic acid (=-78.500 ppm (20°C)).

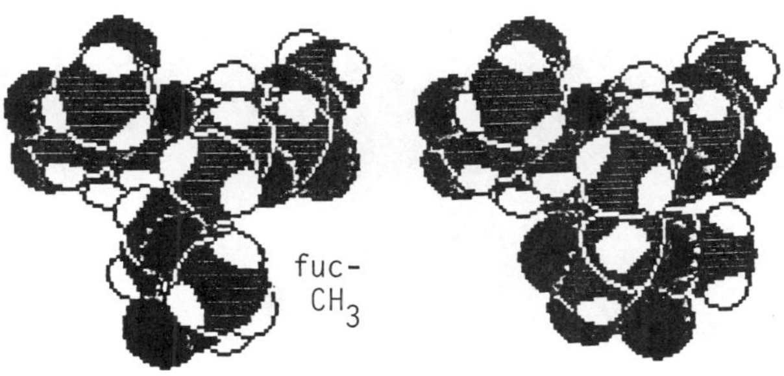

Figure 3. Allowed conformations.

STABLE ISOTOPIC STUDIES ON CHITIN, MEASUREMENTS ON CHITIN/CHITOSAN ISOLATES AND D-GLUCOSAMINE HYDROCHLORIDE FROM CHITIN

A. SCHIMMELMANN and M. J. DeNIRO

Department of Earth and Space Science, University of California
Los Angeles, CA 90024, USA

ABSTRACT

Measurements of the stable isotope ratios $^{13}C/^{12}C$ and $^{15}N/^{14}N$ in chitin and chitosan isolates show large variations which reflect intrinsic compositional and isotopic heterogeneities as well as differences caused by methods of preparation. Analysis of D-glucosamine hydrochloride from chitin hydrolysates eliminates these problems.

INTRODUCTION

Stable isotope ratios of carbon and nitrogen in different types of animals have been used in many ecological, environmental and physiological studies. In general, the substrate used in these studies was either total biomass or a selected tissue such as muscle. Because different biochemical components do not have the same isotopic composition[1,2] this approach suffers the inherent disadvantage that differences among the isotope ratios of different organisms may reflect heterogeneity in biochemical composition. One way to circumvent this problem involves the analysis of a single biochemical component from all organisms analyzed in a study. The use of cellulose rather than total floral biomass brought a substantial improvement in analytical precision and has allowed fruitful applications of isotope analysis in climatology[3]. The widespread taxonomic occurrence of chitin together with the availability of paleontological and archaeological chitin make it a valuable substrate for stable isotope studies. Amino sugar biomass has the advantage of being chemically well defined compared to average biomass or an arbitrarily selected tissue, thus eliminating potential isotopic variabilities due to differing biochemical compositions. In this paper, we report on the isotopic compositions of chitin and chitosan isolates and of D-glucosamine hydrochloride (GlcN.HCl) prepared by a variety of methods. Although the findings and conclusions of this study are based solely upon carbon and nitrogen stable isotope ratios, they have implications for hydrogen and oxygen stable isotope ratios in chitins and chitosans[4,5].

RESULTS AND DISCUSSION

Elemental and isotopic compositions of carbon and nitrogen in the various substrates were determined as described elsewhere[6]. The results are expressed in δ-notation, where

$$\delta^m X = \left[\frac{(^m X/^n X)_{SAMPLE}}{(^m X/^n X)_{STANDARD}} - 1 \right] \cdot 1000 \text{ per mil}$$

$^m X = {}^{13}C$ or ^{15}N and $^n X = {}^{12}C$ or ^{14}N.

The standards are the Peedee belemnite (PDB) carbonate for $\delta^{13}C$ values and atmospheric nitrogen (AIR) for $\delta^{15}N$ values. The precisions for the measurements, given as standard deviations of the means calculated for 20 replicate analysis are ±0.1 per mil for $\delta^{13}C_{PDB}$ and $\delta^{15}N_{AIR}$ values.

Crude chitin powder (Sigma) was used throughout this study. In order to interpret the stable isotope results and the elemental N/C ratios in the various isolated compounds, we needed an information base about the quantitative and qualitative isotopic participation of the various organo-chemical components of crude chitin, which may be regarded as separate isotopic pools. Experiments were designed to measure separately not only the amino-sugar fraction apart from all contaminants, but also to determine the carbon isotopic value of GlcN.HCl separately from the acetyl group. First, crude chitin powder was processed in quantities of 4 to 7 mg for the determination of elemental and stable isotope ratios. Second, crude chitin powder (0.4 g) was hydrolyzed in 10 ml 7N sulfuric acid for 6 hours at 100°C. The chitin acetyl group was liberated as acetic acid, which was then selectively recovered from the hydrolysate by steam distillation. After titration with 0.1N sodium hydroxide solution and freeze-drying, sodium acetate served as substrate for the determination of the $\delta^{13}C$ value in the chitin acetyl group. The steam distillation process, the recovery of the acetic acid in the form of acetic acetate, and the measurement of the $\delta^{13}C$ value of the sodium acetate were shown to be without any significant isotope fractionation effect by performing control experiments on pure acetic acid (Baker). Third, pure GlcN•HCl was prepared from crude chitin using the following procedure: Chitin was soaked in 2N HCl for 24 hours at room temperature, followed by rinsing with distilled water to constant pH. Deproteinization was performed in aqueous 1N NaOH for 1 hour at 100°C with subsequent rinsing in distilled water to constant pH. 100 mg aliquots of freeze-dried deproteinated chitin isolate were first dissolved in 5.6 ml concentrated HCl at 50°C, followed by dilution with 4.4 ml distilled water to yield a 7N hydrochloric acid. The hydrolysis was carried out at 100°C during 4 hours after Fischer and Nebel[7] who observed optimal GlcN•HCl yields under these conditions. The resulting hydrolysate was freeze-dried, then dissolved in 1 ml distilled water and passed through a 1.2 μm Metricel filter membrane. The filtrate was injected onto an ion-exchange column (effective length 17 cm, inner diameter 7.5 mm; Dowex-50 W-X8, minus 400 mesh). Fractions eluted with 1N HCl at room temperature were collected. Ca. 5 μl of each fraction were placed on a TLC plate (Eastman chromatogram sheet #6060, silica gel) sprayed beforehand with 1% methanolic ninhydrin solution. Fractions containing compounds with amino-groups produced red coloration after developing the spotting chromatogram at 150°C for 5 minutes. GlcN•HCl was eluted after neutral compounds, but prior to all amino acids as determined by HPLC analyses of the individual fractions. The fractions containing GlcN•HCl were pooled and freeze-dried. The residual crystalline GlcN•HCl was redissolved in 1 ml distilled water and aliquots of the solution were taken for determination of elemental and stable isotope ratios.

The elemental and isotopic compositions of the fractions prepared as described above are given in Table 1. The isotopic and elemental ratios for the non-amino-sugar contamination are based upon isotope mass balance calculations assuming a 90% degree of acetylation in the original crude chitin.

The overall concentration of ca. 25 weight % contaminants had been determined through GlcN•HCl yields, taking into account a ca. 5% loss of GlcN•HCl during hydrolysis[7]. From the results in Table 1 it is evident that the stable isotope distributions of carbon and nitrogen in crude chitin are highly heterogeneous. Similar isotopic heterogeneities were observed between proteinaceous biomass and GlcN•HCl from arthropods[2]. Consequently, non-amino-sugar contamination in chitin and chitosan isolates will cause overall isotopic shifts that show variability due to the character and amount of the contaminants.

Crude chitin powder was used in attempts to prepare chitin isolates with a minimum of non-amino-sugar contamination. Since the acetyl group has a $\delta^{13}C$ value significantly different from GlcN•HCl, the methods employed must show little or no deacetylation. The results for the unaltered crude chitin are given under symbol ① in Figure 1 and should be compared with the results of the various treatment methods. The theoretical data for the completely decontaminated and fully acetylated "ideal" chitin are indicated by dotted lines. Acid treatments (② , ③ , ⑦) and dissolution in N, N-dimethylacetamide/LiCl[8] (⑥) were not found to insure sufficient decontamination and in many cases yielded poor reproducibility of data. Treatment in sodium hydroxide solution (④) gave better results, but a comparison with the differing results of ⑤ casts doubts on the reliability. Also, sclerotized chitinous structures, especially insect chitins, cannot be sufficiently decontaminated by short-term treatment in NaOH, rendering this approach impractical for many substrates.

Long-term heterogeneous deproteinization/deacetylation of chitin in suspensions were performed with amounts of 1 g crude chitin powder in 300 ml 1N aqueous NaOH solution kept for 0.5, 1, 3, 12, 24, 48, and 120 hours at 100°C. After cooling to room temperature, the chitin isolates were filtered off and rinsed with distilled water to constant pH. The chitins were freeze-dried for one day and processed for the determination of elemental and isotopic ratios. We observed a large variability in $\delta^{13}C$ and $\delta^{15}N$ values which did not permit any conclusions as to the absence or presence of trends through time. The statistically significant trend of the elemental N/C ratios through time gave evidence for progressive deacetylation. A logarithmic regression analysis yielded $N/C = 0.13 + 0.0025 \cdot \ln$ time (hours) with a coefficient of correlation $R^2 = 0.74$ and significance at the 0.99 level (two-tailed Student's t-test). Similar relationships were reported elsewhere[9,10]. Based on these considerations, this approach was not found to yield substrates of reliable reproducibility for stable isotope studies on chitin or chitosan isolates.

Homogeneous deproteinization/deacetylation of chitin in alkaline solution was performed by dissolving crude chitin powder in aqueous sodium hydroxide solution[11] and keeping it for various times at 25°C, before precipitating the chitin/chitosan products in methanol and washing them free of alkali with methanol. The freeze-dried products were processed for the determination of elemental and stable isotope ratios. Results are given in Figure 2. The elemental N/C ratio, the $\delta^{13}C$ and the $\delta^{15}N$ values show distinct dependencies on the length of the reaction time. The differences between the crude chitin and the chitin isolate reprecipitated from alkaline solutions after a reaction time of only 15 minutes indicate rapid deproteinization, and the elimination of nitrogen-rich, ^{15}N-enriched and ^{13}C-depleted contaminating material, thus yielding a chitin isolate with the opposite isotopic properties. An increase in the degree of deacetylation with increasing length of reaction time is consistent with the observation of a trend in the $\delta^{13}C$ values through time that is in agreement with the increasing loss of ^{13}C-depleted acetyl carbon. The prolonged deacetylation of chitin results in a chitosan which shows a sufficient degree of crystallinity to form a gel rather than a solution[11]. This event, which occurred between

347 and 600 hours at room temperature, has a profound impact on all parameters measured in the resulting chitosan isolate. The decreasing elemental N/C ratio, the drastic increase of the $\delta^{15}N$ value, and the slight decrease in $\delta^{13}C$ value all suggest the adsorption/incorporation of non-amino-sugar compounds such as amino acids and possibly some acetate into the precipitated, partly crystalline chitosan. The various chemical and physical-chemical processes occurring in the course of homogeneous deamination/deacetylation, with their dependencies on time, temperature and possibly specific substrate properties, render this method impractical for preparing stable isotope substrates.

The highly variable results obtained from chitin/chitosan isolates led to the conclusion that GlcN•HCl is the best substrate for stable isotope ratio determinations in chitin. We report here on possible isotope fractionation effects that occur when improper methods are applied. Intra-peak stable isotope fractionation effects of several per mil have been observed for amino acid eluted from ion-exchange resin columns in case of carbon[12] and nitrogen stable isotopes[13]. To evaluate the potential fractionation effects within GlcN•HCl peaks, 104 mg of pure GlcN•HCl (Eastman Chemical Co.) were passed through a Dowex-50 column as described above. The eluted peak was separated into five consecutive fractions of 1 ml. The amounts of GlcN •HCl recovered and the stable isotope ratios for each fraction were measured (Table 2). An isotope mass balance calculation for the combined GlcN•HCl of all five fractions yielded the expected values for $\delta^{13}C$ and $\delta^{15}N$ in the bulk original GlcN•HCl. The data demonstrate isotopic heterogeneities of 2.9 per mil for carbon and 6.4 per mil for nitrogen within the peak. Thus, it is mandatory to sample entire peaks in order to avoid severe isotopic artefacts.

Isotope fractionation effects of more than 2 per mil for nitrogen and 0.5 per mil for carbon were observed when crystals of GlcN•HCl were obtained by either concentrating the aqueous solutions until partial precipitation occurred, or by mixing the aqueous solution with organic precipitating agents such as ethanol or acetone. These fractionations are influenced by temperature, time, pH, yield, and the α-β-anomer-mutarotation, many of which are inseparably interwoven in their effects. Consequently, for stable isotope assays, GlcN•HCl should never be handled as a solid, but always in isotopically homogeneous aqueous solution in order to maintain overall isotopic homogeneity of the samples.

The chitin hydrolysis, vacuum evaporation, ion-exchange chromatography, etc. procedures must not show any isotope fractionation effects while at the same time producing close to 100% yields of GlcN•HCl. To test the fulfillment of the requirements, pure, isotopically pre-characterized GlcN•HCl (Eastman; $\delta^{13}C = -17.6°/_{oo}$, $\delta^{15}N = \pm 0.0°/_{oo}$) was mixed with 4 to 11 weight % of isotopically significantly different pure aspartic acid hydrochloride (Baker; $\delta^{13}C = -25.1°/_{oo}$, $\delta^{15}N = +4.6°/_{oo}$), ammonium sulfate (U.S. National Bureau of Standards, #2; $\delta^{15}N = +19.5°/_{oo}$), acetic acid (Baker; $\delta^{13}C = -16.4°/_{oo}$), glucose (Baker; $\delta^{13}C = -9.9°/_{oo}$) and NaCl (Mallinckrodt). Of all amino acids, aspartic acid has its retention time closest to the one of GlcN•HCl on Dowex-50 columns under the conditions used here and is therefore the most likely contaminant in case of incomplete peak separation. Additionally, 35 mg aliquots of GlcN•HCl (Eastman) were subjected (with and without contaminants) to the hydrolytic procedure in 7N HCl at 100°C for 4 hours followed by the regular purification procedure. In all cases, the GlcN•HCl fractions were separated from the contaminants and recovered according to the procedure outlined above. The nitrogen and carbon stable isotope ratios were determined and compared to the values of the original GlcN•HCl pool. The GlcN•HCl recovery yield averaged 97.2 ± 1.6% (n=3) for experiments including hydrolytic conditions and 99.6 ± 1.5% (n=8) for the ion-exchange chromatographical purification and the freeze-drying only. None of the experiments showed any significant stable isotope fractionation effect. The

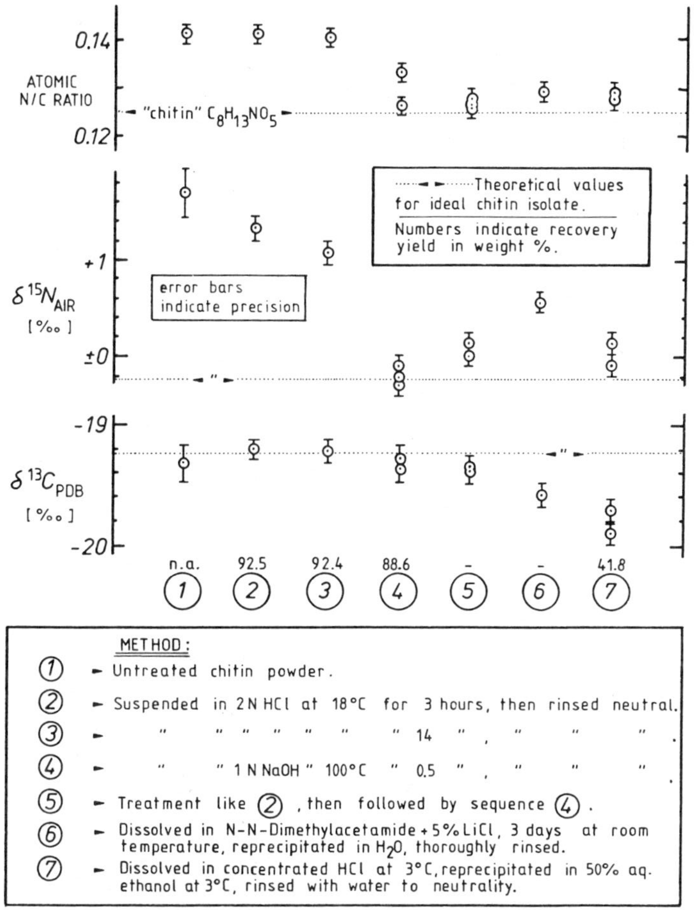

Fig. 1. Atomic N/C ratio, $\delta^{13}C_{PDB}$ and $\delta^{15}N_{AIR}$ in chitin isolates from commercial chitin powder (Sigma Chemical Co.).

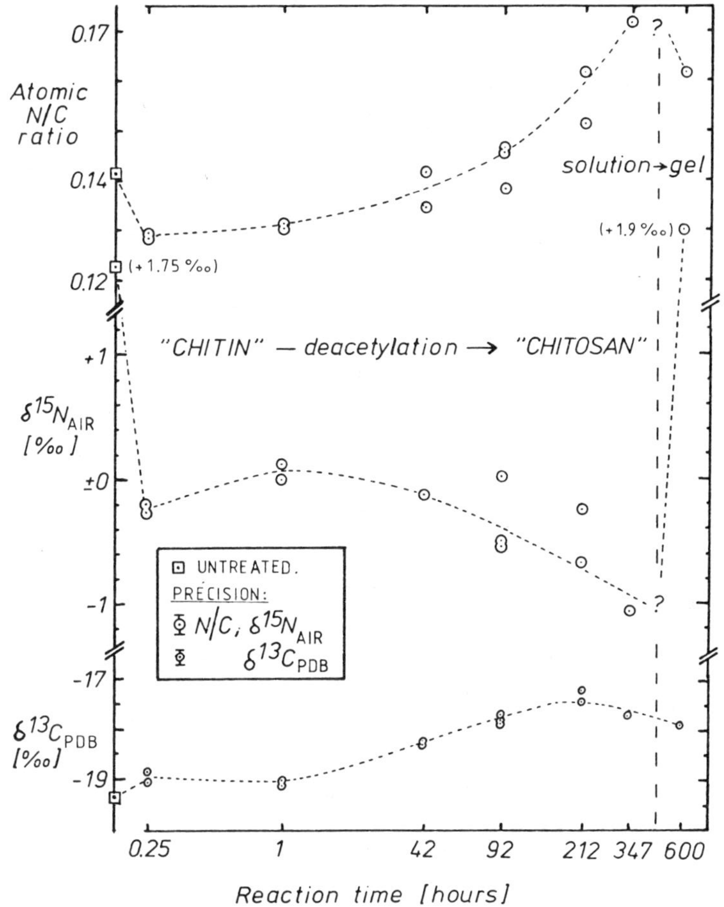

Fig. 2. Homogeneous deacetylation/deproteinization of chitin in alkaline solution at ≈ 25°C.

TABLE 1. Elemental and isotopic compositions of chemical fractions in crude chitin.

Substrate	$\delta^{13}C_{PDB}$ (°/∘∘)	$\delta^{15}N_{AIR}$ (°/∘∘)	Elemental N/C Ratio	N
Crude chitin	-19.3±0.3	-1.7±0.3	0.141±0.003	2
Acetyl group	-23.6±0.1	n.a.	n.a.	2
D-glucosamine hydrochloride	-17.8±0.1	+0.2±0.1	0.167±0.003	3
non-amino-sugar contamination	-19.8±0.3	+7.2±0.1	0.175±0.005	-
calculated: "ideal chitin", $C_8H_{13}NO_5$	∿ -19.3	∿ +0.2	0.125	-

TABLE 2. Peak fractions of GlcN•HCl eluted from Dowex-50 ion exchange resin.

Peak Fraction Number	Recovered GlcN•HCl (mg)	$\delta^{13}C_{PDB}$ (°/∘∘)	$\delta^{15}N_{AIR}$ (°/∘∘)
1	4.0	-16.1	- 4.2
2	27.6	-17.3	- 1.7
3	39.4	-17.8	± 0.0
4	28.1	-18.4	+ 1.0
5	3.0	-19.0	+ 2.2

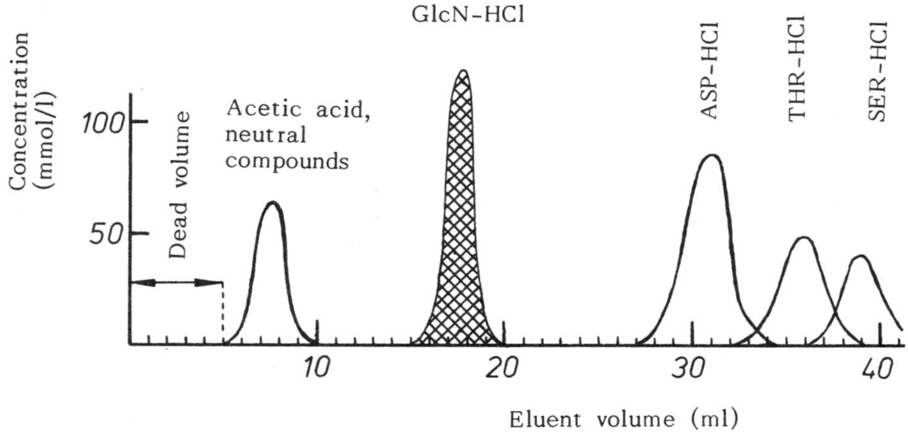

Fig. 3. Elution sequence from Dowex-50 in 1N HCl at ∿ 25°C.

observed average isotopic deviations from the original GlcN•HCl are 0.0 ± 0.0 per mil (n=11) for $\delta^{13}C$ values and 0.0 ± 0.0 per mil (n=11) for $\delta^{15}N$ values. The largest differences observed between original and recovered GlcN•HCl for both $\delta^{13}C$ and $\delta^{15}N$ values were 0.1 per mil.

CONCLUSIONS

Macromolecular chitin and chitosan show significant intrinsic compositional and isotopic heterogeneities. From an isotopic viewpoint, it appears to be difficult to control the chemical kinetics of the deproteinization and deacetylation processes during chitin processing and to achieve reproducibility and comparability of data. These problems may be exacerbated when different chitin sources are concerned. The picture becomes even more complicated when adsorption effects are considered. In particular, the net difference in $\delta^{13}C$ values between the acetyl carbon and the glucosamine carbon casts serious doubts on the usefulness of chitin/chitosan isolates as potential substrates for $\delta^{13}C$ determinations because of the difficulty of avoiding different degrees of deacetylation. Additionally, the large and variable adsorptive capabilities of chitin/chitosan isolates in combination with large nitrogen isotopic differences between the glucosamine units and bulk non-amino-sugar contaminants make a high-precision $\delta^{15}N$-determination impossible.

The use of GlcN•HCl, prepared by methods described here, circumvents all problems caused by deacetylation, deproteinization/decontamination, and adsorption. If care is taken to avoid potential isotope fractionation effects due to peak-splitting and partial GlcN•HCl crystallization, this approach offers a highly sensitive and reliable analytical tool for the determination of carbon and nitrogen stable isotope ratios.

ACKNOWLEDGMENTS

This work was supported by the Lerner-Gray Fund for Marine Research of the American Museum of Natural History, Geological Society of America grant 3216-83, and National Science Foundation grant ATM 79-24591. We thank Ed Hare for helpful discussions and invaluable assistance in developing the ion-exchange chromatographical method and Dave Winter for performing the stable isotope ratio mass spectrometry.

REFERENCES

1. E. M. Galimov, $^{13}C/^{12}C$ in kerogen. in: Kerogen. Insoluble organic matter from sedimentary rocks. B. Durand, ed., Editions Technip, Paris (1980).
2. A. Schimmelmann and M. J. DeNiro, Stable isotopic studies on chitin II. The $^{13}C/^{12}C$ and $^{15}N/^{14}N$ ratios in arthropod chitin. *Contrib. Marine Sci.*, submitted for publication.
3. S. Epstein, C. J. Yapp, and J. Hall. The determination of the D/H ratio of non-exchangeable hydrogen in cellulose extracted from aquatic and land plants. *Earth Planet. Sci. Lett.* 30:241-251 (1976).
4. A. Schimmelmann and M. J. DeNiro, Stable isotopic studies on chitin III. The $^{18}O/^{16}O$ and D/H ratios in arthropod chitin. *Geochim. Cosmochim. Acta*, submitted for publication.
5. A. Schimmelmann , M. J. DeNiro, M. Poulicek, M.-F. Voss-Foucart, and G. Goffinet, Stable isotope ratios and micromorphology of arthropod chitin as paleoenvironmental and paleoclimatic indicators. *Biochem. Syst. Ecol.*, submitted for publication.

6. D. W. Northfelt, M. J. DeNiro, and S. Epstein, Hydrogen and carbon isotopic ratios of the cellulose nitrate and saponifiable lipid fractions prepared from annual growth rings of a California redwood. Geochim. Cosmochim. Acta 45:1895-1898 (1981).
7. F. G. Fischer and H. J. Nebel, Nachweis und Bestimmung von Glucosamin und Galaktosamin auf Papierchromatogrammen. Hoppe-Seyler's Z. Physiol. Chem. (1955).
8. F. A. Rutherford and P. R. Austin, Marine chitin properties and solvents. in: Proceedings First International Conference on Chitin/Chitosan. R. A. A. Muzzarelli and E. R. Pariser, eds., MIT Sea Grant report, MIT-SG-78-7, Cambridge, 182-192.
9. K. Kurita, T. Sannan, and Y. Iwakura, Studies on chitin, 4, Makromol. Chem. 178:3197-3202 (1977).
10. R. A. A. Muzzarelli, F. Tanfani, G. Scarpini, and G. Laterza, The degree of acetylation of chitins by gas chromatography and infrared spectroscopy. J. Biochem. Biophys. Methods 2:299-306.
11. T. Sannan, K. Kurita, and Y. Iwakura, Studies on chitin, 1, Makromol Chem. 176:1191-1195 (1975).
12. P. H. Abelson and T. C. Hoering, Carbon isotope fractionation in formation of amino acids by photosynthetic organisms. Proc. Natl. Acad. Sci. U.S. 47:623-632.
13. S. A. Macko, M. L. F. Estep, P. E. Hare, and T. C. Hoering, Stable nitrogen and carbon isotopic composition of individual amino acids isolated from cultured microorganisms. Ann. Rep. Dir. Geophys. Lab., Carnegie Inst., Washington, D.C., 404-410 (1982).
14. P. A. Levene, Hexosamines and mucoproteins. Longmans, Green and Co., London (1925).
15. M. Stacey and J. M. Webber, 2-Amino-2-deoxy-α-D-glucose (α-D-Glucosamine) from crustacean shell. in: Methods in Carbohydrate Chemistry, J. Whistler, ed., Vol. I, Academic Press, 228-230 (1962).

CHARACTERIZATION OF CHITIN DERIVATIVES BY PYROLYSIS-GAS CHROMATOGRAPHY

D. H. DAVIES, E. R. HAYES* and G. S. LAL◊

Chemistry Dept., Saint Mary's University
Halifax, N.S., B3H 3C3, Canada

*Chemistry Dept., Acadia University
Wolfville, N.S., BOP 1X0, Canada

◊Chemistry Dept., Virginia Polytechnic and State University
Blacksburg, VA 24061, USA

INTRODUCTION

The development of chitin chemistry has been hampered by the lack of convenient analytical techniques for the chitin family of polymers. Desirable features of appropriate techniques are sensitivity to structural variations, simplicity, speed, and low cost. The characterization of chitin and its derivatives by analytical pyrolysis has been investigated as the technique which appears to satisfy these requirements.

The analysis of chitin and chitosan by pyrolytic techniques has already been investigated. Curie point pyrolysis mass spectrometry was employed by Meuzelaar et al. (1) and by Weijam (2) to monitor the quality of chitin extracted from fungi. Probe pyrolysis mass spectrometry was used by Davies and Hayes (3) in a study of the effect of temperature on the pyrolysis of chitin and chitosan. Mattai and Hayes (4) showed by a similar technique that there is a direct correlation between the amine content and the ratio of intensities of certain peaks in the pyrolysis mass spectra of chitosans. Later Lal and Hayes (5) showed by pyrolysis gas chromatography that the amine content of chitosan can be correlated with the ratio of the areas of pyrolysis peaks obtained from glucosamine and N-acetyl-D-glucosamine units. Schlotzhauer et al (6) identified products formed by furnace pyrolysis using pyrolysis gas chromatography/mass spectrometry. In a recent publication Kaaden et al. have used chitin as a model polymer to investigate a number of analytical pyrolysis techniques (7).

This paper describes a rapid and inexpensive pyrolytic method that can distinguish between chitin, chitosan, N-acetyl-D-glucosamine and the hydrochloride salts of glucosamine and chitosan. The technique employes a pulse pyrolyzer interfaced with a gas chromatograph containing a packed column. Major products formed by the pyrolysis of the natural compounds have been identified by mass spectrometry. The pulse pyrolyzer heats the sample very rapidly to any temperature up to 1000°C. The rapid heating and the instantaneous removal of the primary products from the hot zone minimizes secondary reactions and forms pyrolysates that are reproducible and characteristic of the sample. Pyrolytic techniques have been reviewed recently by Lieberman (8).

EXPERIMENTAL

Preparation of materials for analysis

Commercial chitin was obtained from the Kypro Company, Seattle, WA, USA. Chitosan of varying degrees of deacetylation (amine content in the range 3.74 to 7.78 %) were prepared by procedures previously developed (9, 10). N-acetyl-D-glucosamine was purchased from Sigma Chemical CO., St Louis, MO, USA as an Analar-grade product.

Pyrolysis gas chromatography mass spectrometry

All sample tubes were cleaned by heating in a propane-air flame before use. A sample of approximately 0.4 mg was weighed in a thin quartz tube containing a quartz wool plug, and inserted into the interface of the gas chromatograph and purged with helium carrier gas. After the interface had obtained the required temperature, the sample was pyrolyzed. The pyrolysates were swept by the carrier gas into the column of the gas chromatograph. The compounds that emerged from the column were passed through the jet separatory system and the mass spectra of the major chromatographic peaks were obtained.

Pyrolyzer

Pyrolyses were performed in a Pyroprobe 100 pyrolyzer equiped with a coil probe. The samples were brought from ambient to 750°C with the ramp in the off position and the interface at 200°C for the pyrolysis of chitin and chitosan and at 180°C for the pyrolysis N-acetyl-D-glucosamine (thermal decomposition was observed at 200°C). The temperature was maintained at 750°C for 10 sec.

Gas chromatograph

The gas chromatographic separations were performed on a Hewlett Packard Model 5880A instrument equipped with a flame ionization detector and a 3.658 m x 3.175 mm i.d. stainless steel column packed with 10 % SP-1000 on 50/100 Chromosorb W (AW-DMCS). The flow rate of the helium carrier gas was 30 ml/min. The column temperature was held at 75°C for 5 min after injection and then increased to 200°C at a rate of 10°C/min and was maintained at 200°C for 10 minutes.

Mass spectrometer

To identify the compounds in pyrolysate, the gaseous products eluted from the column installed in a Finnegan mod. 9610 gas chromatograph were introduced into the jet separator of a Finnegan model 4000 quadropole mass spectrometer. The injection port and jet separator temperatures were respectively 200°C and 250°C. Analyses were performed with an anode current of 0.35 A and an ionization energy of 70eV. The electron multiplier voltage was maintained at 1700 V.

RESULTS AND DISCUSSION

Typical pyrograms for chitin, chitosan, N-acetyl-D-glucosamine, glucosamine hydrochloride and chitosan hydrochloride are reproduced in Figure 1 and 2. Each pyrogram provides a distinctive fingerprint for the saccharide. When the pyrolyses are conducted under the same conditions, the pyrograms are very reproducible with retention times within 0.04 minutes.

The programs of chitin and its predominant monomer, N-acetyl-D-glu-

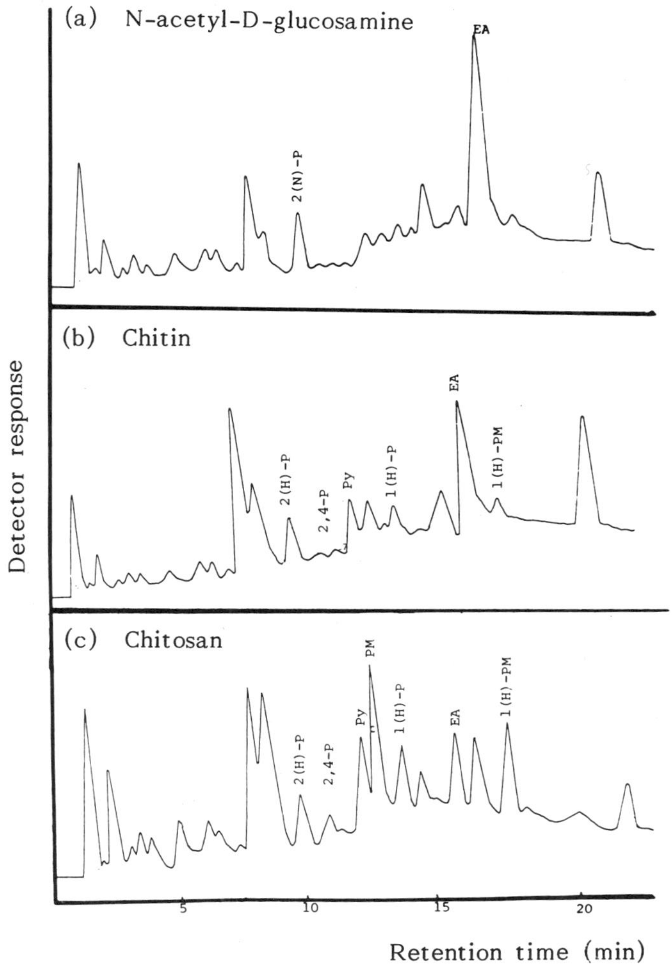

Fig. 1. Typical pyrograms

cosamine, are very similar. Some of the peaks characteristic of N-acetyl--D-glucosamine are also present in the pyrogram of chitosan. However, the pyrogram of chitosan contains a number of major peaks which are not present in that of N-acetyl-D-glucosamine but which are observed as minor peaks in the pyrogram of chitin, presumably because chitin contains a small proportion of free amine.

The gross similarity of the products obtained from chitin and its predominant monomer unit, N-acetyl-D-glucosamine, suggests that extensive decomposition of the polymer chains occurs during the pyrolysis of chitin. This evidence therefore supports the pyrolysis mechanism proposed by Fu and Shafizadeh (11) for 1,4-linked polysaccharides and by Hayes and Davies (3) for chitin. According to this scheme, the β-glycosidic bonds are cleaved to produce monomer-like units which then undergo further degradation. Therefore it would be expected that the pyrolysate of chitin would contain a larger proportion of products derived from N-acetyl-D-glucosamine than from-glucosamine units. This proportion should be reversed for chitosan. This assumption was confirmed when it was shown that there is a direct correlation between the amine content and the ratio of the areas of the peaks derived from glucosamine and N-acetyl-D-glucosamine units (5).

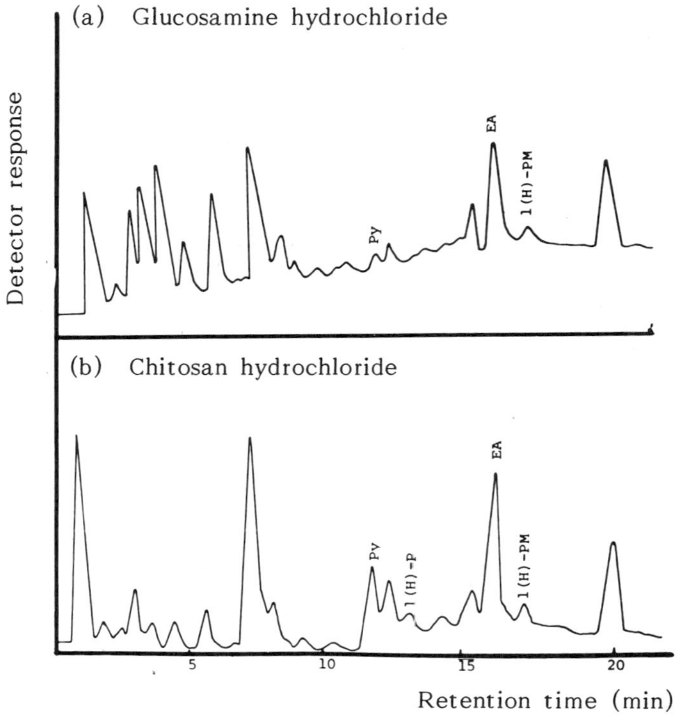

Fig. 2. Typical pyrograms of saccharide salts.

It was not possible to obtain a pyrogram for glucosamine because of its instability. However, glucosamine hydrochloride purchased from Sigma and chitosan hydrochloride prepared in the laboratory were pyrolyzed. The pyrograms of these salts (Fig. 2) are distinctly different from that of chitosan as in the pyrogram of the salt formed between chitosan and nitric acid (not shown). These observations and studies of the pyrolysis of other polysaccharides that contain sulphate and carboxylate groups confirm that acid groups cause significant changes in the thermal decomposition pathways.

The major peaks in the pyrograms of chitin, chitosan, and N-acetyl-D-glucosamine were identified by forwarding the gaseous effluent from the gas chromatograph column via a separator to the ionization chamber (electron impact) of a quadrupole mass spectrometer. The spectra of those peaks whose area was more than 10 % of that of the largest peak in the reinforced ion current were carefully evaluated. Those compounds which could be identified with reasonable certainty are listed in Table I. Identification was based on a number of factors : computer comparison with the National Bureau of Standards library of spectra; Fit, Purity and R-Fit values; comparison of the spectra with the spectra of known compounds; chromatographic consideration of polarity and boiling point with retention times; recurrence of the same compound in different pyrolysates; correlation with compounds reported by other investigators; and, where possible, by comparing retention times with those of known compounds.

Table I lists slightly less than half of the compounds detected by the mass spectrometer. The others have been omitted because they did not satisfy the specified criteria. It is also possible that the observed spectra were not the spectra of the primary pyrolysis products but secondary products formed in the ion source when the pyrolysates were bombarded by the electron beam. It is interesting to note that the mass spectrometer detected several important nitrogen containing compounds that eluted after most chromatographic runs had been terminated. The peaks of these polar compounds on the pyrogram were so broad that they were considered to be unimportant. However, acetamide was common in the pyrolysis of all three saccharides; 2-pyridinamine-N-methyl was detected in the pyrolysates of chitin and chitosan; and 1(H)-pyrrole, 2,5-dimethyl was found in the pyrolysate of chitosan.

Table I. Identity of some products from the pyrolysis of N-acetylglucosamine (I), chitin (II), and chitosan (III).

Retention Time (min)	Compound	I	II	III
1.98	Furon (F)	X		
4.02	Furan, 2,5-dimethyl (FDM)	X		
9.75	2(H)-pyrrol-2-one, 1,5-dihydro-1-methyl (2(H)-P)	X	X	X
10.88	2,4-pentadienonitrile (2,4-P)		X	X
12.12	Pyrimidine (PY)	X	X	X
12.51	Pyrazine, methyl (PYM)			X
13.69	1 (H)-pyrrole (1(H)-P)		X	X
16.45	Ethanoic acid (EA)	X	X	X
17.75	1 (H)-pyrrole, 1-methyl (1(H)-PM)		X	X
24.4	2-pyridinamine, N-methyl		X	X
25.2	1(H)-pyrrole, 2,5-dimethyl			X
30.9	Acetamide	X	X	X

The main objective of this study was achieved when it was demonstrated that chitin and its derivatives can be characterized by a simple and inexpensive analytical pyrolysis apparatus. The system is simple because all of the pyrolysate from the pulse pyrolyzer is transferred to the packed column. With capillary columns a complicated split gas flow injection system is required. Packed columns are less expensive than capillary columns, often have a longer life, and give the experimenter a wide range of stationary phase polarities from which to choose. The method is rapid, reproducible, and can be used for either qualitative or quantitative analysis. Any chromatograph can be converted for pyrolysis work by interfacing it with a suitable pyrolyzer.

While a gas chromatograph with a packed column is satisfactory for routine analysis, a capillary column will provide more structural detail if such information is needed, and will be simpler to interface with a mass spectrometer. The results from the analysis of chitin and some of its derivatives by pyrolysis capillary column gas chromatography will be reported in the near future.

REFERENCES

1. H. L. C. Meuzelaar, P. G. Kistemaker and P. G. Posthumus, Recent advances in pyrolysis mass spectrometry of complex biological materials, Biomed. Mass Spectrum. 1 (1974).
2. A. C. M. Weijman, The application of curie point pyrolysis mass spectrometry in fungal taxonomy, in : "Analytical Pyrolysis", C. A. Jones and C. A. Cramers, eds., Elsevier, Amsterdam (1972).

3. E. R. Hayes and D. H. Davies, Characterization of chitosan, II : The determination of the degree of acetylation of chitosan and chitin, in: "Proceedings of the First International Conference on Chitin/Chitosan", R. A. A. Muzzarelli and E. R. Pariser, eds., MIT Sea Grant Program Cambridge (1978).
4. J. Mattai and E. R. Hayes, Characterization of chitin by pyrolysis mass spectrometry, J. Anal. and Appl. Pyrolysis, 3:327 (1982).
5. G. S. Lal and E. R. Hayes, Determination of the amine content of chitosan by pyrolysis-gas chromatography, J. Anal. and Appl. Pyrolysis, 6:183 (1984).
6. W. S. Schlotzhauer, O. T. Chontyrk and P. R. Austin, Pyrolysis of Chitin. A potential tobacco extender, J. Agric. Food Chem., 24:177 (1976).
7. A. van der Kaaden, J. J. Boon, J. W. de Leeuw, F. de Lange, P. J. W. Schuyl, H. R. Schulten and U. Bahr, Comparison of analytical pyrolysis techniques in the characterization of chitin, Anal. Chem., 56:2160 (1984).
8. S. A. Liebman, T. P. Wampler and E. J. Levy, Developments in pyrolysis capillary GC, High Res. Chrom. & Chrom. Comm., 1:172 (1984).
9. T. Fujita, Japan Patent, 7, 013, 599 (1970).
10. G. W. Rigby, U.S. Patent, 2, 072, 771 (1936).
11. F. Shafizadeh and Y. L. Fu, Pyrolysis of cellulose, Carbohydr. Res., 29:110 (1973).

SYNTHESIS AND PHYSICO-CHEMICAL PROPERTIES OF THE POLYAMPHOLYTE CHITOSAN 6-SULFATE

A. M. NAGGI, G. TORRI, T. COMPAGNONI and B. CASU

Istituto di Chimica e Biochimica "G. Ronzoni"
Via G. Colombo, 81
20133 Milano, Italy

INTRODUCTION

Most of the reactions reported for sulfating chitosan lead to N-sulfation, either exclusively or in addition to O-sulfation.[1] In the framework of a systematic study on the influence of molecular size and sulfation pattern of polysaccharide sulfates, a simple method of sulfation (based on reaction with a 2:1 mixture of sulfuric and chlorosulfonic acids) was developed.[2] By use of this method, polysaccharides preferentially sulfated at the primary hydroxyls were obtained in high yield.[2] This paper reports the synthesis and structural characterization of the product obtained from chitosan.

EXPERIMENTAL

Sulfation of chitosan

1 g of the polysaccharide (ANIC, batch No 116, degree of acetylation 0.2) was added to a mixture of 40 ml H_2SO_4 95% and 20 ml $HClSO_4$ 98% previously cooled at 0-4 °C, and the solution was stirred for 60' at room temperature. The product was precipitated by pouring the reaction mixture on 250 ml of cold diethyl ether. The precipitate was collected on a sintered-glass filter, washed with cold diethyl ether, solubilized in water, neutralized with 0.5M $KHCO_3$, and dialyzed through a 14,000 dalton membrane. A product with essentially the same physico-chemical properties was obtained by performing the reaction with a fourfold concentration of chitosan (1g/24 ml total reagents). With both procedures, yields were higher than 90%. A low-molecular weight chitosan 6-sulfate was obtained with a similar procedure, from chitosan oligosaccharides ($\overline{DP}$~15) obtained by nitrous acid depolymerization of chitosan.[3] The degree of polymerization of the chitosan 6-sulfates was controlled, to some extent, by proper choice of the experimental conditions, especially the reaction time. Most of the products had nonetheless degrees of polymerization higher than 50, as evaluated by neocuproine and ^{13}C-NMR end-group analysis. All products were freely soluble in water (up to 20% w/v). A water-soluble low-molecular weight chitosan (used for NMR titration studies) was kindly provided by Dr. J.R. Vercellotti, V-Labs, USA.

Conductimetric titration

Sulfate groups were determined conductimetrically with 0.1M NaOH, after exchange on a column of IRA-120H$^+$ resin, as reported[4] for heparin and other polysulfates.

IR spectra

The spectra were obtained on a Perkin-Elmer Mod. 137 spectrometer, on films cast from aqueous solutions.

NMR spectra

NMR measurements were performed on 1-10% (w/v) solutions in D_2O, following several exchanges with the same solvent (99.9% D). The spectra were measured at 75 MHz for ^{13}C, and 300 MHz for ^1H, on a Bruker CXP-300 spectrometer, at 20°C. ^{13}C chemical shifts are referred to internal methanol = 51.75 ppm from internal TSP. ^1H chemical shifts are referred to internal $CHCl_3$ = 7.3 ppm from internal TMS.

NMR titrations

^1H-NMR spectra were recorded from solutions of chitosan (or chitosan 6-sulfate) brought to different pD by addition of DCl or NaOD. pD values were obtained by adding 0.4 to pH-meter readings.

RESULTS AND DISCUSSION

Reaction of chitosan with a 2:1 mixture of sulfuric and chlorosulfonic acids afforded products freely soluble in water ($[\alpha]_D^{20}$ = -10.4 at pH 5).

Conductimetric titration of products indicated degrees of sulfation between 0.95 and 1.0. The IR spectra (solid films) displayed the typical bands of sulfate esters, at 1240 and 810 cm^{-1}.

The products were fully characterized by high-field NMR spectra. ^{13}C-NMR spectra are currently used for assessing the homogeneity and major structural features of polysaccharides.[5-7] In their most common (proton-decoupled) form, ^{13}C-NMR spectra of polysaccharides consist of a set of singlets, assigned to specific carbons by correlation criteria, isotopic effects, or heteronuclear decoupling with the more easily assignable signals of the appended hydrogens.[5-7] Whereas recent papers deal with ^{13}C spectra of chitin and chitin derivatives,[8,9] NMR data on chitosan have not been previously reported, probably because of solubility limitations. In this work, NMR spectra for chitosan were obtained over a wide pH (pD) range, using a low-molecular weight chitosan ($\overline{DP}$~30). The ^{13}C chemical shifts of chitosan 6-sulfate are compared in Table 1 with those of the low-molecular weight chitosan and of the β-anomer of 2-amino-2-deoxy-D-glucose (β-glucosamine). Signal assignments, made by comparison with those for β-glucosamine hydrochloride,[10,11] are in accordance with those reported for chitosan in the solid state and in the gel form.[9] Major chemical shift differences between chitosan and β-glucosamine are those expected for carbons involved in the glycosidic bridge,[5-7] i.e., C-1 and C-4 (large downfield shift).

The ^{13}C-NMR spectrum of a typical preparation of chitosan sulfate obtai

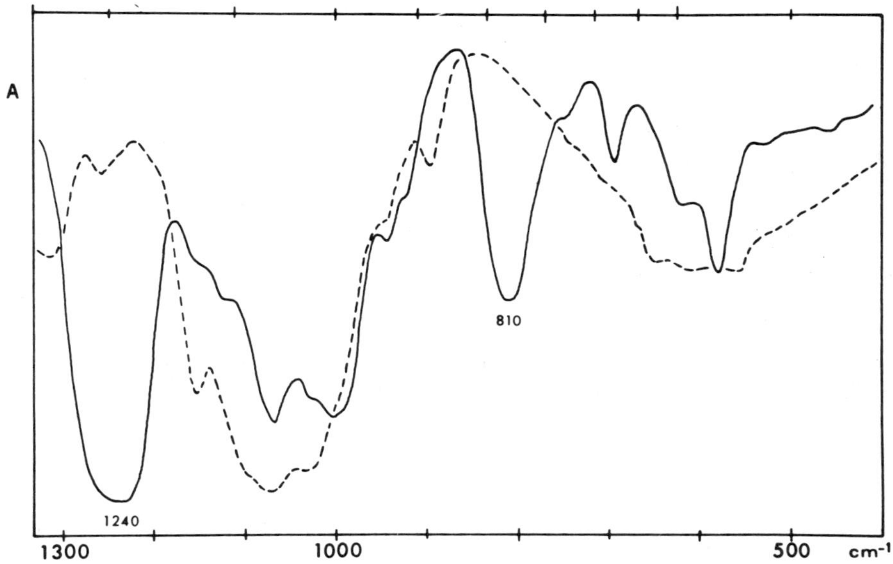

Fig.1. Partial IR spectra (solid films) of chitosan 6-sulfate (full line) and a chitosan oligosaccharide $\overline{DP}$~15 (dashed line).

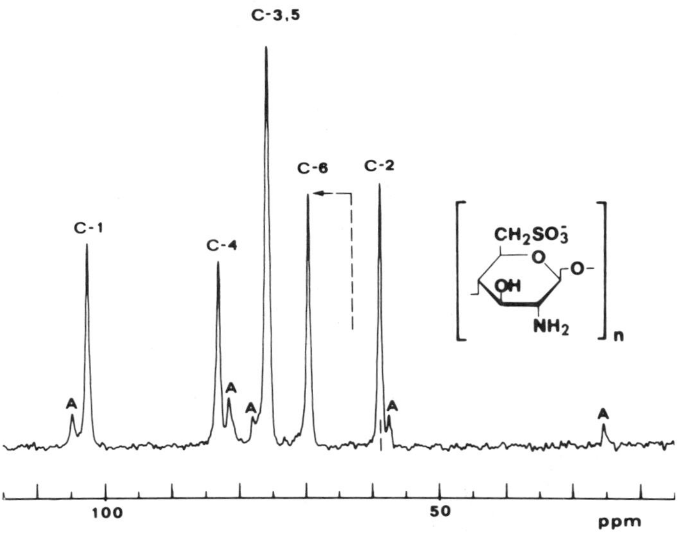

Fig. 2. ^{13}C-NMR spectrum (75 MHz) of chitosan-6-sulfate in D_2O solution (pD 8). Signals labelled A are from N-acetylated residues. The arrow indicates the shift of the C-6 signal from its typical position when bearing an unsubstituted hydroxyl.

Table 1. ^{13}C-NMR chemical shifts (ppm) for β-D-glucosamine, chitosan and chitosan 6-sulfate in D_2O solution (pD 5).

	C-1	C-2	C-3	C-4	C-5	C-6
β-D-glucosamine	95.2	59.5	74.5	72.2	78.6	63.1
chitosan[a]	100.5	58.8	73.1	79.3	77.6	63.0
chitosan 6-sulfate	99.0	57.8	75.3	78.5	75.8	69.2

[a]Low-molecular weight preparation ($\overline{DP}$~30).

ned in the present work (Figure 2) is remarkably simple, essentially consisting of the six major signals given in Table 1. Signals are shifted to different extent with respect to chitosan. However, whereas shifts for carbons 1-to-5 are relatively small, the C-6 signal is dramatically and completely shifted downfield (-6.2 ppm), as expected[6,12] for formation of sulfate esters at the primary hydroxyl. In the absence of minor signals (other than those of N-acetylated residues), the ^{13}C spectrum clearly indicates that sulfation occurred exclusively at the level of O-6. Moreover, the practical absence of C-1 signals of reducing end-groups in the 95 ppm region[6] indicates that the sulfation reaction did not involve a substantial depolymerization of chitosan. Although, in principle, concurrent sulfation and depolymerization can be obtained also using sulfuric acid alone, reaction with sulfuric acid affords products sulfated also at the amino group.[13]

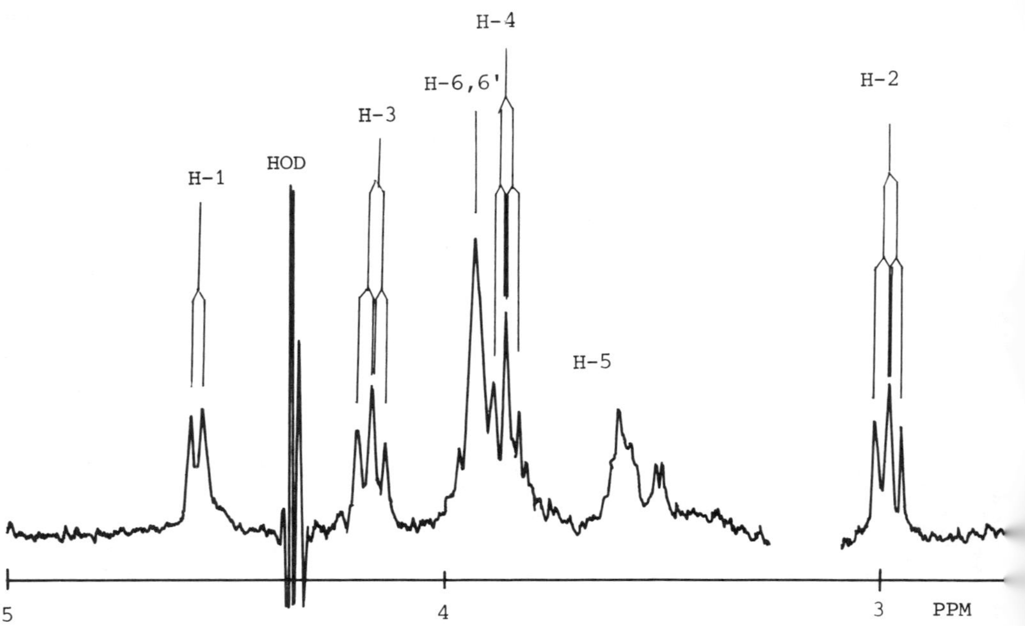

Fig. 3. ^{1}H-NMR spectrum (300 MHz) of chitosan 6-sulfate in D_2O (pD 7). (The signal from residual HOD in the solvent was partially saturated.)

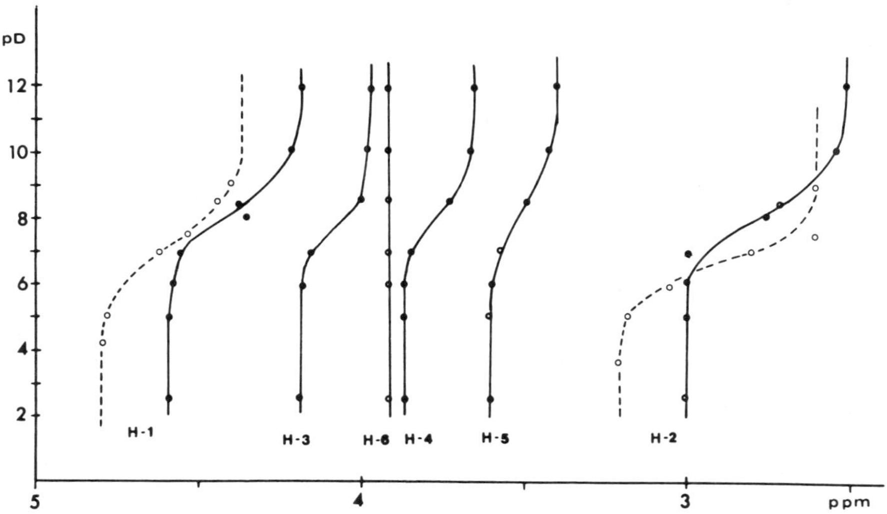

Fig. 4. ^{1}H-NMR titration curves for chitosan 6-sulfate (full lines). Curves for H-1 and H-2 of chitobiose[10] are also given for comparison purposes.

Also the ^{1}H-NMR spectra indicate a remarkable homogeneity of the products. Because of interproton spin-coupling, the ^{1}H spectra of polysaccharides usually consist of doublets and multiplets, the magnitude of signal splitting being related to the dihedral angles between vicinal hydrogens.[6,7] Figure 3 shows the ^{1}H-NMR spectrum of chitosan 6-sulfate, with signal assignments as made by homonuclear spin-decoupling. Similarly to ^{13}C signals (Table 1), ^{1}H signals associated with the anomeric centre and with that bearing the amino group are at the lowest and highest regions of the spectrum, respectively. Coupling constants between vicinal protons (^3J values, deduced from first-order analysis of signal splittings) are all large (7-10 Hz), in accordance with the expected $^4C_1$(D) conformation[6,7] of the glucosamine residues.

When a carbon or a hydrogen are near an ionizable group, there is usually a notable pH-dependence of NMR chemical shifts.[6,7] This effect is clearly shown by data in Table 2, reporting the differences between ^{13}C chemical shifts measured at pD 8 and pD 5 for β-D-glucosamine, chitosan and chitosan 6-sulfate. Most signals shift downfield on increasing pH (negative shifts). However, the pH-dependence is markedly different for different carbons in the same molecule, as well as for the same carbons in the three molecules. In the monomer, the effect of changing the ionization state of the amino group is significantly felt at the α-carbon (C-2) to the same extent as at the β-carbon C-3, the other β-carbon C-1 (probably because of its higher polarizability) being even more dramatically affected. Perhaps surprisingly, C-2 is barely affected by pH changes in both chitosan and chitosan 6-sulfate. On the other hand, in the latter polysaccharide C-3 is not pH-sensitive, the only significant relative shifts being those of carbons involved in the glycosidic bridge, i.e., C-1 (-3.3 ppm) and C-4 (-4.2 ppm).

Table 2. pH-dependence of ^{13}C-NMR chemical shifts for β-D-glucosamine, chitosan and chitosan 6-sulfate, expressed as differences (ppm) between chemical shifts at pD 8 and at pD 5.

	C-1	C-2	C-3	C-4	C-5	C-6
β-D-glucosamine	-9.8	-3.6	-3.6	-1.7	-0.6	-1.1
chitosan[a]	-4.3	-0.4	-4.4	-1.6	0.0	-0.5
chitosan 6-sulfate	-3.3	-1.0	-0.2	-4.2	0.0	-0.4

[a]Low-molecular weight preparation ($\overline{DP}$~30).

All the ^{1}H-NMR signals of chitosan 6-sulfate (except H-6) are pH-dependent in the 2-to-12 pD range. This is illustrated by the "titration curves" of Figure 4, which show the expected inflexions corresponding to the pKa value of the amino group. The signal most affected by changes in ionization i that of H-2. However, similarly to carbons, significant changes also occur for protons across the glycosidic bridge, i.e., H-1 and H-4. Similar change (though different in magnitude) have been observed for H-1 of chitobiose, and associated to changes in the values of torsional angles ϕ and ψ between adjacent glucosamine residues.[14]

Introduction of a sulfate group at position 6 of chitosan while leaving the amino group unsubstituted, is expected to have an important influence on solution and rheological properties of the polysaccharide. In alkali solution, chitosan 6-sulfate should behave like a typical polyanion, and the local conformation of its chains should be essentially similar to that of chain made up of β-1,4-linked glucopyranose residues, where ribbon-like segments ar stabilized by intramolecular hydrogen bonds between OH-3 and the ring oxygen of adjacent residues, as demonstrated also in solution for cellulose[15] and chitin.[16] However, in this conformation a polar interaction between the sul fate groups at O-6 and the amino groups on adjacent residues can contribute to stabilize the chains in a more rigid conformation. At low pH values, whe re a zwitterionic interaction can significantly contribute to stabilize chai as depicted in Figure 5, molecular rigidity is expected to be higher. The different pH-dependence of NMR signals associated with the glycosidic centre (as well as of the C-3 signal) for chitosan and chitosan 6-sulfate might accordingly reflect a re-adjustement of the torsional angles ϕ and ψ driving the chain conformation to achieve a closer proximity of 2-sulfo and 2-amino groups on adjacent residues.

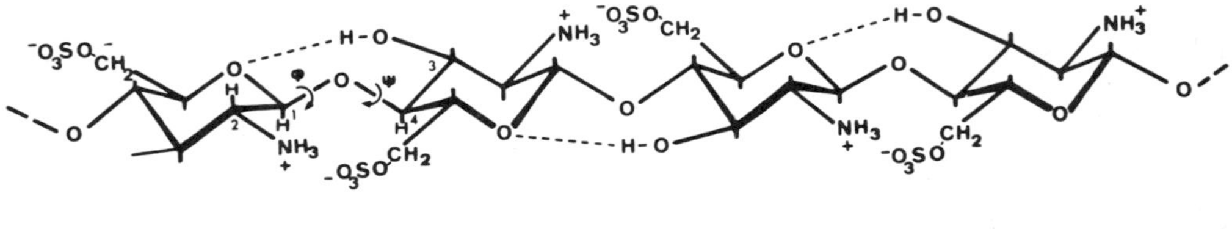

Fig. 5. Molecular model of chitosan 6-sulfate, showing possible concurrenc of intramolecular hydrogen bonding and zwitterionic interaction to increase the rigidity of chain segments at low pH values.

ACKNOWLEDGEMENTS

The authors thank Enrico Pivari for technical assistance. This work was supported in part by the Italian National Research Council, Progetto Finalizzato Chimica Fine e Secondaria, Roma.

REFERENCES

1. R. A. A. Muzzarelli, "Chitin", Pergamon Press, Oxford, (1977).
2. A. Naggi and G. Torri, EP 0 116 251 (1984).
3. F. Yaku, E. Muraki, T. Tsuchiya, Y. Shibata and K. Koshijima, Cellulose Chem. Techn., 11:421, (1977).
4. B. Casu and U. Gennaro, Carbohydr. Res., 39:168, (1975).
5. P. A. J. Gorin, Adv. Carbohydr. Chem. Biochem., 38:13, (1981).
6. A. S. Perlin and B. Casu, in: "The Polysaccharides", G. O. Aspinall, ed., Accademic Press; New York, 1:133, (1982).
7. B. Casu, in: "Polisaccharides", E. D. Atkins, ed., McMillan, London, in press (1985).
8. D. Gagnaire, J. St.-Germain and M. Vincendon, Makromol. Chem., 183:593, (1982).
9. H. Saito, R. Tabeta and S. Hirano, in: "Proceedings 2nd International Conference on Chitin and Chitosan", S. Hirano and S. Tokura, eds., Japanese Soc. Chitin, Tottori (1982).
10. T. E. Walker and R. Barker, Carbohydr. Res., 64:266, (1978).
11. K. Bock and C. Pedersen, Adv. Carbohydr. Chem. Biochem., 41:27, (1983).
12. V. M. Doctor and D. Esho, Carbohydr. Res., 121:312, (1983).
13. K. Nagasawa, Y. Tohira, Y. Inoue and N. Tanoura, Carbohydr. Res., 18:95, (1971).
14. S. Tsukada and Y. Inoue, Carboydr. Res., 88:19, (1981).
15. D. Gagnaire, J. St.-Germain and M. Vincendon, J. Appl. Polym. Sci. Appl. Polymer Symposium, 37:261, (1983).
16. M. Vincendon, this volume.

A SMALL ANGLE X-RAY SCATTERING STUDY OF CHITOSAN 6-SULFATE IN WATER

S. S. STIVALA, A. PATEL and S. PATEL

Department of Chemistry and Chemical Engineering
Stevens Institute of Technology
Hoboken, NJ 07030, USA

INTRODUCTION

It was recently reported that the reaction of chitosan, using a 2:1 mixture of sulfuric and chlorosulfonic acids, results in the preferential sulfation of the primary hydroxyl groups (1). The use of ^{13}C-n.m.r. spectroscopy (2) clearly indicates that the reaction occurs at the hydroxyl group at the O-6 position of the chitosan (see Naggi et al., this Volume).

Earlier studies on the sulfation of chitosan by various investigators (3) using diverse methods produced products which were either exclusively N-sulfated or in addition, O-sulfated. Sulfated chitosan has been demonstrated to possess high blood anticoagulant activity (3) and has often been compared to heparin. Heparin is a highly sulfated polysaccharide used medicinally as an anticoagulant in the treatment of various cardiovascular diseases. It is obtained from the tissue of mammals, specifically bovine lung and porcine intestinal mucosa. Reviews on the chemistry and pharmacology of heparin have been published (4). Though we have reported our work on the physical chemistry of heparin in relation to anticoagulant activity (5), including those by other workers (4), little, if any, studies on the physico-chemical properties of sulfated chitosans have appeared in the open literature. Accordingly, it is the intent of this paper to examine some dilute solution parameters of chitosan 6-sulfate using small angle x-ray scattering (SAXS). This study is compared to the physico-chemical parameters we have reported on heparin.

EXPERIMENTAL

The chitosan 6-sulfate used in the present study is the same described in the previous Chapter of this book. We have found that chitosan 6-sulfate is highly hydroscopic and, accordingly, all measurements were made on samples dried to constant weight, about 48 h at 40° C in a vacuum oven.

Partial specific volume V_1

The V_1 was obtained in water at 25 ± 0.1° C using the digital precision densitometer mod. DMA-02C (Anton-Paar Co., Austria). The value of 0.48 $cm^3\ g^{-1}$ was obtained for the chitosan 6-sulfate.

Small angle x-ray scattering

The procedure used in this work has been described in an earlier work (6). Detailed procedures, theory and treatment of data may be found elsewhere (7). The entrance slit of 100 μm and counter-slit of 250 μm were used in the Kratky camera. The collimation error of the raw data was eliminated on a PDP-10 computer according to the method of Glatter (8). The corresponding corrected data is the desmeared scattering data.

SAXS measurements of chitosan 6-sulfate in water were made at 25 ± 0.2° C on concentrations of 20, 30 and 40 mg/ml. Concentration dependence of the data is eliminated by extrapolation to infinite dilution.

The electron density of the chitosan was calculated from the experimentally determined elemental analysis, Table I. For comparison, Table I also contains the elemental analysis computed from the structural formula assuming 100 % O-6 sulfation and complete deacetylation (chitosan is not totally deacetylated).

Intrinsic viscosity $|\eta|$

The $|\eta|$ of the chitosan 6-sulfate in water at 25.5° C was determined from a Ubbelohde viscometer having flow time for the solvent in excess of 100 seconds. The $|\eta|$ was computed from the plot of $(\eta_{sp}/C)^{-1}$ vs $\sqrt{C}$, according to the Fuoss Equation $(\eta_{sp}/C) = A/(1 + B\sqrt{C})$, where A is the intrinsic viscosity obtained from the intercept at C = O.

RESULTS

Molecular weight, M, and radius of gyration, R_g

The M was obtained according to the equation

$$M = (I_o/P_o)21.0\ a^2/|(z_1 - \bar{v}_1\ \rho_2)^2\ dc| \tag{1}$$

where I_o is the scattering intensity extrapolated to zero scattering angle; a is the distance in cm between the sample and plane of registration; d is the sample thickness (0.1 cm); c is the concentration in g/ml; $\bar{v}_1$ is solute partial specific volume; ρ_2 is the number of mole electrons per ml of the solvent; z_1 is the number of electrons per gram of the solute; and P_o is the energy of the primary beam per cm length in the plane of registration.

The radius of gyration was obtained from the Guinier equation,

$$\ln I = \ln I_o - K\ R_g^2\ (2\theta)^2 \tag{2}$$

or

$$\ln I = \ln I_o - R_g^2\ Q^2/3 \tag{2a}$$

where $Q = (4\pi/\lambda)\sin\theta$ (For λ = 1.54 Å, K = 5.55) and 2θ is the scattering angle in radians. Since decay of the intensity at very small angles (inner portion of the scattering curve) is a Gaussian function of 2θ, a straight line is obtained for the several innermost points according to Eq. (2) or Eq. (2a). The slope of this linear portion yields the radius of gyration whereas the intercept gives the intensity at zero angle. Figure 1 shows the Guinier plot of the chitosan 6-sulfate in water according to Eq. (2a). Since R_g is concentration dependent, the apparent values of R_g for each concentration are extrapolated to zero concentration.

Radius of gyration of the cross section, R_q

The R_q may be obtained from the equation,

$$\ln (I2\theta) = \ln (I2\theta)_o - k' R_q^2 (2\theta)^2 \quad (3)$$

or

$$\ln (IQ) = \ln (IQ)_o - R_q^2 Q^2/2 \quad (3a)$$

where $k' = (2\pi/\lambda)^2/2$, (for $\lambda = 1.54$ Å, $K' = 8.32$).

Mass per unit length, Mu, and the total length, L

The mass per unit length is calculated from the absolute value of the intensity at zero angle of the cross-section factor, $(I2\theta)_o$ using an equation similar to Eq. (1), thus,

$$M_u = |(I2\theta)_o/p_o||27.3\ a^2/(z_1 - \bar{v}_1\ \rho_2)^2\ dc| \quad (4)$$

The intercepts of the cross-sectional plots yield values of $(I2\theta)_o$ for a given concentration from which the value of $(I2\theta)_o/c$ at infinite dilution is obtained. The total length of the molecule is obtained from the division of Eq. (1) by Eq. (4), i.e., $L = M/M_u$.

Table I. Elemental analysis of chitosan 6-sulfate

	Experimental, %	calculated*, %
Carbon	20.30	27.38
Hydrogen	3.24	3.83
Nitrogen	3.70	5.32
Oxygen	44.71	42.55
Sulfur	15.61	12.18
Sodium	9.57	8.73

* Computed from the structure of the monomer, assuming complete O- 6 sulfation only and complete deacetylation (chitosan is not totally deacetylated).

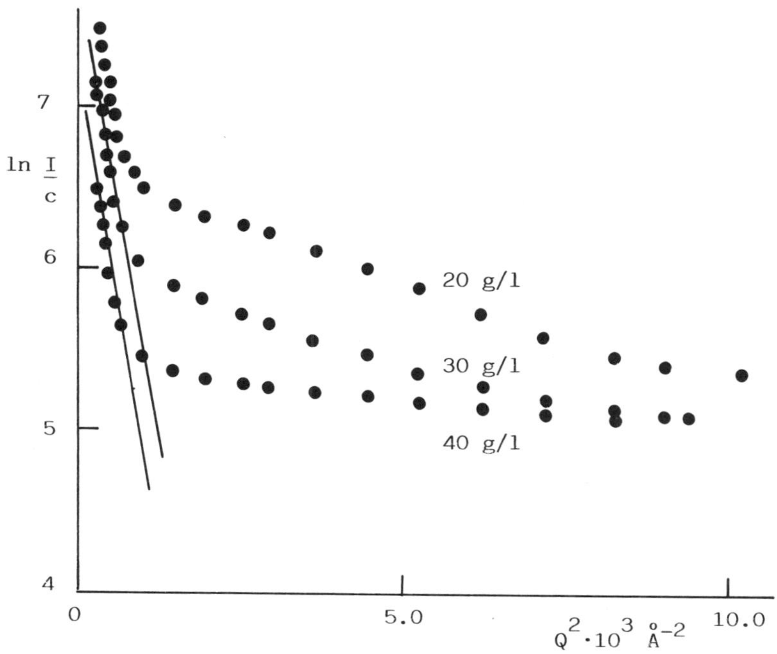

Fig. 1. Guinier plot of chitosan 6-sulfate in water at three concentrations.

Table II. Solution parameters from SAXS in water of chitosan 6-sulfate, including other polysaccharides.

	Chitosan 6-sulfate	Heparin (6)	Dextran (9)	Levan (10)
M_w	31,600	12,900	12,000	27,300
R_g, Å	83.6	35.2	38.0	50.1
L_{exp}, Å	271	237	355	534
L_{calc}	620	260		
a^*, Å	20.3	21.1		17.7
$(M_u)_{exp}$, $(Å)^{-1}$	117	54.5	33.8	45.8
$(M_u)_{calc}$	51.0	52.7	31.5	28.9
$\|\eta\|$, dl/g	0.8	4.17	0.11	0.14
$\bar{v}$, cm^3/g	0.48	0.47	0.61	0.65

Persistence Length, a^*

The persistence length was calculated from the equation below

$$a^* = 2.3\lambda/2\pi(2\theta)^*$$

or alternately,

$$a^* = 2.3/Q^*$$

where $(2\theta)^*$ is the transition between the $I/(2\theta)^2$ and $I/(2\theta)$ portions of the scattering curve. Figure 2 shows the plot of $\ln IQ^2$ vs. Q.

Table II summarizes the solution parameters obtained from the SAXS of the chitosan 6-sulfate. Included in the Table are corresponding values of heparin, and the neutral polysaccharides dextran (9) and levan (10).

DISCUSSION

The molecular weight obtained for the chitosan 6-sulfate is 31,600 daltons. This value compares favorably with the average molecular weight of sulfated chitosan prepared by Nagasawa and co-workers (11), by treating chitosan with concentrated sulfuric acid. The chitosan used by Nagasawa was of 92 % free amine and had an average molecular weight of 120,000. They noted the remarkable depolymerization following reaction with concentrated sulfuric acid.

The SAXS parameters of chitosan 6-sulfate may be compared to those of heparin. Heparin may be considered as a sulfated copolymer of glucosamine and glucuronic acid having a 1-4 linkage sequence. Heparin is heterogeneous in molecular weight. The heparin shown in Table II, is a fractionated sample of narrow molecular weight distribution. It may be noted that, for this sample, $M = 12,900$ and $R_g = 35.2$ compared to $R_g = 83.6$ Å for the chitosan 6-sulfate. The mass per unit length, M_u, of heparin, observed experimentally from SAXS, is 54.5 daltons per Å. The M_u may be computed from the molecular weight of the monomer, M_o, and from its length, h_o i.e., $M_u = M_o/h_o$. For heparin, the value of 52.7 per Å was calculated (6) on the basis of its structure where h_o is identical to half the cell length (5.15 A). Accordingly, based on the observation that $(M_u)_{exp} \simeq (M_u)_{calc}$ it may be concluded that heparin is not branched. On the other hand, for the chitosan 6-sulfate,

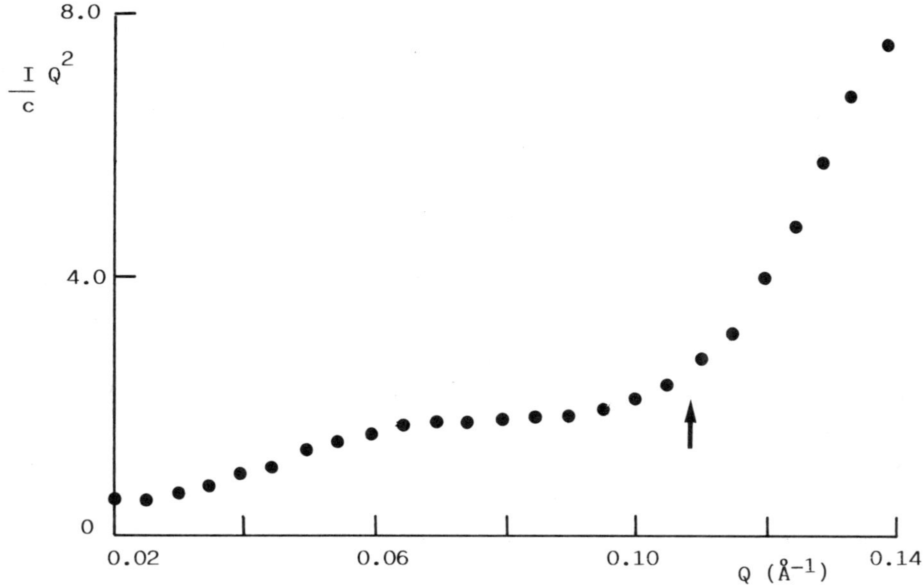

Fig. 2. Plot of IQ^2/C vs. Q of chitosan 6-sulfate in water.

$(M_u)_{exp}$ = 117 per Å whereas $(M_u)_{calc}$ = 51.0 per Å. The latter was computed from M_o = 526 daltons for the disaccharide unit of the sodium salt of O-6 polyglucosamine sulfate and h_o = 10.32 A for chitin (from x-ray diffraction; see Ref. 3). That $(M_u)_{exp} > (M_u)_{calc}$ suggests that the chitosan 6-sulfate, as prepared by the procedure described by Naggi and co-workers (1, 2), is branched. It should be noted that the $(M_u)_{calc}$ is the mass per unit length of the linear (unbranched) structure. These observations were also noted for the case of dextran (9) and levan (10). As would be expected, the values of $(M_u)_{calc}$ for heparin and chitosan 6-sulfate are quite close, i.e., 51.0 vs. 52.7 per Å, respectively. The $L_{calc} > L_{exp}$ for the chitosan, where L_{calc} (i.e., for the linear structure) is obtained from $(DP)h_o$, where DP, the degree of polymerization, is M/M_o, thus (31,600 : 526) .(10.32) = 620 Å. Measurement of chitosan 6-sulfate in 0.15 N LiCl also showed that $(M_u)_{exp} > (M_u)_{calc}$ (12).

The plot of reduced viscosity, $\eta sp/C$ versus C, where ηsp is specific viscosity and C is concentration, showed typical polyelectrolyte behavior, i.e., non-linear curve rising rapidly with decreasing C. However, plotting $C/\eta sp$ against C, according to the Fuoss equation, afforded a linear curve from which the intrinsic viscosity, $|\eta|$, was readily obtained upon extrapolation to c = 0, giving $|\eta|$ = 0.81 dl/g, cf. Table II.

This value is considerably lower than 4.17 dl/g for the polyelectrolyte heparin (13), which has an extended conformation. Since the chitosan 6-sulfate, also a polyelectrolyte, has a molecular weight about 2.5 times that of the heparin, it should have exhibited a higher $|\eta|$ in water than that of heparin. The low value of $|\eta|$ = 0.81 dl/g strongly suggests that the chitosan 6-sulfate is branched (heparin is unbranched). It is interesting to note that the $|\eta|$ in water at 25° C of the dendritic structures of the neutral polysaccharides dextran (9) and levan (14) are also low (Table II).

The premise that chitosan 6-sulfate is branched, may be explained by the fact that some degradation occurs in chitosan during its sulfation, as reported by Nagasawa (11). The -OH in carbon 6 of both the oligomer and the partially degraded chitosan can simultaneously react with sulfuric acid to form the diester linkage, constituting the branch point.

A band at about 1390 cm^{-1} was observed in the i.r. spectra of dimethyl sulfate. The i.r. spectra of the chitosan 6-sulfate shows a weak, but distinct peak at this frequency, together with a relatively stronger band at about 1230 cm^{-1} characteristic of the monosulfate ester.

If three monosaccharide residues are attached to a disaccharide unit on the main chain (3 residues / 2 monosaccharides) we compute $(M_u)_{calc}$ = 121 per A compared to $(M_u)_{exp}$ = 117 per A. Since the chitosan 6-sulfate was shown to be 95 to 100 % sulfated from potentiometric titratrions of the sulfates by Naggi et al. (2) this indicates an oligomer of about at least 15 residues attached to 10 monosaccharides along the chain, corresponding to one non-titratable sulfate to 25 titratable sulfates, thus slightly above 95 % sulfation. Further, branch points separated by about 10 monosaccharides should be just outside the resolution for detection by ^{13}C-n.m.r. (14). Though the possibility of forming dendritic structures in chitosan 6-sulfate exists, this may be ruled out in this case on the basis that lower titratable sulfates and lower $|\eta|$ should have prevailed.

ACKNOWLEDGEMENT

The authors appreciate the preparation and supply of the chitosan 6-sulfate by Naggi, Torri and Casu, Istituto di Chimica e Biochimica "G. Ronzoni", Milano, Italy. S.S.S. is grateful to Drs. Casu, Torri and Focher for their stimulating discussions in connection with this work.

REFERENCES

1. A. Naggi and G. Torri, EPA 00116 251 (26/12/83).
2. A. Naggi, G. Torri, T. Compagnoni and B. Casu, this volume.
3. R. A. A. Muzzarelli, "Natural Chelating Polymers", Pergamon Press, New York (1973).
4. J. Ehrlich and S. S. Stivala, J. Pharm. Sci., 62:517 (1973).
5. S. S. Stivala, Federation Proc., 1:83 (1977).
6. S. S. Stivala, M. Herbst, O. Kratky, and I. Pilz, Arch. Biochem. Biophys., 127:795 (1968).
7. O. Glatter, and O. Kratky, "Small angle X-ray scattering", Academic Press, New York (1982).
8. O. Glatter, J. Appl. Cryst., 1:147 (1974).
9. S. Garg and S. S. Stivala, J. Polymer Sci., (Phys. Ed.), 16:1419 (1978).
10. B. H. Khorramian and S. S. Stivala, Carbohydrate Res., 108:1 (1982).
11. K. Nagasawa, Y. Tohira; Y. Inoue and N. Tanoura, Carbohydr. Res., 181:95 (1971).
12. To be published.
13. P. Liberti and S. S. Stivala, Arch. Biochem. Biophys., 119:510 (1967).
14. S. S. Stivala and J. Zweig, Biopolymers, 20:605 (1981).

THE DETERMINATION OF THE DEGREE OF ACETYLATION OF CHITOSANS BY SPECTROPHOTOMETRY

R.A.A. MUZZARELLI and R. ROCCHETTI

Faculty of Medicine, University of Ancona
I-60100 Ancona, Italy

INTRODUCTION

The current interest (1-4) in highly deacetylated chitosans, useful for the preparation of metal ion chelating agents, textiles and pharmaceutics, has brought forward the need for a rapid and simple instrumental method for the determination of the degree of acetylation. The method sought should provide accurate and precise results, especially for those highly deacetylated chitosans which can be hardly analyzed by existing techniques, namely by infrared spectrometry (5), where the signals recorded are those of the N-acetyl group: when this group is removed by extended deacetylation processes and the degree of acetylation becomes lower than 20%, the errors associated with the graphical reading become exceedingly large. In those cases, therefore, methods based on the reactions of the liberated amino group are preferred, among which the colloid titration method making use of anionic polymers as titrants (6) and various colorimetric methods developed after the Elson-Morgan methods (7). Ultraviolet spectrophotometry so far has not been used to determine chitin: unsuccessful attempts were made (8) to determine chitin in amide and lithium chloride solvents through measuring absorbances. Derivative spectrophotometry is a relatively recent analytical principle; although the uv-vis range offers a very wide field of applications, no publication exists dealing with the determination of the degree of acetylation of chitin/chitosan by this technique. In this work we have evaluated first derivative spectrophotometry as a method for the rapid, precise and non-destructive determination of the acetyl content of chitin / chitosan.

OUTLINE OF THE METHOD

First derivative spectra of acetic acid solutions (0.01, 0.02 and 0.03 M) were recorded in the range 240-190 nm, using water as a reference. Scanning speed was 30 nm/min, chart speed 10 cm/min, response time 4 sec. and scale expansion +0.35 absorbance units. Superimposing of the three spectra permitted the exact individualization of the zero crossing point of the acid. Standard solutions containing 5 - 35 mg/l of N-acetylglucosamine were prepared in 0.01 M acetic acid and first derivative spectra recorded exactly as outlined above. All of the spectra were then superimposed and the height (H) in mm of each standard measured, starting from the zero crossing point of the acid, as shown in Figure 1. A calibration curve of H against N-acetylglucosamine concentration was then drawn.

For the solution, 100 mg of dry chitosan (preferably freeze dried), was dissolved in 10 ml of 0.1 M acetic acid, and diluted to 100 ml with water. Immediately the first derivative spectrum of this solution was recorded, and H measured as before with careful attention to detail, and accounting for the dilution, the concentration of N-acetylglucosamine, obtained from the calibration curve, and expressed in mg/l, should be exactly the weight percentage of N-acetylglucosamine in the polymer. Solutions were also prepared of D-glucosamine hydrochloride, in 0.01 M HCl (120 mg/l), to check the effect of D-glucosamine on the H signal for the N-acetylglucosamine.

RESULTS AND DISCUSSION

The zero order absorption spectra of glucosamine and N-acetylglucosamine show maxima at 197 and 193 nm, respectively. At 199 nm the molar absorptivity, ε, of N-acetylglucosamine is about 130 times higher than for glucosamine. The presence of acetic acid greatly disturbs the determination of both compounds, especially N-acetylglucosamine, its contribution being particularly high around 199 nm. The first derivative spectra for N-acetylglucosamine and acetic acid are in Fig. 1; it is apparent that small amounts of N-acetylglucosamine (2.5 mg per l of 0.01 M acetic acid) give signals whose maxima correspond to the lowest contribution from acetic acid (199 nm); their intensity is much higher than for glucosamine at the same concentration level.

When the first derivative spectra of acetic acid solutions in the range of 0.01 to 0.03 M are recorded against water, it is possible to note that all the acetic acid spectra share a common point at 202 nm, denoted as zero crossing point. The zero crossing point corresponds to the N-acetylglucosamine maximum on the wavelength axis: this makes the N-acetylglucosamine determination independent of the acetic acid concentration range usually encountered in the dilute chitosan solutions.

The detection limit for N-acetylglucosamine at 199 nm is 0.5 mg/l, its signal height, H, being twice as big as the noise signal. To define the effect, if any, of acetic acid, on the determination of N-acetylglucosamine, various aqueous mixtures of N-acetylglucosamine and glucosamine were prepared and read in the first derivative mode.

The data show that, for varying concentrations of N-acetylglucosamine the signal heights at 202 nm are not influenced by the acetic acid present even when the latter is 300 times more concentrated than N-acetylglucosamine. Moreover, the data show that calibration curve can be easily drawn.

Calibration curve for N-acetylglucosamine

To obtain a calibration curve for N-acetylglucosamine, the first derivative spectra of standard N-acetylglucosamine solutions were taken for 8 varied concentrations (0.005 to 0.40 mg/ml) of N-acetylglucosamine in 0.01 M acetic acid. The spectrum was measured five times for each concentration, so that 40 sets of H values (y) and N-acetylglucosamine concentrations (x) were obtained. The correlation coefficients between y and x was calculated to be 0.999 and a plot of y against x yielded a good straight line which passed through the origin. The slope was 3.270 mm. $(mg/l)^{-1}$, with a relative standard deviation $(S/\bar{x})$ of $\pm$ 2.7 %, which corresponded to the confidence limits of 3.27 $\pm$ 0.081 mm.$(mg/l)^{-1}$ at the 95 % level. The N-acetylglucosamine concentration, x, (mg/l), could be calculated from 1/slope.H, that is

$$x = 0.0306\ H \qquad \text{Equation 1}$$

Accuracy of the first derivative method for N-acetylglucosamine was calculated by the following equation, using the data of the calibration curve:

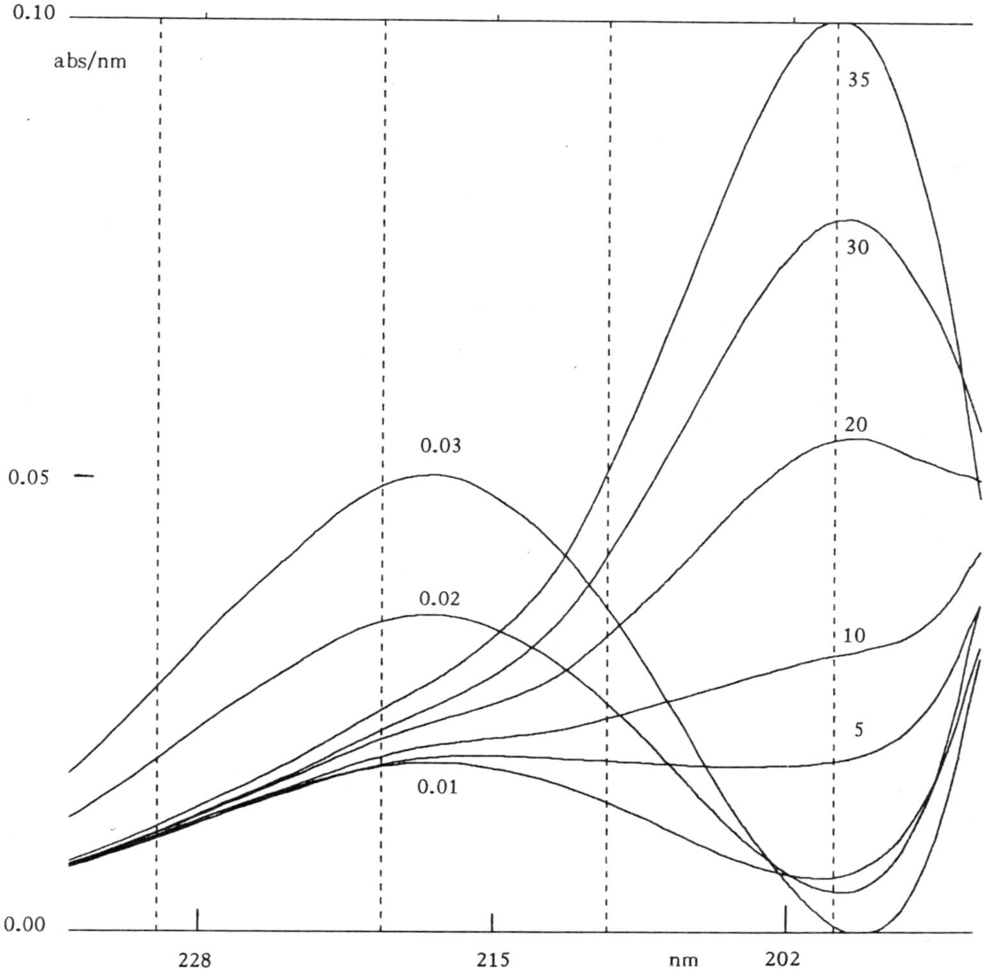

Figure 1. First derivative spectra of 0.01, 0.02 and 0.03 M acetic acid solutions (three lower curves crossing at 202 nm), and of N-acetyl glucosamine at various concentrations (mg/l) in 0.01 M acetic acid. Actual recording with the Kontron Uvikon 680 spectrophotometer.

Table I. Accuracy of the first derivative u.v. spectrophotometric determination of N-acetyl glucosamine.

N-acetyl glucosamine concentration, mg/ml		% Accuracy
Prepared solution, $\hat{x}$	Calculated value, $\bar{x}$ *	$100\ (\bar{x} - \hat{x})/\hat{x}$
10.0	10.1	1.0
15.0	14.9	- 0.7
20.0	20.2	1.8
25.0	24.8	- 0.8
30.0	29.9	- 0.3

* Mean value of five x's calculated from H values using Equation 1.

$$\% \text{ Accuracy} = (\bar{x} - \hat{x}) / \hat{x}.100 \qquad \text{Equation 2}$$

where $\hat{x}$ is the prepared concentration of N-acetylglucosamine in a standard solution and $\bar{x}$ is the mean value of the N-acetylglucosamine concentrations calculated by Equation 1, for each five values of H. As the results in Table I show, good accuracy was proved for the first derivative method at every N-acetylglucosamine concentration.

Effect of glucosamine on H values

The experimental data show that the presence of glucosamine in the N-acetylglucosamine solutions contributes detectably to the H signal, when the N-acetylglucosamine to glucosamine ratio is below 0.11. For the value of 0.01 of the said ratio, the value is twice as high as the one due to N-acetylglucosamine alone. Correction factors can be derived from the reference curve in Fig. 2.

Assay for commercial chitosan powders

Chitosan powders of various origins (0.500 g) were dissolved in 0.1 M acetic acid and brought to 500 ml. These solutions were analyzed as such or after dilution, depending on their N-acetylglucosamine contents. The results are in agreement with those previously obtained by other techniques by ourselves and by other authors. Their precision is much better, for instance, the results for *Euphausia superba* chitosan manufactured by Rybex are given as 42.6 with confidence limits ± 0.7, instead of 42 ± 4 as determined by infrared spectrometry.

CONCLUSIONS

In conclusion, it was demonstrated that first derivative u.v. spectrophotometry permits a simple and time saving assay of N-acetylglucosamine residues in chitosan, which has precision and accuracy. Inherent advantages are: 1) There is no need to know exactly the acetic acid concentration, if a reasonable concentration range is taken. 2) The spectrophotometric determinations can be carried out with water as a blank because the readings made at the zero crossing point make the determinations independent of the acid concentration. The use of water as a reference blank avoids light absorption in the optical system, thus permitting a better signal to noise ratio. 3) Sensitive and precise measurements can be carried out in the simplest way by using commonly available spectrophotometers.

Acknowledgement. This work was carried out with the financial contribution of Consiglio Nazionale delle Ricerche, Progetto Finalizzato Chimica Fine e Secondaria, Roma, Italia.

REFERENCES

1. S. Mima, M. Miya, R. Iwamoto & S. Yoshikawa, *J. Appl. Polymer Sci.*, 28:1909 (1983).
2. A. Domard & M. Rinaudo, *Intl. J. Biol. Macromol.*, 5:49 (1983).
3. S. Hirano & S. Tokura, *(Eds.) Proc. 2nd Intl. Conf. Chitin/Chitosan* Tottori, Japan. Soc. Chitin (1982).
4. R. A. A. Muzzarelli, "Chitin", Oxford, Pergamon Press (1977).
5. R. A. A. Muzzarelli, F. Tanfani, G. Scarpini and G. Laterza *J. Biochem. Biophys. Methods*, 2:299 (1980).
6. K. Toei and T. Kohara, *Anal. Chim. Acta*, 83:59 (1976).
7. G. C. Chen & B. R. Johnson *Appl. Environ. Microbiol*, 46:13 (1983).
8. J.E. Castle, J.R. Deschamps and K. Tice, "Chitin, Chitosan and Related Enzymes", P. Zikakis, ed., New York, Academic Press (1984).

Editor's Report

R. A. A. MUZZARELLI

FILMOGENIC PROPERTIES OF CHITIN / CHITOSAN

based on works by

S. AIBA, Y. FUJIWARA, T. HIDESHIMA,
C. HWANG, M. KAKIZAKI, M. IZUME, N. MINOURA
C.K. RHA, T. SHOIJ, A.J. SINSKEY
and A. TSUTSUMI

INTRODUCTION

The film-forming ability of chitosan has been the object of many studies, some of which have led to industrial uses in various fields. Photographic films, reverse osmosis membranes and cosmetics containing significant amounts of chitosan are now being marketed.

Procedures for preparing photographic images by diffusion transfer include chitosan as a protective layer, together with copper salts and glycerol or gelatin. Relatively thick protective layers can be used to enhance the long term stability and integrity of the films. The chelation of copper(II) ions by chitosan in form of membranes has been studied in detail. Reverse osmosis membranes were found to possses resitance to high alkali concentrations and being durable in some organic solvents. Hollow chitosan fibers were spun with a special spinneret: the hollow fiber was coagulated externally with 1 M NaOH solution and internally with gaseous ammonia.

Chitin too lends itself to membrane formation: chitin membrane can be cast from chitin solutions such as the mixture of formic acid and dichloroacetic acid, trichloroacetic acid and dichloroethane, and N,N-dimethyl acetamide, N-methyl-2-pyrrolidone and lithium chloride, the coagulating agents being 2-propanol and acetone.

ENCAPSULATION WITH CHITOSAN: TRANS-MEMBRANE DIFFUSION OF PROTEINS IN CAPSULES

(C. Hwang, C.K. Rha and A.J. Sinskey)

INTRODUCTION

Our laboratory is establishing principles for encapsulation processes for mammalian cell culture technology and other applications. More specifically, current studies are emphasizing control of capsule membrane porosity in order to design and fabricate capsules to desired specifications.

There are a number of demanding initial requirements for mammalian cell culture. Among them are: (1) a surface for anchorage-dependent cell growth, (2) protection from shear and mechanical damages, and (3) packing of the cells. One of the best methods for meeting these requirements is by encapsulation of the cells in a biocompatible capsule. When the cells are encapsulated, the capsule membrane can, for some cells, provide the surface for growth, protect cells from shear and mechanical impact and allow for compartmentalization that leads to higher cell density. Such high density cell systems may simulate natural tissue conditions and reduce serum usage. Encapsulation offers further advantages because it can serve, at the same time, as a separation aid by providing controlled porosity of the semi-permeable capsule membrane. Control of this important variable allows concentration of the product and removal of metabolites or elimination of contaminants.

In our laboratory we have investigated encapsulation processes which are simple, versatile and utilize low cost materials while allowing for control of the membrane strength and porosity. The encapsulation process involves the forming of capsules (with liquid core) by adding drops of a solution of either an anionic polymer composition or a cationic polymer composition to a solution of an ionic polymer of opposite charge.

The most important aspect of the encapsulation process is the control of the porosity and mechanical strength of the capsule membrane. In experiments reported here we have investigated the diffusional properties of capsule membranes by encapsulating proteins of various molecular weights and determining outward transport rates through the capsule membrane.

The chitosan used in this study was obtained from Madera Products, Albany, Oregon. Its degree of deacetylation was 80% and the protein content less than 1%. The molecular weight estimated from intrinsic viscosity using the Mark-Houwink constant was 130,000.[4] A low-calcium sodium alginate was supplied by Kelco, San Diego, CA (Trade name: Keltone). The proteins chosen for the diffusion experiments are fibrinogen (MW 400,000), γ-globulin (MW 156,000), and BSA (MW 69,000). The medium use for the experiments is deionized water.

Background Information

In our laboratory, we have made an extensive study on the solution behavior, both diluted and concentrated, and the film properties of chitosan.[1,2,3] Such studies have provided the fundamental evidence that the microstructure and mechanical properties of chitosan can be controlled by the manipulation of chitosan solution conditions (i.e., pH, concentration, ionic strength, and solvent) and suggested an approach for controlling the porosity of the capsule membrane.

Our studies showed evidence that the morphology of the chitosan molecules, more specifically the hydrodynamic volume or intrinsic viscosity, plays a significant role in the structure of the capsule membrane. We have shown that the intrinsic viscosity $[\eta]$ is dependent on the ionic strength and the pH of the chitosan solution. As the ionic strength increases, the inter- and intra-molecular electrostatic repulsions between chitosan chains are reduced due to the shielding effect of the counter ions. This will allow the chains to come closer together and thus lowers the intrinsic viscosity or the hydrodynamic volume of the chitosan molecules (Figure 1). This effect will enhance the inter- and intra-chain hydrogen bonding and causes the chains to form a special network structure (Figure 2) that results in the increase of the average pore size in the matrix. Similarly, if the pH of the solution is increased, the data on Figure 1 indicate that the intrinsic viscosity will also decrease and the average pore size in the matrix increase. These behaviors of the chitosan solution provide useful predictive tools for controlling the porosity, and possibly the mechanical strength, of the chitosan-alginate membrane.

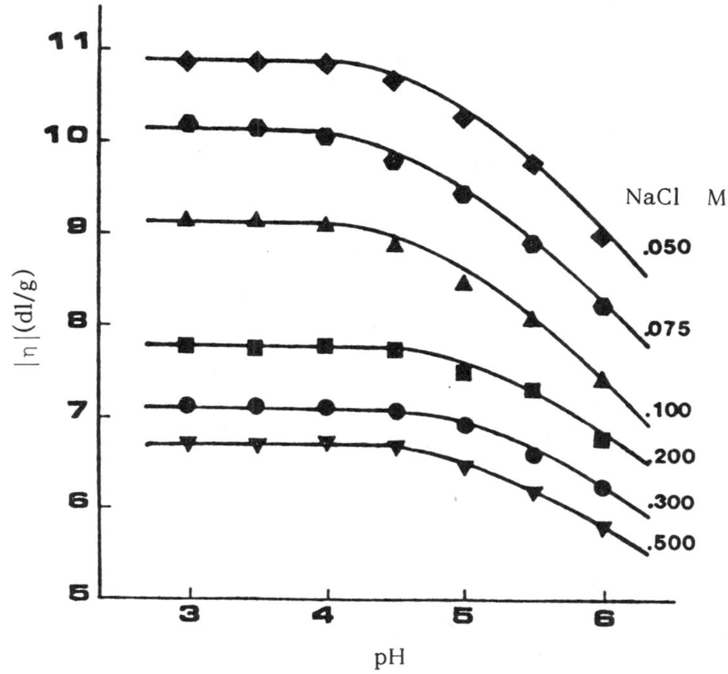

Figure 1. The effects of pH and ionic strength on the intrinsic viscosity |η| of the chitosan solution.

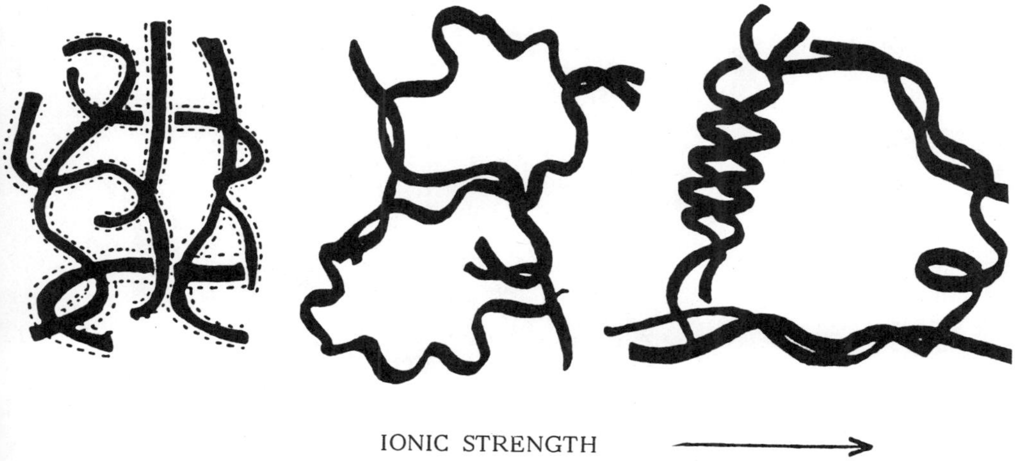

Figure 2. The effect of ionic strength on the network of chitosan chains.

Encapsulation Procedure

The encapsulating process being evaluated is based on the electrostatic interaction between chitosan, a natural polycationic polymer and sodium alginate, an anionic polysaccharide (Figure 3). Initially, chitosan is acid solubilized in 0.01 M HCl and to the solution, varying amounts of salts (NaCl and $CaCl_2$) are added. Afterwards, the chitosan solution is pH adjusted with 1 N NaOH to a desired pH and then the solution is filtered to remove insolubles. Similarly, alginate is completely solubilized in deionized water and NaCl (0.15 M) and protein (0.3% w/v) are added to the solution. When both solutions are prepared, capsules are fabricated by dropping the alginate solution (casting solution) into the chitosan solution (coagulation bath) using the encapsulation device (Figure 4). The device is very easily made, and it simply consists of a plastic cannister with syringe needle outlets (0.9 mm in diameter and 2.5 cm in length) in the bottom and an air inlet near the top. To encapsulate, the alginate solution is added into the cannister and the chitosan solution into the stirred coagulating bath. It is essential to stir the chitosan solution to prevent capsules from sticking together. The process is initiated by applying air pressure (10 psig) to the cannister and to the needles outside the cannister. This back blowing effect for the needles will decrease the drop size and increase the drop rate; this technique also offers a mean to adjust the drop size or drop rate of an individual needle by moving the concentric PVC tube up or down along the needle. When enough capsules are made in the beaker, the capsules are filtered and washed twice with deionized water. Then the capsules are collected and ready to be used.

Diffusion Experiments

Capsules prepared are distributed 0.15 volume fractions into a series of small sample tubes and the outward diffusion of protein from the capsules is initiated by suspending the capsules in water. The diffusion takes place at 37°C with slow mixing. The experiment is monitored at various time intervals by analyzing the protein concentration in the water using the Biorad

Alginate (1.2 %) in Deionized Water
↓ Stir 15 min.
Add NaCl to make 150 mM
↓
Add 0.3 % protein to Alginate solution
↓
Drop into chitosan solution

Chitosan (0.1 %) in in 0.01 MHCl
↓ Stir
Add NaCl to make 117 mM
$CaCl_2$ 5 mM
↓
Adjust pH to 5.9 W/1N NaOH (slowly)
↓
Filter

Filter separate capsules from solution
↓
Wash capsules with deionized water
↓
Collect capsules

Figure 3. Encapsulation procedure.

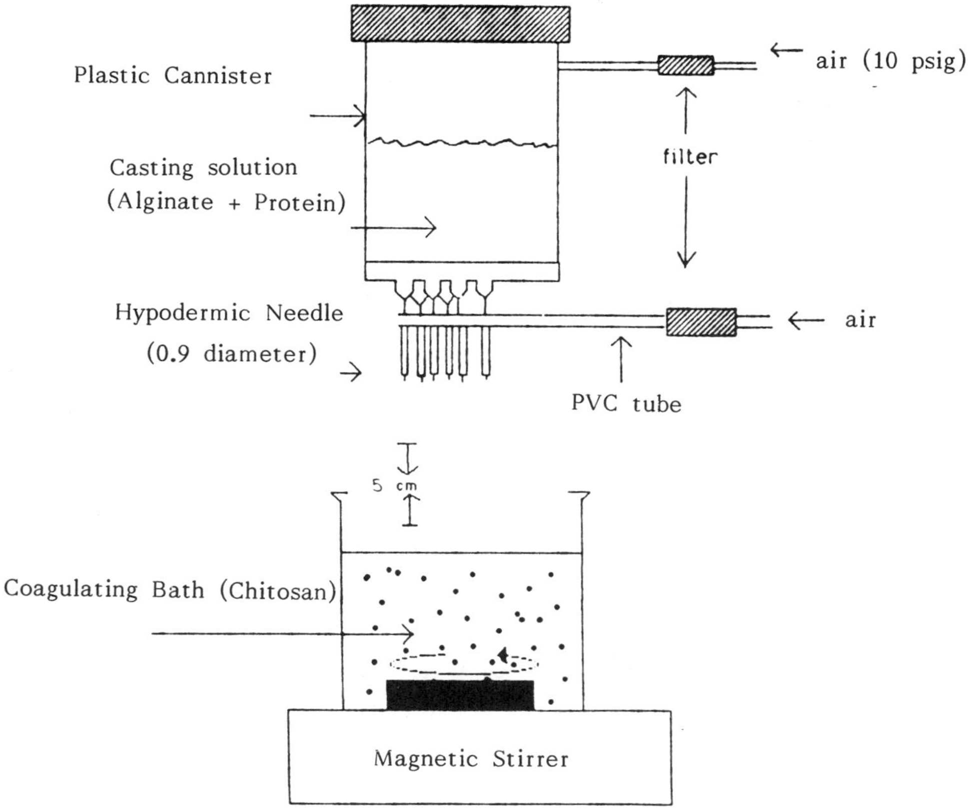

Figure 4. Encapsulation equipment diagram.

protein assay. The experiment is continued until no further increase in the protein concentration outside the capsules was observed.

RESULTS AND DISCUSSION

Initial studies indicate that the thickness of a typical single membrane is approximately 2-5 μms. Depending on the size of the needle used to drop the alginate solution into the chitosan solution, the capsule size obtained ranged from 50-3000 μm in diameters. By blowing air over the needles, capsules of 750-1100 μm in diameter and density of 0.9997 o/ml can be routinely produced. Using this process, 75 ml volume of capsules can be fabricated in less than half-an-hour.

The concentration of the chitosan solution was varied between 0.1% and 0.5% (w/v) and the pH from 5.9 to 3.5. The results for the diffusion of γ-globulin and BSA are shown in Figures 5 and 6. The y axis for both figures is defined as the protein concentration outside the capsules at time t over the equilibrium protein concentration that would have reached if there were no capsules (the fraction of equilibrium reached). For fibrinogen, the largest protein used, there was no detectable diffusion in all four solution conditions that were investigated.

As indicated in the results, the concentration and the pH of the chitosan solution can affect the diffusional rate of proteins which is related to the membrane pore size. By decreasing the pH of the chitosan

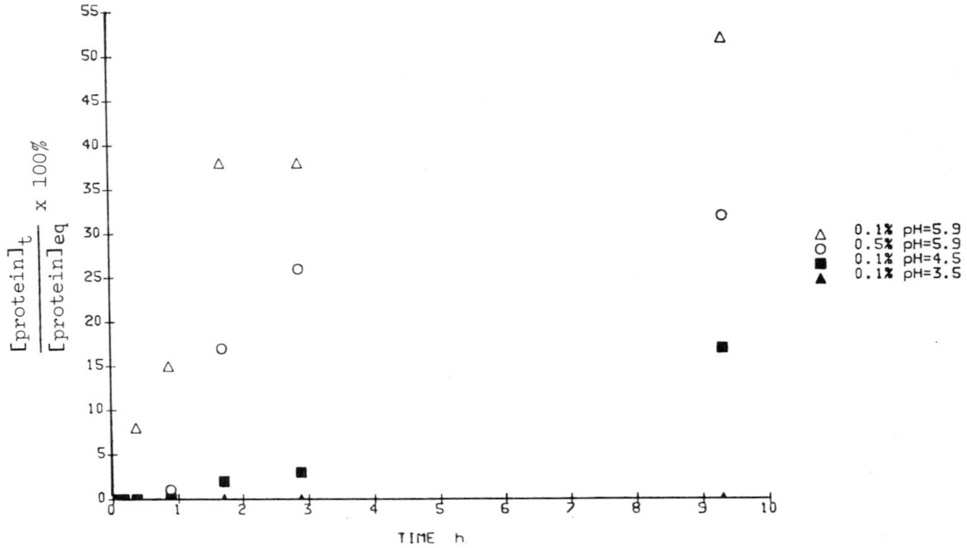

Figure 5. Diffusion of γ-globulin in capsules.

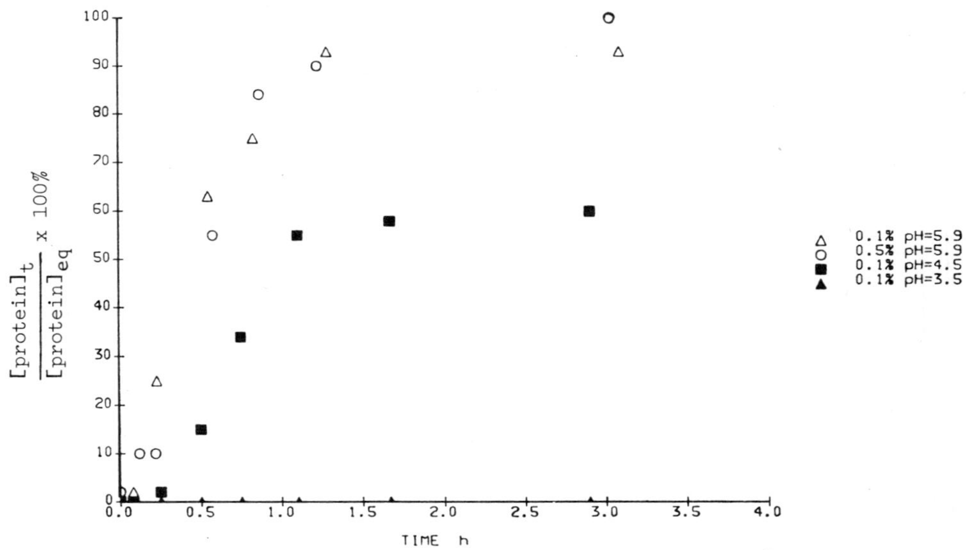

Figure 6. Diffusion of BSA in capsules.

solution from 5.9 to 3.5 at 0.1% concentration, a decrease of the diffusion rate is observed, until zero diffusion occurs for γ-globulin and BSA at pH 3.5. These results illustrate the effect of pH on the membrane pore size as discussed in the background section; the results showed that by decreasing the pH of the chitosan solution at constant concentrations and ionic strength, the rate of diffusion is decreased.

Although not much work is done on the effects of ionic strength and concentration on the diffusional properties of capsule membrane, initial studies have shown the following:

1. As the concentration of $CaCl_2$ and/or NaCl is increased, the protein diffusion rate is decreased and the mechanical strength of the capsule membrane increased.

2. As the pH of the chitosan solution is decreased, the mechanical strength of the capsule membrane is increased.

3. As the concentration of chitosan in the solution is increased to 0.5%, the rate of BSA diffusion is increased while γ-globulin diffusion is decreased.

4. Ionic strength is an important parameter in the formation of a strong capsule.

CONCLUSION

A workable encapsulation process using chitosan has been developed and the effect of pH on the diffusional properties of the capsule membrane has been examined. These experiments demonstrate that we can control and estimate the porosity of capsule membrane from dilute solution behavior of polysaccharides; we have shown that by selecting the appropriate chitosan solution conditions, we can retain or release molecules of different molecular weights.

Although our main objective in this project is to develop an encapsulation system suitable for mammalian cells, we must first determine the quantitative relationship between the casting solution and coagulation bath parameters and the porosity, as well as mechanical strength, of the capsule membrane. These parameters include the type and concentration of membrane-forming polymers, the type and amount of acid used in the chitosan solution, the pH and the ionic strength. After this initial phase is completed, the optimum operating conditions will be established and then the effectiveness of capsules in mammalian cell culture will be evaluated.

Acknowledgements. This work was partially supported by M.I.T. Sea Grant No. 93291 "Chitosan matrix for biotechnology process", from NOAA, Dept. Commerce, the B.F. Goodrich Co., Snow Brand Milk Products Co. Ltd., Tokyo Japan, and Dept. Navy, Office of Naval Research.

REFERENCES

1. C.A. Kienzle-Sterzer, D. Rodriguez-Sanchez and C.K. Rha, "Applied Polymer Science," (1982).

2. E. Robinson-Lang, C.A. Kienzle-Sterzer, D. Rodriguez-Sanchez and C.K. Rha, "Chitin and Chitosan, Proc. 2nd International Conference on Chitin-Chitosan," S. Hirano and S. Tokura, eds., (1982).

3. D. Rodriguez-Sanchez, C.A. Kienzle-Sterzer and C.K. Rha, "Chitin and Chitosan, Proc. 2nd International Conference on Chitin-Chitosan," S. Hirano and S. Tokura, eds., (1982).

4. C.A. Kienzle-Sterzer, D. Rodriguez-Sanchez and C.K. Rha, Characterization of chitosan films, in: "Rheology Applications, Vol. 3," G. Astarita, G. Marrucci and L. Nicolais, eds., Plenum Press, New York, (1982).

PREPARATION AND PROPERTIES OF DIALYSIS MEMBRANES

(S. Aiba, M. Izume, N. Minoura and Y. Fujiwara)

Chitin has potential as a blood dialysis membrane. A number of reports have implied the feasibility of chitin as a dialysis membrane. Tokura et al.[1] prepared a chitin film from a formic acid and dichloroacetic acid (FA-DCA) solution. Kataoka and Ando[2] prepared semipermeable chitin membranes from a trichloroacetic acid and dichloroethane (TCA-DCE) solution. Uragami et al.[3] reported the water flux and poly(ethylene glycol) rejection of chitin membranes cast from a N,N-dimethyl acetamide, N-methyl 2-pyrrolidone and lithium chloride (DMA-NMP-LiCl) solution. Rutherford and Dunson[4] prepared chitin films from DMA-5%LiCl solvent system as reverse osmosis membranes. To the best of our knowledge, the permeation characteristics of chitin as a dialysis membrane have been described in only two patents,[5] in which it was claimed that the dialysis characteristics of the chitin membrane were superior to that of Cuprophane^{R}. There is not much published data about chitin membranes which have been characterized in terms of both dialysis and tensile properties. In this paper we report on a method of preparing a chitin membrane from DMA-NMP-LiCl solvent system (DMA-membrane) and compare its permeation and tensile properties with other membranes prepared by two previously reported methods: from FA-DCA and TCA-DCE solutions (FA and TCA-membranes).

EXPERIMENTAL

The well-characterized starting materials (crab shells) were used throughout the work. A chitin sample (C-1) from the shells of Benizuwaigani (Chionoecetes japonicus) was obtained from Katakura Chikkarin Co. Ltd, Tokyo, Japan. A refined chitin sample (C-5) wasprepared by treating again the C-1 sample in our laboratory. The characteristics of the samples were detailed elsewhere.[6,7] Their molecular weights were over one million. FA and TCA-membranes were prepared in similar manners to the literature.[1,2,6] DMA-membranes were prepared as follows:[6,7] chitin solution in DMA-NMP-LiCl(100:100:5) was prepared by the literature.[3] The solution was cast on a glass plate and immersed in 2-propanol. After gelation the membrane was immersed in water and hot water and dried. The membranes from the C-5 sample were placed in a vacuum oven and annealed. Solute permeation experiments were conducted at 37°C using the permeation apparatus in the similar manner to that of Nakagawa and Fujiwara.[8]

RESULTS AND DISCUSSION

A chitin solution in DMA-NMP-LiCl prepared using the C-1 sample was prepared. The solution (1 %) was very viscous and so it was diluted to

Table 1. Properties of Annealed Chitin Membranes

Heat treatment	Water sorption (%)	Tensile strength at break (MPa)	Elongation (%)	Permeability coefficient to vitamin B_{12} (cm^2s^{-1})
None	220	2.2±0.3	51± 9	8.2 X 10^{-7}
120°C, 2h	170	2.1±0.6	33±12	5.8
120°C, 5h	130	4.6±1.4	46±14	4.6
145°C, 2h	110	5.7±0.4	46± 4	2.8
170°C, 2h	70	4.0±0.9	10± 2	---
Cuprophane	100	29 ±3	38± 7	1.7

Measured at 37°C in water.

0.75 % for the membrane preparation. When the solution cast on a glass plate was immersed in 2-propanol, the solution coagulated. The gel membrane thus obtained was then immersed in water. The membrane shrunk slightly and DMA, NMP and LiCl were washed out. On treatment in hot water the membrane became transparent and DMA, NMP and LiCl were removed. The DMA-membrane was compared with other membranes (FA and TCA-membranes). All the membranes were hard, stiff and brittle in the drystate. The membranes absorbed about three times their weight of water. The DMA-membrane had higher tensile strength than the others in both the dry and the wet states. The use of the solvent system, DMA-NMP-LiCl, seems to be an improvement over other methods of preparation. The three chitin membranes had greater solute permeability than Cuprophane. The order of the permeability of the three membranes was as follows: FA>TCA>DMA.

The chitin membrane was successfully prepared by coagulation of DMA-NMP-LiCl solution of chitin with 2-propanol. The possibility of other coagulants was investigated using the C-5 sample. The solution cast on a glass plate was immersed in methanol, ethanol, acetone, 2-propanol, water and mixtures of 2-propanol and water. The gel membrane obtained in 2-propanol was transparent with minimal contraction in this study. Acetone was not useful because the membrane became opaque. Methanol, ethanol and mixtures of 2-propanol and water were not favorable since the gel membranes obtained in these coagulants shrunk moderately and were not flat. 2-Propanol was favorable as a coagulant as compared with the others. The DMA-membrane was not so strong in the wet state. The weakness seemed to be due to low crystallinity. In order to increase the crystallinity and tensile strength the DMA-membranes from the C-5 sample were annealed under reduced pressure under the conditions cited in Table 1. Annealing made the membranes dense and strong. The water sorption of the membranes decreased and the tensile strength increased with increasing annealing time and temperature. D.s.c. was applied in order to estimate the crys-

Table 2. Effect of γ-ray Irradiation on Tensile Properties of Chitin Membranes

Dose of γ-ray (Mrad)	Tensile strength at break (MPa)	Elongation (%)
0.0	2.5±0.2	53±14
1.0	1.9±0.2	34± 6
4.6	1.4±0.2	14± 3
8.37	0.9±0.3	7± 3

Water-swollen membranes were tested at 25°C.

tallinity of the membranes in terms of the energy of fusion. The annealed membranes had high values for the energy of fusion than the unannealed one. D.s.c. data imply clearly that the annealing caused the membranes to have high crystallinity. Permeability of the membranes to vitamin B_{12} was summarized in Table 1. Even the membrane annealed at 145°C had higher permeability than Cuprophane. Table 2 summarized the effect of γ-ray irradiation in the wet state on the tensile properties of the C-5 chitin membranes. The tensile strength was decreased with the increase in the irradiation dose. In general medical goods are sterilized with the dose of 2.5 or 4.5 Mrad. The water-swollen chitin membrane deteriorated even with the dose of 1.0 Mrad. Other sterilizing methods should be used.

CONCLUSION

The membrane from DMA-NMP-LiCl solution by coagulation with 2-propanol was stronger in the wet state than the other membranes and had higher permeability to solutes than Cuprophane. Annealing made the membrane dense and strong with low permeability to solutes. Even the annealed membranes had higher permeability than Cuprophane. The chitin membrane deteriorated by γ-ray irradiation in the wet state.

REFERENCES

1. S. Tokura, N. Nishi, and J. Noguchi, Studies on chitin.III. Preparation of chitin fibers, Polym. J., **11**:781 (1979).
2. S. Kataoka, and T. Ando, Regenerated chitin film from the solution of trichloroacetic acid systems, Kobunshi-Ron., **36**:175 (1979).
3. T. Uragami, Y. Ohsumi, and M. Sugihara, Studies on syntheses and permeability of special polymer membranes: 35. Preparation and permeation characteristics of chitin membranes, Polymer, **22**:1155 (1981).
4. F.A. Rutherford and W. A. Dunson, The permeability of chitin films to water and solutes, in: "Chitin, Chitosan, and Related Enzymes," J. P. Zikakis, ed., Academic Press, New York (1984).
5. Mitsubishi Rayon Co., Japan Kokai Tokkyo Koho, 79-139294 (1979); 81-38103 (1981).
6. S. Aiba, M. Izume, N. Minoura, and Y. Fujiwara, Studies on chitin: 2. Preparation and properties of chitin membranes, Carbohydr. Polym., in press.
7. S. Aiba, M. Izume, N. Minoura, and Y. Fujiwara, Studies on chitin: 3. Effect of coagulants and annealing on the preparation and the properties of chitin membrane, Br. Polym. J., in press.
8. T. Nakagawa and Y. Fujiwara, Improvement of permeability of low molecular weight substances through poly(L-leucine-co-γ-benzyl L-glutamate) by addition of hydrophilic property, Nippon Kagaku Kaishi, 860 (1983).
9. R.H. Hackman, Chitin. 1. Enzymic degradation of chitin and chitin esters, Aust. J. Biol. Sci., **7**:168 (1954).

MOLECULAR MOTION AND DIELECTRIC RELAXATION IN CHITIN AND ACYLCHITINS

(M. Kakizaki, T. Shoji, A. Tsutsumi and T. Hideshima)

INTRODUCTION

Chitin is not soluble in usual organic solvents and consequently very intractable for processing. Recently a series of acylated chitins having various aliphatic chains has been derived, which are soluble in various organic solvents.[1~3] Introduction of a flexible side chain into chitin molecule modifies the mode of molecular motion and consequently causes considerable change in physical properties as well as in the solubility in the organic solvents. Therefore, it is interesting and important to investigate the effect of the acyl groups on the molecular motion and physical properties. In this report we made broad-line NMR and dielectric measurements on chitin and several acylchitins to investigate the effect of the side-chain length on the molecular motion and dielectric properties.

EXPERIMENTAL

The samples used were chitin, acetylchitin, butyrylchitin, caproylchitin and caprylchitin, in which the acyl group is $-COCH_3$, $-CO(CH_2)_2CH_3$, $-CO(CH_2)_4CH_3$ and $-CO(CH_2)_8CH_3$, respectively. Films of the samples were obtained by casting their solutions on a glass plate. The solvent for acetylchitin was formic acid, while that for other acylchitins was chloroform. The film of chitin was made by pressing powdered sample at room temperature. These films were heat-treated in vacuum at 50°C for 3 days. DSC thermogram suggested that the glass transition temperatures of these samples were higher than 170°C though they were not clearly determined owing to the change in quality. Complex permittivity was measured with Gen Rad capacitance bridge type 1616 in the audiofrequency range at temperatures ranging from ca. -180°C to room temperature. Broad-line NMR measurements were carried out with JNM-W-40 spectrometer (rf. 40 MHz) made by Japan Electron Optics Laboratory Co. in the temperature range from ca. -160°C to ca. 50°C.

RESULTS AND DISCUSSION

Figure 1 summarizes the data on temperature dependence of loss permittivity at 1 kHz for the five samples. It is remarkable that chitin and acetylchitin show only one peak whereas the other samples having longer side chains show two separate peaks. We designate these relaxation processes as β and γ in the order of descending temperature. The α process suggested by

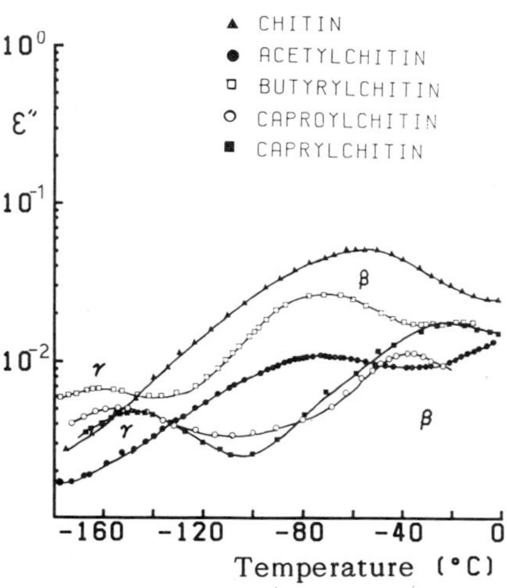

Fig.1. Temperature dependence of loss permittivity at 1 kHz for chitin and acylchitins.

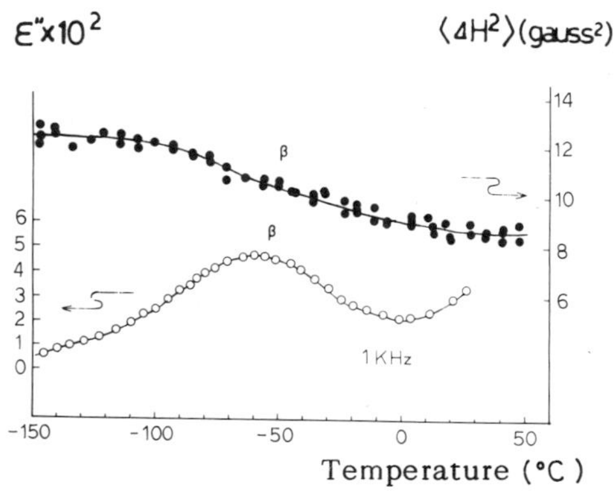

Fig.2. Temperature dependence of second moment and loss permittivity at 1 kHz for chitin.

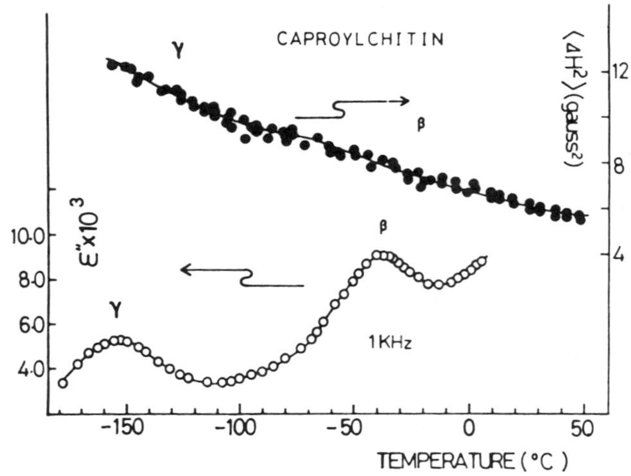

Figure 3. Temperature dependence of second moment and loss permittivity at 1 kHz for caproylchitin.

the upswing of loss permittivity at higher temperatures was hardly observed because the samples change in quality at higher temperatures. It seems that the acyl groups shorter than acetyl shift the β process slightly to lower temperatures, while the longer acyl groups shift it to higher temperatures and besides they cause an additional relaxation called γ in the lower temperature region. The γ process shifts to higher temperatures with the increase of the side-chain length.

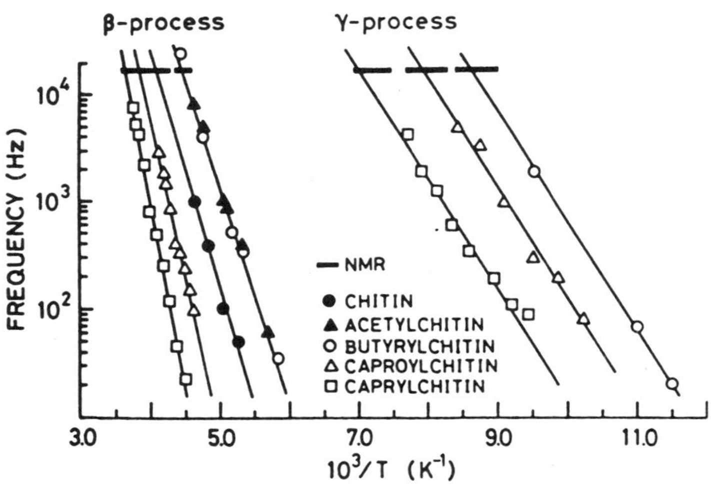

Figure 4. Relaxation map for chitin and acylchitins.

In connection with the loss permittivity data, temperature dependence of second moment of the NMR spectrum shows motional narrowing at ca. -20°C for all the samples and another motional narrowing below -100°C for the samples having longer side chain than acetylchitin, as shown typically in Figs. 2 and 3 for chitin and caproylchitin. In these figures loss permittibity data are also shown and the temperatures for loss peaks are seen to be in good agreement with those for the motional narrowing. Therefore, we designate the narrowing processes also as β and γ. The correspondence between the NMR and dielectric processes are more clearly evidenced on a relaxation map shown in Fig. 4, where frequency-temperature positions of loss permittivity peaks and narrowing regions observed at the effective frequency of ca. 10^4 Hz are plotted against the reciprocal absolute temperature. The narrowing regions represented by bars correspond well to the straight lines for the dielectric relaxation processes. The activation energy for the β relaxation varies from ca. 9 kcal/mol to 15 kcal/mol while that for the γ relaxation is ca. 5 kcal/mol.

Characteristic features of the relaxations given above are reminiscent of the side-chain relaxations in poly(n-alkyl methacrylate)s, PnAMA, which are also designated as β and γ.[4,5] The β relaxation in PnAMA is due to the motion of the side chain as a whole and appears in every member of this series, while the γ relaxation is attributed to the local motion of the side chain and arises in the members having longer side chain. Both relaxations seem to shift to higher temperatures with the increase of the side-chain length. Furthermore, the shapes of the loss permittivity vs. frequency curve for the β and γ relaxations in chitin and acylchitins are independent of the side-chain length as in the case of PnAMA[6] and very similar to those for the corresponding relaxations in PnAMA as shown in Figs. 5(a) and 5(b). The activation energies for the β and γ relaxations in PnAMA are ca. 19 kcal/mol and ca. 6 kcal/mol respectively. Though the value for the β relaxation is somewhat larger, that for the γ relaxation is nearly equal to that in acylchitins. From these results, it is strongly suggested that the β and γ relaxations in chitin and acylchitins are caused by the same motions of the side chain as those in PnAMA.

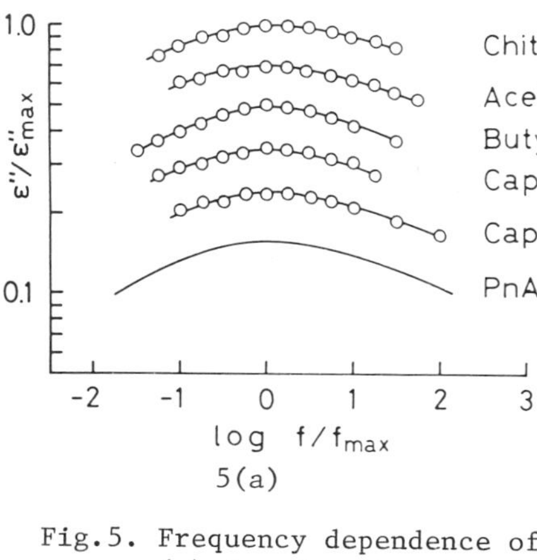

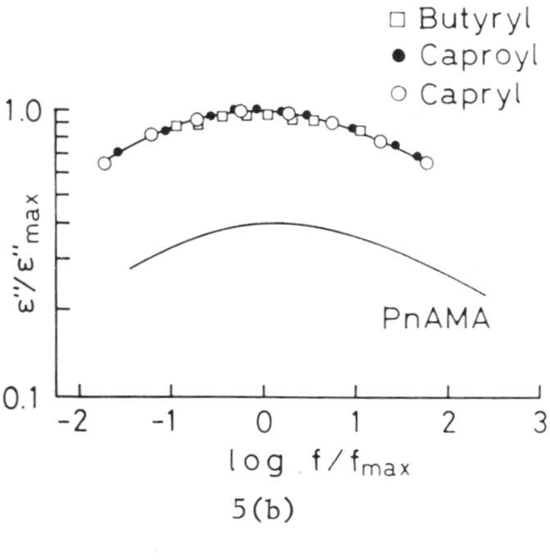

Fig.5. Frequency dependence of loss permittivity in normalized form: (a) for β relaxation in chitin, acylchitins and PnAMA; (b) for γ relaxation in acylchitins and PnAMA; ordinate is correct only for the top curves.

REFERENCES

1. N. Nishi, J. Noguchi, S. Tokura and H. Shiota, Polym. J., 11, 27 (1979).

2. K. Kaifu, N. Nishi, T. Komai, S. Tokura and O. Somorin, Polym. J., 13, 241 (1981).
3. K. Kaifu, N. Nishi and T. Komai, J. Polym. Sci., Polym. Chem. Ed., 19, 2361 (1981).
4. E. A. Hoff, D. W. Robinson and A. H. Willbourn, J. Polym. Sci., 18, 161 (1955).
5. Y. Kawamura, S. Nagai, J. Hirose and Y. Wada, J. Polym. Sci., Part A-2, 7, 1559 (1969).
6. T. Tetsutani, M. Kakizaki and T. Hideshima, Polym. J., 14, 305 (1982).

Affiliations

M. Izume, Division of Research & Development, Katakura Chikkarin Co., Ltd., Ohtemachi, Chiyoda-ku, Tokyo 100, Japan.
S. Aiba, N. Minoura and Y. Fujiwara, Industrial Products Research Institute, 1-1-4 Yatabe-machi Higashi, Tsukuba-gun, Ibaraki 305, Japan.
C. Hwang, C.K. Rha and A.J. Sinskey, Biomaterial Science & Engineering Lab. M.I.T., Dept. Biological Sciences, Cambridge MA 02139, USA.
M. Kakizaki, T. Shoij, A. Tsutsumi and T. Hideshima, Department of Applied Physics, Faculty of Engineering, Hokkaido University, Sapporo 060, Japan.

PHYSICAL MODIFICATIONS OF CHITIN AND CHITOSAN

M. YALPANI

Domtar Inc., Research Centre
P.O. Box 300
Senneville, P.Q. H9X 3L7, Canada

INTRODUCTION

The utility of water soluble polysaccharides as suspending, stabilizing, viscosifying and gelling agents for many industrial applications has been recognized for many years. Products such as alginic acid, cellulose, guar-gum, starch and xanthan gum and their derivatives are being widely used in the food, cosmetic, paint, pharmaceutical, oil and various other industries (1, 2).

In addition to non-fibrous derivatives, fibrous, microcrystalline polysaccharides, such as cellulose, chitin and amylose, are known to display useful properties for a wide spectrum of commercial applications, primarily in the food and pharmaceutical industries, as described by Battista (3) and Austin et al. (4). For example, one of the features of colloidal microcrystalline cellulose gels is their unusual viscosity stability with respect to prolonged standing or elevated temperatures. This property is exploited in food applications in which viscosity-control agents are required to be compatible with sterilization or other types of heat treatments.

Austin et al. (4) have recently described a method for the preparation of microcrystalline chitin which is based on a combined chemical and physical treatment involving an initial acid hydrolysis step, followed by washing, centrifugation, shearing, and lyophilization. However, the molecular weight of the resulting product (M.W. 75,00) was reported to be substantially lower than that of the native polymer (800,000). Other indications of the degradation and modification of the microcrystalline chitin included a reduced acetyl content, limited solubility in the dimethylacetamide-lithium chloride system, as well as low level incorporations of phosphorous (from the hydrolytic treatment). In addition to these disadvantages, the procedure is relatively complex and economically unattractive.

One aspect of the work reported here evolved from previous studies (5, 6) directed at the development of methods which may facilitate a more extensive commercial utilization of chitin and chitosan, and was prompted, in part, by recent reports of the conversion of pulp samples into "microfibrillated" products (7, 8). The main objective here was to design a method for the preparation of stable chitin and chitosan suspensions and gels, whose properties would be amenable to exploitation either directly or via subsequent chemical modifications. The process selected for this purpose is based on a ysical method in which the solid biopolymers are subjected to high shear orces which result in drastic changes in the physical properties of the materials. This modification can be conveniently accomplished by repeatedly passing slurries of the polysaccharide through a commercial homogenizer at pressures of between 4,000 to 15,000 psig.

Another aspect of this work relates to a problem commonly encountered in the preparation of uniform solutions or dispersions of chitin and chitosan, and of many other polysaccharides. The solubilization process may often require special equipment or pretreatment in order to obtain homogeneous solutions free of undesirable clumps, micro gels or "fish eyes". Among the methods commonly employed for good dispersion are high-shear mixing, addition of the solid polymer via an aspirator, and preblending with various ingredients such as non-aqueous or non-miscible liquids. Sandford et al. (9) have reviewed another method which is based on the pretreatment of xanthan gum and other polysaccharides with glyoxal to obtain improved dispersibility. It would be clearly desirable to have available facile, inexpensive and widely applicable methods for this purpose which do not lead to depolymerization or extensive chemical modification or the degradation of the polymer.

EXPERIMENTAL

Materials and equipment

Chitin powder and chitosan flakes (both from crab shells) were obtained from Sigma Chemical Co. and Kypro Co., respectively. The coarse chitosan flakes were ground into a powder to pass a 0.5 mm screen. Sodium alginate and xanthan gum were kindly supplied by Kelco Co. Sodium carboxymethyl cellulose was obtained from Hercules Inc., lactose from Sigma Chemical Co., glucose and sucrose from Fisher, and Emersol 213 from Emery.

A laboratory homogenizer Model 15M (Gaulin Corporation) was employed which had a pumping rate for water of 750 ml/min. Viscosity determinations were performed at 25.0° C using a Haake Rotovisco RV12 viscometer and the MIPSt profiled cup and bob measuring system.

Preparation of polysaccharide suspensions and solutions

Slurries containing 1-5 % by weight of the polysaccharides in either water, dimethyl sulfoxide or other polar solvents were repeatedly passed through the homogenizer for between 3 to 40 volume passes at pressures of 4,000-15,000 psig, until the desired consistencies were attained. Alternatively, the solid biopolymers were slowly added during several volume passes to a circulating solution of the appropriate solvent. The latter method allowed for the processing of coarser materials, such as chitosan flakes, without causing plugging of the homogenizing valve which would otherwise be observed. Unless indicated otherwise, additives were incorporated into the suspensions prior to the homogenization step. Polysaccharide solutions were prepared in essentially the same manner. All samples contained 0.02 % NaN_3 and products were stored at 5° C for periods of up to two years.

RESULTS AND DISCUSSION

Homogeneous and stable chitin suspensions were obtained by homogenizing slurries containing 1-5 % biopolymer in water, dilute aqueous acetic acid, or dimethyl sulfoxide at pressures of between 6,000 to 9,000 psig for five to ten volume passes. The products were lighter in colour than the starting materials and had consistencies which ranged from syrupy to almost gel-like, depending on the concentration of solids and the number of passes employed. Thus, a 2 % chitin slurry was treated at 9,000 psig for five passes resulting in a fully homogeneous, viscous suspension which, after further treatment for five passes, assumed the consistency of a stable, non-flowing gel. Similarly, uniform suspensions of 2 % chitin could be obtained using dilute aqueous acetic acid, dimethyl sulphoxide or other solvents. Interestingly, samples of homogenized chitin suspensions retained ther homogeneity even after processing at 7,000 psi for 15 passes, remained homogeneous after a seven-fold dilution with water.

In addition to the simple biopolymer suspensions, it is also possible to incorporate other components either prior, during, or after processing. Addition of hydrophilic materials, such as carbohydrates, may, for instance be useful for the purpose of tailoring rheological or other properties of the product suspensions or gels. Thus, chitin suspensions were prepared which included varying concentrations of glucose, lactose, sucrose, sodium alginate, and chitosan, without affecting the homogeneity of the resulting suspensions. On the other hand, incorporation of a surfactant (2 % Emersol 213) in a 1.5 % aqueous chitin slurry afforded, after homogenization at 9,000 psi for 10 passes, a stable, homogeneous emulsion.

Similar results were obtained for chitosan, where 1-2% slurries produced thick suspensions typically at pressures of 7,000-9,000 psi and after 10-40 volume passes.

Some of the other interesting features of the products included the following. The chitin and chitosan suspensions and gels were found to be stable over extended periods (about two years) without any significant settling, softening, or other apparent physical changes. It appears that the homogenization process does not lead to depolymerization. Preliminary rheological evaluations of the suspensions have revealed a general pseudoplastic behaviour, i.e. high viscosities at zero or low shear rates which rapidly decline with increasing shear rates. For example, the apparent viscosity of a 5% aqueous chitin suspension (obtained after processing at 9,000 psi for 25 passes) was 923,190 mPa.s at a shear rate of 0.01 s^{-1}, 4,000 mPa.s at 0.64 s^{-1}, and 147 mPa.s at 512 s^{-1}. Similar trends were observed for chitosan, where, for example, a 2% aqueous suspension (9,000 psi, 40 passes) displayed a viscosity of 2,946 mPa.s at 1 s^{-1} and 42 mPa.s at 512 s^{-1}. The apparent viscosity of a 2% chitin suspension in 1% aqueous acetic acid (9,000 psi, 30 passes) declined from 6,440 mPa.s at 1 s^{-1} to 32 mPa.s at 512 s^{-1}. The modification of the biopolymer suspension rheology can be exemplified by a 2% aqueous chitin product, for which incorporation of 1.3% (w/w) sodium alginate resulted in a considerably smaller slope of the pseudoplastic viscosity curve, as compared to an identically processed simple chitin suspension over a similar shear interval.

In addition to facilitating the preparation of biopolymer suspensions as described above, the homogenization process is equally applicable to the preparation of homogeneous solutions. It is, for example, well known that aqueous organic acids, such as formic or acetic acid, at low concentrations are good solvents for chitosan. However, using these solvents and conventional dissolution procedures for chitosan (flakes or powders), various amounts of microgels or undissolved solids often remain, even after prolonged agitation of the solutions. This necessitates filtration or other suitable separation procedures prior to any processing of the solutions for most applications. The said process on the other hand, allows for the preparation of homogeneous chitosan solutions in, e.g., dilute aqueous organic acids, such as 1% acetic acid, in the presence or absence of other ingredients within a matter of a few minutes. The process is similarly applicable to other polysaccharides, such as xanthan gum. It should be noted that the polysaccharides can be processed in various ways, using, for example, fine powders, relatively coarse flakes (to pass mesh size 32) or preswelled or prehydrated materials. It is apparent from the above, that the process described here can contribute to significant reductions in the time and cost of processes involving the preparation and use of chitin and chitosan solutions.

CONCLUSIONS

Although the results presented here are mostly preliminary in nature, they do provide an indication of the utility of the homogenization process for the preparation of chitin and chitosan solutions and new types of suspensions and gels (10). The products obtained by this procedure can be envisaged to have a range of applications. Thus, chitin and chitosan suspensions may be used as inexpensive thickening and suspending agents in pharmaceuticals, foods and cosmetics. The enhanced surface area and chemical reactivity of chitin and chitosan products are also of interest in other areas, such as in metal chelation (11) or catalysis (12).

REFERENCES

1. R. L. Whistler, ed., "Industrial Gums", 2nd edn., Academic Press, New York (1973).
2. G. O. Aspinall, ed., "Polysaccharides", Academic Press, New York (1982).
3. O. A. Battista, "Microcrystal Polymer Science", McGraw Hill, New York (1975).
4. P. R. Austin, C. J. Brine, J. E. Castle, and J. P. Zikatis, Science, 212:729 (1981).
5. M. Yalpani and L. D. Hall, Macromolecules, 17:272 (1984).
6. L. D. Hall and M. Yalpani, U. S. Patent 4,424,346 (1984).
7. A. F. Turbak, F. W. Snyder, and K. R. Sandberg, U.S. Patent 4,341,807 (1982).
8. F. W. Herrick, R. L. Casebier, J. K. Hamilton, and K. R. Sandberg, J. Appl. Polym. Sci., Symp. 37:797 (1983).
9. P. A. Sandford, J. Baird, and I. W. Cottrell, ACS Symp. Ser. No. 150:33 (1981).
10. M. Yalpani, U. S. Patent Application (1985).
11. L. D. Hall and M. Yalpani, Carbohydr. Res., 83, C5 (1980).
12. B. J. Arena, U. S. Patent 4,274,980 (1983).

Editor's Report

R. A. A. MUZZARELLI

WHOLE CELLS AND ENZYMES IMMOBILIZED ON CHITOSAN

based on works by

J. BRAUN, R. CHAMY, E. FLASCHEL, D. KNORR, A. ILLANES, P. LeCHANU, F. LeGOFFIC, J. LEONIL, M. MITSUTOMI, A. OHTAKARA, A. RENKEN, S. SICSIC, J. SYNOWIECKI, Y. UCHIDA, E. VAN GRIETHUYSEN, and M. E. ZUNIGA

INTRODUCTION

A proliferation of carriers for immobilizing enzymes have been reported in the literature over the last twenty five years. Among the polysaccharides cellulose, not surprisingly, has been a popular choice in many investigations and applications. More recently however some attention has been focussed on the possibility of using chitin, or its derivatives, as possible carriers. The high availability and low cost, the low toxicity, the potential for varied derivatization, and the multiple physical forms of this polymer, all combine to make it an interesting compound for study as a suitable support.

Shrimp chitin, crab chitin particles, chitosan flocs, ground chitosan membranes, as well as modified chitosans have all been successfully used in this regard. Various enzymes such as acid phosphatase, α-chymotrypsin, gluco-amylase, glucose oxidase, trypsin, urease, β-glucosidase, β-galactosidase and lysozyme have been immobilized under various conditions on these supports. Table I gives a more complete list of those enzymes which have been immobilized, together with proposed applications.

Enzymes can be immobilized on chitin/chitosan by chemical linking with glutaraldehyde and/or by adsorption. The preferred immobilization technique is based on the use of glutaraldehyde as a crosslinking reagent to form aldimine bonds with chitosan and protein. However, fundamental research on the actual mechanism of combination between chitin and glutaraldehyde is needed, as no scientific basis for choosing a chitin/glutaraldehyde ratio has been established. In some cases chitin can apparently bind enzymes directly, as they contain specific binding sites. The occupation of these sites does not apparently detract from the catalytic effect of the protein. This is the case with chitin-immobilized α-amylase and glucoamylase from the mutant HF-15 of *Aspergillus awamori* var. *kawachi*.

From an industrial point of view, the morphology of the carrier is extremely important with respect to surface area and pore parameters. Ideally the carrier should combine the advantages of an inorganic support such as mechanical stability and appropriate density, with those of organic supports such as flexibility, elasticity and high coupling efficiency. With this in mind, novel organic-inorganic carriers, suitable for use in fluidized bed reactor systems, are being developed. Examples are the coating of silica gel particles

TABLE I. Enzymes and cells immobilized on chitin or chitin derivatives, and proposed uses.

Alkaline phosphatase and acid phosphatase.
AMP-deaminase: deamination of AMP to IMP.
Amylase, diastase and glucoamylase: conversion of potato starch, raw corn starch and glycogen to D-glucose.
α-Chymotrypsin: plastein synthesis from soya and alfa-alfa.
α-Galactosidase.
β-D-Galactosidase: hydrolysis of lactose in milk and whey.
D-Glucose isomerase: isomerization of D-glucose to D-fructose.
D-Glucose oxidase: preparation of D-gluconic acid.
β-D-Glucosidase: hydrolysis of cellobiose.
Invertase: sugar production.
Lysozyme: preparation of pharmaceuticals.
Papain: removal of haze from beer.
Penicillin acylase.
Pepsin: controlled digestion of proteins.
Pronase, subtilisin and trypsin: cosmetics and food proteins.
Urease: conversion of urea to ammonia and carbon dioxide.
Amaranthus tricolor cells.
Asclepias syriaca cells.
Escherichia coli cells: synthesis of L-tryptophan.
Nitrosomonas europaea cells: nitrification of waste water.

---

with chitosan, and their subsequent use for immobilization of the enzymes trypsin, invertase and β-galactosidase.

In recent years the practice of using non-growing whole or lysed microbial cells, rather than purified enzymes, is gaining popularity. Advantages of immobilized cells are simple product isolation, and repeated use in continuous processes. The cells are normally trapped within a network, such as an ionotropic gel. In this case a polyelectrolyte (polyanion or cation) is ionically cross linked with the oppositely charged multivalent counterions. Various authors have used this principle to develop a cell immobilization process to getting spherical chitosan biocatalysts. By dropping a chitosan solution (1-2 %) containing suspended cells to a cross linking solution with anionic counterions (1-4 %) they obtained spherical chitosan cell biocatalysts, suitable for use in a bioprocess.

Immobilization of whole cells is also used in plant biotechnology. Immobilization of culture plant cells has concentrated on the entrapment of cells in various matrices, with gel formation by ionic cross linking of a charged polymer, with alginate being the most widely used polymer. However, plant metabolites of commercial interest are often stored within cellular vacuoles, requiring membrane permeabilization for their release. Increase in cell permeability of cultured soybean cells by chitosan has been reported, and also the increased release of oxalate from cultured amaranth cells by chitosan has been demonstrated. Chitosan gels have been used effectively for entrapment of microbial cells and have been proposed for flavour encapsulation. Recently the feasibility of chitosan gels for entrapment of cultured plant cells has also been shown.

As chitin and chitosan are especially acceptable to the food industry, and with expanding interdisciplinary interest in chitin, it is likely that much more interest will be focussed on possible uses of these materials in biotechnological applications.

IMMOBILIZATION OF THERMOSTABLE ALPHA-GALACTOSIDASE ON CHITIN AND ITS APPLICATION

(A. Ohtakara, M. Mitsutomi, and Y. Uchida)

α-Galactosidase is a useful enzyme not only in the determination of raffinose and the structural studies of carbohydrates, but in the beet sugar industry and for soymilk processing. This paper deals with immobilization of the *P. cinnabarinus* α-galactosidase on chitin and its application for the hydrolysis of galactooligosaccharides in soymilk.

MATERIALS AND METHODS

Pycnoporus cinnabarinus IFO 6139 was cultured according to the procedure described in our previous report (2). α-Galactosidase was purified from the culture broth by precipitation with ammonium sulfate, heat treatment and DEAE-Sephadex A-50 column chromatography, as described previously (2). Immobilized α-galactosidase was prepared by incubating the enzyme (50 units) with colloidal chitin (0.2 g) previously treated with glutaraldehyde (2.5 % at final concentration), in 0.05 M acetate buffer of pH 5.0 at 25° C for 1 h. In the assay of α-galactosidase activities, the reaction mixtures consisting of 0.4 ml of 4 mM p-nitrophenyl-α-D-galactoside (pNPGal), 0.4 ml of McIlvaine buffer (pH 5.0), and 0.2 ml of native enzyme solution or immobilized enzyme suspension, were incubated at 37° C for 10 min with shaking. The reaction was stopped by adding 4 ml of 0.2 M Na_2CO_3, and the p-nitrophenol released was measured by reading the absorbance at 420 nm, after centrifugation in the case of immobilized enzyme. One unit of enzyme activity was defined as the amount of enzyme which released one mol of p-nitrophenol.

RESULTS AND DISCUSSION

Immobilization of α-galactosidase on chitin

Immobilization of this α-galactosidase on chitin, colloidal chitin, chitosan, colloidal chitosan, DEAE-Sephadex A-50 and DEAE-cellulose DE-23, was first tested. The enzyme was well bound to chitin and chitosan, but the activity yield was highest when colloidal chitin was used as a carrier.

The optimal condition for immobilization of the enzyme on colloidal chitin was then examined. The enzymes (various amounts from 10 to 150 units) were incubated with 0.2 g of colloidal chitin at pH 5.0, at 25° C for 1 h (Table I). The activity yield was highest when 50 units of the enzyme was applied. In this condition, the effects of pH and incubation time on the immobilization were investigated (Fig. 1). As pH changed from 3.5 to 7.0, the unbound enzyme activity rapidly increased and the enzyme was well

Table I. Effect of Enzyme Amount on Immobilization of α-Galactosidase on Colloidal Chitin

Added activity (units)	Unbound activity (units)	Immobilized activity (units)	Activity yield (%)
10	1.3	7.0	70.0
25	3.1	19.8	79.2
50	5.6	42.6	85.2
100	28.5	56.1	56.1
150	58.5	68.5	45.7

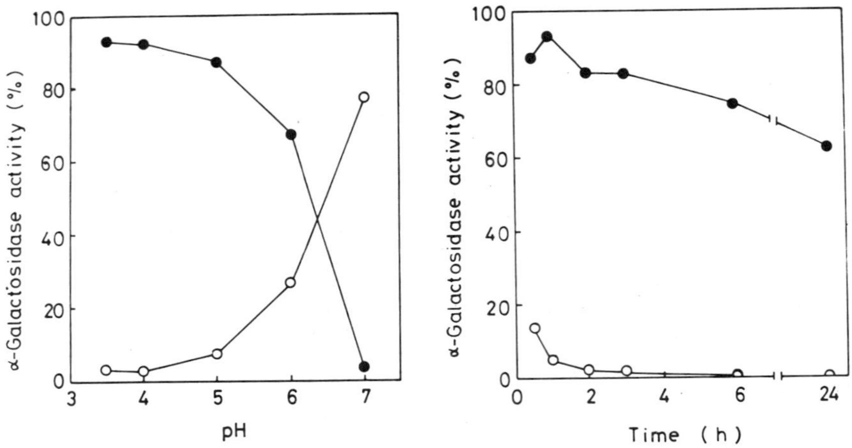

Fig. 1. Effect of pH (left) and incubation time (right) on immobilization of α-galactosidase on colloidal chitin. Enzyme, 50 units; colloidal chitin, 0.2 g.
-○-, unbound enzyme activity; -●-, immobilized enzyme activity.

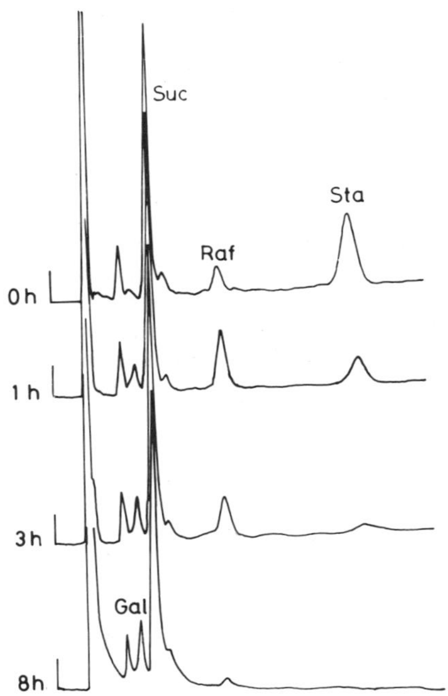

Fig. 2. Hydrolysis of galactooligosaccharides in soymilk by immobilized α-galactosidase. Gal, galactose; Suc, sucrose; Raf, raffinose; Sta, stachyose.

immobilized in the pH range of 3.5 to 5.0. When the enzymes were incubated (various times from 30 min to 24 h) about 90 % of the activity was retained on colloidal chitin at the short incubation times of 30 to 60 min.

Enzymatic properties of immobilized α-galactosidase

The immobilized α-galactosidase was most active around pH 5.0, and was stable between pH 3 and 9 after incubation for 2 h at 37° C. The enzyme had maximum activity at 75° C, and was stable below 80° C after incubation for 15 min at pH 5.0. The enzyme was strongly inhibited by Ag^{+} and Hg^{++} at concentration of 0.1 mM. The enzyme was inhibited powerfully by galactose and melibiose, and markedly by glucose, maltose and stachyose at the concentration of 10 mM. The Km value of the enzyme for pNPGal was 0.31 mM. The enzymatic properties of the immobilized α-galactosidase are very similar to those of native enzyme (1). These results strongly suggest that the catalytic site of the enzyme is not affected by immobilization. The immobilized α-galactosidase retained about 90 % of the initial activity even after repeated uses for 20 times. Such stability is advantageous to continuous use of this α-galactosidase.

Application of immobilized α-galactosidase

The soymilk tested contained 0.68 % sucrose, 0.10 % raffinose and 0.79 % stachyose. To 4 ml of soymilk adjusted to pH 6.2, 1 ml of the enzyme (1 unit) was added, and the mixture was incubated for 1, 3 and 8 h at 60° C. The mixture was then deproteinized, centrifuged, treated with Dowex 50 and Amberlite CG4B, and subjected to HPLC. As shown in Fig. 2, raffinose and stachyose in soymilk were decomposed into galactose and sucrose by the immobilized enzyme. Thus, this α-galactosidase can be used to remove galacto-oligosaccharides in soymilk.

REFERENCES

1. A. Ohtakara, M. Mitsutomi and Y. Uchida, Purification and enzymatic properties of α-galactosidase from Pycnoporus cinnabarinus, Agric. Biol. Chem., 48:1319 (1984).
2. M. Mitsutomi and A. Ohtakara, A simplified procedure for purification and crystallization of thermostable α-galactosidase from Pycnoporus cinnabarinus, Agric. Biol. Chem., 48:3153 (1984).

IMMOBILIZATION OF INVERTASE ON CROSS LINKED CHITIN

(A. Illanes, R. Chamy, and M. E. Zuniga)

Invertase (E.C.3.2.1.26, β-D-fructofuranosidase) is a commercial enzyme being used mainly in soluble form for the production of invert sugar and confections. We have attempted to develop an insoluble invertase catalyst suitable for the continuous production of invert sugar. Chitin was used in powder and flake form and three immobilization procedures followed, two of them involving a crosslinking step with glutaraldehyde. The immobilization process was assessed in terms of efficiency (percentage of loaded enzyme expressed in the solid catalyst) and operational stability (half-life at optimal temperature for analysis), and optimized in terms of invertase to chitin, and glutaraldehyde to chitin ratios, selected as relevant variables of the process.

MATERIALS AND METHODS

Chitin, in flake and powder form, was obtained from waste shrimp shells by a procedure similar to that reported by Muzzarelli (1). Results obtained with this material were always similar to those obtained with the corresponding products sold by Sigma Chemical Co. (St. Louis, Mo, U.S.A.). Partially purified invertase from baker's yeast (851 I.U./mg protein; 698 I.U./mg of solid enzyme product) was purchased from Sigma or elsewhere, a crude extract (102 I.U./mg protein) was produced from autolyzed baker's yeast cells by a procedure developed by Illanes and Gorgollon (unpublished results). Glutaraldehyde, in a 25 % solution, was purchased from Sigma. Other reagents were all analytical grade and purchased from Merck.

Invertase activity was expressed in international units (I.U.). The unit of activity was defined as the amount of enzyme that hydrolyzes one micromol of sucrose per minute at 53° C and pH 4.5 from a 125 g/l sucrose solution in 0.1 M acetate buffer. Reducing sugars were determined according to Folin-Wu (2), to calculate the amount of sucrose hydrolyzed. Protein was determined according to the procedure of Lowry et al. (1951). Immobilization was carried out using three different procedures: Procedure 1 involved no glutaraldehyde; the enzyme was contacted with chitin for 12 hours at 4° C with occasional stirring, immobilization being produced solely by adsorption. Procedure 2 involved chitin pretreatment with 4 % glutaraldehyde solution for 24 hours at 18° C and then contacting with the enzyme for 12 hours at 4° C, stirring occasionally. Procedure 3 involved pretreatment of the enzyme with 4 % glutaraldehyde solution for 45 minutes at 4° C and then contacting with chitin for 12 hours at 4° C with occasional stirring. In all cases, the catalyst was washed four times with 0.1 M acetate buffer pH 4.5 and once with 2 M NaCl solution. Immobilization was always completed after about 10 hours of contacting the enzyme with the support.

Table I. Immobilization efficiency and operational stability of invertase catalysts

Support	Proc.	Retained Activity, %	Residual Activity, % *	Activity Lost, %	Half-Life at 53° C, days **
Activated carbon					
Bone	1	15.6	28.7	44.3	
	2	12.6	31.1	56.3	
	3	20.2	18.8	61.0	
Vegetal	1	13.0	12.0	75.0	0.9
	2	26.5	8.0	65.5	1.1
	3	34.7	6.0	59.3	0.7
Coffee pulp					
	1	0.8	51.4	47.8	
	2	2.1	47.8	50.1	
	3	1.2	43.4	55.4	
Chitin					
Flake	1	30.0	51.0	19.0	
	2	71.1	21.0	7.9	26
	3	53.0	31.0	16.0	5.7
Powder	1	25.0	62.1	12.9	
	2	44.0	33.4	22.6	
	3	39.0	39.2	21.8	

* Refers to the activity remaining in solution after immobilization
** Under substrate protection (5 % sucrose)

RESULTS AND DISCUSSION

Results of immobilization efficiency and operational stability are summarized on Table I for different invertase catalysts produced under the procedures described. Reported values are in all cases optimal in terms of the variables studied.

It is clear from these results that only chitin represents a support of potential interest. Immobilization on coffee pulp was extremely low and activity losses on activated carbon were high, presumably because of internal diffusional restrictions owing to its microporous structure.

Best results were obtained with chitin flakes under procedure 2; therefore, all subsequent results are referred to this catalyst. An efficiency of 71 % was obtained with only 8 % loss of the loaded activity, which implies a potential of 92 % of enzyme utilization. Immobilization efficiency decreased only to 61 % when a crude invertase preparation (102 I.U./mg protein) was used. The same results were obtained with the partially purified invertase, artificially contaminated with albumin. In this case efficiency of 71 % remained unaltered down to a specific activity of 350 I.U./mg protein. Chitin could be recovered from the exhausted catalyst by alkaline washing with 0.1 M NaOH, protein being completely desorbed. Immobilization capacity was lost at a rate of only 3 to 4 % per regeneration cycle. The catalyst remained fully active after six months of storage at 4° C. Operational half-life at 53° C under substrate protection was 26 days, assuming first-order kinetics of deactivation as shown in Fig. 1. Data however, tend to fit more closely a two-phase first order kinetics, as described by Dagys et al. (3) for other immobilized enzyme systems, but more experimental data are required. Half-life for the soluble enzyme at 53° C was only 65 minutes. Immobilization produced therefore a 576 fold increase in enzyme operational stability.

These values compare favourably with those reported for invertase bound to different supports (4-6) and different enzymes bound to chitin (7-10), and were considered satisfactory for reactor operation.

The immobilization process was optimized in terms of invertase to chitin and glutaraldehyde to chitin ratios. Results are presented in Figure 2. Optimal values were 0.2 mg. (140 I.U.) of invertase preparation and 8 ml of 4 % glutaraldehyde solution per gram of dry chitin. Figure 2.a represents the

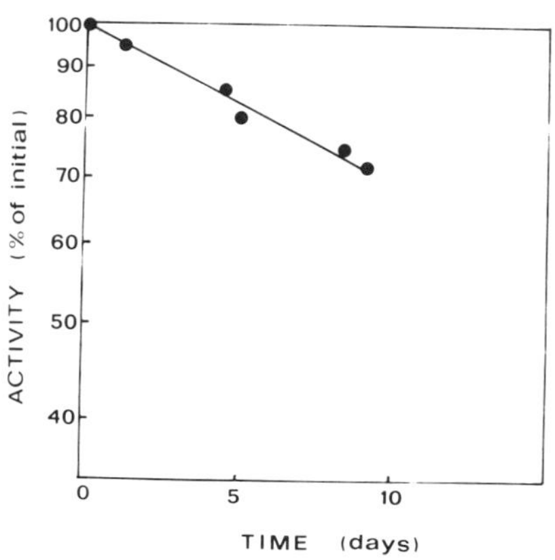

Fig. 1. Kinetics of deactivation of chitin-immobilized invertase at 53° C under substrate protection (5 % sucrose).

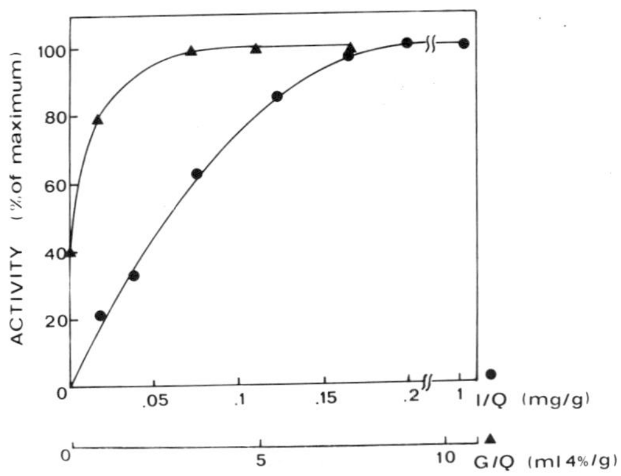

Fig. 2. Enzyme activity expressed in the solid catalyst as a function of: a) contacted invertase (I) to chitin (Q) ratio, at optimal glutaraldehyde to chitin ratio; b) contacted gluraraldehyde (G) to chitin ratio, at optimal invertase to chitin ratio.

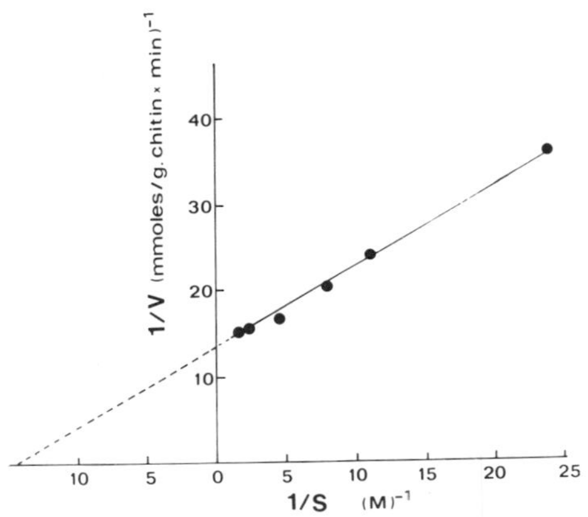

Fig. 3. Lineweaver-Burke plot for chitin-immobilized invertase.

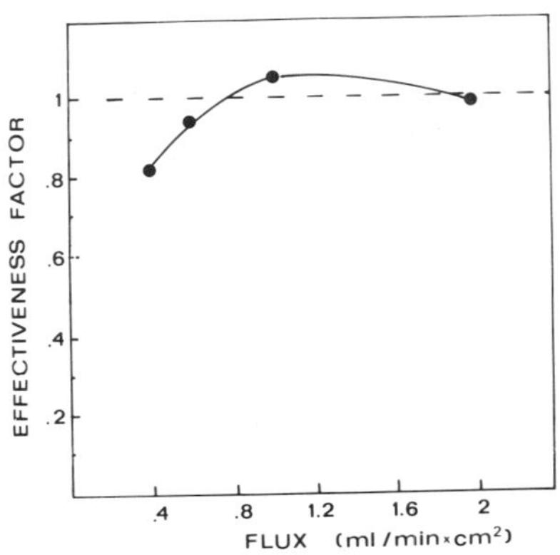

Fig. 4. Effectiveness factor as a function of column flux, at feed substrate concentration of 75 g/l of sucrose and 53° C.

saturation of chitin by enzyme protein. The curve could not be fitted however to a sorption isotherm model, implying a difference between protein binding and enzyme activity expression in the support.

Chitin-immobilized invertase followed simple Michaelis-Menten kinetics (Figure 3), with no substrate nor product inhibition up to the upper levels tested of 200 g/l of sucrose and 50 g/l of glucose respectively. Kinetic parameters, as determined by initial rate studies, are presented on Table II. The apparent Km value for the chitin-based catalyst, as opposed to that for the carbon-based catalyst, is very similar to the Km value for the soluble enzyme, indicating the low incidences of diffusional restrictions.

Temperature and pH activity profiles, as determined by initial rates, gave optimal values of 53° C and 4.3 respectively. Optimal pH was displaced 0.2 units to the acid side with respect to the soluble enzyme, as expected for a positively charged support.

The selected catalyst was tested in continuous fixed-bed column operation at 53° C and 75 g/l of sucrose. Results were in good agreement with a simple steady state mathematical model based on piston flow, showing negligible external diffusional restrictions for fluxes over 0.5 $ml{\cdot}min^{-1}cm^{-2}$ as shown by the values of the effectiveness factor in Fig. 4.

In summary, a chitin-based catalyst has been developed for the continuous inversion of sucrose, and the immobilization process has been optimized. Its characteristics compare favourably with those reported in the literature for similar systems. Chitin has proven to be an excellent support both from enzyme kinetics and physicochemical considerations. At present, catalyst deactivation is studied in more detail to develop a sound mathematical model for optimal policy of reactor operational temperature. Besides, this immobilization scheme is being extended to lactase catalysts which should be relevant to the dairy industry in Chile.

Table II. Kinetic Parameters of Invertase Catalysts

	Km	Vmax
Soluble enzyme	3.92×10^{-2}M	1.77 mmoles/min.l
Carbon (proc. 3)	1.85×10^{-1}M	5.81×10^{-2} mmoles/min.g support
Chitin (proc. 2)	6.85×10^{-2}M	7.44×10^{-2} mmoles/min.g support

REFERENCES

1. R. Muzzarelli, "Chitin", Pergamon Press, Oxford, (1977).
2. W. Horwitz, "Official Methods of Analysis of the Association of Official Analytical Chemists", ed., A.O.A.C., Washington, (1975).
3. R. Dagys, A. Pauliukonis, and D. Kazlauskas, Biotechnol. Bioeng., 26:620 (1984).
4. J. Hradil, and F. Svec, Enzymc & Microbial Technol., 3:331 (1981).
5. J. Synowiecki, Z. Sikorski, and M. Naczk, Biotechnol. Bioeng., 23:231 (1981).
6. A. Wiseman, New and Modified Invertases and their Applications, in: "Topics in Enzyme and Fermentation Biotechnology 3", A. Wiseman, ed., Ellis Horwood Ltd., Chichester (1979).
7. O. Lowry, N. Rosenbrough, F. Lewis and R. Randall, J. Biol. Chem., 19:265 (1951).
8. R. Muzzarelli, Enzyme Microbial Technol., 2:177 (1980).
9. W. Stanley, G. Watters, B. Chan, and J. Mercer, Biotechnol. Bioeng., 17:315 (1975).
10. W. Stanley, G. Watters, S. Kelly, and A. Olson, Biotechnol. Bioeng., 20:135 (1978).

IMMOBILIZATION OF PENICILLIN ACYLASE ON CHITOSAN

(J. Braun, P. LeChanu and F. LeGoffic)

The immobilization of penicillin acylase on chitosan by a simple method, produces an interesting biocatalyst. Different forms of support, powder, particles and beads were prepared and the enzymes was immobilized either by adsorption reticulation by the glutaraldehyde, or by covalent linkage to glutaraldehyde-pretreated chitosan. Moreover, a column reactor controlled by a micro-computer was developed to test the activity over a long period of time and the production of the immobilized enzyme.

MATERIALS

For these two methods of immobilization, the preparation of different supports is the same. By using a chitosan solution in 5 % acetic acid, either powder, particles or beads were prepared. The powder was obtained by precipitation of dissolved chitosan in 0.5 M sodium hydroxide under strong agitation; the particles were prepared by addition of 1 M sodium hydroxide to an emulsion of chitosan acetic solution in cyclohexane; the beads were prepared by precipitation with the aid of a syringe, from the chitosan acetic solution containing 0.67 % powdered glyoxal in 1.0 M sodium hydroxide.

The activity tests for the long time periods were carried out in a fixed bed reactor without a buffer; the regulation of the pH level was a means to control and to follow the advancement of the hydrolysis of the penicillin G. This method was chosen after having verified its validity in comparison with the spectrophotometric methods. The reactor included one or more immobilized enzyme columns, supplied with a penicillin G solution, contained in a central vessel where the pH was maintained at 8 by 0.1 - 0.05 M sodium hydroxide. The pH electrode, the sodium hydroxide valve and the pump were connected to a micro-computer which also calculated the transformation level and regulated the transfer of the initial penicillin G solutions and the hydrolyzed solutions.

Adsorption and cross-linking

The supports were impregnated with a penicillin acylase solution in 0.05 M phosphate buffer, pH = 7.5 for 1 - 2 h at 25° C. The solid phase was separated by centrifugation at 2000 rpm for the powder, and by decantation for the particles and the beads; then it was immersed in a minimal amount of 25 % glutaraldehyde solution, pH = 6 at 25° C for 12 h.

Table I. Yields of immobilization

Support *		Beads	Particles	Powder
Initial Enzyme	mg	100	80	80
	UE	22	17,5	17,5
Immobilized	UE	3,3	9,4	15
Activity	UE/g	33	94	150
Fixed Proteins	%	38	57	82
Activity Yield	%	15	54	85

* from 100 mg dry chitosan

Table II. Maximum specific activities

Support	Size (mm)	Specific activity (EU/g)
Powder	0.1	210
Particles	0.5 - 1	112
Beads	2 - 3	37

The yield of immobilization ranged from 15 % to 85 % (see Table I); the impregnating enzymatic solutions could be loaded again with enzyme and reused, the denaturation of the enzyme being negligible. The maximum specific activities ranged from 37 to 210 enzymatic units by gram (see Table II).

In the powder form, the immobilized enzyme was very stable; after a year with weekly testing, 90 % of the initial activity was maintained; a test lasts 1.5 h and results in 80 % hydrolysis of the Penicillin G solution used (ca. 20 mg/ml). The beads and the particles were tested in a batch reactor and in a column reactor; in a semi-continuous process for more than 350 h, they lost less than 10 % of their activity.

Immobilization by covalent linkage

The supports were treated with a 5 % glutaraldehyde solution in 0.1 M phosphate buffer, pH = 8 for 30 minutes at 25° C. The activated chitosan was washed to eliminate the glutaraldehyde and immerged in a penicillin acylase solution in the same buffer for one night at 25° C. The results obtained with this method are nearly the same as with adsorption reticulation: the enzymatic solutions could also be reused after readjusting their concentrations.

The maximal specific activities for the powder and for the beads were respectively 194 and 60 enzymatic units by gram of dry support. The activities were measured by cycle, each cycle being continued until 70-75 % transformation level of penicillin G to 6-APA was obtained, after the 20th cycle, the loss of activity was around 2-5 %.

The activity tests for the long time periods were carried out in a fixed bed reactor without a buffer; the regulation of the pH level was a means to control and to follow the advancement of the hydrolysis of the penicillin G. This method was chosen after having verified its validity in comparison with the spectrophotometric method.

The reactor includes one or more immobilized enzyme columns, supplied with a penicillin G solution, contained in a central vessel where the pH is maintained at 8 by 0.1 - 0.05 M sodium hydroxide. The pH electrode, the sodium hydroxide valve and the pump are connected to a microcomputer which also calculates the transformation level and regulates the transfer of the initial penicillin G solution and the hydrolyzed solutions.

USE OF KRILL CHITIN AS AN ENZYME SUPPORT

(J. Synowiecki)

Chitin from krill offals was obtained by demineralization with HCl and by deproteinization with aqueous KOH, washing with water to pH 7.0, twofold washing with methanol, drying at 105° C, and screening to separate the fraction 22 - 52 mesh. Some chitin preparations were modified with formaldehyde or carbon disulphide, as described elsewhere (1, 2). The availability of free amino groups of chitin was determined by electrometric titration to pH 7.0. The dithiocarbamino groups of modified chitin were assayed after Ellman (3).

For immobilizing of α-amylase, β-amylase, diastase, invertase, and amyloglucosidase two procedures were applied: 1) crosslinking by glutaraldehyde, and 2) adsorption on unmodified chitin, or on chitin modified by carbon disulphide or by formaldehyde. In all investigations 50 mg chitin or chitin modified with formaldehyde or carbon disulphide were mixed with 5 ml of the enzyme solution and 5 ml of 0.02 M acetate or phosphate buffer, which in the case of crosslinking of the enzyme contained glutaraldehyde. The mixture was left overnight, filtered, and washed on a sintered glass funnel with 0.8 l distilled water. The activity of invertase was determined according to Somogyi (4) and the activity of the other investigated enzymes was assayed as described elsewhere (5).

The suitability of krill chitin as an enzyme support

The activity of enzymes immobilized on chitin obtained at different parameters of demineralization and deproteinization depends on the degree of deproteination of chitin, the content of minerals, availability of amino groups, particle size, structure of the surface, and conformation of the chitin molecules. It was found that increase in the content of minerals or proteins in chitin prepared at not too rigorous conditions decreased the enzyme activity (6). The correlation between the number of available amino groups and adsorbed α -amylase activity, $r = 0.94$ and $\alpha = 0.05$, suggests the participation of these groups in binding of the enzyme (Table I).

Table I. The activity of enzymes adsorbed on chitin containing different number of amino groups.

Amino groups in chitin	Rate of enzymatic reaction mg/min		
µmol/g	α-amylase	diastase	invertase
190	2.9	1.6	19
232	3.7	1.9	21
408	6.8	2.2	21

The best chitin preparation for immobilizing all investigated enzymes contained 6.23 % nitrogen, only 0.5 % proteins or residual amino acids, traces of minerals, and had very low powder density, about 0.03 g/ml. The insolubility of this preparation in 10 % acetic acid, and the presence of amide bands at 1655, 1555, and 1313 cm^{-1} in the infrared spectra indicate that the conditions used for deproteinization did not convert chitin into chitosan. In comparison with chitins from other sources, krill chitin has a porosity more suitable for adsorption of the enzymes. The best results for immobilizing of enzymes were obtained on chitin fractions of 22 - 52 mesh.

METHODS

Glutaraldehyde used as a crosslinking agent reacted with the free amino groups of chitin and the enzyme to form Schiff's bases. However, glutaraldehyde brought about a loss in activity of the immobilized enzymes (Table II).

Table II. The activity of immobilized enzymes, %, relative to the activity of enzymes adsorbed on unmodified chitin.

Enzyme	Crosslinked by glutar-aldehyde	Immobilized on chitin pretreated with:		
		HCHO	CS_2	$NaIO_4$
α-amylase	61	114	200	-
β-amylase	69	143	240	-
amyloglucosidase	92	94	200	-
diastase	57	-	278	85

The suitability of aldehyde derivatives prepared from chitin by periodate oxidation was also investigated. Diastase immobilized on such support had lower activity than the enzyme adsorbed on control chitin. The gradual desorption of the enzyme by washing with KCl solutions of increasing ionic strength indicated that no covalent linkages were formed.

The most active preparations were obtained by immobilizing enzymes on krill chitin treated with carbon disulphide (3). In carbon disulphide treated chitin the concentration of amino groups decreased from 296 μmol/g in the control sample to 194 μmol/g. Probably free amino groups on chitin condense with carbon disulphide forming dithiocarbamino groups. Because chitin pretreated with carbon disulphide has both positively charged residual amino groups and negatively charged dithiocarbamino groups it can form ionic linkages with different charged amino acid residues in the enzymes.

Formaldehyde can react with the amino groups of chitin to form Schiff's bases and with various functional groups of the enzymes, creating methylene bridges. However, enzymes immobilized on chitin pretreated with formaldehyde are not stable at pH lower than 5.5.

The activity of diastase adsorbed on chitin pretreated with succinic or acetic anhydride at pH 7.5 was higher than that of the enzyme adsorbed on control chitin by 481 and 365 % respectively. However, the acylation of chitin in anhydrous acetic anhydride and perchloric acid caused a loss in concentration of amino groups in the support from 408 to 67 μmol/g and a loss in activity of the immobilized enzyme by about 93 %.

Effects of immobilization on the operating parameters of the enzymes

The ionic complexes of enzymes and chitin or chitin pretreated with carbon disulfide are stable from pH 5.0 to 8.0 and at low ionic strength. Increasing the ionic strength above 0.01 caused a gradual desorption of the enzymes. The influence of the ionic strength on the stability of chitin-enzyme complexes was not observed for enzymes fixed on chitin by methylene bridges. After washing with 5 % KCl solution, the diastase immobilized on chitin pretreated with formaldehyde had about three times higher activity than diastase adsorbed on unmodified chitin.

After immobilization, the pH-activity profile of all enzymes shifted towards the acid side by about 0.8 pH units. The immobilization on krill chitin caused no significant increase in the thermal stability of the investigated enzymes. The moist preparations of immobilized diastase had good storage stability. The activity of these preparations decreased only by about 15 % after 20 days at 4° C.

REFERENCES

1. J. Synowiecki, Z. E. Sikorski, M. Naczk, Immobilizing of enzymes on krill chitin activated by formaldehyde, Biotechnol. Bioeng., 24:1871 (1982).
2. J. Synowiecki, A. Sikorska-Siondalska, A. F. El-Bedawey, Adsorption of enzymes on krill chitin modified with CS , Biotechnol. Bioeng., in press.
3. R.A.A. Muzzarelli, F. Tanfani and M. Emanuelli, Preparation and characteristic properties of dithiocarbamate chitosan, Carbohydr. Res., 104:235 (1982).
4. M. Somogyi, Notes on sugar determination, J. Biol. Chem., 19:195 (1952).
5. J. Synowiecki, Z. E. Sikorski, M. Naczk, Immobilization of amylases on krill chitin, Food Chem., 8:239 (1982).
6. J. Synowiecki, Z. E. Sikorski, M. Naczk, The activity of immobilized enzymes on different krill chitin preparations, Biotechnol. Bioeng., 23:2211 (1981).

PHOTOACTIVATABLE CHITOSAN FOR PROTEIN IMMOBILIZATION

(S. Sicsic, J. Léonil, J. Braun and F. LeGoffic)

New photoactivatable chitosans were synthesized with 4-azidopyridine derivatives, leading to insoluble or soluble supports usable for protein immobilization.

Photochemically binding proteins on a support through an arylazido group can be a particularly attractive method since the binding reaction is rapid and does not require any reactive group at the protein level (1, 2). The yields of photochemical fixations are generally low however, so it was necessary to have a support rich in photoactivatable groups. For this reason, chitosan was chosen. Two photoactivatable reagents were used: 4-azido-2,3,5,-6-tetrafluoropyridine and 4-azido-3,5-dichloro-2,6-difluoropyridine. These reagents were bound to the chitosan either directly or through a spacer arm.

MATERIALS AND METHODS

4-Azido-2,3,5,6-tetrafluoropyridine and 4-azido-3,5-dichloro-2,6'-difluoropyridine were synthesized as previously described (3). Crab shell chitosan was from Rousselot-France. It was ground in an electric blender and the resulting beads passed through different sieves. Before the coupling reaction with the photoactivatable reagents, the chitosan was successively washed with 1 M ammonia, water, methanol and then dried.

Preparation of 4-azido-2-(4-hydroxybutyl)-3,5,6trifluoropyridine, I

A mixture of 3 g of 4-azido-2,3,5,6-tetrafluoropyridine, 1.59 g of 4-aminobutanol and 1.32 ml of triethylamine in 30 ml of dimethylformamide (DMF) was heated at 60° during four hours in darkness. Then DMF was evaporated, and the residue was dissolved in methylene chloride. The solution thus obtained was washed with 1 M HCl, dried with magnesium sulfate and evaporated. The crude product was poured on a silica gel chromatography column. The pure product I (2.15 g, 55 % yield) was eluted with pentane-ethylacetate (8.5:1.5). It was characterized by i.r., ^{1}H-n.m.r., ^{19}F-n.m.r. and u.v. spectroscopy.

Preparation of the tosylate, II

It was prepared from alcohol I using p-toluenesulfonyl chloride, and purified by liquid chromatography on silica gel using the same eluent system as the above one. The product was characterized by ^{1}H -n.m.r., ^{19}F-n.m.r. and u.v. spectroscopy.

Preparation of photoactivatable chitosan, III

A mixture containing 0.415 g of the tosylate II, 0.158 g of chitosan and 0.049 ml of triethylamine in 10 ml of DMF, was heated for 20 hours at 60°C. The mixture was then cooled and centrifuged. The separated resin was extensively washed with methanol and dried under vacuum; 0.182 g of resin was obtained.

Enzyme immobilization

The irradiation reaction was promoted by a Philips high pressure mercury lamp SW 500 W using a glass filter. The reactor was kept in a jacketed pyrex vessel maintained at 8° C. The vessel was shaken and flushed with nitrogen during irradiation. a) Immobilization on modified chitosan III: 2 ml of the solution being irradiated contained 100 mg of resin II, 1 mg of beta-galactosidase and 0.01 M phosphate buffer pH 7.3; b) Immobilization on modified chitosan IV: 50 mg of modified chitosan IV dissolved in 2 ml of 0.3 M HCl. The solution was adjusted to pH 6 with NaOH and 1 ml of a solution containing 0.9 mg of beta-galactosidase was added. The mixtures were irradiated for five minutes.

RESULTS AND DISCUSSION

Synthesis of photoactivatable chitosan

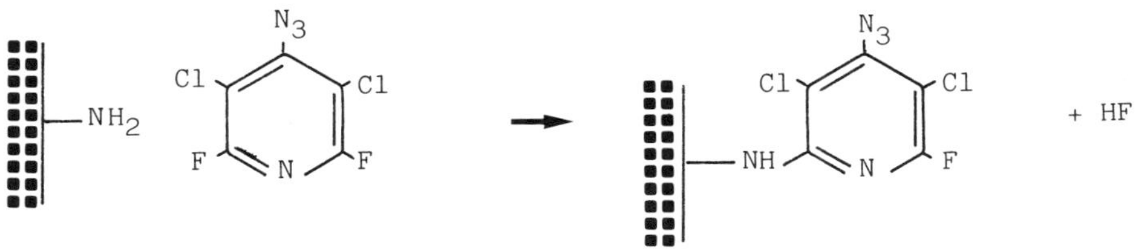

The control of the direct fixation of the photoactivatable reagent I on chitosan, is very easy to achieve. As the results show, it is possible to yield different loading of the photoreactive group, by modifying reaction time and temperature. At 80° C and for 72 h up to 90 % of the amino groups are substituted. The resins thus obtained are insoluble in aqueous media and in organic solvent. As a photochemical reaction with a solid support can only occur at the surface of the support, we have also prepared photoactivatable chitosan soluble in acidic aqueous media. The indirect fixation through a spacer arm leads to such resins.

Tosylate II is not stable and its degradation occurs even during the purification step, probably due to an intramolecular reaction between the nucleophilic nitrogen and the tosylate leading to a favoured five membered ring. In the conditions described, 0.6 meq of II are bound per g of chitosan, giving a polymer IV soluble in acidic solution.

If the u.v. spectra of chitosan, modified chitosan IV and irradiated modified chitosan IV are compared, it is possible to notice that the absorption peak at 318 nm disappears as a consequence of the irradiation.

Immobilization of beta-galactosidase

Table 1 shows the results of the immobilization of beta-galactosidase on different sized modified chitosans III. The importance of granulometry is

Table I. Binding of β-galactosidase on modified chitosans.

Granulometry mm diameter	Bound protein mg/g matrix	Activity U/mg protein	Relative specific activity (%)
< 0.160	7.5	205	21.2
0.250-0.300	2.3	35	3.5
> 0.500	1.7	12	1.2

Azide loading of the matrix: 3.5 meq/g
Initial protein: 7.5 mg/g matrix.

pointed out: the lower the granulometry, the higher the ability to bind beta-galactosidase. The specific activity of the bound enzyme is also the best for the lowest granulometry. With a bead size of 0.160 mm the maximum of bound protein was 18.4 mg (97 % of the total protein) per g of matrix with a retained specific activity of 27 %. One of these matrices has been used in a packed bed reactor with D-lactose as a feed substrate.

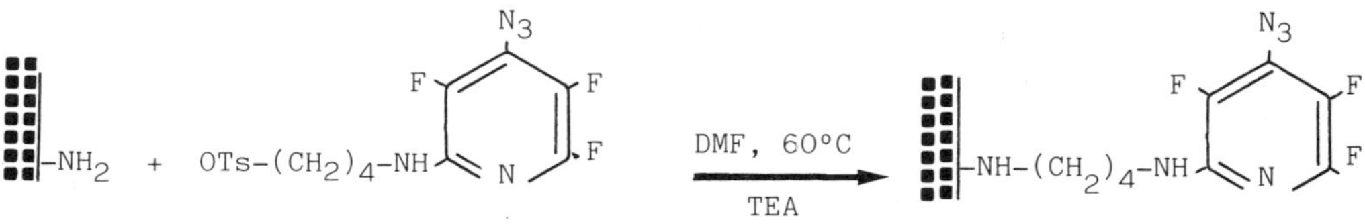

Immobilization of beta-galactosidase was also performed with soluble chitosan IV: 12.4 mg (71 % of the total protein) were bound per g of matrix with a retained specific activity of 13 %. This result is a preliminary one, and was not optimized, but it shows that introduction of a spacer arm of 4 carbons between the matrix and the photoactivator does not improve the specific activity of the bound enzyme.

REFERENCES

1. P. Guire, Methods Enzymol., 44:280 (1976).
2. P. Guire, D. Fliger, and J. Hodgson, J. Pharm. Res. Commun., 9:131 (1977).
3. R. L. Banks, and G. R. Sparkes, J. Chem. Soc. Perkin Trans., 1:2964 (1972).
4. J. Léonil, S. Sicsic, J. Braun, and F. Le Goffic, Enzyme Microb. Technol., 6:517 (1984).

IMMOBILIZATION OF LACTASE ON CHITOSAN-COATED SILICA GEL PARTICLES

(E. Van Griethuysen, E. Flaschel and A. Renken)

With a view to using immobilized lactase in fluidized bed reactors, it was necessary to develop a carrier with suitable fluidization behavior. An immobilization technique for A. niger lactase was suggested using chitosan flakes activated with glutaraldehyde (2), but their low density and their tendency to agglomerate prevented its use in such types of reactors. Moreover, a further disadvantage of this carrier was seen in the difficulty to obtain and work with small particles, because it tended to form compressible beds. Therefore, a method was developed in order to render the support more rigid and to increase its density. Porous silica beads of defined particle size, used as inner core, were coated with chitosan. The chitosan layer then was cross-linked by means of glutaraldehyde. This treatment simultaneously gave an activated support for enzyme immobilization.

MATERIALS AND METHODS

Enzymes and chemicals

The lactase of A. niger was purchased from Rapidase (Séclin, France) and the lactase of A. oryzae was the technical grade Lactosin AO from Yakult Biochemicals (Nishinomiya, Japan). Glutaraldehyde was purchased from Merck (Darmstadt, W. Germany Cat. No. 4239) as was lactose (Cat. No. 7656). Chitosan was purchased from a Sigma (St. Louis, U.S.A. Cat. No. C-2263). Silicagel type 432 was purchased from Grace (Worms, W. Germany). The particles had an apparent density in water of 1340 $kg.m^{-3}$ and in air of 380 $kg.m^{-3}$. The specific surface was 344 $m^2.g^{-1}$.

Immobilization procedure

A chitosan solution was prepared by dissolving 3 g chitosan in 250 ml of 20 % (v/v) aqueous acetic acid at 40° C. For the first tests with A. niger lactase, a sieved fraction of silica gel was used without any pretreatment. Later on, it was also fluidized to eliminate fines and dried under vacuum before adding it to the filtered chitosan solution under gentle stirring. The concentration of dry silica gel in the chitosan solution was 170 g/l in the case of A. niger lactase immobilization and 600 ml/l in the case of the A. oryzae lactase. The suspension was placed under water jet vacuum to evacuate the air present in the pores of the silica gel. The chitosan coated silica gel was filtered off and washed extensively with water. A sample was placed in an aqueous glutaraldehyde solution (75 g/l in a solution containing 3.1 % glutaraldehyde for A. niger lactase immobilization, 300 ml/l in a solution containing 12.5 % glutaraldehyde for A. oryzae lactase). The suspension was gently stirred for 2 h. The activated support was filtered off and washed extensively with water. An enzyme solution was prepared and added to the activated support. The immobilization was performed in stirred tanks, in vials agitated by rotation, in a rotary evaporator as well as in a fluidized bed with continuous recycling of the enzyme solution. This last method was a particularly gentle operation and well suited for large-scale immobilization. The immobilized lactase was filtered off and washed with water. With a view to using the immobilized lactase with substrates having a high ionic strength like whey, the catalyst was retreated with an aqueous glutaraldehyde solution. A. niger lactase was retreated with a solution containing 5 % glutaraldehyde. The influence of the glutaraldehyde concentration on the A. oryzae lactase was tested. Since no significant difference in the final activity was observed, a concentration of 25 % glutaraldehyde was adopted. The retreated immobilized lactase was filtered off and washed with water. It could be utilized immediately or stored wet.

Analytical methods

The activity of free and immobilized lactase preparations was measured using lactose as a substrate. The amount of glucose released was analysed with a glucose analyzer, type 23, from Yellow Springs Instruments (Ohio, U.S.A.). The data of lactose conversion as a function of time was used to estimate the actual activity according to the kinetics of the A. niger lactase (9) or the kinetics of the A. oryzae lactase at a lactose concentration of 150 mmol/l (10), respectively.

RESULTS AND DISCUSSION

The immobilization procedure has been studied in detail for the A. niger lactase as well as for the A. oryzae lactase. The activity yield of the immobilization was a function of several factors which were studied separately in order to optimize the procedure.

Two types of activity yields might be distinguished, the activity yield of the immobilized lactase corresponded to the ratio of the immobilized activity and the total activity initially present. The total residual activity (sum of carrier and filtrate) was defined as the amount of the immobilized activity and the activity remaining in the liquid phase referred to the initial activity.

The activity yield of the immobilized enzyme depended strongly on the pH during the contact period of the activated carrier and the enzyme. As shown in Fig. 1, the best activity yield was obtained for A. niger lactase at pH 4, while the total residual activity was nearly constant in the range of pH 3 to 6. Further immobilizations were carried out at pH 4 which approaches the optimum pH of 3.5 for the activity of this enzyme. Subsequently, this pH was also used for the immobilization of the A. oryzae lactase which exhibited an activity maximum at pH 4.

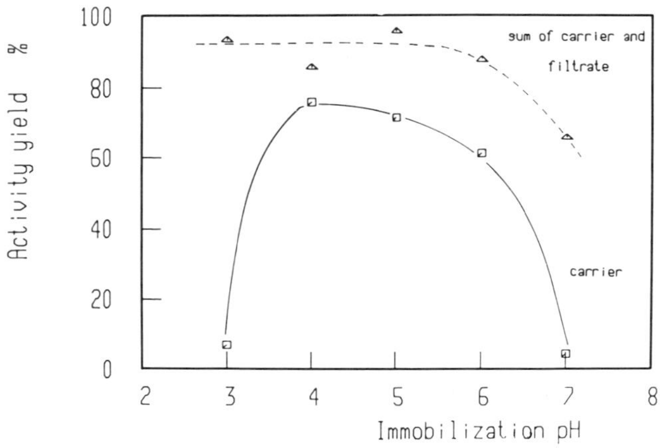

Fig. 1 pH-profile for the immobilization of A. niger lactase (immobilization in a stirred tank during 2 h, mean particle size of silica gel d_p = 0.100-0.125 mm)

The immobilization was carried out at room temperature, since no influence on the activity yield of a temperature variation between 20 and 50° C was observed for the A. niger lactase.

In order to define the most favorable enzyme/support ratio, several immobilization tests were carried out for both enzymes. The results obtained for the A. oryzae lactase are given in Figures 2 and 3.

As Fig. 2 shows, the activity yield decreased strongly in the range of low enzyme concentrations and were less pronounced for higher concentrations, while the total residual activity remained approximately constant. However, the specific activity of the immobilized enzyme increased considerably with

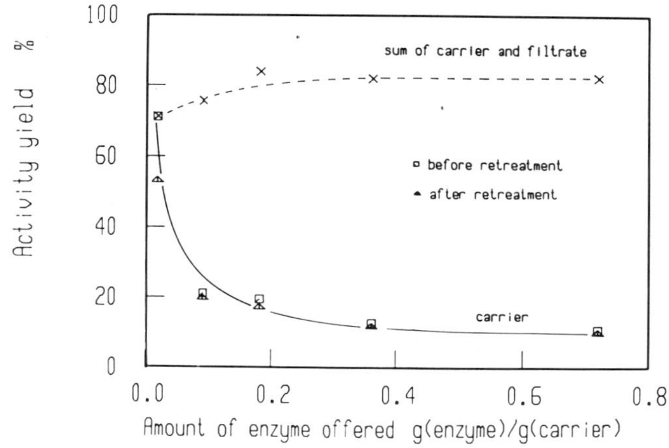

Fig. 2 Activity yield of A. oryzae lactase (immobilization in reciprocally shaken vials during 2 h, d_p = 0.125-0.160 mm)

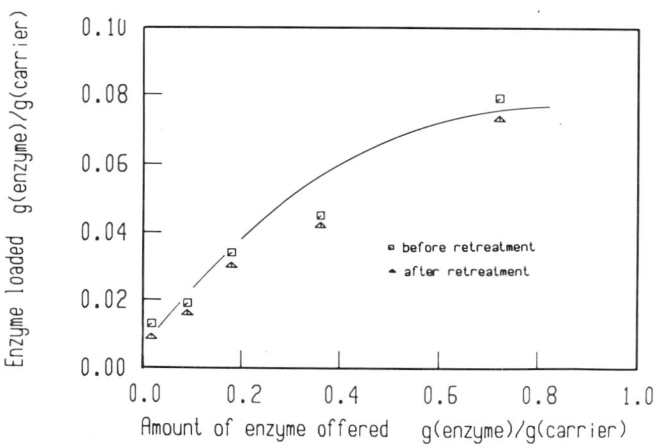

Fig. 3 Total activity of immobilized A. oryzae lactase (immobilization in reciprocally shaken vials during 2 h, d_p = 0.125-0.160 mm)

the amount of enzyme offered, as shown in Fig. 3. The most favorable enzyme/support ratios were estimated for the A. oryzae lactase to represent 0.18 g/g and 0.1 g/g for the A. niger lactase. Nevertheless, the activity yield corresponding to the enzyme/support ratio of 0.18 g/g for the A. oryzae lactase was only 20 %.

In order to improve this result, the contact time of the enzyme with the support was extended. Fig. 4 gives the immobilization kinetics for A. oryzae lactase and shows that the activity yield was 50 % after 25 hours, while the same value was obtained for the A. niger lactase in only 4 hours (Fig. 5). Generally, it seemed that the A. oryzae lactase was more sensitive to the immobilization procedure since the losses were higher than for the A. niger lactase. As shown in Figures 2 and 4, the activity loss was in the order of 20 - 30 % compared with 10 - 20 % for the A. niger lactase (Figures 1 and 5).

The retreatment of the catalyst with glutaraldehyde was introduced after having observed a rapid decrease in the activity of the immobilized lactase during continuous operation with whey as the substrate (Fig. 6). The aim of this supplementary step was to stabilize the immobilized enzyme by crosslinking of the enzyme immobilized by adsorption in order to avoid the competition of ions with the enzyme. The retreatment step did not significantly reduce the activity yield (Figures 2 and 3), but considerably improved the enzyme stability against substrates with high ionic strength.

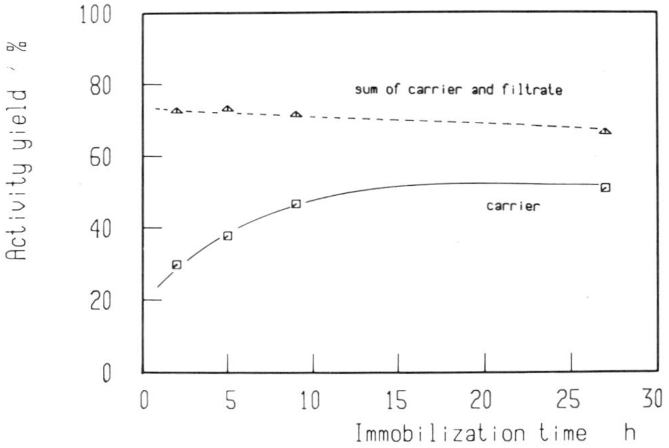

Fig. 4 Immobilization kinetics of A. oryzae lactase (enzyme/carrier ratio 0.18 g/g immobilization in reciprocally shaken vials d_p = 0.125-0.160 mm)

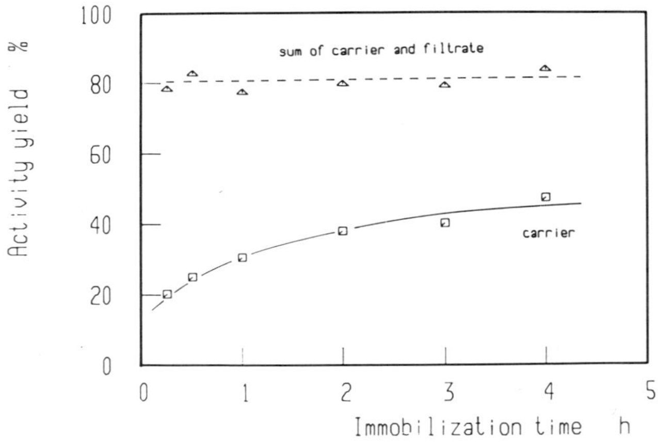

Fig. 5 Immobilization kinetics of A. niger lactase (enzyme/carrier ratio 0.1 g/g immobilization in a fluidized bed d_p = 0.100-0.125 mm)

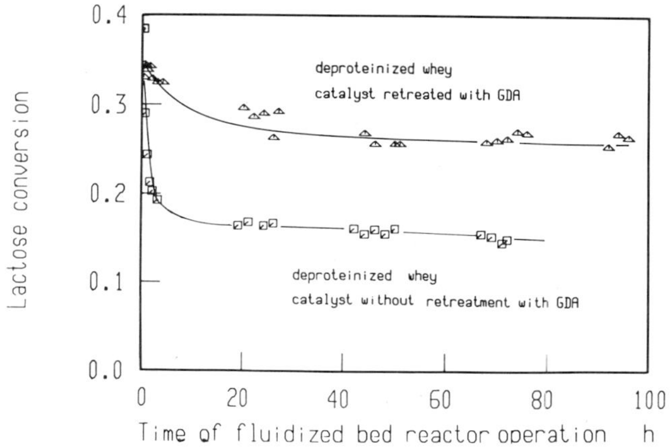

Fig. 6 Operational stability of immobilized *A. niger* lactase (substrate: deproteinized whey)

CONCLUSION

The developed procedure showed that an inorganic support can easily be coated with chitosan, activated with glutaraldehyde and then used for the immobilization of enzymes. The lactases from *A. niger* and *A. oryzae* were successfully immobilized on chitosan-coated silica gel particles with an activity yield of about 50 %. The deleterious effect of whey salts could be avoided by retreatment with the coupling reagent. This immobilization process combines the different advantages of an inorganic support, such as appropriate density and mechanical stability, with the coupling efficiency of the coating polymer which is compatible for utilization in the food industry. In addition, the catalyst is well adapted for its utilization in fluidized bed reactors and in fixed bed reactors.

REFERENCES

1. R. A. A. Muzzarelli, Immobilization of Enzymes on Chitin and Chitosan *Enzyme Microb. Technol.*, 2:177 (1980).
2. J. L. Leuba and F. Widmer, Immobilization of the β-Galactosidase from A. niger on Chitosan, *J. Solid-Phase Biochem.*, 2(3):257 (1977).
3. R. A. A. Muzzarelli, G. Barontini, and R. Rocchetti, Immobilized Enzymes on Chitosan Columns α-Chymotrypsin and Acid-phophatase, *Biotechnol. Bioeng.*, 18:1445 (1976).
4. T. Kasumi, M. Tsuji, K. Hayashi, and N. Tsumura, Preparation and Some Properties of Chitosan Bound Enzymes, *Agric. Biol. Chem.*, 41:1865 (1977).
5. F. Bissett, and D. Sternberg, Immobilization of *Aspergillus* β Glucosidase on Chitosan, Appl. *Environ. Microbiol.*, 35:750 (1978).
6. W. L. Stanley, G. G. Watters, B. Cham, and J. M. Mercer, Lactase and Other Enzymes Bound to Chitin with Glutaraldehyde, *Biotechnol. Bioeng.*, 17:315 (1975).
7. R. A. A. Muzzarelli, G. Barontini, and R. Rocchetti, Isolation of Lysozyme on Chitosan, *Biotechnol. Bioeng.*, 20:87 (1978).
8. R. W. Coughlin, and M. Charles, Applications of Lactose and Immobilized Lactase, *in*: Immobilized Enzymes for Food Processing, W. H. Pitcher ed., Boca Raton, CRC Press (1980).
9. E. Flaschel, E. Raetz, and A. Renken, The Kinetics of Lactose Hydrolysis

for the β-Galactosidase from A. niger, Biotechnol. Bioeng., 24:2499 (1982).

10. E. van Griethuysen, E. Flaschel, and A. Renken, The Influence of the Ion Content of Whey on the pH-Activity Profile of the β-Galactosidase from A. oryzae, J. Chem. Technol. Biotechnol. (in press).

CHITOSAN GELS FOR ENTRAPMENT OF CULTURED PLANT CELLS

(D. Knorr)

Based on the findings that chitosan has a permeabilizing effect on cultured plant cells and can also serve as an immobilizing agent for them, it seemed of interest to examine the potential of chitosan to serve as both a concomitant immobilizing and permeabilizing agent for cultured plant cells (1-13).

Cell suspensions of amaranth (Amaranthus tricolor) and common milkweed (Asclepias syriaca L.) were used to examine the release of oxalate and protein respectively (8, 15). Commercially available chitosan (Bioshell Inc., Albany, OR) dissolved in ascorbic acid (16) was used as the polyelectrolyte for gel formation and sodium triphosphate as the multivalent counterion (8, 11, 15). In the case of alginate gel entrapped cells, sodium alginate and calcium chloride served as polyanion and counterion respectively. A cross sectional view of a chitosan capsule used for entrapment of plant cells (14) is shown in Fig. 1.

A comparison of oxalate release from cultured A. tricolor cells (8) subjected to various treatments (Fig. 2) resulted in significantly ($P< 0.01$) lower amounts of oxalate released from freely suspended cells and calcium-alginate entrapped cells than from treatments where chitosan was present. No differences in oxalate release were observed between chitosan gel entrapped cells and freely suspended cells with chitosan present in the medium (treatment D or E).

Protein concentrations of medium filtrates of Asclepias syriaca L. cells entrapped in chitosan gels or calcium-alginate gels are presented in Fig. 3. Protein concentrations in the medium 96 hours after entrapment in calcium-alginate gels (2.5 μg/ml medium filtrate) were significantly lower ($P<0.05$) than with chitosan entrapped samples (5.4 μg/ml medium filtrate) (15).

These data indicate that chitosan was effective in increasing the release of oxalate, a low molecular weight plant metabolite (17), as well as proteins, high molecular weight substances. Chitosan binds to polygalacturonate in plant cell walls (18), so the increase in cell permeability may be due to disruption of the intermolecular bonding responsible for maintaining an intact membrane, changes in membrane fluidity, or effects on the components associated with membrane transport (19). Recently, labeled fragments of chitosan were detected within plant cytoplasm when high-molecular weight tritiated chitosan was applied to the surface of pea tissue (20).

Release of oxalate, as well as protein, from calcium-alginate gel entrapped cells was significantly lower than from chitosan gel entrapped cells. This is of interest because it indicates a permeabilizing effect of chitosan and also a potential inhibitory effect of calcium-alginate gels in the case of milkweed (Asclepias syriaca L.) cells because proteins such as asclepain, a sulfhydryl protease, are usually extracellular products (21). Recent findings (22) refer to the importance of calcium ions for the production of secondary metabolites from calcium-alginate gel entrapped plant cells.

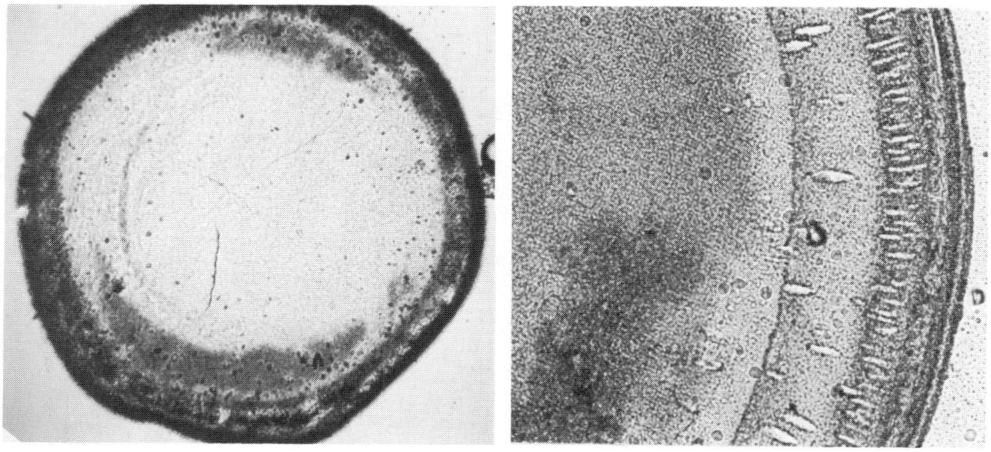

Figure 1: Cross-sectional view of a chitosan capsule and detail of membrane

Overall these data suggest that chitosan concurrently can serve as an effective immobilizing and permeabilizing agent for cultured plant cells. This provides an effective tool for scale-up and commercial production of secondary plant metabolites.

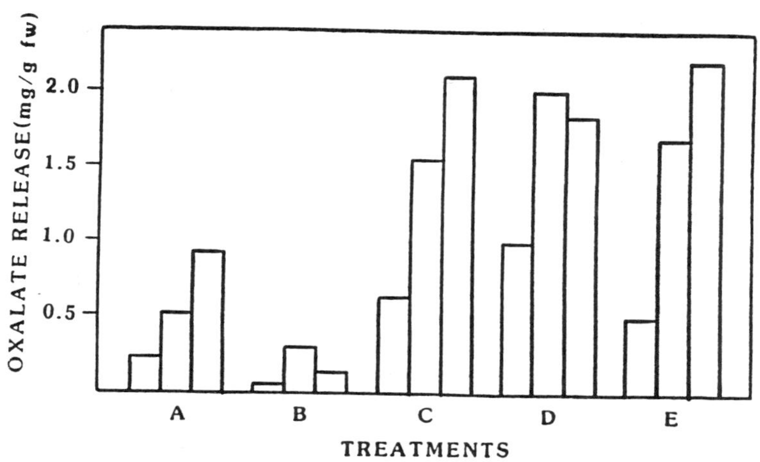

Figure 2: Release of oxalate from cultured A. tricolor cells at 1,3,and 5 days of treatment. A: freely suspended cells, B: calcium-alginate gel entrapped cells, C: chitosan gel entrapped cells, D: freely suspended cells with 5 mg chitosan/100 ml medium, E: freely suspended cells with 10 mg chitosan/100 ml medium.

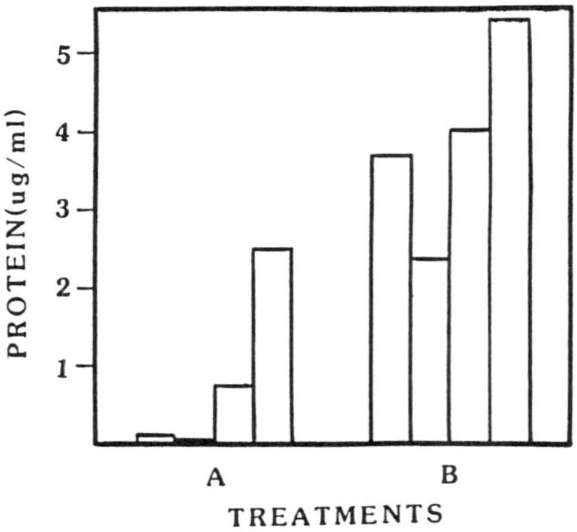

Figure 3: Protein concentration of medium filtrates of Asclepias syriaca L.cells entrapped in chitosan or calcium alginate gels 24, 48, 72 and 96 hours after immobilization

REFERENCES

1. A. I. Laskin, G. T. Stao, and L. B. Wingard, Enzyme Engineering 7, New York Academy Sciences, Vol. 434, New York, NY (1984).
2. K. Venkatsubramanian, Immobilized Microbial Cells, ACS Symposium Series 106, Am. Chem. Society, Washington, DC (1979).
3. M. L. Shuler, O. P. Sahai, and G. A. Hallsby, Entrapped Plant Cell Tissue Cultures, in: "Biochemical Engineering III"; K. Venkatasubramanian, A. Constantinides, and W. R. Vieth, eds., New York Academy of Sciences, Vol. 412, New York, NY (1983).
4. P. Brodelius, Production of Biochemicals with Immobilized Plant Cells: Possibilities and Problems; in: "Biochemical Engineering III", K. Venkatasubramanian, A. Constantinides, and W. R. Vieth, eds., New York Academy of Sciences, Vol. 413, New York, NY (1983).
5. P. Brodelius, Immobilized Viable Plan Cells, in: "Enzyme Engineering 7", A. I. Laskin, G. T. Stao, and L. B. Wingard, New York Academy of Sciences, Vol. 434, New York, NY (1984).
6. H. R. Lernes, D. Ben-Bassat, L. Reinhold, and A. Poljakoff-Mayber, Introduction of "pore" formation in plant cell membranes by toluene, Plant Physiol., 61:213 (1978).
7. H. Köhle, D. H. Young, and H. Kauss, Physiological changes in suspension-cultured soybean cells elicited by treatment with chitosan, Plant Sci. Lett., 33:221 (1984).
8. D. Knorr, and R. A. Teutonico, Chitosan immobilization and permeabilization of Amaranthus tricolor cells, J. Agric. Food Chem. (submitted).
9. D. Knorr, Use of chitinous polymers in food - a challenge for food research and development, Food Technol., 38(1):85 (1984).
10. D. Knorr, Utilization of chitinous polymers in food processing and biomass recovery, in: "Biotechnology of Marine Polysaccharides", R. R. Colwell, E. R. Pariser and A. J. Sinskey, eds., Hemisphere Publishing Co., Washington, DC (1985).

11. K. D. Vorlop, and J. Klein, Fermentation of spherical biocatalysts by ionotropic gelation, Biotechnol. Lett., 3:9 (1981).
12. J. Klein, and K. D. Vorlop, Immobilized Cells Catalysts Preparation and Reaction Performance, in: "Foundations of Biochemical Engineering: Kinetics and Thermodynamics in Biological Systems", H. W. Blanch, E. T. Papoutsakis, and G. Stephanopoulas eds., Am. Chemical Soc., Washington, DC (1983).
13. C. K. Rha, D. Rodriguez-Sanchez and C. Kienzle-Sterzer, Novel applications of chitosan, in: "Biotechnology of Marine Polysaccharides", R. R. Colwell, E. R. Pariser, and A. J. Sinskey, eds., Hemisphere Publishing Co., Washington, DC (1985).
14. E. R. DeBernardez, D. Knorr, and M. Beaumont, Use of chitosan for plant biotechnology processes (unpublished).
15. S. Miazga, and D. Knorr, Use of cultured common milkweed (Asclepias syriaca L.) cells for protease production, Enzyme, Microbiol. Technol. (submitted).
16. R. A. A. Muzzarelli, F. Tanfani, and M. Emanuelli, Chelating derivatives of chitosan obtained by reaction with ascorbic acid, Carbohydr. Polym., 4:137 (1984).
17. R. A. Teutonico and D. Knorr, Plant tissue culture: Food application and the potential reduction of nutritional stress factors, Food Technol. 38(2):120 (1984).
18. D. H. Young, H. Köhle, and H. Kauss, Effects of chitosan on membrane permeability of suspension-cultured Glycine max and Phaseolus vulgaris cells, Plant Physiol., 70:1449 (1982).
19. D. H. Young, and H. Kauss, Release of calcium from suspension-cultured Glycine max cells by chitosan, other polycations and polyanions in relation to effects on membrane permeability, Plant Physiol., 73:698 (1983).
20. D. Kendra, and L. A. Hadwiger, Characteristics of the smallest chitosan oligomer that is maximally antifungal to Fusarium solani and elicits pisatin formation in Pisum sativum, Exper. Mycol., 8:276 (1984).
21. S. Miazga, Determination of proteolytic activity in freely suspended and immobilized cell cultures of Asclepias syriaca L. MS Thesis, University of Delaware, Newark, DE (1985).
22. P. K. Bajpai, and A. Margaritis, Effect of calcium chloride on the growth and ethanol production by free cells of Zymomonas mobilis, Biotechnol. Lett., 6:673 (1984).

Affiliations

A. Ohtakara, M. Mitsutomi and Y. Uchida, Dept. Agricultural Chemistry, Fac. Agriculture, Saga University, Saga 840, Japan.

A. Illanes, R. Chamzy and M. E. Zuniga, School of Biochemical Engineering, Universidad Catolica de Valparaiso, Casilla 4059, Valparaiso, Chile.

J. Braun, CERCOA-CNRS, 2 rue H. Dunant, F-94320 Thiais, France.

P. Le Chanu, F. Le Goffic, S. Sicsic and J. Léonil, Laboratoire de Bioinorganique & Biotechnologie, Ecole Nationale Supérieure de Paris, 11 rue Curie, F-75231 Paris cedex, France.

E. Van Griethuisen, E. Flaschel and A. Renken, Institute of Chemical Engin., Swiss Federal Institute of Technology, EPFL, Ecublens, CH-1015 Lausanne Switzerland.

D. Knorr, Dept. Food Science & Human Nutrition, University Delaware, 203 Alison Hall, Newark, DE 19716, USA.

Section VI

MEDICAL USES OF CHITIN

CHITOSAN EFFECT IN VASCULAR SURGERY, TISSUE CULTURE AND TISSUE REGENERATION

W. G. MALETTE, H. J. QUIGLEY Jr. and E. D. ADICKES

Veterans Administration Medical Center
Omaha, NE, USA

INTRODUCTION

In our previous studies of hemostasis utilizing Chitosan solution (2 mgm/ml.pH4) [1,2] we found that the Chitosan solution formed a coagulum in contact with whole blood. No such reaction was obtained with albumen, globulin, white cells, or platelets. Further investigation showed that the reaction did occur with heparinized blood, defibrinated blood, washed red cells and pure red cell membranes. Therefore, the mechanism for the coagulation appears to be a reaction between the red cell membrane and the Chitosan solution leading to crosslinkage or re-polymerization. (Fig 1)

TISSUE REGENERATION

Vascular Grafts

Our investigation using porous knitted Dacron grafts in heparinized dogs resulted in the ingrowth of vascularized smooth muscle and a complete endothelial intima. A series of these grafts, with and without heparin, was undertaken to clarify this observation. These grafts were harvested at forty-eight hour intervals for the first thirty days and at thirty day intervals to six months. The Chitosan coagulum remained in place until replaced from the outside of the graft to the intima. The coagulum was initially invaded by vascular buds that seemed to contribute rapidly growing smooth muscle cells as they moved from the outside toward the lumen. The muscle cells were initially rapidly dividing and random in arrangement. As they matured they assumed a circumferential pattern within and without the graft as one would expect in a laminar reaction to stress. The entire graft-vessel wall matured in thirty days. Electron microscopy also showed some myelinated nerve fibers invading the wall. The grafts remained stable when examined at six months. In control grafts implanted in a like manner without Chitosan extensive fibroplasia was seen.

Skin and Subcutaneous Tissue

Scalpel incisions were made in skin and subcutaneous tissue in the abdominal surface of dogs. Sixteen matched wounds were made with each animal serving as its own control. It became necessary to make a tattoo at each end of the incisions to identify the incised area. These wounds were harvested at twenty-four hours for the first five days and forty-eight hours thereafter. The Chitosan treated wounds simply had the solution applied to them with a surgical sponge soaked in the solution. The control wounds were treated with a similar sponge soaked in saline solution. The control series exhibited the classic wound healing sequence leading to the deposition of collagen by fibroplasia. The Chitosan treated wounds displayed no such sequence, but, slightly slower, simply regenerated the normal tissue elements. There was no visible scar at the surface. (Figure 2)

Bone

As a tissue with a different healing mechanism a pilot study was done in bone. Identical burr wounds were made in the radii of dogs using a five millimeter burr. The wound was made through the cortex and into the marrow cavity. The right radius was treated with Chitosan solution by application, the left was treated with a saline sponge. The animals served as their own controls. The wounds were followed radiographically each week until regeneration appeared to be taking place. The control wounds exhibited the classic callus osteoblastic-osteoclastic sequence with periosteal bone formation and bone formation in the marrow cavity. The Chitosan treated bone simply reformed cortical bone without going through this sequence. (Figure 3)

TISSUE CULTURE

With the above observations in vivo, we undertook to study the phenomenon in tissue culture. Standard tissue culture methods were used to study several cell types.[3] We had access to rat myocardial cell cultures that exhibited cell contraction as a measure of viability and activity. We also felt it necessary to study other cells that seemed affected in vivo endothelium and fibroblasts.

Myocardial Cells

Neonatal Sprague Dawley rat pups were sacrificed in the usual fashion. The heart muscle was prepared, trypsinized, and grown in a Dulbecco's Modified Eagle Medium (DMEM) with additives. As usual the fibroblasts were removed by allowing them to lay down first, then the myocardial cells were replated. The control cultures were grown in the above fashion. The Chitosan treated cultures were grown in the same medium to which Chitosan solution at a 10% concentration was added. At forty-eight hours the control cultures had layed down in a monolayer and some contraction was seen. The Chitosan cultures grew much faster. They grew in large masses in three dimensions and were beating synchronously. (Figure 4) During this experiment a culture was contaminated with mycoplasma. On the chance that Chitosan might have an effect, the 10% Chitosan feeding solution was added to the plate. The myoplasma disappeared and the myocardial cells grew as before.

Fibroblasts

Since our in vivo observations indicated a specific effect on the fibroblast, control and treated cultures of human fibroblasts were undertaken. The control cultures grew in disorganized fashion with the extension of long processes throughout the culture. An attempt to grow them in the medium with Chitosan resulted in a rounding up of the cells, the extension of many short (exploratory?) processes, an inability to lay down, and death of the fibroblast. (Figure 5)

Endothelial Cells

In the same manner human endothelial cells were grown. The control cultures eventually assumed the desired "cobblestone" appearance, however

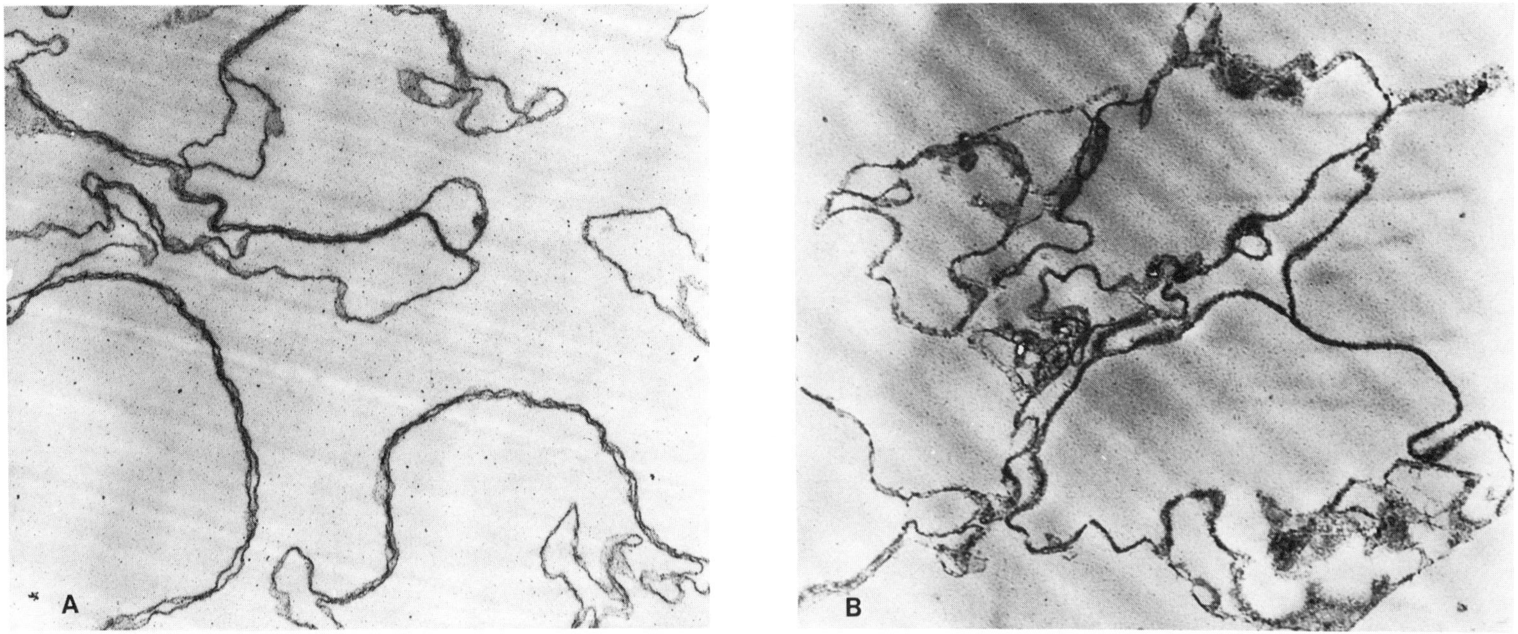

Figure 1

A. Red cell "ghosts" electron microscopy (x 10,000) without Chitosan.

B. Red cell "ghosts" in a Chitosan coagulum (x 10,000) with "reaction sites".

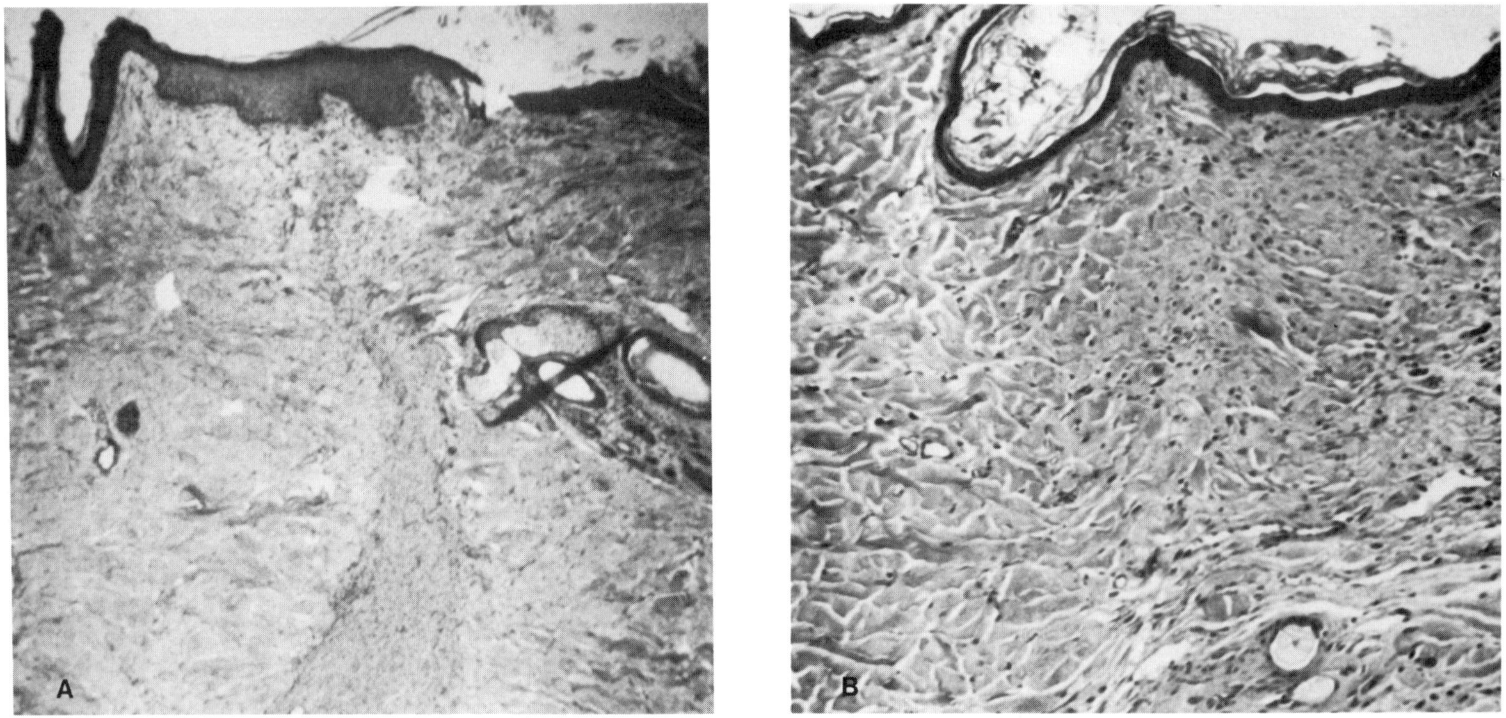

Figure 2

A. Skin and subcutaneous wound without Chitosan showing fibroplasia.

B. Skin and subcutaneous wound with Chitosan showing normal tissue regeneration without fibroplasia.

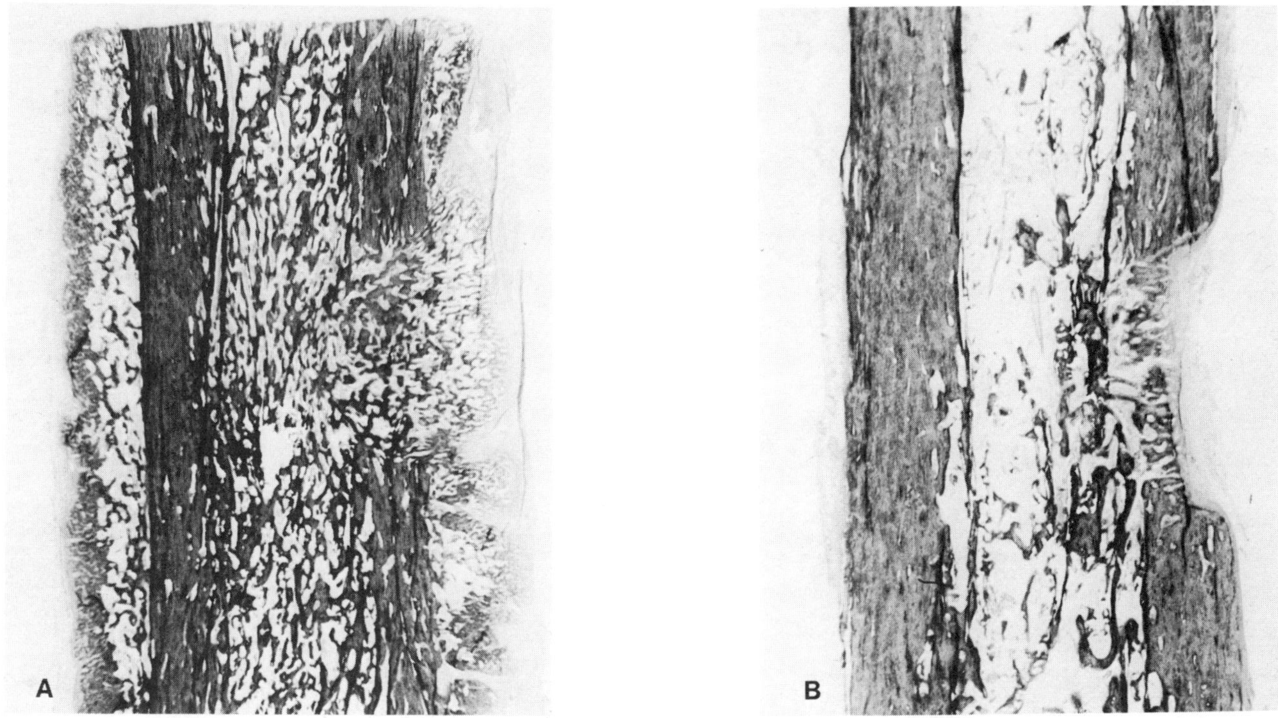

Figure 3

Radius

A. Bone burr wound at six weeks without Chitosan showing periosteal bone formation and bone formation in the marrow cavity.

Radius

B. Bone burr wound at six weeks (same dog) with Chitosan showing cortical bone formation only.

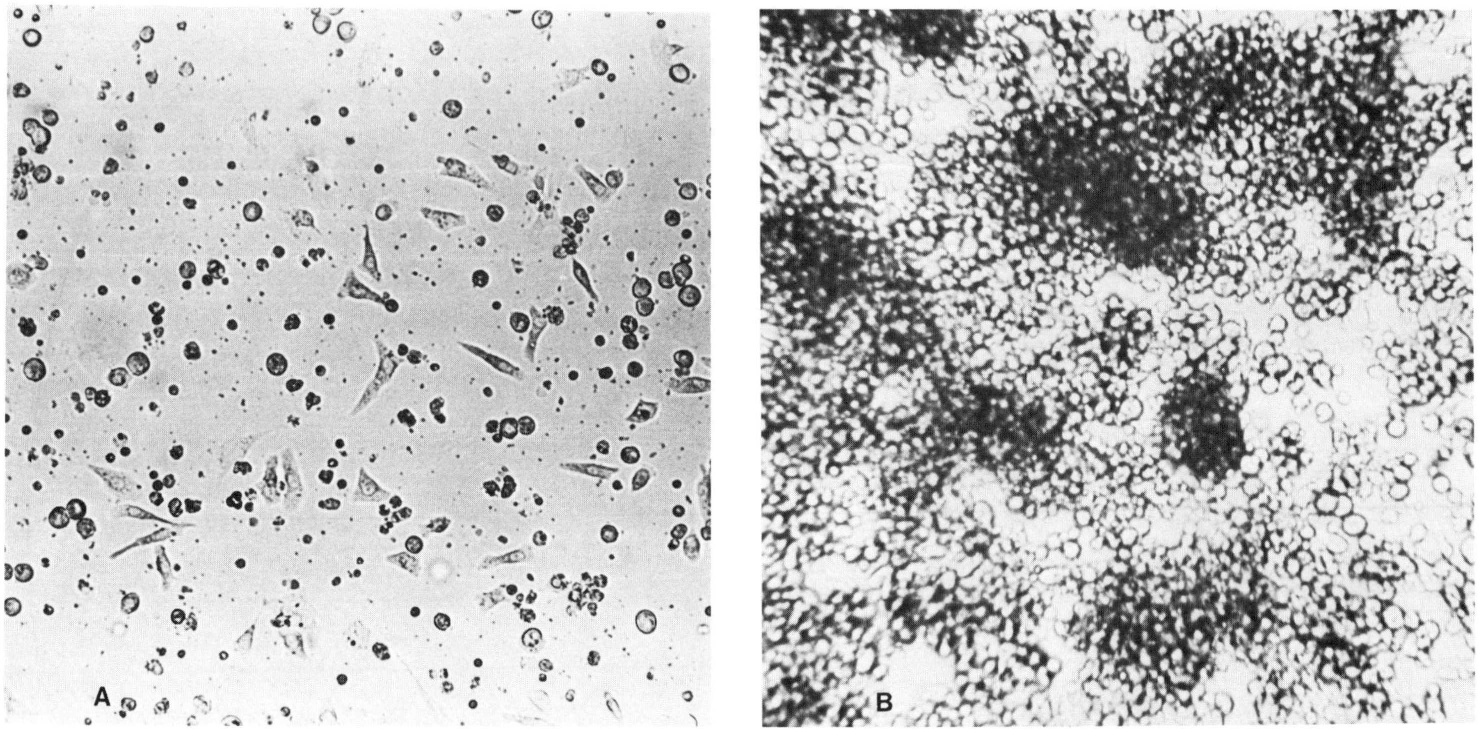

Figure 4

A. Rat pup myocytes in tissue culture at 48 hours without Chitosan showing slow growth and mono-layer formation.

B. Rat pup myocytes in tissue culture at 48 hours with Chitosan showing rapid three dimensional growth with synchronous contraction.

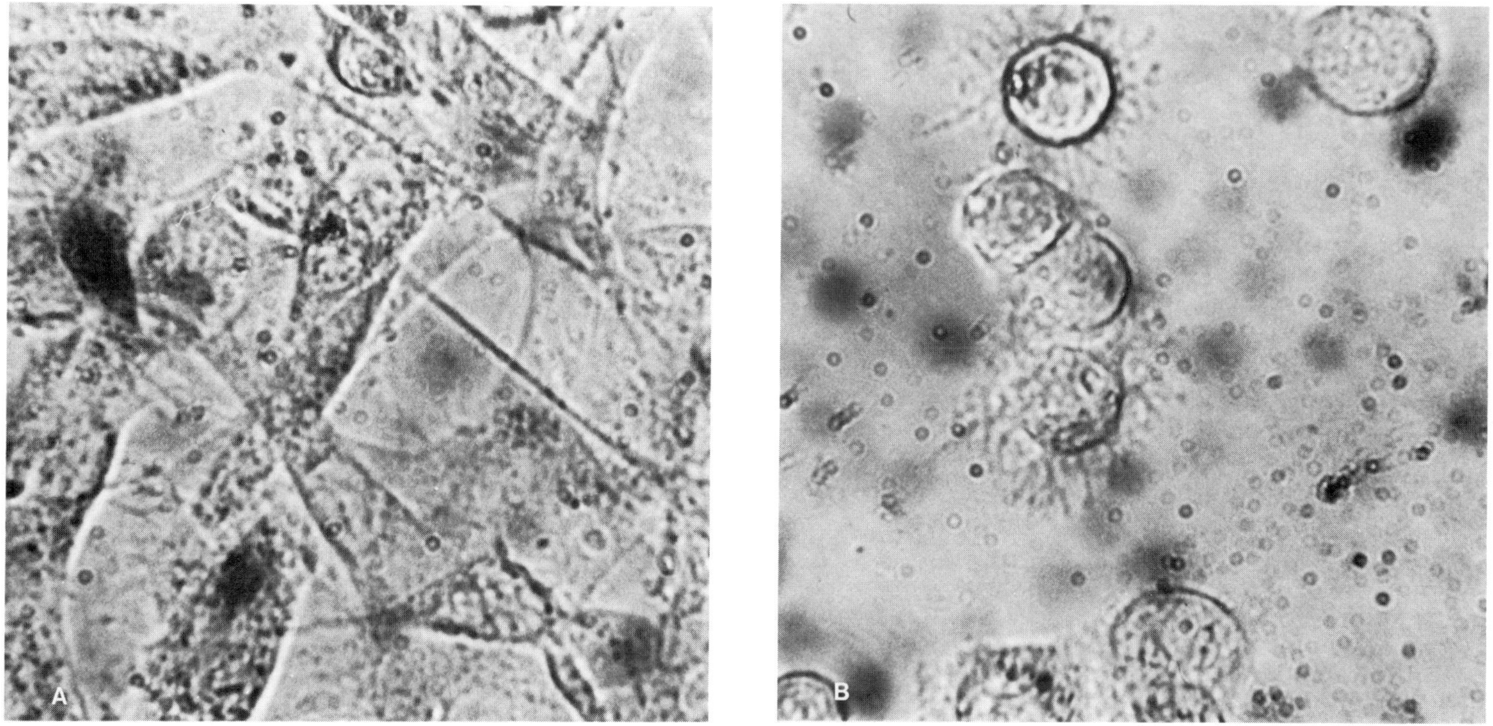

Figure 5

A. Fibroblast tissue culture without Chitosan showing rapid growth and long processes.

B. An attempt to grow fibroblasts with Chitosan results in a "rounding-up" of the cell, the formation of many short processes, failure to "lay down" and cell death.

the Chitosan treated cultures grew faster and assumed the configuration in much less time.

DISCUSSION

The observations in the preceeding experiments leads us to a hypothesis that the Chitosan molecule in solution, as we have used it, has a specific cell wall connection. Coagulation occurs with red cell membrane contact. Three dimensional growth of myocardial cells indicates that the need for "contact" to grow and divide is satisfied. Mycoplasma is inhibited. Finally the inhibition of the fibroblast to prevent the production of collagen and allow normal tissue to regenerate suggests such a mechanism. Whether this phenomenon is a receptor site, several receptor sites, or whether there are multiple and different mechanisms at play are as yet unknown and must await further study.

CONCLUSIONS

1. Chitosan solution in contact with red cell membranes forms a hemostatic coagulum that withstands arterial pressure in vascular grafts and leads to the growth of a vascularized smooth muscle vessel wall with an endothelial intima.

2. Chitosan solution used to treat various tissues results in the inhibition of fibroplasia and the regeneration of normal tissue elements.

3. The addition of Chitosan solution to tissue culture methods results in the inhibition of fibroblasts and mycoplasma organisms, three dimensional growth of myocardial cells, and accelerated growth.

REFERENCES

1. W. G. Malette, H. J. Quigley, Jr., R. D. Gaines, N. D. Johnson, and W. G. Rainer: Chitosan, A New Hemostatic, Ann Th Surg. 36:55 (1983)

2. G. Brandenberg, L. G. Leibrock, R. Shuman, W. G. Malette, and H. J. Quigley Jr.: Chitosan, A New Hemostatic Agent for Diffuse Capillary Bleeding in Brain Tissue, Neurosurgery: 15:9 (1984)

3. D. Yaffe, Rat Skeletal Muscle Cells, in Tissue Culture, Methods and Applications, (Kruse and Patterson eds) Academic Press. (1972)

EVALUATION OF CHITOSAN AS A NEW HEMOSTATIC AGENT: IN VITRO AND IN VIVO EXPERIMENTS

G. FRADET, S. BRISTER, D. S. MULDER, J. LOUGH
and B. L. AVERBACH*

Department of Cardiovascular Surgery and Pathology
McGill University
Montreal, Quebec H3G 1A4, Canada
*Department of Material Science and Engineering
Massachusetts Institute of Technology
Cambridge, MA 02139, USA

INTRODUCTION

Life-threatening bleeding through the pores of vascular grafts is a well-recognized complication in major cardiac and thoracic aortic surgery despite the use of low porosity dacron grafts (1, 2). There have been a number of investigations in which this situation is improved by the preclotting of the grafts, and Bethea and Reemtsmak (1) have described a process wherein the graft was autoclaved after immersion of the graft into heparinized blood. Cooley (3) has since modified the technique by immersion of the grafts in plasma rather than blood, and then autoclaving. He found that this improved the handling characteristics of the grafts. More recently, Malette et al. (4) described the use of a new hemostatic agent, chitosan. They replaced the abdominal aorta of dogs by grafts which had been treated in chitosan and found, at autopsy, that the grafts were encased in smooth muscle, with a living endothelial lining and an abundant vasa-vasorum. The control grafts showed the usual fibrous healing. They also found the chitosan-treated grafts to be impermeable to blood. Following these observations, we decided to investigate the use of chitosan as a hemostatic agent in critical situations, mimicking in vivo cardio-pulmonary bypass conditions where patients are given high doses of heparin systemically. We also looked at the effect of chitosan on graft healing and pseudointima formation. Several techniques for the treatment of the grafts with chitosan were used.

In this study we infiltrated and coated grafts with chitosan in several forms. The film-forming characteristics of chitosan served to seal the grafts, thus acting as a hemostatic agent. The in vitro and in vivo experiments indicate that this approach is promising.

MATERIALS AND METHODS

Graft Hemostasis: In Vitro Study

A perfusion system with systemic pressure maintained at 100 mm Hg by a roller pump (Med Science electronic pump) and a flow of 165 ml/min was used. Different groups of 6 cm long vascular low-porosity Cooley dacron vascular grafts (n = 5 per group) were perfused for 5 minutes with human packed red blood cells diluted in 200 ml of a 0.9% saline solution with 2100 u of heparin added per 500 ml of solution (equivalent of 300 μ/kg in a 70 kg man). The blood losses were collected for the 5 minutes of the perfusion and weighed.

The chitosan used in these experiments was specially prepared from a single species of crab (blue crab) and had a molecular weight of about 900,000. A solution of chitosan acetate was prepared, at a concentration of 8 wt pct. The grafts were dipped into chitosan acetate solutions for fifteen minutes and allowed to dry in air. At this point, the grafts were coated with chitosan acetate. The coated grafts were then treated in 1 molar sodium hydroxide for 30 minutes, rinsed in distilled water, and then dried in air. This treatment converted chitosan acetate to chitosan. In one experiment, the grafts were dipped into chitosan acetate and then perfused wet. Since the blood was slightly alkaline, the acetate was probably converted to chitosan in situ, in the form of a gel.

Six different groups of grafts were perfused.

I. Control: grafts without preclotting.
II. Grafts preclotted with non-heparinized human blood.
III. Grafts autoclaved after immersion in heparinized human blood.
IV. Grafts treated with chitosan solution, used in the air-dried conditions.
V. Grafts treated with chitosan, air dried and then autoclaved at 130°C.
VI. Grafts immersed in chitosan solution and perfused wet (chitosan gel).

Graft Hemostasis and Healing: In Vivo Experiment

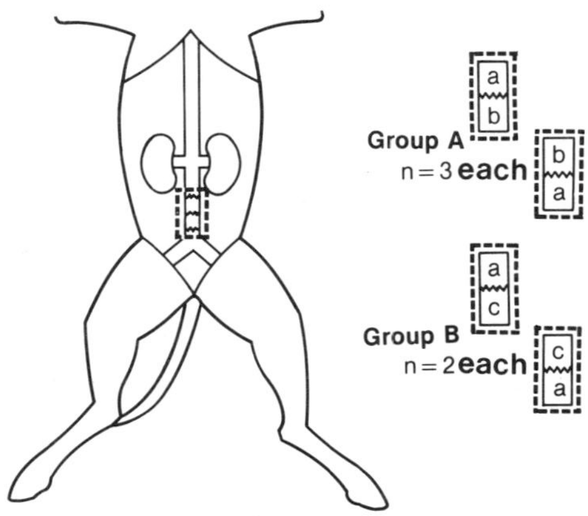

Figure 1. Graft hemostasis and healing: in vivo experiment.

Ten mongrel dogs weighing from 20 to 30 kg underwent a midline laparotomy; the infrarenal aorta was dissected free from the surrounding tissue and the collateral branches were tied. Two 3 cm. lengths of lowporosity Cooley vascular dacron grafts were inserted into place after resection of a suitable length of abdominal aorta. All anastomoses were end-to-end (Figure 1). The pulsating grafts were observed for 5 min in order to observe hemostasis after the clamps were removed, and finally, the retroperitoneum and the abdominal incision were closed. The dogs were divided into two groups:

Group A -	grafts a (n = 6):	autoclaved grafts after immersion in animal blood (control)
	grafts b (n = 6):	grafts treated with chitosan and then air-dried for 24 hrs under sterile conditions
Group B -	grafts a (n = 4):	autoclaved grafts after immersion in animal blood (control)
	grafts c (n = 4):	grafts immersed into chitosan acetate solution and then inserted wet (chitosan gel)

We used 6 animals in groups A and 4 in group B. The graft segments were alternately inserted proximally and distally in the animals in order to eliminate the effects of flow in the evaluation of their performance. (See Figure 1). A total of 6 cm of graft (2 segments of 3 cm end-to-end) was inserted into each animal, using prophylactic antibiotic therapy, penicillin G, 10,000 u/kg. Heparin 50 u/kg was given before clamping of the aorta, and 1/2 amp. (20 meq) of $NaHCO_3$ was given before unclamping. A running of 5-0 polypropylene suture was used for the graft anastomosis. The average cross-clamp time was 35 min. The animals were sacrificed at 6 to 8 weeks after the operation and all specimens were examined macroscopically as well as by light microscopy (LM) and electron microscopy (EM).

Graft Hemostasis: In Vitro Study

The blood losses were collected in containers and then weighed. Except for the control group, the losses occurred mainly during the first 30 sec,

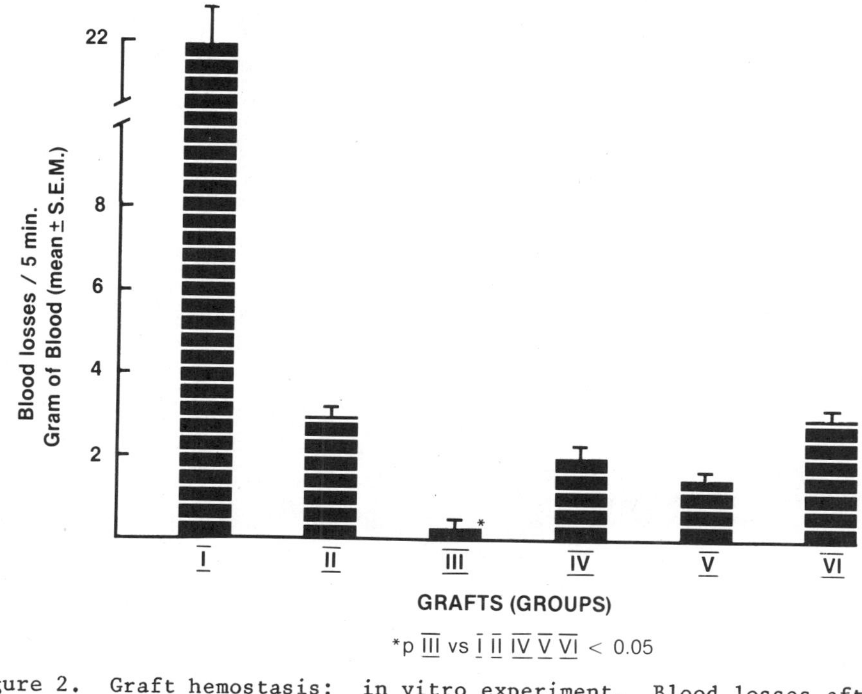

Figure 2. Graft hemostasis: in vitro experiment. Blood losses after 5 min of perfusion (Mean ± S.E.M.).

stabilizing thereafter without any gross differences. The leakage of blood was expressed as grams of blood loss per 5 min of perfusion per group (n = 5 each), and the results are reported in Figure 2. It is evident that all of the groups performed better than the controls, which were non-preclotted grafts ($p < 0.05$). The chitosan coated grafts (groups IV to VI) performed as well as the blood-clotted grafts (group II). The blood-autoclaved grafts (group III) performed the best ($p < 0.05$), but this is explained by a difference in leakage which occurred immediately after the beginning of the perfusion; there is a formation of a coagulum between the chitosan-coated grafts and the blood at this point. This difference becomes clinically insignificant once the coagulum is formed, after 30 to 60 sec.

Graft Hemostasis and Healing: In Vivo Experiment

In both groups (A and B), the macroscopic blood losses after unclamping of the aorta were similar and not significant. In none of the cases was it necessary to reclamp the proximal aorta or to use suction. One animal in group B, the first of our series, had pre- and intraoperative cardiorespiratory arrests because of inadequate intubation. He became paraplegic with cold lower extremities on the first day after the operation and was sacrificed.

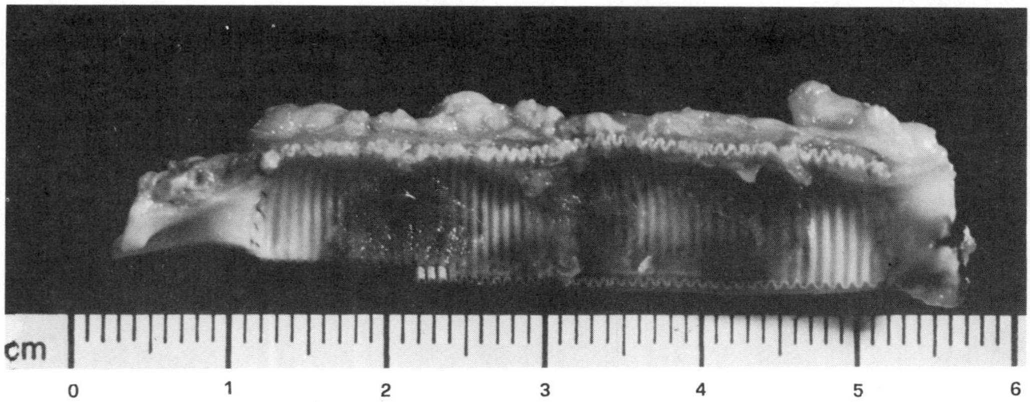

Figure 3. Platelet-fibrin aggregates and organizing thrombus are shown in the center of the graft.

Table 1. Pseudointimal Thickness (mm) -- Group A (n = 6) (Mean ± S.D.)
a. Control grafts (blood autoclaved)
b. Grafts with chitosan films

	Grafts a		Grafts b	
n	Mean	S.D.	Mean	S.D.
1	0.0526	0.0091	0.1382	0.0391
2	0.1343	0.0674	0.0711	0.0079
3	0.1158	0.0319	0.0750	0.0167
4	0.0158	0.0000	0.2396	0.0633
5	1.8960	0.6703	0.1237	0.0405
6	0.4239	0.0935	0.0057	0.0800

Table 2. Pseudointimal Thickness (mm -- Group B (n = 3) (Mean ± S.D.)
a. Control grafts (blood autoclaved)
c. Grafts with chitosan gel

	Grafts a		Grafts c	
n	Mean	S.D.	Mean	S.D.
1	0.2027	0.0987	0.1659	0.0961
2	0.0895	0.0120	0.1027	0.0158
3	0.0289	0.0091	0.0289	0.0091

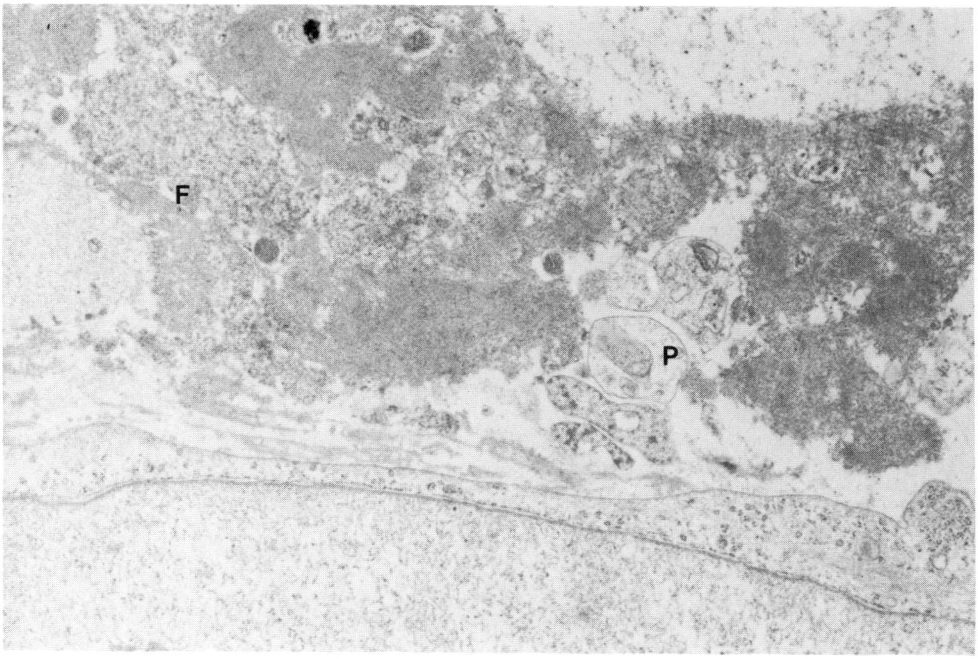

Figure 4. Undifferentiated multipotential mononuclear cell is shown adjacent to fibrin (F) and degranulated platelets (P). Uranyl acetate, lead citrate -- 12,000 X.

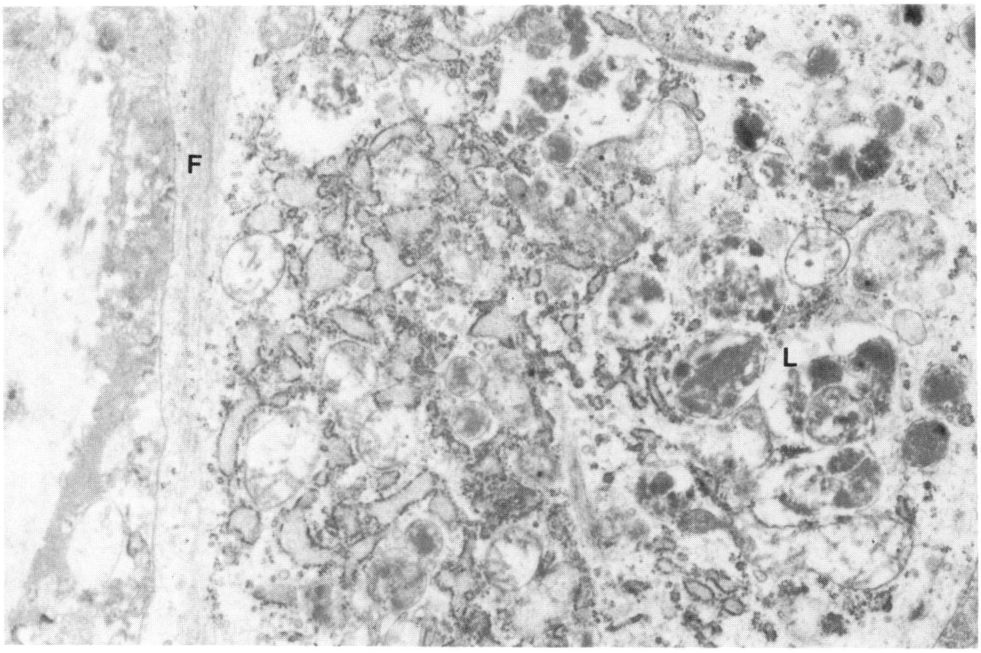

Figure 5. A mononuclear macrophage is shown phagocytosing platelet-fibrin thrombus. Peripheral filaments adjacent to the plasma membrane suggest that this cell may be differentiating to a fibroblast. Lysosomes (L) containing fibrin (F) filaments. UA, LC -- 16,000 X.

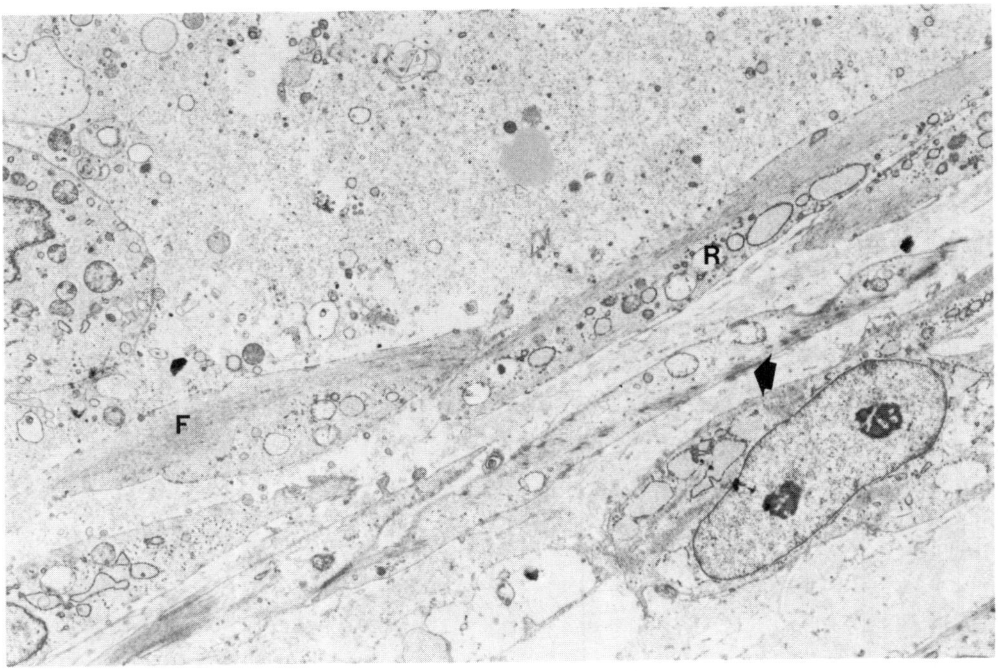

Figure 6. Spindle-shaped mononuclear cells are shown infiltrating thrombus. Rudimentary rough endoplasmic reticulum (R), filaments (F), and attachments zones (arrow) are characteristic of myofibroblasts. UA, LC -- 5680 X.

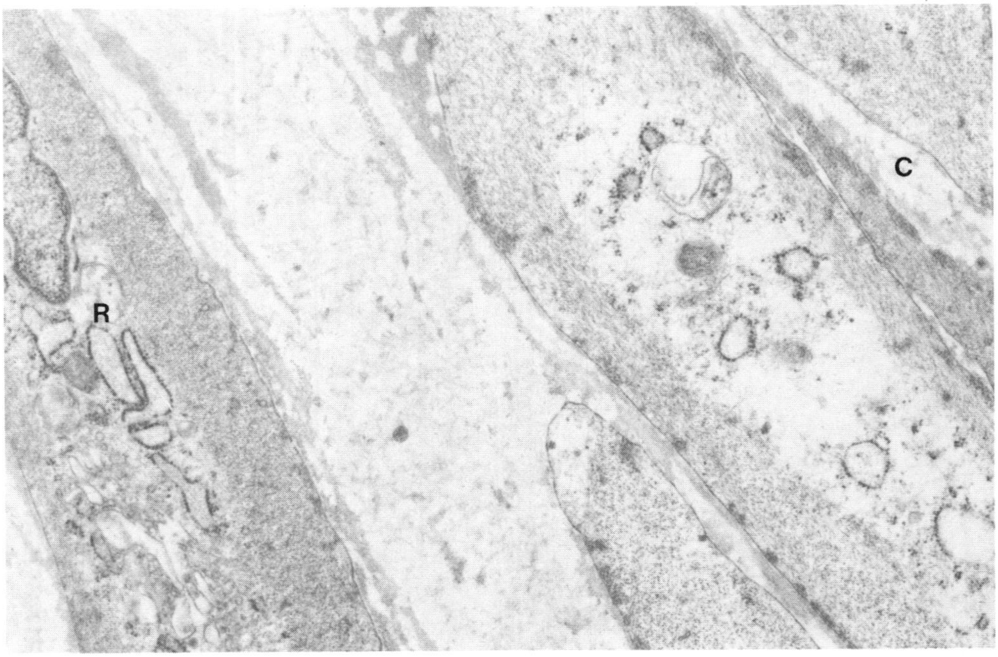

Figure 7. More mature myofibroblasts with a well-developed complex R.E.R. (R) and peripheral longitudinal myofilaments. These cells appear to be synthesizing collagen (C). UA, LC -- 7520 X.

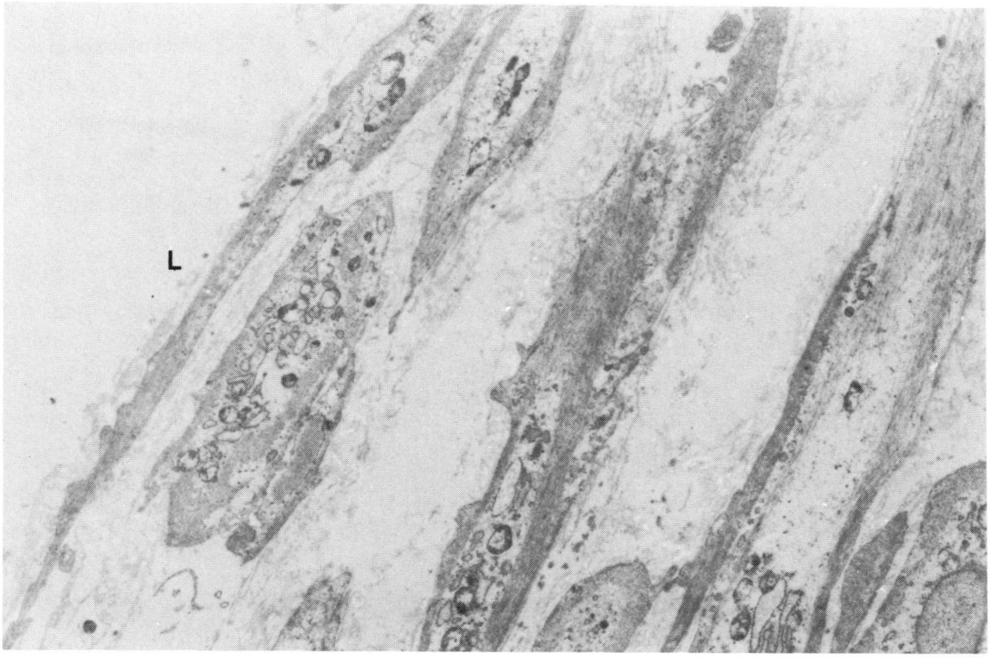

Figure 8. Mature neo-intima with a surface pseudo-endothelial cell adjacent to the lumen (L). Many myointimal cells are shown in the collagenized collagenizing stroma. UA, LC -- 6000 X.

The autopsy findings showed a clotted abdominal aorta from 2 cm above the proximal anastomosis down to both iliac arteries, with necrosis of the bladder, uterus, and tubes. All suture lines were intact and there were no hematoma around the graft. Microscopy corroborated those findings, showing intact anastomosis with intraluminal thrombosis above and below the grafts.

Macroscopic and LM examinations showed no differences in groups A and B between the chitosan-treated grafts (chitosan-gel and air-dried chitosan) and the control grafts (autoclave blood). The grafts were found to be encased by an outer fibroblastic cast with the presence of vasa-vasorum in close proximity to the dacron fibers but without any permeation through the graft. The pseudointimal lining was composed of nonuniform layers of fibroblasts with the intraluminal cells appearing more flattened and elongated. In the center portion of the grafts, a small organizing thrombus or a quantity of platelet-fibrin aggregate was found (Figure 3). We observed a marked discrepancy between the thickness of the pseudointima at the convexity and concavity of the crimped grafts, and we thus limited our measurements of pseudointimal thickness to the convexity, obtaining three random measures per specimen (Tables 1 and 2). In both groups, the pseudointima was less than 1 mm thick, in both the controls and the chitosan-treated grafts. Statistical analysis was not performed because of the small population studied.

E M examination of neo intima revealed no differences when the chitosan grafts were compared with blood treated grafts. The sequence of neo-intima formation is shown in the electron micrographs (fig. 4-8). Initially, undifferentiated multipotential mononuclear cells were found infiltrating and phagocytozing platelet fibrin thrombus. These cells often have peripheral longitudinal filaments which suggests that they gradually differentiate into fibroblasts (Fig. 5, 6). The well developed rough endoplasmic reticulum also suggests that they are involved in collagen synthesis. Mature myofibroblasts with myofilaments and peripheral pinocytic vesicles were the predominant cells in the neo-intima (Fig. 8). Re-endothelialization was seen to progress from both proximal and distal anastomosis. Preliminary pseudo-endothelialization is shown in Fig. 8. Pseudo-endothelial cells were see on the surface and differ from true endothelial cells because of the presence of peripheral myofilaments suggesting that they are adapted from the myofibroblasts.

DISCUSSION

Our data on the efficacy of chitosan as a hemostatic agent correlate well with the ones of Malette et al. (4). The chitosan-treated grafts (both chitosan gel and chitosan film) performed as well as the blood-clotted ones in extreme in vitro conditions of anticoagulation. Indeed, this was corroborated in our in vivo study where no clinical differences in hemostasis were found between the chitosan and blood coagulated grafts.

Our observations on the role of chitosan in the graft healing process, however, do not corroborate the report by Malette, who described the growth of vascularized smooth muscle in their chitosan-treated grafts. They found muscle cells penetrating the graft and linking the outer layers to the inner subendothelial layer, with abundant vessels penetrating both layers. Our observations correlate more with the work of Sottiurai et al. (11) and favor the theory of the monocyte being a multipotential cell in the circulating blood, and acting as one of the major components of pseudointimal formation. Indeed, we found both chitosan and blood treated grafts to be encased in a fibroblastic cast where nerves and vasa-vasorum are found, but we could not demonstrate the permeation of vasa-vasorum through the graft pores, although the passage of escaped cells from the vessels is quite possible. As for pseudointimal healing, ingrowth of endothelium from the aortic end was found at both proximal and distal anastomotic sites. At some distance from the anastomosis, the pseudoindima was composed mostly of fibroblasts in the deepest layers close to the dacron fibers. Closer to the intraluminal surface, the cell population consisted mainly of myofibroblastic cells, and finally, of pseudoendothelial cells, which differ from myofibroblastic cells only by the appearance of peripheral intracytoplasmic vesicles, making them more closely related to true endothelial cells, the only difference being the presence of the peripheral myofilaments.

These observations seem to support Mackenzie's (6) hypotheses of graft healing and pseudointimal formation. The initial event of healing involves platelet and fibrin aggregates, followed by red blood, white blood cells, and finally by monocytes that differentiate into fibroblasts, then myofibroblasts, and finally pseudoendothelial cells. The various stimuli of differentiation and maturation are unknown, but these could be a consequence of the blood flow and the different cell contacts as we progress toward the intraluminal surface. This mechanism could play a major role in graft healing along with endothelium ingrowth and possibly permeation of capillaries or escaping cells and nutrients through the pores of the graft.

CONCLUSION

We have demonstrated that chitosan is an effective and adequate hemostatic agent even under the most severe conditions of anticoagulation. It is, apparently, a safe agent which did not adversely affect graft healing. Both chitosan- and blood-treated grafts presented the same light microscopy and electron microscopy features at 6 weeks after insertion. We conclude from these observations that chitosan is a good substitute for blood for graft hemostasis.

REFERENCES

1. M. C. Bethea and K. Reemtsmak, Graft hemostasis: an alternative to preclotting, Ann. Thor. Surg., 27:374 (1979).
2. R. L. Therer, J. M. Hauer, and R. M. Weintraub, A comparison of preclotting techniques for prosthetic aortic replacement, Circulation, 66:1.143 (1982).
3. D. A. Cooley, A. Romagnoli, J. D. Milam, and M. I. Bossart, A method of preparing woven Dacron aortic grafts to prevent interstitial hemorrhage, Cardiovasc. Dis. Bull. Tex. Heart Inst., 8:48 (1981).
4. W. G. Malette, H. J. Quigley, R. D. Gaines, N. D. Johnson, and W. G. Rainer, Chitosan: a new hemostatic, Ann. Thor. Surg., 36:55 (1983).
5. V. S. Sottiurai and R. C. Batson, Role of myofibroblasts in pseudointima formation, Surgery, 94:792 (1983).
6. J. R. Mackenzie, M. Hackett, C. Topuzlu, and D. J. Tibbs, Origin of arterial prosthesis lining from circulating blood cells, Arch. Surg., 97:879 (1968).

INHIBITORY ACTION OF SULPHATED CHITIN DERIVATIVES ON THE HYDROLYTIC ACTIVITY OF THROMBIN

W. OKIEI, S. NISHIMURA, O. SOMORIN,* N. NISHI and S. TOKURA

*Chemistry Department, University of Lagos, Lagos, Nigeria
Department of Polymer Science, Faculty of Science, Hokkaido University
Sapporo 060, Japan

INTRODUCTION

Heparin, an anticoagulant polysaccharide, inhibits thrombin activity which plays the most important role in the blood clotting process. It is known to accelerate neutralization of thrombin activity instantaneously in the presence of antithrombin III (AT-III) (1). The thrombin-AT-III complex also inhibits Factor Xa moiety in the blood clotting system (2). In order to understand the exact mechanism of heparin action, the stepwise preparation of heparin-like materials should be investigated from the point of view of inhibition mechanism.

The biomedical properties of chitin have been receiving a great deal of attention recently, since it has a similar sugar skeleton as that of heparin and is biodegradable to some extent in the animal body. Chitosan has been used to prepare a heparin-like derivative owing to its similar chemical structure and to the easiness of obtaining a homogeneous chemical modification (3, 4). Although there are many investigations on the preparation of chitosan heparinoids, including the latest work by Muzzarelli et al. (5), their anticoagulant activity and their toxicity in the human body have not been fully investigated. Although chitin is poorly reactive because of its rigid crystalline structure (6), the crystalline structure has been found to be broken down by the freezing procedure during alkylation reaction (7). The alkylated chitin, which was converted to a hydrophilic one, has some lysozyme accessibility, even at low degrees of substitution. The lysozyme accessibility and water solubility were markedly enhanced by the increase of the degree of carboxymethylation, probably owing to the destruction of hydrogen bonds (8). Carboxymethyl-chitin (CM-chitin) has been reported to adsorb blood proteins with effective contribution of calcium ion (9). As CM-chitin can be prepared by a simple procedure and the substitution site is suggested to be only the C-6 position in the GlcNAc residue when the degree of substitution is less than 0.8 (10), the preparation of chitin heparinoid was started from CM-chitin in this study. The contribution of the carboxymethyl group to the heparin like property was investigated in the case of dextran heparinoid (11). A water soluble sulphated-chitin (S-chitin) has also been prepared by the direct sulphation of chitin to investigate the influence of the sulphonyl group. The type of inhibition on the thrombin activity was suggested to be a non-competitive one with use of synthetic chromogenic substrates, and the fibrinogen transformation to fibrin was investigated in the absence or presence of AT-III to demonstrate the influence of chitin heparinoids clearly.

EXPERIMENTAL

Materials

Chitin was prepared from Queen Crab shells according to the method of Hackman (12) and powdered to 45-60 mesh before use. Monochloroacetic acid and other reagents of reagent grade were obtained from Wako Pure Chemical Industries Ltd. and used without further purification. Fibrinogen 95 % clottable (Lot. 100) was purchased from Seikagaku Kogyo Co. Ltd. Thrombin was supplied from Mochida Pharmaceutical Co. Ltd. with activity of 1,000 NIH Units/vial (Lot. 3B428). Human antithrombin III of 25 Units/vial was obtained from AB KABI Co. Ltd. (Lot. 74868). Heparin sodium salt of 159 I.U./mg (Lot. M9R8164) was purchased from Nakarai Chemical Industries Ltd. L-BAPA HCl (Benzoyl-L-Arg-p-nitroanilide hydrochloride) was synthesized according to the method of Nishi et al. (13). Chromogenic Substrates, "KABI TESTZYME", such as BPVAPA HCl (Benzoyl-L-Phe-L-Val-L-Arg-p-nitroanilide hydrochloride) and BIGGAPA HCl (Benzoyl-L-Ile-L-Glu(γ-OCH_3) -Gly-L-Arg-p-nitroanilide hydrochloride) were obtained from AB KABI Co. Ltd.

Preparation of chitin derivatives

CM-chitins were prepared from chitin powder according to the method reported previously (8). A 35 or 50 % NaOH solution (w/w) was used to prepare CM-chitin with DS 0.56 and 0.80, respectively. The sulphation of chitin and CM-chitin was carried out according to the method of Horton and Just (14). In general, 2 g of chitin or CM-chitin was suspended in 70 ml of distilled pyridine and separated by centrifugation after standing for 1 h at room temperature. This procedure was repeated three times to remove water. A chlorosulfonic acid-pyridine mixture, prepared by the slow addition of $ClSO_3H$ (10 ml) to 60 ml of pyridine at ice cold temperature, was added, dropwise, to the above suspension in pyridine for 40 min at below 10°C. The reaction mixture was refluxed for 1.5 h with stirring. The reaction mixture was poured into 200 ml of ice-water and adjusted to pH 10 with 2M NaOH. The precipitate obtained by the addition of 2 ml of ethanol was redissolved in water and dialyzed against deionized water. The product was lyophilized. Yield 1.3-2.1 g.

Estimation of the degree of substitution

The degree of carboxymethylation was estimated by elemental analyses with Yanagimoto CHN Corder MT-2 and by i.r. spectrometry with JASCO i.r. spectrophotometer A-302. The sulphur content was determined by combustion method.

Estimation of molecular weight of the heparinoids

The apparent molecular weights of chitin heparinoids were estimated from viscosity measurements, using an Ubbelohde viscometer, by applying the equation for heparin (17). The molecular weight of CM-chitin was also estimated by applying the viscosity equation proposed by Kaneko et al. (18).

Fluorescence titration of thrombin by sulphated chitin derivatives

Fluorescence titration was carried out with a Hitachi fluorescence spectrophotometer Model 204 by excitation at 280 nm and recorded at 330 nm at pH 7.4 (10 mM Tris-HCl buffer containing 5 mM $CaCl_2$ and 0.1 M KCl) at 25°C. Double reciprocal plots of fluorescence yield ($\Delta F/F°$) and heparinoid concentration were made to calculate the binding constant and binding free energy.

Kinetics for chromogenic substrates

The rate of hydrolysis of the chromogenic susbstrate by thrombin was measured at pH 7.4 (Tris-HCl buffer) and 37°C in the presence of 3.3 NIH Units/ml of thrombin. 0.2 mg/ml of heparinoid was applied to inhibit the thrombin activity. The concentrations of thrombin and heparinoid were reduced to 1/20 in the case of BPVAPA to prevent precipitation.

Measurements of the clottability

Fibrinogen was dissolved in deionized water and adjusted to 0.5 mg/ml as determined from the optical density at 280 nm according to Mihalyi (15). Thrombin was dissolved in the buffer to give a concentration of 0.5 mg/ml (100 NIH Units/ml). 5 mg of AT-III was dissolved in 10 ml of buffer solution to give a solution of 2.5 I.U./ml; 3.7 ml of reaction mixture contains 0.1 ml of thrombin solution and 215 μl of AT-III (molar ratio of thrombin to AT-III was 5.0). Firstly thrombin or AT-III was treated with heparin or heparinoids of various concentration for 5 min at 25°C and pH 7.4 (10 mM Tris-HCl buffer containing 5 mM $CaCl_2$ and 0.1 M KCl) prior to starting the clotting. Thrombin was treated with heparin or heparinoid-AT-III complex for 5 min under the same conditions as above. Then 1.5 ml of fibrinogen solution was combined with the above reaction mixture and the mixture was agitated gently with glass hook until the first appearance of fibrin threads, as shown in text (16). The clotting time was converted into thrombin activity. The thrombin activity was standardized by the rate of TAME HCl hydrolysis according to the text (16).

RESULTS AND DISCUSSION

Preparation of sulphated chitin derivatives

The sulphation of chitin and its derivatives was carried out successfully in the chlorosulphonic acid-pyridine complex according to reference 14. A sulphur trioxide-DMF system was also applied but it was too drastic to maintain the molecular weight of products. The chitin heparinoids were easily soluble in water. As deacetylation was reported to be less than 5 % during the carboxymethylation from potentiometric titration (10), the amount of N-sulphonyl group (would be negligibly small on the sulphated CM-chitin) as shown in Fig. 1. From the i.r. measurements, the characteristic absorptions due to sulphonyl group have been shown at 820, 980 and 1240 cm^{-1} respectively. The peak, due to the crystalline structure of chitin at 1620 cm^{-1}, has disappeared after the sulphation (as well as) in CM-chitin. The results of elemental analyses of chitin derivatives are listed in Table I with apparent molecular weights. The sulphur content of S-chitin was controlled up to 0.7 to prevent substitution on C-3 hydroxyl group, and to maintain a similar level of molecular weight for the chitin derivatives. The apparent molecular weights were almost the same among our chitin derivatives.

Fluorescence titration of thrombin by chitin derivatives

Fluorescence spectra of thrombin were observed at various heparinoid concentrations by the excitation wave length of 280 nm at pH 7.4 of Tris-HCl buffer. The peak of the fluorescence spectrum tended to shift slightly to shorter wave lengths with increase in heparinoid concentration. The fluorescence yields were noticeably increased by the addition of S-CM-chitins and slightly by CM-chitin, in the presence of calcium ion as shown in Fig. 2. The highest activation among our samples was observed with S-CM-chitin II. The degree of activation seems to depend on the degree of carboxymethylation. Though the decrease in activation seems also to be due to the decrease in the sulphonyl groups, sulphur content alone does not seem to have any active action on the fluorescence yield because of little effect by S-chitin.

The binding constants and free energy changes to form the complex between thrombin and chitin derivatives are listed in Table II. They were calculated by the double reciprocal plots of the fluorescence yields and the concentration of the chitin derivatives. The highest binding constant was shown by S-CM-chitin II and the resulting complex was suggested to be the most stable among other complexes. However, the fluorescence data alone does not give direct information on the denaturation near to the thrombin catalytic site.

Table I. Analytical data for chitin heparinoids.

Sample	S, %	SO_3Na*	COONa*	M.W.**
CM-chitin	0	0	0.80	$6.3x10^4$
S-CM-chitin I	1.48	0.14	0.80	$1.8x10^4$
S-CM-chitin II	7.66	0.81	0.56	$2.4x10^4$
S-chitin	7.07	0.69	0	$2.0x10^4$

* moles of functional group/GlcNAc residue.

** average molecular weight determined by the viscosity equation for heparin.

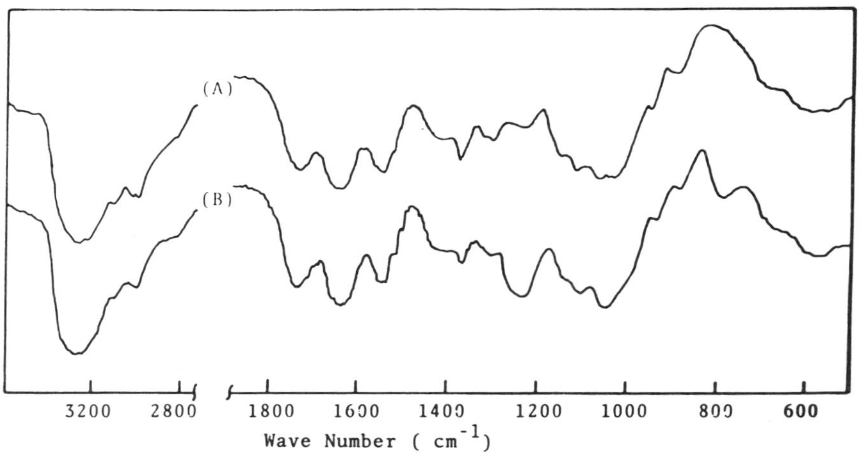

Fig. 1. Infrared absorption spectra of chitin derivatives(KBr). (A): CM(0.80)-chitin. (B): S(0.14)-CM(0.8)-chitin.

Inhibition of thrombin activity by chitin derivatives on the chromogenic substrates.

The hydrolytic activity of thrombin toward chromogenic substrates was studied in the presence of S-CM-chitin II to show the type of inhibition. The double reciprocal plots of the rate of hydrolysis and the substrate concentration would suggest a non-competitive inhibition on the substrates used in these experiments. The results of L-BAPA hydrolysis by thrombin are shown in Fig. 3 in the presence of rather higher S-CM-chitin II concentration.

Fig. 2. Dependence of fluorescence yield of thrombin on the concentration of chitin derivatives at pH 7.4 (10 mM Tris-HCl buffer containing 5 mM $CaCl_2$ and 0.1 M KCl) at 25°C.

--◑--; S-CM-chitin I
--◐--; S-CM-chitin II
--○--; S-chitin
--●--; CM-chitin

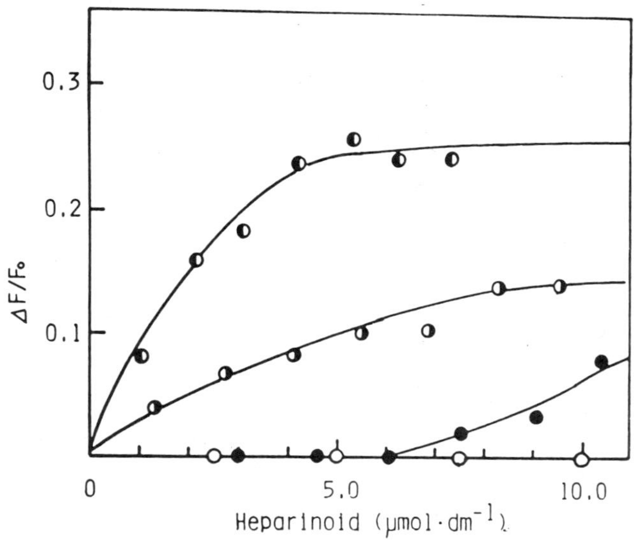

But S-chitin hardly demonstrated any inhibition on the thrombin activity with single peptide substrates, and clear inhibition with tri or tetra-peptide substrates. Kinetic parameters for the hydrolysis of various chromogenic substrates which were given by applying the equation for the non-competitive inhibition are listed in Table III. As can be seen from Table III, lower Ki values are shown for S-CM-chitin than for S-chitin. Abnormally low values were obtained in the case of BPVAPA.HCl, probably due to the formation of tight complexes between thrombin and substrates or chitin derivatives. Thus the hydrophobic group in the thrombin molecule might contribute to form the complex between thrombin and S-CM-chtin, in addition to the basic aminoacid residues.

Neutralization of thrombin activity on the fibrin clotting

The profiles for the neutralization of thrombin activity by chitin heparinoids are shown in Fig. 4 and 5 together with that by heparin (Fig. 6) in the absence and presence of AT-III. The gradual depression of thrombin activity was observed to start at 0.2 μg/ml of S-chitin in the absence of AT-III as seen in Fig. 4. S-chitin showed a little activation effect for AT-III, although it was shown to combine with AT-III. The ratios of remaining thrombin activity to that in heparin at 0.1 μ g/ml of inhibitor concentration are 100 in the absence of AT-III and 86 % with AT-III influence on S-chitin.On the other hand, S-CM-chitin II became effective at lower concentration than S-chitin even in the absence of AT-III. Thrombin activity was indicated to remain ca. 70 % without AT-III and 43 % with AT-III at 0.1 μ g/ml compared

Table II. Binding constants and free energies on the adsorption of heparinoids to thrombin.

Sample	$K_a(M^{-1})$	$-\Delta G^o(K_{cal}/mol)$	$-\Delta G_u(K_{cal}/mol)$
CM-chitin	$7.9x10^3$	5.3	7.7
S-CM-chitin I	$2.1x10^4$	5.9	8.2
S-CM-chitin II	$7.9x10^4$	6.6	9.0
S-chitin	n.d.	---	---

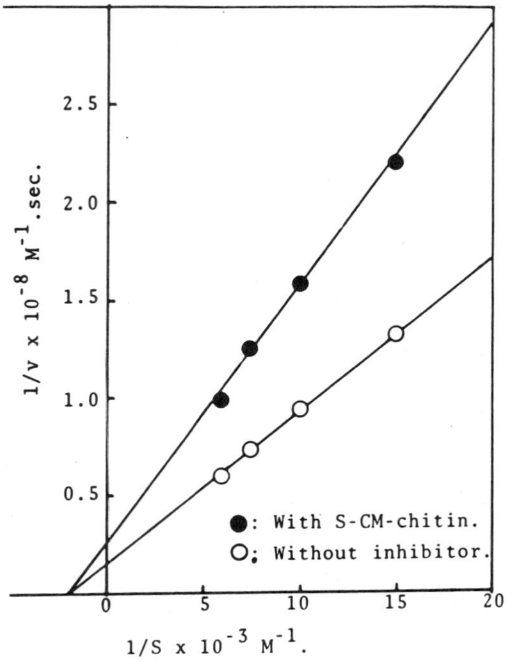

Fig.3. Double reciprocal plots of L-BAPA HCl hydrolytic reaction by thrombin in the absence and presence of S-CM-chitin II at pH 7.4 and 37°C.

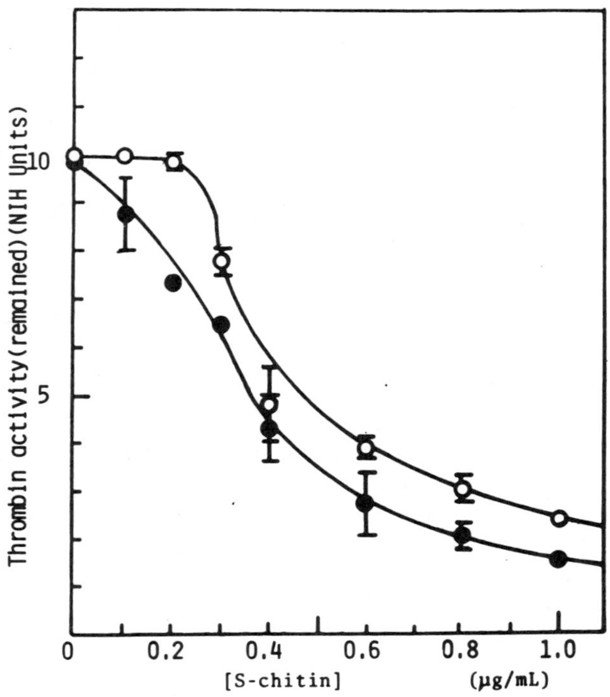

Fig. 4. Inhibition of thrombin activity on the fibrin clotting by S-chitin in the absence and presence of AT-III at pH 7.4 and 25° C. Incubation time was 5 min in each case.

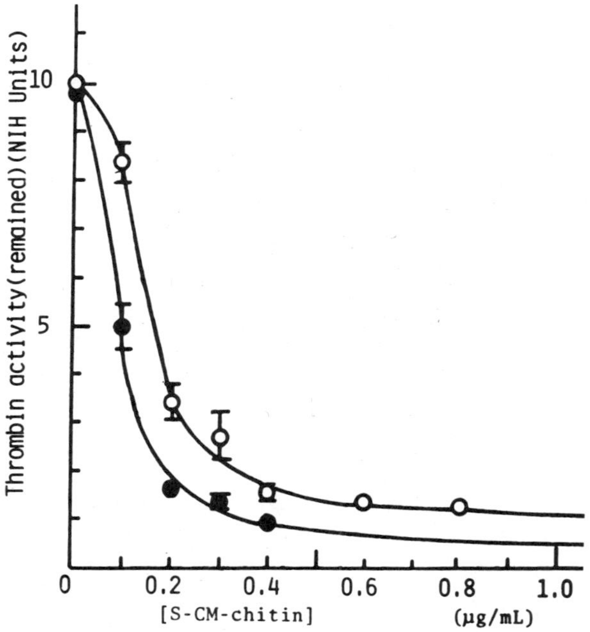

Fig. 5. Inhibition of thrombin activity on the fibrin clotting by S-CM-chitin II in the absence and presence of a AT-III at pH 7.4 and 25°C.

--o--; withour AT-III
--●--; with AT-III

Incubation time at 25°C was 5 min each case.

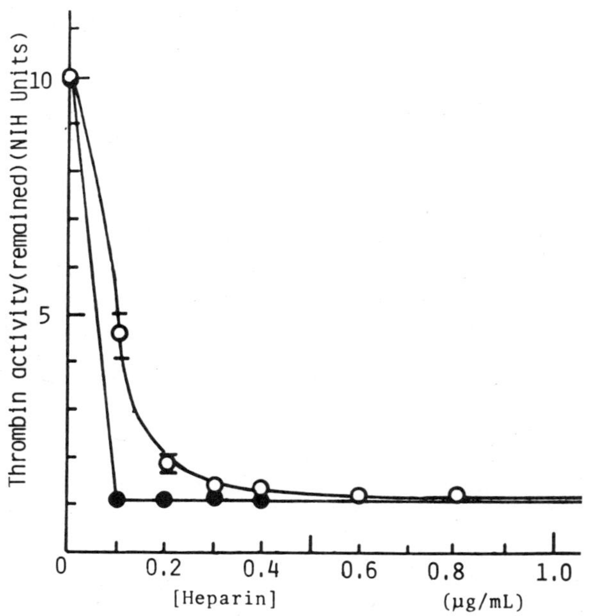

Fig. 6. Inhibition of thrombin activity on the fibrin clotting by heparin in the absence and presence of a AT-III at pH 7.4 and 25°C.

--o--; without AT-III
--●--; with AT-III

Incubation time at 25°C was 5 min each case.

to those by heparin. It is not clear yet if the lower sulfur content of the S-CM-chitin I is responsible for the lower effect than by S-CM-chitin II on the inhibition of thrombin activity, though S-CM-chitin I was less effective than S-chitin. It seems to be more important to regulate the molar ratio between sulphonyl and carboxyl groups than to increase the sulfur content. Since heparinoids used in this study have molecular weights higher than that of heparin, it will be necessary, in the future, to investigate the relationship between molecular weights of heparinoids and their neutralization activity on thrombin.

Table III. Kinetic constants for the hydrolysis of chromogenic substrates by thrombin.

Substrate	Inhibitor	Km, M	Ki, M	Vm, M/sec
BAPA.HCl	S-CM-chitin II	5.0×10^{-4}	3.8×10^{-5}	6.7×10^{-8}
BPVAPA.HCl	S-CM-chitin II	2.6×10^{-5}	3.7×10^{-7}	2.6×10^{-8}
	S-chitin	2.6×10^{-5}	2.1×10^{-6}	2.6×10^{-8}
BIGGAPA.HCl	S-CM-chitin II	3.3×10^{-4}	2.7×10^{-5}	1.8×10^{-7}
	S-chitin	3.3×10^{-4}	1.3×10^{-4}	1.8×10^{-7}

Kinetics were carried out at pH 7.4 (Tris-HCl buffer containing 5 mM $CaCl_2$ and 0.1 M KCl) and 37° C. Hydrolytic reaction was started by the addition of substrate after 30 min incubation of thrombin-inhibitor complex.

REFERENCES

1. R.D. Rosenberg, The effect of Heparin on Factor XIa and Plasmin, Thr. Diathes. Haemorrh., 33:51(1974).
2. O.R. Φdegard, M. Lie and U. Abildgaard, Antifactor Xa activity measured with amidolytic methods, Haemostasis, 5:265(1976).
3. D.T. Warner and L.L. Coleman, Selective sulfonation of amino groups in amino alcohols. J. Org. Chem., 23:1133(1958).
4. K. Nagasawa, Anticoagulant sulfonated chitosan. Japan, 76.06720(1976)
5. R.A.A. Muzzarelli, F. Tanfani and M. Emanuelli, Sulfated N-carboxymethyl chitosans: Novel blood anticoagulants, Carbohyd. Res., 126:225(1984).
6. K.H. Gardner and J. Blackwell, Refinement of the structure of β-chitin. Biopolymers, 14:1581(1975).
7. S. Tokura, J. Yoshida, N. Nishi and T. Hiraoki, Studies on chitin VI. Preparation and properties of alkyl-chitin fibers, Polym. J.,14:527(1982)
8. S. Tokura, N. Nishi, A. Tsustumi and O. Somorin, Studies on Chitin VIII. Some properties of water soluble chitin derivatives, Polym.J. 15:485(1983)
9. S. Nishimura, Y. Ikeuchi and S. Tokura, The adsorption of bovine blood proteins onto the surface of O-(carboxymethyl)chitin. Carbohyd. Res. 134:305(1984).
10. S. Tokura, S. Nishimura and N. Nishi, Studies on chitin IX. Specific binding of calcium ions by carboxymethyl-chitin. Polym. J. 15:597(1983).
11. Y. Kikuchi and A. Noda, Polyelectrolyte complexes of heparin with chitosan. J. Appl. Polymer Sci., 20:2561(1976).
12. R.H. Hackman, Chitin I. Enzymic degradation of chitin and chitin esters. Austr. J. Biol. Sci., 7:168(1954).
13. N. Nishi, S. Tokura and J. Noguchi, Synthesis of benzoyl-L-arginine-p-nitroanilide. Bull. Chem. Soc. Jpn., 43:2900(1970).
14. D. Horton and K. Just, Preparation from chitin of (1→4)-2-amino-2-deoxy-D-glucopyranuronan and its 2-sulfoamino analog having blood-anticoagulant properties. Carbohyd. Res., 29:173(1973).
15. E. Mihalyi, Physicochemical studies of bovine fibrinogen IV. Ultraviolet absorption and its relation to the structure of the molecule. Biochemistry, 7:208(1968).
16. R.L. Lundbrand, H.S. Kingdon and K.G. Mann, Thrombin, in:"Methods in Enzymology vol. XLV", L. Lorand, ed., Academic Press, New York(1976).
17. S.E. Lasker and S.S. Stivala, Physicochemical studies of fractionated bovine heparin. Arch. Biochem. Biophys., 115:360(1966).
18. Y. Inoue, M. Kaneko and S. Tokura, The viscosity relationship of water soluble carboxymethyl-chitin. Rept. Prog. polym. Phys. Jpn. XXV:759(1982)

SULFATED DERIVATIVES OF CHITOSAN AND THEIR CHARACTERIZATION WITH RESPECT TO BIOLOGICAL ACTIVITY

S. HIRANO, J. KINUGAWA and A. NISHIOKA

Department of Agricultural Biochemistry
Tottori University
Tottori 680, Japan

INTRODUCTION

Chitosan is easily prepared by N-deacetylation of chitin, an unbranched (1-4)-linked 2-acetamido-2-deoxy-β-D-glucan. Sulfated derivatives of chitosan possess blood anticoagulant activity (1). Conversion of position 6 into a carboxyl group in N-sulfated chitosan gives a product with 23 % of the activity of heparin (2), and its O-sulfation exhibits 45 % activity (3). O-sulfated N-carboxymethyl (CM) chitosan exhibits 45 % activity (4). Chitosan is almost non-toxic LD_{50} 16g/Kg (5), and N,O-sulfated chitosan has a low toxicity in mice LD_{50} 1.25-3.25 g/Kg (6), cf. LD_{50} 1.50-2.00 g/Kg for heparin (7). When position 6 of N-sulfated chitosan is converted to carboxyl, the product is much more toxic LD_{50} 0.237 g/Kg (3). On the other hand, little is known about the activation activity of the sulfated derivatives of chitosan for lipoprotein lipase (LPL).

We now report on the preparation of some sulfated derivatives of chitosan, the induced Cotton effects of their Methylene blue (MB) complexes, and ^{13}C-n.m.r. analysis, with respect to a dual biological function as an inhibitor of blood coagulant factors and as an activator of LPL.

EXPERIMENTAL

Materials

Purified chitosan, $|\alpha|_D^{20}$ -7° (c 1 %, CH_3COOH, 2 % aqueous); d.s. for N-Ac 0.01-0.10/GlcN, was prepared from Flonac-N (commercial chitosan of crab shell, Kyowa Yushi Co., Ltd.). O-CM chitosan was prepared from O-CM N-acetylchitosan (8). N-Acylation of chitosan was performed with carboxylic anhydride in aqueous 2 % acetic acid-methanol (9). Sulfation was carried out with N,N-dimethylformamide-sulfur trioxide (10) and the resulting product was isolated as a sodium salt (Tables I and II).

Methods

^{13}C-N.m.r. spectra were recorded with a JEOL JEM-FX 200 FT NMR spectrometer, ^{1}H-n.m.r. spectra with a Hitachi R-24 spectrometer, i.r. spectra with a Hitachi 215 grating spectrometer, specific rotations with a JASCO DIP-180 polarimeter, and optical absorptions with a Hitachi 100-50 spectrometer. C.d. spectra were recorded with a JASCO ORD/UV-5

Table 1. Data for the Yields and Some Properties of the Sulfated Derivatives of Chitosan

Chitosan(Compound)	Yield (%)	Mol. wt.[a] (x 10^3)	D.s. for sulfate	$[\alpha]_D$(degree) (c, solvent, temp.)
N,O-Sulfated(1)	79	12	2.6	-11(0.5, water, 14)
(2)	n.d.[b]	22	1.7	-11(0.5, water, 17)
O-Sulfated N-acetyl(3)	86	26	2.0	-31(1.2, water, 18)
(4)	75	150	2.0	-16(0.8, water, 17)
O-Sulfated N-hexanoyl(5)	83	27	1.8	-32(0.8, water, 21)
O-Sulfated(6)	83	22	0.7	-9(0.8, water, 17)
Sulfated O-CM(7)	54	245	1.8(0.5)[c]	-6.7(0.8, aqueous 5% NaOH, 17)
(8)	75	540	0.6(0.9)[c]	-13(0.8, aqueous 5% NaOH, 14)

[a]Estimated by gel filtration(see Experimental).
[b]Not determined.
[c]D.s. for CM.

Table 2. Data for the Elemental Analysis of the Sulfated Derivatives of Chitosan

Compound	Formula	Calc.(%) C	H	N	S	Found(%) C	H	N	S
1	$[C_6H_8NO_4(SO_3Na)_{2.63}(H)_{0.37}\cdot 1.60H_2O]_n$	15.72	2.54	3.06	18.4	15.44	2.69	3.10	18.7
2	$[C_6H_9NO_4(C_2H_3O)_{0.08}(SO_3Na)_{1.65}(H)_{1.27}\cdot 2.89H_2O]_n$[a]	19.22	4.00	3.64	13.7	19.16	3.91	3.70	13.7
3	$[C_8H_{11}NO_5(SO_3Na)_{1.55}(SO_3H)_{0.45}\cdot 2.20H_2O]_n$	22.00	3.66	3.21	14.7	22.02	3.98	3.18	14.7
4	$[C_8H_{11}NO_5(SO_3Na)_{2.00}\cdot 2.75H_2O]_n$	21.03	3.64	3.07	14.0	21.04	3.71	2.93	14.0
5	$[C_{12}H_{19}NO_5(SO_3Na)_{1.80}(H)_{0.20}\cdot 2.56H_2O]_n$	29.47	5.01	2.86	11.8	29.48	5.07	2.70	11.8
6	$[C_6H_{10}NO_4(C_2H_3O)_{0.01}(SO_3Na)_{0.74}(H)_{0.25}\cdot 0.89H_2O]_n$[a]	28.57	4.80	5.53	9.4	28.44	4.79	5.85	9.4
7	$[C_6H_9NO_4(C_2H_3O)_{0.01}(C_2H_2O_2Na)_{0.54}(SO_3Na)_{1.00}(H)_{0.36}\cdot 2.80H_2O]_n$[a,b]	24.22	4.56	3.88	8.9	24.22	4.43	3.88	8.8
8	$[C_6H_9NO_4(C_2H_2O_2Na)_{0.88}(SO_3Na)_{0.58}(H)_{0.54}\cdot 1.49H_2O]_n$[b]	30.71	4.74	4.61	6.1	30.90	5.04	4.65	6.4

[a]C_2H_3O :acetyl group.

[b]$C_2H_2O_2Na$: sodium carboxymethyl group.

recorder, equipped with a CD apparatus. The absorption and induced Cotton effects were analyzed(500-650 nm), and molar ellipticities[θ] are expressed on the basis of MB concentration. Sulfur content was calculated from the sulfate analysis by the barium chloroanilate method (11).

Molecular Weight. The molecular weight of the sulfated derivatives was estimated by gel filtration on columns of Bio-Gel P-300(fine; 1.6 x 46 cm), Toyopearl HW-65(fine; 1.6 x 74 cm) and HW-50(fine; 1.6 x 65 cm), which were eluted with 0.3M NaCl. A calibration curve was obtained for each column by analysis of relationship between Ve/Vo and the log of the molecular weights of standard pullulans and dextrans.

Blood Anticoagulant Activity. The anticoagulant activity of the sulfated derivatives was determined for human plasma, with respect to activated partial thromboplastin time APTT (12) thrombin time, TT (13), antithrombin activity, AA (14), and anti(factor Xa) activity (15). These activities were expressed as units/mg in reference to that of porcine intestinal heparin (174 units/mg) as standard.

LPL Activity. The LPL activity of the sulfated derivatives was determined by titration with 0.05M NaOH of fatty acids, which were released from plasma lipoproteins after injection of the sulfated derivatives into rabbit's vein at a dosage 0.1 or 1.0 mg/Kg body weight (16, 17). The activity was expressed as Δmol/L, and compared with that of heparin.

RESULTS

Data for yield , molecular weight, specific rotation and elemental analysis of the sulfated derivatives of chitosan are shown in Tables 1 and 2. The proposed structure of the repeating units of these sulfated derivatives is shown in Figure 1.

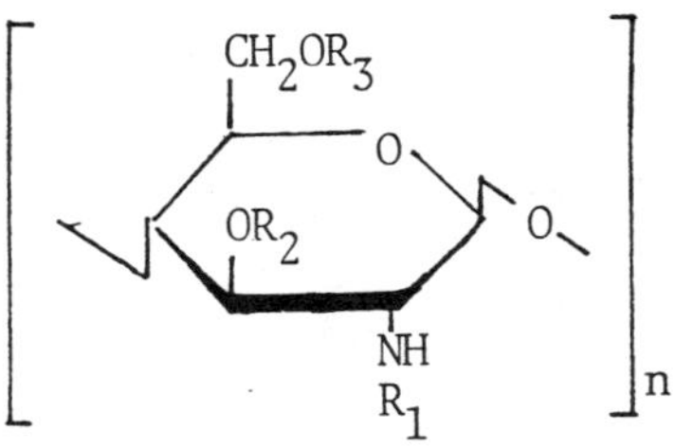

Fig. 1. Structure of the repeating units of the sulfated derivatives of chitosan

Compound	R_1	R_2	R_3
1 and 2	SO_3Na	H, SO_3Na	SO_3Na
3 and 4	$COCH_3$	SO_3Na	SO_3Na
5	$CO(CH_2)_4CH_3$	H, SO_3Na	SO_3Na
6	H	H	H, SO_3Na
7 and 8	SO_3Na	H	H, CH_2CO_2Na

Table 3. Metachromatic Bands and Induced Cotton Effects of the Methylene Blue Complexes of the Sulfated Derivatives of Chitosan

Compound	Procedure[a]	P/D[b]	Metachromatic band(λ_{max}^{water}, nm)	Induced Cotton Effects[c]	
				λ (nm)	$[\theta] \times 10^3$
2	A	1.5	554	555	+17.8
	B	1.2	557	557	-7.5
	C	1.2	557	554	+2.6
4	A	1.2	556	556	-27.4
	B	1.2	559	553	-12.5
	C	1.2	558	554	+110

[a]A, Both the sample and MB solutions were mixed at room temperature; B, the sample solution was heated at ∿80° for two min., then the MB solution was added, and the mixture solution was allowed to cool at room temperature; C, the mixture solution A was heated at ∿80° for two min., and the mixture solution was allowed to cool at room temperature.
[b]Molar ratio of the sulfated derivative(P) to Methylene Blue(D).
[c]Molar ellipticity based on Methylene Blue.

Table 4. Data for ^{13}C Chemical Shifts of N-Acetylchitosan, Chitosan, N,O-Sulfated and O-Sulfated N-acetyl chitosans

Carbon	N-Acetylchitosan[a]	Chitosan[a]	1	3
C=O	174.3	---	---	175.3
C-1	104.3	105.5,107.49	97.3	100.3
C-4	83.7	81.6, 86.0	76.4	78.8
C-5	75.5	76.0	74.0	75.9
C-3	75.5	76.0	73.5	73.6
C-6	61.4	61.0	67.5	67.6
C-2	55.7	57.6	55.9	55.5

[a]CP/MAS n.m.r. spectra

Each sulfated derivative was soluble in water, and had i.r. absorptions at 1240-1250(S=O) and 800-810 cm^{-1}(eq. C-O-S).

Metachromatic Bands and Induced Cotton Effects of the Methylene Blue Complexes

Each sulfated derivative exhibited its unique metachromacy with MB at ∿ 557 nm, and the maximum absorbance appeared at polyanionic site(p)/cationic dye(D) 1.2-1.5. The sign of the induced Cotton effects in the 500-650 nm range was dependent on the procedure for the preparation of the MB complexes: A, both the sample and MB solutions were mixed at room temperature; B, the sample solution was heated at ∿80° for two minutes, the MB solution was added, and the mixture solution was allowed to cool at room temperature; C, the mixture solution A was heated at ∿80° for two minutes, and the solution was allowed to cool at room temperature. As shown in Table 3, the induced Cotton effects were negative by procedure B, positive by procedure C, and both negative and positive by procedure A at the same metachromatic bands with slight experimental deviation (< 5 nm). The induced Cotton effects produced by procedures B and C were unchangeable upon heating and cooling. Fig. 2 shows the induced Cotton effects of the MB complexes of 4 by the three procedures, indica-

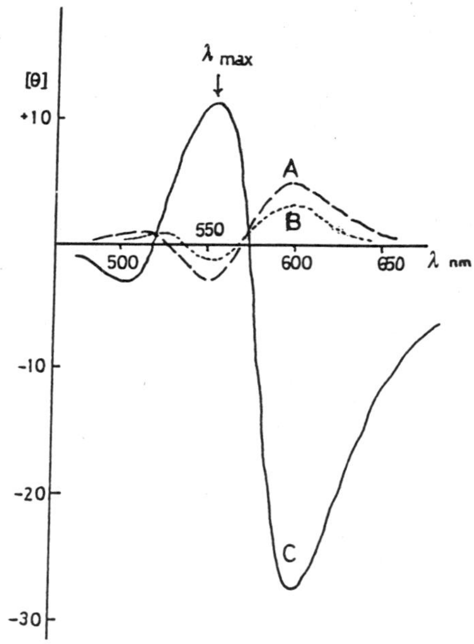

Fig. 2. Induced Cotton effects of the MB complexes of O-sulfated N-acetylchitosan(4) produced by the three procedures: A, both the sample and MB solutions were mixed at room temperature; B, the sample solution was heated at ∿80° for two min., the MB solution was added, and the mixture solution was allowed to cool at room temperature; C, the mixture solution A was heated at ~80° for two min., and the solution was allowed to cool at room temperature.

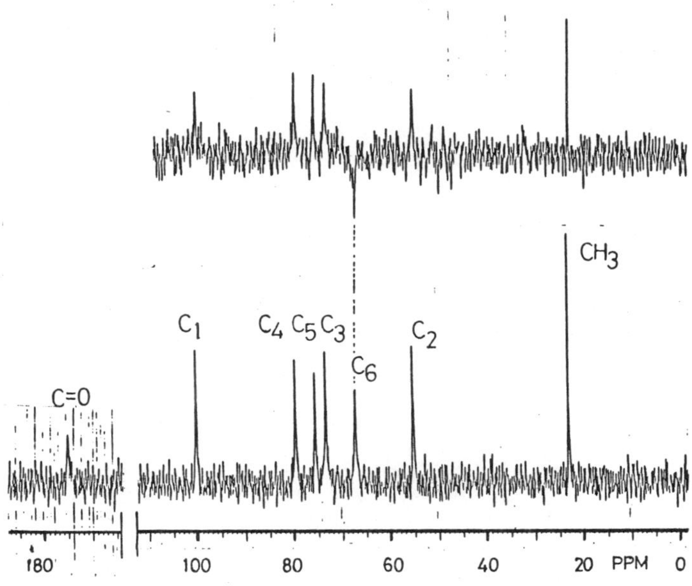

Fig. 3. ^{13}C-N.m.r. spectrum of O-sulfated N-acetylchitosan(3)

Table 5. Dual Biological Function of the Sulfated Derivatives of Chitosan

Compound	APTT[a] (units/mg)	LPL($\Delta\mu$mol/L)[b] 15 min	30 min.
Heparin(control)	174(1.0)	360(1.0)	950(1.0)
N,O-Sulfated(1)	190-287(1.1-1.6)	n.d.[c]	
(2)	n.d.	840(2.3)	3,200(3.4)
O-Sulfated N-acetyl(3)	331-379(1.9-2.2)	20(0.1)	100(0.1)
	331-379(1.9-2.2)	380(1.1)[e]	1350(1.4)[e]
4	n.d.	0	0
O-Sulfated N-hexanoyl(5)	n.d.	0	0
O-Sulfated(6)	n.d.	0	0
Sulfated O-CM(7)	n.d.	200(0.6)	700(0.7)
8	21-31(0.2-0.3)	n.d.	

[a]Activated partial thromboplastin time. The bracket indicates its relative activity to that of heparin.
[b]Lipoprotein lipase activity for 15 and 30-min incubations at a dosage 0.1 mg/Kg body weight. The bracket indicates its relative activity to that of heparin.
[c]Not determined.
[e]At a dosage 1.0 mg/Kg body weight.

ting its conformational change with respect to the helical array of bound dye.

^{13}C-n.m.r. spectra

Figure 3 shows ^{13}C-n.m.r. spectrum of 3. The ^{13}C signals were assigned on the basis of ^{13}C shifts of N-acetylchito-oligosaccharides (18), N-acetylchitosan (chitin), and chitosan (19, 20). The ^{13}C chemical shifts in solid state and in aqueous solution were in good agreement with slight experimental deviation (< 2 ppm), except for the C-1 and C-4 signals. CH_3 and C=O signals of N-Ac were characteristic of 3, but absent in 1. Table 4 summarizes the ^{13}C chemical shifts of N-acetylchitosan, chitosan, 1 and 3. It is obvious that sulfation brought about a down-field shift of C-6 signal (+ 6.5 for 1 , and + 6.2 for 3) because of the O-sulfation of position 6 (21) and up-field shifts of C-4 signal (-9.6 for 1, and -4.9 for 3), probably because of conformational change.

A Dual Biological Function as an Inhibitor of Blood Coagulant Factors and as an Activator of LPL

The sulfated derivatives of chitosan had a dual biological function. As shown in Table 5, the anticoagulant activity(units/mg) for human plasma was in the sequence 3(mol. wt. 26,000)> 1(12,000)> heparin(21,000)> 8(540,000) with respect to APTT at 5-400 mg/ml, TT at 5-60 g/ml, and AA at 10-1000 mg/ml. Compounds 1, 3, and 8 did not inhibit factor Xa activity at 10-1000 mg/ml. The activation activity of LPL was in the sequence 2(mol. wt. 22,000)> heparin(21,000) >7(245,000)> 3(26,000), but the LPL activity was not observed with 4-6 at a dosage 0.1 mg/Kg body weight. The maximum LPL activity($\Delta\mu$mol/L) was observed with heparin(950) and with 7 (700) for 30-min. incubation, and with 2(6,100) for 60-min. incubation at a dosage 0.1 mg/Kg body weight. It was observed with 3(2,600) for 60-min. incubation at a dosage 1.0 mg/Kg body weight.

DISCUSSION

Antithrombin III (22) and heparin cofactor II (23, 24) are known as heparin-dependent inhibitors for blood coagulant factors in human plasma by forming a 1:1 complex with heparin. Heparin cofactor II inhibits thrombin activity but not factor Xa activity, while antithrombin III inhibits both of these factors as well as other coagulant factors. Heparin is also known as an *in vivo* activator of LPL (16) to humans. As demonstrated in the present study, the sulfated derivatives of chitosan had a dual biological function as an inhibitor of coagulant factors and as an activator of LPL. A 6-sulfate group in the hexosaminyl moiety of 1-5 was confirmed by a down-field shift of C-6 in ^{13}C-n.m.r. spectra. The 6-sulfate group may be a main active or binding site because of no activity when it is absent (1); the activity is maintained more or less when position 6 is CM. A 3-sulfate group in the hexosaminyl moiety promotes the activities when position 6 has an acidic group, but it is not essential. The important contribution of the N-sulfate to the anticoagulant activity has been pointed out (6, 25), but its presence is not a prerequisite for the activities, as demonstrated in the present study with respect to APTT and LPL. In fact, dextran sulfate, which has no hexosaminyl moiety, has anticoagulant activity (26) and LPL activity (27), and the N-acetyl group is a minor component of heparin (28). Molecular weight is also an important factor for these biological activities, as reflected by the sequence 3 (26,000) > 1 (12,000) > heparin > 7 (540,000) for APTT, and 2 (22,000) > heparin (21,000). However, it is interesting to note that 7 (240,000) had an activation activity of LPL, and 8 (540,000) had an anticoagulant activity, in spite of their high molecular weights.

No apparent correlation between the induced Cotton effect and the biological activities was found. This indicates that the biological activities are not due to the conformation of whole sulfates polymer, but are due to the stereospecific array of sulfate or acidic groups in a small region of the polymer having some limited molecular weights.

Acknowledgements. This work was supported by grants from the NATO Science Foundation and the Grant-in-Aid for Scientific Research, Ministry of Education, Japan.

REFERENCES

1. D. T. Warner and L. L. Coleman, J. Org. Chem., 23:1133 (1958).
2. D. Horton and E. K. Just, Carbohydr. Res., 28:173 (1973)
3. R. J. Whistler and M. Kosik, Arch. Biochem. Biophys., 142:106 (1971).
4. R. A. A. Muzzarelli, F. Tanfani, M. Emanuelli, D. P. Pace, E. Chiurazzi and M. Piani, Carbohydr. Res., 126:225 (1984).
6. J. Doczi, A. Fishman and J. A. King, J. Am. Chem. Soc., 75:1512 (1953).
5. K. Arai, T. Kinumaki and T. Fujita, Bull. Tokai Reg. Fish. Res. Lab., 56:89 (1968).
7. J. Seifter and A. J. Begany, Am. J. Med. Sci., 216:234 (1948).
8. R. Trujillo, Carbohydr. Res., 7:483 (1968).
9. S. Hirano, Y. Ohe and H. Ono, Carbohydr. Res., 47:315 (1976).
10. R. G. Schweiger, Carbohydr. Res., 21:219 (1972).
11. J. E. Hodge and B. T. Hofreiter, Methods Carbohydr. Chem., 1:388.
12. R. R. Proctor and S. I. Rapaport, Am. J. Clin. Pathol., 36:212 (1961).
13. R. W. Bensnes and W. J. Sweeney, Am. J. Obstet. Gynecol., 70:334 (1955).
14. U. Abildgaad, M. Lie and O. R. Oedegard, Thromb. Res., 11:549 (1977).
15. A.N. Teien and L. Mettle, Thromb. Res., 10:399 (1977).

16. E. D. Korn, Methods Biochem. Anal., 7:145 (1959).
17. V. P. Dole, J. Clin. Inves., 35:150 (1956).
18. H. Saito, T. Mamizuka, R. Tabeta and S. Hirano, Chem. Lett., 1483 (1981).
19. H. Saito, R. Tabeta and S. Hirano, Chem. Lett., 1474 (1981).
20. M. G. Peter, L. Grün and H. Förster, Angew. Chem. Int. Ed. Engl., 23: 638 (1984).
21. B. Casu, P. Oreste, G. Torri, G. Zoppetti, J. Choay, J. Lormeau, M. Petitou and P. Sinaÿ, Biochem. J., 197:599 (1981).
22. R. D. Rosenberg and P. S. Damus, J. Biol. Chem., 248:6490 (1973).
23. G. F. Briginshaw and J. N. Shanberg, Arch. Biochem. Biophys., 161:683 (1974).
24. D. M. Tollefesen, D. W. Majerus and M. K. Blank, J. Biol. Chem., 257: 2162 (1982).
25. M. L. Wolfrom, T. M. Shen and C. G. Summners, J. Am. Chem. Soc., 75: 1519 (1953).
26. C. R. Ricketts and K. W. Walton, Chem. Ind. (London), 869 (1952).
27. D. S. Robinson, P. M. Harris and C. R. Ricketts, Biochem. J., 69:110 (1958).
28. U. Lindahl and M. Höök, Ann. Rev. Biochem., 47:385 (1978).

SULFATED N-CARBOXYMETHYL CHITOSANS AS BLOOD ANTICOAGULANTS

R. A. A. MUZZARELLI, F. TANFANI, M. EMANUELLI, E. CHIURAZZI [◊] and M. PIANI [◊]

Faculty of Medicine, University of Ancona
I-60100 Ancona, Italy
and [◊]Department of Hematology, Regional Hospital
I-60020 Ancona Torrette, Italy

INTRODUCTION

Heparin has several biological activities, mainly associated with interactions with plasma proteins and/or components of the endothelium. Due to incomplete biosynthesis and depending on the preparation, heparin is heterogeneous as regards sequences and molecular weight. Different heparins contain different proportions of regular regions made up of sequences of the trisulfated disaccharide α-L-iduronic acid 2-sulfate → 2-deoxy-2-sulfamino α-D-glucose 6-sulfate, as well as undersulfated regions, including a specific pentasaccharide segment representing the binding site for antithrombin. Different sequences play various roles in modulating the biological activities of heparin. Several further features of the heparin structure may be summarized as follows. The bulk of the polysaccharide chain consists of alternating units of uronic acid and glucosamine, bound in 1-4 linkages. Most of the uronic acid residues are L-iduronic, the remainder D-glucuronic acid. Most of the amino groups are sulfated, but a minor part is acetylated and yet another small fraction is unsubstituted. The glucuronic acid units are β-linked, whilst the iduronic acid units are α-linked. Sulfate esters occur on C-6 and to some extent on C-3. To produce heparin-like substances, efforts have been made toward the introduction of sulfate and carboxyl groups to the chitosan macromolecule. Early reports on this topic deal with the sulfation of chitosan (1), the selective formation of sulfamido groups (2), the sulfation and depolymerization of chitosan (3), and the preparation of formyl chitosan, a remarkably potent inhibitor of aldosterone production (4). Among a variety of polysalts of chitosan, those with dextran sulfate have been reported to inhibit clot formation in vitro (5). The antithrombogenicity of polysalts of chitosan and carboxymethyl dextrans was also studied (6). In addition to these polysalt complexes, chitosan sulfate (27,000 dalton) was studied, and evidence of its anticoagulant activity was given (7). Other polymers studied were sulfated aminodeoxycellulose (8), alginic acid (9), cellulose sulfate (10), synthetic polymers (11, 12), and sulfated pectins and dextrans (13). Chitosan is particularly well suited as a raw material from which heparin-like substances can be obtained, because it has 1-4 anhydroglycosidic linkages and linear chains and already carries amino and acetamido groups, whose introduction in a polysaccharide chains has been found to be most difficult.

The present study is therefore based on the following considerations:

(a) the heparin-like substances should possess simultaneously O-sulfated, acetamido and carboxyl groups, which are recognized to be effective in anticoagulant action; (b) carboxyl groups should preferably be in the form of carboxymethyl groups or, even better, in the form of α-amino acid groups. It has been recently reported (14) that dextrans bearing carboxymethyl, sulfate, benzyl sulfate and α-aminoacid groups possess high antithrombogenic activity; (c) the introduction of the said groups should be as mild as possible, to avoid side reactions and alterations whose consequences are difficult to foresee and which prevent the correct description of the product; (d) the ionic groups should be in optimum ratios and (e) should be distributed regularly along the chain; (f) the chain length should be tailored to impart optimum blood anticoagulant properties.

For this purpose, we have used Euphausia superba chitosan (15) to prepare N-carboxymethyl chitosan, which we then submitted to sulfation (16). In this study, N-carboxymethyl chitosan sulfates have been prepared to define the chemical characteristics which impart blood anticoagulant activity.

EXPERIMENTAL

Three different sulfated N-carboxymethyl chitosans were prepared in order to evaluate their anticoagulant capacities for human blood. The N-carboxymethyl chitosan 6-sulfate (NCMC monosulfate) and the N-carboxymethyl chitosan 3,6,N-trisulfate (NCMC trisulfate) were obtained from N-carboxymethyl chitosan having a N-carboxymethylation degree of 50 % and the remaining amino groups in free form. The third derivative, namely N-carboxymethyl chitosan 3,6-disulfate (NCMC disulfate) was obtained from NCMC having degree of N-carboxymethylation of 58 % and degree of N-acetylation of 42 %. The reactions were carried out on freeze dried NCMC, using dried glassware under exclusion of moisture.

Preparation of N-carboxymethyl chitosan 6-sulfate (NCMC monosulfate)

The NCMC monosulfate was obtained with a mixture of sulfuric acid and chlorosulfonic acid in the ratio of 1:1 as described by Naggi et al. (this Volume). The sulfur content at the elemental analysis was found to be 12.2 % by weight. The product was white.

Preparation of N-carboxymethyl chitosan 3,6-disulfate (NCMC disulfate)

N-Carboxymethyl chitosan (degree of acetylation, 42 %; degree of N-carboxymethylation, 58 %) was stirred in DMF overnight. The DMF/SO_3 complex (ca. 20 g) was added at 0° C with stirring and was left at room temperature overnight. Then it was neutralized by adding a saturated solution of sodium bicarbonate, at 0° C and finally the pH was raised to 10 - 11 with NaOH. It was dialyzed and freeze dried. The product had a yellow color. The sulfur content was 12.4 by weight.

Preparation of N-carboxymethyl chitosan 3,6,N-trisulfate (NCMC trisulfate)

This derivative was obtained by treating partially hydrolyzed N-carboxymethyl chitosan (0.5 g) with a mixture of pyridine (30 ml) and chlorosulfonic acid (3 ml) added dropwise at -10° C. The temperature raised to 60° C and stirring was continued overnight. A dark brown colored mixture was obtained. This was cooled to -10° C and the polymer was separated as described above. The sulfur content was 11.8. The product was yellow.

For the introduction of the sulfur groups in position 3 of the glycosidic ring it was first necessary to hydrolyze the NCMC to avoid obtaining products with a low percentage of sulfur. The hydrolysis was carried out in the follo-

Table I. Chemical characteristics of the sulfated N-carboxymethyl chitosans.

NCMC monosulfate:	Groups: 6-O sulfate, N-carboxymethyl, amino; sulfur, 10.2 %; degree of substitution 1.00.
NCMC disulfate:	Groups: 6-O sulfate, 3-O sulfate, N-carboxymethyl, acetamido; sulfur, 12.4 %; degree of substitution, 1.8.
NCMC trisulfate:	Groups: 6-O sulfate, 3-O sulfate, N-sulfate N-carboxymethyl; sulfur, 11.8 %; degree of substitution 1.8. When partially depolymerized, sulfur, 15.8 %; degree of substitution 2.47.

wing manner: NCMC (1 g) was boiled for no less than 10 min with 6 M HCl; then, after cooling, acetone was added and the mixture was filtered.

Hematological tests

Anticoagulant activity of the sulfated NCMC's. To test tubes containing venous blood (3 ml) provided by healthy donors, one of the following was added: NCMC monosulfate (polydisperse), NCMC disulfate (M.W. fractions of 45,000, 75,000 and 100,000) or NCMC trisulfate (polydisperse) to obtain a concentration of 0.17 mg/ml. The test tubes were kept at room temperature and observations were made after 2 and 24 h.

Action of the sulfated NCMC's on the coagulation. These tests were made: thromboelastogram (TEG), platelet aggregation and Factor Xa inhibition test (Heparin Coatest by Ortho Diagnostic System). The measurements of PT, PTT and the Factor Xa inhibition test sought information on the steps of the blood coagulation process affected by the action of the sulfated polysaccharides. The TEG gave the instrumental confirmation of the anticoagulant activity, previously observed visually. The platelet aggregation test provided information on the activity of the platelets in the presence of the compounds studied.

Conservation state of red blood cells. The following tests were carried out: sedimentation rate of the red blood cells (VES), rouleaux formation, hemolysis, and a test for the conservation state of the erythrocytes. The determinations of the conservation state of the red blood cells and the extent of rouleaux formation were made during the optic microscopic observation of blood smears treated with the various sulfated NCMC's. To determine the degree of hemolysis, the blood was repeatedly stirred, left to settle and examined for reddening of the plasma. Before proceeding with this test an amount of blood necessary to fill a Westergreen pipette for the determination of the VES was removed from each test tube. The aim of these tests was to see if the sulfated NCMC would adversely affect the red blood cells. The determination of VES in blood samples with the sulfated NCMC permits evaluation of the influence of polysaccharides on the test itself compared with heparin and citrate controls.

RESULTS AND DISCUSSION

Chemical characterization of sulfated N-carboxymethyl chitosans

Evidence of the introduction of sulfate groups in NCMC's was obtained by i.r. spectrometry, ^{13}C-n.m.r. spectrometry and spectropolarimetry. The i.r. spectra of NCMC sulfates showed evident absorption bands at 1230 and 800 cm^{-1} due to the presence of the sulfate group in all of the products obtained.

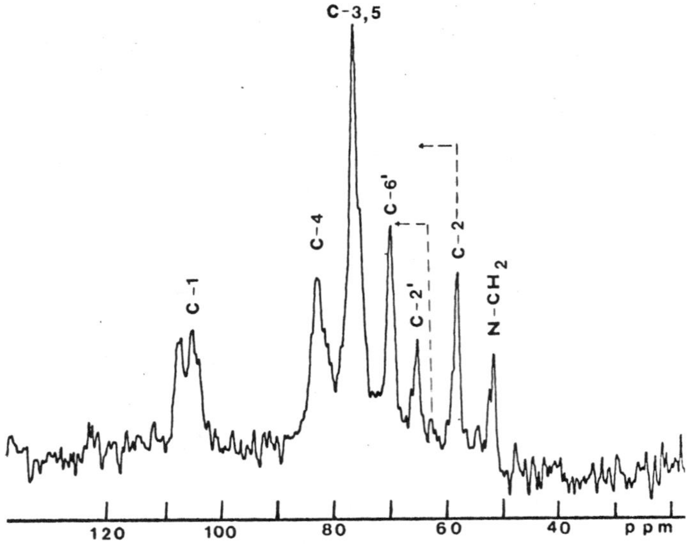

Figure 1. ^{13}C-n.m.r. spectrum of NCMC-6-sulfate (10.8 % S) obtained with $H_2SO_4/ClSO_3H$ sulfating mixture. The chemical shifts of the C-2 and C-6 are respectively due to the presence of carboxymethyl and sulfate groups.

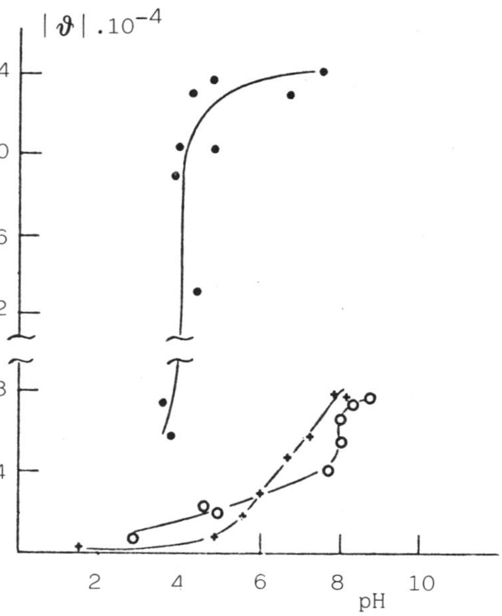

Figure 2. pH-Dependence of the molecular ellipticity, $|\theta|$, degree. $cm^2.decimole^{-1}$, for (•) NCMC-3,6,N-trisulfate (S 15.8 %) obtained from 20 min hydrolysed NCMC, (o) NCMC-3,6,N-trisulfate (S 11.8 %) obtained from 10 minutes hydrolysed NCMC and (+) NCMC 6-sulfate (S 10.2 %).

Sulfate groups also affected the ^{13}C-n.m.r. spectra by introducing evident chemical shifts. Figure 1 shows the n.m.r. spectrum of NCMC monosulfate (S, 10.8 %) where the C-6 signal has been shifted as a consequence of the sulfation (see also Fig. 2 in the chapter by Naggi et al., this Volume). This spectrum also contains the signals due to the presence of the N-carboxymethyl group (see also Fig. 1 in the chapter by Kurita, this Volume).

All of the circular dichroism spectra for the sulfated derivatives were pH-dependent. Fig. 2 shows the $|\theta|$ values of monosulfate, trisulfate (hydrolysed for 10 min), and trisulfate (hydrolysed for 20 min), calculated respectively at 192, 210 and 230 nm. The latter shows an impressive increase of the $|\theta|$ value in a narrow pH interval centered at pH 4, probably due to easier conformational changes of the polymer.

The circular dichroism spectra of NCMC monosulfate and NCMC disulfate show one negative Cotton band at 192 and 203 nm, respectively. The spectrum of trisulfate shows negative and positive Cotton bands which depend on the extent of hydrolysis to which the starting NCMC was submitted (Figures 3 and 4). In fact, higher degrees of sulfation were obtained with low molecular weight NCMC's. This trend, which is common to other substitution reactions of chitosan, (see Kurita, this Volume) was confirmed by gel chromatography.

The NCMC disulfate was subjected to chromatography on Bio-gel P-100 in 0.1 M acetate buffer at pH 4.5 (column 82 x 2 cm, flow-rate 0.25 ml/min). NCMC disulfate was separated into 45,000, 55,000, 65,000, 75,000 and 100,000 dalton fractions for which the respective sulfur percentages were: 12.4, 12.3, 11.6, 11.8 and 9.5 %. It is possible to notice that the sulfur content was lower for the higher M.W. fractions. Therefore, the choice of the molecular dimensions of the chitosan or chitosan derivative to be submitted to sulfation influences the degree of sulfation attainable.

While the NCMC disulfate was fractionated before biochemical characterization, NCMC monosulfate and NCMC trisulfate were used as polydisperse products with M.W. between 40,000 and 100,000.

Biochemical characterization of sulfated N-carboxymethyl chitosans

The analysis of the data obtained shows that the anticoagulant activity of the sulfated NCMC's at the preferred concentration of 0.17 mg/ml is excellent. There are small quantitative differences in the anticoagulant activity of the three derivatives studied. The thromboelastograms show how, after the recalcification of the samples, an indefinite extension of R occurred in all cases. This indicates that no signs of coagulation appear in a reasonable time period.

The Factor Xa inhibition test demonstrated that these compounds are capable of inhibiting the activated Factor Xa even if the inhibition extent varies from sample to sample. When it was possible to perform the platelet aggregation test, it demonstrated the existence of a slight alteration of the platelet function. The data show that all of the examined compounds, with the exception of NCMC trisulfate, increased the velocity of erythrocyte sedimentation (VES) with respect to the two controls and this increase tended to be higher as the M.W. of the fractions increased. Some correlations may exist between VES and rouleaux formation because both increased with increasing M.W.

NCMC monosulfate and NCMC trisulfate, at the concentration of 0.17 mg/ml did not show any hemolytic action after 24 h and the conservation of the red blood cells after this period remained good. In the case of NCMC disulfate the 45,000 and 55,000 M.W. fractions gave the same good results.

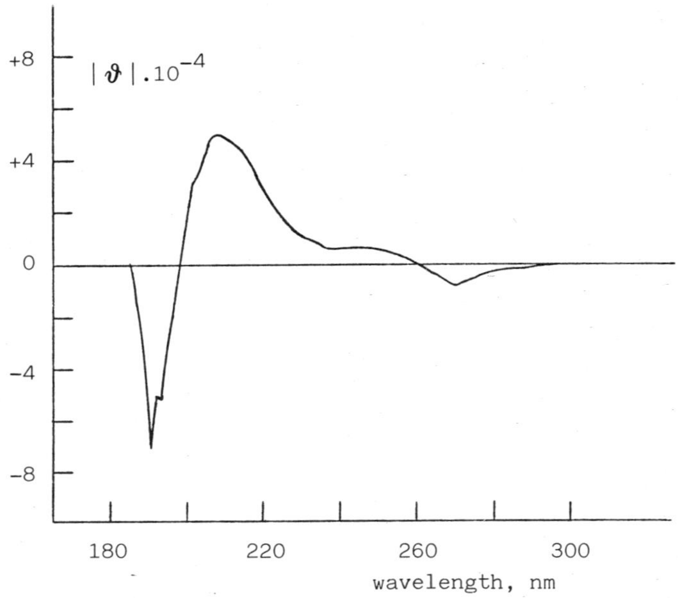

Figure 3. Circular dichroism spectrum of NCMC-3,6,N-trisulfate (S 11,8 %) at pH 6.7 obtained with pyridine/$ClSO_3H$, sulfating mixture. The NCMC (50 % NH_2; 50 % NH-CH_2COO^-) was previously hydrolysed for 10 min.

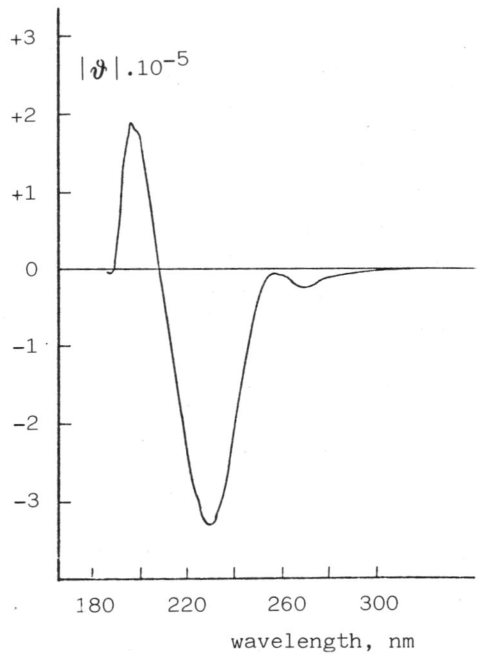

Figure 4. Circular dichroism spectrum of NCMC-3,6,N-trisulfate (S 15.8 %) at pH 6.7 obtained with pyridine/$ClSO_3H$ sulfating mixture. The NCMC (50 % NH_2; 50 % NH-CH_2COO^-) was previously hydrolysed for 20 minutes.

CONCLUSIONS

The NCMC sulfates, obtained according to three different sulfation reactions, are blood anticoagulants, the best of them being N-carboxymethyl chitosan 3,6-disulfate of low molecular weight. The said compound exhibited the same blood anticoagulant activity as heparin, and showed no adverse effects on the cellular structures, when added to blood to be used *in vitro*.

The N-carboxymethyl chitosan 3,6-disulfate carries both acetamido and glycine groups at C-2. A large proportion of acetamido groups favors the blood anticoagulant action of the sulfated chitosans (see Hirano et al., this Volume), and the introduction of the carboxyl groups *via* Schiff reaction is valid for the purpose of obtaining blood anticoagulants which could replace heparin in blood samples to be used *in vitro*.

The reduction of the molecular size, besides making easier the sulfation reaction, improves the blood anticoagulant activity of the products while depressing adverse effects on cellular structures. The knowledge of the molecular weight is a prerequisite for the preparation of products with reproducible degree of sulfation and consequent biochemical characteristic properties. The assessment of the applicability of these substances *in vivo* will be only possible when the chemistry necessary to produce them will be fully understood.

ACKNOWLEDGEMENT

This work was carried out with the financial contribution of the Italian National Research Council, Progetto Finalizzato Chimica Fine e Secondaria, Roma, Italy.

REFERENCES

1. M. L. Wolfrom and T. M. Shen Han, The sulfonation of chitosan, *J. Am. Chem. Soc.*, 81:1764 (1959).
2. D. T. Warner and L. L. Coleman, Selective sulfonation of amino groups in amino alcohols. *J. Organ. Chem.* 23:1133 (1958).
3. K. Nagasawa and N. Tanoura, Depolymerization and sulfation of chitosan by sulfuric acid. *Chem. Pharm. Bull.* 20:157 (1972).
4. E. Glaz and P. Vecsei, "Aldosterone", Pergamon Press, Oxford (1975).
5. H. Fukuda and Y. Kikuchi, In vitro clot formation on the polyelectrolyte complexes of sodium dextran sulfate with chitosan. *J. Biomed. Mat. Res.* 12:531 (1972).
6. Y. Kikuchi and A. Noda, Polyelectrolyte complexes of heparin and chitosan, *J. Appl. Polymer Sci.*, 20:2561 (1976).
7. K. Nagasawa, Anticoagulant sulfonated chitosan. *Japan Patent* 76.06720 (1976).
8. T. Teshirogi, H. Yamamoto, M. Sakamoto and H. Tonami, Preparation of sulfated aminodeoxycelluloses. *Sen-I Gakkaishi*, 36:78 (1980).
9. O. Larm, K. Larsson, E. Scholander, L.O. Andersson, E. Holmer and G. Söderström, The preparation of a heparin analog from alginic acid. *Carbohydr. Res.*, 73: 332 (1979).
10. G. Kindness, W. F. Long and F. B. Williamson, Evidence for antithrombin III involvement in the anticoagulant acitivity of cellulose sulfate. *Br. J. Pharmacol.* 68:645 (1980).
11. M. Okada, H. Sumimoto, M. Hasegawa and H. Komada, Sulfated synthetic polysaccharides having physiological acitivity. *Macromol. Chem.*, 180:813 (1979).
12. H. Komada, M. Okeda and H. Sumimoto, Synthetic polysaccharides containing amino groups. *Macromol. Chem.*, 181:2305 (1980) .
13. L. B. Jaques, Heparins: anionic polyelectrolyte drugs. *Pharm. Review*, 31: 99 (1979).

14. M. Jozefowicz, J. Jozefowicz, C. Fugnot and D. Labarre, New heparin-like insoluble materials, In: "Chemistry and Biology of Heparin", R. L. Lundblad, W. V. Brown, K. G. Mann and H. R. Roberts, eds., Elsevier-North Holland, Amsterdam (1981).

15. R. A. A. Muzzarelli, F. Tanfani, M. Emanuelli, M. G. Muzzarelli and G. Celia, The production of chitosan of superior quality. J. Appl. Biochem., 3:316 (1981).

16. R. A. A. Muzzarelli and F. Tanfani, N-Carboxymehtyl chitosans and N-carboxybenzyl chitosans: novel chelating polyampholytes. In: "Proc. 2nd Intl. Conference on Chitin / Chitosan", S. Hirano and S. Tokura, eds., Japanese Soc. Chitin, Tottori, Japan (1982).

17. R. A. A. Muzzarelli, F. Tanfani, M. Emanuelli and S. Mariotti, N-Carboxymethylidene chitosans and N-carboxymethyl chitosans: novel chelating polyampholytes obtained from chitosan. Carbohydr. Res., 107:199 (1982).

18. R. A. A. Muzzarelli, F. Tanfani, E. Emanuelli, D. P. Pace, E. Chiurazzi and M. Piani, Sulfated N-carboxymethyl chitosans: novel blood anticoagulants. Carbohydr. Res., 126:225 (1984).

19. R. A. A. Muzzarelli, Heparin-like susbtances and blood compatible polymers obtained from chitin and chitosan. In: "Polymers in Medicine", E. Chiellini and P. Giusti, eds., Plenum Press, London (1983).

IMMUNOLOGICAL ACTIVITY OF CHITIN DERIVATIVES

K. NISHIMURA, S. NISHIMURA◊, N. NISHI◊, S . TOKURA◊ and I. AZUMA

Institute of Immunological Sciences
and ◊Department of Polymer Science
Faculty of Science, Hokkaido University
Sapporo 060, Japan

INTRODUCTION

Previously, we have reported the immunological activities of chitin derivatives for the stimulation of non-specific host resistance in mice.[1] Among derivatives of chitin tested, deacetylated chtin derivatives such as 70% deacetylated chitin (DAC-70) and 30% deacetylated chitin (DAC-30) were shown to have potent immunological activities for activation of peritoneal macrophages *in vivo*, suppression of Meth-A tumor cells in syngeneic BALB/c mice and stimulation of non-specific host resistance against *Escherichia coli* infection in mice. Recently, Suzuki et al.[2,3] have reported that chitin and chitosan were effective for the protection of host against infection with *Candida albicans* and *Staphylococcus aureus* and against growth of Ehrlich and Sarcoma 180 ascites tumor.

The chemically modified chitin derivatives are characteristic of their susceptibility to lysozyme *in vivo* (biodegradability), and of their reconstructive easiness into fibers and films,[4,5] and beads (unpublished data). The susceptibility to lysozyme of chitin derivatives was controled by the degree of deacetylation of C_2-position and/or by the introduction of various substituents at the 6-0-position of the N-acetyglucosamine residue. The purpose of this study was to investigate the immunological properties of chitin derivatives as immunoadjuvants in comparison with well-known adjuvants such as bacterial cell wall skeletons and chemically-defined synthesized agents, and discuss the possibility of chitin derivatives as new biomedical materials.

EXPERIMENTAL

Animals

Inbred C57BL/6, DBA/2 and BALB/c female mice, seven to ten weeks of age, and outbred STD-ddY male mice, five weeks of age, and female Hartley strain guinea pigs weighing 350 to 400g were obtained from Shizuoka Agricultural Co-operation for Experimental Animals, Hamamatsu, Japan.

Chitin Derivatives and Other Adjuvants

The chemical structures and properties of chitin derivatives used in this study were reported previously.[1] Chitin derivatives were dissolved or suspended in physiological saline and sterilized by heating at 60°C for 30 min just before use. Maleic anhydride divinylether (MVE-2) of molecular weight 15,500 was kindly supplied from Dr. Michael Chirigos, NCI-FCRF, Frederick. The cell-wall skeletons of *Mycobacterium bovis* BCG (BCG-CWS) were prepared by methods described previously.[6] Synthesis and properties of N-acetylmuramyl-L-alanyl-D-isoglutamine (MDP), the minimum adjuvant-active subunit of BCG-CWS, were reported in previous paper.[7]

Determination of Circulating-Antibody Formation to Bacterial α-amylase (BαA)

Each Hartley strain guinea pig was immunized in its four footpads with a total of 200μg of BαA with or without 100μg of chitin derivatives or MDP in Freund incomplete adjuvant (FIA)[a mixture of liquid paraffin and Aracel A (85:15)]as water-in-oil emulsion. After 4 weeks, each guinea pig was bled. Anti-BαA antibody in serum was measured by the method described by Okada et al.[8]

Determination of Adjuvant Activity on the Induction of Delayed-Type Hypersensitivity to Azobenzenearsonate-N-Acetyl-L-Tyrosine (ABA-Tyr)

Guinea pigs were immunized in four footpads with a total of 50 μg of ABA-Tyr with or without 100μg of various adjuvants per animal in FIA. Two weeks later, skin tests were carried out with 50 and 100μg of ABA-bovine serum alubmin, and the reactions were measured at 24 h after intradermal injection of test antigen.

Assay for Allogeneic Cell-Mediated Cytotoxicity

C57BL/6 mice ($H\text{-}2^b$) were immunized intraperitoneally (i.p.) with 5 $\times 10^5$ viable mastocytoma P815-X2 cells, chemically induced in DBA/2 mice ($H\text{-}2^d$), with or without adjuvants. On day 14 after immunization, the cell-mediated cytotoxicity assay was carried out by using the spleen cells (1×10^6) of immunized mice and ^{51}Cr-labeled mastocytoma P815-X2 cells (1×10^4). The culture fluids were harvested after 4 h, and the radioactivity in the culture supernatant was determined by γ-counting to estimate the specific target cell lysis by the formula:[(release of effector cells - spontaneous release)/(maximum release - spontaneous release)]x 100. Maximum chromium release was determined by freezing and thawing labeled target cells three times.

Assay for Natural Killer Cell-Mediated Cytotoxicity

C57BL/6 mice were injected i.p. with 500μg of chitin derivatives or MVE-2. On day 3 after injection, the natural killer (NK) cell-mediated cytotoxicity assay was performed by using the spleen cells (1×10^6) of treated or untreated mice and ^{51}Cr-labeled NK sensitive YAC-1 cells, molony virus-induced T cell lymphoma of A/Sn mice origin, or NK resistant P815-X2 cells (1×10^4). The cultures were incubated for 4 h, and the specific target cell lysis was determined as described above.

Determination of Mitogenic Activity on Normal Mouse Spleen Cells

Spleen cells (5×10^5) of C57BL/6 or BALB/c mice were cultured with or without chitin derivatives or mitogens, such as Concanavalin A (Con A; Pharmacia Fine Chemicals, Uppsala, Sweden) and lipopolysaccharide B (LPS) from *Escherichia coli* 055:B5 (Difco Lab., Detroit, Mich., USA), in 96-well micro tissue culture plates for 72 h at 37°C. At 24 h before the end of culture, 0.5 μCi of ^{3}H-thymidine (The Radiochemical Centre, Amersham, Buckinghamshire, UK) was added, and the spleen cells were harvested on a glass filter by using a multiple cell harvester. The incorporation of ^{3}H-thymidine was measured by a liquid scintilation method.

Assay for Interleukin 1 (IL 1) Activity

C57BL/6 mice were injected with 1 ml of 3% thioglycolate broth i.p. and 4-5 days later the peritoneal exudate cells were collected. The cells were plated in 24-well culture dishes and non-adherent cells were washed out after incubation for 2 h at 37 °C (final concentration was 2×10^{6} cells/well). The adherent cells, displayed morphological characteristics of macrophages, were cultured with or without adjuvants for 24 h. IL 1 activity was measured as the ability of a test supernatant to augment the phytohemagglutinin (PHA)-induced proliferation of C57BL/6 thymocytes, assesed by uptake of ^{3}H-thymidine.[9] In brief, thymocytes were cultured with dilutions of macrophage culture supernatants in the presence of 1μg of PHA per ml for 3 days at 1.5×10^{6} cells/well in 96-well micro plates. Cultures were pulsed for the final 4 to 5 h with 0.5μCi ^{3}H-thymidine, and harvested onto glass filters. The radioactivity was counted by a liquid scintillation counter.

Table 1. Adjuvant activity of chitin derivatives on the circulating-antibody formation to bacterial α-amylase and on the induction of delayed-type hypersensitivity to ABA-N-acetyl-L-tyrosine in guinea pigs

Adjuvant[a]	Dose (μg)	Anti-BαA titer[b] (units ± S.E.)	skin reaction with[c] ABA-BSA (mm ± S.E.)
water-insoluble derivatives			
chitin	100	462 ± 11 (1.2)	(5.0 ± 0.3)
chitosan	100	483 ± 32 (1.2)	
DHP-chitin	100	459 ± 43 (1.1)	
DHP-chitosan	100	490 ± 12 (1.2)	
water-soluble derivatives			
DAC-30	100	1393 ± 119 (3.5)	13.4 ± 1.7
DAC-70	100	1216 ± 135 (3.0)	12.4 ± 1.2
CM-chitin	100	251 ± 8 (0.6)	
CM-chitosan	100	835 ± 41 (2.1)	
P-chitin	100	362 ± 6 (0.9)	(6.5 ± 0.2)
S-chitin	100	1192 ± 178 (3.0)	(2.3 ± 0.9)
MDP	100	1638 ± 50 (4.1)	22.5 ± 1.4
control (saline + FIA)		401 ± 16 (1.0)	(4.3 ± 0.9)

[a] DHP-chitin: dihydroxypropyl- chitin; DAC-30: 30% deacetylated chitin; DAC-70: 70% deacetylated chitin; CM-chitin: carboxymethyl-chitin; P-chitin: phosphorylated-chitin; S-chitin: sulphonated-chitin.

[b] Hartley guinea pigs were immunized in each footpad with 200 μg of BαA in FIA with or without adjuvants. The data are expressed as mean ± standard error. The numbers in parentheses are stimulation index, as calculated with the control.

[c] Hartley guinea pigs were immunized with 50μg of ABA-N-acetyl-L-tyrosine in FIA with or without adjuvants. After 2 weeks, skin test was performed with 50 μg of ABA-BSA, and skin reaction was measured 24 h after the intradermal injection of test antigen. The data are expressed as mean ± standard error of the skin reaction for four guinea pigs. The numbers in parentheses indicate the size of faint erythema.

Table 2. Effect of chitin derivatives on the induction of allogeneic cell-mediated cytotoxicity in mice

Mice immunized with[a] (on day 0)	Dose (μg)	Specific target cell lysis (%)
Expt. 1		
Mastocytoma P815 (5 x 10^5)		
+ chitin	200	25.6 ± 1.7
+ DAC-30	200	27.6 ± 2.7
+ DAC-70	200	37.2 ± 1.2
+ chitosan	200	16.8 ± 1.4
+ CM-chitin	200	13.4 ± 1.2
+ BCG-CWS	100	46.3 ± 0.8
Mastocytoma P815 alone (5 x 10^5)		12.3 ± 1.5
Mastocytoma P815 alone (1 x 10^7)		43.0 ± 1.4
DAC-70 alone	200	3.8 ± 1.8
Expt. 2		
Mastocytoma P815 (5 x 10^5)		
+ DAC-70 (on -7 days)	200	31.3 ± 3.2
+ DAC-70 (on 0 day)	200	36.1 ± 2.1
+ DAC-70 (on 7 days)	200	27.5 ± 2.5
+ BCG-CWS (on 0 day)	100	36.6 ± 4.8
Mastocytoma P815 alone (5 x 10^5)		9.7 ± 1.5
Mastocytoma P815 alone (1 x 10^7)		52.9 ± 5.2

[a] C57BL/6 mice were immunized intraperitoneally (i.p.) with mastocytoma P815-X2 cells 7 days before or after or on the same day as the administration of each adjuvant (i.p.). Two weeks later, cell-mediated cytotoxicity was determined by the method described in EXPERIMENTAL by the incubation of spleen cells obtained from immunized mice and ^{51}Cr-labeled P815-X2 cells at a ratio of 100:1 for 4 h.

RESULTS AND DISCUSSION

Chitin and three water-insoluble and six water-soluble derivatives were tested for their activities on the circulating-antibody formation to BαA compared with MDP (Table 1). Among water-soluble derivatives of chitin, DAC-30, DAC-70 and S-chitin showed potent adjuvant activity. Other chitin derivatives were not or hardly effective. DAC-30 and DAC-70 also showed significant but less adjuvant activity in comparison with MDP for the development of delayed-type hypersensitivity (DTH) to ABA-Tyr in guinea pigs (Table 1). S-chitin, which was effective for the circulating-antibody formation to BαA, was not active at all on the induction of DTH.

Table 2 shows the effect of chitin derivatives on the generation of alloreactive cytotoxic T lymphocytes (CTL) in allogeneic C57BL/6 mice. DAC-70, in particular, showed the most adjuvant activity among the derivatives of chitin. No effect was observed when DAC-70 was administered without P815-X2 cells on the induction of CTL activity, indicating that DAC-70 enhanced antigen (P815-X2 cells)-specific immune responses. DAC-70 enhanced the CTL activity when administered 7 days before and after as well as on the same day of the immunization of P815-X2 cells. For the *in vitro* generation of alloreactive CTL,[10] the enhancing effect of DAC-70 was observed, however, the effect of DAC-70 was much lower than that of BCG-CWS *in vitro* (not shown). Effect of chitin derivatives on the stimulation of helper function of carrier-primed T cells was determined in C57BL/6 mice by using hapten-carrier (dinitrophenyl-ovalbumin) as antigen. DAC-70 enhanced the helper T cell function most effectively among derivatives of chitin tested, and the effect of DAC-70 was comparable to that of BCG-CWS. DAC-30 was less effective and other derivatives were much less effective than DAC-70.

As shown in Table 3, all derivatives of chitin tested enhanced the activity of NK cells. Previously, we have reported that DAC-70 augmented the cytolytic activity of murine peritoneal macrophages *in vivo* most effectively, and DAC-30 was as or less effective, and chitosan was less effective than DAC-70, and chitin was hardly effective. In this study also similar structure-activity relationships were observed.

Mitogenic activites of chitin derivatives were examined by the use of normal spleen cells of BALB/c or C57BL/6 mice (Table 4). CM-chitin, HE-chitin and DHP-chitin showed weak mitogenic activities over a dose of 50μg per ml. Other derivatives were inactive as mitogens on normal mouse spleen cells. This indicated that adjuvant activity and mitogenic activity of chitin derivatives were not correlative.

It has been reported by Oppenheim et al.[9] that MDP stimulated human peripheral and mouse peritoneal adherent cells to produce a factor which was mitogenic for mouse thymocytes (IL 1). As shown in Table 5, DAC-70 stimulated the production of IL 1 by thioglycolate-induced mouse peritoneal macrophages. We also examined the effect of DAC-70 on the production of interleukin 2 (IL 2), which stimulates proliferation and differentiation of T cells and activity of NK cells, by T cells. DAC-70 did not stimulate T cells to produce IL 2 at all (data not shown). Furthermore, the effect of DAC-70 on the production of other factors, such as macrophage activating factor and colony-stimulating factor, is now under investigation.

Table 3. Effect of chitin derivatives on the activation of NK cells in mice[a]

Adjuvant	Dose (μg)	Timing (days)	Specific target lysis (%)	
			YAC-1	P815
chitin	500	-3	18.9 ± 2.4	-1.1 ± 2.6
DAC-30	500	-3	19.7 ± 4.1	1.3 ± 3.1
DAC-70	500	-3	23.0 ± 1.9	-1.0 ± 2.8
chitosan	500	-3	19.7 ± 2.8	2.8 ± 2.2
MVE-2	500	-3	30.0 ± 2.9	5.5 ± 1.9
control	—	—	12.0 ± 1.4	-0.9 ± 1.0

[a]C57BL/6 mice were injected i.p. with each adjuvant on 3 days before the assay of NK cell-mediated cytotoxicity. Spleen cells (1 x 10^6) obtained from treated or untreated (control) mice were incubated with ^{51}Cr-labeled target cells (1 x 10^4) for 4 h.

Table 4. Mitogenic activity of chitin derivatives on normal spleen cells of BALB/c or C57BL/6 mice[a]

Mitogen	Dose (μg/ml)	^{3}H-thymidine incorporation (cpm) BALB/c		^{3}H-thymidine incorporation (cpm) C57BL/6	
Expt. 1					
CM-chitin	250	6,065 ± 208	(3.4)[b]		
	50	3,822 ± 280	(2.2)		
	10	3,478 ± 296	(2.0)		
	2	2,957 ± 219	(1.7)		
Con A	5	126,058 ± 2247	(71.2)		
LPS	10	10,466 ± 190	(5.9)		
control	—	1,771 ± 169	(1.0)		
Expt.2					
HE-chitin	250	3,386 ± 158	(2.2)	5, 953 ± 205	(1.6)
	50	1,855 ± 165	(1.2)	4,086 ± 112	(1.1)
	10	2,016 ± 101	(1.3)	3,760 ± 70	(1.0)
	2	1,665 ± 54	(1.1)	3,575 ± 162	(1.0)
DHP-chitin	250	5,507 ± 393	(3.5)	6,703 ± 279	(1.8)
	50	3,522 ± 266	(2.2)	4,219 ± 237	(1.1)
	10	2,517 ± 133	(1.6)	3,384 ± 186	(0.9)
	2	2,376 ± 223	(1.5)	2,869 ± 87	(0.8)
DAC-70	250			3,519 ± 303	(0.9)
	50			4,150 ± 96	(1.1)
	10			3,563 ± 316	(1.0)
	2			3,325 ± 210	(0.9)
Con A	5	123,835 ± 1818	(79.1)	88,301 ± 4633	(23.6)
LPS	10	8,901 ± 366	(5.7)	15,183 ± 1113	(4.1)
control	—	1,566 ± 117	(1.0)	3,744 ± 279	(1.0)

[a]Normal spleen cells (5 x 10^5) of BALB/c or C57BL/6 mice were incubated with mitogen or medium alone (control) for 72 h. ^{3}H-thymidine (0.5μCi) was added from 24 h until the end of culture.

[b]Numbers in parentheses are stimulation ratios.

We have found that the solubility in water and the susceptibility to lysozyme of CM-chitin were parallel with the degree of substitution at 6-O hydroxyl groups with CM groups (Nishimura et al., submitted). This increment of the solubility in water and the susceptibility to lysozyme digestion of CM-chitins were also correlated to their immunological activities for the activation of peritoneal macrophages *in vivo*. With a view to utilizing the advantages of CM-chitins, we synthesized antitumor drug (adriamycin)-combined CM-chitin (D.S. 0.8), and tested its efficacy against tumor growth *in vitro*. Adriamycin-combined CM-chitin was less effective than free adriamycin against *in vitro* tumor growth, however, it was expected to exhibit reduction of toxicity and prolonged retention of adriamycin *in vivo*.

The above results suggest it may be possible to develop an adjuvant having reasonable solubility, biodegradability, and unique pharmacodynamics by chemical modifications of chitin.

Table 5. Effect of DAC-70 on the *in vitro* production of IL 1 by thioglycolate-induced peritoneal macrophages

Adjuvant added to macrophages	Dose (μg/ml)	Incorporation of ^{3}H-thymidine by thymocytes (mean cpm ± S.E.)	
		Expt. 1	Expt. 2
DAC-70	1	7349 ± 484	N.D.[b]
	10	5030 ± 76	11794 ± 795
MDP	10	11437 ± 154	3554 ± 449
None	—	1392 ± 122	2264 ± 333
medium control		1259 ± 5	977 ± 76

[a]Thioglycolate-induced peritoneal macrophages were incubated with or without adjuvant for 24 h. Supernatant fluids were removed and assayed at 1/2 dilution (Expt. 1) or at 1/4 dilution (Expt. 2) in medium for effect on thymocyte incorporation of ^{3}H-thymidine.

[b]N.D., not done.

REFERENCES

1. K. Nishimura, S. Nishimura, N. Nishi, I. Saiki, S. Tokura, and I. Azuma, Immunological activity of chitin and its derivatives, Vaccine 2: 93 (1984).
2. S. Suzuki, Y. Ogawa, Y. Okura, K. Hashimoto, and M. Suzuki, Proceedings of the Second International Conference on Chitin and Chitosan, Sapporo, Japan, pp.210-212 (1982).
3. K. Suzuki, Y. Ogawa, K. Hashimoto, S. Suzuki, and M. Suzuki, Protecting effect of chitin and chitosan on experimentally induced murine candidasis, Microbiol. Immunol. 28: 903 (1984).
4. S. Tokura, N. Nishi, and J. Noguchi, Studies on chitin. III. Preparation of chitin fibers, Polymer J. 11: 781 (1979).
5. S. Tokura, J. Yoshida, N. Nishi, and T. Hiraoki, Studies on chitin. VI. Preparation and properties of alkyl-chitin fibers, Polymer J. 14: 527 (1982).
6. I. Azuma, F. Kanetsuna, T. Taniyama, Y. Yamamura, M. Hori, and Y. Tanaka, Adjuvant activity of mycobacterial fractions. I. Purification and *in vivo* adjuvant activity of cell wall skeletons of *Mycobacterium bovis* BCG, *Nocardia asteroides* 131 and *Corynebacterium diphtheriae* PW8, Biken J. 18: 1 (1975).
7. P. Lefrancier, J. Choay, M. Perrier, and I. Lederman, Synthesis of N-acetylmuramyl-L-alanyl-D-isoglutamine, an adjuvant of the immune response and some N-acetyl-muramyl peptide analogs, Int. J. Peptide Protein Res. 9: 249 (1977).
8. Y. Okada, K. Onoue, S. Nakashima, and Y. Yamamura, Studies on enzyme-antienzyme systems. II. N-bromosuccineimide modified bacterial α-amylase, J. Biochem. (Tokyo) 3: 113 (1963).
9. J.J. Oppenheim, A. Togawa, L. Chedid, and S. Mizel, Components of mycobacteria and muramyl dipeptide with adjuvant activity induced lymphocyte activating factor, Cell. Immunol. 50: 71 (1980).
10. K. Nishimura, S. Nishimura, N. Nishi, F. Numata, Y. Tone, S. Tokura, and I. Azuma, Adjuvant activity of chitin derivatives in mice and guinea pigs, Vaccine in press (1985).

IMMUNOPOTENTIATING EFFECT OF N-ACETYL-CHITO-OLIGOSACCHARIDES

S. SUZUKI, K. SUZUKI, A. TOKORO, Y. OKAWA and M. SUZUKI

Department of Microbiology and The Second Department of Hygienic Chemistry, Tohoku College of Pharmacy, Komatsushima 4-4-1, Senday Miyagi 983, Japan

INTRODUCTION

In our previous papers on this subject (1, 2), we revealed that chitin and chitosan were able to display growth-inhibitory effects against tumor cells or pathogenic microbes inoculated in the peritoneal cavity of mice. The action mechanism of both polysaccharides was assumed to involve the activation of phagocytic cells attracted to the peritoneal cavity from blood and lymphoid fluid through the peritoneal membrane by intraperitoneal administration of these polysaccharides.

However, water-insolubility of chitin and chitosan is disadvantageous to their use as immunotherapeutic agents. Therefore, the investigation of immunopotentiating effect of water-soluble, lower analogues of chitin and chitosan such as N-acetyl-chito-oligosaccharides (NACOS) and chito-oligosaccharides (COS) was attempted. Quite recently, we published a preliminary paper (3) stating that NACOS, from N-acetyl-chitotetraose (Tetraose) to N-acetyl-chitohexaose (Hexaose), were able to increase the number of mouse peritoneal exudate cells by intraperitoneal administration and that they displayed higher active oxygen-generating and candidacidal effects than the cells from mice of non-treated control did. In the present paper, details of immunological properties of NACOS, especially those of N-acetyl-chitohexaose which demonstrated the strongest effects among NACOS, are reported.

MATERIALS AND METHODS

Animals. BALB/c male mice, 4- to 6-weeks old, were obtained from Shizuoka Agricultural Cooperation for Experimental Animals (Hamamatsu, Japan). These mice were housed 5/cage in air-conditioned quarters and were provided food and water *ad libitum.*

Chemicals. NACOS, i.e., di-N-acetyl-chitobiose (Biose), tri-N-acetyl-chitotriose (Triose), tetra-N-acetyl-chitotetraose (Tetraose), penta-N-acetyl-chitopentaose (Pentaose), hexa-N-acetyl-chitohexaose (Hexaose), and hepta-N-acetyl-chitoheptaose (Heptaose) were supplied by Ihara Chemical Industry Co. Ltd., Tokyo, Japan. All NACOS except for Heptaose were found to be chromatographically homogeneous when these materials were separately examined on a column of Bio-Gel P-4 (2.5 x 90 cm) connected with a refractive index monitor using water as the eluant. Heptaose contained approximately 20 %

of Hexaose, and was used without further fractionation. The crude chitin was obtained from Wako Pure Chemical Industies, Ltd., Tokyo, Japan, and purified by the method of Hackman (4). Casein and muramyl dipeptide (MDP), used as the positive controls, were purchased from Junsei Chemical Co. Ltd., Tokyo, Japan, and Peptide Research Institute Inc., Osaka, Japan, respectively. Biose, Triose, Tetraose, Pentaose, Hexaose, casein, and MPD were separately dissolved in saline, while Heptaose and chitin were suspended in the same sterilized saline.

Assay of number of peritoneal exudate cells (PEC) from mice administered with NACCS, casein, and chitin. This was performed in accordance with the method of our previous descriptions (1, 5). Namely, seven groups of mice (16 per group) were given intraperitoneal (ip) injections of Tetraose, Pentaose, Hexaose, Heptaose, chitin, MDP, and casein, each 50 mg/kg, 3 hr before the assay. Then the mice were sacrificed by bloodshed followed by leaving the peritoneal cavity with Hanks' balanced salts solution (HBSS). The fluid was centrifuged at 500 x g for 15 min, and the cells were washed 3 times with the same medium by centrifugation. The cells were resuspended in HBSS, counted by means of a hemocytometer, and used immediately.

Dose-Response and time-course studies on the attractive effect of Hexaose on mouse peritoneal exudate cells. For the dose-response study, BALB/c mice (6 per group) were injected intraperitoneally with Hexaose in the doses of 1, 10, 25, 50, and 100 mg/kg. After the 3 hr administration, PEC were collected, and the number of cells per mouse was counted. For time-course study, 7 groups of BALB/c mice, each 6 per group, were administered with 50 mg/kg of Hexaose. After 1, 2, 3, 6, 9, 12, and 24 hr, the mice of each group were sacrificed, the peritoneal exudate cells were collected, and the number of the cells was counted by the same method as described above.

Separation of leukocytes from the circulating blood of normal BALB/c mice. This was conducted according to the method described by West et al. (6): Blood obtained from a normal mouse heart containing 100 units of heparin (Nakarai Chemicals, Ltd., Kyoto, Japan) per ml was mixed with a half volume of 6 % dextran T-500 (Pharmacia, Uppsala, Sweden) in saline and allowed to sediment at room temperature for 30 min. The leukocyte-rich layer was collected into another test tube by pipetting. The cell suspension was diluted with HBSS and centrifuged at 250 x g for 10 min. The number of cells was counted by using hemocytometer and adjusted to 1 x 10^7 cells/ml. The cell suspension was used immediately for the *in vitro* chemotaxis assay.

In vitro chemotactic assay of Hexaose on leukocytes from normal BALB/c mice. This was conducted by essentially the same method as Falk et al. (7) using a 48-well multi-chemotaxis chamber (Neuro Probe Bethesda, Maryland). Briefly, HBSS solutions of Hexaose, from 10^{-6} mg/ml to 10^{-1} mg/ml, were added in the bottom of the chemotaxis chamber, and a membrane (pore size, 0.3 μm) was placed on top of the cell layer. Then 5 x 10^5 cells of leukocytes in 50μl of HBSS were added to the upper side of the polycarbonate membrane. The chamber was incubated in air containing 5 % CO_2 at 37°C for 3 hr, then the cells passed through the membrane were counted microscopically. MDP in HBSS at the same concentrations as those of Hexaose were used as the positive control.

Chemiluminescence measurement. PEC, 1 x 10^6 cells, and PMN from blood, 1 x 10^5 cells, were separately suspended in 1 ml of HBSS, and those were preincubated with 5 mM D-glucose and 50 μl of luminol solution (2 mg/ml in dimethyl sulfoxide) for 10 min, then 5 μl of dimethyl sulfoxide solution of phorbol myristate acetate (Sigma Chemical Co., St. Louis, U.S.A.), 10 μg/ml, was added to each vial, and chemiluminescence was measured in a

Biolumat LB 9500 (Berthold Co., Wildbad, West Germany) according to the previous description (1, 5). Chemiluminescence response was expressed as counts per minute (cpm).

In vitro activation of mouse PMN by tratment with Hexaose. For this purpose, leukocytes separated from a normal mouse, as described above were further fractionated by the following method. Namely, leukocyte-rich layer obtained by fractionating with dextran T-500 solution was added to Lympholyte-M^{R} (Wako Pure Chemical Industries, Ltd, Tokyo, Japan), and centrifuged at 500 x g for 30 min. After centrifugation, the lymphocyte-rich fraction in the middle layer was removed, and the precipitate was treated with sterilized water for lysing contaminated red blood cells, and centrifuged again at 250 x g for 5 min. The cells were washed twice with HBSS, then resuspended in HBSS. The number of PMN in the suspension was counted by means of a hemocytometer. Homogeneity of the PMN preparation was more than 95 % as determined by May-Giemsa stain. PMN, 1.0 to 1.3 x 10^5 cells, were collected from one mouse, and 10 to 12 mice were necessary for one assay. PMN from these mice were mixed and adjusted 1 x 10^6 cells per ml of HBSS. Then 0.1 ml of this suspension, containing 1 x 10^5 cells of PMN, was added in each chemiluminescence vial. Subsequently, a solution of 100 μg of Hexaose in 100 μl of HBSS was added in this vial, and total volume was made up to 1.0 ml with HBSS. After addition of luminol solution, the tubes were incubated at 37°C for 10 min, and chemiluminescence response was determined by the above method. As a positive control, an HBSS solution of MDP was run simultaneously.

Candidacidal activity of PEC in mice administered with Hexaose. Candidacidal activity of PEC from mice treated with NACOS was assayed by the same method as that of our previous report (1). Briefly, a suspension of viable cells of *Candida albicans* NIH A-207 strain, 200 in 10 μl of Hanks' BSS containing 10 % normal mouse serum, was added to a suspension of PEC, 1 x 10^5 cells in 100 μl of the same medium, and the mixture was incubated at 37°C in the 96-well trays. After 3-hr incubation, PEC were scraped off with a rubber policeman, and the cell suspension was placed on Sabouraud agar. The number of colonies was counted after incubation for 24 hr at 28°C, and the killing activity index was calculated by the following formula: Killing activity index = 200 - Average number of colony/200 x 100. As positive controls, MDP, casein, and chitin were also assayed under the same conditions.

RESULTS

Effect of intraperitoneal administration of NACOS on number of peritone exudate cells of BALB/c mice. Fig. 1 shows comparison of the attractive effects of PEC in BALB/c mice with Tetraose, Pentaose, Hexaose, and Heptaose, and Heptaose, each 50 mg/kg, after 3 hr ip administration. These conditions were established based on the fact that casein showed a maximum of neutrophil-attracting response after 3 to 6 hr ip administration (8). A time-course study with Hexaose showed that the peak number of PEC was 3 hr after the administration (Fig. 2). As can be seen in Fig. 1, PEC-attracting effects evoked by Hexaose and Heptaose were almost the same, and were stronger than Tetraose and Pentaose. These effects were also identical to those displayed by chitin, the water-insoluble parent polysaccharide, and were significantly higher than those of two positive controls, MDP and casein. The fact that the PEC-attracting effects of Pentaose, Hexaose, and Heptaose were higher than those of MDP seems to account for the necessity of consecutive β-(1→4)-linked GlcNAc residues to manifest this effect. Lower analogues of NACOS, N-acetyl-D-glucosamine, Biose, and Triose, were not effective.Chito-oligosaccharide analogues, D-glucosamine, chitobiose, chitotriose, chitotetraose, chitopentaose, and chitohexaose, were also inactive under the same assay conditions, demonstrating the importance of N-acetyl groups for manifestation of the attracting effect to PEC.

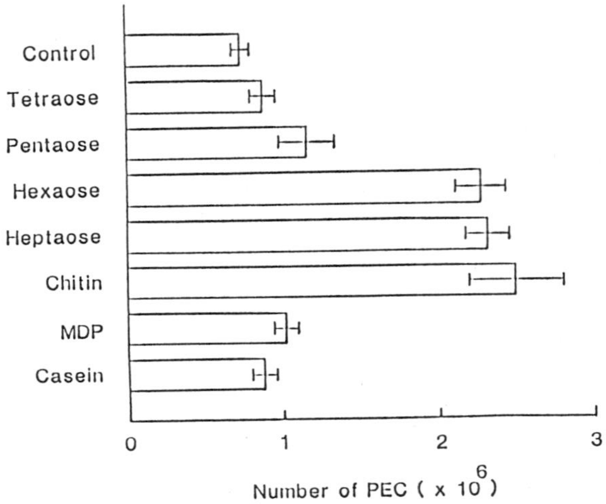

Figure 1. Number of PEC in mice given NACOS, chitin, MDP and casein intraperitoneally.

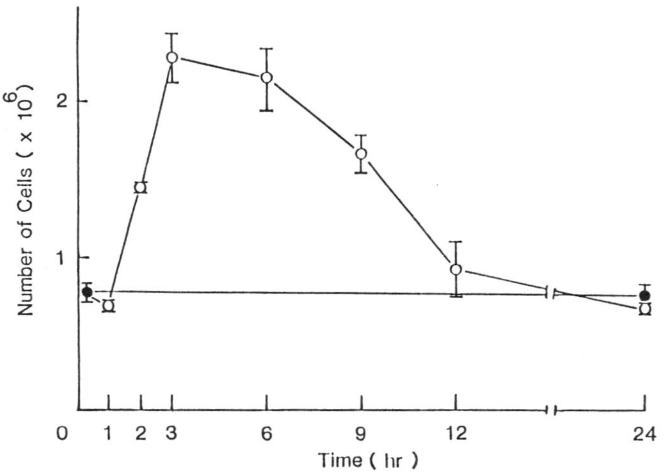

Figure 2. Kinetic study of PEC in mice given Hexaose intraperitoneally.

Dose response study between number of attracted PEC and amounts of Hexaose administered ip in mice. The results of the attraction assay of PEC with varying amounts of Hexaose, from 10 to 100 mg/kg, indicate that a proportional relationship between dose of Hexaose and PEC response existed in the range from 10 to 50 mg/kg, while a high dose, 100 mg, showed a decrease of the response.

Effect of Hexaose on chemotaxis of normal mouse leukocytes in vitro. In order to analyze the mechanism of the rapid attraction of PEC, Hexaose was investigated for its accelerating effect of chemotaxis on normal leukocyte from BALB/c mice in vitro using MDP a positive control. It was found that Hexaose exhibited a higher chemotaxis-accelerating effect than that of MDP in the same weight basis, demonstrating that a fragment of chitin consisting solely of consecutive β-(1→4)-linked D-GlcNAc residues is able to serve as a strong attractant of normal leukocytes.

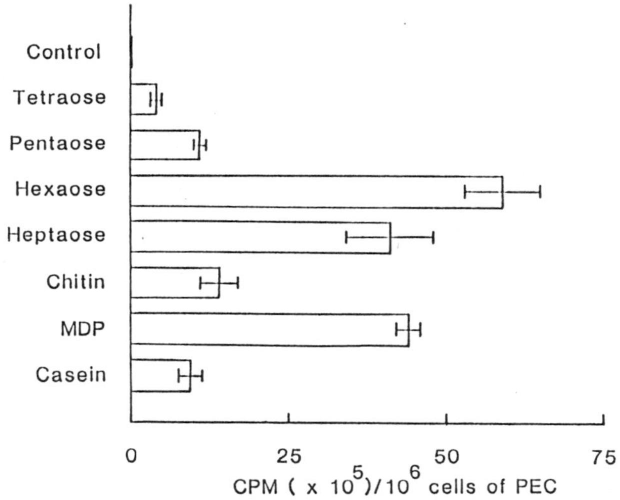

Fig. 3. Chemiluminescence Response of PEC in Mice Administered with NACOS, Chitin, MDP, and Casein.

Chemiluminescense response of PEC attracted by NACOS in mouse peritoneal cavity. Fig. 3 depicts the results of the active oxygen-generation assay of PEC, each 1 x 10^6 cells, attracted by the ip administration of Pentaose, Hexaose, and Heptaose, 50 mg/kg. It is obvious that the PEC attracted by Hexaose showed the strongest chemiluminescence response among all NACOS, chitin, MDP, and casein. The above findings demonstrate that the size of the NACOS receptor locating the surface of PEC might be quite large, corresponding up to 6 consecutive β-(1→4)-linked GLcNAc residues, and that contribution of N-D-lactoyl-L-alanyl-D-isoglutamine moiety in MDP to this effect is considerably large. It is also presumable that the low effects of Heptaose and chitin are attributable to their water-insolubility.

Effect of Hexaose on active oxygen generation of normal mouse PMN in vitro. Normal leukocytes obtained from mouse peripheral blood were further fractionated by means of Lympholyte-M, a fractionating reagent of mouse lymphocytes and centrifugated. Then PMN with a high homogeneity, more than 95 %, obtained by this method were assayed for their active oxygen-generating activity after treatment with Hexaose in vitro. The results indicated that PMN treated with Hexaose showed a higher active oxygen-generating effect than those treated with MDP.

In vitro candidacidal effect of PEC from mice administered ip with NACOS, chitin, MDP, and casein. In the previous study (16), we demonstrated that PEC from mice treated either with chitin or with chitosan acquired the killing effect on viable cells of *Candida albicans.* Therefore, PEC from mice pretreated with NACOS were investigated for their candidacidal effect by the same method as that of the previous study (16). As shown in Fig. 4, PEC treated with Hexaose showed the strongest effect among materials including Teatrose, Pentaose, chitin, MDP, and casein in the same dose level. This tendency is similar to that of the number of attracted PEC in mice given intraperitoneally the same materials shown in Fig. 1, and is significantly different from that of the chemiluminescence response of PEC treated with these materials, especially of MDP (Fig. 2). Therefore, the candidacidal effect by PEC is not simply proportional to the amounts of active oxygens.

DISCUSSION

In the previous study it was reported that chitin was able to exhibit an immunopotentiating action (1). Namely, mice treated with chitin showed a longer life span against the callenge of C. albicans cells via iv route. A series of kinetic studies of a number of PMN from blood and PEC in chitin-treated mice were conducted in order to resolve the infection-defense mechanism. The results indicated that chitin was able to increase the number of circulating PMN, and the amounts of generated active oxygens from these phagocytes increased by either treatment with chitin or with chitosan. These results suggest that the activated PMN in chitin-treated mice play an important role in the defense mechanism against the iv challenge of C. albicans.

In this decade, many papers have been published on animal lectins specifically binding to N-acetyl-D-glucosamine, D-glucosamine, D-glucosamine, and D-mannose residues on the surface of phagocytic cells (9, 11). According to the description by Oda et al. (12), a significant suppression of the phagocytic effect against fungal cells was observed on macrophage treated with D-mannose, D-glucosamine, and N-acetyl-D-glucosamine in vitro.

It is, therefore, reasonable to presume that the active sites of the phagocytic effect of macrophage include the receptor sites for N-acetyl-D-glucosamine and D-glucosamine residues, and that the recognition process of these sugar residues serves as the strong trigger of the proliferation process of phagocytes participating in the generation of active oxygens to kill the parasitic cells.

This paper can, therefore, be regarded as the first one dealing with immunological properties of NACOS, the lower analogues of chitin. The most striking finding obtained in the present study was that NACOS were able to attract PEC in the mouse peritoneal cavity 3 hr after ip administration, and that PEC collected from the peritoneal cavity of NACOS-administered mice showed elevated active oxygen-generating and candidacidal effects. The extents of both effects depended on the molecular weights of NACOS, and Hexaose was found to show the strongest effect among all materials so far assayed. It was also evident that the PEC-attracting effect of Hexaose was dose-dependent, and was stronger than that of MDP or casein. Because MDP is considered as the minimum active unit of the immunoadjuvant of microbial cell wall peptidoglycan (13 - 16), and casein as a representative proteinous attractant of PEC (8), it is of interest that NACOS, consisting solely of GlcNAc residues, are able to display stronger attracting and activating effects of PEC than those displayed by MDP and casein. The fact that Hexaose shows the strongest biological effects among NACOS, chitin, and MDP seems to imply that size of the receptor sites involving GlcAc units corresponds to this magnitude. The additional fact that Hexaose attracted larger numbers of mouse PMN than MDP or casein containing proteinous components also indicate the presence of a different activation mechanism of PMN initiated by interacting with ClcNAc-containing material. It is well-known that formyl-methionylleucyl-phenylalanine (f-MLP), a peptidic PMN attractant, which stimulates superoxide generation activity, (17, 18), has its specific receptor on the PMN membrane (19). After binding of f-MLP to the receptor, activation of phospholipase A_2 takes place to cause the degradation of methylated phospholipids (20[2]), thus releasing 5-lipoxygenase products such as leukotriene B_4 (LTB_4). Then LTB_4 enhance chemotactic activity and chemiluminescence response of PMN in vitro in short term incubation (18, 21).

In the present study, Hexaose attracted PMN within 3 hr after ip administration in mice, enhacing chemiluminescence response. These results suggest that Hexaose may activate phospholipase A_2, and release LTB_4 enhancing generation of active oxygens. Therefore, the existence of different recognition sites from peptidic materials such as f-MLP of MDP on PMN membrane is highly likely.

All findings obtained in the present study seem to indicate that the existence of a defense mechanism triggered by the interaction between cell surface receptor(s) of lectin-like properties and NACOS consisting solely of GlcNAc residues is obvious, and that development of some immunotherapeutic agents can be expected by investigating the properties of Hexaose in detail.

REFERENCES

1. K. Suzuki, Y. Okawa, H. Hashimoto, S. Suzuki, and M. Suzuki, Protecting effect of chitin and chitosan on experimentally induced murine candidiasis., Microb. Immunol., 28:903 (1984).
2. S. Suzuki, Y. Okawa, Y. Okura, K. Hashimoto, and M. Suzuki, Immunoadjuvant effect of chitin and chitosan. In chitin and chitosan, S. Hirano, S. Tokura, eds., The Japanese Society of Chitin and Chitosan, Tottori Univ., Tottori, 210 (1982.
3. K. Suzuki, A. Tokoro, Y. Okawa, S. Suzuki, and M. Suzuki, Enhancing effect of N-acetyl-chito-oligosaccharides on active oxygen generating and microbicidal activities of peritoneal exudate cells, Chem. Pharm. Bull., 33:886 (1985).
4. R. H. Hackman, Studies on chitin, Austr. J. Biol. Sci., 15:526 (1962).
5. K. Hashimoto, Y. Okawa, Y. Okura, K. Suzuki, S. Suzuki, and M. Suzuki, Antitumor activity of acidic mannan fraction from bakers' yeast, J. Pharm. Dyn., 6:668 (1983).
6. B. C. West, A. S. Rosental, N. A. Gleb, and H. R. Kimball, Separation and characterization of human neutrophil granules, Am. J. Pathol., 77:41 (1974).
7. W. Falk, R. H. Goodwin, and E. J. Lenonard, A 48-well micro chemotaxis assembly for rapid and accurate measurement of leukocyte migration, J. Immunol. Methods, 33:239 (1980).
8. M. Yoshinaga, K. Nishime, S. Nakamura, and F. Goto, A PMN-derived factor that enhances DNA-synthesis in PHA or Antigen-stimulated Lymphocytes, J. Immunol., 124:94 (1980).
9. Z. Bar-Shavit, I. Ofek, R. Goldman, D. Mireman, and N. Sharon, Mannose residues on phagocytes as receptors for attachment of E. coli and Salmonella typhimurium, Biochem. Biophys. Res. Comm., 78:455 (1977).
10. Y. Fukazawa, K. Kagaya, and H. Miura, Biological and biochemical characterization of macrophage-activating factor (MAF) in murine lymphocytes: Role of mannopyranosyl residue of the MAF molecule in macrophage activation, Microbiol. Immunol., 25:1163 (1981).
11. G. A. Warr, A macrophage receptor for (mannose-glucosamine)-glycoproteins of potential importance in phagocytic activity, Biochem. Biophys. Res. Comm., 93:737 (1980).
12. L. M. Oda, C. F. Kubelka, C. S. Alviano, and L. Travassos, Ingestion of yeast forms of Sporothrix schenckii by mouse peritoneal macrophages, Infect. Immun., 39:497 (1983).
13. I. Azuma, K. Sugimura, T. Taniyama, M. Yamawaki, Y. Yamamura, S. Kusumoto, S. Okada, and T. Shiba, Adjuvant activity of mycobacterial fractions: Adjuvant activity of synthetic N-acetylmuramyl-dipeptide and the related compounds, Infect. Immun., 14:18 (1976).
14. E. B. Fraser-Smith, D. A. Eppstein, M. A. Larsen, and T. R. Mathews, Protective effect of muramyl-dipeptide analogue encapsulated in or mixed with liposomes against Candida albicans infection, Infect. Immun., 39:172 (1983).
15. K. N. Mashi, I. Azuma, W. Brehmer, and W. Lange, Stimulation of chemiluminescence by synthetic muramyl dipeptide and analogs, Infect. Immun., 40:16 (1984).

16. Y. Osada, M. Mitsuyama, T. Une, K. Matsumoto, T. Otani, M. Satoh, H. Ogawa, and K. Nomoto, Effect of L-18MDP(Ala) a synthetic derivative of muramyl-dipeptide, on nonspecific resistance of mice to microbial infection, Infect. Immun., 37:292 (1982).
17. L. Haravath, and R. R. Aksamit, Oxidized N-formyl-methionyl-leucyl-phenylalanine: effect on the activation of human monocytes and neutrophil chemotaxis and superoxide production, J. Immunol., 84:1471 (1984).
18. J. Palmblad, H. Gyllenhammar, J. A. Lindgren, and L. Malmsten, Effect of leukotriens and f-Met-Leu-Phe on oxidative methabolism of neutrophils and eosinophils, J. Immunol., 132:3041 (1984).
19. J. S. Niedel, Wilkinson, and P. Cuatrecasas, Receptor-mediated uptake and degradation of 125 I-chemotactic peptide by human neutrophils, J. Biol. Chem., 254:10700 (1979).
20. F. Hirata, B. A. Corcoran, K. Venkatasubramanian, E. Schiffmann, and J. Axelrod, Chemoattractants stimulate degradation of methylated phospholipids and release of arachidonic acid, Pro. Natl. Acad. Sci., USA, 76:2640 (1979).
21. C. Dahlgren, and O. Stendahl, Effect of in vitro preincubation of polymorphonuclear leukocytes on formylmethionyl-leucyl-phenylalanine-induced chemiluminescence, Infect. Immun., 37:34 (1982).

ADHESION PROPERTIES OF MOUSE SPLEEN LYMPHOCYTES ON CHITIN AND ITS DERIVATIVES

S. NISHIMURA, N. NISHI, M. KANEKO and S. TOKURA

Department of Polymer Science, Faculty of Science, Hokkaido University
Kita-Ku, Sapporo 060, Japan

INTRODUCTION

In the course of our adsorption chromatographic studies of metal ions (1), amino acids and bovine blood proteins (2) on chitin and its derivatives, it was suggested that the interactions between chitin derivatives and ligand molecules were significantly regulated by the content of the substituted functional groups on the sugar skeleton. It is interesting to investigate the relationship between the recognition property of lymphocytes and the chemical structures of chitin derivatives on the molecular and morphological aspects, since the macrophage activations have been demonstrated on the peritoneal injections of chitin derivatives such as O-carboxymethyl-chitin (O-CM-chitin) and chitosans (30% and 70% deacetylated chitins) (3).

In this presentation, adhesion behavior of murine spleen lymphocytes onto chitin and its derivatives was investigated to obtain fundamental information on the interactions between biologically active cells and chitin derivatives, which will accelerate the evaluation of chitin as a biomedical material. The fibrous chitin derivatives were applied to investigate the adhesion behavior of lymphocytes according to the method applied for some synthetic polymers containing micro-phase-separated structure (4).

EXPERIMENTAL

Mice

Inbred C57BL/6 female mice, eight to ten weeks of age, were obtained from Shizuoka Agricultural Cooperative for Experimental Animals, Hamamatsu, Japan.

Chitin derivatives and their fibers

Preparations of chitin derivatives used in this study were carried out by the method described in previous papers (5). In brief, O-CM-chitin dihydroxypropyl-chitin (DHP-chitin) were prepared from chitin by carboxymethylation and dihydroxypropylation of C-6 position of GlcNAc residues. The degrees of substitution were around 0.4. Diacetyl-chitin was prepared from chitin by acetylation at C-3 and C-6 positions in methanesulfonic acid.

Preparation of lymphocyte suspension

Lymphocytes were prepared from mouse spleens and suspended in Hanks balanced salt solution (HBSS) (pH 7.4) after thorough washing with Tris-NH_4Cl solution in order to remove erythrocytes. The cell concentration was controlled to 10 cells/mL and their viability was more than 95% as determined by the method of trypanblue indicator.

Estimation of adhered lymphocytes and identification of B lymphocytes

Fibrous samples were packed into small glass tubings (6.0 mm x 30 mm). Columns were saturated with HBSS and then lymphocytes suspension was loaded into the column. The number of lymphocytes retained on the fiber surface was estimated by the microscopic observation of total effluent at the equilibration. B lymphocytes (IgG positive cells) were identified by the reaction of fluorescein isothiocyanate (FITC)-conjugated rabbit antibody to mouse IgG. B lymphocytes conjugated with FITC-labelled anti-mouse IgG were counted by means of fluorescence microscope. The IgG negative fraction was assumed to be T lymphocytes, though it contained a small amount of macrophages and granule cells.

Scanning electron microscopic observations of adhered lymphocytes

Fibrous samples adhered by lymphocytes were lyophilized after rinsing with the buffer solution and observed by a scanning electron microscope (SEM) following the Au-Pd spattering.

RESULTS AND DISCUSSION

Estimation of differential retention of lymphocyte subpopulations in the column

Adhesivities of lymphocyte are expressed by the percentage of lymphocytes trapped in the column, which has been estimated from the difference between the number of applied lymphocytes and the number of unadsorbed ones washing from the column.

Table I. Percentages of adhered lymphocytes.

Polymers	D.S.[a]	Adhesion of lymphocytes (%)[b]	B (%)[b] (IgG pos.)	T' (%)[c] (IgG neg.)	Pos./Neg.[d]
Chitin	==	69.1 $\pm$ 4.3	70.6 $\pm$ 2.5	66.3 $\pm$ 6.6	1.1
Chitosan	0.84[e]	69.2 $\pm$ 5.0	93.8 $\pm$ 6.2	57.5 $\pm$ 4.3	1.6
DHP-chitin	0.4	54.3 $\pm$ 9.7	76.4 $\pm$ 5.1	40.6 $\pm$ 1.6	1.9
CM-chitin	0.4	27.7 $\pm$ 6.7	64.7 $\pm$ 8.0	20.0 $\pm$ 7.4	3.2
Acetyl-chitin	2.0	47.4 $\pm$ 9.3	77.8 $\pm$ 12.2	34.0 $\pm$ 9.4	2.3

4 Mice were used to prepare the suspension of spleen lymphocytes and the concentration of each suspension was adjusted to $1x10^8$ /ml of Hanks balanced salt solution. The initial content of B lymphocytes (IgG positive cells) was 34.4 $\pm$ 7.1 (%).

a) Degree of substitution.

b) $\frac{\text{Number of B lymphocytes retained in column}}{\text{Number of B lymphocytes loaded into column}} \times 100$

c) $\frac{\text{Number of IgG negative cells retained in column}}{\text{Number of IgG negative cells loaded into column}} \times 100$

d) Ratio of adhered B lymphocytes (%) and IgG negative cells (%).

e) Degree of deacetylation. The value 0.84 means 84% deacetylated chitin.

B (IgG positive cells) (%) or IgG negative cells (%) in the Table expresses each percentage of adhered lymphocytes calculated according to the equations described in the Table. The pos./neg. will indicate an apparent selectivity in the lymphocyte adhesion. Although the adhesion of lymphocytes was shown to be significantly suppressed by the introduction of carboxyl groups into chitin molecules, the apparent selectivity for B lymphocytes was the highest among other chitin derivatives as shown in the Table. The order of apparent selectivity for B lymphocytes was O-CM-chitin < diacetylchitin < DHP-chitin < chitosan < chitin.

The role of the degree of deacetylation on the adhesivity of lymphocytes

The relationship between the degree of deacetylation and the adhesivity of lymphocytes was investigated to estimate the effect of the electrostatic interaction between the cell surface and the primary amino groups of partially deacetylated chitin as shown in Figure 1. The total amount of adhered cells was not affected much by the increase of the degree of deacetylation. On the other hand, the adhesivity of B lymphocytes showed a clear tendency to increase as the extent of deacetylation increased. As a result, the selectivity for B lymphocyte seems to depend on the content of primary aminogroups on the fiber surface.

SEM observation of adhered cells

SEM pictures of adhered cells on the surface of chitin, O-CM-chitin, and chitosan are shown in Figures 2, 3 and 4.

The adhered cells on the chitin surface remained without any morphological deformation, which was in significant contrast to those on the chitosan surface. Though the B lymphocytes showed a similar affinity both to O-CM-chitin and chitosan, the stimulation property of chitosan on the lymphocytes

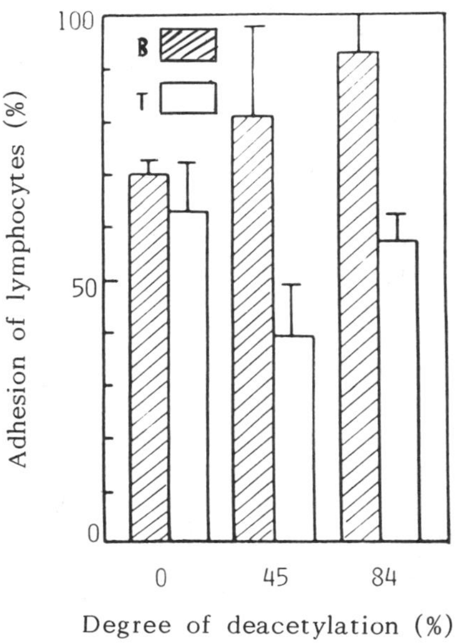

Figure 1. The effect of the degree of deacetylation on the cell adhesion.

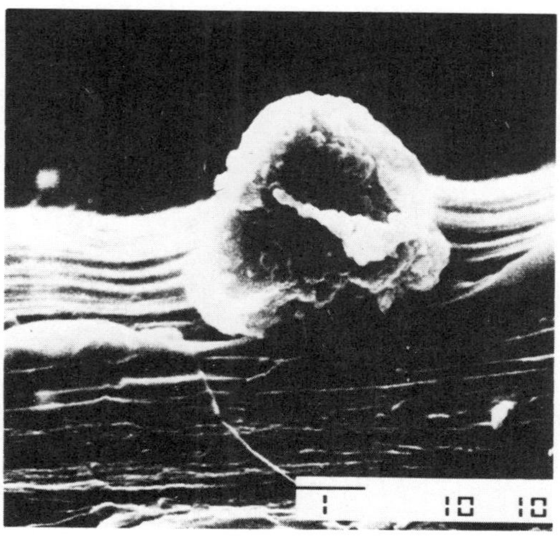

Figure 2. SEM picture of adhered lymphocyte (x 8000) on the surface of chitin fiber

seems to be stronger than that of O-CM-chitin. As the activation of immune systems has been observed especially in the presence on chitosans in vivo (3, 6) and the toxicity of chitosans has not been found, probably because of its biodegradation by lysozyme in plasma or by some lysosomal enzymes in macrophages, the morphological deformations might suggest the future importance of chitosans as biomedical materials for the delayed release of drugs.

From the structural aspects of these derivatives, the presence of acetamide groups at C-2 position in GlcNAc residues seems to be important for the interaction of polysaccharides with lymphocytes. The charge effect of amino groups should be considered for the deformation of cells in the adhesion process onto chitosan surface, because chitin and O-CM-chitin charged negatively or neutral were accepted by the cells without shape deformation.

Although the present investigation provides some valuable information for the biomedical evaluation of chitin and its derivatives, futher detailed study should be performed to clarify the exact relationship between the structures of chitin derivatives and their ability to recognise the biologically active cells.

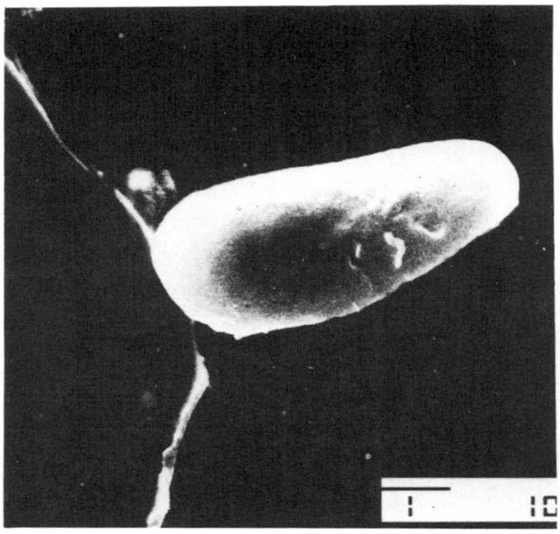

Figure 3. SEM picture (O-CM-chitin)

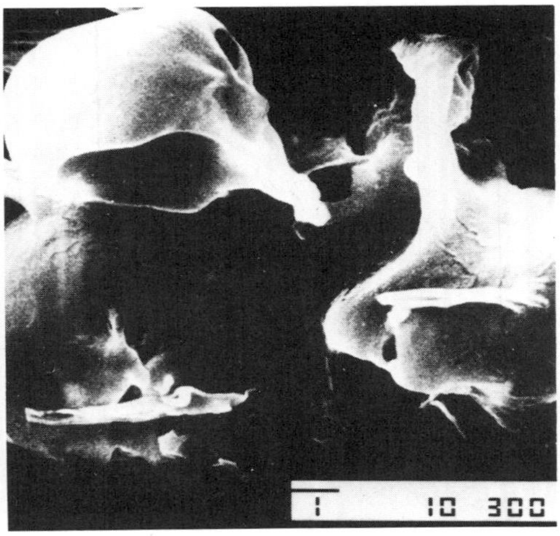

Figure 4. SEM picture (chitosan)

REFERENCES

1. S. Tokura, S. Nishimura, and N. Nishi, Studies on Chitin IX. Specific Binding of Calcium Ions by Carboxymethyl-Chitin, Polym. J., 15, 597(1983)
2. S. Nishimura, Y. Ikeuchi, and S. Tokura, The Adsorption of Bovine Blood proteins onto The Surface of O-(Carboxymethyl)Chitin, Carbohydr. Res., 134, 305(1984).
3. K. Nishimura, S. Nishimura, N. Nishi, I. Saiki, S. Tokura, and I. Azuma, Immunological Activity of Chitin and Its Derivatives, Vaccine, 2, 93(1984).
4. K. Kataoka et al., Differential Retention of Lymphocyte Subpopulations (B and T Cells) on The Microphase Separated Surface of Polystyrene/Polyamine Graft Copolymers, Eur. Polym. J., 19, 979(1983).
5. S. Tokura et al., Studies on Chitin I,III,IV,VI,VIII., Polym. J., 11, 27(1979), 11, 781(1979), 12, 695(1980), 14, 527(1982), 15, 485(1983).
6. K. Nishimura, S. Nishimura, N. Nishi, F. Numata, Y. Tone, S. Tokura, and I. Azuma, Adjuvant Activity of Chitin Derivatives in Mice and Guinea Pigs, Vaccine (1985) in press.

BIOMEDICAL EVALUATION OF ACYLATED CHITINS AS COATING MATERIALS

T. KOMAI, K. KAIFU, M. MATSUSHITA[◇],
I. KOSHINO[◇] and T. KON[◇]
Department of Polymer Science, Faculty of Science
and [◇] Department of Surgery, Medical School
Hokkaido University
Sapporo 060, Japan

INTRODUCTION

The chemically modified materials arising from natural tissue or organs sometimes show excellent blood-compatibility and can be utilized in reconstructive surgery. Hundreds of glutaraldehyde-treated human umbilical cord vessels (GA.UCV) have been implanted in humans as an artificial blood vessel and acetylated bovine pericardium (AC.PC) appears to be very promising as a material for the substitution of bladder tissue. (1-3). Both GA.UCV and AC.PC are well known to be composed of collagen, polysaccharide and lipid, and their surfaces are comparatively hydrophobic. In spite of the successful utilization of these materials in therapeutic reconstructive surgery, the characteristics and functional properties of these conjugated materials still remain to be clarified, mainly because of their complicated chemical and physical structures. An examination of model compounds simulating the GA.UCV or AC.PC surfaces chemically, should provide some understanding of this. (4-6). Accordingly, we prepared acylated chitins with various lengths of acyl groups in side chain hydroxy groups as model compounds simulating lipid polysaccharide conjugates and investigated their practical use as materials for blood contacting surfaces. (7). In this study, we have attempted to elucidate the relationship between the various types of acylated chitin and the thromboresistant properties and wetting characteristics of the respective materials.

EXPERIMENTAL

Materials

According to the method previously reported, (8-10) acetyl, propionyl, butyrylchitin were prepared by the reaction of chitin (from Alaska king crab shell purchased from Nippon Suisan Co., Ltd.) with the appropriate carboxylic anhydride in methanesulfonic acid, and degrees of substitution were 2.0, 1.9, and 1.8 acylgroups per N-acetylglucosamine residue, respectively. Caproyl, capryl, lauroyl, and benzoylchitin were prepared using the appropriate acyl chloride in methanesulfonic acid, and their degrees of substitution were 2.0, 1.8, 1.9, and 1.9, respectively. Formylchitin was prepared using formic acid and methanesulfonic acid and degree of substitution was 1.4. The degrees of substitution were determined by elemental analyses. Henceforth, the degree of acyl group substitution per N-acetylglucosamine residue will appear before the chitin derivative, i.e., 1.9 propionylchitin or 1.8 butyrylchitin. Table I illustrates the structural formulas of these acylchitins.

Table I. Formulae of acyl chitins

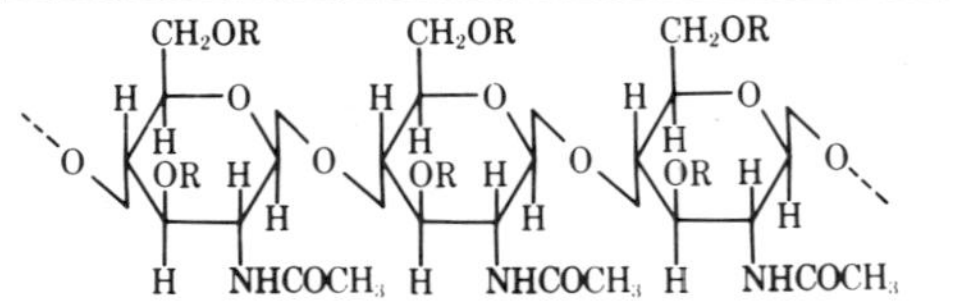

Chitin derivative	Acyl Group (R)
Formylchitin	- COH
Acetylchitin	- $COCH_3$
Propionylchitin	- $COCH_2CH_3$
Butyrylchitin	- $CO.(CH_2)_2.CH_3$
Caproylchitin	- $CO.(CH_2)_4.CH_3$
Caprylchitin	- $CO.(CH_2)_8.CH_3$
Lauroylchitin	- $CO.(CH_2)_{10}.CH_3$
Benzoylchitin	- COC_6H_5

Contact angle measurements

An Elma Model M-2010 goniometer was used for the contact angle measurements. The sessile drop method using water ethanol mixture of various proportions was utilized at 25°C. The surface tension values for the water ethanol mixtures were obtained from the literature. (11). The relationships $\cos\theta$ vs. γ_L were plotted, and the critical surface tension (γ_c) was determined for each acylated chitin surface. (12). Furthermore, the dispersive and nondispersive contributions to the surface energy of each acylchitin were calculated according to the following equation (13):

$$\frac{\gamma_L(1+\cos\theta)}{2(\gamma_L^d)^{1/2}} = (\gamma_s^d)^{1/2} + (\gamma_s^p)^{1/2}(\gamma_L^p/\gamma_L^d)^{1/2} \qquad (1)$$

As the values γ_L, γ_L^d, and γ_L^p were obtained from the literature and θ was a measured values, eq. 1 could be treated as a linear equation. When the measured values of $\gamma_L(1+\cos\theta)/2(\gamma_L^d)^{1/2}$ and $(\gamma_L^p/\gamma_L^d)^{1/2}$ were plotted, a straight upward line was obtained. The intercept$(\gamma_s^d)^{1/2}$ and slope$(\gamma_s^p)^{1/2}$ were obtained using a least-squares fit.

Clotting time measurements

In vitro thromboresistant properties were measured with human ACD blood using the kinetic method investigated by Imai and Nose (14) and improved by Komai and Nose (15). Onto the watch glasses coated with the acyl chitins, ACD blood (0.2 ml) was pipetted and 0.2 M calcium chloride (0.02 ml) was added simultaneously to initiate blood clotting. At predetermined times, the contents of each watch glass were emptied into a beaker containing distilled water (100 ml). The red blood cells not entrapped in the formed thrombus were emolyzed, and the free emoglobin was dispersed in solution. The concentration of the free hemoglobin in the resulting solution was colorimetrically measured at 540 nm. The variation of optical density with time was plotted, and the clotting times were estimated for each acyl chitin specimen, on glass and siliconized glass as control materials (watch glass treated with chromic acid, siliconized watch glass with Pierce dry film SC-87). Two or three acyl chitin samples were tested at each predetermined time.

Plasma protein absorption

The amounts of absorbed plasma proteins onto the acyl chitin film surface were measured by the following procedure. Blood plasma was obtained by centrifugal separation (3000 rpm) of fresh human blood containing ACD solution (1/10 volume). The Petri dish (85 mm diameter) coated with the acyl chitin was placed in water bath regulated at 37° C. A sample of fresh human plasma (5 ml) was added to the Petri dish and was incubated for 10 min. Then the plasma was discarded and sample surface was rinsed with 4 portions (50 ml) of 1 % sodium dodecylsulfate (SDS) with occasional shaking for 2 days. Each 0.1 ml of SDS solution containing plasma proteins which were released from the sample surface was applied to polyacrylamide gels, which were cast from 7.5 % acrylamide, 0.2 % methylene bis acrylamide, 0.07 % ammonium persulfate, 0.03 % N,N,N',N'-tetramethylenediamine, 0.38 M Tris, and 0.06 M HCl. Electrophoresis was performed in a disc gel apparatus at 3 mA per tube until the Bromophenol Blue had traveled about 90 % of the gel length. After marking the tracking dye position, the gels were stained for 1 h in 0.025 % Coomassie Brilliant Blue G-250 in methanol, acetic acid and water (5:1:5). Gels were stored in 7 % acetic acid with several solvent changes during storage in order to remove the excess dye distributed in gels. The stained proteins were then characterized and quantitated at 550 nm.

In vitro absorbability test on 2.0 acetyl chitin (2.0 A. Ch) coated XAD-4 resin

XAD-4 Resin coated with 2.0 A. Ch was evaluated for its blood compatibility and detoxifying compatibility to be applied for a hepatic support.

To 2% solution of 2.0 A. Ch in formic acid/dichloroacetic acid (50/2 V/V), 18.2g of XAD-4 resin was suspended for 2h with gentle stirring, filtered dried under IR irradiation, and stored in ethanol for 48h. Final product was stored in water ethanol mixture (50/50 V/V).

To a Pentobarbital solution (100 ml, 200 mg/l, M.W. 225) in phosphate buffer saline, coated or non-coated XAD-4 (1.5 g) was suspended for 2 h at 37° C with gentle shaking. The concentration of Pentobarbital in the mother liquid was measured by fluorescence spectrometry at predetermined times. In the case of Indocyanine Green (M.W. 924), absorption was tested as described above; concentration was 10 ml/l.

In vivo hemoperfusion studies

Closed circuit direct hemoperfusion in dogs with obstructive juandice were performed. The time course of absorbed component were obtained by occasional blood sampling. Bilirubin contents, platelet count, and white blood cell count were measured at the clinical analysis center of Hokkaido University Hospital.

RESULTS AND DISCUSSION

Surface properties of acyl chitins

Figures 1(a) and 1(b) show the relationship between the surface tension of the liquid (γ_L) and the cos θ values.

The critical surface tension values of the acylated chitins were all in the range of 20-30 dyn/cm. The Zisman plots were not linear because of the high nondispersive force contribution of the water-ethanol solvent mixtures. Akers et al. (13) have suggested that surfaces with critical surface tension

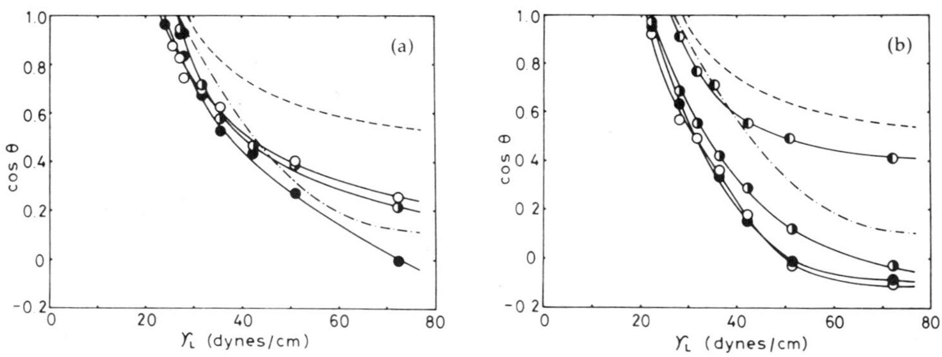

Figure 1. (a) The Zisman plot of the contact angle data: formyl(◑);propionyl(●);butyryl(○);chitin(---);acetyl(-·-). (b) caproyl(◑);capryl(●); lauroyl(○);benzoyl(◐);chitin(---);acetyl(-·-).

between 20 and 30 dyn/cm should exhibit good biocompatibility thromboresistance. In the measurement of the respective contact angles for the acylchitin, the contact angle values gradually changed because of evaporation of the liquid drop on the acylchitin surface and its solvation into the samples. In the case of the water-ethanol solvent mixture, the solvent ratio of the drop changed by evaporation and solvation, and therefore it is expected that the surface tension of the liquid would change. To better appreciate this phenomena and obtain a more accurate contact value, the time course of the contact angle values was measured and the values at zero time were determined by extrapolation. The dispersion or nondispersion force contributions of the surface tension values were calculated according to eq. 1 , and the various values derived from the contact angle measurement are summarized in Table II. As can be seen in Table II, acylation reduced the critical surface tension of chitin, and the nondispersion component (γ_s^d) decreased with increasing length of the acyl group. This effect, however, was also dependent on the degree of substitution.

Figure 2 shows the relationship between the nondispersion (γ_s^p) and dispersion (γ_s^d) force components of the respective surface tension value acylchitin surfaces. The surface tension of chitin had a large nondispersion force contribution, while the surface tension of acylchitins had large dispersion force contributions. The dispersive component value of 2.0 acetylchitin was 20 dyn/cm, which is the largest dispersive component of the surface tension in this series of acylchitins.

Introduction of the longer aliphatic side chain into the chitin molecule resulted in the regular decrease in the nondispersion force component. The dispersion force component value was irregularly ordered with the chain length of the acylchitins. In Figure 2, the large dashed line circle is illustrated as the zone of good thromboresistance suggested by Akers et al.(13). This zone was calculated using the GA.UCV and Dardik Biografts, which have been used widely in vascular allograft substitution and show excellent thromboresistant properties. The γ_s^d and γ_s^p values of 2.0 acetylchitin place this material nearest to the zone of good thromboresistance, and this material is expected to exibit a longer clotting time. The 1.8 butyrylchitin showed unusual behavior in both dispersion and nondispersion force components of the surface energy.

Anticoagulant properties

The clotting time on the respective acylchitin surfaces were compared to those of the control surfaces (glass and siliconized glass) and these values are summarized in Table III.

Table II. Surface tension values for acyl chitin films

Sample	Contact angle with water, degrees	γ_s^d (a) dyn cm-1	γ_s^p dyn cm-1	γ_s dyn cm-1	γ_c (b) dyn cm-1
Chitin	57	4.6	29.1	33.7	29
1.4 Formylchitin	77	14.4	9.9	24.3	27
2.0 Acetylchitin	83	20.3	6.3	27.3	27
1.9 Propionylchitin	93	18.2	5.6	23.8	24
1.8 Butyrylchitin	75	9.0	14.6	23.6	23
2.0 Caproylchitin	92	16.2	4.6	20.8	22
1.8 Caprylchitin	95	17.5	2.3	19.8	21
1.9 Lauroylchitin	96	17.6	2.5	20.1	21
1.9 Benzoylchitin	66	8.3	19.4	27.6	27

(a) γ_s^d and γ_s^p are the dispersion and nondispersion force components of the surface tension on the acylated chitin surfaces, respectively.
(b) γ_c is the critical surface tension from the Zisman plot.

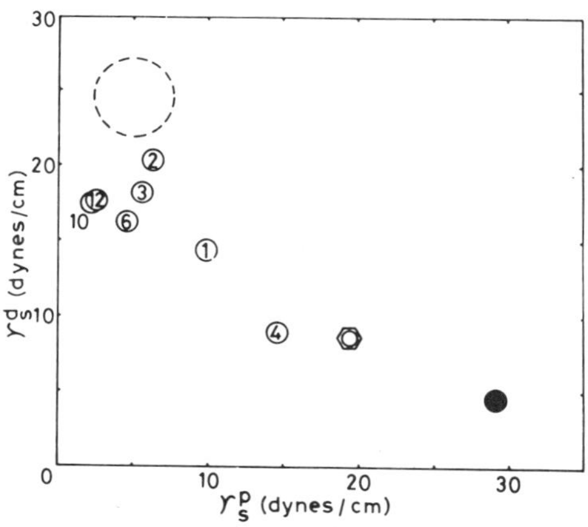

Figure 2. Relationship between (γ_s^d) and (γ_s^p):chitin(●);formyl(①);acetyl(②);propionyl(③);butyryl(④);caproyl(⑥);capryl(⑩);lauroyl(⑫);benzoyl(⬡).

All acylated chitins showed better anticoagulant properties than the original chitin. Formyl, propionyl, butyryl, caproyl, capryl, lauroyl and benzoylchitin showed similar clotting time to siliconized glass suggesting that those materials are considered to have good thromboresistant properties. The 2.0 acetylchitin was particularly interesting as its clotting time was twice as long as siliconized glass and three times as long as the original chitin. Blood coagulation on the surfaces of materials depends on several factors. Surface characteristics of the material itself and the blood-material interface reaction are primarily important. Contact angle and blood-clotting time measurements were applied for the screening of the thromboresistant surfaces; combination of these methods was quite effective for that purpose, and we found that the 2.0 acetylchitin showed promise for use as a blood contact surface. As one of the characteristics of the materials, molecular motion in acylchitin films were investigated by ESR measurement with 2,2,6,6-teatramethyl-4-hydroxypiperidine-1-oxyl(TENPO) as spin-probe. Introduction of a longer side chain into the chitin molecule resulted in a remarkable lowering of the temperature at which the separation between the extremes of the ESR spectrum decreases abruptly (T_N). Below and above T_N , two kinds of motional process were observed, one at a lower temperature having a smaller activation energy and the other at a higer temperature with a higher activation energy. The activation energy of the latter process (large scale motion: rotation) decreased with increasing chain length, except for 2.0 acetylchitin. Fully acetylated chitin films were anomalous with respect to the value of the activation energy. Corresponding to this, an anomaly was found in the wetting chracteristics and blood clotting time of 2.0 acetylchitin mentioned above. At present, no adequate explanation of these anomalies is available. The 2.0 acetylchitin contains no free hydroxy groups, and the hydrophobic interaction among molecules would therefore be smaller than the other acylchitin having longer aliphatic side chains. This fact may lead to the formation of a specific structure on a solid surface. Anomalies found in this work may be attributed to this specific surface structure.

Plasma proteins adsorption

The electrophoresis patterns after the extraction of adsorbed plasma proteins with SDS are shown in Figure 3. The peaks of albumin were clearly isolated and showed remarkable difference depending upon the acylation. The amount of adsorbed albumin was found to be increased according to the length of acyl group of chitin derivatives.

Table III. Relative clotting time index

Sample	Clotting time ratio	
	Sample/Glass	Sample/Siliconized glass
Chitin	1.3	0.7
1.4 Formylchitin	1.6	1.0
2.0 Acetylchitin	3.8	1.9
1.9 Propionylchitin	2.4	0.9
1.8 Butyrylchitin	2.0	1.2
2.0 Caproylchitin	2.5	1.1
1.8 Caprylchitin	2.5	1.1
1.9 Lauroulchitin	2.5	1.1
1.9 Benzoylchitin	2.6	1.0
Glass	1.0	--
Siliconized Glass	--	1.0

Also, the peak which presumed to be transferred appeared in the pattern of butyrlyl, caproyl, capryl, lauroyl, and benzoylchitins. The amounts of adsorbed albumin which were estimated from electrophoresis patterns were summarized in Table IV.

The amount for lauroylchitin (0.11 $\mu g/cm^2$) corresponds to half the amount which is calculated for a monolayer in side-on adsorption of albumin (0.25 $\mu g/cm^2$), though the consideration of other proteins would be necessary. A 0.02 $\mu g/cm^2$ for 2.0 acetylchitin seems to be appreciably small. Any other proteins could not be detected from the electrophoresis pattern carried out by low gel concentration even though many peaks of the other proteins appeared in the pattern for lauroylchitin. These results suggested that the plasma

Table IV. Adsorption of albumin onto acyl chitins

Sample	Surface Concentration, $\mu g\ cm^{-2}$
Chitin	0.02
1.4 Formylchitin	0.01
2.0 Acetylchitin	0.02
1.8 Butyrylchitin	0.06
2.0 Caproylchitin	0.05
1.8 Caprylchitin	0.06
1.9 Lauroylchitin	0.11
1.9 Benzoylchitin	0.04
Glass	0.07

proteins adsorbed onto 2.0 acetylchitin film surface at a remarkably low level. Normally, blood clotting on material surface starts with plasma protein adsorption in a few seconds after the contact of the surface with blood. In the next few minutes, components of adsorbed proteins are converted to make a stable interface between blood and material surface. Activation of the blood clotting system is initiated by the contact of F-XII with this surface and adhesion or aggregation of platelet takes place at the next step. In this experiment, we found such plasma protein adsorption could not be initiated on the 2.0 acetylchitin film surface. It is suggested that the activation of the material surface should be done by direct contact of blood components with material film surface.

In vitro adsorbability test

The time course of pentabarbital adsorption by the XAD-4 derivatives are shown in Figure 4(a). Adsorption profiles of uncoated and 2.0 acetylchitin coated resins were found to be very similar indicating that the adsorbability cannot be interfered by the 2.0 acetylchitin coating. As can be seen in Figure 4(b), remarkable differences between types of coating materials and the adsorbability of indocyanine green are noted. Remaining indocyanine green contents in the mother liquid indicate that the adsorbability of 2.0 acetylchitin is quite similar to that of uncoated XAS-4. It should be quite an interesting finding that difference of molecular weights in the reagents (Pentbarbital and Indocyanine green) influenced adsorption profiles to such an extent.

In vivo hemoperfusion studies

Reduction profiles of serum bilirubin during hemoperfusion in dogs with obstructive jaundice is illustrated in Figure 5; 33% of bilirubin was removed by 2.0 acetylchitin coated and uncoated XAD-4 after 2h. PBLG coated resin adsorbed bilirubin half as much as 2.0 acetylchitin or uncoated resin as in the case of the in vitro indocyanin adsorption test. The serum bilirubin is one of the typical toxic materials in the blood of patients which hepatic failure and well known to be combined with carrier protein in the blood. Comparing with adsorbability test in vitro, the reduction of serum bilirubin (M.W. 585) in dogs with obstructive jaundice looks very low. As the molecular weight of bilirubinprotein conjugate is considered to be around 70,000 (17), efficiency of the adsorption could not be regarded as low. Fig. 6(a) shows the platelet counts in the XAD-4 hemoperfusion. As can be seen in the Figure specific recovery of the platelet count during the hemoperfusion with 2.0 acetylchitin coated XAD-4 was observed. In this case, platelet count decre-

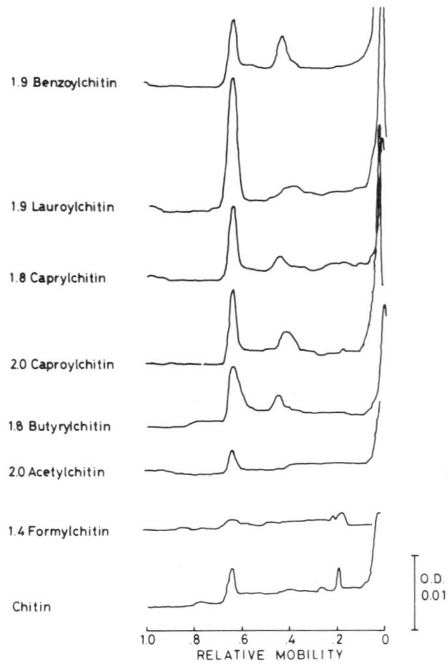

Figure 3. Electrophoresis patterns after the extraction of adsorbed plasma proteins with SDS.

ased and showed a minimum 30 minutes after the hemoperfusion started. PBLG and cellulose acetate coated resin also showed a minimum after the same period and similar recovery profiles in platelet count. Total recovery ratios for both materials in 2h hemoperfusion were 60.5% $\pm$ 8.7 for PBLG, cellulose acetate and uncoated, and 112.1% $\pm$ 13.1 for 2.0 acetylchitin respectively. Figure 6(b) shows the white blood cell count (WBC) in the XAD-4 hemoperfusion. Depressions of the WCB count were found in all of the materials showing a minimum at the 1h hemoperfusion. Final adsorption ratios were 55.6%, 58.3%, 43.5% and 23.3% for PBLG, 2.0 acetylchitin, cellulose acetate respectively. Only 2.0 acetylchitin showed recovery of the WBS count after 1 h hemoperfusion.

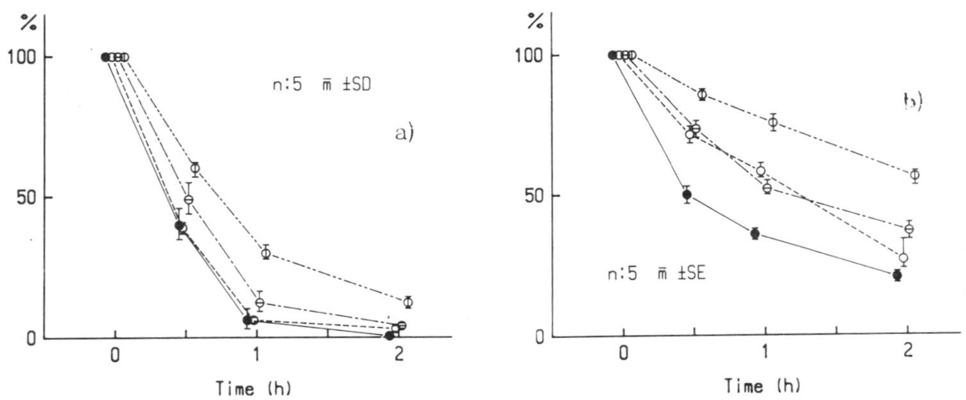

Figure 4. in vitro Adsorption profile: (a) Pentbarbital. (b) Indocyanine Green. PBLG(-Φ-);Cellulose Acetate(-θ-);Acetylchitin(-O-);Uncoated(-●-)

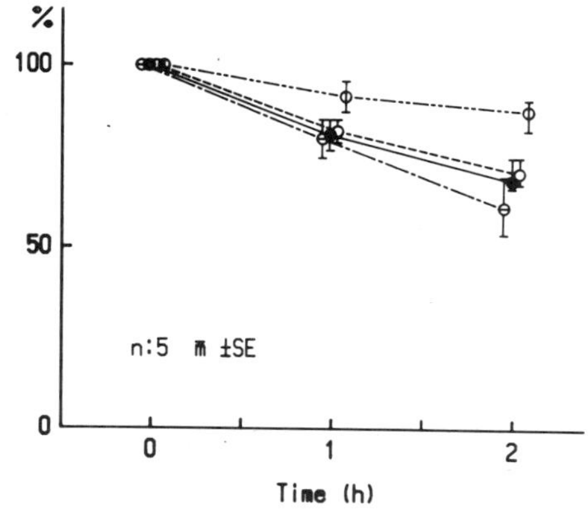

Figure 5. In vivo adsorption of bilirubin: acetylchitin(○); cellulose acetate(⊖); PBLG(⏀); uncoated(●).

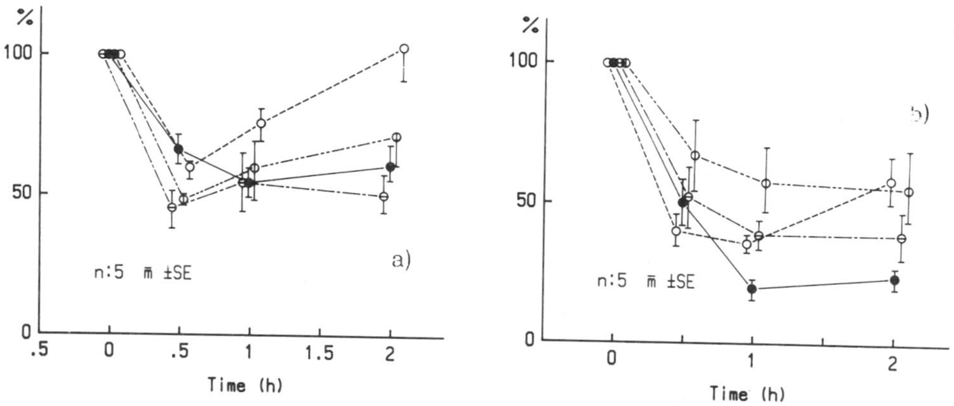

Figure 6. The time course of platelet (a) and WBC (b) counts during hemoperfusion. Cellulose acetate(⊖); Acetylchitin(○); PBLG(⏀); Uncoated(●).

From scanning electromicroscopic observation the surface of 2.0 acetylchitin, coated on XAD-4, was found to be a sponge like and porous structure, and the diameter of an opening was estimated as 0.2-0.3 mm. This structure seems to prevent the direct contact of cell fraction in blood with XAD-4 resin surface. As in vitro tests proved that the small M.W. compounds could be adsorbed with 2.0 acetylchitin coated resin to the same extent as uncoated resin, the porosity of the 2.0 acetylchitin should make it suitable as coating materials for XAD-4 hemoperfusion. Regarding the recovery of the platelet adsorption during hemoperfusion, the following mechanism would be proposed: First, platelets contact with 2.0 acetylchitin surface directly, without any competitive adhesion with plasma proteins, and in the next step, fixation of the platelets, adsorbed on the material, should occur without any progressive adhesion aggregation and release reaction of platelets. Then, fixed platelets start to be released 30 minutes after the first contact on the material. At this moment, homeostasis, (thermodynamical equilibrium in biology) would be recovered, and blood components should not recognize the material surface as foreign. Thus we can investigate the best material to regulate platelet adhesion and prevent acute reaction of blood with foreign material, and hope to apply this in clinical trials.

REFERENCES

1. T. Komai, Y. Sawada, Y. Onda, I. Koshino, Y. Nakanishi, A. Nashimura, and Y. Kasai, Umbilical Cord Vessels as a Bioprostheses for Vascular Substitute, Artif. Organs., 2 (Suppl):155 (1978).
2. T. Komai, K. Kaifu, and K. Nakamura, Biomedical Evaluation of Glutaraldehyde Treated Umbilical Cord Vessels, J. Jap. Soc. Strength Fracture Mater., 15:109 (1980).
3. M. Nakazono, T. Komai, K. Koiso, G. Picha, R. Kiraly, and Y. Nose, New. Approach for Experimental Substitution of Bladder Tissue, Trans. Am. Soc. Artif. Organs., 16:376 (1973).
4. K. Koiso, T. Komai, and T. Niijima, Experimental Urinary Bladder Reconstruction Using a Synthetic Poly (α-Amino Acids) Membrane, Artif. Organs., 7:232 (1983).
5. T. Komai, S. Yamada, T. Kawasaki, E. Yano, K. Kaifu, and K. Itouama, Wetting Characteristics and Blood Clotting on Surfaces of Immobilized Hyalunoic Acid, Heparin and Chondroitin Sulfuric Acid with Chitosan, Artif. Organs., 8: (1984) in press.
6. T. Komai, E. Yano, T. Kawasaki, and S. Haschizume Poly(O-Acyl-Hydroxy-L-Proline) (IV), Polym. J., 16:313 (1984).
7. I. Koshino, M. Matsushita, H. Sakamoto, M. Kondo, T. Kawamata, T. Kon, Y. Kasai, and T. Komai, Biphasic Approach to an Artificial Liver, Artif. Organs., 3 (Suppl):262 (1979).
8. K. Kaifu, N. Nishi, T. Komai, S. Tokura, and O. Somorin, Studies on Chitin. V. Formylation, Propionylation and Butylation of Chitin, Polym. J., 13:241 (1981).
9. K. Kaifu, N. Nishi, and T. Komai, Preparation of Hexanoyl, Decanoyl and Dodecanoylchitin, J. Polym. Sci., Polym. Chem. Ed., 19:2361 (1981).
10. N. Nishi, J. Noguchi, S. Tokura, and H. Shiota, Studies on Chitin. I, Acetylation of Chitin, Polym. J., 11:27 (1979).
11. J. R. Dann, Forces Involved in the Adhesive Process, I. Critical Surface Tension of Polymeric Solids ad Determined with Polar Liquids, J. Colloid Interface Sci., 32:302 (1970).
12. W. A. Zisman, Relation of the Equilibrium Contact Angle and Solid Constitution, in: "Contact Angle, Wettability, and Adhesion", F. M. Fowkes, Ed., Adv. Chem. Ser., No. 43, Am. Chem. Soc., Washington, (1964).
13. C. K. Akers, I. Dardik, H. Dardik, and M. Wodka, Computational Method Comparing the Surface Properties of the Inner Walls of Isolated Human Veins and Synthetic Biomaterials, J. Colloid Interface Sci., 59:461 (1977).
14. Y. Imai, and Y. Nose, A New Method for Evaluation of Antithrombogenicity of Materials, J. Biomed. Mater. Res., 6:165 (1972).
15. T. Komai, and Y. Nose, Thromboresistant Properties of Poly(α-Amino Acids), Artif. Org. Jap., 4:114 (1975).
16. K. Kaifu, T. Komai, and A. Tsutsumi, Studies of Molecular Motions in Acylchitin Films by Means of Spin Probe Technique, Polym. J., 14:803 (1982).
17. L. Zieve, Metabolic Abnormalities in Hepatic Coma and Potential Toxins to be removed, in: "Artificial Liver Support", R. Williams and I. Murray-Lyon, eds., Pitman Med., Kent, (1974).

CHITOSAN IN DENTISTRY

P. L. SAPELLI, V. BALDASSARRE,
R. A. A. MUZZARELLI and M. EMANUELLI

Faculty of Medicine, University of Ancona
I-60100 Ancona, Italy

INTRODUCTION

While the proposed uses of chitosan in medicine are numerous (1 - 7), no applications in dentistry have been reported so far. This might be due to the characteristics of the oral cavity, where various bacteria are constantly present and where lysozyme, a chitin degrading enzyme, occurs in the salivary secretion.

In parodontology, remarkable advantages attainable with the use of chitosan, either in the form of a membrane or as powder, are: decreased subjective symptomatology, good hemostatic action, delayed release of antibiotics, wound healing acceleration and better conditions for asepsis.

The present paper reports some clinical applications of chitosan, specially on the palatal donor site of the free flap, on the dental alveoli, on the large periodontal intrabone pockets, in various surgical mucogingival operations and after tooth extraction.

EXPERIMENTAL

Chitosan powder and membranes

Crab chitosan with a degree of acetylation of 27.5 % with low inorganic content and free from proteins was used in the form of a powder milled to pass 100-mesh and 200-mesh sieves. One hour before use, the chitosan powder was suspended in an antibiotic solution selected according to the patient's anamnesis and kept in a syringe (Figure 1). Just before use, the liquid's excess was expelled from the syringe, the piston removed and the sterile chitosan paste collected with the aid of a spatula for immediate application on the wounds. Chitosan membranes were cast from a 2 % solution in acetic acid (1 %). After filtration on a fritted glass filter and deaeration, the solution (15 ml) in a Petri dish was evaporated (35° C, 60 h) and the membrane was neutralized with sodium hydroxide (1 M) and with distilled water. This flexible and transparent membrane was easily applicable when soaked in the antibiotic solution. Membranes with various thicknesses (50, 100 and 200 μm) were prepared.

Figure 1. Chitosan powder suspended in an antibiotic solution and kept in a syringe before use. The excess liquid will be expelled through the needle and the paste taken with a spatula after removal of the piston.

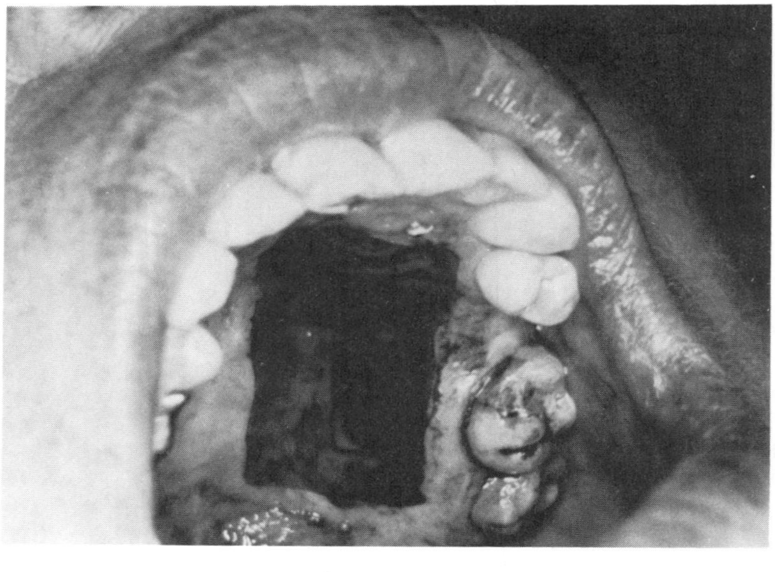

Figure 2. A chitosan membrane applied to the donor site. The membrane was preliminarily soaked in an antibiotic solution.

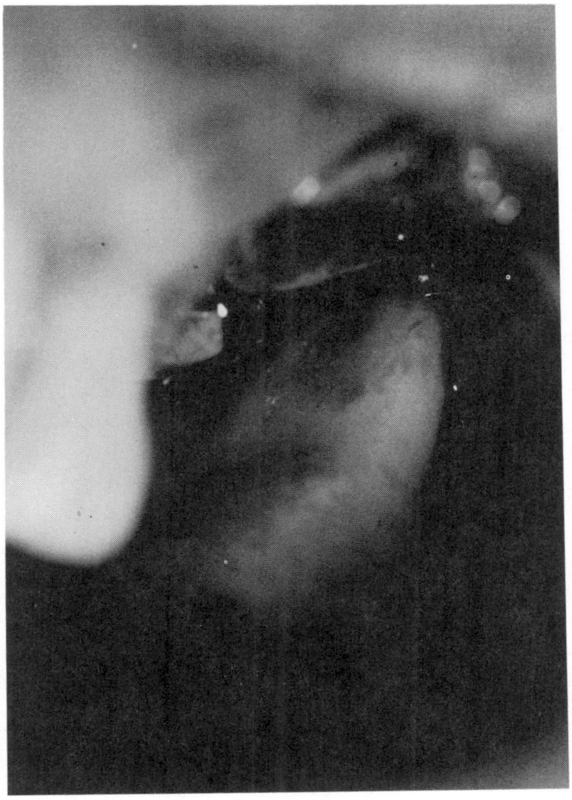

Figure 3. Periodontal intrabone pocket of large dimensions.

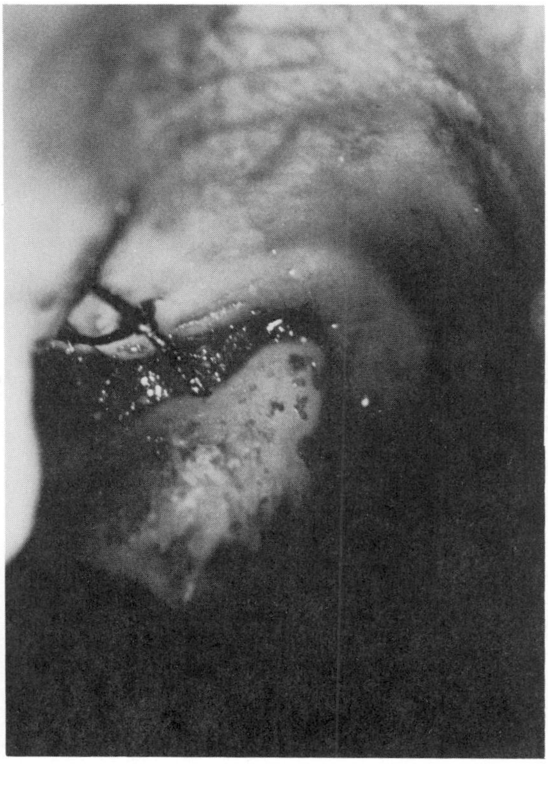

Figure 4. Chitosan powder applied within the intrabone pocket, with sutured flaps.

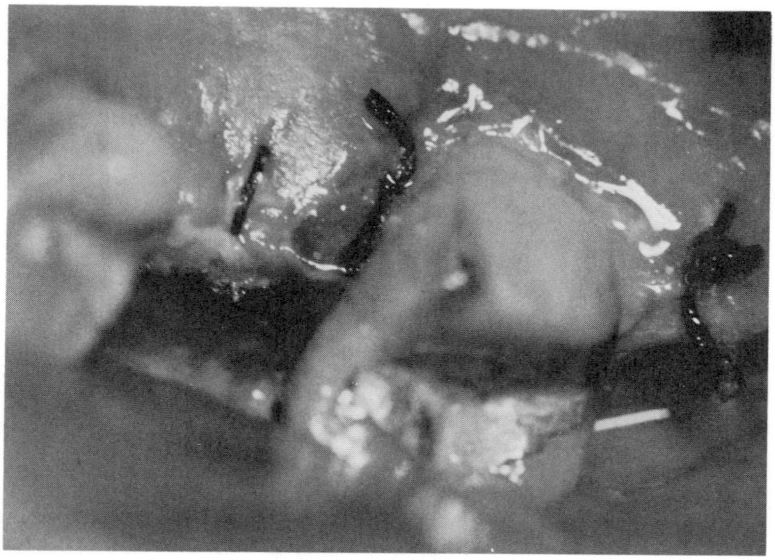

Figure 5. The wound after removal of the surgical paste.

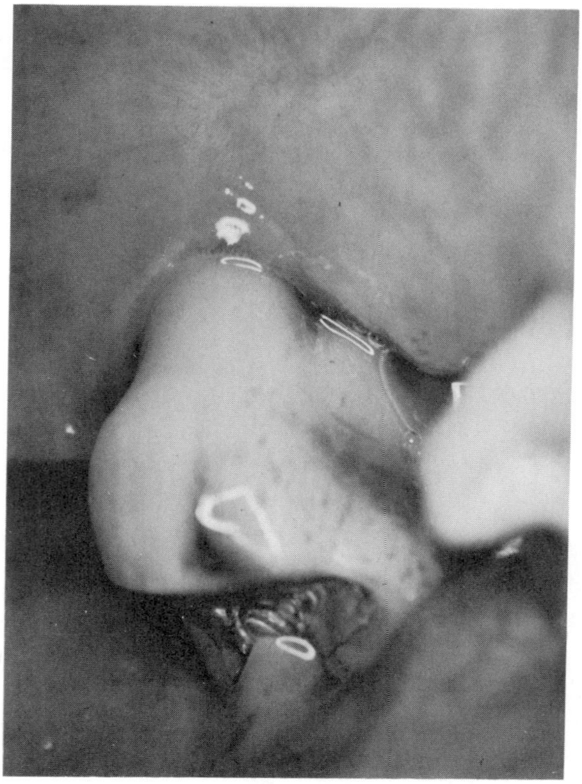

Figure 6. The healed surgical wound, showing a well-structured epithelial attachment.

RESULTS AND DISCUSSION

Chitosan membranes

Four patients (aged 20 - 50) who needed palatal grafts to treat gingival recessions were medicated with chitosan membranes (100 μm). After removal of the free flap from the donor site, the surgical wound was covered with the membrane, which immediately promoted hemostasis, perfectly visible due to the transparency of the membrane itself. A conventional surgical paste was then applied. At the time of removal of the paste, an uniform fibrin layer was visible; the peripheral part of the membrane, exceeding the area of the wound had disappeared, the membrane exactly matching the size of the wound. Due to its transparency, the healing of the wound could be observed directly (Figure 2). Even in the absence of the surgical paste, the wound remained isolated and protected, thus permitting the patient to eat normally. In no cases infections occurred. Complete healing was obtained within ca. 13 days.

Thicker chitosan membranes (200 μm) were too rigid to adapt themselves to the wound, therefore they were kept in place with sutures or with cyanoacrylate (Histoacril) or with fibrin (Tissucol). These thicker membranes, however, had the tendency to become more rigid, with the consequence that the fibrin underneath did not adhere to the chitosan. They could not be kept in place long enough to permit healing. By the other hand 50 μm membranes were too thin and were rapidly degraded; therefore, 100 μm membranes were preferred.

Chitosan powder

Twenty patients with various affections were treated with chitosan powder; they were daily examined until complete recovery. All of them recovered in spite of the serious initial conditions. In the case of tooth extractions, the chitosan powder deposited into the dental alveolus promoted clot formation without retraction. The epithelialization of the gingival margins occurred within a short period of time and maintained the original size of the clot. Chitosan powder was similarly deposited into periodontal intrabone pockets of remarkable sizes (Figures 3, 4 and 5). Complete recoveries were obtained, with reduced dental mobility and with the formation of an epithelial attachement at the same level at which the marginal gingiva was positioned (Figure 6). The healing of the sutured gingival flap had evidently been favored by chitosan. No allergic reactions were observed in the patients. Future observations will eventually give evidence of the possible formation of bone tissue rather than a deep epithelial attachment.

CONCLUSIONS

Chitosan has been used for the first time in dentistry, on a total of 24 patients, all of which recovered completely. No allergic reactions nor infections took place. Chitosan could be used as a transparent membrane or, preferably, as a thin powder, soaked in antibiotic solution; it accelerated wound healing, promoted regular fibrin formation and favored the epithelialization.

REFERENCES

1. R. A. A. Muzzarelli, "Chitin", Pergamon Press, Oxford (1977).
2. R. A. A. Muzzarelli, Heparin-like substances and blood-compatible polymers obtained from chitin and chitosan,In: "Polymers in Medicine", E. Chiellini and P. Giusti, eds., Plenum Press, London (1984).
3. R. A. A. Muzzarelli, Chitosan and its derivatives: new trends of applied research, Carbohydr. Polymers, 3:53 (1983).

4. L. L. Balassa and J. F. Prudden, Chitosan, a wound healing accelerator, In: "Proc. 1st Intl. Conf. Chitin/Chitosan", R. A. A. Muzzarelli and E. R. Pariser, eds., M. I. T., Cambridge, USA (1978).
5. W. G. Malette, H. J. Quigley, Jr., R. D. Gaines, N. D. Johnson and W. G. Rainer, Chitosan, a new hemostatic, Ann. Thor. Surg., 36:55 (1983).
6. G. Brandenberg, L. G. Leibrock, R. Shuman, W.G. Malette and H. J. Quigley, Jr., Chitosan, a new hemostatic agent for diffuse capillary bleeding in brain tissue, Neurosurgery 15:9 (1984).
7. W. G. Malette, H. J. Quigley, Jr., and E. D. Adickes, this Volume.

Section VII

CHITIN ECOLOGY

CHITIN PRODUCTION BY ANIMALS AND NATURAL COMMUNITIES IN MARINE ENVIRONMENT

C. JEUNIAUX, J. C. BUSSERS, M. F. VOSS-FOUCART and M. POULICEK

Oceanographic Reseach Station STARESO,
Calvi, Corsica

Laboratory of Animal Morphology, Systematics and Ecology
Zoological Institute, University of Liège
B-4020 Liège, Belgium

INTRODUCTION

Since the discovery of the remarkable properties of chitin and of its derivative, chitosan, offering a wide range of possible industrial and agricultural applications [1,2,3], the need of new and trusting sources of chitin for its isolation on an industrial scale was claimed.

At present, chitin is usually isolated from crab and lobster shells, owing to their high chitin content and to their availability in relatively high amounts from fisheries and tinned food industries. However, this kind of raw material offers some disadvantages and the exploitation of alternative sources of chitin has been proposed and investigated. These sources could be, for instance, marine plankton or biological cover of marine rocky shores. As with any prospect of exploitation, these potential sources of chitin require preliminary estimations of chitin availability and renewal, i.e. chitin biomass and chitin production in marine environment.

On a more theoretical and fundamental background, chitin is a highly chemically stable biopolymer widely distributed in animals and fungi, which must certainly play an important part in biogeochemical cycles of carbon and nitrogen. The importance of chitin from this ecochemical point of view relies on its production by natural biocenoses (or communities), on one hand, and on its biodegradation and bioaccumulation in sediments on the other.

Our knowledge of chitin biodegradation and accumulation in sediments has increased considerably, thanks to the works of authors such as Bolin [4], Goffinet [5,6], Poulicek [6-9], Seki [10], Smucker [4,11], Warnes [12], Wirsen [13], Voss-Foucart [15], some of them being able to present the most recent conclusions of their investigations in the present Volume. On the contrary, very little quantitative and experimental data on chitin production are available as yet [14,15].

The aim of the present paper is to try to estimate the biomass and production of chitin at the level of several biological communities living in a single, relatively well defined marine area, namely the bay of Calvi (Isle Corsica), in the Mediterranean Sea.

POSITION, TOPOGRAPHY AND MAIN ECOLOGICAL CHARACTERISTICS OF THE BAY OF CALVI

The bay of Calvi, situated at the north-west end of the Isle Corsica, in the Mediterranean Sea, is roughly semi-circular in shape, with a maximum width of 6 km. Widely open towards the North, the bay is delimited by a relatively abrupt rocky coast (mainly granitic and granulitic in nature [16]), interrupted near Calvi by a sand beach extending over about 3 km. The bay extends over a surface of about 2200 ha (5500 acres). With a mean slope of 2 %, the depth increases rapidly up to 100 meters.

Tidal movements are insignificant. Salinity varies from 37.5 ‰ in winter to 38.1 ‰ in autumn; surface water temperatures fluctuate from 13°C in february to 25°C in summer, with a thermocline located at 20 m in May and 50 m in September [17, 18].

Around the bay, the rocky benthic communities occupy a fringe from 0 to 27 m depth, estimated approximately to 145 ha, which represent an actual available surface of at least 190 ha. The biological benthic communities growing on the infralittoral rocky shore, (identified according to Peres and Picard, [19]) are mainly :

1) photophilous algae communities characterized by the brown-algae species *Cystoseira stricta*, *C. crinita* and *C. balearica*, in well enlighted conditions and
2) sciaphilous communities in crevices and semi-dark caves.

An important surface (about 44 %) of the bottom of the bay is taken up by a *Posidonia oceanica* meadow, from 5 m down to 38 m depth [17,18]. The remaining surface (about 50 %) is covered by different kinds of sandy, terrigenous and organo-detritic (bioclastic) sediments [16].

Owing to the poor development of human habitat, industrial and agricultural activities on the coasts of this part of the isle, the bay of Calvi is considered as only slightly polluted (oligotrophic waters), and especially suitable for approaching ecology and ecobiochemistry of non perturbated marine systems.

METHODS FOR QUANTITATIVE ESTIMATION OF CHITIN

In order to estimate the amount of chitin in crude, highly composite materials, a very specific method is needed. The unique highly specific technique allowing a quantitative estimation is the enzymatic method based on the use of purified chitinases [20,21,22]. Powerful purified chitinases can be prepared from cultures of bacteria (*Streptomyces antibioticus*, *S. griseus*, *Serratia marcescens*) according to the procedures of Berger et al.[23], Jeuniaux [24, 25] or Roberts et al. [26].

The enzymatic method for chitin determination [21], was applied as summarized as follows. The analysed material was washed, weighed ("calcified weight"), then decalcified by HCl 0.5 N at room temperature. After washing and drying ("decalcified weight"), the material was treated by 1N NaOH at 100°C during 3 h, in order to remove proteins, to liberate most of the bound chitin and thus to facilitate its enzymatic hydrolysis. After washing, the chitin in the insoluble residue was hydrolysed by a solution of purified chitinase (Koch-Light, 1 mg/ml, or Serva, 3 mg/ml) at pH 5.2 and 37°C, during 6 h, then incubated in dilute buffer at pH 5.2. These incubations in chitinase and buffer were repeated if necessary until "complete" hydrolysis was achieved. After addition of a solution of N-acetyl-D-glucosaminidase (or chitobiase) from lobster serum, the total N-acetyl-D-glucosamine liberated during the successive incubations was measured by colorimetric method [27].

The values given by this enzymatic method are systematically lower than the actual ones, due to the facts, that some chitin, strongly associated to other components, is able to withstand prolonged enzymatic action and that some acetylglucosamine residues in the chitin chain are replaced by non-acetylated residues, which are not taken into account in the colorimetric measure of the final hydrolytic products. The mean error is estimated to about 10 %.

When the studied material is made of Crustaceans only, chitin may be more readily estimated by weighing the residue after decalcification and hot alkali treatment. The values given by this simpler procedure are somewhat higher than the actual chitin amounts.

RESULTS

Chitin Biomass in Rocky Benthic Communities

In infralittoral communities of photophilous algae characterized by *Cystoseira spp*., the chitin biomass is principally due to sessile and incrusting colonies of Bryozoa and Hydrozoa, living on brown algae, on one hand, to vagile and crawling species of Crustaceans on the other hand. The whole benthic biological cover, excluding Decapod Crustaceans, contributes to a chitin biomass of 0.71 $g.m^{-2}$ (± 0.28). Decapod Crustaceans of medium size (taller than 5 mm. long) contribute to more variable chitin biomass values (from 0 to 1.8 $g.m^{-2}$; average 0.38 $g.m^{-2}$ ± 0.51) [28].

In sciaphilous communities inhabiting semi-dark caves [28], the chitin biomass of encrusting colonies is lower and amounts to 0.27 $g.m.^{-2}$ ± 0.1. The contribution of Decapod Crustaceans is very variable and may be sometimes very high (big species inhabiting holes and crevices).

Chitin Production by Infralittoral Rocky Benthic Communities of Photophilous Algae, estimated by a Study of Pioneering Communities Growing on Naked Substrates.

The rate of chitin production by infralittoral communities living on the rocky shore was estimated by measuring the amount of chitin accumulated after given periods of time by pioneering communities allowed to settle and grow on naked substrates submerged in sea water in the vicinity of the coast of the bay of Calvi.

Several sets of rectangular plates of granite, baked clay, glass and PVC, of the same size (10 x 20 cm) were submerged and fastened by Scuba divers at different depths (6 and 18 meters) and maintained in natural conditions. At given intervals (from 1 to 15 months), sets of plates were removed, and the whole biological cover of each face of each plate was separately scraped, dried, weighed and powdered, if necessary, by mechanical grinding ("Ika-werk" grinder). Aliquots of the powder were decalcified and used for determination of total organic matter and total chitin. The results are given in table I, in which individual values obtained for both faces of the same plates are pooled, and are calculated per square meter.

The species composition of the biological cover corresponds roughly, after a few months, to that of an infralittoral community of photophilous algae.

At a depth of 6 meters, the development of benthic pioneering communities (total dry matter production) was relatively similar on the different types of experimental substrates during the first 6-10 months, but was found to be much better on granite and baked clay thereafter.

Table I. Chitin production by pioneering communities growing on naked substrates, at 6 m depth (from June 1981 forward).

Substrate	chitin biomass, mg. m^{-2}				Calculated chitin production mg. m^{-2}. $month^{-1}$	
	Sept. 81 3 months	Jan. 82 7½ months	May 82 11 months	Sept. 82 15 months	winter	spring-summer
PVC	2.5	33.9	36.7	46.7	-	-
Glass	1.3	-	7.0	7.4	-	-
Baked Clay	1.7	15.8	19.2	136.	0.97	29.2
Granite (1)	2.3	12.8	22.1	99.0	2.65	19.2
(2)	2.3	12.8	22.1	298.0	2.65	69.0

(1) whole biological cover, Decapod Crustaceans excluded.
(2) whole biological cover, including Decapod Crustaceans.

As far as chitin is concerned, the quantitative data in Table I indicate that, when the naked substrates were immerged in June at a depth of 6 meters, chitin production proceeded slowly during the first 10 months, then increased considerably on granite and clay substrates. This difference may be explained either by season influence, or by the progressive constitution of the biological cover, including enhancement of colonization rate by the presence of a pioneering community. It is likely that the faster chitin production rate measured during the last 6 months in this experiment is a better estimation of the actual chitin production of a well settled community. As granite is a natural substrate more similar to those building the rocky shore of the bay than the other experimental substrates, chitin production by photophilous communities of the first 12 meters depth can be evaluated to 19.2 mg.m^{-2}. $month^{-1}$ from May to September, and to 69 mg. m^{-2}. $month^{-1}$ when Decapod Crustaceans are taken into account (table 1). The chitin production of pioneering communities varies to some extent with the season and seems to be more important from september to december. This point will be discussed more fully elsewhere.

Values for plates immerged at a depth of 18 meters are so far incomplete, but seem to indicate a higher chitin production during winter season, around 33.1 mg. m^{-2} $month^{-1}$ on baked clay and upto 51.7 mg. m^{-2}. $month^{-1}$ on granite.

Chitin Biomass and Production by Epiphytic Communities growing on Posidonia Leaves.

The primary and secondary production of epiphytic communities growing on the leaves of *Posidonia oceanica* in the *Posidonia* meadow of the bay of Calvi was studied by Mesureur [29,30], at a depth of 10 meters.

A study of the seasonal foliar renovation allowed consideration of leaf as a new substrate available for epiphytic communities, the age of which was identified by its position in the foliar bundle, and confirmed by experimental markings. The epiphytic biological cover of leaves was scraped, pooled according to age, weighed and powdered. Besides carbonates, organic matter and proteins, chitin was identified and measured using our enzymatic method.

The main producers of chitin were found to be Hydrozoa and Bryozoa. The chitin production by the epiphytic biological cover was estimated to 1.25 mg per m^2 of leaf surface per month during autumn and winter, and 0.5 mg per m^2 of leaf surface per month during spring and Summer. According to the numerical density of Posidonia foliar bundles (407 bundles per m^2 at a depth of 10 meters) [17,18] and to the number and age of leaves in each bundle, epiphytic chitin production was calculated with respect to the surface of sediment occupied by the *Posidonia* meadow on the ground of the bay. The annual chitin production was thus estimated to 75 mg. m^{-2}. year $^{-1}$.

Chitin Production by Zooplankton

The biomass and production of zooplankton in the bay of Calvi was studied by Dauby [31,32] during two annual cycles, and the results expressed with respect to a square meter of surface water, for the whole water column. The most stricking characteristic is that Copepod Crustaceans contributed to more than 90 % of the total zooplancton biomass, except during short periods, when Pteropod Mollusks and larvae of benthic Crustaceans became abundant.

The Copepod fauna is dominated by five species or groups of species belonging to the genus *Clausocalanus, Centropages, Oithona, Acartia* and *Calanus*. The Copepod production was estimated to about 20 g (dry matter). m^{-2}.year $^{-1}$.

On the basis of a mean proportion of chitin of about 5 % with respect to Copepod dry matter, the chitin production was roughly evaluated during a period of twelve months from March 1983 to February 1984 (Dauby, 32). The Copepod chitin production was low (2 mg. m^{-2}. day $^{-1}$) at the end of the winter, than increased considerably in spring (up to 17 mg.m^{-2}. day $^{-1}$); a further but lower peak (7 mg. m^{-2}. day $^{-1}$) was observed in July; the production was low (from 0.5 to 2 mg. m^{-2}. day^{-1}) during the rest of the year. The average annual chitin production was thus estimated to 1 g.m^{-2}. year^{-1}.

DISCUSSION AND CONCLUSION

The bay of Calvi can be taken as an example of a relatively well delimited and explored marine area from an ecochemical point of view. The contribution of different biocenotic compartments to chitin production can be tentatively calculated on the basis of the data reported or gathered in this paper, as summarized in Table II.

Crustaceans, benthic Decapod species as well as planctonic Copepods, appear as the main contributors to chitin production per surface unit and per year. Moreover, it must be recalled that larger species inhabiting holes and crevices, such as lobsters and big crabs, have not been taken into account, nor nectonic Crustaceans such as shrimps. Despite the abundance of other animal groups with chitinous exoskeletons, such as Hydrozoa, Bryozoa, Annelids and Mollusks, most of the chitin production in this marine system is accounted for by Crustaceans.

This is even more evident when chitin production is expressed for a whole marine area, such as the bay of Calvi in its entirety. In this case (table II), planctonic and benthic Crustaceans account for about 95 % of total chitin production in the bay, which is estimated to 24.8 Tons per year.

Table II. Chitin production by several communities in the bay of Calvi.

Community	chitin production $g.m^{-2}.\ year^{-1}$	Surface covered ha	Chitin production (whole bay) kg. year -1
Benthic			
- photophilous algae on rocky shores			
* 0-12 m (1)	0.23	76	175
*12-27 m (2)	0.62	114	707
* Small Decapod Crustaceans	0.60	190	1 140
- *Posidonia* meadow (epiphytes)	0.075	1 056	792
Planctonic			
- Zooplancton (Copepods)	1.0	2 200	22 000
Total	-	2 200	24 814

(1) On basis of values obtained for pioneering communities growing on granite plates in spring and summer, after 1 year's immersion at a depth of 6 m.

(2) idem, for communities growing on granite plates in autumn and winter, after a 6 months' immersion at a depth of 16 m.

Acknowledgments

Thanks are due to Miss Toussaint for technical assistance. This work was supported by a grant n° 2.4550.80 of the "Recherche Collective" of the "Fonds National Belge de la Recherche Scientifique".

BIBLIOGRAPHY

1. R.A.A. Muzzarelli, "Chitin", Pergamon Press, Oxford (1977).
2. R.A.A. Muzzarelli and E.R. Pariser, "Proceedings of First Internat. Conference on Chitin/Chitosan, MIT, Cambridge (1978).
3. S. Hirano and S. Tokura, "Proceedings Second International Conference on Chitin/Chitosan", Japanese Society for Chitin, Tottori (1982).
4. J.T. Bolin, R.A. Smucker and C.E. Warnes, Differential chitin/chitosan degradation in aquatic habitats. Abstr. Ann. Meet. Soc. Microb., N_1 (1982).
5. M.F. Voss-Foucart, J.C. Bussers, G. Goffinet, M. Poulicek, Cl. Toussaint and Ch. Jeuniaux, Etude préliminaire de la diagenèse précoce des carapaces de *Carcinus maenas* dans un sédiment marin. Altération ultrastructurale et chimique .Ann. Soc. Roy. Zool. Belg., 114, suppl. 1, 145 (1984).
6. M. Poulicek, M.F. Jaspar-Versali and G. Goffinet, Etude expérimentale de la dégradation des coquilles de Mollusques au niveau des sédiments marins .Bull. Soc. Roy. Sc. Liège, 50, 513 (1981).
7. M. Poulicek and Ch. Jeuniaux, Biomass and biodegradation of Mollusk shells chitin in some marine sediments, in "Proceedings of the 2nd International Conference on Chitin/Chitosan", S.Hirano and S. Tokura, eds., Japanese Soc. Chitin, Tottori (1982).

8. M. Poulicek, Patterns of Mollusk shell biodegradation in bathyal and abyssal sediments. J. Moll. Stud., Suppl. 12A, 136 (1983).
9. M. Poulicek, Shell matrix weathering in marine sediments . Malacologia, in the press.
10. H. Seki and N. Taga, Microbiological studies on the decomposition of chitin in marine environment. I-V. J. Oceanogr. Soc. Japan, 19, 27 (1963).
11. R.A. Smucker, Determination of chitinase hydrolytic potential in an estuary. in "Chitin and Chitosan", Hirano S. and Tokura S. ed., Japan. Soc. Chitin Chitosan, Sapporo, 135 (1982).
12. C.E. Warnes and C.I. Randles, Preliminary studies on chitin decomposition in lake Erie sediments. Ohio J. Sci., 77, 224 (1977).
13. C.O. Wirsen and H.W. Jannasch, Decomposition of solid organic materials in the Deep Sea. Environ. Sc. Techn., 10, 880 (1976).
14. H. Yamamoto and H. Seki, Impact of nutrient enrichment in a water-chestnut ecosystem at Takahama-Iri bay lake Kasumigaura, Japan. IV. Population Dynamics of Secondary Producers as Indicated by Chitin.. Water Air Soil Pollution, 12, 519 (1979).
15. P.F. Berry and M.J. Smale, An estimate of production and consumption rates in the Spiny Lobster *Panulirus homarus* on a shallow littoral reef off the Natal Coast, South Africa. Mar. Ecol. Prog. Ser., 2, 337 (1980).
16. M. Burhenne, Faciès sédimentaire du précontinent calvais. Bull. Soc. Roy. Sc. Liège, 50, 387 (1981).
17. D. Bay, "Etude de la production primaire d'un herbier de posidonies (*Posidonia oceanica* L.,Delile), baie de Calvi, Corse". Doctorat Thesis, Liège University, 251 p. (1978).
18. D. Bay, A field study of the growth dynamics and productivity of *Posidonia oceanica* (L.) Delile in Calvi bay, Corsica. Aquatic Botany, 20, 43 (1984).
19. J.M. Peres and J. Picard, "Nouveau manuel de biònomie benthique de la mer Méditerranée".Rec. Trav. Sta. Mar. Endoume, 31, 137 p. (1964).
20. Ch. Jeuniaux, "Chitine et Chitinolyse", Paris, Masson ed., (1963).
21. Ch. Jeuniaux, Chitine et phylogénie : application d'une méthode enzymatique de dosage de la chitine. Bull. Soc. Chim. Biol., 47, 2267 (1965).
22. Ch. Jeuniaux, Chitinases in "Methods in Enzymology, VIII, Complex carbohydrates", Neufeld E.F. and Ginsburg V. ed., Academ. Press, N.Y.,644 (1966).
23. L. R. Berger and D.M. Reynolds, The chitinase system of a strain of *Streptomyces griseus* . Bioch. Bioph. Acta, 29, 522 (1958).
24. Ch. Jeuniaux, Recherches sur les chitinases. I. Dosage néphélométrique et production de chitinase par des Streptomycètes. Arch. internat. Physiol. Bioch., 66, 408 (1958).
25. Ch. Jeuniaux, Recherches sur les chitinases. II. Purification de la chitinase d'un Streptomycète et séparation électrophorétique de principes chitinolytiques distincts. Arch. internat. Physiol. Bioch., 67, 597 (1959).
26. R.L. Roberts and E. Cabib, *Serratia marcescens* chitinase : one-step purification and use for the determination of chitin. Analyt. Biochem., 127, 402 (1982).
27. J.L. Reissig, J.L. Strominger and L.F. Leloir, A modified colorimetric method for the estimation of N-acetylamino sugars. J. Biol. Chem., 217, 959 (1955).
28. Ch. Jeuniaux, M.F. Voss-Foucart and J.C. Bussers, Preliminary results on chitin biomass in some benthic marine biocenoses. In: "Chitin and Chitosan", Proceed. 2d Intern. Conf. Chitin and Chitosan, Hirano and Tokura ed., Sapporo, 200 (1982).

29. B. Mesureur, "Un aspect écologique de l'herbier de Posidonies dans la baie de Calvi : productions primaire et secondaire de la couverture épiphytique". Thèse Licence Sc. Zoologiques, Univ. de Liège, 55 p. (1981).
30. B. Mesureur, Contribution à l'étude de l'herbier de Posidonies dans la baie de Calvi : quantification de la production épiphytique de la frondaison. Oceanologica Acta, submitted for publication (1985).
31. P. Dauby, Cycle annuel du zooplancton de surface de la baie de Calvi (Corse). Biomasse totale et plancton copépodien. Oceanol. Acta, 3, 403 (1980).
32. P. Dauby, "Dynamique et productivité de l'écosystème planctonique du golfe de Calvi-Corse". Thèse Doctorat, Univ. Liège, 277 p. (1985).

CHITIN DIAGENESIS IN DEEP-WATER SEDIMENTS

M. POULICEK, R. MACHIROUX* and C. TOUSSAINT

Laboratory of Animal Morphology, Systematics and Ecology,
University of Liège, 22, Quai Van Beneden,
B-4020 Liège, Belgium and

*Laboratory of Analytical Chemistry, University of Liège
Bat. B 16 Radiochemistry
B-4000 Sart-Tilman Liège, Belgium

INTRODUCTION

In the oceanic sphere, complete systems (producers, consumers and decomposers) exist almost exclusively in the epipelagic zone. It has been proposed for some time that food-chains on the deep-sea floor are sustained by the "organic snow" coming from the surface layers, where phytoplanktonic producers are living. Recent investigations (1) indicate that it is the relatively swift fall of faecal material together with molted cuticules (particularly from copepods), rather than the slow "perpetual snowfall of organic flakes" that provides most of the energy to the deep-sea benthos (with the possible exception of hydrothermal "oases" around tectonic spreading centers where chemolithotrophic microorganisms, similar to primary producers, are used as food source by suspension feeders).

Considering the high chitin content and production of planctonic communities (2), it was interesting to investigate the importance of chitin as nutrient and energy source for deep-water organisms. Moreover, considering the well-known binding properties of chitin (3), it seemed fruitful to approach chitin diagenesis in an oceanographic perspective.

CHITIN CONTENT OF DEEP-SEA SEDIMENTS

Over most of the oceans (97.6 % of the total amount of seawater lies over depths exceding 1000 meters), the deep-sea floor is covered, and indeed formed, by terrigenous or hemipelagic muds, the color of which is due to varying degrees of oxidation of the organic matter. Our estimations of chitin contents (by the enzymatic method of Jeuniaux (4)) of deep-sea muds are summarized in Table 1.

Except in the case of the bathylittoral sediments (5), the chitin content of the deep-water sediments is low, of the same order of magnitude as for shallow water sediments already examined (5, 6). Nevertheless, ZoBell and Rittenberg (7) estimated that several million tons of chitinous material

Table 1 : $CaCO_3$ and chitin content of some deep-sea sediments

Type of sediment	Depth (m)	$CaCO_3$ content (g/100 g sed)	Chitin content (µg/g decalcif. sediment)
BATHYLITTORAL FRINGE			
- Muddy sand	175 - 190	60.3 - 90.4	2808
BATHYAL AREA			
- Blue mud	440 - 1150	39.5 - 60.4	61.2 - 127.4
- Yellow sandy mud	1150 - 1600	34.7 - 58.2	92.7 - 157.3
ABYSSAL AREA			
- Black mud	2800 - 3600	21.1 - 46.5	63.0 - 134.8
- Red clay	4800	7.6	65.4

settle every year onto the bottom, mainly originating from copepod exoskeletons. In the bay of Calvi (Corsica, Mediterranean Sea), more than 90 % of the chitin yearly produced came from planktonic producers (2).

Most of this organic matter produced in the epipelagic zone, including most of the chitin input, is actually turned over in the water column (8,9), through bacteria adsorbed on sedimenting particulate organic matter. Some implications for chemical oceanography of this turnover in the water column will be discussed later.

A further biodegradation step takes place in the uppermost layers of the bottom sediment, namely the decomposition of skeletal pieces, heavier or more refractory to diagenesis.

CHITIN DIAGENESIS IN DEEP-WATER SEDIMENTS

A huge bulk of mineralized skeletal structures sheltering chitinoproteic complexes settle onto deep-sea sediments. In some areas, tests, shells,...can accumulate to a large extent (deep calcareous oozes).

Extraction of organic matter from calcareous skeletons, for example Pteropod or other Mollusk shells, occurs at any depth in marine sediments (10). Microborers are ubiquitous (6,10,11,12,13). In aphotic environment, Fungi (Oomycotina, Ascomycotina, Yeasts,...), colorless blue-green algae, bacteria and Streptomycotina...(Fig.1) are supposed to function heterotrophycally, and thus use the organic matrix of skeletal remains as nutrients and energy sources (11). Most of those endolithic organisms are living in close contact with the organic sheaths around crystallites (6). Extracellularly secreted enzymes would thus be able to hydrolyse the organic compounds of those sheaths, as shown in shallow depths (6,13). In that perspective, Figure 3 shows that there is a direct relation between the decreasing chitinoproteic content of weathering shells (for example Propeamussium meridionale shells dredged at 4830 meters deep in the "Mombasa trench"

Indian ocean) and the microborers density (up to 300.000 borings/cm^2) with a very good correlation coefficient (R = 0.95186, signification level sup. 0.01 %).

Despite the importance of the deep-sea ecosystem, the data about enzymatic degradation processes under high hydrostatic pressure are scarce. Kim and ZoBell (14) showed that chitinase was active and very stable under deep-sea conditions, and, recently, Koch and Disteche found it even slightly activated by pressure elevation (pers.comm.). These observations can explain the low chitin content of deep-sea sediments.

From an oceanographical point of view, the biodegradation processes of calcified skeletons, and thus the breakdown of chitin is also responsible for an increased dissolution rate of the biogenic carbonates, and thus has repercussions on chemical equilibriums of the carbonate phase : the organic sheaths around $CaCO_3$ crystallites partly protect them from dissolution in thermodynamically undersaturated media (6). After extraction of the organic matter, the calcified skeletons, less "protected" from inside, are more rapidly dissolved (6,10,11). Moreover, through their boring activity, microborers increase the surface accessible to dissolution : the ratio surface to volume (S/V) rapidly increases, up to 7 to 10 fold increase (Fig.4). This ratio is directly proportional to dissolution rates and thus its estimation will affect the calculated dissolution rates based on thermodynamical laboratory constants and sedimentation rates.

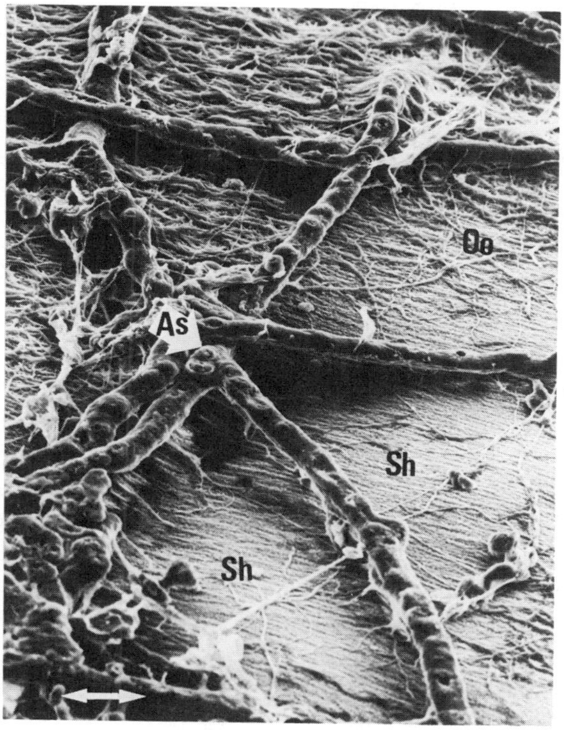

Fig. 1 : Scanning Electron Microscopy of preserved microorganisms boring into the shell of *Propeamussium meridionale* (4830m) As:Ascomycotina;Oo:Oomycotina;Sh:Shell. (Scale : 10 μm)

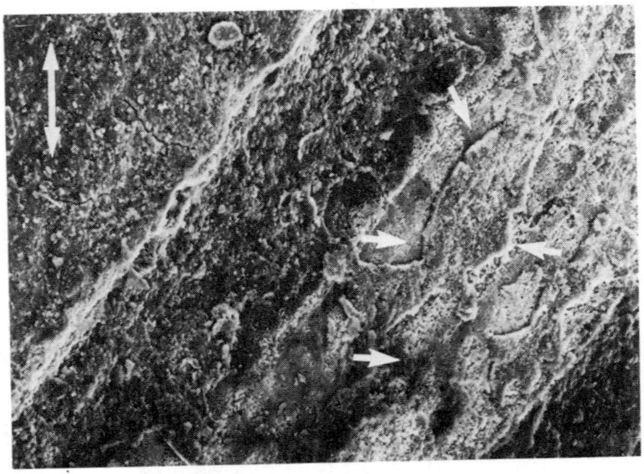

Fig. 2 : Continuous layer of aggregate metallic crystals on a *Katadesmia kolthoffi* shell (arrows). (Scale : 50 μm)

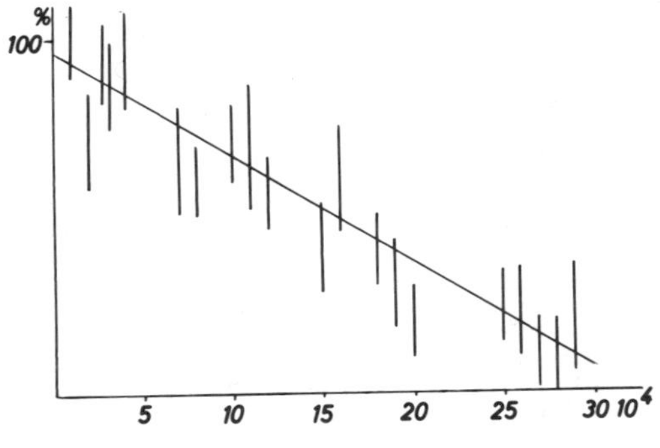

Fig. 3 : Relation between mean number of microborers (N/cm^2) and organic matter content of detritic abyssal shells (*Propeamussium meridionale*)

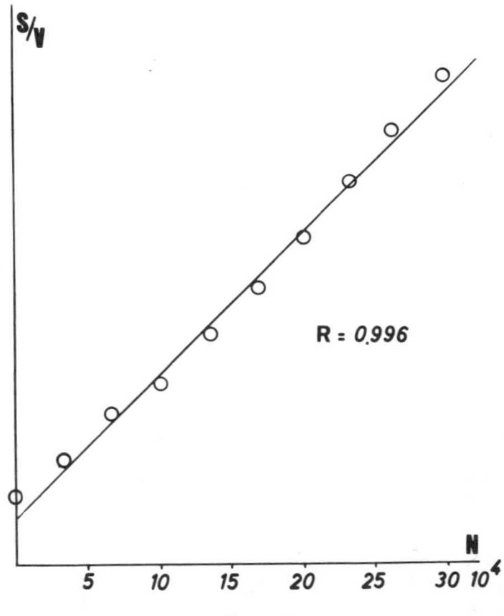

Fig. 4 : Relation between mean number of microborings and surface to volume ratio (S/V) in abyssal detritic shells (*Limopsis pelagica*)

METAL BINDING AND CHITIN DIAGENESIS : OCEANOGRAPHICAL IMPLICATIONS IN DEEP-SEA SEDIMENTS

In addition to the role played by chitin in the input of nutrients and energy to the deep-sea food-webs and in carbonate equilibrium, the chelating properties of chitin are of great oceanographical interest. As stressed for long by Muzzarelli and coworkers (3,15) and Hunt (16), chitin is presumably

Table 2 : Experimental metal binding by chitin and natural chitino-proteic complexes revealed by histochemical methods (5 mg of material was incubated at 4°C in 500 ml of 10 ppm solution in distilled water, pH 7.5, washed several times by centrifugation and resuspended).

+, ++, +++ : increasing intensities of the reaction.
0 : no reaction observed
(?) : doubtful

Element introduced	CP(1)	MM(2)	CM(3)	CC(4)	P(5)
Al^{+++}	+	+	0	0	+
V^{5+}	0	++	++	++	0
Mn^{++}	++	++	+++	++	+
Fe^{+++}	++	+++	+++	++	++
Co^{++}	++	++	++	+	+
Ni^{++}	++	+	+++	+	++
Cu^{++}	++	+++	+++	++	+
Zn^{++}	+	+	++	+	0
As^{+++}	+	+(?)	0	0	+(?)
Rb^{++}	+	+(?)	0	0	+
Sr^{++}	+	++	0	0	+
Ru^{+++}	++	++	++	+	+
Ag^{++}	+++	+++	++	++	+++
Sn^{++}	0(?)	+	0(?)	0	+
Ba^{++}	+	+	+(?)	0	0
Au^{+++}	++	+++	+++	++	++
Hg^{++}	+	++	+	+	+
Tl^{++}	+	+	0	0	0(?)
Pb^{++}	+	+	++	+	+
U^{6+}	++	+++	+++	+++	++

(1) CP : crude cuticles from planctonic crustaceans

(2) MM : decalcified, finely dispersed matrix of Mollusk shell (*Nautilus*)

(3) CM : decalcified matrix of Mollusk shell (*Nautilus*) partly weathered by 3 hours boiling with 0.1 N NaOH

(4) CC : chitin isolated from Mollusk shell (*Nautilus*) (="native chitin")

(5) P : proteic matrix (collagen-like) without chitin, extracted by decalcification of sea-urchin test (*Sphaerechinus granularis*)

responsible for the adsorption of metal ions on organic residues which fall to the bottom of the sea.

By histochemical methods, we pointed out (Table 2) that purified chitin (CC) and even more "natural" chitinoproteic complexes partly weathered (MM, CM) were able to firmly bind several elements. Our results are in very good agreement with quantitative data from the literature (3). The chelating properties of partly weathered chitinoproteic complexes isolated from Mollusk shells are better than those of "fresh" skeletons (CP) or purified chitin (CC) from the same shell. The non chitinous matrix (P) of a sea-urchin test (Sphaerechinus granularis) had very poor chelating properties. These properties enable chitin to collect metals from very diluted solutions (less than 1 ppm Fe^{+++}, Mn^{++} , U^{5+} ,...) close to their natural concentration in sea-water.

The bound metals are released during chitin hydrolysis, either in sea-water or in sediments, as shown by the following experiments : pure chitin or chitinoproteic complexes with or without adsorbed metals (Mn, Fe) were incubated with purified chitinases (Serva, 0.5 mg/ml) and N-acetylglucosamine (NAG) and metal release was monitored as a funtion of time.

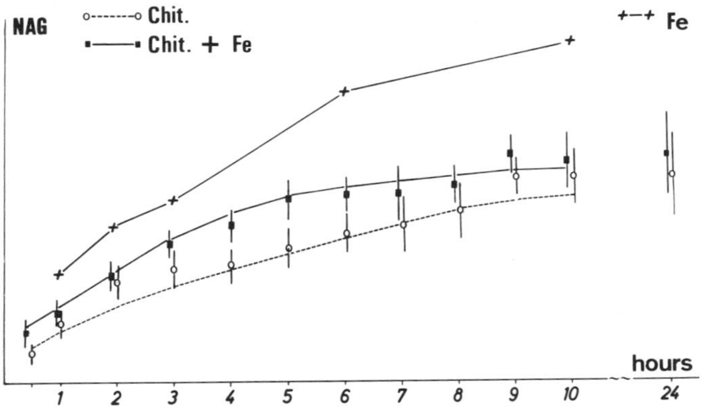

Fig. 5 : NAG and Fe^{+++} release during enzymatic hydrolysis of pure and metal adsorbed chitin (mean ± extreme values and standard error of the method)

Figure 5 is an example of such an experiment using pure chitin and Fe adsorbed chitin.

The metal ions previously adsorbed on chitin and chitinoproteic complexes are freed at a rate proportional to NAG release in the media : one mole of Fe^{+++} for every two moles of NAG and/or one mole of Mn^{++} for every ten moles of NAG. Moreover, the released metal ions have a slight but significative activation effect on chitinases.

From these observations, it appears likely that the removal of some elements from the water column, and the subsequent enrichment of sediments and interstitial water (17,18) may be associated with sinking detritic chitin and its weathering during settling or in the uppermost layers of the sediment

As a result of this enrichment, partly weathered chitinoproteic sheaths outcropping from empty microborings walls in dead shells can act as active sites for adsorption of heavy metals and subsequent nucleation and growth of metallic deposits on detritic material (Fig. 2). Some metal oxidising bacteria (and other microorganisms) may be associated with this metallic growth (19, 20). As further evidence of these phenomena, dead shells recovered from abyssal biocenoses were highly enriched in Mn and Fe (factor 20 to 300) when compared to the live shells of the same species, coming from the same locality (Limopsis pelagica from the Bengal gulf, 2800 meters deep) (10).

In such a perspective, chitin diagenesis in deep-water sediments may help to solve one of the most intriguing problems in marine geochemistry : the formation and the growth of polymetallic nodules. Some authors proposed a biological origin of nodules, like Monty (21) who considered them as stromatolites. However, the most widely held theory of nodule growth is that nodules form by purely physicochemical accretion of manganese on iron oxides. Nevertheless, the binding properties of chitin and chitinoproteic complexes, the microorganisms associated with their degradation (including Metallogenium-like microorganisms very close to the descriptions of Burnett (22, 23) of organic films and microorganisms associated with manganese nodules), and the presence of organic or calcified nuclei within nodules should be carefully considered in relation to that problem. Beyond purely fundamental results, these observations, if confirmed, could have some important economic repercussions.

REFERENCES

1. N.B. Marshall, "Developments in Deep-Sea Biology", Blanford Press Ltd, Poole (Dorset), (1979).
2. Ch. Jeuniaux, J.C. Bussers, M.F. Voss-Foucart and M. Poulicek, Chitin production by animals and natural environment, IIId Intern.Conf. Chitin and Chitosan (1985).
3. R.A.A. Muzzarelli, "Chitin", Pergamon Press, Oxford (1977).
4. Ch. Jeuniaux, Chitine et phylogénie : application d'une méthode enzymatique de dosage de la chitine, Bull.Soc.Chim.Biol., 47:2267 (1965).
5. M. Poulicek and Ch. Jeuniaux, Biomass and biodegradation of Mollusk shell chitin in some marine sediments, IId Internat.Conf.Chitin Chitosan, S. Hirano and S. Tokura ed., Sapporo (1982).
6. M. Poulicek, "Coquilles et autres structures squelettiques des Mollusques: composition chimique, biomasse et biodégradation en milieu marin", Ph.D. Thesis, University of Liège, 180 p. (unpublished) (1982).
7. C. ZoBell and S.C. Rittenberg, The occurence and characteristics of chitinoclastic bacteria in the sea, J.Bact., 35 : 275 (1938).
8. F. Azam, T. Fenchel, J.G. Field, J.S. Gray, L.A. Meyer-Reil and F. Thingstad, The ecological role of water-column microbes in the sea, Mar.Ecol.Prog.Ser., 10 : 257 (1983).
9. R. Iturriaga, Bacterial activity related to sedimenting particulate matter, Marine Biology, 55 : 157 (1979).
10. M. Poulicek, Shell matrix weathering in marine sediments, Malacologia, in press (1985).
11. M. Poulicek, Patterns of Mollusk shell biodegradation in bathyal and abyssal sediments, J.Mollusc.Stud., Suppl.12 A : 136 (1983).
12. M.L. Zeff and R.D. Perkins, Microbial alteration of bahamian deep-sea carbonates, Sedimentology, 26 : 175 (1979).
13. M. Poulicek and M.F. Jaspar-Versali, Biodégradation de la trame organique des coquilles de Mollusques en milieu marin : action des microorganismes endolithes, Bull.Soc.r.Sci.Liège, 53(2) : 114 (1985).
14. J. Kim and C.E. ZoBell, Agarase, amylase, cellulase and chitinase acti-

vity at deep-sea pressure, J.Oceanog.Soc.Japan, 28 : 131 (1972).

15. R.A.A. Muzzarelli and O. Tubertini, Chitin and chitosan as chromatographic supports and adsorbents for collection of metal ions from organic and aqueous solutions and sea-water, Talanta, 16 : 1571 (1969).

16. C.D. Hunt, Incorporation and deposition of Mn and other trace metals by floculent organic matter in a controlled marine ecosystem, Limnol. Oceanog., 28(2) : 302 (1983).

17. P. Duchart, S.E. Calvert and N.B. Price, Distribution of trace metals in the pore waters of shallow water marine sediments, Limnol.Oceanog., 18(4) : 605 (1973).

18. B.J. Prestley, R.R. Brooks and I.R. Kaplan, Manganese and related elements in the interstitial water of marine sediments, Science, 158 : 906 (1967).

19. H.L. Ehrlich, Conditions for bacterial participation in the initiation of manganese deposition around marine sediment particles, in "Environmental biogeochemistry and geomicrobiology", Vol.3, Chap.65 W.E. Krumbein ed., Oldenburg (Germany) (1978).

20. H.W. Jannasch, Microbes in the oceanic environment, in "S.G.M. Symposium N°36", Part 11, D.P. Kelly and N.G. Carr ed., Pitman Press, NBath, (1984).

21. C.L.V. Monty, Les nodules de manganèse sont des stromatolites océaniques, C.R. Acad.Sci.Paris, 276 : 3285 (1973).

22. B.R. Burnett, Quantitative sampling of microbiota of the deep-sea benthos-1.Sampling techniques and some data from the abyssal central North Pacific, Deep-sea Research, 24 : 781 (1977).

23. B.R. Burnett and K.H. Nealson, Organic films and microorganisms associated with manganese nodules, Deep-sea research, 28A : 637 (1981).

ANAEROBIC DECOMPOSITION OF CHITIN IN FRESHWATER SEDIMENTS

H. STURZ and J. ROBINSON

School of Biological Sciences, Queen Mary College, University of London
Mile End Road
London E1 4NS, England

INTRODUCTION

It has been estimated that 10^{11} tonnes of chitin are produced annually in the aquatic biosphere (Tracey 1957) in such forms as the exoskeletons of crustaceans, molluscs, coelenterates, in protozoa and in the cell walls of fungi and algae. The decomposition of chitin is therefore an important component in the recycling of both carbon and nitrogen in aquatic environments.

In the aquatic biosphere, a significant proportion of carbon cycling is anaerobic. It is known that in the organically enriched sediments of coastal waters and eutrophic lakes, the oxygen concentration is depleted by microbial activity and anaerobic decomposition of polymeric compounds is an important process. In these environments, organic compounds are decomposed initially by fermentation and the organic end-products of fermentation are then oxidised further by anaerobic respiration in the presence of NO_3^-, SO_4^{2-}, or CO_2. In the absence of $SO_4^{2-}$and NO_3^- polymers are fermented to methane and carbon dioxide by a complex community of bacteria.

Under these conditions organic matter in sediments is fermented to methane and carbon dioxide according to the stoichiometry given by the Buswell equation,

$$C_nH_aO_b + (n-a/4-b/2)H_2O \longrightarrow (n/2-a/8+b/4)CH_4 + (n/2-a/8 + b/4)CO_2$$

For chitin, ignoring the contribution of nitrogen, the proportions would be,

$$8C_8H_{13}(N)O_5 \quad + \quad 18H_2O \longrightarrow 29CO_2 \quad + \quad 35CH_4$$

Although chitin is resistant to chemical attack, little is found to accumulate in sediment (Brewer and Pfaender 1979). Aerobic microbial degradation has been extensively studied in marine (Seki and Taga 1963a, 1965; Campbell and Williams 1951; Hood and Meyers 1977), freshwater (Warnes and Randles 1977, 1980; Aumen 1981) and soil (Okafor 1966, Baxby and Gray 1968) environments.

It is also known that the polymer is decomposed anaerobically in freshwater (Warnes and Randles 1977, 1980) and salt marsh sediments.

Chitin degrading microorganisms representing a variety of genera have been isolated from a wide range of habitats.

Chitinolytic bacteria have been isolated from the gastro intestinal tracts of whales (Seki and Taga 1965a) and chitinolytic *Vibrio*, *Aeromonas*, and *Alginomonas* spp have been isolated from the gut of marine fish (Okutani 1966).

As well as commensal relationships between chitinolytic bacteria and animals (Iverson et al 1984. Buck and Barbaree 1971; Seki and Taga 1963a), many chitinolytic bacteria are associated with the disease states of crustaceans (Rosen 1970).

Free living chitinolytic bacteria have been described in the marine environment (Seki and Taga 1963a) and in freshwater lakes, aerobic chitinolytic actinomycetes have frequently been observed (Veldkamp 1955; Johnston 1972; Willoughby 1969). Other commonly isolated genera of chitinolytic microorganisms from freshwater lakes include *Myxobacteria spp*, *Pseudomonas spp*. (Warnes and Rux 1983) and *Serratia spp* (Aumen 1980; Warnes and Rux 1983).

In the soil chitinolytic fungi (Okafor 1966) and aerobic bacteria of the genera *Bacillus*, *Pseudomonas*, *Zymononas*, *Serratia*, *Aeromonas*, *Alcaligenes*, *Flavobacterium* and *Cytophaga* have been isolated. (Veldkamp 1955. Srikantaiah and Mohankumar 1980).

Most of the bacteria described above have been isolated using aerobic techniques. However, many of the predominant isolates have been shown to be facultative anaerobes.

Willoughby (1969) isolated genera of *Nocardia* and *Micromonospora* capable of withstanding anaerobic conditions. Stanier (1947), isolated a strain of *Cytophaga johnsonii var denitrificans* capable of growing anaerobically in the presence of NO_3^- and a number of isolates obtained by Warnes and Rux (1983) were capable of fermentative growth.

Decomposition of chitin under strictly anaerobic conditions has been less well investigated. Boyer and Wolfe (1983) isolated chitinolytic bacteria from salt marsh sediment, and Warnes and Randles (1977, 1980) using slides coated with reprecipitated chitin found that anaerobic degradation appeared to be initiated by rod shaped bacteria which were succeeded by vibrioid or spiral bacteria.

Most estimations of chitin decomposition have been determined from *in vitro* studies using water and sediment from marine and estuarine environments (Seki 1963b, 1965b; Chan 1970; Hood and Meyers 1977). *In situ* degradation rates have also been determined in the water column of a freshwater lake (Warnes and Rux 1983).

Although enumeration of chitinolytic microorganisms has been carried out aerobically in seawater and marine sediments and in both the sediments and water columns of freshwater lakes (Warnes and Randles 1977; Okutani 1975) little work has been carried out on the decomposition of chitin in freshwater sediment under anaerobic conditions. This paper describes an investigation of the relationship between the *in situ* decomposition rates and viable counts of cellulolytic and chitinolytic microorganisms in freshwater sediment.

METHODS AND MATERIALS

Study site

Blelham Tarn is a small freshwater eutrophic lake situated close to Lake Windermere in the English Lake District. It is 0.67 km long, contains 0.69 x $10^6 m^3$ of water and has a maximum depth of 14.5 m (Jones and Simon 1980).

The Tarn becomes thermally stratified in the summer and the hypolimnion is deoxygenated usually to within 6 m of the surface (Jones 1976).

Collection and storage of sediment

Samples of sediment were taken from the littoral zone of the Lake using a Jenkins surface mud sampler (Macan 1970). Cores were extruded by the method of Jones (1976). Mud samples from different depths were dispensed under nitrogen into boiling tubes, stoppered with butyl rubber bungs and stored in the dark at 4°C. All samples were used within 14 days of collection.

Media and Isolation Techniques

Media were prepared using pebble-milled, deproteinated Nephrops chitin or pebble-milled cellulose prepared from Whatman number 1 filter paper. Anaerobic media were prepared under a nitrogen atmosphere using the Hungate technique (1969) and reduced by the addition of $5gl^{-1}$ of cysteine hydrochloride and $5gl^{-1}$ sodium sulphide.

Mineral salts medium consisting of per litre of distilled water; $KHPO_4$ 0.4g; $MgCl_2$ 0.1g; $NaCO_3$ 1.5g; Vitamins (Wolin et al 1963) 1ml; pebble-milled cellulose, 100ml of 3% suspension or pebble-milled chitin, 250ml of 1% suspension: 1.5% agar, was used (for solid medium). This medium was supplemented with 0.1% yeast extract for the anaerobic viable counts and $0.75gl^{-1}$ NH_4Cl for cellulose medium.

Bacteria were isolated directly from sediment by dilution into cellulose or chitin medium without enrichment.

Anaerobic and Aerobic viable counts

Samples of anaerobic sediment taken from different depths were serially diluted through tubes of anaerobically prepared lake water. For aerobic or anaerobic counts, aliquots from selected tubes were either plated aerobically or inoculated into anaerobic roll tubes of cellulose or chitin agar. Viable counts for each depth were carried out in triplicate. Tubes and plates were incubated at 30°C. Chitinase or cellulase producing colonies were indicated by surrounding zones of clearing. Plates and tubes were examined periodically until the maximum number of clearing colonies developed.

Preparation of chitin

Purified chitin was prepared by soaking Nephrops carapaces in 0.3M HCl to remove $CaCO_3$. Protein was removed by extraction in hot 1.8M KOH followed by treatment in 0.06M $KMnO_4$ and 20% sodium metabisulphite for the removal of recalcitrant molecules and pigments. The chitin was stored dry. Chitin was added to agar media as a suspension prepared by pebble-milling $10gl^{-1}$ at 4°C for 48 hours.

Preparation of cellulose

$30gl^{-1}$ Whatman number one cellulose filter paper was pebble-milled for 24 hours at 4°C for the preparation of cellulose agar.

Cellophane strips for the *in situ* cellulolysis studies were dyed with Remazol Blue by the method of Poincelot and Day (1972).

In situ studies

Weighed chitin strips of approximately equal dimensions and dyed cellophane strips were sewn into nylon bags. The bags containing the samples were placed in metal frames which were embedded in the sediment in the littoral zone of Blelham Tarn. Replicate bags were removed from the sediment of each sample over a period of 12 weeks.

After removal from the sediment at 2,4,8,12 week intervals, the nylon mesh bags containing the partially mineralised chitin and cellulose were stored frozen at -20°C. The bags containing the chitin samples were washed with distilled water to remove loose organic debris and dried. Dried chitin samples were weighed and the percentage weight loss calculated.

Remazol blue dye from the control and exposed cellophane strips was extracted by heating in 0.06M KOH at 120°C for 10 minutes. The absorbance was measured at 595nm. The percentage weight loss of cellulose was determined by using the relationship;

$$W = \frac{\text{Abs Control} - \text{Abs Exposed}}{\text{Abs Control}} \times 100$$

Preparation of ^{14}C labelled chitin

Wild type and dumpy mutant 3rd instar Drosophila larvae were suspended in 16ml distilled water containing 4.4 µCi/ml U ^{14}C-glucose in a petri dish.

After incubation for three hours at 25°C the larvae were filtered out of the liquid and stored frozen at -20°C until treated. Whole larvae were heated for seven hours at 100°C in $100gl^{-1}$ NaOH with three changes of solvent. The treated larval chitin was then washed several times in distilled water to remove any traces of alkali. After washing, small aliquots of larval chitin were homogenised in a tissue grinder until the particles were fine enough to be drawn through a 23 gauge needle. The chitin suspension was gassed out with nitrogen, reduced with $5gl^{-1}$ sodium sulphide, and stored under nitrogen at 4°C until used.

Determination of ^{14}C labelled chitin decomposition in vitro

Triplicate slurries of Blelham Tarn sediment from different depths were prepared in stoppered bellco tubes under a nitrogen atmosphere using anaerobic Blelham Tarn water. To each tube 0.5ml of anaerobic ^{14}C- labelled chitin was added. A control tube containing sediment slurry and ^{14}C labelled chitin which had been autoclaved was prepared.

Measurement of total ^{14}C labelled gas

The total ^{14}C labelled gas released as the result of decomposition was determined by the combustion of gas samples removed from the headspace of experimental tubes in a furnace at 750°C. The combusted gas was collected as $^{14}CO_2$ by absorption into carbosorb. Scintillant was added and the samples counted in a liquid scintillation counter.

DISCUSSION

There was a marked decrease in the viable count of anaerobic chitinolytic and cellulolytic bacteria below the 0-1cm horizon (Fig.1). The number of anaerobic chitinolytic bacteria at each depth was greater than the number of anaerobic celluloytic bacteria, which suggests that the anaerobic chitin degradation observed in these sediments is carried out by a chitinolytic population distinct from the cellulolytic population. The observation that the dominant chitinolytic anaerobe was a small gram negative rod which was unable to utilise cellulose is consistent with this hypothesis. The number of anaerobic chitinolytic bacteria in the surface sediment, 10^6/g dry weight sediment, is higher than reported by Warnes and Randles (1977) who found 10^5/g dry weight of freshwater sediment, but the numbers at greater depth are comparable. Okutani (1975) found 10^3 to 10^4 chitinolytic microorganisms /g wet weight in the 0-3cm horizon of freshwater sediment and 10^3 and 10^5 - 10^6 in marine and estuarine sediments respectively. The numbers of aerobic

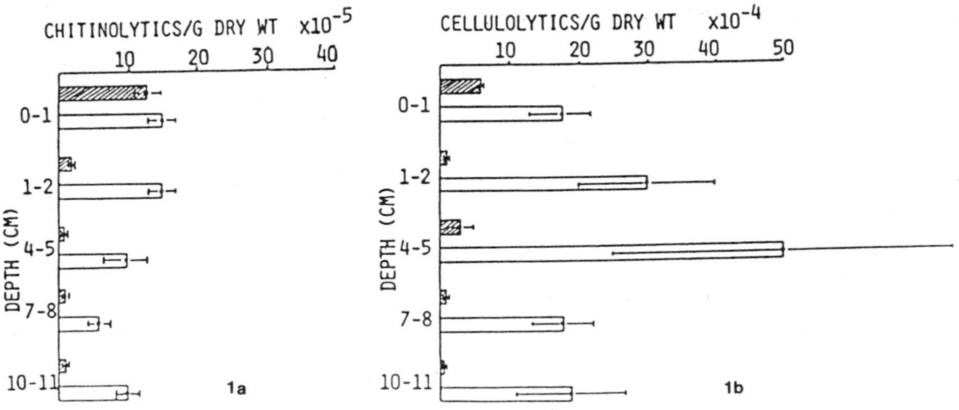

Fig.1. Viable counts of chitinolytic (1a) and cellulolytic (1b) microorganisms on anaerobic ▨ and aerobic □ medium from various depths below the sediment: water interface of Blelham Tarn.

cellulolytic and chitinolytic microorganisms were greater than the numbers of anaerobes at all depths and did not show the same marked decrease below the 0-1cm horizon.

The in vitro measurement of anaerobic degradation (fig 2) suggests that there is greater anaerobic chitinolytic activity in the 0-1cm horizon than in the 10-11cm horizon. This is consistent with the distribution of anaerobic chitin degraders (Fig.1). Degradation of cellulose and chitin buried in the littoral sediment is shown in Fig.3. The initial rate of degradation of buried chitin was slightly faster in the surface sediment than at depth being 10.7mg/g/day and 9.3mg/g/day in the 0-6cm and 42-48cm horizons respectively. This represents aerobic and anaerobic decomposition and thus cannot be directly compared with the data shown in Figs 1 and 2. The lack of a marked effect of depth is consistent with the distribution of aerobic counts.

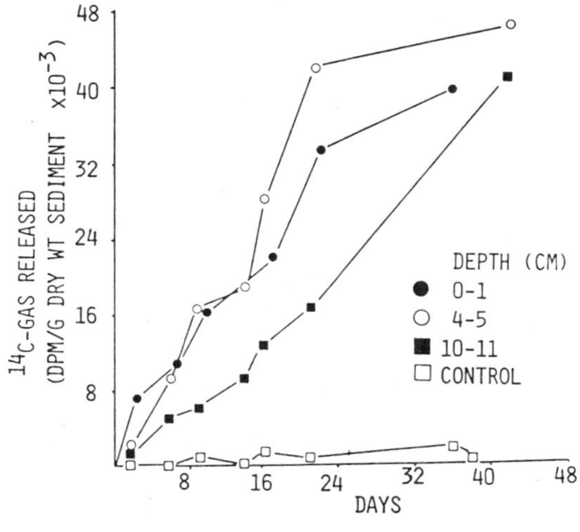

Fig.2. Release of ^{14}C- labelled gas from ^{14}C-labelled chitin incubated in sediment from various depths below the sediment water interface of Blelham Tarn.

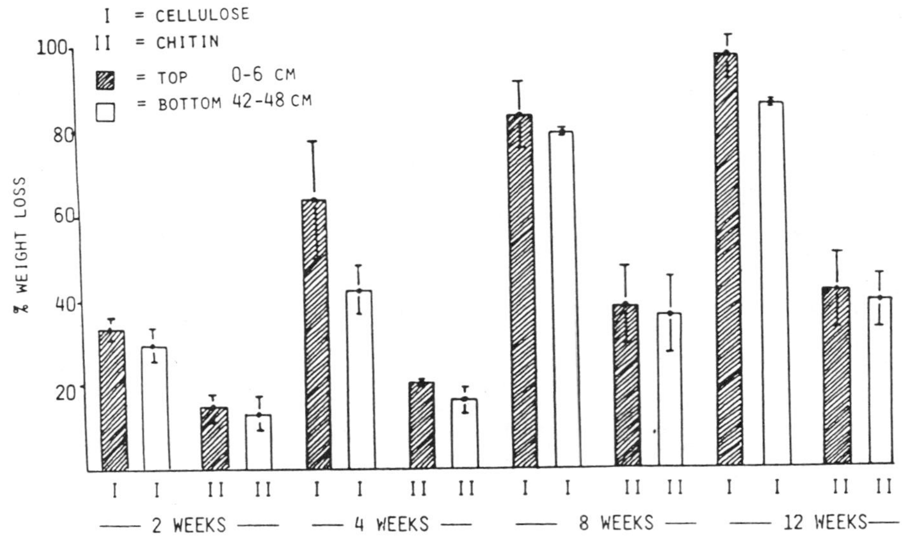

Fig.3. Percentage weight loss from chitin and cellulose samples buried in littoral zone sediment of Blelham Tarn.

Most estimations of chitin decomposition have been based on data from *in vitro* studies using water and sediment from marine and estuarine sources. Rates equivalent to 27mg/g/day (Seki and Taga 1963b), 7mg/g/day (Hood and Meyers 1977) and 8.3 mg/g/day (Seki 1965b) have been reported in marine and estuarine water columns and 16mg/g/day in sediments (Hood and Meyers 1977; Seki 1965b). The *in situ* rates reported are generally higher than those obtained *in vitro*. Hood and Meyers (1977) found decomposition rates of 118mg/g/day (1977) and 440mg/g/day (1979) in estuarine water columns and Warnes and Rux (1983) reported a rate of 43.6 mg/g/day at the sediment/water interface of a freshwater lake.

The maximum *in situ* rate of chitin decomposition measured in the 0-6cm horizon in this study was 10.7mg/g/day, similar to the rates reported for *in vitro* studies on marine sediments. The average rate of decomposition in the 0-5cm horizon measured with the ^{14}C assay was 10.5mg/g/day similar to that obtained *in situ*. Since the *in vitro* temperature of incubation, 28°C, is higher than the average temperature of the sediment, 6-8°C, this similarity is probably fortuitous. Particle size is also known to affect the rate of chitin decomposition (Hood and Meyers 1977; Warnes and Rux 1983. Seki and Taga 1963b). Since the radioactive chitin had been finely divided by homogenisation, *in situ* and *in vitro* results are not directly comparable.

CONCLUSIONS

Chitin and cellulose placed in the littoral sediment of a freshwater lake were degraded slightly faster in the 0-6cm horizon than at greater depth and cellulose was decomposed faster than chitin under these conditions. *In vitro*, the fastest rates of ^{14}C- labelled gas release from ^{14}C-labelled chitin was associated with the 0-1cm and 4-5cm horizons.

A marked decrease in the viable count of anaerobic but not aerobic chitinolytic and cellulolytic microorganisms was observed below the 0-1cm horizon. The data suggest that the sediment surface is the most active region of anaerobic chitin decomposition.

REFERENCES

Aumen, N.G., 1981, Microbial succession on a chitinous substrate in a woodland stream. Microb. Ecol. 6(4) : 317-328.

Baxby, P., and Gray, T.R.G., 1968, Chitin decomposition in soil. I. Media for isolation of chitinoclastic microorganisms from soil. Trans. Br. Mycol. Soc. 51:287-292.

Buck, S.K., and Barbaree, J.M., 1971, Chitinolytic bacteria associated with marine copepods, Bact. Proc. 71:G148.

Boyer, J.N., and Wolfe, R.S., 1984, Anaerobic chitin degradation as a carbon and hydrogen source for sulfate reduction and methanogenesis in salt marsh bacteria. Biological Bulletin, 165.N2:505-506.

Brewer, W.S., and Pfaender, P.K., 1979, The distribution of selected organic molecules in freshwater sediment. Water Research, 13(3) 237-240.

Campbell, L.L., and Williams, O.B., 1951, A study of chitin decomposing microorganisms of marine origin, J.Gen.Microbiol. 5:894-905.

Chan, J.C., 1970, The occurrence, taxonomy and activity of chitinoclastic bacteria from sediment, water and fauna of Puget Sound, Dissertation (PhD) University of Washington. Seattle. p.1-312.

Hood, M.A., and Meyers, S.P. 1977. Rates of chitin degradation in an estuarine environment. J.Oceanog. Soc. Japan, 33:328-334.

Hood, M.A., and Meyers, S.P. 1979. Chitin degradation in estuarine environments and implications in crustacean biology. In "Proceedings of the 1st International Conference on Chitin/Chitosan." Eds R.A.A. Muzzarelli and E.R. Pariser.p.563-576. M.I.T. Cambridge USA (1978).

Hungate, R.E., 1969, A roll tube method for the cultivation of strict anaerobes, in: "Methods in Microbiology, Volume 3B", J.R. Norris and D.W. Ribbons, eds., pp. 117-132, Academic Press, London.

Iverson, K.L. Bromel, M.C., Anderson, A.W., and Freeman, T.P., 1984, Bacterial symbionts in the sugar beet root maggot, *Tetanops myopaeformis* (von Roder), Appl. Environ. Microbiol. 47(1):22-27.

Johnston, D.W. 1972, Actinomyces in aquatic habitats. Ph.D. Thesis, University of Bradford.

Jones, J.G., 1976, The microbiology and decomposition of seston in open water and experimental enclosures in a productive lake, J. Ecology 64: 241:278.

Macan, T.T., 1970, "Biological Studies of the English Lakes", Longmans, London.

Okafor, N., 1966, Ecology of microorganisms on chitin buried in soil, J.Gen.Microbiol. 44:311-327.

Okutani, K.,1966, Studies of chitinolytic systems in the digestive tract of *Lateolabrax japonicus*, Bull.Misaki Mar.Biol.Inst. 19:1-47.

Okutani, K.1975. Microorganisms related to the mineralisation of chitin in aquatic environments. In "Nitrogen fixation and Nitrogen cycle. Eds. H.Takahashi." p.147-154.

Poincelot, R.P. and Day, P.R. 1972. Simple dye release assay for determining cellulolytic activity of fungi. Appl.Microbiol. 23(5): 875-879.

Rosen, B., 1970, Shell diseases of aquatic crustaceans, in "Symposium on Disease of Fishes and Shellfishes", S.Snieszko, ed., Spec Pub. No.5, Amer. Fish. Soc., Washington D.C. p409-15.

Seki, H., and Taga, N. 1963a. Microbiological studies on the decomposition of chitin in marine environment I and II. J.Oceanog. Soc. Japan 19(2) 27-37.

Seki, H., and Taga, N., 1963b, Microbiological Studies on the decomposition of chitin. J. Oceanog. Soc. Japan, 19(3),23-31.

Seki, H., and Taga, N., 1965a, Microbiological studies on the decomposition of chitin in marine environment. VI. J.Oceanog. Soc. Japan,20(6):18-23.

Seki, H., 1965b, Microbiological studies on the decomposition of chitin in marine environment IX, and X. J.Oceanog. Soc. Japan 21(6):17-34.

Srikantaiah, M., and K.C. Mohankumar, 1981. Isolation and characterisation of chitin degrading bacteria. Indian J. Microbiol. 20(3): 216-9.

Stanier, R.Y. 1947, Studies on non fruiting Myxobacteria. I. Cytophaga johnsonae, n. sp, a chitin decomposing Myxobacterium.J.Bacteriol. 53:297.

Tracey, M.V., 1957, Chitin. Rev. Pure Appl. Chem. 7:1-14.

Veldkamp, H. 1955, A study of aerobic decomposition of chitin by microorganisms. Mededelingen van de Landbouwhogeschool te Wageningen Nederland, 55(3): 127-174.

Warnes, C.E., and Randles, C.I., 1977, Preliminary studies on chitin decomposition in Lake Erie sediments, Ohio J. Sci. 77(5):224-230.

Warnes, C.E., and Randles, C.I., 1980, Succession in a microbial community associated with chitin in Lake Erie sediment and water, Ohio J. Sci. 80(6):250-255.

Warnes, C.E., and Rux, T.P., 1983, Chitin mineralisation in a freshwater habitat, in: "Proceedings of the Second International Conference on Chitin and Chitosan", The Japanese Society of Chitin and Chitosan, July 12-14, 1982. Sapporo Japan, S. Hirano and S. Tokura, eds.

Willoughby, L.G. 1969, A study of the aquatic actinomyces of Blelham Tarn. Hydrobiologica 34:465-483.

CHITINOLYTIC COMMUNITIES FROM AN ANAEROBIC ESTUARINE ENVIRONMENT

R. PEL and J. C. GOTTSCHAL

Department of Microbiology, University of Groningen
Haren, The Netherlands

INTRODUCTION

Mineralization of organic matter in anaerobic environments is accomplished by a community of physiologically different bacteria.[1] Biogenic polymers such as carbohydrates and proteins are hydrolyzed and the monomers produced are subsequently converted to a variety of products by fermentative bacteria. Their fermentation products are further oxidized by sulfate-reducing bacteria or by syntrophic consortia of acetogenic and methanogenic bacteria. Our knowledge of anaerobic polysaccharide degrading communities mainly originates from the cellulolytic system in the rumen[2,3,4] Much less is known about the microbiology of anaerobic polysaccharide composition in marine environments. This is particularly true for the structure of these communities and the possible interspecies interaction involved (i.e. the primary hydrolysing bacteria and those using hydrolysis and fermentation products).

For the inner part of the Ems-Dollard estuary, largely consisting of tidal flats, anaerobic mineralization (measured as the rate of sulfate reduction) has been estimated at 90 g $C.m^{-2}.y^{-1}$.[5]. This is 30% of the total mineralization. Unfortunately no data exists on the precise identity and quantity of the polysaccharides entering the sediments. As copepods and shrimps constitute both numerically and by biomass the largest populations of the zoo-plankton and macrofauna, respectively,[6] chitin can be expected to be an important substrate for a considerable part of the microbial sediment population. For this reason, and because of its well defined structure, chitin has been chosen as a model-substrate in studies on the structure of anaerobic polysaccharide degrading communities in marine environments.

So far very little work has been done on the isolation and characterization of anaerobic chitinolytic bacteria. To our knowledge there is only one report on the isolation in pure culture of marine bacteria capable of strictly anaerobic chitin degradation.[7] Unfortunately this report offers only fragmentary information on the growth characteristics and the chitinolytic mechanism of the isolates. Aerobic chitinolytic microorganisms, on the other hand, have been isolated repeatedly and their growth and hydrolysing capabilities have been studied quite extensively in the past.[8]

In this report results on the growth characteristics of pure and mixed chitinolytic populations isolated from intertidal sediments will be presented.

MATERIALS AND METHODS

Cultivation conditions

For growth of mixed and pure cultures, the following basal medium was used (components in g/liter): KH_2PO_4, 0.2; Na_2SO_4, 3.0; $MgCl_2.6H_2O$, 2.0; NaCl, 10.0; NH_4Cl, 0.3; KCl, 0.3; $CaCl_2.2H_2O$, 0.15; $NaHCO_3$, 2.52; $Na_2S.9H_2O$, 0.4; yeast extract, 0.1; resarurin, 0.001; trace elements solution (1 ml/l) and vitamin solution (1 ml/l). The trace elements solution contained per liter: 12.5 M HCl, 4.0 mg; $FeCl_2.4H_2O$, 2.000 mg; $ZnCl_2$, 70 mg; $MnCl_2.4H_2O$, 100 mg; $COCl_2.6H_2O$, 190 mg; $CuCl_2.2H_2O$, 17 mg; $NiCl_2.6H_2O$, 24 mg; $Na_2MoO_4.2H_2O$, 36 mg; $Na_2SeO_3.5H_2O$, 39 mg; $Na_2WO_4.2H_2O$, 49 mg. The vitamin solution contained (mg/l): biotin, 10; nicotinic acid, 100; ß-aminobenzoic acid, 50; thiamin, 100; panthotenic acid, 50; pyroxamin, 250; cobalamin, 50. The following stock solutions were sterilized separately and added aseptically to the basal medium: vitamins, trace elements, $NaHCo_3$, $Na_2S.9H_2O$ and yeast extract. The vitamin and trace solutions were filter sterilized (0.2 µm), the other components were autoclaved. All soluble substrates were added aseptically from stock solutions which had been filter sterilized. Chitin was added from an autoclaved stock suspension (10 mg/l) or sterilized together with the basal medium. Stock suspensions were prepared by ball-milling for about 4 h. Chitin oligosaccharides were prepared from chitin by acid-hydrolysis and subsequent fractionation on a polyacrylamide gel column according to Guire.[9] (80 x 2,8 cm column packed with Bio-Gel P-2, Bio-Rad Lab., Calif.). Fractions containing the dimer were combined and the aminosugar concentration estimated by comparing the organic carbon content with a NAG-standard using a total carbon analyser. Chitobiose stock solution thus obtained, contained approximately 20 mg NAG-equivalents per ml. Column fractions containing the trimer and higher oligosaccharides were combined too. In this way a mixture of oligosaccharides was obtained containing approximately 20 mg N-acetylglucosamine-units per ml. Final pH of the medium was 7.0 - 7.2.

Cultures were grown in screw-cap bottles with a head-space flushed with oxygen free gas composed of 80% N_2 and 20% CO_2. Incubations were done in the dark at 25°C.

Chemical analyses

Both volatile and non-volatile short chain fatty acids were analyzed with a Packard 437 gaschromatograph equipped with a flame ionization detector, connected to a Packard 604 integrator. Glass columns (2 m long; 2 mm inner diameter) contained Chromosorb WAW, 100 - 120 mesh, coated with 10% SP-1000 and 3% H_3PO_4 (Chrompack Nederland B.V. Middelburg, The Netherlands). The flow rate of the carrier gas (N_2) was 50 ml/min. The temperature of injection port, column and detector were 175, 120 and 175°C respectively. The flow rates of H_2 and air were 30 and 200 ml/min, respectively. Volatile fatty acids were determined after diethyl ether extraction with iso-valerate as described by Laanbroek et al.[10]

Non-volatile fatty acids were determined after methylation and chloroform extraction using malonic acid as interval standard.[10] Formate was determined in the same way as these non-volatile acids but with a column temperature of 50°C.

H_2, CH_4 and CO_2 in the gasphase of the cultures were analyzed on a

Pye Unicam 104 gaschromatograph equipped with a katharometer detector.[10] Methanol and ethanol were analysed with a Packard 427 gaschromatograph,[11] using 2 m glass columns (2 mm inner diameter), packed with Poropack Q (Waters Ass. Inc., Milford, Mass.) mesh 100 - 120.

Sulphide was determined by the method of Packmayer as described by Trüper and Schlegel.[12] Organic carbon was determined using a Beckman Total Carbon Analyser (Model 915A). For determination of chitin the polymer was completely hydrolyzed to its deacetylated monomers (HCl-glucosamine) which were subsequently quantified colorimetrically.[13]

N-acetylglucosamine (NAG) was determined either colorimetrically according to Tracey[14] or by high pressure liquid chromatography according to Blumberg et al.[15] The equipment used was a Waters Chromatography 6000A pump/injector, and a Waters variable wavelength LC-Spectrophotometer model 480 monitoring at 190 nm. For quantification of the peaks a Waters integrator, model 730 Data Module, was used. The reverse-phase column (250 x 4.6 mm) employed was CP tm Sphere C_{18} (Chrompack). Chitobiose was determined according to Blumberg et al.[14]

Chemicals

Chitin powder (practical grade), glucosamine and N-acetyl glucosamine were purchased from Sigma.

Isolation of chitin hydrolysing bacteria

From mixed populations pure chitinolytic bacteria were isolated by serial dilution in agar shake tubes with a chitin containing medium. This medium contained 0.5% chitin and 0.5% agar, thus allowing clearing-zones to form. To permit simple observation of clearings, glass tubes with an inner diameter of 6 mm were used. Both ends of these tubes were closed with rubber stoppers, permitting easy isolation of colonies using a syringe punched through the rubber stopper.

RESULTS

Mixed culture studies

Anaerobic chitinolytic mixed populations were isolated from the sediment of an intertidal mud flat in the brackish water of the Ems-Dollard estuary. A mud sample was taken immediately below the upper grey (oxic) layer, at a depth of approximately 2 cm (redox potential (Eh) ca. +50 mV). This sample was diluted 1 : 50 and was used to inoculate bottles containing strictly anaerobic chitin containing liquid medium. Growth was observed in all incubations within 4 days and active cultures could be maintained by serial transfer in chitin containing liquid medium every 3 - 4 weeks. As far as could be judged microscopically, chitin decomposition showed a similar pattern of events in all mixed cultures thus obtained: within a day following inoculation chitin particles became colonized by slender rods (4 - 7 µm long), often containing terminal or subterminal spores. In some areas large numbers of these bacteria were observed to lie perpendicular to the chitin surface with the spore-bearing end of the cell away from the chitin particle. After about one week of incubation these cultures enter a distinctly new phase marked by the appearance of a conspicious bacterium. This organism (10 µm long - 4 µm thick) forms refractive cells of variable morphology which tend to cluster. These events coincided with acetate accumulation in the culture fluid and subsequent consumption of the fatty acid. Figure 1 shows a typical picture of the transient accumulation of acetate during the time of incubation. In the

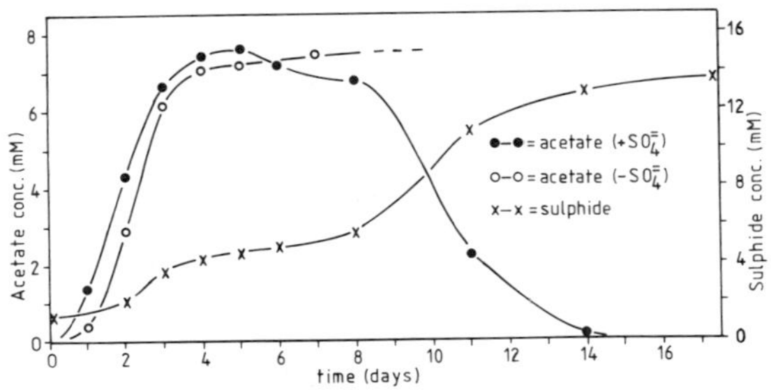

Fig. 1. Chitin fermentation (acetate formation) in mixed culture in the absence and in the presence of sulphate (sulphide formation).

presence of sulphate, chitin fermentation (acetate formation) proceeded with a doubling time (t_d) of approximately 8 h. The fermentation products formed were acetate, CO_2 and ammonia. Up to about 5% of the acetate produced is converted to CH_4. However, the bulk of this fatty acid is consumed by a sulphate-reducing population, as indicated by the sulphide formation during the second week of incubation (Fig. 1). In an acetate/sulfate medium the acetate oxidizing sulfate reducing population could be easily enriched yielding mainly the refractive cells. This organism is morphologically distinct from any sulfate reducing bacterium described sofar. Our attempts, however, to isolate the bacterium in pure culture from enrichments on acetate/sulfate medium, were as yet unsuccessful.

When sulphate was omitted from the growth medium the rate of chitin fermentation was not different from that in its presence (Fig. 1). The number of fermentation products now detectable in the culture fluid was much greater. Besides acetate, formation of H_2, formate, ethanol, propionate and butyrate could be observed (not shown). H_2, formate and ethanol were accumulating only transiently. Conversion of acetate to CH_4 started after ca. 2.5 weeks and was completed within 6 weeks. Propionate and butyrate were not used at all within a period of 8 weeks.

<u>Pure cultures studies</u>

From the type of mixed populations as described above, eight chitinolytic organisms have been isolated and studied in pure culture. The isolates are straight or slightly curved motile rods (3-7 µm long) which tend to form non-motile filaments when associated with chitin particles. Five out of eight of these isolates form terminal spores. They are strict anaerobes and show similar fermentation patterns. Acetate, ethanol, H_2, formate, CO_2 and ammonia are the fermentation products formed in these cultures. Only one of these isolates produces propionate, butyrate and propanol in addition. Acetamide, a possible degradation product from chitin as suggested by some workers in the past,[8] could not be detected in the culture fluids.

Compared with mixed populations, growth on chitin (acetate formation) in the pure cultures, was much slower (t_d = 3 days versus 8h) and lag-phases of approximately 2 weeks were observed before degradation started. Furthermore, chitin was used incompletely in these cultures: 40% remains unhydrolyzed. Supplementation of the medium with dithionite, lowering the E_h from ca. −160 mV to ca. −400 mV, markedly stimulated growth on chitin of all isolates, reducing lag phases to 2-4 days and decreasing t_d to 19-24 h (Fig. 2A).

Glusose, glucosamine (GA) and N-acetyl-glucosamine (NAG) do not, or only marginally support growth: 2 out of 8 strains are capable of slow

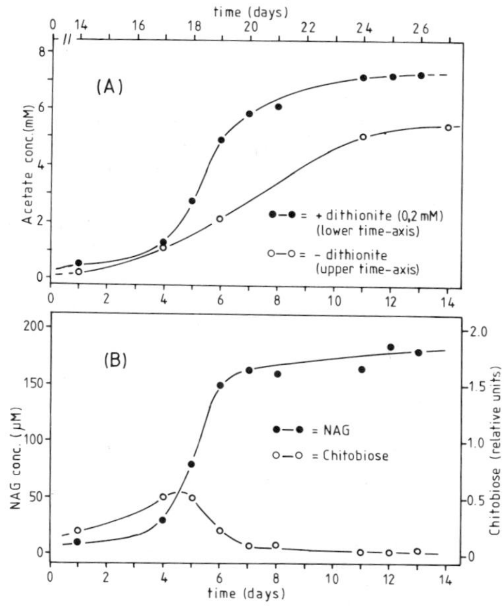

Fig. 2. (A) Chitin fermentation (acetate formation) by strain 9.1 in the absence ((o); upper time axis) and in the presence ((●); lower time axis) of 0.2 mM dithionite. (B) Accumulation of NAG and chitobiose in the culture fluid of strain 9.1 during chitin fermentation in the presence of dithionite. (As no pure standard was availabe, chitobiose concentration is presented on a relative scale).

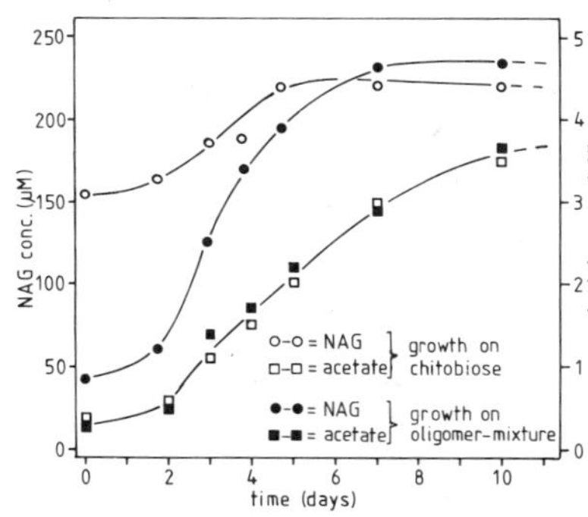

Fig. 3. Growth of chitinolytic strain 9.1 (measured as acetate formed) on chitobiose and a soluble chitin-oligomer mixture (NAG unit > 2).

growth on the amino-sugars (t_d = 6-7 days). During growth on chitin NAG gradually accumulated in the culture fluids and was not further consumed, whereas in the beginning of the active growth phase a transient accumulation of chitobiose could be detected (Fig. 2B). As these observations suggested a preference of the chitinolytic organisms for a hydrolysis product other than NAG, growth experiments on NAG-oligomers were initiated (Fig. 3). For these experiments with soluble oligomers, growth of one of our isolates (strain 9.1) was studied in more detail. The extent of NAG accumulation during growth on the mixture of oligomers (number of NAG-units > 2) is comparable to that during growth on chitin. However, with the chitobiose preparation the amino-sugar accumulation was less pronounced. The high initial amount of NAG in this experiment originated from NAG present in the chitobiose preparation.

Apparently the soluble oligomers do not offer more favourable growth conditions to this chitinolytic strain. This was indicated by somewhat larger doubling times (t_d ca. 30 h) and smaller final quantities of acetate.

DISCUSSION

Although at present no detailed knowledge exists on the quantitative importance of chitin in the flux of organic carbon in anaerobic tidal estuarine sediments, it could represent a key substrate in such environments. This is substantiated by the ease of isolating strictly anaerobic chitin degrading populations. The primary chitin hydrolyzing group of bacteria in the isolated mixed chitinolytic cultures appeared to consist of species sharing some interesting growth characteristics. In pure cultue the bacteria require very low redox potential (< -160 mV) for optimal growth, yet in the mixed chitinolytic culture E_h never drops significantly below the redox potential of the couple S^{2-}/S^0 (ca. -160 mV). It may be noted that the cellulolytic systems of for example Clostridium thermocellum and Bacteroides succinogenes is very easily inactivated by SH-groups oxidizing agents, which can be reversed by strongly reducing agents.[16,17,18] Retarded growth of the chitinolytic species when dithionite supplementation is omitted, could therefore be due to traces of oxygen entering the culture during inoculation. However, the situation is probably more complicated, as results from preliminary experiments (not shown) indicated that continued addition of dithionite resulted in an even further increase in the rate of growth. For as yet unexplained reasons this additional stimulation seems connected to the fact that the redox potential is thus maintained at a value below -160 mV.

The inability of the pure chitinolytic cultures to ferment aminosugars and the accumulation of NAG during growth on chitin indicated the preference for hydrolysis products other than the monomers. The results from experiments with NAG-oligomers make it likely that growth occurs at the expense of the dimer. NAG is accumulating in the culture as a result of the random cleavage of chitin fragments by the chitinases. The relatively small amount of NAG accumulating during growth on chitobiose probably originates from some higher oligomers (trimer) still present in our dimer preparation. A strict dependency of growth of oligomers has been reported for other anaerobic species by hydrolyzing cellulose, polyethylene glycol and polygalacturonic acid.[19,20,21] The possible advantage of the uptake of the NAD-dimer for chitinolytic organisms still has to be investigated. Chitobiose (NAG-dimer) uptake would be particulalry favourable in chitobiose-phosphorylase activity is present in the cells. Such a situation was shown to exist in for example R. flavefaciens and C. thermocellum.[22,23] Thus one ATP would be gained per two NAG fermented. Furthermore, dimer uptake may be a suitable mechanism for relieving the competitive pressure exerted by non-chitinolytic NAG fermenting bacteria, which were found to be present in numbers (data not shown) comparable to the chitinolytic species.

In the absence of both sulfate reduction and methanogenesis the rate of chitin fermentation was the same as observed during active sulfate reduction. This indicates the absence of a strict dependency of the primary chitin hydrolyzing population on the removal of fermentation products. Yet, in pure culture, chitin hydrolysis was incomplete and much slower than in the mixed culture. This could point to a similar situation as found with cellulolytic pure cultures, since several anaerobic cellulose degrading species were shown to be inhibited by glucose and cellobiose.[2,23] From our present observations (Fig. 3) it seems likely that NAG does not normally inhibit chitin fermentation at concentrations found in the pure cultures. This is concluded from the fact that growth on chitobiose (initial NAG concentration = ca. 150 μM) proceeded at the same rate as growth on oligomers in the presence of an initial NAG concentration of ca. 40 μM and because in pure chitinolytic cultures NAG was never found at concentrations higher than 200 μM.

Future research will focus on (i) the precise mechanism of the growth stimulating effects of very low redox potentials observed with the chitinolytic strains, (ii) the influence of non-chitinolytic organisms on the process of polymer hydrolysis/fermentation with particular reference to consumption of chitin hydrolysis products.

REFERENCES

1. H. J. Laanbroek and H. Veldkamp, Microbial interactions in sediment communities. Phil. Trans. R. Soc. Lond. B 297:550 (1982).
2. R. E. Hungate, "The rumen and its microbes", Academic Press, New York (1966).
3. H. D. Peck Jr. and M. Odom, Anaerobic fermentations of cellulose to methane. in: "Basic Life Sciences" vol. 18, Trends in the biology of fermentations for fuels and chemicals. A. Hollaender, ed., Plenum Press (1981).
4. N. O. van Gylswyk and H. M. Schwarz, Microbial ecology of cellulose and hemicellulose metabolism in gastro-intestinal ecosystems, in: "Current perspectives in microbial ecology", M. J. Klug and C. A. Reddy, eds., American Society for Microbiology, Washington (1983).
5. F. Colijn, Primary production in the Ems-Dollard estuary, Thesis University of Groningen, The Netherlands (1983).
6. Invertebrates of the Wadden Sea, Final report of the section "Marine Zoology" of the Wadden Sea working group, N. Dankers, H. Kühl and W. J. Wolff, eds., (1981).
7. K. Timmis, G. Hobbs, and R. C. W. Berkeley, Chitinolytic clostridia isolated from mud. Can. J. Microbiol. 20:1284 (1974).
8. R. C. W. Berkeley, in: "Microbial polysaccharides and polysaccharases", R. C. W. Berkeley, G. W. Gooday and D. C. Ellwood, eds., Special Publ. SGM: 3, Academic Press, (1979).
9. P. E. Guire, Fractionation of oligosaccharides by polyacrylamide gel filtration, Proc. Okla. Acad. Sci. 51:63 (1971).
10. H. J. Laanbroek, H. J. Geerligs, A. A. C. M. Peijnenburg, and J. Siesling, Competition for L-lactate between *Desulfovibrio*, *Veillonella* and *Acetobacterium* species isolated from anaerobic intertidal sediments, Microb. Ecol. 9:341 (1983).
11. H. J. Laanbroek, H. J. Geerligs, L. Sijtsma, and H. Veldkamp, Competition for sulfate and ethanol among *Desulfobacter*, *Desulfobulbus* and *Desulfovibrio* species isolated from intertidal sediments, Appl. Environm. Microbiol. 47:329 (1984).
12. H. G. Trüper and H. G. Schlegel, Sulphur metabolism in Thiorhodacea. I. Quantitative measurements on growing cells of *Chromatium okenii*. Antonie van Leeuwenhoek 30:225 (1964).
13. G. C. Chen and B. R. Johnson, Improved colorimetric determination of cell wall chitin in wood decay fungi, Appl. Environ. Microbiol. 46:13 (1983).
14. M. V. Tracey, Chitin, in: "Modern methods of plant analysis", K. Paech and M. V. Tracey, eds., Springer, Berlin (1955).
15. K. Blumberg, F. Liniere, L. Prestilnik, and C. A. Bush, Fractionation of oligosaccharides containing N-acetylamino sugars by reverse-phase high-pressure liquid chromatography. Anal. Biochem. 119:407-412 (1982)

16. N. Aït, N. Creuzet, and J. Cattaneo, Properties of ß-glucosidase purified from Clostridium thermocellum, J. Gen. Microbiol. 128:569 (1982).
17. E. A. Johnson and A. L. Demain, Probable involvement of sulfhydryl groups and a metal as essential components of the cellulase of Clostridium thermorellum, Arch. Microbiol. 137:135 (1984).
18. C. W. Forsberg and D. Grolean, Stability of the endo-ß-1,4-gluconase and ß-1,4-glucosidase from Bacteroides succinogenes, Can. J. Microbiol. 28:144 (1982).
19. C. Giuliano and A. W. Khan, Cellulase and sugar formation by Bacteroides cellulosolvers, a newly isolated cellulolytic anaerobe, Appl. Environ. Microbiol. 48:446. (1984).
20. B. Schink and M. Stieb, Fermentative degradation of polyethylene glycol by a strictly anaerobic, gram-negative, nonspore-forming bacterium, Pelobacter venetianus sp. nov., Appl. Environ. Microbiol. 45:1905 (1983).
21. B. Schink and J. G. Zeikus, Clostridium thermosulfurogenes sp. nov., a new thermophile that produces elemental sulphur from thiosulphate, J. Gen. Microbiol. 129:1149 (1983).
22. T. K. Ng and J. G. Zeikus, Differential metabolism of cellobiose and glucose by Clostridium thermohydrosulfuricum, J. Bacteriol. 150:1391 (1982).
23. M. C. Fasee and J. M. Leatherwood, Regulation of cellulase from Ruminococcus, Can. J. Microbiol. 18:347 (1972).

CHITIN DEGRADATION IN NATURAL ENVIRONMENT (MOLLUSK SHELLS AND CRAB CARAPACES)

M. POULICEK, G. GOFFINET, M. F. VOSS-FOUCART, J. C. BUSSERS,
M. F. JASPAR-VERSALI and C. TOUSSAINT

Laboratory of Animal Morphology, Systematics and Ecology
University of Liège, Zoological Institute, 22, Quai Van Beneden
B-4020 Liège, Belgium

INTRODUCTION

Most marine sediments are organodetritic in nature, and many particles of the coarser fraction are constituted of more or less stable remains of biological origin (skeletal pieces of benthic or planctonic animals, plant remains,...)(1). In this respect, dead crustacean cuticles and mollusk shells introduce into the sediments the chitinoproteic compounds of their matrix, closely associated to calcified crystallites.

All the material settled down is not actually embedded in the sedimentary layers since the chemical composition of the skeletal pieces is more or less affected by several diagenetic events during their transit through the upper aerobic layers of the sediments (1) :

- Where hydrodynamism is high, mechanical effects (waves, currents, turbidity floods,...) brake and erase the skeletal pieces (2).
- In undersaturated sea-water (high latitudes, deep waters,...), purely chemical dissolution processes affect the carbonates in different ways following their mineralogy (1, 2).
- Everywhere, the biological weathering of skeletal pieces by microorganisms, microborers and/or meiofauna is obvious. One of us (MP) showed this is, by far, the most important way of degradation of skeletal pieces when experimental material is laid down on shallow-water sediments (- 37 meters) in aerobic conditions.

The scope of this paper is precisely to enlighten some aspects of crustacean cuticles and mollusk shells organic matrices biodegradation (chitin was estimated by the method of Jeuniaux (3) and proteins by the method of Lowry (4)). We shall discuss some aspects of the results of a long term experiment of biological skeletons weathering in the Calvi bay (Corsica, Mediterranean sea).

MOLLUSK SHELLS BIODEGRADATION

Various communities of endolithic organisms, together with chasmolithic bacteria, diatoms, yeasts,...and meiofauna of the upper layers of the sediments are accountable for the biodegradation of the chitinoproteic matrices of mollusk shells (1). Extracellular hydrolytic enzymes (chitinase, protease) are produced by microborers in close contact with the organic sheaths around calcified crystallites (5,6).

Fig. 1 shows the kinetics of chitin and protein extraction in three isolated layers of shells, mother of pearl (A), prisms (B) and crossed-lamellar microstructure (C).

In mother of pearl, the degradation of chitin proceeds very fast during the first months after settling of a dead shell onto the sediment (1,7). This first step, similar to the degradation of crustacean cuticles, is tentatively attributed to the hydrolytic activity of "free enzymes" in the inter-

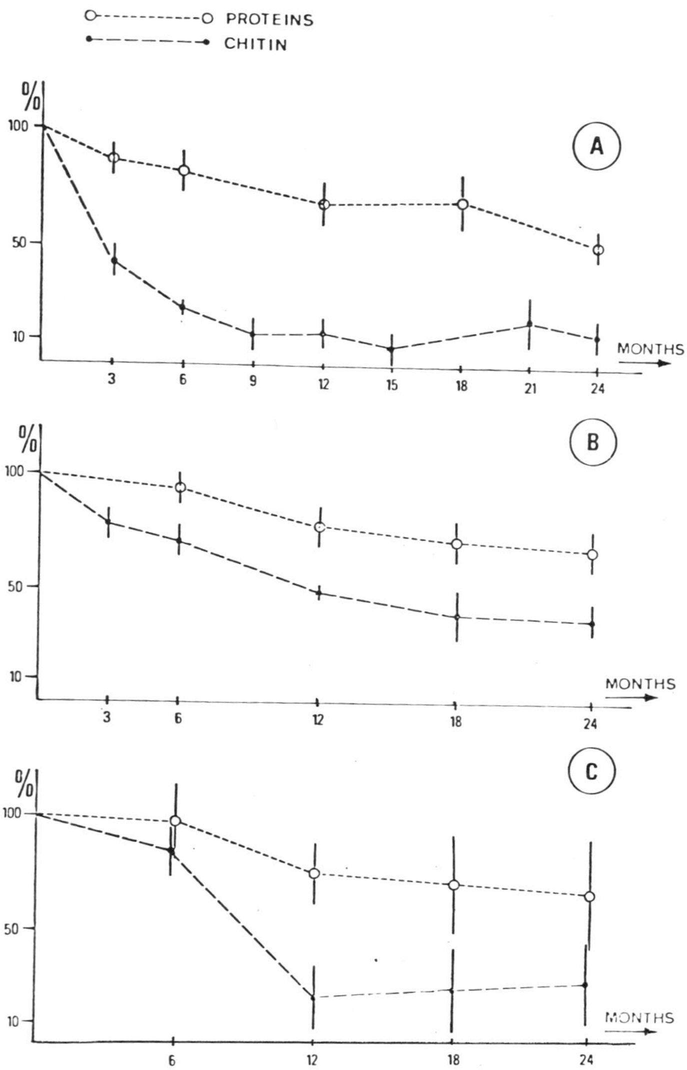

Fig. 1 : Decrease of the organic content of isolated layers of Mollusk shells in the course of early diagenesis. (A): Mother of pearl of *Mytilus edulis* ; (B): prismatic layer of *Pinna nobilis* ; (C): crossed-lamellar layer from *Tridacna gigas*. (Reproduced from (8)).

stitial waters, or enzymes adsorbed to sediment particles (1). In all three layers (A,B,C) proteins are extracted at a constant rate (8). Chitin, even thought more resistant, appears more unstable than proteins, probably due to its external location around calcified crystallites, whereas proteins can be bound to or sheltered within calcified crystallites (9).

CRUSTACEAN CUTICLES BIODEGRADATION

The cuticle of crustacea shows drastic ultrastructural alteration as early as 11 days after calcified fragments (cephalonotal shields of *Carcinus maenas*) were laid down onto an organodetritic sand.

As a general rule, the fragments are devoid of the membranous layer which splitts very soon from the overlayered calcified principal layer. Progressively, the chitin-protein macrofibers of the principal layer and the chitin-protein reticulate arrangement of the pigmented layer become loose and more or less disorganized according to the fragment and the cuticle level examined. In contrast, the non chitinous epicuticle (tanned lipoproteins)

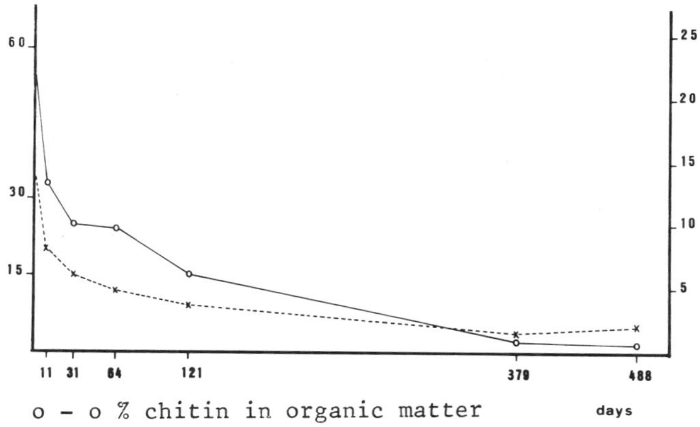

o - o % chitin in organic matter

x - x % organic matter in residual mat.

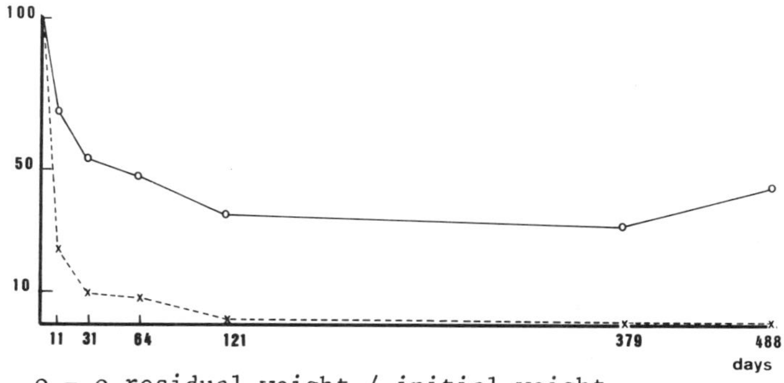

o - o residual weight / initial weight

x - x chitin : % of initial chitin

Fig.2 . Chemical changes in *Carcinus maenas* cuticles during aerobic weathering in a marine sand.

exhibits a higher resistance to the weathering factors since the first morphological alterations appear only two months after fragments immersion.

From a chemical point of view, one can observe a gradual loss of the organic compounds so that, four months after immersion, the organic matter and chitin constitute respectively only 4.3 % and about 1.0 % when compared with their initial amounts.(Fig.2)

In contrast with the shells of mollusks, the activity of endolithic organisms, at least under our experimental conditions, seems to be little concerned with the aerobic weathering of the chitinoproteic cuticles of crustacea.

REFERENCES

1. M. Poulicek, "Coquilles et autres structures squelettiques des Mollusques Composition chimique, biomasse et biodégradation en milieu marin", Ph.D. Thesis, University of Liège, 180 p. (unpublished) (1982).
2. J.D. Milliman, "Marine carbonates", Springer Verlag, Berlin (1974).
3. Ch. Jeuniaux,"Chitine et chitinolyse", Masson ed., Paris (1963).
4. O.H. Lowry, N.J. Rosebrough, A.L. Farr and R.J. Randall, Protein measurements with the Folin phenol reagent, J.Mar.Res.,33 : 265 (1951).
5. M. Poulicek, M.F. Jaspar-Versali and G. Goffinet, Etude expérimentale de la dégradation des coquilles de Mollusques au niveau des sédiments marins, Bull.Soc.r.Sci.Liège, 50(11-12) : 513 (1981).
6. M. Poulicek and M.F. Jaspar-Versali, Biodégradation de la trame organique des coquilles de Mollusques en milieu marin : action des microorganismes endolithes, Bull.Soc.r.Sci.Liège, 53(2) : 114 (1985).
7. M. Poulicek and Ch. Jeuniaux, Biomass and biodegradation of Mollusk shell chitin in some marine sediments, Proc.IId.Intern.Conf.Chitin Chitosan, S. Hirano and S. Tokura ed., Sapporo (1982).
8. M. Poulicek, Shell matrix weathering in marine sediments, Malacologia, in press (1985)
9. M. Poulicek, M.F. Voss-Foucart and Ch. Jeuniaux, Chitinoproteic complexes in Mollusk shells, this Volume.

Editor's Report

Ch. JEUNIAUX

CHITOSAN AS A TOOL FOR THE PURIFICATION OF WATERS

based on works by

K. A. ALMAS, A. BARADARAJAN, H. BLAIR, P. DECOCK,
S. DEIANA, A. DESSI, B. DUBOIS, A. FINDON,
H. KOZLOWSKI, J. McKAY, G. MICERA,
C. A. SASTRY, C. SENSTAD, J. P. THOME
J. VAN DAELE, and B. VENKATRAO

INTRODUCTION

It is well known that surface waters are polluted not only by inorganic chemicals such as heavy metals, but also by organic compounds. Owing to the high sorption capacities of chitosan this polymer can be utilized as a tool for the epuration of wastewaters on polluted stream waters. Some typical examples of the application of chitosan to this kind of environmental problems are given here. They include the prevention of water pollution by higly toxic chlorinated aromatic compounds, metal ions, and dyes which are amply used in the textile industry. The recovery of proteins from wastewaters is an example of pollution control utilized in other industries, including shrimp processing, as well.

ADSORPTION OF POLYCHLORINATED BIPHENYLS (PCB) ON CHITOSAN AND APPLICATION TO DECONTAMINATION OF POLLUTED STREAM WATERS

(J. P. Thomé and Y. Van Daele)

In spite of the widespread contamination of aquatic environment by PCB (1), it is surprising that the ability of chitosan to eliminate these xeno-biotics from natural waters has not been researched. However, standard methods for water purification used in water softening plants remain largely ineffective to remove PCB (2).

The present study deals with adsorption capabilities of chitosan towards PCB's and especially the most chlorinated ones which are the most frequently detected in aquatic ecosystems. Chitosan, manufactured from shrimp shells (*Crangon crangon*) according to the procedure described by Asano et al. (3), has been used in laboratory experiments to filter low and highly polluted water by PCB (4). The impact of such a treatment on the aquatic environment was tested by following PCB accumulation in young specimens of the sensitive teleost *Barbus barbus* kept in PCB contaminated water filtered or not through chitosan.

PCB adsorption capabilities

Adsorption capabilities of chitosan and other adsorbing agents (C_{18}, activated charcoal, sand and chitin) was tested using the following procedure:

100 ml distilled water spiked with 0.5 ppb PCB (Aroclor 1260) were filtered through cartridges packed with the material tested (100 to 200 mg). PCB remaining in water after filtration was extracted and analyzed by GLC as previously described (5).

According to Table I, sorption abilities of chitosan were more efficient than those of other tested organic substances. Indeed, up to 84% of the PCB present in water was adsorbed by 100 mg of chitosan. Chitin, the structure of which is close to that of chitosan, also had relatively good adsorption properties towards PCB but the reproducibility was poor. Furthermore, the efficiency of chitin was not significantly different from that of activated charcoal and sand ($P < 0.01$).

The concentration of PCB on chitosan was higher that that measured on the other adsorbing materials. C_{18} could be used as an efficient PCB adsorbing medium, but its high cost makes it unsuitable for application on an industrial scale.

<u>Decontamination of PCB polluted stream water and impact on a test fish species</u>

The ability of chitosan to remove PCB from contaminated stream water was tested using a dynamic flow through a system composed of two 50 liter glass tanks. At the beginning of the experiment, tanks were filled with natural pond water free of PCB. Polluted water (0.5 ppb Aroclor 1260) filtered or not through chromatographic columns packed with 1 to 10 g chitosan enriched, at a constant rate (36 l/day), each of experimental tanks. Chitosan was replaced every day. Thirty barbels (one year old, mean weight: 1.365 ± 0.55 g), originating from experimental fish farms, were placed in each of the experimental tanks. Analyses of PCB concentrations in water and in whole fish were carried out as a function of time as previously described (5).

The evolution of PCB concentration in water (Bob), without filtration on chitosan, was quite similar to the theoretical curve (Bth) quoted in Figure 1 (χ^2 test: $P < 0.01$).

The amount of PCB in fish linearly increased during 120 h of progressive contamination, reaching a maximum mean value (1.6 ppm) after 120 h and then slowly resuming a lower level.

Table I. Efficiency of various materials to remove PCB from distilled water (100 ml) spiked with 0.5 ppb PCB

adsorbing material		mean percentage of adsorption ± S.D.	Concentration of PCB on adsorbent ± S.D. (ng/g)
chitosan	(100 mg)	83.3 ± 7.5	416 ± 35
chitin	(100 mg)	66.1 ± 33	330 ± 153
activated charcoal	(200 mg)	66.6 ± 1.2	166.4 ± 3
sand	(200 mg)	59.2 ± 3.6	148 ± 9
C_{18}	(150 mg)	97 ± 4.1	323 ± 14

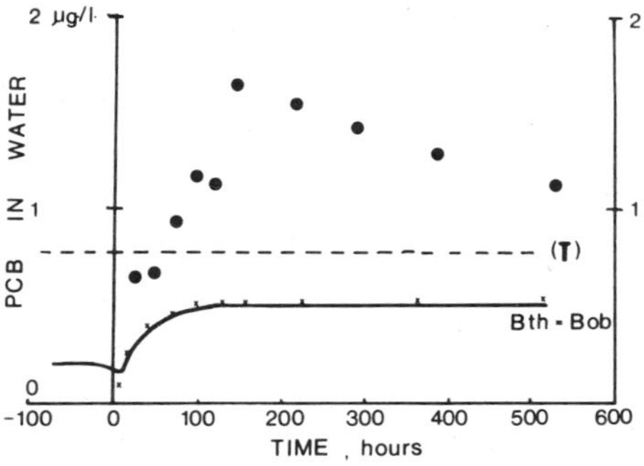

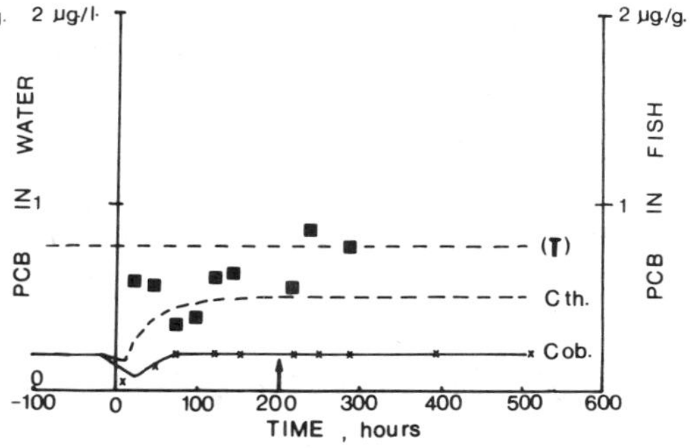

Figure 1 : Evolution of the PCB concentrations in water and in fish (●) in the (left) experimental tank receiving contaminated water without filtration on chitosan. Level of contamination of control fish (---T).

Figure 2 : Evolution of PCB concentrations in water (Cob.) and in fish (■) (right) after filtration through chitosan of PCB polluted stream water. Level of contamination of control fish (---T).

The results quoted in Figure 2 were obtained after the filtration of the PCB contaminated stream water on chitosan. The concentrations of PCB in filtered water reached only 0.2 ppb instead of 0.5 ppb and remained unchanged during the 250 h of the experiment. One gram chitosan is thus sufficient to eliminate more than 60 % of PCB from 0.5 ppb polluted stream water (36 liters). Efficiency of such colums allowed the maintainance of contamination in fish at a non toxic level (<1 ppm) (6). Fish contamination in such a water was quite similar to that of controls. Experience has proved that larger amounts of chitosan (10 g) did not result in further purification. As a consequence, while the efficiency of chitosan was not sufficient to completely eliminate PCB from water, it however appeared to be quite sufficient to protect the animals against any further intoxication.

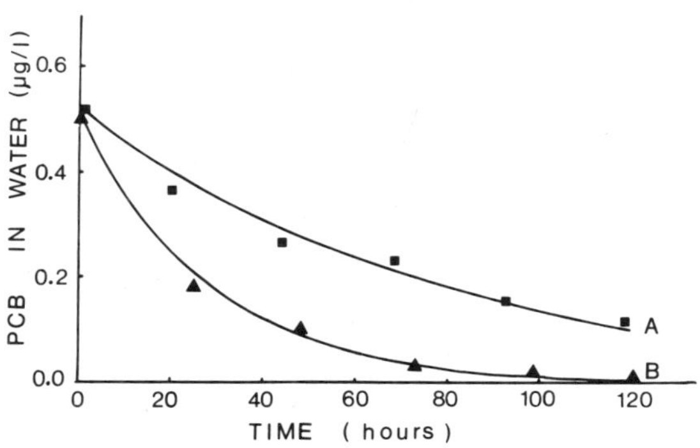

Figure 3 : Evolution of PCB concentrations in water containing 0.5 ppb Aroclor 1260 during filtration in a closed loop system through activated charcoal (1.5 kg)(A curve) and through activated charcoal (1.5 kg) + chitosan (75g)(B curve).

Elimination of PCB from contaminated water in steady state

The procedure of filtration is suitable to protect "clean" natural water from contamination by PCB polluted effluents. However its efficiency could be very poor to eliminate PCB from water already contamined by these toxics. With this in mind, we filtered 50 liters of PCB contaminated water (0.5 ppb) in a closed loop system using a chitosan filter (75 g) provided with an activated charcoal filter (1.5 kg). Water circulation was achieved by a pump at a rate of 180 l/h.

Complete removal of PCB from water was rapidly obtained, after less than 120 hours (B curve in Fig. 3). With an activated charcoal filter, without chitosan, concentrations decreased only to 0.2 ppb after the same time (A curve). Activated charcoal would work as an adsorbent of a fraction of PCB present in water but principally as a filter towards particles in suspension able to saturate adsorption sites of chitosan.

The highly efficient purification of PCB contaminated water using a chitosan filter could find favourable application to environment preservation. Chitosan appears to be much more efficient than activated charcoal and chitin for a nearly identical cost. Its powerful adsorption capabilities towards most of the persistent toxics found in aquatic environment make chitosan one of the most suitable polymers to be used for elimination of non biodegradable toxics.

REFERENCES

1. P. Davar, and J. P. Wightman, Interaction of pesticides with chitosan, in: "Adsorption from Aqueous Solutions", P. H. Tewari, ed., Plenum Publishing Corporation, New York (1981).
2. E. S. Tücker, and H. Saeger, Activated sludge primary degradation of PCB, Bull. Environ. Contam. Toxicol., 14:705 (1975).
3. T. Asano, and N. Hamakawa, Chitosan applications in waste water sludge treatment, in:"Proceedings 1st International Conference Chitin/ Chitosan", R. A. A. Muzzarelli and E. R. Pariser, eds., MIT, Cambridge (1978).
4. A. Lafontaine, and S. Delforge, "Evaluation de l'impact des PCB et PCT sur l'environnement aquatique", CEE Env/178/80, IHE Bruxelles (1979).
5. Y. Van Daele, and J. P. Thomé, Purification of PCB contaminated water by chitosan: a biological test of efficiency using the common Barbel (Barbus barbus), Bull. Environ. Contam. Toxicol., under press (1985).
6. C. L. M. Poels, M. A. Van Der Gaag, and J. F. J. Van De Kerkhoff, An investigation into the long-term effects of Rhine water on rainbow trout, Water Res., 14:1029 (1979).

ADSORPTION OF DYESTUFFS ON CHITOSAN

(B. Venkatrao, A. Baradarajan and C. A. Sastry)

Effluents from dyehouses, when discharged into any watercourse, at once become obvious on account of the high tinctorial value of many dyes currently used for dyeing a variety of natural and synthetic fibres. Removal of colour from dye waste waters by adsorption has been studied extensively in the recent years. A wide variety of adsorbents have been employed (1-7). More recently, chitin has been used in the removal of colour from direct and acid dyes (8, 9). However, there has only been a limited mention of the adsorption characteristics of chitosan on dyes (10).

In this paper, adsorptive behaviour of chitosan with respect to two dyes (C.I. Reactive Red 73 and C.I. Direct Red 31) has been studied. Information on removal of reactive dyes in particular, has been very limited, despite being one of the most popular brands of dyes currently employed in the textile industry for dyeing cotton, protein and polyamide fibres (11).

EXPERIMENTAL

All experiments to study the influence of process parameters on the dye uptake were carried out using a cylindrical vessel, 110.8 mm in diameter. Four baffles were fixed, evenly spaced around the circumference. Agitation was achieved using a flat paddle stirrer driven by a 1/20 h.p. (Remi) fractional motor. In all experiments the stirrer was maintained at 800 rpm. The various parameters studied and their ranges were: initial dye concentration, 80 - 500 ppm; initial pH of the dye solution, 2 - 12; temperature, 30 - 70°C; particle size, 150 - 1000 microns.

In preparation for each run, 1.314 litres of 250 ppm of respective dye solution was placed in the vessel and 4 g of chitosan were added. Samples were withdrawn, at varying intervals of time and dye concentrations were determined using ECIL spectrophotometer. Transmittance measurements were made at λ_{max} corresponding to each dye (537 nm for C.I. Reactive Red 73 and 526 nm for C.I. Direct Red 31).

RESULTS AND DISCUSSION

In diffusion studies rate processes are usually expressed (12) in terms of the square root of time. Fractional uptake is generally plotted against $t^{0.5}$ to study the effect of intraparticle diffusion on the rate of dye removal. The diffusion process is generally described by Fick's Law with various restrictive assumptions exercised in showing the diffusion equations. Often, a constant diffusivity is assumed throughout the adsorption process and the sorbent particle. Since the particle size of the adsorbent during any run remains constant the dye uptake is linearly dependent on the square root of time. Consequently a rate parameter, K, has been defined as the slope of a graph of the amount of dye adsorbed per gram of adsorbent, against $t^{0.5}$. This is given by

$$\frac{X}{M} = \frac{C_o - C_t}{C_o} \cdot \frac{V}{M}$$

where X/M is the amount of dye adsorbed/g of adsorbent. V is the volume of solution, M is the mass of adsorbent added. C_o and C_t are initial and instantaneous concentrations respectively. So,

$$K = \frac{X/M}{t^{0.5}}$$

K has dimensions of mg/g/min $^{0.5}$. Although this is not a normal dimension of rate, the relative values of K for similar conditions should be significant as rate parameters.

Effect of initial dye concentration

Whereas, in the case of C.I. direct red 31 (Fig. 1), it is observed that the time required to reach equilibrium increases considerably (more than 15 hours for C_o = 500 ppm) as the initial dye concentration is increased, no such trend is evident in the adsorption of C.I. Reactive Red 73. A similar trend has been observed by Ramaprasad (13) for adsorption of methylene blue and Safranine T dyes on white garlic onion skin.

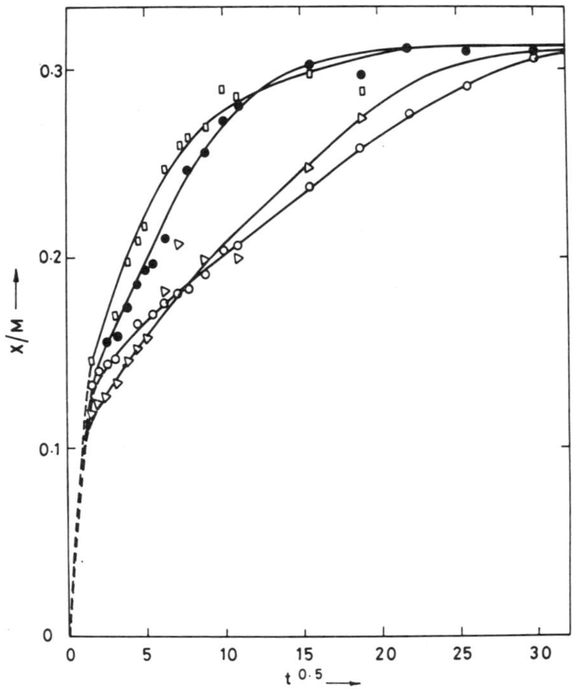

Fig. 1. C.I. Direct Red 31 uptake on chitosan.

Rate parameter data for both the systems are determined from the linear portions of the plots and are given in Table 1.

The trend shows that for the direct red dye, K values are almost constant for a wide range of initial dye concentration, whereas those of reactive red dye suggest a proportionality.

The correlations obtained for the two dyes are as follows :

Direct Red dye (C.I. 31) $k = 0.014\ C_o^{0.88}$

Reactive Red (C.I. 73) $k = 0.074\ C_o^{0.14}$

Considering the linearity of the plots and the power constant values obtained, it appears that while intraparticle diffusion is the main rate controlling step for most part of the process, a boundary layer resistance is experienced (14). Then the early onset of diffusional process in case of direct red dye is probably indicative of a strong affinity between the sorbent and sorbate. This would explain the steepness in the initial region of the plots which is due to surface adsorption followed by diffusional constraints in the later stages. The increasing steepness of approach to equilibrium, with increasing initial dye concentration is due to an increase in the driving force, forcing the dye from solution upto and onto the sorbent particle.

Effect of Temperature

An initial increase in rate with increase in temperature is observed for both the systems but with increase of time, the dye uptake falls off to an equilibrium value. This is expected in systems where physical adsorption takes place in which the increase in rate of adsorption in the early stages of the process is offset by increasing rate of desorption as time progresses,

Table I. Rate parameter data.

Initial dye concentrations	K, mg/g/min$^{0.5}$	
	C.I. Direct Red 31	C.I. Reactive Red 73
500	0.095	0.058
375	0.078	0.044
250	0.063	0.031
125	0.097	--
80	0.086	--

at high temperatures. In this context it is surprising to note the large initial uptake of the Reactive Red dye, at 40°C. Also noteworthy is the trend of the plot, showing a "step", at the lower temperatures.

The rate parameter for the adsorption of two dyes on chitosan is presented in Table II.

Table II. Rate parameter data.

Temperature, ° C	K, mg/g/min$^{0.5}$	
	C.I. Direct Red 31	C.I. Reactive Red 73
30	0.059	0.021
40	0.063	0.029
50	0.064	0.038
60	0.079	--
70	0.091	0.045

$$K = K_o\ 2.303.\ e^{-E/RT} \qquad \text{or} \qquad \log K = \frac{E}{RT} + \log\ (2.303\ K_o)$$

From the plot of log K vs. 1/T, the activation energy for adsorption is determined to be - 2.59 for C.I. Direct Red 31 and - 5.78 Kcal/mol for C.I. Reactive Red 73.

Thus it is seen that the activation energy required for the adsorption of Reactive Red 73 on chitosan is slightly more than double the activation energy of Direct Red 31 onto the adsorbent. Since both these dyes indicate increasing adsorption rate with decreasing temperature, its seems that physical forces such as molecular orientation or intermolecular ionic interactions are responsible for the large increase in the activation energy when passing from one dye system to the other. Also, many dyes are known to form agglomerates and these clusters are known to behave as single molecular species. In this study, this point seems to have some weight as indicated by a higher activation energy for reactive red dye as compared to the Direct red dye. This is because the "larger" dye molecule of C.I. Reactive Red 73, if it were to form agglomerates, would face increased diffusional constraints as compared with that for simple molecules.

Effect of particle size of chitosan

It was proposed (12) that if the mechanism of uptake is one of adsorption on the external sites of a non porous adsorbent the rate of adsorption and hence the parameter 'K' should vary reciprocally with the first power of the particle diameter. The same is proposed to hold for a porous adsorbent as well, when mass transfer to particle is controlled by an external resistance.

This is, film diffusion (or boundary layer resistance) is significant. On the other hand, if intra-particle diffusion is the sole rate determining step, then the rate parameter should vary as the square root of the particle diameter. As can be seen from the correlations obtained for the two dyes studied, this is approximately the case :

Direct Red (C.I. 31) $k = 0.02\ dp^{0.43}$

Reactive Red (C.I. 73) $k = 0.025\ dp^{0.62}$

Thus in the case of C.I. Direct 31 adsorption on chitosan, intraparticle diffusion is predominantly rate controlling whereas, for C.I. Reactive Red 73, more than one process is involved in the rate controlling step. While intra-particle diffusion is important, there is some contribution of surface mass transfer process as well to the overall rate of dye uptake (see Table III).

Table III. Rate parameter data.

Particle size, microns	K, mg/g/min$^{0.5}$ C.I. Direct Red 31	C.I. Reactive Red 73
150 - 300	0.0716	0.0043
300 - 500	0.0625	0.0559
500 - 700	0.0625	0.0345
700 - 1000	0.452	0.0277

Effect of pH of the dye solution

Any solid surface in contact with a liquid is known to have a surface charge and this plays a dominant role in the adsorption of solutes from solution. Thus, at low pH values, adsorbent surfaces that are negatively charged are neutralised and hence attract solute molecules much better. The nature of the solute is also important in determining the rate of its adsorption on the sorbent surface. For C.I. Direct Red 31 dye, adsorption is considerably enhanced around pH 2.0.

CONCLUSIONS

The kinetics of sorption of C.I. Direct Red 31 and C.I. Reactive Red 73 on chitosan are affected markedly by the particle size of chitosan, and temperature. The influence of the initial dye concentration on the uptake of reactive red dye and direct dye is complex in that boundary layer resistance must also be considered, apart from intra-particle diffusion, in any description of the sorption process. Temperature effects indicate that physical adsorption is taking place. However, McKay (14) has pointed out that in certain situations, it is difficult to distinguish between physical adsorption and chemisorption of dyestuffs on sorbent surfaces. However, intra-particle diffusion is a major rate controlling step in the adsorption of Direct and Reactive dyes from aqueous solutions. A rate parameter, K, has been defined and used to correlate experimental data against the process parameters in the following form :

$$K = P\ (\text{Parameter})^{Q}$$

Nomenclature

C_o : Initial dye concentration in solution (mg/l); dp : particle diameter (cm); K : Rate parameter (mg/g/min$^{0.5}$); M : Mass of chitosan (g); P : Constant in the correlation of K with process parameter; Q : Power constant value in the correlation; X : mass of dye adsorbed (mg).

Acknowledgement. The grant of a fellowship to B.V.R. by the Indian Institute of Technology Madras, is gratefully acknowledged. The authors also thank M/S Amar Dye Chem. Ltd., Madras for supplying the dyes.

REFERENCES

1. Mantell, "Adsorption", McGraw-Hill, London (1945).
2. V. J. P. Poots, G. McKay and J.J. Healy, Sci. Proc. Roy. Dub. Ser., A, 6:61 (1976).
3. D. H. Sharp and A. E. Lambden, Chemistry and Industry, London, 39:1207 (1955).
4. M. Dohanyos, V. Madera and M. Sedlacek, Prog. Wat. Tech., 10:559 (1978).
5. R. A. A. Muzzarelli, "Natural Chelating Polymers", Pergamon Press, New York (1973).
6. R. A. A. Muzzarelli, "Chitin", Pergamon, Oxford (1977).
7. R. A. A. Muzzarelli and O. Tubertini, Talanta, 16:1571 (1969).
8. G. McKay, H. S. Blair and J. Gardner, J. Appl. Polym. Sci., 27:3043 (1982).
9. G. McKay, H. S. Blair and J. Gardner, J. Appl. Polym. Sci., 27:4251 (1982).
10. H. Yamamoto, Makromol. Chem., 185:1613 (1984).
11. F. Sadov, M. Korchagin and A. Matetsky, "Chemical Techology of fibrous materials", English Translation, Mir Publishers (1978).
12. W. J. Weber, and J. C. Morris, Advances in Water Pollution Research, Pergamon, Oxford (1962).
13. G. Ramprasad, Unpublished data.
14. G. McKay, J. Chem. Tech. Biotechnol., 33A:196 (1983).

KINETICS OF COPPER UPTAKE ON CHITOSAN

(G. McKay, H. Blair and A. Findon)

Adsorption may be used as a method of metal recovery, it has long been known that chitosan due to its high nitrogen content (7.2%) is an excellent metal adsorbent.[1]

The aim of the present work is to determine equilibrium isotherms for particle size and solution temperature variation, to determine the saturation capacity. Batch adsorbers were used to measure the effect of copper removal, with variation with mass and particle size and the effect of initial copper concentration. Using single resistance models the external film transfer coefficient (βL) and internal mass transfer coefficient (K) were determined. The chitosan used in this study was derived from prawn shell (nephrops) from Portavogie Sea Food Factory, N.I., where in 1981 3,752 tonnes of nephrops were landed.

EXPERIMENTAL

Preparation of Chitosan

Chitin was prepared from prawn shells, it was deacetylated by heating in a 1:1 (w/w) NaOH solution for two, 30 minute periods at 110°C in a nitrogen atmosphere. The chitosan was then ground to the required particle sizes.

Equilibrium Studies

Equilibrium isotherms were determined by contacting a constant mass of chitosan of known particle size range with a series of different concentrations of copper solutions. The chitosan and 0.050 dm^3 Cu(II) solutions were agitated for seven days at room temperature.

Contact Studies

The influence of the previously mentioned variables were studied in batch adsorption vessels. To obtain good mixing the vessel dimensions were designed according to standard specifications[2]. A 2 dm^3 cylinderical glass vessel was used, with 1.7 dm^3 liquid volume. Mixing was provided using a 6 bladed glass, flat-bladed impeller having a 0.075 m diameter and blade height 0.01 m. A heidolph type 5011 variable speed motor was used to drive the impeller using a 0.013 m diameter glass shaft. 8 perspex baffles were spaced evenly around the vessel circumference at 45° intervals and was 0.013 m wide.[3]

RESULTS AND DISCUSSION

Adsorption Isotherms

Isotherms were produced by plotting amount of copper adsorbed by a given mass of chitosan Vs the final concentration of copper left in solution at equilibrium (Fig. 1). The plateau of each isotherm corresponds to monolayer coverage of chitosan by copper ions. Langmuir[4] and Freundlich[5] equations were used to describe each isotherm and the respective constants, b, Q and p, n were calculated (Table 1). The Langmuir equation best described the isotherms within the concentration range tested. The Freundlich equation, however, can only be used to describe the isotherm at low concentrations. The results indicate only a small variation in adsorption capacity per unit mass of chitosan with particle size, but a decrease with increasing temperature. The influence of isotherm shape may be used to predict the favourability of an adsorption system[6] by using the Langmuir constant 'b'. All isotherms indicate a favourable adsorption process.

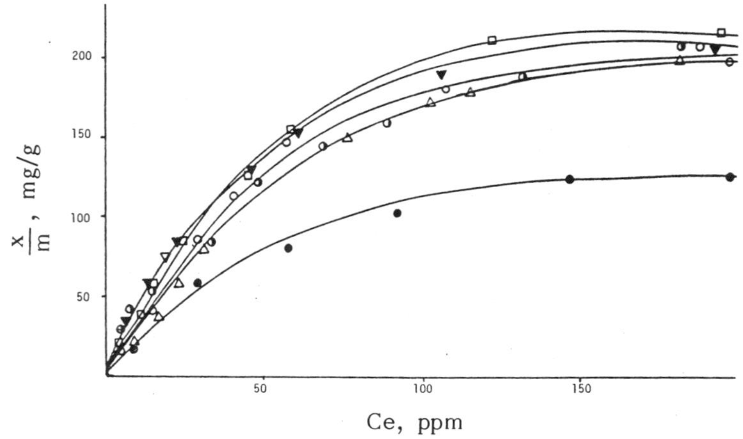

Fig. 1. Isotherm for the Adsorption of Cu II Ions on Various Particle Sizes of Chitosan, and Different Solution Temperatures, Amount Adsorbed, $\frac{x}{m}$, vs Ce

d_p (μm) (o) 710-1000, (□) 500-710, (Δ) 355-500 (◑) 210-355, at 20°C (▼) 40°C, (●) 60°C, at d_p 710-1000 μm.

Table I. Langmuir and Freudlich constants for the adsorption of Copper(II) ions of chitosan.

Particle Size, μm	Temp. °C	Langmuir Constants Q, mg/g	b, mg^{-1} dM^{-1}	Freudlich Constants p, mg/g	n, g/1
805	25	245.0	0.024	4.57	1.22
605	25	252.0	0.022	5.24	1.14
427	25	228.0	0.023	4.57	1.19
282	25	240.0	0.023	5.25	1.11
805	40	230.0	0.024	4.78	1.14
805	60	128.0	0.036	5.0	1.5

Kinetics and Rate Determining Steps of Adsorption

In most treatments in a well agitated adsorption vessel, 3 consecutive steps in the adsorption of the solute are visualized.[2]

(1) Mass transfer of solute from the bulk solution to the particle surface (external mass transfer βL).

(2) Intraparticle diffusion.

(3) Adsorption at an interior site.

The 3rd step is assumed rapid whilst steps 1 and 2 are the rate determining steps, either singly or in combination whether step 1 or 2 offers the most resistance to adsorption, depends to a great extent upon the method of contact of the solution with the adsorbent particle.

A simple model was used to calculate the external film mass transfer coefficient (βL). The model assumes that the surface concentration is negligible at t = 0, and as a consequence, intraparticle diffusion is also negligible.

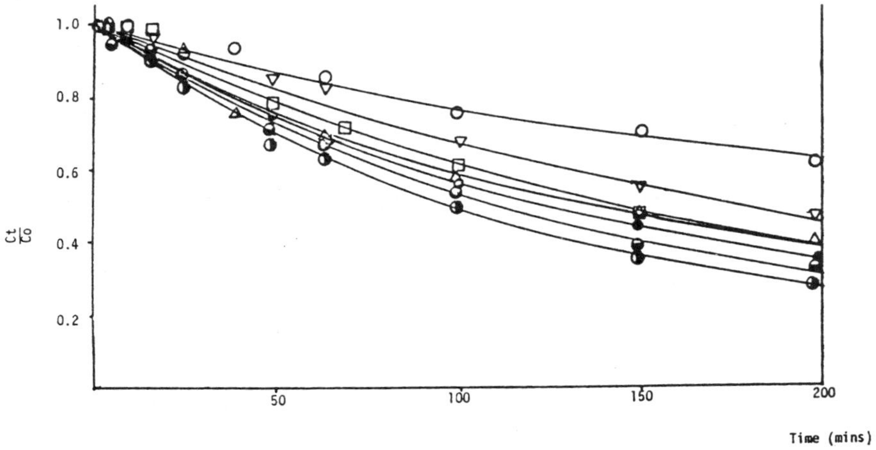

Fig. 2. Effect of agitation for the adsorption of CuII ions on chitosan. Co = 100 ppm. M = 1.7g, dp = 805 x 10^{-4} cm. (O) 50 rpm, (∇) 100, (□) 200, (Δ) 300, (●) 400, (◒) 500, (◑) 600.

In a well agitated system the solute and adsorbent particle concentrations in the liquid phase is assumed to be uniform and 'm' (solid concentration) is therefore determined from the measured mass of chitosan, w, and the volume of particle free liquid, v, according to the equation:

$$m = \frac{w}{v} \tag{1}$$

The change in solute concentration with respect to time is related to the liquid-solid film mass transfer coefficient, βL, by the equation:

$$\frac{dct}{dt} = - \beta L \quad S(Ct - Cs) \tag{2}$$

where Ct = Co at t = 0.

Assuming smooth, spherical particles, the surface area for mass transfer, S, can be obtained from 'm' using the equation:

$$S = \frac{6m}{dp\ \rho(1 - \varepsilon)} \tag{3}$$

If effective diffusion coefficients are neglected, then since Cs → 0 and Ct → C_o at t → 0 then equation 2 becomes:

$$\left.\frac{d\frac{ct}{co}}{dt}\right|_{t \to 0} = - \beta L.S. \tag{4}$$

Hence, a plot of $\frac{ct}{C_o}$ vs time will yield an initial slope of

$$\left.\frac{d\frac{ct}{co}}{dt}\right|_{t \to 0} = \beta L.S \tag{5}$$

βL was calculated for each variable (Table 2) $\frac{ct}{co}$ vs t was plotted for agitation data (Fig. 2).

Intraparticle Diffusion

Having crossed the boundary layer the copper ions may diffuse into the particles. Ficks 2nd Law[7] describes diffusion in an isentropic medium where there is a concentration gradient along the x axis.

Table II. External film mass transfer (βL, cm^2/sec) for each variable.

Variable	βL, $cm^2/sec \cdot 10^{-3}$	Variable	βL
concentration, mg/l		agitation, rpm	
47	63.6	100	7.7
137	29.8	200	10.7
282	15.8	400	12.2
366	7.9	500	11.9
temperature, °C		particle size, μm	
20	20.0	855	12.6
40	31.3	605	14.0
60	56.7	427	10.8
70	102.3	202	5.7

$$\frac{dc}{dt} = D \frac{d^2C}{dx^2} \qquad (6)$$

This may be modified for spherical co-ordinates (spherical co-ordinates r, θ and ϕ) by the equation,

$$\frac{dc}{dt} = \frac{1}{r^2} \left(\frac{\delta}{\delta r} Dr^2 \frac{\sigma c}{\sigma r} + \frac{1}{\sin\theta} \frac{\delta}{\delta\theta} D\sin\theta \frac{\delta c}{\delta\theta} + \frac{D}{\sin^2\theta} \frac{\delta^2 c}{\delta\phi^2} \right) \qquad (7)$$

The mathematical solution to unsteady state diffusion for spheres has been solved taking into account the particular boundary equations.

The model assumes:

(1) The concentration of solute is uniform at initial concentrations and is zero in the adsorbent particle at t = 0.

(2) The diffusion is radial and has no variation in concentration with angular position.

(3) Resistance to transfer in the medium surrounding the particle is significant only in the initial stages of diffusion (i.e. due to external film mass transfer).

One solution for diffusion (unsteady state) in a sphere has found a linear relationship between,

$$\frac{Co - Ct}{Co - Ce} \quad \text{and} \quad \left(\frac{Dt}{r^2}\right)^{0.5} \qquad (8)$$

for much of the adsorption process.

Since D and r are constant for a given system, then the amount adsorbed, $\frac{x}{m}$, will be proportional to

$$\frac{Co - Ct}{Co - Ce}$$

and will thus be linearly dependent on $t^{0.5}$.

The boundary condition (3) explains the initial curved portion of the $\frac{x}{m}$ vs $t^{0.5}$ plot, after which intraparticle diffusion will predominate over a 10-60% adsorption range, with the decreasing rate of adsorption producing another curved portion leading to an equilibrium plateau.

In a plot of $\frac{x}{m}$ vs $t^{0.5}$ the slope of the linear section is designated 'K' the intraparticle diffusion coefficient (cm sec^{-1}) Table III.

Batch Systems

It was found that the βL increased with agitation from 50 to 200 rpm after which it remained constant with increasing agitation, which suggests at low agitation the boundary layer resistance inhibits adsorption, and in a well agitated adsorber vessel this problem would be minimized Weber and Morris[10] stated that if adsorption is one of adsorption onto the external sites of a non-porous adsorbent the rate of adsorption (k) should vary reciprocally with the first power of the diameter of the particle, this relationship holds for the porous adsorbents were the rate of transport to internal surface areas is controlled by external resistance, i.e. film transport. However no such relationship could be found in this

Table III. Intraparticle diffusion coefficient (K, cm/sec) determined for each variable

Variable	K, cm/sec	Variable	K, cm/sec
concentration, mg/l		agitation, rpm	
47	2.6	100	2.5
137	4.0	200	2.9
244	5.4	400	3.0
366	5.5	600	3.5
temperature, °C		particle size, μm	
20	3.6	855	3.6
40	3.4	605	3.6
60	9.7	427	2.9
70	10.6	202	4.1

work suggesting intraparticle diffusion is involved in the rate limiting step. Theoretical equations for intraparticle diffusion indicate that the concentration dependence of a diffusion-adsorption process will vary depending on the characteristics of the adsorption isotherm and on the solute adsorbed at equilibrium. Where intraparticle diffusion is the only rate determining step Weber and Morris [8] found that K varied with the square root of the initial concentrations used. The results for chitosan were found to conform to the following equation $K = 1.7\ Co^{0.5}$ with an initial copper concentration up to 200 mg L^{-1}. It is apparent that the adsorption of copper on chitosan is complex involving boundary layer adsorption and intraparticle diffusion in the rate controlling step.

NOMENCLATURE

- A = correlation coefficient for external film mass transfer
- a = dimensionless separation factor
- B = correlation coefficient for external film mass transfer
- b = Langmuir constant $mg^{-1}\ dm^3$
- Ce = equilibrium liquid phase solute concentration
- Co = initial liquid phase solute concentration
- Cs = liquid phase solute concentration at particle surface
- Ct = liquid phase solute concentration at time, t
- dp = chitosan particle diameter, μm, cm
- Dt = total effective diffusivity
- K = intraparticle diffusion coefficient (cm sec^{-1})
- KL = Langmuir equilibrium constant, $dm^3\ g^{-1}$
- m = mass of chitosan per unit volume of Cu(II) soln., g cm^{-3}
- n = exponent in Freundlich isotherm
- P = Freundlich isotherm constant, mg. g^{-1}
- Q = surface coverage of particle mg g^{-1}
- r = radial distance within particle, μm, cm
- S = specific particle surface, cm^{-1} time, min, sec
- t = time, min, sec.
- v = volume, cm^3, dm^3
- w = mass of chitosan, g
- $\frac{x}{m}$ = mass Cu(II) adsorbed per g chitosan mg g^{-1}
- ε = voidage of chitosan particle
- ρ = solid density, g cm^{-3}
- βL = external film mass transfer, $cm^2\ sec^{-1}$

REFERENCES

1. R. A. A. Muzzarelli, "Natural Chelating Polymers", Pergamon, Oxford, (1973).
2. T. Furusawa and J. M. Smith, Fluid-Particle and Intraparticle Mass Transport Rate in Slurries, Ind. Eng. Chem. Fundam., 12:197 (1973).
3. G. McGray, S. J. Allen, I. F. McConvey, and M. S. Otterburn, Transport Processes in Sorption of Coloured Ions by Peat Particles, J. Colloid Interface Sci., 80:323 (1980).
4. I. Langmuir, The Adsorption of Gases on Plane Surfaces of Glass, Mica and Platinum, J. Amer. Chem. Soc., 40:1361 (1918).
5. H. Freundlich, Z. Phys. Chem., 57:385 (1906).
6. T. W. Weber, R. J. Chakravorti, Pore and Solid Diffusion Models for Fixed Bed Adsorbes, A.I.Ch.E. Journal, 20:228 No. 2 (1974).
7. J. Crank, in: "The Mathematics of Diffusion", Oxford Clarendon Press (1956).
8. W. J. Weber and J. C. Morris, Equilibria and Capacities for Adsorption on Carbon, J. Sanit. Eng. Div. Proc. Am. Soc. Civ. Engr., 50:79 (1974).

COPPER AND VANADIUM COMPLEXES OF CHITOSAN

(G. Micera, S. Deiana, A. Dessì, P. Decock, B. Dubois and H. Kozłowski)

Chitin and its congeners are particularly effective in chelating metal ions (1-4) and thus may play a significant role in controlling and cycling trace elements in the environment.

We report here the results of a study on the complexes formed by interaction of chitosan with aqueous solutions of copper(II) (sulphate, chloride, acetate or nitrate) and oxovanadium(IV) (sulphate). E.s.r. and electronic absorption spectroscopy have been employed to characterize the solid samples obtained at various pH values.

As can be seen in the e.s.r. spectra shown in Fig. 1, at least two kinds of copper(II) complexes (I and II) are present in the samples obtained at pH below 6. One of them (I), characterized by $g_{\parallel} = 2.267$ and $A_{\parallel} = 175 \cdot 10^{-4}$ cm^{-1}, is the major species in the sample obtained by exchange from a copper(II) acetate solution at pH 5.0. Coordination in this complex should involve the nitrogen atom of an amino group and, probably, three oxygen atoms from hydroxyl groups and water molecules. Complex II, $g_{\parallel} = 2.251$ and $A_{\parallel} = 184 \cdot 10^{-4}$ cm^{-1}, predominates over the pH range 6 - 7 and most likely involves a chromophore of the CuN_2O_2-type.

The changes of the $g_{\parallel}$ and $A_{\parallel}$ values observed at higher pH values substantiate the formation of complex species where stronger metal-ligand interaction takes place. Complex III, $g_{\parallel} = 2.222$ and $A_{\parallel} = 202 \cdot 10^{-4}$ cm^{-1}, is the major species in the pH range 7.5 - 9. At higher pH values the signals of this species overlap with weaker resonances due to at least another species (IV). Based on the low-field shift of the parallel absorptions, it can be supposed that the complex species IV should involve metal ions with stronger axial interaction or weaker in-plane coordination bonds than in III.

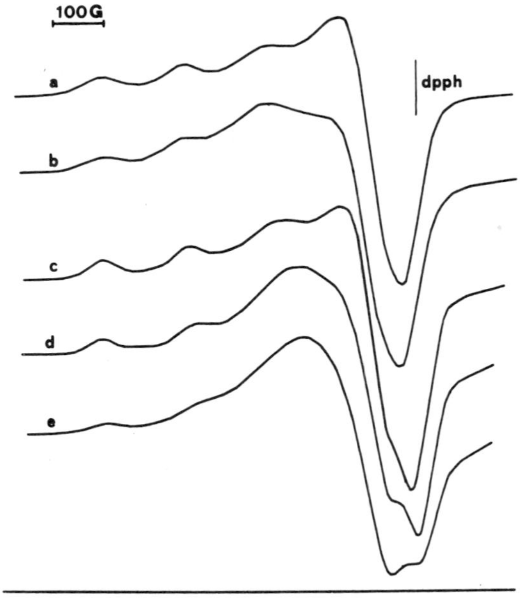

Fig. 1. X-Band ESR spectra of Cu(II)-chitosan samples obtained by exchange from a solution of copper(II) acetate at pH 5.0 (a) and from solutions of copper(II) sulphate at pH 4.6 (b), 6.6 (c), 7.5 (d) and 11.5 (e).

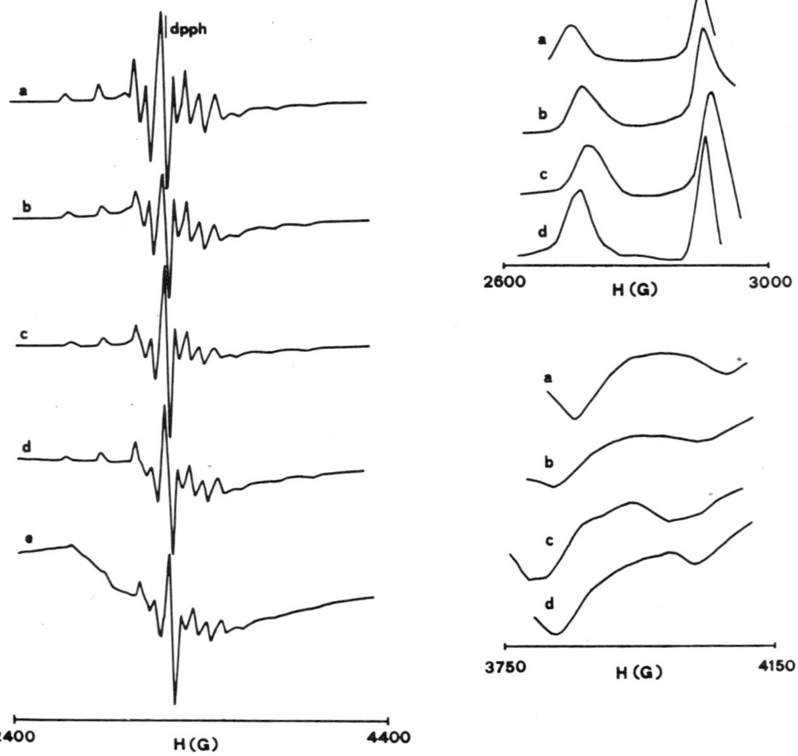

Fig. 2. X-Band ESR spectra of VO(IV)-chitosan samples obtained at pH 5.3 (a), 6.4 (b), 7.0 (c), 9.0 (d) and 10.0 (e).

The electronic spectra of the Cu(II) complexes display a broad d-d absorption maximum around 675 nm, with an evident shoulder at about 760 nm, for the samples obtained at pH below 6. With increasing pH, the absorption maximum shifts to higher energy values and the shoulder is no longer observed. The spectra of the samples obtained at pH > 9 display a maximum around 635 nm and a shoulder at about 500 nm.

On the whole, according to literature data (5) which substantiate that in the Cu(II)-chitosan system the NH_3^+ groups dissociate with pKa = 6.9, while the deprotonation of the hydroxyl groups occurs over the pH range 7 - 9, and by analogy with the trend observed for the Cu(II)-D-glucosamine system (6), it can be suggested that: i) the $CuNO_3$ species formed at pH 5 - 6 changes into a CuN_2O_2 complex (II), involving two amino and two undissociated hydroxyl groups (or two water molecules), as more free amino groups are made available; ii) CuN_2O_2 chelates III may arise from the coordination of deprotonated hydroxyl groups; and iii) complex species like IV may be formed by hydrolysis of III or may involve ions coordinated to units containing acetylated amino groups and thus acting as O-donors. These findings are in substantial agreement with a previous study by Muzzarelli et al. on the chelation of cupric ions by chitosan membranes (7).

The behaviour of the oxovanadium(IV)-chitosan system is similar to that observed for the copper(II) samples. Indeed, the species with only one nitrogen atom in metal coordination ($g_{\parallel}$ = 1.935, $g_{\perp}$ = 1.973, $A_{\parallel}$ = $197 \cdot 10^{-4}$ cm^{-1}, $A_{\perp}$ = $72 \cdot 10^{-4}$ cm^{-1}) is observed in the e.s.r. spectra of the samples obtained at pH 5 - 6 (Fig. 2). The species involving two nitrogen atoms and, probably, two undissociated hydroxyl groups is revealed by the spectra of the samples at pH 6 - 7 ($g_{\parallel}$ = 1.939, $g_{\perp}$ = 1.974, $A_{\parallel}$ = $191 \cdot 10^{-4}$ cm^{-1}, $A_{\perp}$ = $68 \cdot 10^{-4}$ cm^{-1}). A complex analogous to III could be suggested by resonances which show up at pH about 7 ($g_{\parallel}$ = 1.95, $g_{\perp}$ = 1.97, $A_{\parallel}$ = $180 \cdot 10^{-4}$ cm^{-1}, $A_{\perp}$ = $69 \cdot 10^{-4}$ cm^{-1}). At higher pH values broad resonances, indicative of magnetic interaction between neighbouring ions, are detected. The dilute species which display resolved hyperfine features ($g_{\parallel}$ = 1.94, $g_{\perp}$ = 1.97, $A_{\parallel}$ = $190 \cdot 10^{-4}$ cm^{-1}, $A_{\perp}$ = $69 \cdot 10^{-4}$ cm^{-1}) may be complexes analogous to IV.

REFERENCES

1. R.A.A. Muzzarelli, "Natural Chelating Polymers", Pergamon, Oxford (1973).
2. R.A.A. Muzzarelli, "Chitin", Pergamon, Oxford (1977).
3. R.A.A. Muzzarelli and L. Sipos, Chitosan for the collection from sea water of naturally occurring zinc, cadmium, lead, and copper, Talanta, 18:853 (1971).
4. T. Yoshinari, and V. Subramanian, Adsorption of metals by chitin, Environ. Biogeochem., 2:541 (1976).
5. Y. Inaki, M. Otsuru, and K. Takemoto, Vinyl polymerization by metal complexes. XXXI. Initiation by chitosan-copper (II) complex, J. Macromol. Sci. Chem., A12:953 (1978).
6. G. Micera, S. Deiana, A. Dessi, P. Decock, B. Dubois, and H. Kozlowski, Copper (II) complexation by D-glucosamine. Spectroscopic and potentiometric studies, Inorg. Chim. Acta, 107:45 (1985).

7. R.A.A. Muzzarelli, F. Tanfani, M. Emanuelli, and S. Gentile, The chelation of cupric ions by chitosan membranes, J. Appl. Biochem., 2:380 (1980).

USE OF CHITOSAN IN THE RECOVERY OF PROTEIN FROM SHRIMP PROCESSING WASTEWATER

(C. Senstad and K. A. Almas)

Of the total Norwegian catch of shrimp, Pandalus borealis, (approximately 60,000 tons), more than 95 %, is peeled in machines. While the theoretical yield of shrimp meat is 35 %, the practical yield is only 25 %. Shrimp meat is lost partly because of mechanical disruption of the raw material in the machines, but also simply by washing out during the processing. Approximately 50 tons of water is used for processing 1 ton of raw material.

The purpose of the present work is: 1) To test chitosan as a coagulating agent in shrimp processing wastewater, determining how pH and the initial ratio of chitosan to protein affect the precipitation of organic material. 2) To recover organic material lost during shrimp processing.

Chitosan as a coagulating agent has been discussed by several authors (1 - 4). Detailed information about the effect of pH on the destabilization of charged colloids is essential for practical applications of chitosan for this purpose. Another parameter is the ratio of positive charged chitosan molecules to negative charged colloids. Too high a ratio will inhibit destabilization. Recalling that the charge ratio depends on both the pH and the molecular number ratio, the optimization of the coagulation has to be done with respect to both the pH and the ratio of chitosan to protein. It should be noted that pI for chitosan is 6.3, while the majority of the watersoluble proteins have their pI below 5.0. The theory and principle of particle destabilization is fully described (5, 6).

EXPERIMENTAL

The protein concentrations of the wastewater varied from 0.1 to 6.0 g/l at different steps along the shrimp processing line. The highest concentration was found in the steamcooker, and wastewater samples for experiments were withdrawn at this step. The concentrations of protein varied from 1 to 6 g/l. Coagulation tests were carried out both for these concentrations and for diluted samples. Tests were also carried out on ultra- and diafiltered samples (Ultrafilter, Romicon PM10, molecular cut-off 10.000 D) to see how minerals and organics with molecular weights less than 10.000 D affected the coagulation.

Chitosan solutions (1.0 g/l) were made by dissolving Flonac N in 5 % acetic acid. A Jar-test apparatus (6 x 1 l) was applied for coagulation experiments. The pH in both the wastewater and chitosan solutions was adjusted prior to mixing at 100 rpm for 20 minutes. The flocculation proceeded for 20 minutes at 20 rpm. The coagulation and flocculation was carried out according to Black and Harris (7). After settling for 1 h, the chitosan-treated wastewater was decanted and centrifuged at 20,000 g for 10 min. Protein concentrations were determined colorimetrically by the method of Folin-Ciocalteu.

RESULTS AND DISCUSSION

The following results (Fig. 1) explain how the pH and the ratio of chitosan to protein affect the percent of protein removal.

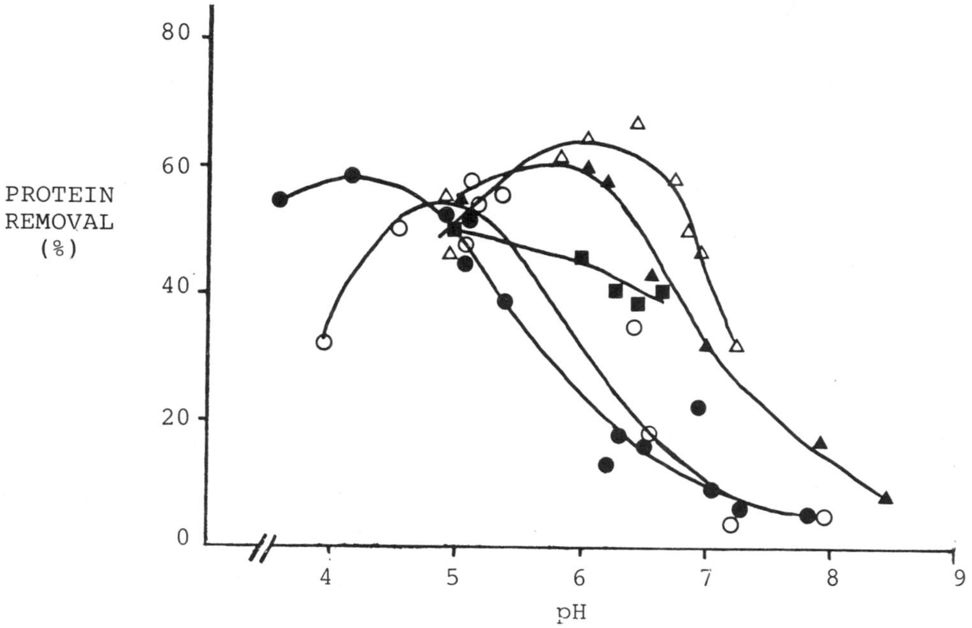

Figure 1. Removal of proteins from shrimp processing wastewater. The ratio of chitosan to protein (weight %) is calculated on the basis of the protein content prior to coagulation.
Ratio: ●—● = 0, ○—○ = 0,5, ▲—▲ = 6, △—△ = 12, ■—■ = 25 .

By adding 6 % chitosan (as weightpercent protein in solution before coagulation) at pH 5.5 - 6.0, the protein removal was about 65 %. A lower chitosan content lead to reduced protein removal at this pH. The pH alone caused the protein removal to be 25 - 30 %. The destabilization and coagulation of these proteins arose from operating in a pH region where some proteins (25 - 30 % of the total) have their pI. An interesting observation was that, when increasing the ratio of chitosan/protein up to 25 %, the result was, in fact, a decreased protein removal.

At lower pH values, the addition of chitosan had a negative effect as a coagulating agent; 60 % of the protein was removed at pH 4.5 without adding chitosan. Increasing the ratio of chitosan/protein caused the proteins to be less destabilized, due to an excess of positive charged chitosan molecules.

The opposite observation was found in the higher pH range (6.5 - 8.0). Here, the protein removal increased as the ratio of chitosan to protein was increased. The chitosan molecule had no net charge, but the amino groups were free to form chemical bridges with reactive sites on the colloids. As the pH was increased, a larger portion of the proteins assumed a net negative charge, and more chitosan was needed to form bridges. Sweep-floc (due to precipitation of chitosan) is an alternative mechanism for the removal of protein in this pH range.

These data show that one can remove ca. 65 % of the protein present in the shrimp processing wastewater. The turbidity removal (at 660 nm) was found to be 85 %. The highest removal of proteins can be achieved by altering the pH and the chitosan/protein ratio, as given in Fig. 1. No difference in protein removal was observed for the ultrafiltrated samples, meaning that the salt content did not have any significant effect on the destabilization of the proteins in these experiments.

The settling time, the sludge volume and the mechanical strength of the flocs were dependent upon the pH and the chitosan/protein ratio. Futher investgations have to be carried out to obtain well separatable flocs.

REFERENCES

1. W. A. Bough, Chitosan - a polymer from seafood waste, for use in treatment of food processing wastes and activated sludge, Process Biochem., 11:13 (1976).
2. W. A. Bough, and D. R. Landes, Treatment of food-processing wastes with chitosan and nutritional evaluation of coagulated by-products, in: "Proceedings of the 1st International Conference on Chitin/Chitosan", R. A. A. Muzzarelli and E. R. Pariser, eds., MIT, Cambridge (1978).
3. R. A. Johnson, and S. M. Gallanger, Use of coagulants to treat seafood processing wastewaters, J. of Water Poll. Contr. Fed., 56:970 (1984).
4. K. L. Simpson, The recovery of proteins and pigments from the shrimp and crab meals and their use in salmonid pigmentation, in: "Proceedings of the 1st International Conference on Chitin/Chitosan", 253 (1978).
5. T. D. Reynolds, Unit Operations and Processes in Enviromental Engineering, Brooks/Cole Engineering Division, Wadsworth Inc., California, 15 (1982).
6. V. T. Chow, R. Eliassen, and R. K. Linsley, "Wastewater Engineering: Treatment, Disposal, Reuse", 2nd ed., McGraw-Hill (1979).
7. A. P. Black, and R. H. Harris, New dimensions of the old jar test, Water Wastes Engin., Dec.: 49 (1969).

Affiliations

J. P. Thomé and Y. Van Daele, Institut de Zoologie "Van Beneden", 22 quai Van Beneden, B-4020 Liège, Belgium.

B. Venkatrao, A. Baradarajan and C. A. Sastry, Dept. Chemical Engineering, Indian Institute of Technology, Madras 600036, India.

G. McKay, H. Blair, A. Findon, Dept. Chemical Engineering, Queen's University Belfast, Belfast BT9 5DL, North Ireland.

G. Micera, Istituto di Chimica Generale, Università, Sassari I-07100, Italy.

S. Deiana and A. Dessì, Istituto Applicazione Tecniche Chimiche Avanzate ai Problemi Agrobiologici, CNR, Sassari I-07100, Italy.

P. Decock and B. Dubois,Laboratoire de Chimie Minérale & Méthodologie Analytique, Université de Lille, F-59655 Villeneuve d'Ascq, France.

H. Kozłowski, Institute of Chemistry, University of Wroclaw, PL-50383 Poland.

C. Senstad and K. A. Almas, Institute of Fisheries, Technol. Research, P. O. Box 677, N-9001 Tromso, Norway.

AUTHOR INDEX

ORGANISMS INDEX

SUBJECT INDEX